AF724909

Peptides for Youth

ADVANCES IN EXPERIMENTAL MEDICINE AND BIOLOGY

Editorial Board:

NATHAN BACK, *State University of New York at Buffalo*
IRUN R. COHEN, *The Weizmann Institute of Science*
ABEL LAJTHA, *N.S. Kline Institute for Psychiatric Research*
JOHN D. LAMBRIS, *University of Pennsylvania*
RODOLFO PAOLETTI, *University of Milan*

For other titles published in this series, go to
www.springer.com/series/5584

Susan Del Valle • Emanuel Escher
William D. Lubell
Editors

Peptides for Youth

The Proceedings of the 20th American Peptide Symposium

Springer

Editors
Susan Del Valle
Department of Chemistry & Biochemistry
New Mexico State University
1175 N. Horseshoe Drive
Las Cruces, NM 88003-8001
USA

Emanuel Escher
Department of Pharmacology
University of Sherbrooke Medical School
Sherbrooke, QC J1H 5N4
Canada

William D. Lubell
Département de Chimie
Université de Montréal
C.P. 6128, Succursale Centre Ville
Montréal, QC H3C 3J7
Canada

ISBN: 978-0-387-73656-3 e-ISBN: 978-0-387-73657-0

Library of Congress Control Number: 2008931302

© Springer Science+Business Media, LLC 2009
All rights reserved. This work may not be translated or copied in whole or in part without the written permission of the publisher (Springer Science+Business Media, LLC, 233 Spring Street, New York, NY 10013, USA), except for brief excerpts in connection with reviews or scholarly analysis. Use in connection with any form of information storage and retrieval, electronic adaptation, computer software, or by similar or dissimilar methodology now known or hereafter developed is forbidden.
The use in this publication of trade names, trademarks, service marks, and similar terms, even if they are not identified as such, is not to be taken as an expression of opinion as to whether or not they are subject to proprietary rights.

Printed on acid-free paper

springer.com

Contents

Preface

The Twentieth American Peptide Society Symposium, *Peptides for Youth*, was held on June 26–30, 2007, at the Palais de Congrès and the Hyatt Regency Hotel in Montréal, Québec, Canada. The third time the APS meeting has come to Canada, and the first time in Montréal, *Peptides for Youth* celebrated the society, its history and traditions, and launched many new initiatives for enhancing member participation at the meeting. Attended by over 1,000 participants, *Peptides for Youth* was a success in scientific content, balance of industrial and academic interests, speaker diversity (scientific, gender and geographic), speaker and poster quality and quantity, the high turnout at the lectures, the venues, the entertainment, the dinners, the celebration of past pioneers, the recruitment, education and encouragement of young investigators, and the sense of freshness.

The Symposium was filled with excellent lecture sessions and workshops, organized by the Scientific Advisory Board and often dedicated to the Pioneers of the field (with grateful sponsorship): *Folding, Recognition, and Catalysis: A session dedicated to Elkan Blout* Sponsored by Wiley, *The Makineni Lecture* endowed by Polypeptide Laboratories, *Peptide, Protein, and Peptidomimetic Synthesis honoring Ralph Hirshmann* sponsored by Merck and Organic Letters, and *Peptide and Membrane Proteins: dedicated to Miklos Bodansky* sponsored by IPSEN, Abbott Laboratories, and Synthetech. Portraying the valued history of our Society, ten Portraits of the Pioneers of the American Peptide Society were prepared by volunteer students from the Universities of Montréal, Sherbrooke and Montpellier, and displayed at the entrance to the Exposition. These posters have now been reformatted for redisplay in this publication. The student volunteers also prepared and performed on the middle day of the meeting a historical slide show scored to the Cole Porter song "Experiment", the lyrics of which in this publication, preface a reproduction of this retrospective with North American bias of the growth of our field from the first peptide synthesis of Fischer more than a hundred years ago to *Peptides for Youth.*

With a nod of gratitude and respect to the pioneers of the Society, the stage was set for the main objective of *Peptides for Youth*, to encourage and recruit the next generation of peptide scientists. Peptide Idol, a new format for selecting the winner of the Young Investigator poster competition, was a hit, so was placing the Young Investigator Oral Presentations into the heart of the meeting. We are grateful to ESCOM, CS Bio, and Amgen for providing sponsorship for these events, which illustrated the strength of the Society's future. Moreover, the Molecules of Life Project was launched with the debut of the Peptides of Life in which art and science were employed together to teach a syllabus of twelve peptides to elementary school students, as presented in three posters at the meeting and discussed in more detail in a proceeding in this publication.

Other new additions to the meeting format included the General Assembly, in which members had an open forum to voice their opinion on Society activities; the Networking Lunch, in which Young Investigators were given opportunity to network with senior members, and the Session on Complementary and Alternative Medicine, featuring new avenues for developing peptide science. In addition, a morning Yoga Session on the roof of the Hyatt hotel and an evening gazing at fireworks from Club 737 at the top of one of Montréal's tallest buildings, both provided complementary social and professional settings for the exchange of ideas, to complement the traditional venues of the Opening Reception, Merrifield and Speakers Dinners and Closing Banquet.

Peptides for Youth was driven with heart and volunteer students rose to the occasion to deliver. The students prepared slides to advertise the sponsors for all of the events of the meeting, which were effectively shown on the third screen in parallel to the talks. They created an up-to-date information board, which provided highlights on the many events of the meeting. The presence of helpful student volunteers addressed many of the meeting needs. Our students made us extremely proud as they proved to be a group of very high quality colleagues who in the face of a formidable challenge worked undaunted side-by-side in a valiant effort. For this, we will be forever grateful. Particular thanks goes to head student volunteers, Ph.D. students Carnie Bourguet and Danny Fillion, as well as the new co-chairs of the Student Affairs Committee, Dr. Audrey Kelleman and Professor Jung-Mo Ahn.

This book captures about a half of the total number of oral and poster contributions from the 20th APS meeting, including the memorable Merrifield Award Lecture by Dr. Isabella Karle and the Makinini Award Lecture by Professor Ronald Raines. We are pleased that this volume is included in the *Advances in Experimental Medicine and Biology*, a highly-regarded and ISI-ranked Springer book series. As per the meeting, advances in the synthesis, production, delivery, characterization and application of peptide products particularly in the fields of medicine and materials science are well represented here.

As the American Peptide Society now looks toward the 21st Symposium, *Breaking Away*, momentum and knowledge exist for creating a meeting that addresses more effectively the interests of our younger membership from both the academic and industrial sectors.

Message from the American Peptide Society President

The American Peptide Society continues it's commitment to advancing and promoting the knowledge of the chemistry and biology of peptides and proteins. I invite you to visit our web site at www.ampepsoc.org to learn more about the Society and the benefits of membership. The Society has a dedicated council of 15 individuals elected by Society members at large. There are a number of important active committees which focus on areas vital to the Society including Awards, Membership, Publications, and Student Affairs. The American Peptide Society also actively participates in activities of the broader research community through our associate membership in the Federation of American Societies for Experimental Biology (FASEB). The official journal of the Society is *Peptide Science* (through *Biopolymers*) and all members receive a subscription. The journal publishes original research and review articles. The editor of *Peptide Science*, Dr. Lila Gierasch, and I strongly encourage you to submit your best work to our journal.

The 20th American Peptide Symposium, *Peptides for Youth*, was held on June 20–25 in Montreal Canada and co-chaired by William Lubell, Université de Montréal and Emanuel Escher, Université de Sherbrooke. The consistently outstanding program of scientific lectures and poster presentations is documented in this volume. Combined with the cultural and social events associated with the meeting, attended by over 900 registrants, this Symposium was both a scientific and social success. We all congratulate co-chairs William Lubell and Emanuel Escher along with their dedicated and hardworking staff in addition to the organizing and program committees. Thanks and gratitude must also be extended to all of our very generous sponsors, as well as and to the exhibitors for their far reaching and informative displays.

The American Peptide Society is actively planning for future symposia. The 21st American Peptide Symposium, with its theme *Breaking Away*, will be held in Bloomington Indiana on June 2009. It will be co-chaired by Richard DiMarchi, Indiana University and Henry Mosberg, University of Michigan. The 22nd American Peptide Symposium, will be held in San Diego on June 2011 with Phil Dawson, Scripps Research institute and Joel Schneider, University of Delaware as co-chairs.

My term as President of the Society runs from June 2007 to June 2009. During this period my priorities are to convey to all the importance of peptides as not only essential biological entities, but as highly relevant and focused therapeutics agents (there is an ever increasing number of peptides moving through the clinical trial process). Additionally I hope to increase not only our membership, but also the value of the Society to its members, especially our younger scientists. Finally, I hope to be able to continue to work to improve the Symposia both scientifically and socially, and to establish the locations of our 23rd and 24th bi-annual meetings, as I anticipate strong growth of our society over the next ten years with the increasing importance of peptide science as a rich field for discovery and innovation.

I look forward to working with all of the members of the American Peptide Society. My best wishes for your continued success in your scientific activities.

Richard A. Houghten
Torrey Pines Institute for Molecular Studies
January 2008

20th American Peptide Society Symposium

June 26 – 30, 2007
Montréal, Québec, Canada

Co-Chairs

Emanuel Escher, Ph.D.
Université de Sherbrooke
Department of Pharmacology
Faculty of Medicine
Sherbrooke, Québec J1H 5N4 Canada
Tel: 1-819-564-5346
Fax: 1-819-564-5400

William D. Lubell, Ph.D.
Université de Montréal
Département de Chimie
C.P. 6128, Succursale Centre Ville
Montréal, Québec H3C 3J7 Canada
Tel: 1-514-343-7339
Fax: 1-514-343-7586

Scientific Program Committee

Annette G. Beck-Sickinger, Ph.D., ETH Zurich
Waleed Danho, Ph.D., Hoffmann-La Roche, Inc.
Charles Deber, Ph.D., Hospital for Sick Children
Richard DiMarchi, Ph.D., The Lilly Research Laboratories
Samuel H. Gellman, Ph.D., University of Wisconsin – Madison
Pierre Lavigne, Ph.D., Université de Sherbrooke
Robin Offord, Ph.D., University of Geneva
Peter W. Schiller, O.Q., Ph.D., FRSCC, Clinical Research Institute of Montréal
Joel Schneider, Ph.D., University of Delaware

Symposium Organizing Committee

Jane Aldrich, Ph.D., University of Kansas
Susan Del Valle, Ph.D., New Mexico State University
Emanuel Escher, Ph.D., Université de Sherbrooke
Renée Landry, Université de Sherbrooke
Tom Lobl, Ph.D., Neurosystec Corporation
William Lubell, Ph.D., Université de Montréal
Becci Totzke, American Peptide Society

Student Affairs Committee

Jung-Mo Ahn, Ph.D., University of Texas
Audrey Kelleman, Ph.D., Senn Chemicals
Florence Brunel, Ph.D., The Scripps Research Institute

Teresa Lama, Ph.D., Université de Montréal
Krista Wilson, University of Florida
Carine Bourguet, Université de Montréal
Fernanda Marques Burke, University of Michigan
James Patterson, University of Indiana
Eunice Murage, University of Texas at Dallas

Young Investigators' Poster Judges

Lila Gierasch, Ph.D.
Ernest Giralt, Ph.D.
Mibel Aguilar, Ph.D.
Chris Soares, Ph.D.
Neal Zondlo, Ph.D.
Jesse Dong, Ph.D.
Oliver Zerbe, Ph.D.
Helma Wennemers, Ph.D.
Robert Ben, Ph.D.
Paolo Grieco, Ph.D.
Laszlo Otvos, Ph.D.
John Tolle, Ph.D.
Anna Diez, Ph.D.
Jeff Bode, Ph.D.
Carsten Baldauf, Ph.D.
Yvonne Angell, Ph.D.
Hee-Seung Lee, Ph.D.
Annelise Barron, Ph.D.
Ajoy Basak, Ph.D.
Jeff Becker, Ph.D.
Matt Bogyo, Ph.D.
Ale Narvanen, Ph.D.
Robert Hondal, Ph.D.
Kenneth Rotondi, Ph.D.
Fernando Formaggio, Ph.D.
Gilles Guichard, Ph.D.
Mark McLaughlin, Ph.D.
Phil Dawson, Ph.D.
Claudio Mapelli, Ph.D.
Chris Schafmeister, Ph.D.
John Mayer, Ph.D.
Wolfgang Maison, Ph.D.
Felix Polyak, Ph.D.
Pierre Lavigne, Ph.D.
Anna Maria Papini, Ph.D.
Mark Jarosinski, Ph.D.
Jesse Dong, Ph.D.
Michael Carrasco, Ph.D.
David Perrin, Ph.D.

Peptide Idol Hosts and Judges

Charles Deber, Ph.D., Hospital for Sick Kids
Jean Martinez, Ph.D., Université de Montpellier
Felicia Etzkorn, Ph.D., Virginia Tech
Lyne Regan, Ph.D., Yal University
Anna Maria Papini, Ph.D., University of Florence
Victor Hruby, Ph.D., University of Arizona
Laslo Otvos, Ph.D., Temple University
Luis Moroder, Ph.D., Max Plank Institute for Biochemistry
Ian Smith, Ph.D., Monash University

Travel Grant Committee

Pierrette Gaudreau, Ph.D., Université de Montréal
Anna Diez, Ph.D., Universitat de Barcelona
Mark Jarosinski, Ph.D., Midwest Bio-Tech Inc.
John Mayer, Ph.D., Lilly Corporate Center

Fundraising Committee

Dr. Roger Bossé, Ph.D., Perkin Elmer Biosignal
Robert Furic, Matrix Innovation
Witold Neugebauer, Ph.D., D.Sc., Université de Sherbrooke
Felix Polyak, Ph.D., Theratechnologies Inc.
Hossain H. Saneii, Ph.D., AAPPTEC

20th American Peptide Society Symposium Sponsors

Gold
IPSEN
American Peptide Company
PolyPeptide Laboratories, Inc.

Silver
Amylin
CS Bio
NeoMPS
Organic Letters (ACS)

Mercury
Aaptec
Amgen
AnaSpec
Bachem Americas
EKA Chemicals (Kromasil)/ Akzo Nobel
Eli Lilly
Hoffman-La Roche
Merck
Midwest Bio-Tech
Novabiochem, Merck Biosciences
Senn Chemicals
Theretechnologies, Peptide Science
Wiley

Copper
Abbott Laboratories
Cambridge Research Biochemicals
ESCOM
Genentech
New England Biolabs
RSP Amino Acids, LLC
Synthetech
Transzyme
Université de Montréal
Université de Sherbrooke
Wyeth

Contributors
Boehringer- Ingelheim
PHARMACHEM
RSC Pubishing
Thieme Chemistry

20th American Peptide Society Symposium Exhibitors

Aaptec
Activotec
Alta
American Peptide Company
American Peptide Society
ANASPEC
Bachem
BCN Peptides
BioSynthesis
Cambridge Research Biochemicals
C. A. T. CmgH & Co. KG
CBL Patras
CEM Corporation
CPC Scientific Inc.
Creosalus
EMD
EKA Chemicals/ Kromosil
Genzyme
GL Biochem (Shanghai) Ltd.
Intavis
IRIS Biotech GmbH
MultisynTech GmbH
NeoMPS (Groupe SNPE)
Novasep
Peptides International Inc.
PeptiSyntha
Phenomenex
PolyPeptide Laboratories
Protein Technologies
RAPP Polymere
Roche Colorado Corporation
Senn Chemicals
Sichuan Goldentree Bio-Pharmaceutical Co.,Ltd
Teldyne Isco Inc.
Tianjin Nankai Hecheng S&T Co, Ltd.
Varian Polymer Laboratories
Wiley
YMC

American Peptide Symposia

Symposium	Year	Chairs	Location
First	1968	***Saul Landa*** **Yale University** ***Boris Weinstein*** **University of Washington, Seattle**	Yale University New Haven, CT
Second	1970	***F. Merlin Bumpus*** Cleveland Clinic	Cleveland Clinic Cleveland, OH
Third	1972	***Johannes Meinhofer*** Harvard Medical School	Children's Cancer Research Foundation Boston, MA
Fourth	1975	***Roderich Walter*** University of Illinois Medical Center-Chicago	The Rockfeller University and Barbizon Plaza Hotel New York, NY
Fifth	1977	***Murray Goodman*** University of California-San Diego	University of California-San Diego San Diego, CA
Sixth	1979	***Erhard Gross*** National Institutes of Health	Georgetown University Washington, DC
Seventh	1981	***Daniel H. Rich*** University of Wisconsin-Madison	University of Wisconsin-Madison Madison, WI
Eighth	1983	***Victor J. Hruby*** University of Arizona	University of Arizona Tucson, AZ
Ninth	1985	***Kenneth D. Kopple*** Illinois Institute of Technology ***Charles M. Deber*** University of Toronto	University of Toronto Toronto, Ontario, Canada
Tenth	1987	***Garland R. Marshall*** Washington University School of Medicine, St. Louis	Washington University St. Louis, MO
Eleventh	1989	***Jean E. Rivier*** The Salk Institute of Biological Studies, La Jolla	University of California-San Diego San Diego, CA
Twelfth	1991	***John A. Smith*** Massachusetts General Hospital	Massachusetts Institute of Technology Cambridge, MA
Thirteenth	1993	***Robert S. Hodges*** University of Alberta-Edmonton	Edmonton Convention Center Edmonton, Alberta, Canada

Fourteenth	**1995**	***Pravin T .P. Kaumaya*** The Ohio State University	The Ohio State University Columbus, OH
Fifteenth	**1997**	***James P. Tam*** Vanderbilt University	Nashville Convention Center Nashville, TN
Sixteenth	**1999**	***George Barany*** University of Minnesota-Minneapolis ***Gregg B. Fields*** Florida Atlantic University	Minneapolis Convention Center Minneapolis, MN
Seventeenth	**2001**	***Richard A. Houghten*** Torrey Pines Institute for Molecular Studies ***Michal Lebl*** Illumina, Inc., CA	Town and Country Resort Hotel San Diego, CA
Eighteenth	**2003**	***Michael Chorev*** Beth Israel Deaconess Medical & Harvard Medical School ***Tomi K. Sawyer*** ARIAD Pharmaceuticals Inc., MA	Boston Marriott Copley Place Boston, MA
Nineteenth	**2005**	***Jeffery W. Kelly*** Scripps Research Institute, CA ***Tom W. Muir*** Rockefeller University, NY	Town and Country Resort Hotel San Diego, CA
Twentieth	**2007**	***Emanuel Escher*** **Université de Sherbrooke** ***William D. Lubell*** **Université de Montréal**	Palais des Congrès and Hyatt Regency Hotel Montréal, Québec, Canada

"Experiment", a pictorial retrospective from the first peptide bond to the 20th American Peptide Society Meeting

Caroline Proulx, David Sabatino, Tanya Godina and William D. Lubell

Département de Chimie, Université de Montréal, Montréal, Québec, Canada H3C 3J7

The 20th American Peptide Society Meeting celebrated the birth and adolescence of the field of peptide science by paying tribute to the pioneers that established the scientific and social foundation on which our membership stands today. Challenged by the task of creating a moment to give their own congratulations to the society, and armed with photos of past meetings and events thanks to those who contributed to the "Meetings to Remember" part of the 20th APS web site (www.20thAPS.org), the student volunteers produced and performed the following historical, pictorial retrospective of the history of the American Peptide Society to the Cole Porter song "Experiment". On the middle day, after the morning lectures before leaving for the free afternoon, participants were cajoled back to their seats into staying a moment longer and then pleasantly surprised by this sign of appreciation exhibited by the students for the behalf of the society.

Experiment by Cole Porter

Before you leave these portals
To meet less fortunate mortals
There's just one final message
I would give to you
You all have learned reliance
On the sacred teachings of science
So I hope, through life you never will
decline
In spite of philistine
Defiance
Do what all good scientists do
Experiment
Make it your motto day and night
Experiment
And it will lead you to the light
The apple on the top of the tree
Is never too high to achieve
So take an example from Eve
Experiment
Be curious
Though interfering friends may
frown,
Get furious
At each attempt to hold you down
If this advice you'll always employ
The future can offer you infinite joy
And merriment
Experiment
And you'll see

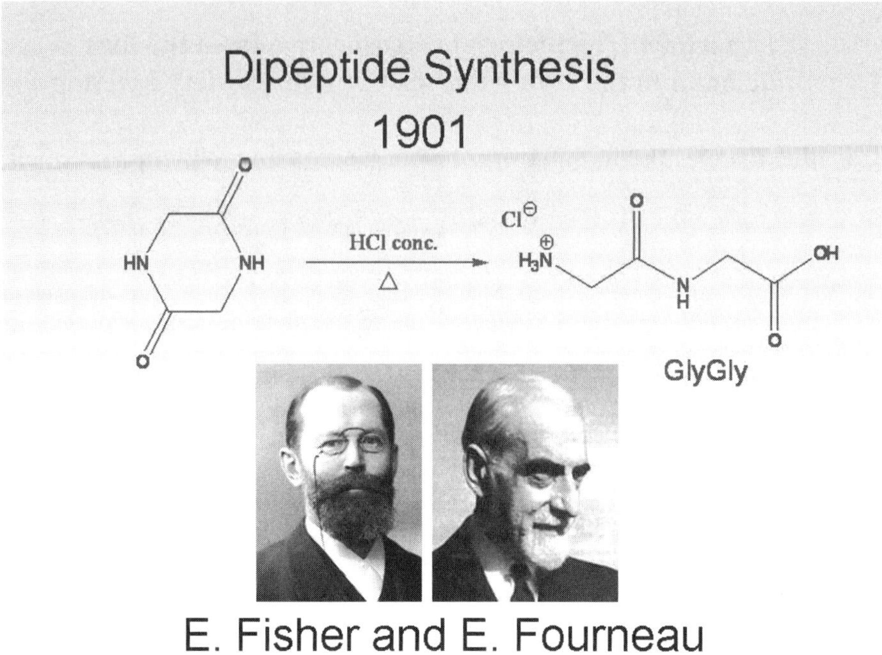
Dipeptide Synthesis
1901
HCl conc.
Δ
HN
NH
Cl⊖
H3N⊕
OH
GlyGly
E. Fisher and E. Fourneau

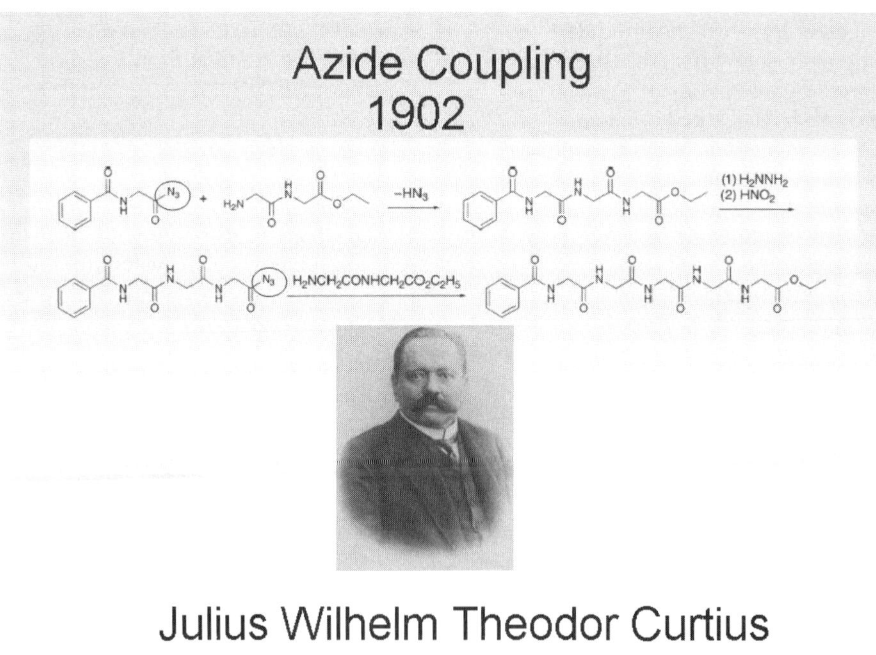
Azide Coupling
1902
N3
H2N
−HN3
(1) H2NNH2
(2) HNO2
H2NCH2CONHCH2CO2C2H5
Julius Wilhelm Theodor Curtius

Isolation of Insulin from Dog Pancreas 1922

THE CANADIAN MEDICAL ASSOCIATION JOURNAL 1 4

PANCREATIC EXTRACTS IN THE TREATMENT OF DIABETES MELLITUS

PRELIMINARY REPORT BY F. G. BANTING AND C. H. BEST, *Dept. of Physiology*

J. B. COLLIP, *Dept. of Path. Chemistry*

W. R. CAMPBELL AND A. A. FLETCHER, *Dept. of Medicine, University of Toronto, and Toronto General Hospital*

SINCE the year 1889, when von Mering and Minkowski (1) produced severe and fatal diabetes by total removal of the pancreas in dogs, many investigators have endeavoured to obtain some beneficial effect in diabetes mellitus, either by feeding pancreas, or by administration of pancreatic extracts.

Minkowski, Sandmeyer (2), Pflüger (3) and others found that feeding pancreas was followed by negative or even harmful results. More recently, Murlin (4), Kleiner (5) and Paulesco (6) have tried the effects of aqueous extracts of the pancreas intravenously, on depancreatized animals and have found transitory reduction in the percentage of blood sugar and in the sugar excreted in the urine.

In 1907, Rennie and Fraser (7), recognizing the possibility that pancreatic enzymes might have harmful effects on the internal secretions, secured islet tissue from teleosteal fishes, where it exists separately from the rest of the pancreas, and fed it to human diabetics. Their studies demonstrated no beneficial influence on the condition of the patient. E. L. Scott (8) in 1912 sought to eliminate the influence of proteolytic enzymes by using alcoholic extracts of the pancreas. He did not find, however, that such extracts caused as marked a reduction in the urinary sugar or in the G-N. ratio as when extracts were made with acidulated water. The whole question has been reviewed recently by Allen (9), by him, and, indeed, by the majority of recent writers, it is usually stated that pancreatic extracts have no clinical value whatsoever. During the past ten months, two of us (F. G. B. and C. H. B.), working in the Department of Physiology of the University of Toronto, have reinvestigated the problem. Certain of the results obtained have already been published, (10) others are now in press. These may be briefly reviewed here.

Believing that extracts of the pancreas, as usually prepared, did not satisfactorily demonstrate the presence of an internal secretion acting on carbohydrate metabolism, because the active principle was destroyed by the digestive enzymes also present in such extracts, attempts were made to eliminate these enzymes. In the first experiments, this was done by taking advantage of the fact that the acinous tissue (from which the digestive enzymes are derived) but not the insular tissue of the pancreas degenerates in seven to ten weeks after ligation of the pancreatic ducts. Extracts were therefore made with ice-cold Ringer's solution, of degenerated pancreatic tissue removed ten weeks after the ligation of the ducts. The extract obtained by this procedure, when injected intravenously or subcutaneously into diabetic dogs, invariably caused a marked reduction in blood sugar and in the amount of sugar excreted in the urine. It also enabled a diabetic dog to retain a much higher percentage of injected sugar than it otherwise would. Extracts of liver or spleen, prepared in the same manner as the extracts of degenerated pancreas, were found to have neither of these effects. The active principle of the extract of degenerated pancreas was destroyed by boiling in neutral or acid solution or by incubating for two hours at body temperature with pancreatic juice.

In later experiments, it was found that the pancreas of foetal calves of under five months development did not contain proteolytic enzymes, thus confirming the observations of Ibrahim (11). By extracting such foetal pancreatic tissue, a highly potent and readily procurable preparation was obtained. Besides affording a much more practicable method for securing larger quantities of extracts, this result demonstrated that the active principle is essentially the same from whatever animal it is prepared. A method

Frederick Grant Banting and Charles Best

Benzyloxycarbonyl Protection 1932

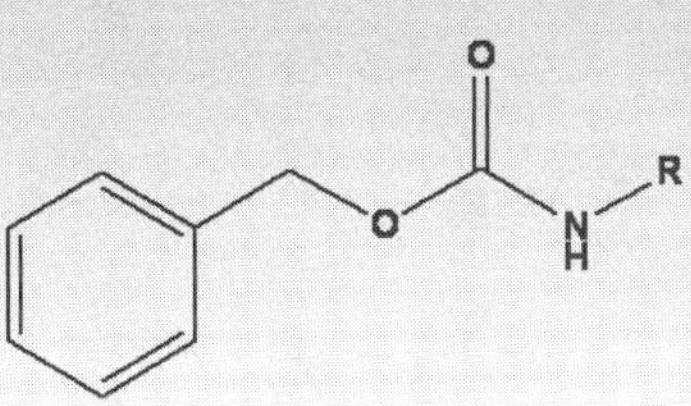

R = amino acid

Bergmann

$$C_6H_5CH_2OCOCl + H_2N\underset{|\atop R}{C}HCOOR' \xrightarrow[(H_2O)]{NaOH^*,\ 0\text{-}20\,^\circ C} C_6H_5CH_2OCONH\underset{|\atop R}{C}HCOOR'$$

* $CHCl + (C_2H_5)_3N$, $CH_3COOC_2H_5 + MgO$, $CHCl_3 + H_2O + Na_2CO_3$

Zervas

Max Bergmann and Leonidas Zervas

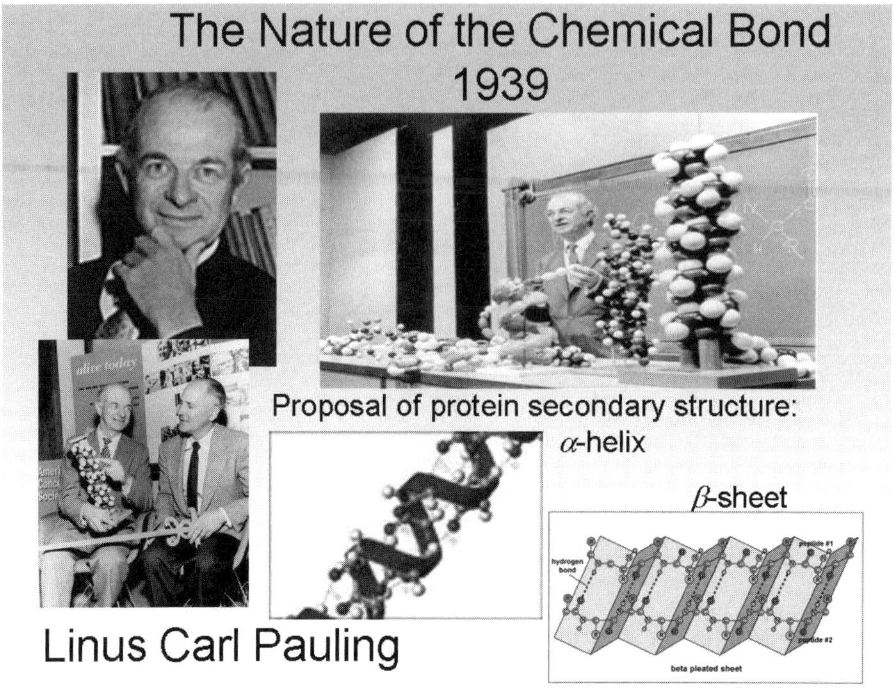
The Nature of the Chemical Bond
1939
Proposal of protein secondary structure:
α-helix
β-sheet
hydrogen bond
beta pleated sheet
Linus Carl Pauling

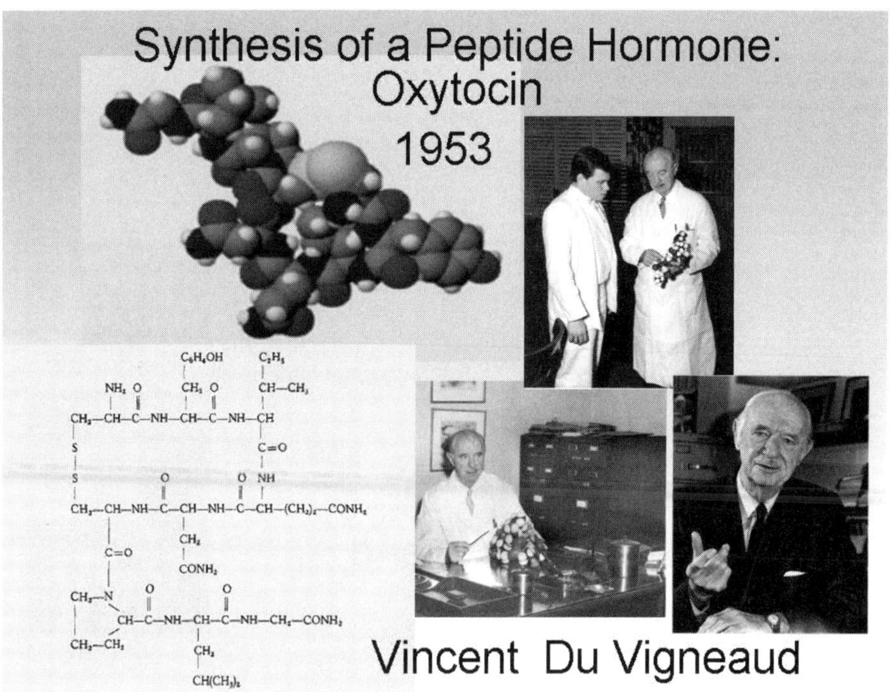
Synthesis of a Peptide Hormone:
Oxytocin
1953
Vincent Du Vigneaud

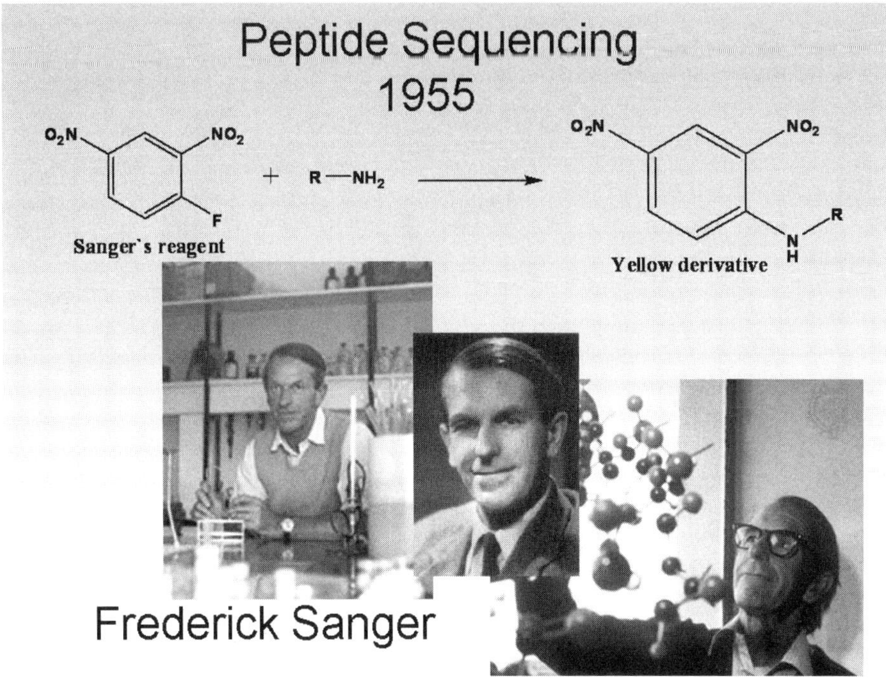
Peptide Sequencing
1955
O₂N
NO₂
F
+ R—NH₂
Sanger's reagent
O₂N
NO₂
N
H
R
Yellow derivative
Frederick Sanger

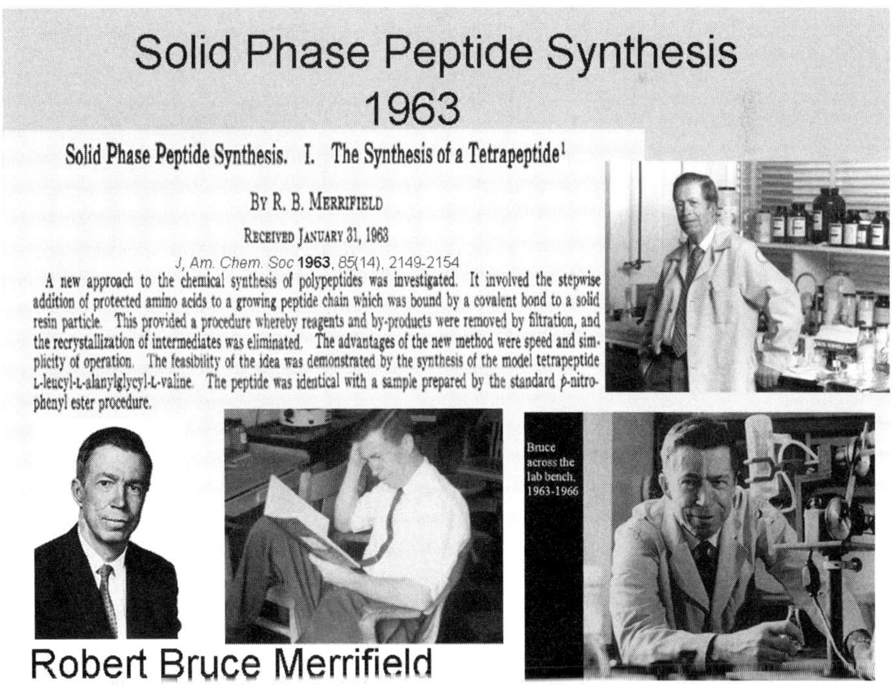
Solid Phase Peptide Synthesis
1963
Solid Phase Peptide Synthesis. I. The Synthesis of a Tetrapeptide[1]
By R. B. Merrifield
Received January 31, 1963
J, Am. Chem. Soc 1963, 85(14), 2149-2154
A new approach to the chemical synthesis of polypeptides was investigated. It involved the stepwise addition of protected amino acids to a growing peptide chain which was bound by a covalent bond to a solid resin particle. This provided a procedure whereby reagents and by-products were removed by filtration, and the recrystallization of intermediates was eliminated. The advantages of the new method were speed and simplicity of operation. The feasibility of the idea was demonstrated by the synthesis of the model tetrapeptide L-leucyl-L-alanylglycyl-L-valine. The peptide was identical with a sample prepared by the standard p-nitrophenyl ester procedure.
Bruce across the lab bench, 1963-1966
Robert Bruce Merrifield

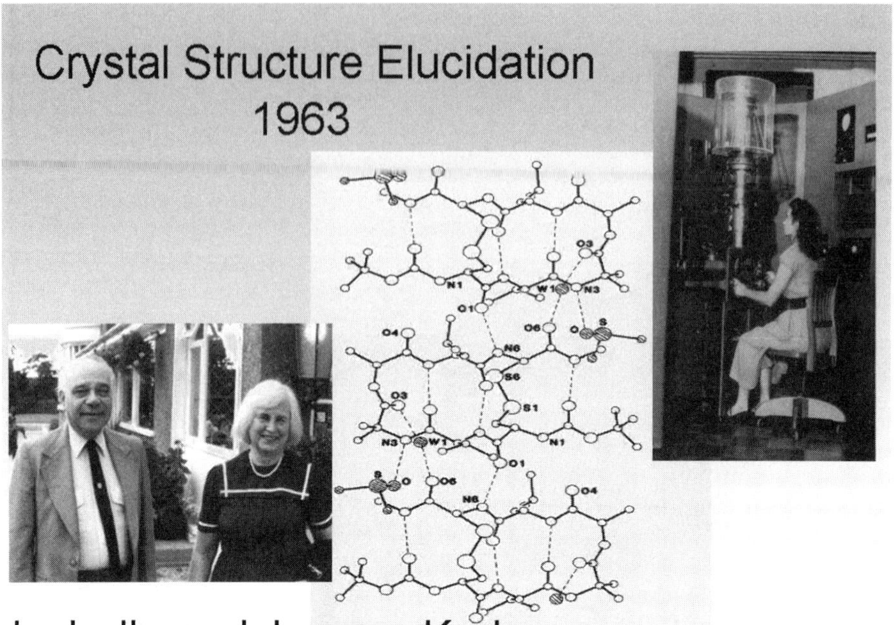

Isabella and Jerome Karle

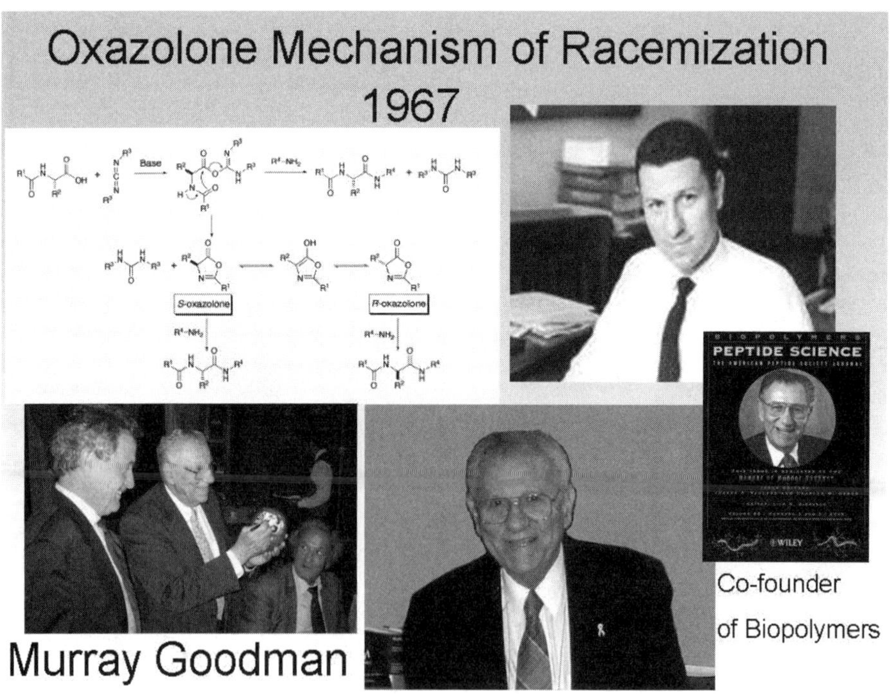

Birth of the American Peptide Society
Symposium 1968
Yale University
Chairs:
Saul Lande
Yale University
Boris Weinstein
U. Washington-Seattle

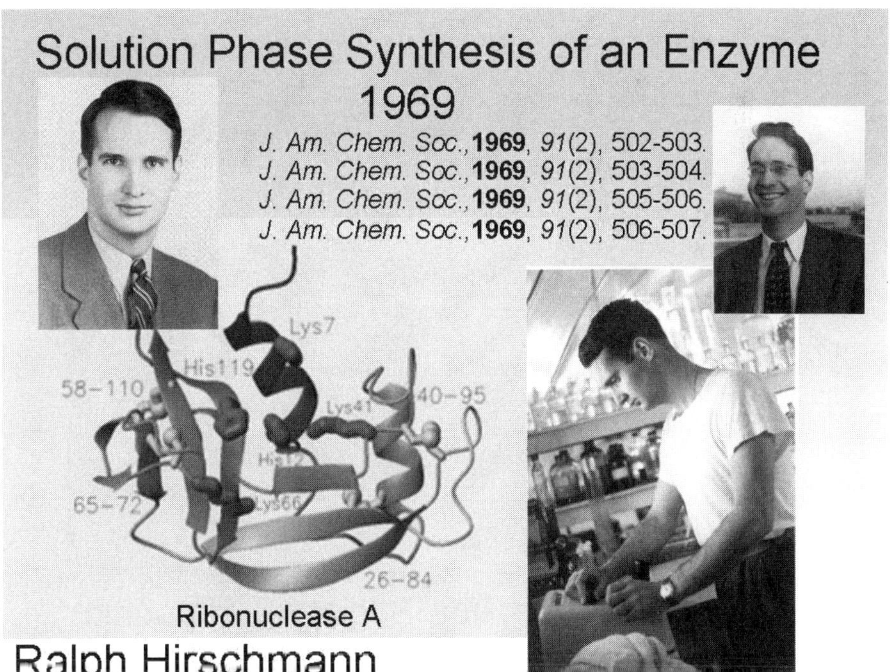
Solution Phase Synthesis of an Enzyme
1969
J. Am. Chem. Soc., 1969, 91(2), 502-503.
J. Am. Chem. Soc., 1969, 91(2), 503-504.
J. Am. Chem. Soc., 1969, 91(2), 505-506.
J. Am. Chem. Soc., 1969, 91(2), 506-507.
Lys7
His119
58–110
Lys41
40–95
His12
65–72
Lys66
26–84
Ribonuclease A
Ralph Hirschmann

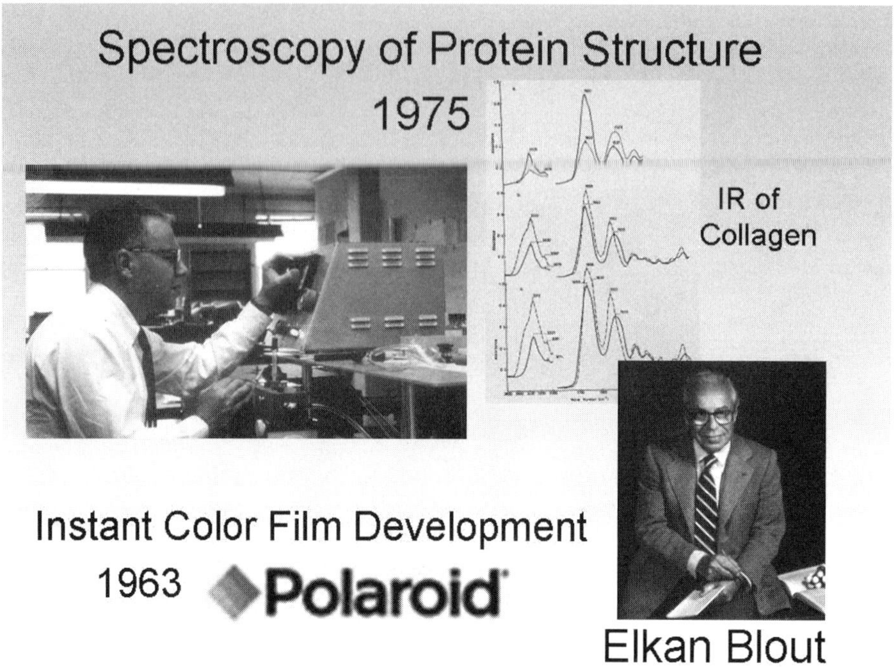
Spectroscopy of Protein Structure
1975
IR of
Collagen
Instant Color Film Development
1963
Polaroid
Elkan Blout

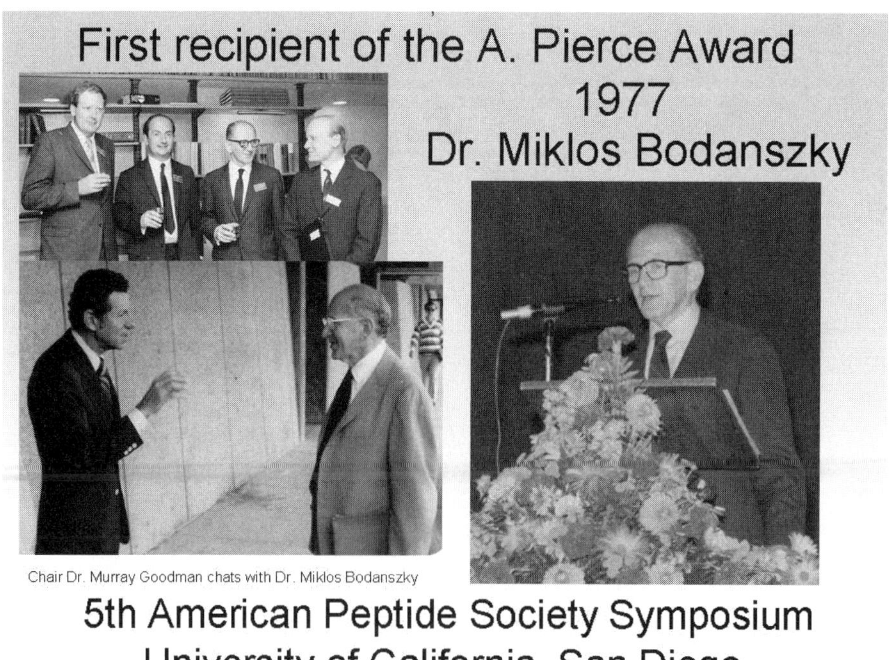
First recipient of the A. Pierce Award
1977
Dr. Miklos Bodanszky
Chair Dr. Murray Goodman chats with Dr. Miklos Bodanszky
5th American Peptide Society Symposium
University of California, San Diego

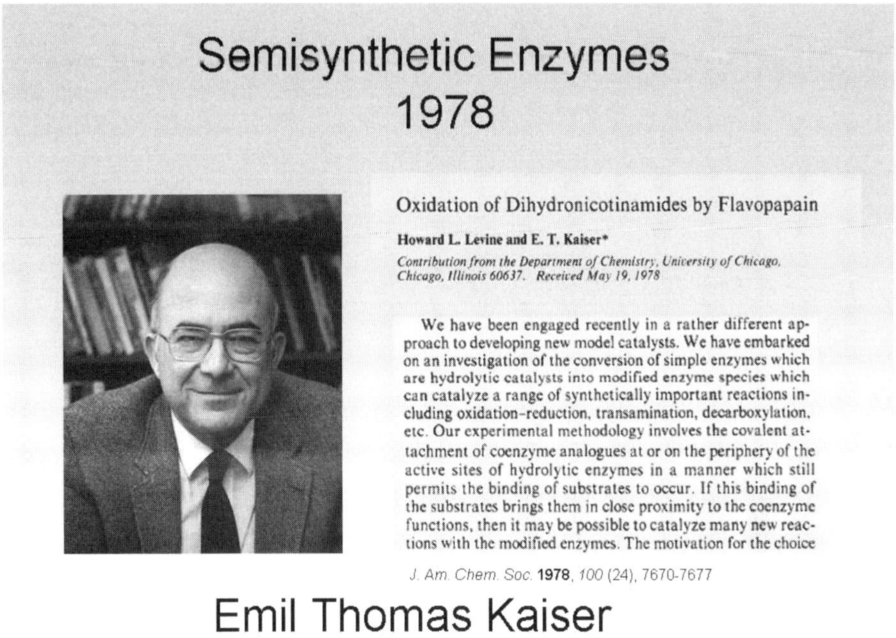
Semisynthetic Enzymes
1978
Oxidation of Dihydronicotinamides by Flavopapain
Howard L. Levine and E. T. Kaiser*
Contribution from the Department of Chemistry, University of Chicago, Chicago, Illinois 60637. Received May 19, 1978
We have been engaged recently in a rather different approach to developing new model catalysts. We have embarked on an investigation of the conversion of simple enzymes which are hydrolytic catalysts into modified enzyme species which can catalyze a range of synthetically important reactions including oxidation-reduction, transamination, decarboxylation, etc. Our experimental methodology involves the covalent attachment of coenzyme analogues at or on the periphery of the active sites of hydrolytic enzymes in a manner which still permits the binding of substrates to occur. If this binding of the substrates brings them in close proximity to the coenzyme functions, then it may be possible to catalyze many new reactions with the modified enzymes. The motivation for the choice
J. Am. Chem. Soc. 1978, 100 (24), 7670-7677
Emil Thomas Kaiser

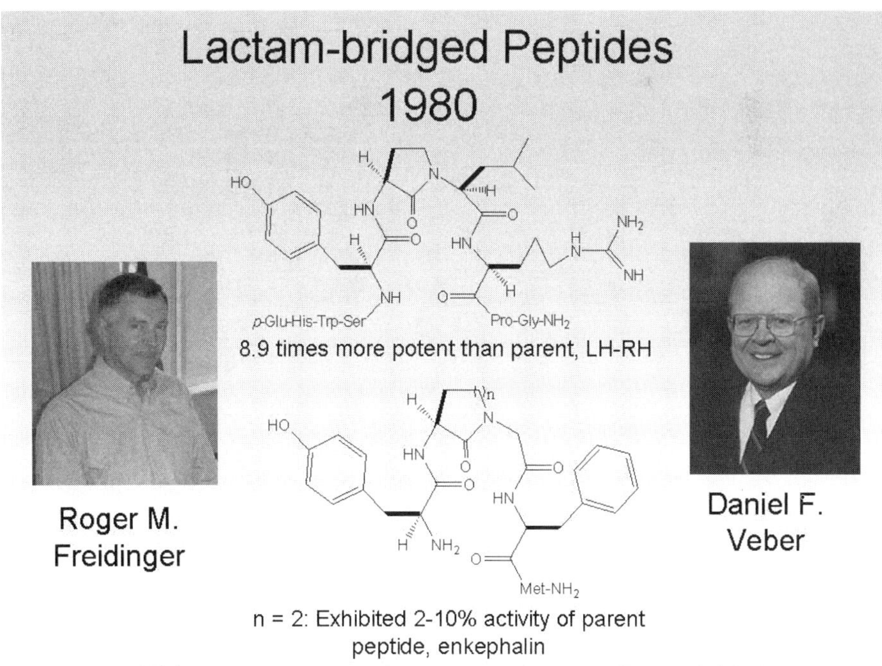
Lactam-bridged Peptides
1980
HO
HN
NH
p-Glu-His-Trp-Ser
Pro-Gly-NH2
8.9 times more potent than parent, LH-RH
n
Met-NH2
Roger M. Freidinger
Daniel F. Veber
n = 2: Exhibited 2-10% activity of parent peptide, enkephalin
Freidinger R.M.; Veber, D.F.; Perlow, D.S.; Brooks, J.R.; Saperstein, R.; Science. 1980 Nov 7;210(4470):656-8.

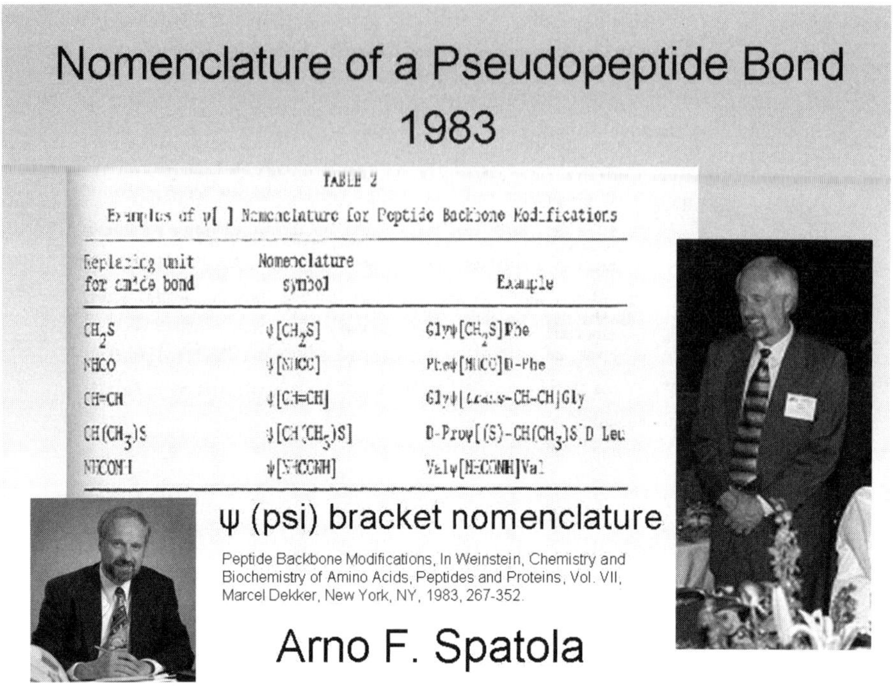
Nomenclature of a Pseudopeptide Bond
1983
TABLE 2
Examples of ψ[ ] Nomenclature for Peptide Backbone Modifications
Replacing unit for amide bond | Nomenclature symbol | Example
CH2S | ψ[CH2S] | Glyψ[CH2S]Phe
NHCO | ψ[NHCO] | Pheψ[NHCO]D-Phe
CH=CH | ψ[CH=CH] | Glyψ[trans-CH=CH]Gly
CH(CH3)S | ψ[CH(CH3)S] | D-Proψ[(S)-CH(CH3)S]D Leu
NHCONH | ψ[NHCONH] | Valψ[NHCONH]Val
ψ (psi) bracket nomenclature
Peptide Backbone Modifications, In Weinstein, Chemistry and Biochemistry of Amino Acids, Peptides and Proteins, Vol. VII, Marcel Dekker, New York, NY, 1983, 267-352.
Arno F. Spatola

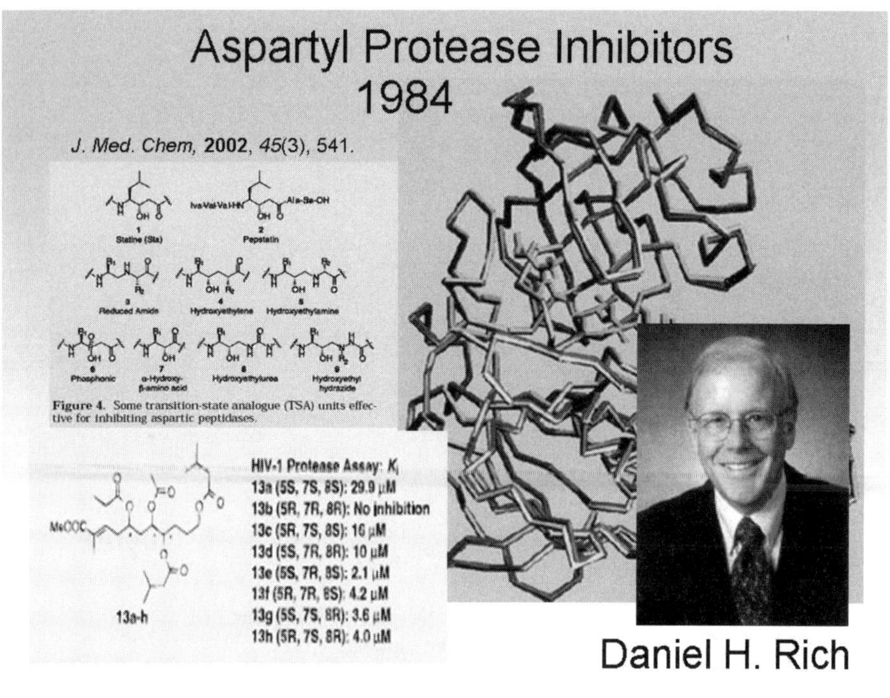
Aspartyl Protease Inhibitors
1984
J. Med. Chem, 2002, 45(3), 541.
1 Statine (Sta)
2 Pepstatin
3 Reduced Amide
4 Hydroxyethylene
5 Hydroxyethylamine
6 Phosphonic
7 α-Hydroxy-β-amino acid
8 Hydroxyethylurea
9 Hydroxyethyl hydrazide
Figure 4. Some transition-state analogue (TSA) units effective for inhibiting aspartic peptidases.
MeOOC
13a-h
HIV-1 Protease Assay: K_i
13a (5S, 7S, 8S): 29.9 μM
13b (5R, 7R, 8R): No inhibition
13c (5R, 7S, 8S): 16 μM
13d (5S, 7R, 8R): 10 μM
13e (5S, 7R, 8S): 2.1 μM
13f (5R, 7R, 8S): 4.2 μM
13g (5S, 7S, 8R): 3.6 μM
13h (5R, 7S, 8R): 4.0 μM
Daniel H. Rich

APS MEETING IN CANADA

1985

Mark Riemen, Maurice Manning, Bruce Merrifield, Cecilia Unson

Co-chairs: Charles Deber, U. of Toronto and Dr. Kenneth D. Kopple, Illinois Institute of Technology

9th American Peptide Society Symposium, Toronto, Ontario, Canada

The "T-BAG TECHNIQUE"

1985

George B. Kauffman (left) and R.A. Houghten (right)

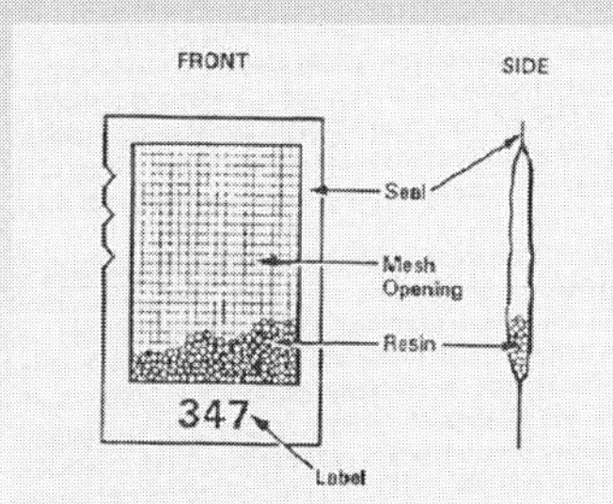

FIG. 1. Illustration of a mesh packet containing resin.

synthesis of up to billions of compounds in mixture-based formats

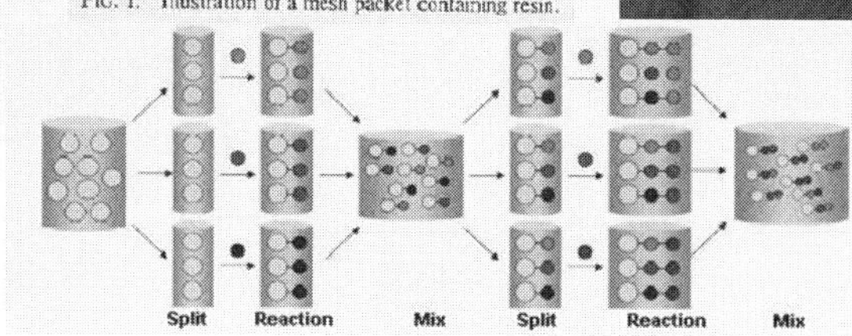

PNAS 1985, 82, 5131-5135

Richard A. Houghten

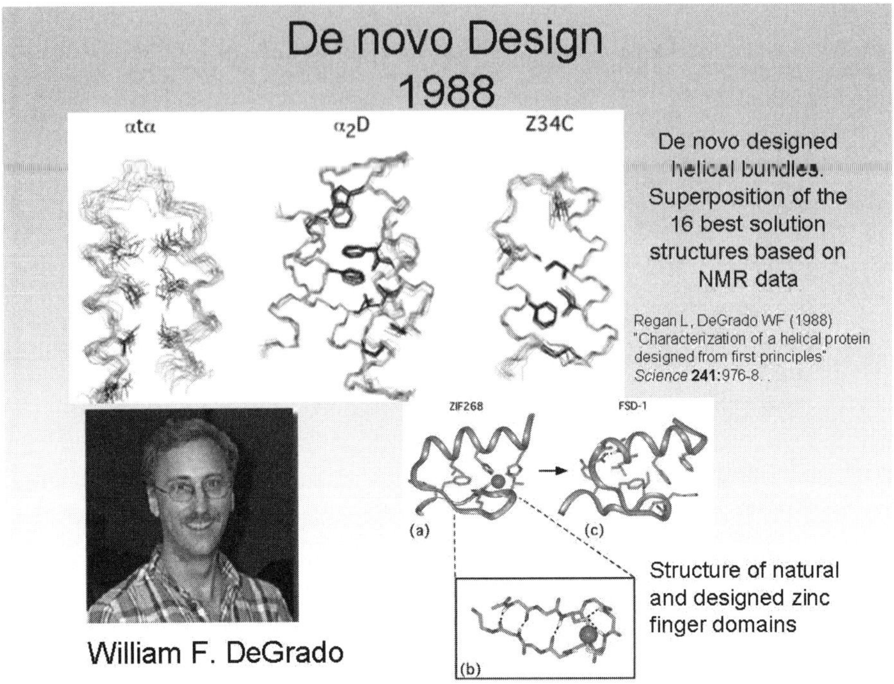
De novo Design
1988
αtα
α2D
Z34C
De novo designed helical bundles. Superposition of the 16 best solution structures based on NMR data
Regan L, DeGrado WF (1988) "Characterization of a helical protein designed from first principles" Science 241:976-8. .
ZIF268
FSD-1
(a)
(c)
(b)
Structure of natural and designed zinc finger domains
William F. DeGrado

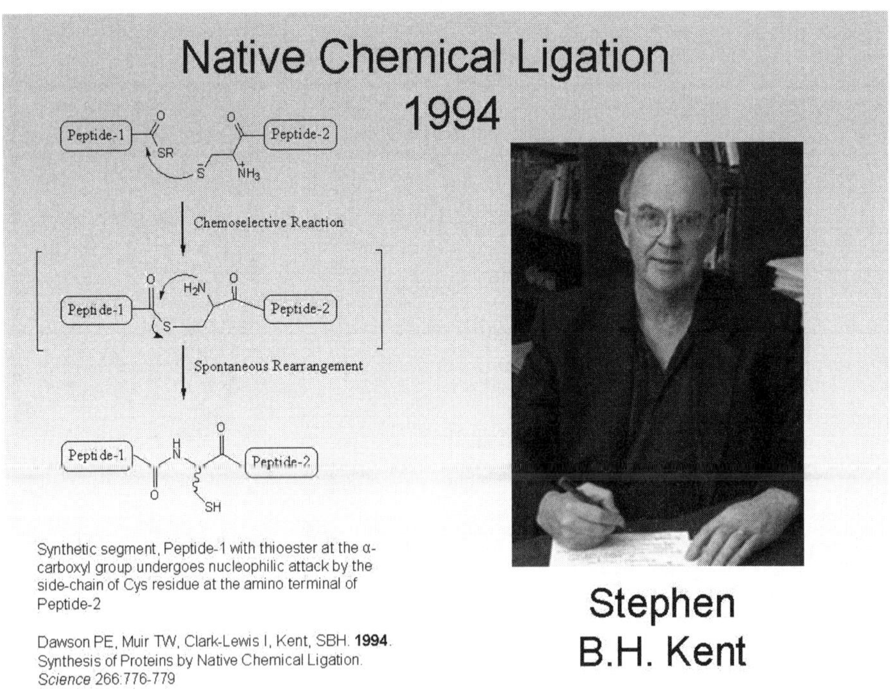
Native Chemical Ligation
1994
Peptide-1
SR
Peptide-2
S
NH3
Chemoselective Reaction
H2N
Spontaneous Rearrangement
SH
Synthetic segment, Peptide-1 with thioester at the α-carboxyl group undergoes nucleophilic attack by the side-chain of Cys residue at the amino terminal of Peptide-2
Dawson PE, Muir TW, Clark-Lewis I, Kent, SBH. 1994. Synthesis of Proteins by Native Chemical Ligation. Science 266:776-779
Stephen B.H. Kent

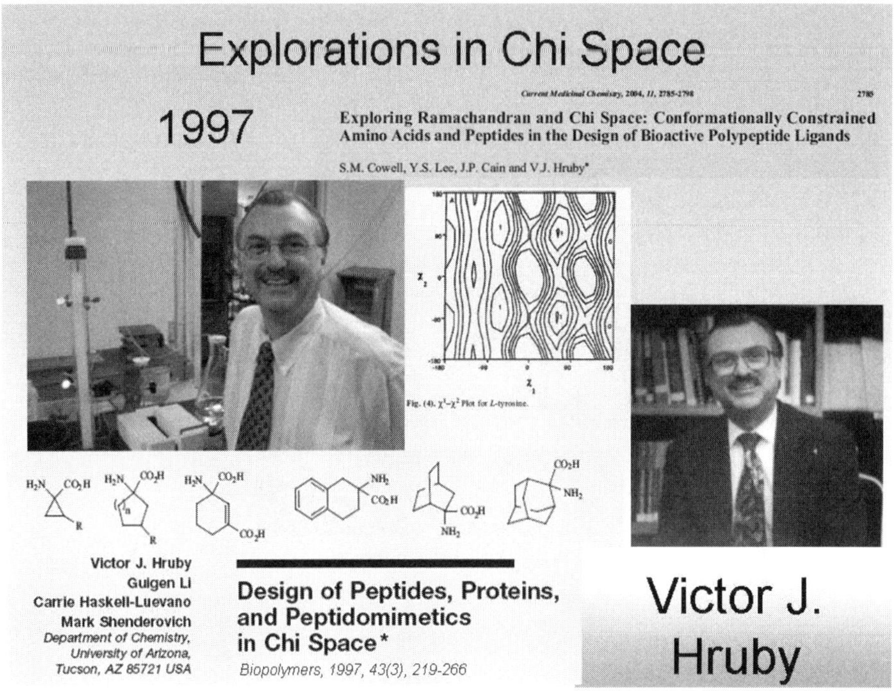
Explorations in Chi Space
1997
Exploring Ramachandran and Chi Space: Conformationally Constrained Amino Acids and Peptides in the Design of Bioactive Polypeptide Ligands
S.M. Cowell, Y.S. Lee, J.P. Cain and V.J. Hruby*
Victor J. Hruby
Guigen Li
Carrie Haskell-Luevano
Mark Shenderovich
Department of Chemistry, University of Arizona, Tucson, AZ 85721 USA
Design of Peptides, Proteins, and Peptidomimetics in Chi Space*
Biopolymers, 1997, 43(3), 219-266
Victor J. Hruby

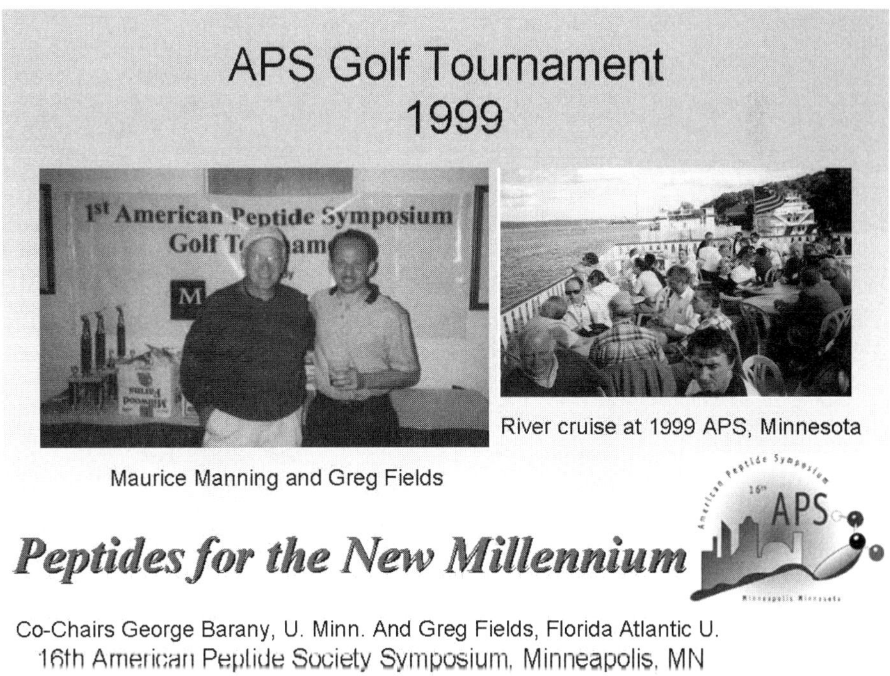
APS Golf Tournament
1999
River cruise at 1999 APS, Minnesota
Maurice Manning and Greg Fields
Peptides for the New Millennium
Co-Chairs George Barany, U. Minn. And Greg Fields, Florida Atlantic U.
16th American Peptide Society Symposium, Minneapolis, MN

Inaugural Makineni Lecture

2003

Prof. James P. Tam

"Peptide Revolution: Genomics, Proteomics & Therapeutics"

Murray Goodman, James Tam

and co-chairs Tomi Sawyer and Michael Chorev

18th American Peptide Society Symposium Boston, Mass.

Female President

2005

Jane Aldrich

President Elect Jane Aldrich and President Roger Freidinger in Boston 2003

Understanding Biology Using Peptides

Co-chairs Jeff Kelley, Scripps Res. Inst. and Tom Muir, Rockefeller U.

19th American Peptide Society Symposium, San Diego, CA

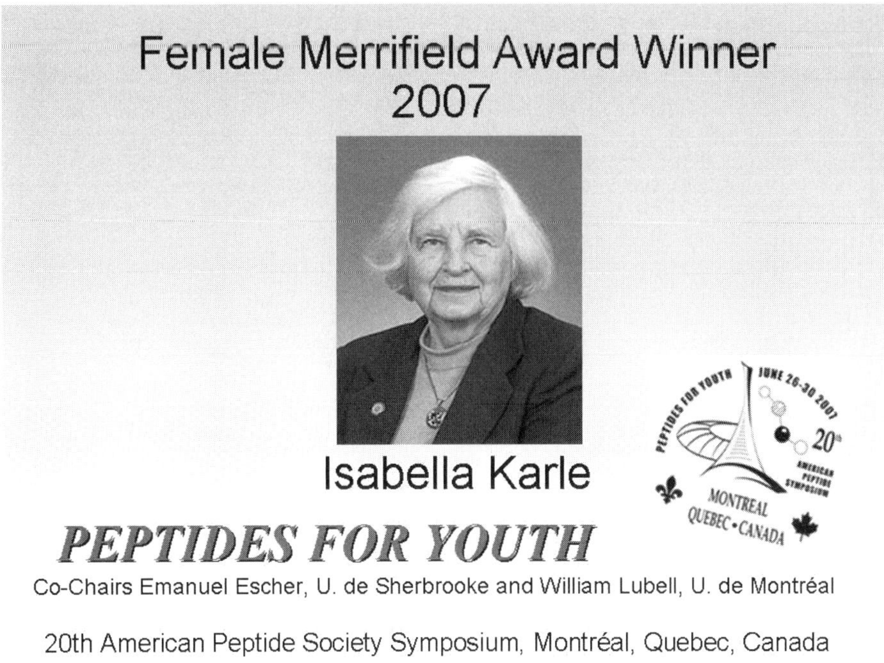
Female Merrifield Award Winner
2007
Isabella Karle
PEPTIDES FOR YOUTH
JUNE 26-30 2007
20th
MONTREAL
QUEBEC • CANADA
PEPTIDES FOR YOUTH
Co-Chairs Emanuel Escher, U. de Sherbrooke and William Lubell, U. de Montréal
20th American Peptide Society Symposium, Montréal, Quebec, Canada

Peptides for Youth
2007
Molecules of Life
http://smu.ca/~mlp/index.html
expanding minds through art and science

20th APS Student Volunteers

Université de Montréal

Carine Bourguet (PhD)
Wajih Ben Tahar (MSc)
Caroline Proulx (MSc)
Philippe Deaudelin (MSc)
Gil Fridkin (PostDoc)
Nicolas Boutard (PostDoc)
Hassan Iden (MSc)
Tanya Godina (MSc)
Aurélie Dörr (PhD)
Nicolas Genest (MSc)
David Sabatino (PostDoc)
Tarek Kassem (PostDoc)
Ali Salhi (MSc)

MIMETOGEN PHARMACEUTICALS

Teresa Lama PhD

UNIVERSITÉ DE SHERBROOKE

Dany Fillion (PhD)
Brian Holleran (PhD)
Martin Clément (PhD)
Jason Arsenault (PhD)
Raymond Hamel Jr. (PhD)
Ivana Domazet (MSc)
Ewa Wieczerzak (PostDoc)
Beata Jastrzebska (PostDoc)
Luta-Luse Basambombo (BSc)
Anne-Marie Bellefleur (BSc)

IBMM
Institut des Biomolécules Max Mousseron

Yann Brouillette (PhD)
Luisa Ronga (PhD)

Portraits of the Pioneers of the American Peptide Society

Carine B. Bourguet[1], Aurélie Dörr[1], Tanya Godina[1], Caroline Proulx[1], Gil Fridkin[1], Andrew Jamieson[1], David Sabatino[1], Tarek Kassem[1], Nicolas Boutard[1], Jason Arsenault[2], Luisa Ronga[3], Yann Brouillette[3], Maria Bednarek,[4] John C. Tolle[5] and William D. Lubell[1]

[1]*Département de Chimie, Université de Montréal, Montréal, Québec H3C 3J7, Canada;*
[2]*Department de Chimie, Université de Sherbrooke, Sherbrooke, Québec, Canada;*
[3]*Département de Chimie, Université Montpellier I*
[4]*Merck Research Laboratories, R50G-140, PO Box 2000, Rahway, NJ 07067*
[5] *Global Pharmaceutical Products Division Process R&D R450 Abbott Laboratories. Abbott Park, 1401 Sheridan Road North Chicago, IL 60064*

Introduction

"Peptides for Youth", the theme of the 20th American Peptide Symposium sought to celebrate our Society by recognizing the achievements of its pioneers and by highlighting the promise of its young investigators. With the goal of achieving the former effectively, the idea of a series of posters entitled "The Portraits of the Pioneers" was conceived. Volunteers from the academic and industrial sectors of our society prepared ten portraits to celebrate leaders in the field of peptide science and their accomplishments. Employing relevant photos, images and text, these posters elaborated on the discoveries, activities, scientific education, awards and distinctions of a representative ten pioneers of our society.

The choices of the representative pioneers of the APS came naturally from the volunteers, and included two contemporary ground-breakers (Isabella Karle and Ralph Hirschmann), five recently departed leaders (Miklos Bodansky, Elkan Blout, Murray Goodman, Bruce Merrifield and Arno Spatola), as well as three legends (Frederick Grant Banting, Thomas Kaiser and Vincent Du Vigneaud). A head volunteer provided instructions, coordinated the received submissions and unified the poster format such that each portrait background was given a different color to provide a rainbow spectrum of the history of the APS. The portraits were arranged in front of the multi-coloured stain glass windows of the entrance hall of the Palais de Congrès and served as a leader toward the archway entrance of the exhibition hall. Throughout the meeting, participants gathered in this historical gallery of portraits of our Society's mentors, and the past was retold as former students of these leaders related, to their own students, memories of the early days of peptide science.

The essence of this display in the portrait gallery of the society's pioneers has now been distilled into this special proceeding in which the ten posters have been edited and reformatted. The spirit of the posters has been maintained to savour the memory of their display at the 20th American Peptide Symposium.

Major Sir Frederick Grant Banting

1891 - 1941

Pioneering Discoveries and Activities

Dr. Banting participated in WWI as a medical officer who was awarded a Military Cross for his act of bravery under fire. Upon return from the war, this young Physician did not have many patients and turned to painting to overcome his financial difficulties. He studied orthopedic medicine and started his residency while teaching orthopedic medicine part-time before becoming a Lecturer in Pharmacology for extra money as his attempt to sell his art-work failed. This came as no surprise as Banting failed his first year in Arts before turning to medicine. Teaching led to a greater income than selling paintings and art became a pass-time.

Banting in the Canadian Army Medical Cops

Banting was a medical practitioner and teacher, who is best known for his work on insulin. He devised a procedure to isolate the active compound from the pancreas for the control of blood sugar levels in diabetics, an idea he got during a sleepless night of reading medical journals. Banting began to work on this idea at the University of Toronto in Professor J.J.R. Macleod's laboratory. Macleod, who was skeptic about the project, appointed the young Dr Charles Best, a biochemist, to work with Banting. The Nobel Prize was awarded to Banting and Macleod for the discovery of insulin; later, Banting shared his half of the prize money with Best. In his time, Dr. Banting was the most famous Canadian. He kept in touch with his original patients and with doctors of many whose lives have been saved by his discovery. He also returned to his passion for the arts by joining A.Y. Jackson (a Group of Seven artist) on an Arctic voyage.

Banting and Best with a diabetic dog

When WWII came, Banting served by studying military research and bacterial warfare for the Canadian Forces. During a voyage to Great Britain for a mission, his life was abruptly ended in a plane crash.

Scientific Education

Honorary Ph.D. convocation from McGill University

- M.B. from University of Toronto (1916)
- M.D. and gold medal U. of Toronto (1922)
- L.L.D. Queen's University (1923)
- D.Sc. University of Toronto (1923)
- Honorary Ph.D. McGill University (1939)

Awards and Distinctions

- Military Cross from the Canadian Army Medical Corps during WWI, appointed in 1919,
- Appointed Senior Demonstrator in Medicine at the University of Toronto in 1922,
- Reeve Prize of the University of Toronto 1922,
- Elected to the Banting and Best Chair of Medical Research in 1923 (endowed by the Legislature of the Province of Ontario),
- Appointed Honorary Consulting Physician to the Toronto General Hospital, the Hospital for Sick Children, and the Toronto Western Hospital in 1923,
- Nobel Prize in Physiology or Medicine for 1923 (shared with Macleod),
- Life Annuity of $7,500 appointed by the Canadian Parliament in 1923,
- Knighted Knight Commander of the Order of the British Empire by King George V in 1934.

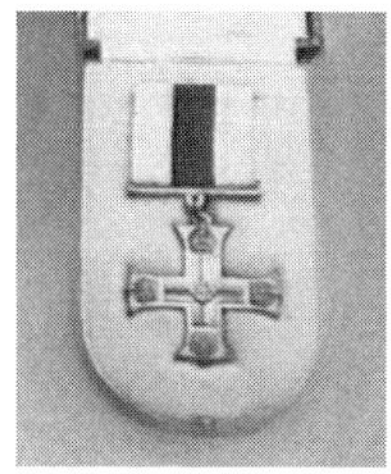

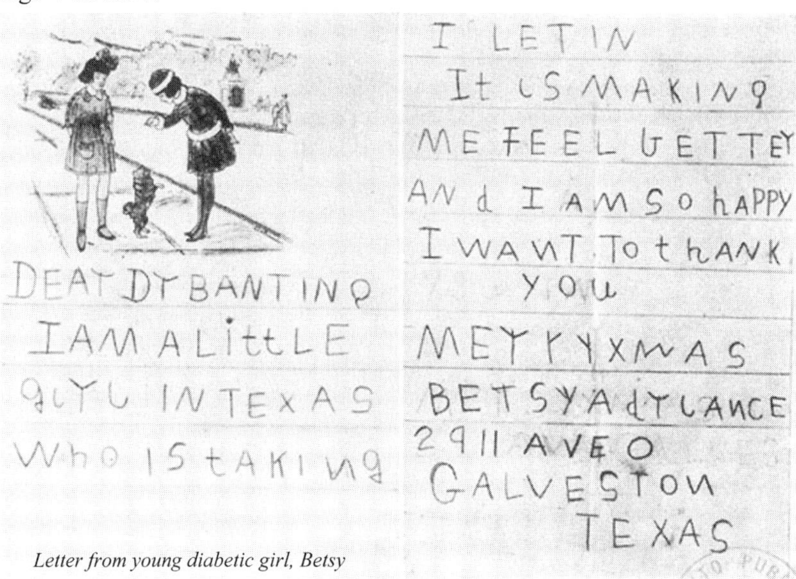

DEAr Dr BANTING
IAM A LittLE
girL IN TEXAS
Who IS tAKING
I LET IN
It IS MAKING
ME FEEL bETTEr
ANd I AM So hAPPY
I WANT To thANK
YOU
MErry XMAS
BETSY AdrLANCE
2911 AVE O
GALVESTON
TEXAS

Letter from young diabetic girl, Betsy

Professor Vincent Du Vigneaud

1901 - 1978

Pioneering Discoveries and Activities

Early work on the structure determination of the hormone Insulin (1930s), nurtured Du Vigneaud to become interested in biologically active sulfur containing compounds. In particular, he continued studies on sulfur containing peptides and completed the first total syntheses of biologically active, oxytocin and vasopressin (1953). He opened the door for synthetic peptide use in medicine and pioneered the use of SAR studies with the discovery that small changes in peptide structure may lead to dramatic differences in bioactivity.

"A scientific problem worth pursuing is worth pursuing for a lifetime."

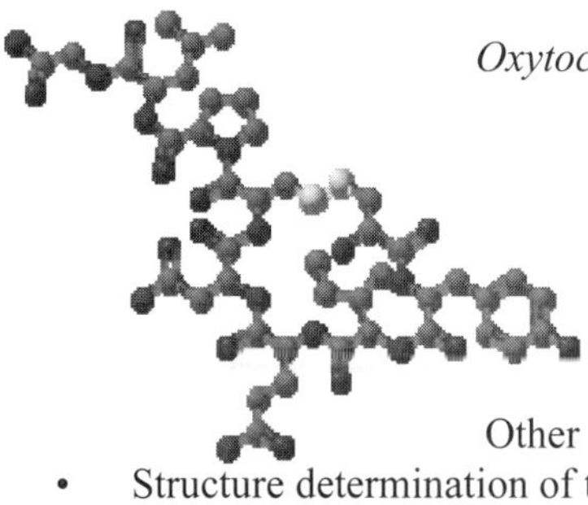

Oxytocin Structure

Other Research Interests:

- Structure determination of the sulfur containing vitamin, biotin (1940s).
- Synthesis of penicillin (1946).
- Intermediary metabolism, trans-methylation and trans-sulfuration.

Scientific Education

Professor DuVigneaud and his Wife Zelia

- Received his B.Sc. (1923) and M.Sc. (1924) from the Univ. of Illinois under the supervision of Professor C. S. Marvel.
- Performed doctorate studies with Professor J. R. Murlin at the School of Medicine at Rochester University and received his Ph.D. in 1927.
- Postdoctoral studies at Johns Hopkins University Medical School, Edinburgh University Medical School and London University College Hospital.
- He began his academic career at the University of Illinois under Professor W. C. Rose, and then in 1932 became head of the Biochemistry department at George Washington University School of Medicine. In 1938 he moved to Cornell University Medical School where he became head of the Biochemistry Department.

Awards and Distinctions

1955 Nobel Prize Winners: Du Vigneaud (Chemistry), Polykarp Kusch(physics), Willis Lamb (physics), Hugo Therell (med), Halldor Kiljan Laxness (lit).

Du Vigneaud received the Noble Prize in Chemistry in 1955 "for his work on biochemically important sulphur compounds, especially for the synthesis of a polypeptide hormone." Among honorary fellowships bestowed on Du Vigneaud include the Royal Society of Edinburgh, the Chemical Society and the Royal Institute of Chemistry, London. He was elected to the board of trustees of the Rockefeller Institute and the National Institute of Arthritis and Metabolic Diseases.

Acknowledgements

- Prof. Victor J. Hruby for quotes and information regarding du Vigneaud's work on oxytocin and vasoprossin.
- Photographs Courtesy of Medical Center Archives of NewYork-Presbyterian/Weill Cornell

Professor Miklos Bodanszky

1915 -2007

Pioneering Discoveries and Activities

Author of more than 250 scientific publications, inventor on about 60 patents, invited speaker at numerous national and international meetings, Professor Bodanszky authored several of the principal textbooks in the peptide field: "Peptide Synthesis", "Principles of Peptide Synthesis, The practice of Peptide Synthesis, "Peptide Chemistry" and "The World of Peptides".

His research led to many important contributions in the field of peptide science including:

- Methodology for peptide synthesis: coupling reagents, protecting groups, techniques to minimize side reactions and racemization, the stepwise approach to peptide chain elongation and the principle of excess
- Peptide hormones : structure elucidation, synthesis, function and conformation (oxytocin, vasopressin, bradykinin, cholecystokinin secretin, VIP)
- Structure elucidation and synthesis of (peptide) antibiotics
- Process research : new methods for the synthesis of known medicines, new pathways for the isolation of streptomycin and oxytetracyclin

Scientific Education

- Diploma of Chemical Engineering, Technical University of Budapest, Hungary
- Technical Director and Research Chemist, Dr. Egger Pharm. Manuf.
- Doctorate from Technical University of Budapest, Hungary

- Department Head, Research Institute for Medicinal Chemistry, Budapest, Hungary
- In 1956, left Hungary, immigrated to the U.S.
- Research Associate with Vincent du Vigneau (Nobel Laureate), Cornell University Medical College, New York, 1957 - 1959
- Head of Peptide Group, The Squibb Institute for Medical Research, New Jersey, 1959 - 1966
- Mabery Professor of Research in Chemistry and Professor of Biochemistry, Case Western Reserve University and CWRU Medical School, Cleveland, 1966 - 1963

Prizes

Recognized for Excellence in Teaching by his fellow colleagues and his many students at Case Western Reserve University.

Recipient of several prizes :

- The Pierce (Merrifield) Award
- The Morley Medal, Cleveland Section of the American Chemical Society
- A. von Humboldt Award (Germany)
- Member of NIH Study Sections
- Foreign Member of the Hungarian Academy of Sciences

Professor Elkan R. Blout

1919-2006

Pioneering Discoveries and Activities

"His path-breaking work in chemistry helped illuminate the large, complex structure of proteins, critical to understanding their function.

It is for this work, as well as his devotion to the scientific enterprise of this nation that he received the National Medal of Science, the nation's highest scientific honor."- President George H.W. Bush on Nov. 13, 1990.

- ➢ Professor Blout pioneered the use of spectroscopy for elucidation of the structure and conformation of complex protein/peptide motifs, including:
- analysis of the UV absorption and optical rotatory dispersion of apomyoglobin and myoglobin

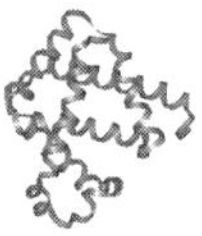

Ribbon Structure of Apomyoglobin

SC Harrison, ER Blout *Journal of Biological Chemistry,* **1965**, *240*(1), 299-303.

- study of *β*-turn conformation in cyclic hexapeptides using NMR and CD spectroscopy.

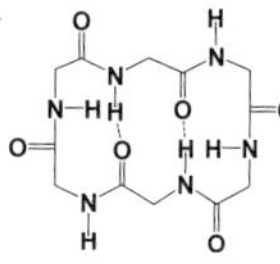

LM Gierasch, CM Deber, V Madison, N-H Niu, ER Blout *Biochemistry,* **1981**, *20*(1), 4730-4738.

- Conformation of Gramicidin

WR Veatch, ET Fossel, ER Blout *Biochemistry*, **1974**, *13*(26), 5249-5226.

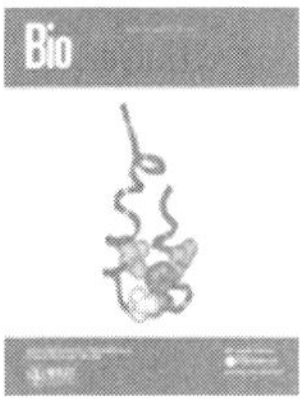

- Blout helped develop instant color film which was introduced by Polaroid Inc. in 1963.
- ER Blout was the co-founder and editor of the journal *Biopolymers*

Scientific Education

Born on July 2, 1919, Blout grew up in Manhattan as an only child. Blout attended DeWitt Clinton High School in the Bronx and earned marks that were high enough to skip three grades. Too young to attend college when he graduated, he enrolled in the Philips Exeter Academy, and after a year, Blout attended Princeton University, becoming one of only twelve Jewish students accepted in 1935, at a time of discrimination from both the University and the students. He graduated in 1939, and in 1942, Blout received his Ph.D. in chemistry from Columbia University.

After completing a research stage as a post-doctoral fellow with Louis Feiser and R.B. Woodward at Harvard, Blout began work with Polaroid Co. in 1943. He joined Harvard University Medical School in 1961, where he stayed until 1991. That year, Blout began work at the Food and Drug Administration (FDA) until 1998. He was active as consultant and board member of various profit and non-profit organizations.

Awards and Distinctions

1942: National Research Council Fellow, Harvard University,

1962: A.M. (honorary), Harvard University,

1976: D.Sc. (honorary), Loyola University,

1982: Honor Scroll Award, Massachusetts Institute of Chemists, Division of the Am. Inst. of Chemists,

1990: National Medal of Science

1991: Ralph F. Hirschmann Award in Peptide Chemistry, ACS

Now protein folding and protein aggregation are considered biologically crucial problems. But nearly fifty years ago, Blout was working at the biology/chemistry interface before it was popular. It was a really pioneering approach."

- Lila Gierasch, Blout's former graduate student at Harvard, and now professor at the University of Massachusetts on Jan. 23, 2007.

Dr. Blout has been the spirit and chief advocate of scientific excellence in our Agency, "More than anyone else, he has laid sound foundations for the strengthening of FDA's scientific potential that is essential for the continued fulfilment of our public health mission. We'll miss his congenial personality, but his innovative ideas and guidance will remain always a significant part of this Agency."

- Commissioner of the FDA Jane E. Henney, M.D. on Sept. 7, 1999.

Dr. Isabella Karle

1921-

Pioneering Discoveries and Activities

- Karle has published over 200 papers throughout her career.
- In 1963, the "symbolic addition method", which uses X-ray and electron diffraction to determine equal-atom crystal and molecular structures, became one of the most important contributions of this renowned chemist and physicist. Aided by her work, the field of molecular analysis was revolutionized and number of new structures published annually increased rapidly from about 150 to over 10,000.
- Karle was first to determine the structure of many complex organic and inorganic substances, including steroids, alkaloids, toxins, ionophores, and peptides such as: valinomycin (1975), antamanide (1979) and enkephalin.
- In solving high-resolution crystal structures of cyclic and linear peptides, Karle helped to identify the fundamental principles governing peptide conformation.

Isabella and Jerome Karle

Scientific Education

Isabella Karle with original electron diffraction apparatus

- By the age of 22, Isabella Karle had received her B.Sc (1941), M.S (1942) and Ph.D. (1944) degrees in physical chemistry from the University of Michigan. Although teaching assistant positions were exclusively allocated to male doctoral students at the time, Karle was able to pursue her doctoral studies as a recipient of the *American Association of University Women Fellowship.*
- In 1943, Karle became one of the few women to work on the Manhattan Project at the University of Chicago, where she worked on plutonium chemistry.
- In 1946, she began work at the Naval Research Laboratory where she remained head of the NRL X-ray Diffraction Section for the Structure of Matter from 1959 to the 1990s.

Awards and Distinctions

- The National Medal of Science (1995).
- Two of her publications (1963, 1966) receive special recognition by Citation Index Classics for the largest number of citations and greatest influence on life sciences.
- The Annual Achievement Award from Women in Science and Engineering (1968).
- The Federal Woman's Award (1973).
- The Garvan Medal from the American Chemical Society (1976).
- The Chemical Pioneer Award by the American Institute of Chemists (1985).
- The Lifetime Achievement Award from Women in Science and Engineering (1986).
- The Franklin Institute's Bower Award (1993).
- The Hirshman Award (1998).
- The Merrifield Award (2007).

Honorary degree recipients, University of Pennsylvania (1999)

Isabella Karle (1993) receives the Bower Award

Professor Bruce Merrifield

1921-2006

Pioneering Discoveries and Activities

May 26, 1959, an entry in one of Merrifield's lab notebooks describes a revolutionary idea that would create new possibilities in the field of peptide and protein chemistry and greatly stimulated progress in biochemistry, molecular biology, pharmacology and medicine. This entry says:

"There is a need for a ***rapid, quantitative, automatic*** *method for synthesis of* ***long-chain peptides****. A possible approach may be the use of chromatographic columns where* ***the peptide is attached to the polymeric packing*** *and added to by an activated amino acid, followed by removal of the protecting group and with repetition of the process until the desired peptide is built up. Finally, the peptide must be removed from the supporting medium."*

In 1963, he was sole author of the JACS classic paper[1] (the fifth most cited paper in the journal's history) in which he first reported this method and called it "**solid phase peptide synthesis**" (SPPS).

In 1965, with John M. Stewart and Niels Jernberg he started designing the first prototype of an automated solid-phase peptide synthesizer (now belonging to the Smithsonian Institute). "Merrifield never patented the machine and instead preferred to share the concept widely, the sign of a true gentleman." *James Tam*

In the early days of SPPS, Pr. Merrifield's laboratory synthesized **bradykinin**[2], **angiotensin**[3], **desamino-oxytocin**[4], **insulin**[5] and **ribonuclease A**[6], which still represents a challenge ("124 amino acids long, requiring 369 chemical reactions and 11,931 steps of the automated peptide synthesis machine without any intermediate isolation steps"). This work proved that medium size proteins can be synthesized by SPPS, and that the primary sequences of proteins determine their tertiary structures and ultimately their biological activities.

Scientific Education

Bruce Merrifield earned his bachelor's degree in chemistry in 1943 from UCLA,

After his graduation, he worked for a year at the Philip R. Park Research Foundation assisting with growth experiments on synthetic amino acid diets,

He returned to graduate school at the UCLA chemistry department with professor of biochemistry M.S. Dunn to develop microbiological methods for the quantitation of the pyrimidines. He completed his PhD under the guidance of Pr. Dunn in 1949,

He then left for New York City and the Rockefeller University, then known as the Rockefeller Institute for Medical Research. He first worked as an assistant of Dr. D.W. Woolley on peptide and nucleic acid growth factors,

Merrifield became a full professor at Rockefeller University in 1966 and was named John D. Rockefeller Jr. Professor in 1983. He became professor emeritus in 1992 and retired the same year.

Prizes

Receiving Nobel Prize

Pr. Merrifield received the 1984 Nobel Prize in Chemistry. He also received numerous awards for his oustanding contribution in peptide synthesis, including the Lasker Award for Basic Medical Research (1969), the American Chemical Society Award for Creative Work in Synthetic Organic Chemistry (1972), the Nichols Medal (1973), the 2nd Alan E. Pierce Award of the American Peptide Symposium (1979), the Ralph F. Hirschmann Award in Peptide Chemistry from the American Chemical Society, the Josef Rudinger Award (1990) and the Seaborg Medal (1993). He received honorary degrees from the University of Colorado (1969), Uppsala University (1970), Yale University (1971), Newark College of Engineering (1972), the Medical College of Ohio (1972), Colgate University (1977), and Boston College (1984).

Merrifield family Stockholm 1984

"Bruce Merrifield was not only a great scientist, but also a wonderful colleague and mentor. He left a legacy in the scientists he trained who have gone on to become leaders in the peptide and protein science fields. As his former students and associates noted, he was known for his modesty, integrity, and courage; and all who met him were changed" *Jane Aldrich, president of the APS.*

1) R.B. Merrifield *J. Am. Chem. Soc.,* **1963**, 85: 2149. 2) R.B. Merrifield *Biochemistry*, **1964**, 3:1385. 3) G.R. Marshall, R.B. Merrifield *Biochemistry*, **1965**, 4:2394. 4) H. Takashima, V. Du Vigneaud, R.B. Merrifield *J. Am. Chem. Soc.*, **1968**, 90:1323. 5) A. Marglin, R.B. Merrifield *J. Am. Chem. Soc.*, **1966**, 88:5051. 6) B. Gutte, R.B. Merrifield *J. Am. Chem. Soc.*, **1969**, 91 501

Professor Ralf F. Hirshman

1922-

Pioneering Discoveries and Activities

Hirschmann has authored more than 160 papers and holds 100 patents. He is particularly well known for his research on:

- The N-carboxy-anhydride method for peptide coupling
- The successful solution-phase synthesis of ribonuclease in 1969
- The design of potent angiotensin converting enzyme inhibitors
- The rational design of somatostatin analogues
- Hirschmann's research while at Merck & Co., Inc., led to a host of new medicines, such as the anti-parasitic drug Ivermec that is helping to eradicate river blindness in the Third World.
- His early work at Merck led to the discovery of stereoselective control of chemical transformations, an important concept in organic chemistry.

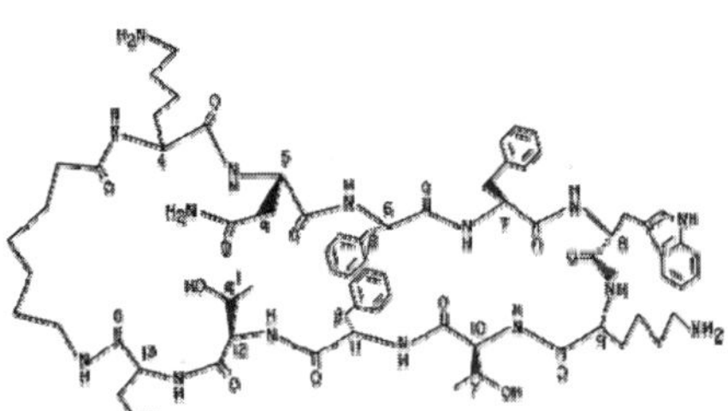

Somatostatine Analog

Scientific Education

- Bachelor's degree from Oberlin College in 1943,
- Doctorate degree in organic chemistry from University of Wisconsin in 1950 under Professor William S. Johnson ,
- He began his career as a steroid chemist in the research laboratories of Merck, where he rose to the position of Senior Vice President of Chemistry.
- He held a concurrent appointment as professor of biomedical research at the Medical University of South Carolina in Charleston
- Executive Director of Medicinal Chemistry at Merck, West Point (PA) in 1974 and Vice President/Senior Vice President of Basic Research (Rahway) in 1976/1978, respectively
- In 1988, he was invited to join the Department of Chemistry of the University of Pennsylvania as the first Research Professor in Chemistry. In 1994 he was appointed the Rao Makineni Professor of Bioorganic Chemistry
- Professional activities included serving on the NIH Medicinal Chemistry A Study Section, on several committees of the NSF, NRC and ACS, on over a dozen Editorial and Advisory Boards, and on the Boards of Trustees of Oberlin College, the Gordon Research Conferences, and others

Awards and Distinctions

Dr. Hirschmann has been recognized with three honorary degrees and three endowed lectureships; three chairs are linked to his name. In addition, the American Chemical Society awards annually the Ralph F. Hirschmann Award in Peptide Chemistry to leaders in peptide science.

- Alan E. Pierce Award—American Peptide Society (1983)
- ACS-Medical Chemistry award (1986)
- Nichols Medal of its New York Section (1988)
- Microbial Chemistry Medal (1993) Carothers Award (1994)
- ACS Alfred Burger Award (1994)
- Philadelphia Organic Chemists' Club Award (1996)
- Edward E. Smissman Bristol-Myers Squibb Award (1999) and the division
- American Chemical Society's Arthur C. Cope Medal (1999)
- National Academy of Sciences' Award for the Industrial Application of Science (1999)
- National Medal of Science (2000)
- Willard Gibbs Medal of the Chicago Section (2002)
- AIC Gold Medal (2003)
- Mendel Medal Recipient (2004)

Professor Murray Goodman

1928 - 2004

Pioneering Discoveries and Activities

Author of nearly 500 journal articles, Professor Goodman served as the Editor-in-Chief of the published 5 volume compendium entitled "*Synthesis of Peptides and Peptidomimetics*", and was Founding Editor of Biopolymers in 1963 and the *Journal of Peptide Science.*
Research Interests, and most referenced publications cited over 1000 times:

- Poly-α-amino acids and oligopeptides as models for understanding fundamental aspects of protein secondary structure,
- Critical chain lengths for alpha-helix formation of a variety of amino acids,
- Fundamental work on racemization during coupling reactions and on the mechanism of N-carboxyanhydride polymerization,
- Use of circular dichroism for elucidating structural aspects of peptides,
- Partial retro-inverso modification of the peptide bond and application on biological target,
- Peptide syntheses via Amino Acid Esters.

Scientific Education

Dr Murray Goodman

- Bachelor's degree from Brooklyn College in 1950.
- Doctorate from University of California, Berkeley, in 1953, in the laboratory of the Nobel Laureate, Professor Melvin Calvin.
- Postdoctoral studies at the Massachusettes Institute of Technology (MIT), in Cambridge, Massachusetts and the University of Cambridge, England.
- In 1956, he began his career at the Polytechnic Institute of Brooklyn, where he became full Professor in 1964 and the Director of the Polymer Research Institute in 1967.
- In 1970, Professor of Chemistry at the University of San Diego, California, (serving as chairman of the Chemistry Department for six years and as acting Provost of Revelle College from 1972-1974).

Awards and Distinctions

Murray, Zelda and Students

Recognized for his excellent pedagogy, he received the UCSD's Chancellor's Associates Recognition Award for Excellence in Graduate Teaching.

Served on and chaired NIH Medicinal Study Sections and was a member of many national and international review panels including IUPAC, AAAS, and the World Health Organization.

Recipient of numerous prizes:

- The Scoffone Medal, University of Padova (1980),
- The Humboldt Professorship, Germany (1986),
- The Pierce (Merrifield) Award (1989),
- The Max Bergmann Medal (1991),
- The Ralph Hirshmann Prize in Peptide Chemistry (1997),
- The Herman F. Mark Polymer Chemistry Award (2000),
- The Arthur C. Cope Scholar Award (2001),
- Foreign member of the Russian Academy of Sciences in 1999.

More than a Peptide Ambassador, a husband, a father, a great teacher

Zelda and Murr

Murray and Sons

Professor Thomas Emile Kaiser

1938-1988

Pioneering Discoveries and Activities

Emil Thomas Kaiser was born in Hungary in 1938. When he was 2 years old, the family moved to US where his father worked in the Armour Pharmaceutical Company. With both of his parents PhD chemists, Professor Kaiser had chemistry in his blood, and with little surprise his scientific career began at an early age. Launching a career in the then novel field of bioorganic chemistry, he started his career at the University of Washington and later moved to University of Chicago. In 1982, he accepted a professorship at Rockefeller University and became Editor of *Bioorganic Chemistry*.

Among other advances that would have distinguished the career of a lesser scientist, including early attempts at site directed mutagenesis in enzymology (1987) and the development of oxime resin, Kaiser was particularly distinguished by two seminal contributions: the development of semisynthetic enzymes and the application of amphiphilic helices.

Semisynthetic Enzymes:

- A strategy for making useful new catalytic systems was developed by the novel combination of the binding properties of one enzyme with the catalytic activity of an unrelated coenzyme. For example, he converted a hydrolytic enzyme into a chimeric one for oxidation-reduction chemistry by attaching a flavin coenzyme at the active site of Papaïn. Kaiser and his coworkers employed semi-synthetic enzymes to couple large peptides such as thiosubtilisin, as well as to ligate activated unprotected peptide segments.

Amphiphilic Proteins

- In pioneering studies on peptide helices, Kaiser demonstrated the importance of the amphiphilic nature for biological activity.

Scientific Education

- He graduated from the University of Chicago at the age of eighteen.
- Received his doctorate when he was only twenty-one from Harvard University, working on cyclic sulfate esters
- In postdoctoral research with Professor E. J. Corey, he proved that sulfone anions can briefly retain their chirality and with Professor Bender, investigated the cinnamoyl intermediates formed on ester hydrolysis by trypsin and chymotrypsin, ,
- Kaiser accepted an assistant professorship at Washington University in St. Louis, and the University of Chicago offered him an assistant professorship in 1963, when he was twenty-five, and a professorship in 1970.
- In 1982 he accepted a professorship at Rockefeller University.

Distinctions

- «He was friendly and smiled easily and often. He was devoted to his wife Bonnie and their children; admired his parents; and valued his graduate students. He was perceived as fair. He had no disputes other than friendly scientific ones. People trusted him; there was never a doubt that he would protect information given to him in confidence or give credit where credit was due.» - F. H. WESTHEIMER (1912-2007).
- The Protein Society sponsors presently the Emil Thomas Kaiser Award which recognizes significant contributions in applying chemistry to the study of proteins.

Professor Arno F. Spatola

1944-2003

Pioneering Discoveries and Activities

- Prof. Spatola authored over 150 papers and 7 patents. He directed numerous graduate and undergraduate students.
- Professor Spatola's major research interests centered around the organic, structural and biological chemistry of peptides, pseudopeptides, cyclic peptides, and their application to the design of biological active ligands with novel properties. In these studies he made critical contributions to cyclic peptide and pseudopeptide synthetic methodology that are used worldwide.
- His contributions to the area of amide bond replacements and pseudopeptides greatly stimulated its development to its current role as a major area in medicinal chemistry, drug development, and peptide and protein science.
- Prof. Spatola originated and promulgated the "psi-bracket (Ψ)" pseudopeptide nomenclature.
- He pioneered studies of metalloprotease inhibitors and inhibitors of nuclear hormone receptor proteins.
- Professor Spatola's ammonium formate catalytic transfer hydrogenation method is used worldwide.

Scientific Education

- **1966 :** Cornell University Ithaca (New York): A.B. degree in Chemistry (major) and Political Science (minor).
- **1968 :** University of Michigan (Ann Arbor): Organic Chemistry M.S.

Prof. Claudio Toniolo
Jackie and Arno Spatola

- **1971 :** University of Michigan (Ann Arbor): Ph.D. "Inert free radicals: ferrocene/thiazole polymers" (Mentor: Prof. Daniel T. Longone).
- **1973 :** University of Arizona (Tucson): Postdoctoral Associate "Peptide synthesis and structure-function studies" (Mentor: Prof. Victor J. Hruby).
- **1973-1978 :** University of Louisville (Kentucky), Department of Chemistry: Assistant Professor.
- **1978-1983 :** University of Louisville (Kentucky), Department of Chemistry: Associate Professor.

Arno and Kimberly Spatola

- **1984-2003 :** University of Louisville (Kentucky), Department of Chemistry: Full Professor.
- **1982 & 1990 :** University of Padova (Italy): Visiting Professor (Prof. Claudio Toniolo).

Awards and Distinctions

- Dr Spatola was a co-founder of the American Peptide Society, served as Secretary of the Society, worked on the editorial board of the Journal of Peptide Research and Letters in Peptide Science, and was an *ad hoc* reviewer for the National Institutes of Health.

27th EPS, 2002

- In **1990** he was a co-chairman of the Gordon Research Conference on Chemistry and Biology of Peptides.
- **2002** Kentucky Science and Engineering Foundation Research and Development Award.
- **2003** President's Award for Outstanding Scholarship, Research and Creativity from the University of Louisville.
- In addition to Dr Spatola's academic career, in **1983** he founded Peptides International, Inc. as an outgrowth of his research interests. This company continues to specialize in fine biochemicals, peptide-related products, and custom synthesis.

In **1996**, Dr Spatola founded a multidisciplinary research and training program; the Institute for Molecular Diversity and Drug Design.

Spatola Research Group 2003

Peptides of Life
Teaching Peptide Science and Art in Elementary School

Teresa Lama,[a] Robert D. Singer[b] and William D. Lubell[a]

[a] *Département de chimie, Université de Montréal, C.P. 6128, Succursale Centre-Ville, Montréal, Québec, Canada, H3C 3J7; lubell@chimie.umontreal.ca;* [b] *Department of Chemistry, Saint Mary's University, Halifax, Nova Scotia Canada B3H 3C3*

Introduction

Elementary school students represent one of the largest and most important populations to which the arts and sciences could be more effectively disseminated. The Molecules of Life Project was conceived to provide a resource for young minds to explore art and science together. In a proof-of-concept demonstration of the power of this combination for teaching elementary students, the "Peptides of Life" project was launched in a second grade class at Oxford Elementary School in Halifax, N.S in Autumn 2006 [1].

A syllabus was prepared consisting of twelve peptides important for basic life functions (Table). Twenty-four university students from departments in the arts and sciences were recruited from Dalhousie, Saint Mary's and Nova Scotia College of Art and Design Universities and challenged with the task of working in teams to prepare and deliver captivating and informative 45-60 min courses on specific peptides. Employing age appropriate discussion, experiments and art projects, the courses inspired and informed the second graders, who became thrilled with learning about molecules in this way that linked science and art together.

Table: Peptides of Life Syllabus

Peptide "etymology"	physiological relevance
1) Gastrin	stomach acid/mucus, digestion
2) Motilin	gastrointestinal motility, bowelmouvement
3) Insulin "island"	blood sugar, body energy
4) Angiotensin "vein tension"	blood pressure
5) Vasopressin	antidiuretic hormone, urination
6) Enkephalin "from the brain"	analgesia, pain relief
7) Bradykinin "slow acting"	sensitizer to inflammation, coughing
8) Melanocite stimulating hormone (MSH)	melanin production, skin pigmentation
9) Growth Hormone Releasing Hormone (GHRH)	growth
10) Collagen "glue producer"	connective tissue, tensile strength
11) Oxytocin "breath of life"	uterine contraction & lactation, birth
12) Prolactin	milk production, natal nutrition

Results and Discussion

Based on the syllabus, twelve descriptions of each of the peptides were prepared to provide the university students with some basic knowledge. A meeting between the university students and the professor was held to discuss the peptide. Ideas were generated on experiments to captivate the student's attention, on questions and teaching aids such as posters to help create dialogue with the elementary students to introduce the topic, and on art projects to assist students to imagine the forms of the molecules and the lives of the people who work with them. A script was then prepared for the course plan.

Fig. top left, learning about peristalsis studying motilin; top right, listening to heart beat studying angiotensin; middle left, constructing a model of the kidney studying vasopressin; middle right, showing off an oxytocin-inspired lasso-shaped bracelet; bottom left, braiding collagen-inspired coiled-coiled tendons; bottom right, painting with photo-reactive dyes studying melanocite stimulating hormone.

One approach often taken to gain the second graders' appreciation for each peptide was to describe the molecules in a metaphorical way based on function: i.e. gastrin was presented as the traffic cop that controls the flow of acid and mucus into the stomach. Although we realized that such anthropomorphism as calling MSH "the painter of skin pigmentation" might lead to some confusion later in life, the use of such depictions provided a palatable means for some children to better grasp the concept of molecules in general. Some may disagree with this method of teaching; however, our goal was to raise enthusiasm for the arts and sciences by focusing on general concepts rather than specific details. Although it may be many years before one realizes the impact of such pedagogy, when evaluated at the end of the project, all of the students expressed interest in doing more art and science projects of this kind in the future.

Blending art and science together, innovative ideas were explored for teaching challenging subjects to elementary school students (Figure). For example, the influences of acid and mucus were depicted by an experiment employing a bottle of diet Pepsi as a model of the stomach into which were placed pennies coated in wax as a model of mucus and uncovered pennies. In the class on motilin, a plastic-walled tube was used as a model of the intestine through which the students passed clay versions of their favorite nutritious food by the squeezing of their hands in peristalsis motions. In a demonstration of peptide folding in the making of insulin, students rolled-out snake-like cookie dough rods, bent them into U-shapes onto which rungs were attached, before cutting open the bend and preparing insulin-inspired ladder-shaped cookies. Employing cardboard cutouts to block the sun's effect on bandanas painted with photo-reactive dyes, students were taught how MSH causes pigment production to block the effects of sun on our skin.

A new method for teaching art and science to elementary school students was successfully developed using a curriculum designed around 12 peptides important for life functions. Working in teams bachelor and master-level university students from the arts and sciences responded effectively to the challenge of teaching elementary school education, and inspired a class of second graders to think about peptides. Considering the enthusiasm for this project shown by the students, teachers and other participants, as illustrated in the Figure above and in final evaluations provided by the elementary school students, the combination of art and science education provides an effective means for engaging young students to begin to learn challenging subjects.

References

1. Molecules of Life Project web site: http://smu.ca/~mlp/index.html

The Merrifield Award

This award is generously endowed by Rao Makineni.

The American Peptide Society's most prestigious award recognizes the lifetime achievement of a peptide scientist whose work exemplifies the highest level of scientific creativity. The American Peptide Society presented the R. Bruce Merrifield Award for outstanding career achievements in peptide research to Isabella Karle at the 20th American Peptide Symposium, on the afternoon of Tuesday, June 26, 2007.

Isabella Karle (nee Lugoski) was born in Detroit, Michigan. After attending the public schools in Detroit, she was awarded a scholarship to the University of Michigan where she earned the B.S. Chem, M.S. and Ph.D. degrees with a speciality in physical chemistry, under the directorship of L. O. Brockway, who had been Linus Pauling's first graduate student. She was an Instructor in Chemistry at the University of Michigan. In 1946, she joined the Naval Research Laboratory in Washington where she still maintains an active research program as a member of the Laboratory for the Structure of Matter.

Dr. Isabella Karle's early research concerned the structure analysis of molecules in the vapor state by electron diffraction. She was instrumental in the development of a quantitative procedure by which vibrational motion as well as bond lengths and bond angles in molecules can be determined accurately. In the fifties, her research was directed toward crystal structure analysis. She developed practical procedures based on the theoretical work developed by her husband in the Laboratory for the Structure of Matter at NRL for the determination of phases directly from the measured intensities of x-ray reflections. These practical procedures have become adopted world-wide and have been essential to the explosive output of crystal structure determinations that are indispensable to the solution of problems in a number of scientific disciplines: chemistry, biochemistry, biophysics, mineralogy, material science, pharmaceuticals, drug design and medicinal chemistry, for example. There are now in excess of 10,000 published analyses per year, as compared to about 150 per year in the early 1960s. Isabella Karle personally has applied the direct method of phase determination to the pioneering elucidation of molecular formulae and determination of conformations for steroids, alkaloids, frog toxins, photorearrangement products caused by radiation, nanotubes and particularly peptides. For peptides she has established the precise or preferred structures at the atomic level, provided needed parameters for computer drug design programs, useful information for conformation analysis and for synthesizing new analogs with more desirable properties and a basis toward understanding physiological processes. She has published more than 380 papers.

The work of Dr. Karle has been recognized by a number of awards and honors. Among them have been election to the National Academy of Sciences, the American Academy of Arts and Sciences, and the American Philosophical Society. She has received the Garvan Award of the American Chemical Society, the Hillebrand Award, the WISE Lifetime Achievement Award, the Gregori Aminoff Prize from the Royal Swedish Academy of Sciences, the Bijvoet Medal from the Netherlands, Robert Dexter Conrad Award (ONR), and eight honorary doctorate degrees, among them one from the University of Athens (Greece). Her first award, however, was presented by the Society of Women Engineers. She has served as President of the American Crystallographic Association, on several editorial boards of journals and a number of national committees concerned with various aspects of chemistry and crystallography. In 1993, Dr. Karle was awarded the prestigious Bower Award and Prize for Achievement in Science (Franklin Institute), and in 1995 she received the National Academy of Sciences Award in Chemical Sciences and the National Medal of Science from President Clinton. Other recognitions include her biography in "Women in Chemistry and Physics" and in "The Door and the Dream".

The Merrifield Award Winners

2007	**Isabella Karle**	**Navel Research Laboratory, D.C.**
2005	Richard A. Houghten	Torrey Pines Institute for Molecular Studies
2003	William F. DeGrado	University of Pennsylvania
2001	Garland R. Marshall	Washington University Medical School, St. Louis
1999	Daniel H. Rich	University of Wisconsin, Madison
1997	Shumpei Sakakibara	Peptide Institute, Inc.
1995	John M. Stewart	University of Colorado, Denver
1993	Victor J. Hruby	University of Arizona
1991	Daniel F. Veber	Merck Sharp & Dohme
1989	Murray Goodman	University of California, San Diego
1987	Choh Hao Li	University of California, San Francisco
1985	Robert Schwyzer	Swiss Federal Institute of Technology
1983	Ralph F. Hirschmann	Merck Sharp & Dohme
1981	Klaus Hofmann	University of Pittsburgh, School of Medicine
1979	Bruce Merrifield	The Rockefeller University
1977	Miklos Bodanszky	Case Western Reserve University

* Previously, the Alan E. Pierce Award sponsored by the Pierce Chemical Company (1977–1995). The Merrifield Award was established in 1997 by an endowment from Rao Makineni.

Meandering through 50 Years of Peptide Folds and Twists

Isabella L. Karle

Laboratory for the Structure of Matter, Naval Research Laboratory, Washington, D.C. 20375, USA

Introduction

I feel very honored to have been selected as the recipient of this years' Merrifield Award and to join the prestigious list of previous winners, all of whom have been pioneers in important areas of peptide research. I am deeply indebted to my nominators and the selection committee. I thank Rao Makineni for his generous support that makes this award possible.

Many scientific careers take a circuitous path, a path that may be influenced by new developments in scientific and technical areas, or perhaps by chance association with scientists from other continents, or even by direction by a family member. One factor is for certain, however. Bernard Baruch, the financier, summarized it by the statement that "no amount of careful planning can substitute for "dumb luck"". My own career has been dependent largely on the activities of other individuals and serendipity, but coupled with a large amount of persistence and stubbornness on my part.

My interest in the conformation of molecules began in graduate school at the University of Michigan where the subject of my thesis was the determination of the structure of molecules in the gaseous state by the use of electron diffraction. I determined that molecules like biphenyl had librations about the single bond connecting the phenyl rings, and that ortho terphenyl and cyclic tetraphenylene [1] were three dimensional molecules, not planar because of steric hindrance. After World War II, my husband (Jerome Karle) and I joined the Naval Research Laboratory where we still are located. Our first project was to design and have constructed a new electron diffraction apparatus, during which time my husband became aware of and immensely interested in the phase problem that existed in the analysis of crystal structures by X-ray diffraction.

I continued electron diffraction research on our home-built apparatus in the late 1940's while my husband and a colleague embarked on solving the 'phase problem' encountered in the X-ray diffraction of single crystals. The position of the sine wave (the phase of the X-ray beam when the diffraction occurred for a particular reflection) appeared to be lost in the experiment. It was generally assumed, and stated in the literature, that the phase values could not be determined experimentally. Therefore, the determination of the crystal structure could not be accomplished, except by devious procedures which were successful if the crystal contained a sufficiently heavy atom. For crystals containing only light atoms, such as C, H, N and O, there were very few successful solutions of the structure.

Jerome Karle realized that from a mathematical point of view the number of data (that is, the diffraction spots) far exceeded the number of unknowns (that is, the x, y, z coordinates for each atom in the molecule). It was a highly overdetermined system and therefore a mathematical solution to the phase problem had to exist. He realized further, that a strong limiting condition was the fact that the electron density in a crystal must be zero or positive throughout the crystal. A crystal does not contain negative electron density. This concept was carried over from our earlier joint work in the electron diffraction analyses of molecules in the vapor state [2]. The result of the mathematical manipulation was indeed the solution to the phase problem, that is, the unknown phases were shown to be related to and derivable from the measured intensities. The answer to the phase problem lies in the infinite set of inequalities displayed in Figure 1. Thirty-five years later he was awarded the Nobel Prize in Chemistry, essentially for the mathematical relationship between unknown phases and measured intensities. However, in the 1950's and for some time to come, the implications contained in the inequalities were not appreciated.

$$\begin{vmatrix} F_{000} & F_{-k_1} & F_{-k_2} & \cdots & F_{-h} \\ F_{k_1} & F_{000} & F_{k_1-k_2} & \cdots & F_{k_1-h} \\ F_{k_2} & F_{k_2-k_1} & F_{000} & \cdots & F_{k_2-h} \\ \cdots & \cdots & \cdots & \cdots & \cdots \\ F_h & F_{h-k_1} & F_{h-k_2} & \cdots & F_{000} \end{vmatrix} \geq 0$$

Fig. 1. Infinite set of inequalities relating unknown phases and observed intensities.

Part of the difficulty was that the solution is in the form of an infinite set of <u>inequalities</u>, not equations. Each of the F values (square roots of measured intensities) are complex numbers that contain the phase values inside of sine and cosine terms. For strong reflections, the third order inequality could be approximated by the function shown in Figure 2. This function expresses a relationship between selected phase angles that are associated with large $|F|$ values and most probably is close to an equality.

$$\begin{vmatrix} F_{000} & F_{-k} & F_{-h} \\ F_k & F_{000} & F_{k-h} \\ F_h & F_{h-k} & F_{000} \end{vmatrix} \geq 0 \quad , \qquad \phi_h \sim \phi_k + \phi_{h-k}$$

Fig. 2. Σ_2 formula.

At this point, let me quote an observation by the pundit Yogi Berra:
"In theory: theory and practice are the same.
In practice: they ain't."

Other X-ray practitioners were very slow to embrace the new relationships. After a number of years passed, my husband strongly suggested that I set up an X-ray laboratory and demonstrate that his formulas were correct. The mathematical problem had been solved. There remained the problem of applying the inequalities to real crystals where the experimentally measured intensities always contained some errors (crystal imperfection, measuring devices) and the number of observable data was limited by the experimental set-up. A number of other problems had to be addressed. Some of them were:

1. Inequalities lead to approximate values. Can not use simultaneous equations.
2. Need for incorporating probability measures. (very important)

3. Fuzzy logic.
4. Where to start the phase determination and how to proceed.
5. Initial assignment of phase values by use of a limited number of symbols.
6. How to refine approximate values and to extend the procedure to determine additional phases? (very important)
7. **50** years ago computers were rather primitive and slow. There was a need to simplify for manual operation.

By the late 1950's, I was successful in formulating a procedure for solving the structures of a number of non-trivial organic compounds by manual means.

Cyclic Backbones

I have not yet mentioned peptides and their conformations. In 1962, Prof. Isidor Fankuchen of Brooklyn Polytechnic Institute, who was a prominent crystallographer, challenged me to solve the structure of a cyclic compound that was intended to be cyclic triglycyl but that could have been cyclic hexaglycyl. The material was synthesized at MIT by John Sheehan and the crystals had already been examined by several different laboratories without yielding a solution to the structure. I invested the necessary time to measure the thousands of X-ray reflections that were available. The results of the analysis, shown in Figure 3, were four different conformers of cyclic hexaglycyl occurring side-by-side in the same crystal. The paper describing the structure [3] became a Citation Classic (most cited 1966-1975). The paper contained various types of information, as for example:

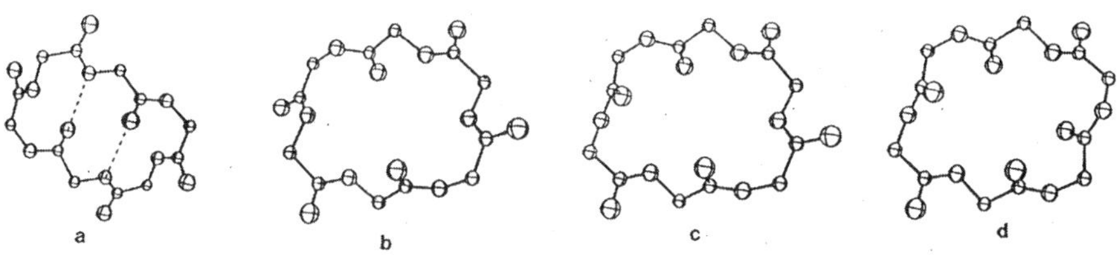

Fig. 3. Four conformers of cyclohexaglycyl in ratio 4:2:1:1.

1. A procedure for obtaining crystal structures directly from the intensities of the scattered reflections.
2. Showed the molecule to be a hexamer, not a trimer.
3. The peptide molecule occurred with four different conformations, side-by-side in the crystal (unexpected), and presumably with the same energy and
4. Provided "accurate" bond lengths, bond angles and torsion angles for a hairpin type beta-bend in a peptide. Later, Ramachandran and Sasisekharan labelled this type of beta-bend as Type I [4].

All this happened because someone gave me crystals with a challenge.

The next few years I spent determining the structures of a variety of natural products including the chemical identification of potent alkaloids from Columbian poison arrow frogs, in collaboration with John Daly and Bernhard Witkop at the NIH, as well as radiation damaged nucleic acid bases and the photocyclizations and photorearrangements of pharmacodynamic amines.

About 1969, through the aegis of Bernhard Witkop, I came into possession of crystals of cylic peptides that were isolated and/or synthesized in the laboratories of Theodor Wieland of Heidelberg and Urry Ovchinnikov of Moscow, both of whom I later met in person at the 1973 meeting of the predecessor of the American Peptide Society. One of the cyclic peptides from Ovchinnikov's laboratory with the sequence: c-(Gly-Gly-D-Ala-D-Ala-Gly-Gly) crystallized with only one conformation [5]. It was very similar to the most prevalent conformation for c-$(Gly)_6$, described above, with the same kind of intramolecular NH···OC hydrogen bonding. Another cyclic peptide from Ovchinnikov's laboratory c(L-Leu-L-Tyr-δAvaler-δAvaler), although a tetrapeptide with two delta amino acid residues, had the same number of atoms in the cyclic backbone as the cyclic hexapeptides with all alpha-amino acids [6], Figure 4. The overall shape of the backbone is similar to another one of the c-$(Gly)_6$ conformers. There are no intra-ring hydrogen bonds, even though an NH and a CO moiety are available for an NH···OC hydrogen bond on the upper and lower strands of the backbone as seen in Figure 4. Conformations such as seen in the above molecules have served as standards for testing force fields. (M. Karplus, private communication).

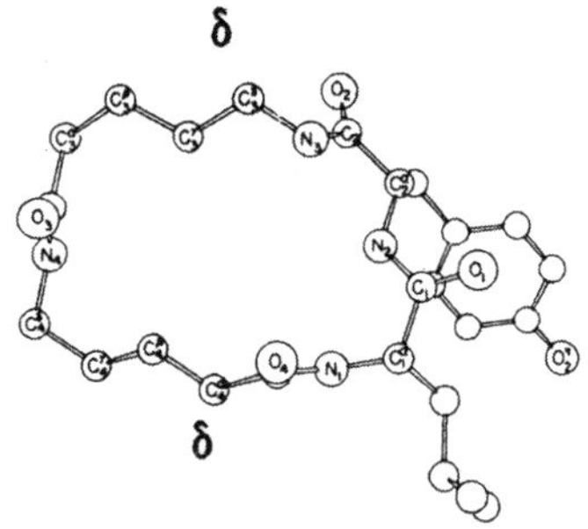

Fig. 4. cyclo-(L-Leu-L-Tyr-δ-Avaler-δ-Avaler) lacking internal hydrogen bonds.

My association with Theodor Wieland (Heidelberg) was very exciting and fruitful. I became acquainted with his work on antamanide, a cyclic decapeptide that was present in and served as an antidote to the potent toxins found in the Amanita Phalloides mushrooms, when Wieland and his collaborators provided me with numerous crystals of the natural antamanide,

```
Pro⁸-Phe⁹-Phe¹⁰-Val¹-Pro²
 |                     |
Pro⁷-Phe⁶-Phe⁵-Ala⁴-Pro³,
```

Na^+ and Li^+ complexes of antamanide, and synthetic analogues in which the residues in antamanide were systematically replaced with other similar residues in order to discover clues to their biological activity. Further, they had established that complexation of antamanide with Na^+ and Li^+ (but not K^+) was a necessary condition for antitoxic activity [7]. The analogs that did not form complexes were inactive. The analogs that did form complexes were active except for one type. When Phe residues were perhydrogenated, one at a time, the complex lost activity progressively, until there was none.

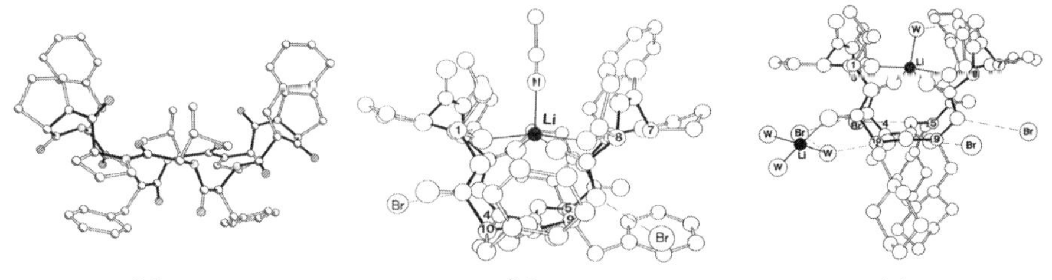

(a) (b) (c)

Fig. 5. (a) Uncomplexed antamanide, (b) Active Li^+ complex, (c) Inactive perhydroantamanide complexed with Li^+.

The crystal structures associated with the perhydrogenation are shown in Figure 5 where (a) shows the relatively flat uncomplexed antamanide, [8] (b) shows the backbone folded to partially encapsulate a Li^+ ion, the encapsulation being completed by a solvent molecule, and the phenyl groups in the side chains are folded against the backbone so that a hydrophobic surface [9] is created for the complex and (c) shows that when the phenyl groups are perhydrogenated to become cyclic hexyl groups. In the complex the bulky side chains hang down away from the globular folded backbone [10]. Perhydrogenation has had the effect of unshielding the middle of the complex and exposing the polar areas to the environment. One result is the attraction of an additional Li^+ ion that is surrounded by water molecules. Hence, the surface of the perhydro complex is very different and much more polar than the surface of the active native antamanide complex [10].

By the time of the 4th American Peptide Symposium (1975), a sufficient number of peptide structures in crystals had been established in laboratories around the world that certain generalizations could be made with considerable certainty. For example, similar conformations do occur for similar atomic groupings, conformational parameters for internal hydrogen bonds are remarkably constant, *cis* conformations for peptide units are fairly common but appear only for peptide units with N-substitution (proline). And, of course, that the peptide bond should be planar, either *cis* or *trans* with torsional angles $\omega \cong 0°$ or $\cong 180°$. However, there always is the possibility that a peptide will assume unusual or unexpected conformational features.

Up to this point in time (1976) cyclic tetrapeptide rings had been observed to have *cis-trans-cis-trans* conformations for the four peptide groups. In fact, it had been shown by theoretical computation that four planar peptide groups with all *trans* conformations cannot be closed into a ring [11]. Nevertheless, crystal structure determinations have demonstrated that a very few cyclic tetrapeptides do exist in the all-*trans* conformation. The crystal structure of dihydrochlamydocin [12], shown in Figure 6, a natural product isolated from the spores of Dihederospora chlamydosporia, is a classic case. In order to achieve ring closure, each peptide bond deviates significantly from planarity with ω values as low as 156°, as compared to 180°. Further, the cyclic tetrapeptide ring provides parameters for two internal 3→1 hydrogen bonds. An all-*trans* conformation for a cyclic tetrapeptide with 3→1 type intramolecular hydrogen bonds had been postulated for cyclic tetraglycyl by Schwyzer [13] already in 1956. Interestingly, these features were found in chlamydocin, a microbial product that was isolated in Switzerland, near Schwyzer's home.

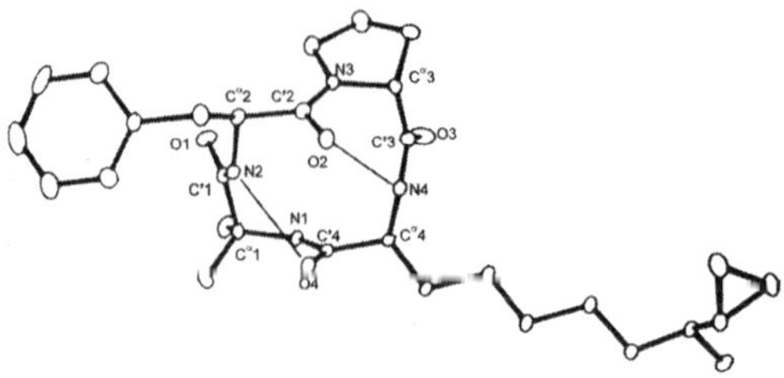

Fig. 6. Dihydrochlamydocin. The omega values in the cyclic tetrapeptide ring for the four nonplanar peptide bonds are +162°, -166°, +157° and -164°.

Lila Gierasch designed and synthesized a series of cyclic pentapeptides that have both a 4→1 and a 3→1 type hydrogen bond. One of the five all-*trans* peptide bonds, the one that is associated with the 3→1 hydrogen bond, is also a twisted C-N bond where the ω value is near 160°, rather than 180°, quite the same as in the all-*trans* cyclic tetrapeptide. An example from a series of cyclic pentapeptide structures is shown in Figure 7 [14].

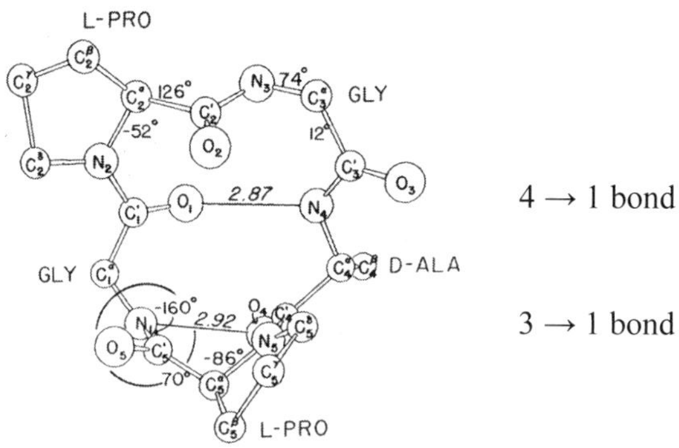

Fig. 7. An all trans cyclic pentapeptide with one nonplanar amide bond, ω_5 = 160°.

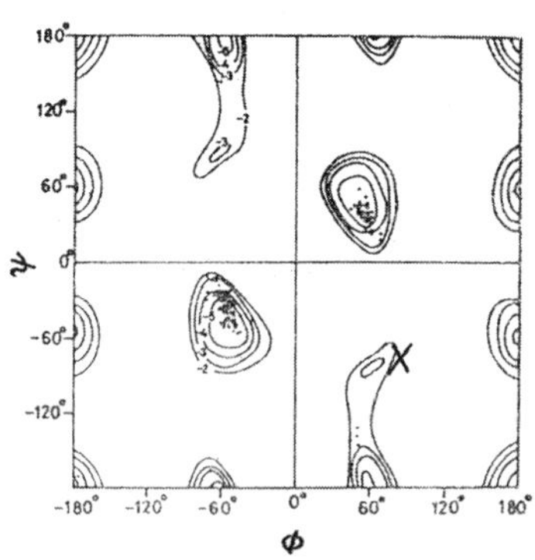

Fig. 8. Crystallographically observed φ, ψ values of 154 Aib residues from 65 crystal structures placed on the theoretically computed potential energy map for Ac-Aib-NHMe [15]. X markes the position of the Aib in dihydrochlamydocin.

The dihydrochlamycin molecule also contains two amino acid residues not found among the 20 amino acids present in the higher forms of life. The structure analysis in Figure 6 may have been the first crystal structure of an Aib (α-aminoisobutyric acid) residue. Curiously, this early example of the Aib residue was found in one of the very few structures in which the Aib residue is not contained in a helix and therefore its intense inclination for helix initiation was not apparent in this structure [12]. Its position on a Ramachandran map of allowable ϕ,φ values for Aib residues is indicated by an X in Figure 8 [15] where it barely abuts a weakly acceptable contour, is far from the helical regions, and would not be expected to exist, although it obviously does exist in a stable naturally occurring compound.

Before we leave cyclic backbones, let us examine an analog of elastin, the load-bearing elastic fiber in animal connective tissue and arteries. The protein polymer has rubber-like extensible properties when in the presence of water. The synthetic analog, prepared by Dan Urry, is the cyclic $(APGVGV)_2$ that assembles in corrugated β-sheets in the crystal [16]. Models of the crystal structure can demonstrate its facile extensibility and retractability. The 12-residue cyclic backbone does not assume the regular β-sheet type motif shown by 6- and 10-residue peptides that have cross-strand N⋯OC hydrogen bonds. Rather, the 12-residue cyclic backbone has a bifurcated acceptor geometry, as illustrated by the N(4)H⋯O(1) and N(5)H⋯O(1) hydrogen bond pair where both N(4)H AND N(5)H are donors to the same O(1) [16], Figure 9. The side view in Figure 9 shows the dome-shaped outline of the backbone with a shape and size that could accommodate a somewhat elusive water molecule that can participate in several different hydrogen bonds with NH and/or C=O moieties as it moves up and down. Models show that a vertical motion of X1 can be coordinated with a horizontal motion of the backbone (like opening and closing an umbrella) and may suggest a mechanism for the extensibility property of the elastin analog.

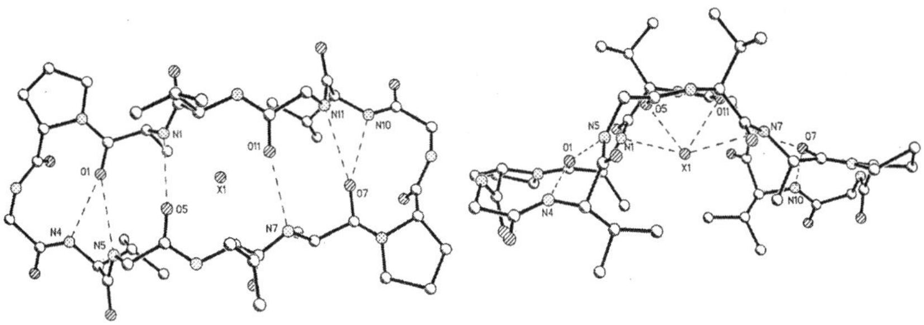

Fig. 9. Cyclic $(APGVGV)_2$ (a) View into the ring. (b) Side view showing a dome-shaped backbone.

Linear Backbones

Crystal structures of peptides with cyclic backbones were generally determined earlier than those with linear backbones, probably because cyclic backbones have additional constraints on the number of degrees of freedom as compared to linear backbones that allow crystallization to be achieved more readily.

The demonstration that the Aib residue (alpha amino isobutyric acid or dimethyl glycine) has an extremely restricted acceptable ϕ, φ space in conformational energy diagrams by Marshall [17], Burgess and Leach [18], and later by others [15] led to the realization that the Aib residue is a helix inducer. Furthermore, the rigidity of the

Aib residue stiffened the entire molecule and encouraged crystal growth. A review in 1983 by Toniolo and associates [19] of structures of linear peptides containing 5 or fewer residues with at least one Aib residue showed a 3_{10}-helix formation or an incipient 3_{10}-helix formation, except for the cyclic dihyrochlamydocin, [12] already discussed.

In 1986, I received a letter from Balaram asking if I would be interested in doing crystal structure analyses of zervamicin and analogs for which he had crystals. Thus began a long-time, long-distance collaboration. The crystals were mailed from India in sealed glass tubes in their mother liquor by ordinary air mail. None were ever lost and almost all the crystals survived the journey.

By 1990, a sufficient number of crystal structures of helical peptides consisting of 7 to 19 residues and containing one or more Aib residues were determined to merit a review of longer peptide sequences, this time by Karle and Balaram [20]. The broad conclusion was that the Aib residue initiates 3_{10}-helix type formations for the shorter residues and α-helix type for the longer residues, tempered by the number of Aib residues or other amino acids with a double substitution on the C^{α} atom. Figure 10 shows a 3_{10}-helix for a 7-residue peptide, [21] Boc-VAL-Dbg-VAL-OMe, (where Dbg is dibutylglycine and induces helices as well as Aib) and an alpha-helix for a 15-residue peptide [22], with three Aib residues in the sequence, Boc-VALUVAL(VALU)$_2$-OMe, where U ≡ Aib.

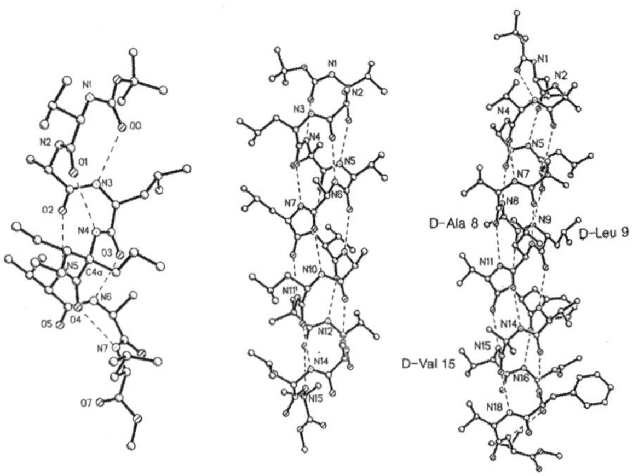

Fig. 10. Helical peptides. (a) Seven-residue 3_{10}-helix induced by dibutyl glycine, b) Fifteen-residue alpha-helix containing three Aib residues, and (c) 19-residue alpha helix containing three D-residues randomly placed in the sequence.

Helices promoted by Aib residues, even one Aib in a sequence, are very robust. Unsuccessful attempts to unwind such helices included the substitution of the CH_3 groups in Aib by longer hydrocarbon chains [23], or by cyclic hydrocarbons that contain the C^{α} atom in the cycle [24,25], by the insertion of additional atoms into the backbone by the use of a dipeptide ββ segment [26] or βγ segment, [27] or by the exchange of several L-residues for D-residues [28]. For example, the insertion of the sequence β-Val-β-Phe into the backbone merely caused an increase in the size of the hydrogen bonded NH···OC loops and a small bulge in the helix. The presence of three D residues in the 19-residue sequence Boc-LUVALUV-^{D}A-^{D}L-VFVU-^{D}V-LFVV-OMe is unnoticed in the continuous winding of the backbone [28]. Compare

Figure 10c with 10b. These structures reinforce the observation made by Seebach and Matthews [29] that "the expectation of many a colleague and protein specialist was that insertion of a CH_2 group into a residue in a peptide backbone would lead to conformational chaos" was not so. Obviously, additional methylene groups inserted into a helical backbone can be readily accommodated [30].

The helical peptides described above were synthetic molecules designed to illustrate and establish geometric parameters for various conformational features. These features have also been found in naturally occurring peptides, such as zervamicin [31] and the related antiamoebin [32] that are produced by microbes in the soil and that act as ionophores for transporting K^+ ions across cell membranes. The sequences are characterized by a large number of Aib and Hyp (hydroxyproline) residues:

AcPheAibAibAibIva^{5}GlyLeuAibAibHyp^{10}GlnIvaHypAibPro^{15}Phol [32]

and

AcLeuIleGlnIvaIle^{5}ThrAibLeuAibHyp^{10}GlnAibHypAibPro^{15}Phol [31].

The backbone in each molecule folds in an identical manner. Despite the large difference in the number of Aib (plus Iva) residues, each molecule is completely helical and folds into a regular 5→1 alpha helix up to the ninth residue. At Aib^9 there is a helix reversal, a severe bend in the helix axis, and the helix becomes a twisted β-bend helix that accommodates the three hydroxyproline and proline residues, Figs. 11a and b. The severely bent composite helix in both molecules has hydrophobic side-chains covering the concave side and hydrophilic side chains coverning the convex side.

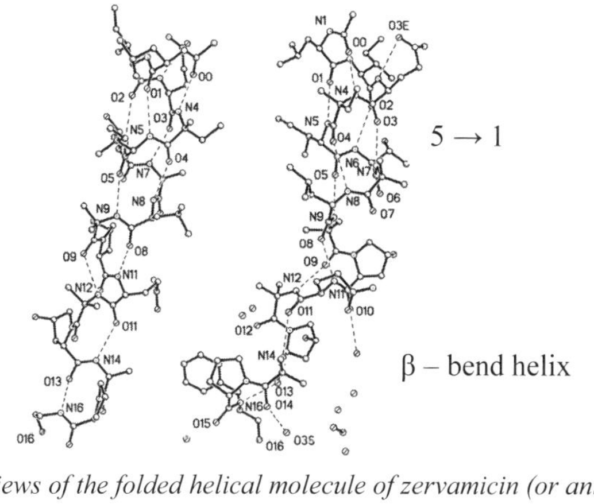

Fig. 11. Two views of the folded helical molecule of zervamicin (or antiamoebin).

Both L-zervamicin and antiamoebin molecules assemble in the crystal in the same manner with their hydrophilic sides facing each other and forming an hour-glass shaped channel. At the constriction, the channel contains a double-gating passage [31,32] for K^+ ions to pass through one at a time. Antiamoebin was used to display the channel in Fig.12 since n-octanol was the chosen crystallizing agent that cocrystallized around the columns of antiamoebin in a fashion that suggests a membrane-mimetic environment for the ion channels. In other words, the arrangement in the crystal may have some relationship to ion channels penetrating cell walls for the purpose of transporting charged ions across non-polar cell membranes, complete with a double-gating mechanism.

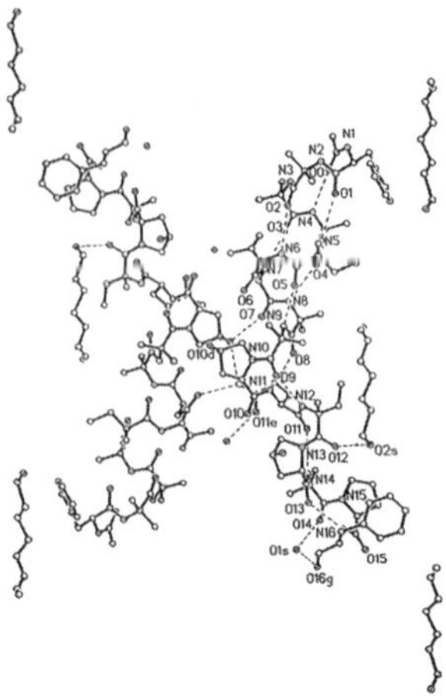

Fig. 12. Hour-glass shaped channel for K^+ ion passage. The peptide walls are surrounded by n-octanol molecules.

Hairpin folding in beta-sheets

The other important folding motif is the formation of beta-sheets by peptides with linear backbones that are constrained to make hairpin turns and support antiparallel strands with cross-strand hydrogen bonds. Balaram had chosen the doublet D-Pro-Gly(Ala) for the hairpin folding constraint. The structure for one of these peptides, shown in the central diagram of Fig. 13, has the sequence (b) where all the residues are alpha amino acids:

BocLeuVal*β-Val*^{D}ProGly*β-Leu*ValValOMe (a) [34]
BocLeuValVal^{D}ProGlyLeuValValOMe (b) [33]
BocLeuValVal^{D}Pro*δ-Ava*LeuValValOMe (c) [35]

(where δ-Ava ≡ -NH(CH_2)$_4$-C(O)-).

For the normal beta hairpin like (b) the cross-strand NH···O=C hydrogen bonds alternate direction and result in a neutral molecule. The β-, γ-, and δ-amino acid residues that contain one, two or three additional CH_2 moieties in their backbone have been used to lengthen the backbone of peptides.

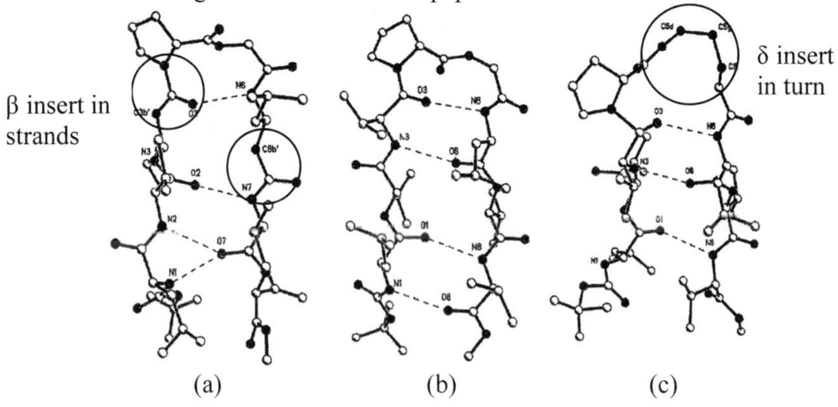

Fig. 13. Hairpin folded peptides that have β-amino acid inserts in strands in (a) and δ-amino acid insert in turn region in (c).

In molecule (a), for example, the substitution of β-Val for Val and β-Leu for Leu (as a pair placed opposite each other) caused no difficulty in the folding of the backbone or in forming cross-strand hydrogen bonds. A major difference in the structures (a) and (b) is the direction of adjacent NH···OC bonds, where in (a) the two upper NH···OC bonds point in the same direction. In fact, additional pairs of β-amino acids inserted into the structure can endow a hairpin molecule with polarity caused by the unidirectional alignment of the cross-strand NH···OC bonds [36]. In molecule (c), the lengthening of the backbone has taken place in the turn region by the substitution of Gly with δ-Ava. The extra three CH_2 groups form a bulge in the turn with little disturbance in the strands and the NH···OC bonds. Only the two termini have been raveled a bit. A virtue of inserting -CH_2- groups, whether in the strands or in the turn, is to provide scaffolds for additional modifications.

Coexisting multiple types of folding

The design of the peptide shown in Fig. 14 that contains an extended backbone, followed by a helix and ending with a beta-sheet [37] was based on the conformational information described in this paper.

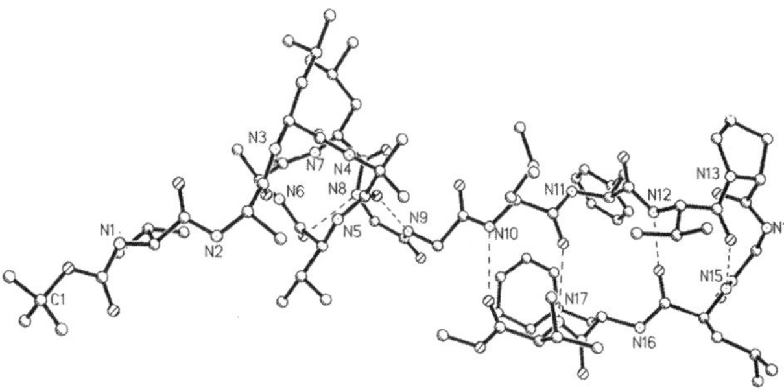

Fig. 14. Three independent folding domains for Boc-VALUVALGGLFV-^{D}Pro-GLFV-OMe.

Conclusion

The structures of selected peptides presented in this paper were chosen to demonstrate the typical conformations for peptides with cyclic backbones, helical backbones, and beta-turn backbones and to show the persistency of these conformations under unusual circumstances, such as insertions and/or substitutions in backbones and side chains, or changing the hand of individual residues. Also highlighted were a number of conformations found in naturally occurring, as well as synthetic peptides that deviated significantly from the established norms. Particularly gratifying to the author were the structures whose folding characteristics could be directly connected with biochemical or physical properties.

References

1. Karle, Isabella L. and Brockway, L.O. *J. Amer. Chem. Soc.* **66**, 1975-1979 (1944).
2. Karle, Isabella and Karle, Jerome *Struc. Chem.* **16**, 5-16 (2005).
3. Karle, I. L. and Karle, J. Acta Cryst. **16**, 969-975 (1963).
4. Ramachandran, G. N. and Sasisekharan, V. *"Conformation of Polypeptides and Proteins"* in Advances in Protein Chemistry, eds. C. B. Anfinson, Jr., M. L. Anson, J. T. Edsall and F. M. Richards (1968) Academic Press, N.Y.
5. Karle, I. L., Gibson, J. W. and Karle, J. *J. Amer. Chem. Soc.* **92**, 3755-3760 (1970).

6. Karle, I. L. *Macromolecules* **9**, 61-66 (1976).
7. Wieland, T. in *Chemistry and Biology of Peptides,* Ed. Meienhofer, J., Ann Arbor Sci., Ann Arbor, MI (1972) pp. 377-396.
8. Karle, I. L. *J. Amer. Chem. Soc.* **99**, 5152-5157 (1977).
9. Karle, I. L. *J. Amer. Chem. Soc.* **96**, 4000-4006 (1974).
10. Karle, I. L. *Proc. Natl. Acad. Sci. USA* **82**, 7155-7159 (1985).
11. Go, N. and Scheraga, H. A. *Macromolecules* **3**, 178-187 (1970).
12. Flippen, J. and Karle, I. L. *Biopolymers* **15**, 1081-1092 (1976).
13. Schwyzer, R., Iselin, B., Rittel, W. & Sieber, P. *Helv. Chim. Acta* **34**, 872-883 (1956).
14. Karle, I. L. *J. Amer. Chem. Soc.* **100**, 1276-1289 (1978).
15. Uma, K. and Balaram, P. *Indian J. Chem.* **28B**, 705-710 (1989).
16. Karle, I. L. and Urry, D. W. *Biopolymers* **77**, 198-204 (2005).
17. Marshall, G. R. and Bosshard, H. E. *Circ. Res. (Suppl. II)*, **30/31,** 143-150 (1972).
18. Burgess, A. W. and Leach, S. J. *Biopolymers* **12**, 2599-2605 (1973).
19. Toniolo, C., Bonora, G. M. Bavoso, A., Benedetti, E., di Blasio, B., Pavoni, V. and Pedone, C. *Biopolymers* **22**, 205-210 (1983).
20. Karle, I. L. and Balaram, P. *Biochemistry* **29**, 6747-6756 (1990.
21. Vijayalakshmi. S., Balaji Rao, R. Karle, I. L. and Balaram, P. *Biopolymers* **53**, 84-98 (2000).
22. Karle, I. L., Flippen-Anderson, J. L., Uma, K., Sukumar, M. and Balaram, P. *J. Amer. Chem. Soc.* **112**, 9350-9356 (1990).
23. Karle, I. L., Gurunath, R., Prasad, S., Balaji Rao, R and Balaram, P. *Inter. J. Pept. Prot. Res.* **47**, 376-382 (1996).
24. Toniolo, C., Crisma, M., Formaggio, F., Benedetti, E., Santini, A., Iacovino, R., Saviano, M., Di Blasio, B., Pedone C. and Kamphuis, J. *Biopolymers* **40**, 519-522 (1996).
25. Karle, I. L., Prasad, S. and Balaram, P. *J. Peptide Res.* **63**, 175-180 (2004).
26. Roy, R. S., Karle, I. L., Raghothama, S. and Balaram, P. *Proc. Natl. Acad, Sci. USA* **101**, 16478-16482 (2004).
27. Karle, I. L., Pramanik, A., Banerjee, A., Bhattacharjya, S. and Balaram, P. *J. Amer. Chem. Soc.* **119**, 9087-9095 (1997).
28. Karle, I. L., Gopi, H. N and Balaram, P. *Proc. Natl. Acad. Sci. USA,* **100**, 13946-13951 (2003).
29. Seebach, D. and Mathews, J. L. *Chem. Commun.* 2015-2022 (1997).
30. Appella, D. H., Christianson, L. A., Karle, I. L., Powell, D. R. and Gellman, S. H. *J. Amer. Chem. Soc.* **118**, 13071-13072 (1996).
31. Karle, I. L., Flippen-Anderson, J. L., Agarwalla, S. and Balaram, P. *Biopolymers* **34**, 721-735 (1994).
32. Karle, I. L., Perozzo, M. A., Mishra, V. K. and Balaram, P. Proc. Natl. Acad. Sci. USA **95**, 5501-5504 (1998).
33. Karle, I. L., Awasthi, S. K. and Balaram, P. *Proc. Natl. Acad. Sci. USA* **93**, 8189-8193 (1996).
34. Gopi, H. N., Roy, R. S., Raghothama, S. R., Karle, I. L. and Balaram, P. *Helv. Chim. Acta* **85**, 3313-3330 (2002).
35. Rai, R., Vasudev, P. G., Ananda, K., Raghothama, S., Shamala, N., Karle, I. L. and Balaram, P. Chem. Eur. J. on line (2007).
36. Karle, I. L., Gopi, H. N. and Balaram, P. *Proc. Natl. Acad. Sci. USA* **99**, 5160-5164 (2002).
37. Karle, I. L., Das, C. and Balaram, P. *Proc. Natl. Acad. Sci. USA* **93**, 3034-3037 (2000).

Makineni Lecturer

Endowed by PolyPeptide Laboratories and Murray and Zelda Goodman (2003)

The Lectureship honors Rao Makineni, a long time supporter of peptide science, peptide scientists, and the American Peptide Society. Presented at the biennial symposia, the Makineni Lectureship recognizes an individual who has made a recent contribution of unusual merit to research in the field of peptide science. The award is intended to recognize original and singular discoveries rather than cumulative or lifetime contributions.

Ron Raines is the Henry Lardy Professor of Biochemistry and a Professor of Chemistry at the University of Wisconsin–Madison.

Ron received Sc.B. degrees in chemistry and biology at MIT, and A.M. and Ph.D. degrees in organic chemistry at Harvard University. He was a Helen Hay Whitney postdoctoral fellow in the Department of Biochemistry and Biophysics at the University of California, San Francisco. He joined the faculty at the University of Wisconsin-Madison in 1989.

Ron has contributed broadly to our understanding of the chemistry and biology of peptides and proteins. He discovered fundamental attributes of the collagen triple helix, enabling him to assemble triple helices that are stronger and longer than any found in nature. Ron demonstrated that mammalian ribonucleases can become potent cytotoxins and potential cancer chemotherapeutics. He developed the traceless Staudinger ligation as a means to couple synthetic peptides and thus synthesize proteins.

Ron has trained more than 70 doctoral and postdoctoral students. He is an author of more than 200 research papers, and an inventor on more than 10 US patents.

Citation for Makineni Lecture:
"Self-Assembly of Synthetic Collagen Triple Helices"
Kotch, F. W.; Raines, R. T. *Proc. Natl. Acad. Sci. U.S.A.* **2006**, *103*, 3028–3033.

2007	**Ronald T. Raines**	**University of Wisconsin - Madison**
2005	Robin E. Offord	Centre Medical Universitaire, Switzerland
2003	James P. Tam	Vanderbilt University

Stronger and (now) Longer Synthetic Collagen

Ronald T. Raines

Department of Biochemistry and Department of Chemistry
University of Wisconsin–Madison, Madison, WI 53706, USA

Introduction

Prevalence of Collagen. Collagen is the most abundant protein in modern animals, including humans.[1,2] The natural selection of collagen as an anatomical scaffold was made long ago. Collagen has been isolated from soft tissue in the fossilized bones of a 68-million-year-old *Tyrannosaurus rex* and a half-million-year-old mastodon.[3,4] No isolated DNA is that old.[5,6] The archaeological longevity of collagen is even more remarkable considering that a peptide bond has a half-life of 400 years, whereas a phosphodiester bond has a half-life of 30 million years.[7-10]

Collagen comprises $^1/_3$ of the human proteome and $^3/_4$ of the dry weight of human skin.[1,2] To date, 28 different types of collagen have been found in humans, and many other human proteins are known to contain collagenous domains.[2] A defining feature of all types of collagen is a unique tertiary structure in which three parallel strands, each in a polyproline II-type (PPII) helix (which is left-handed), are wound around a common axis to form a triple helix (which is right-handed) (Figure 1A). The packing of this coiled-coil structure requires that every third residue be glycine (Gly), resulting in a repeating Xaa–Yaa–Gly sequence. The residue in the Xaa position of these triplets is often 2*S*-proline (L-proline or Pro), and the residue in the Yaa position is often (2*S*,4*R*)-4-hydroxyproline (Hyp). In human type I collagen, which is the most abundant form, 28% of the Xaa residues are Pro, 38% of the Yaa residues are Hyp, and Pro–Hyp–Gly is the most common triplet.[11]

Individual triple helices of collagen are organized into fibrils of great tensile strength and flexibility. These fibrils can be arranged and cross-linked so as to support stress efficiently in one, two, or three dimensions in tissues such as tendon, skin, and cartilage, respectively. Abnormalities in collagen structure are associated with connective tissue diseases, such as osteogenesis imperfecta, Ehlers–Danlos syndrome, and some types of osteoporosis and arthritis.[1,12-15] A complete understanding of the basis for collagen stability (and instability) could lead to effective therapies for these and other disorders.

Collagen is also an important biomaterial.[16,17] For example, natural collagen is the principal component of biodegradable sutures and artificial heart valves. Obtaining natural collagen in high purity without degrading its structural integrity is difficult. Moreover, the natural collagen that is most readily available is bovine collagen, which can engender allergic and immunological side effects in humans.[18] Despite numerous studies on synthetic collagen, few have been tested as biomaterials.[19]

Role of Hyp Residues. The hydroxyl groups of the prevalent Hyp residues have an important role in collagen stability. Hyp residues are not incorporated into collagen by ribosomes.[20] Instead, this difficult post-translational modification is mediated by the enzyme prolyl 4-hydroxylase[21,22] after the strands are biosynthesized but before they form a triple helix. The prevalence of this post-translational modification is extraordinary: the abundance of Hyp in humans is 4.2%, a value calculated from the abundance of collagen in humans ($^1/_3$) and Hyp in collagen (38% × $^1/_3$). Thus, the abundance of Hyp exceeds that of seven "common" amino acids:

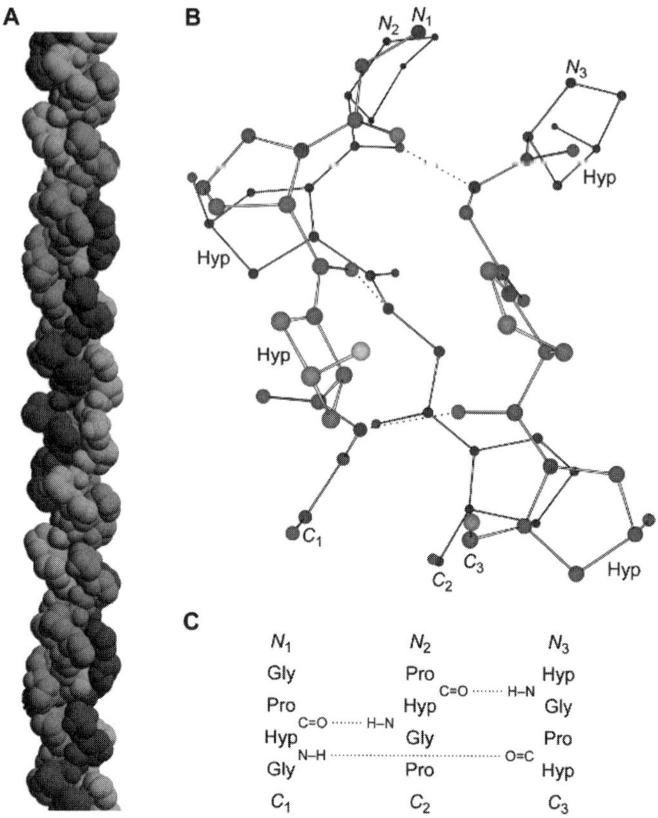

Fig. 1. Structure of a (Pro–Hyp–Gly)$_n$ triple helix. (A) Space-filling model of a long segment. (B) Ball-and-stick model of a short segment indicating Hyp residues and XaaC=O···H–NGly hydrogen bonds. (C) Register of the residues in the three strands of panel B. Atomic coordinates are from PDB entry 1CAG.[23]

Cys, Gln, His, Met, Phe, Trp, and Tyr.[24] Hydroxylation is critical for the folding of collagen, its secretion to the extracellular matrix, and its further processing and incorporation into fibrils or other structures.[25-28] In 2000, we discovered that the absence of prolyl 4-hydroxylase (and hence Hyp residues) is lethal to an animal—the nematode *Caenorhabditis elegans*.[29]

In type I collagen, each strand consists of ~300 Xaa–Yaa–Gly triplets. Discerning the chemical basis for the conformational stability of such a large molecule is difficult. Hence, a reductionist approach using peptide models has played a pivotal role and contributed much insight,[30-33] and led to the following landmark discoveries.

In 1973, Prockop and coworkers used synthetic peptides to demonstrate that the hydroxyl group of Hyp residues dramatically increases the thermal stability of triple-helical collagen.[34] The two decades after 1973 were dominated by the notion that the stability of collagen relies on water molecules that form bridges between the hydroxyl group of its Hyp residues and a main-chain oxygen. In 1994, X-ray diffraction analysis revealed the first truly high-resolution three-dimensional structure of a collagen triple helix.[23] In that triple helix (Figures 1A and 1B), the Hyp residues do indeed have water molecules bound to their hydroxyl groups. These water molecules form bridges between the Hyp residues in one chain and a main-chain oxygen in another chain.[35]

Results and Discussion

Stronger Synthetic Collagen—Stereoelectronic Effects. We were skeptical of the "water-bridge" hypothesis, believing that the bridges were artifactual rather than meaningful. In 1998, we tested this hypothesis by replacing Hyp with (2*S*,4*R*)-4-fluoroproline (Flp). We found that Flp confers unprecedented stability upon a collagen triple helix.[36,37] Because a fluoro group exerts strong inductive effects but does not form strong hydrogen bonds, our result called into question the 25-year-old paradigm.

In 2001, we demonstrated that replacing Hyp with (2*S*,4*S*)-4-fluoroproline (flp), which is a diastereomer of Flp, does not allow for triple helix formation (Table 1).[38] This result indicated that the fluoro group was manifesting a stereoelectronic effect, rather than merely an inductive effect. In 2002, we discovered that the key attribute of the fluoro group (as well as the natural hydroxyl group) was its imposition of a C^{γ}-*exo* pucker on the pyrrolidine ring via the *gauche* effect (Figure 2).[39] This pucker preorganizes the main-chain dihedral angles (ϕ, ψ, and ω) to be those in the triple helix. This work not only demonstrated the role of the hydroxyl group of Hyp residues, but was the first example of a stereoelectronic effect conferring stability upon a protein.

In 2003, we attempted to enlist stereoelectronic effects in the Xaa position to enhance collagen stability. Again, we used a fluoro group, demonstrating that flp but not Flp enables the formation of a stable triple helix (Table 1).[40,41] Thus, stereoelectronic effects can operate adventitiously (or deleteriously) in the Xaa position of collagen. The stereochemical dichotomy (Xaa prefers flp, whereas Yaa prefers Flp) results from the preference for a C^{γ}-*endo* pucker in the Xaa position. In 2006, we showed how to fix the pyrrolidine ring pucker with reciprocal steric effects, thereby reiterating the conformational stability conferred by stereoelectronic effects.[42]

A conundrum led us to another insight. We could not explain why the pucker imposed by the 4-substituent had a marked effect on ω, which refers to the dihedral angle of the peptide bond. In AcFlpOMe, the amide bond has $K_{trans/cis}$ = 6.7; in its diastereomer, AcflpOMe, $K_{trans/cis}$ = 2.5.[38]

In 2002, we discovered that the explanation arose from another unappreciated stereoelectronic effect—an $n \rightarrow \pi^*$ interaction.[39,43] In an $n \rightarrow \pi^*$ interaction (which is not to be confused with an $n \rightarrow \pi^*$ electronic transition) the oxygen of a peptide bond

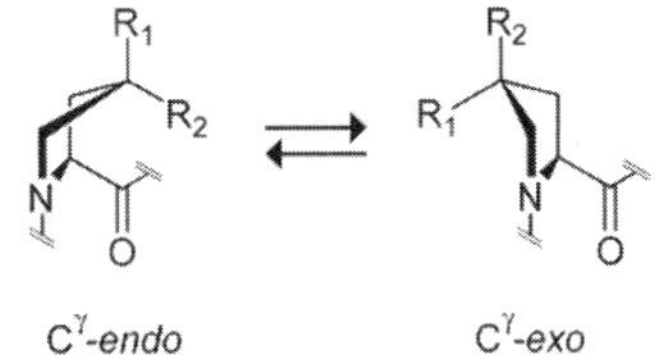

Fig. 2. Ring conformations of 4-substituted proline residues. The C^{γ}-endo conformation is favored strongly when R_1 = H and R_2 = F (as in flp). The C^{γ}-exo conformation is favored strongly when R_1 = OH (Hyp) or F (Flp) and R_2 = H. The C^{γ}-exo:C^{γ}-endo ratio is ~1:2 when R_1 = R_2 = H (Pro).[39]

(Oi–1) donates electron density from one of its lone pairs into the antibonding orbital of the carbon in the subsequent peptide bond (Ci ′=Oi) (Figure 3A). The C$^\gamma$-exo pucker of a proline residue provides a more favorable Oi–1···Ci ′=Oi distance and angle for an $n \rightarrow \pi^*$ interaction than does the C$^\gamma$-*endo* pucker.[39] Because the $n \rightarrow \pi^*$ interaction can occur only if the peptide bond containing Oi–1 is trans (*i.e.*, *Z*), as opposed to cis (*E*), the $n \rightarrow \pi^*$ interaction has an impact on the trans/cis ratio. Thus, the two stereoelectronic effects influence peptide conformation as in the scheme:

gauche effect ⇒ ring pucker ⇒ $n \rightarrow \pi^*$ interaction ⇒ trans/cis ratio

The $n \rightarrow \pi^*$ interaction resembles the approach of a nucleophile to the electrophilic carbon of an acyl group. Accordingly, the $n \rightarrow \pi^*$ interaction is strongest when Oi–1 is proximal to Ci ′and along the Bürgi–Dunitz trajectory to Ci ′=Oi. In polypeptides, this geometry occurs in the right-and left-handed α-helix and the PPII helix (Figure 3B), which is the conformation assumed by the strands of the collagen triple helix (Figure 1A).

Table 1. Correlation of ring pucker with collagen triple-helix stability.[38-40,44] *In a crystalline triple helix, proline residues have $\phi = -73°$, $\psi = 164°$ in the Xaa position and $\phi = -60°$, $\psi = 150°$ in the Yaa position.*[23]

		Residue Ring Pucker		Triple Helix T_m	
		C$^\gamma$-*endo*	C$^\gamma$-*exo*	$(XaaProGly)_7$	$(ProYaaGly)_7$
Flp	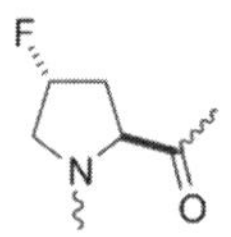	14%	86% $\phi = -55°$ $\psi = 140°$	no helix	45 °C
Hyp (natural)	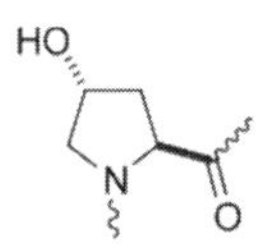			no helix	36°C
Pro (natural)	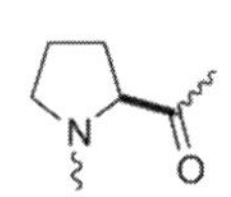	66%	34%	6 °C	6 °C
hyp	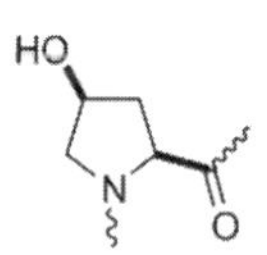			no helix	no helix
flp	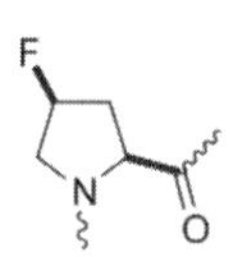	95% $\phi = -76°$ $\psi = 172°$	5%	33 °C	no helix

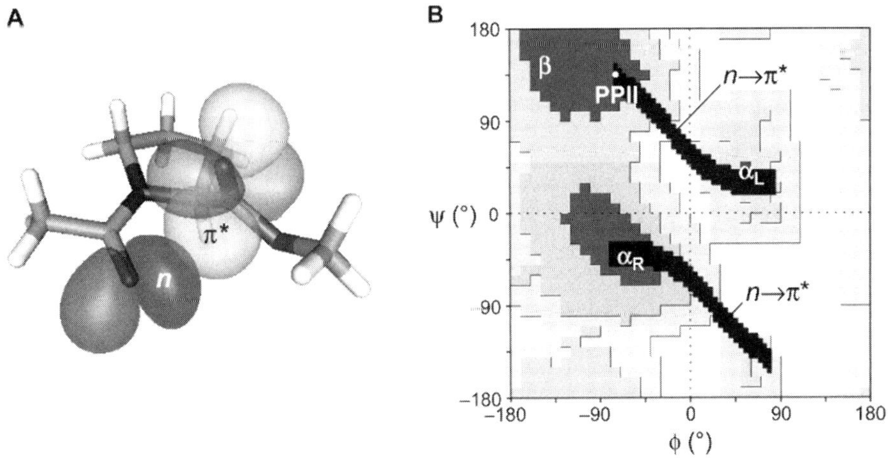

Fig. 3. The n→π interaction. (A) Natural bond orbitals depicting the n→π* interaction between O_{i-1} and C_i'' in AcProOMe with a trans peptide bond and $C^{\#}$-exo ring pucker.*[43] *(B) Ramachandran plot showing the regions where the n→π* interaction will be strongest. In these regions, the $O_{i-1}\cdots C_i'$ ≤ distance is 3.2 Å and the $O_{i-1}\cdots C_i'{=}O_i$ angle is ≥ 99° and ≤ 119° The white dot indicates the ϕ and ψ angles for the PPII helix assumed by the strands of a collagen triple helix.*[43]

The $n{\rightarrow}\pi^*$ interaction can occur only if the peptide bond containing O_{i-1} is trans (*i.e.*, *Z*), as opposed to cis (*E*). Accordingly, the trans/cis ratio of an Xaa_{i-1}–Pro_i peptide bond can report on the strength of an $n{\rightarrow}\pi^*$ interaction. We have used the trans/cis ratio of a peptide bond to quantify the energetics of an $n{\rightarrow}\pi^*$ interaction.

We found that the trans/cis ratio of the amide bond in *N*-formylproline phenylesters correlates with electron-withdrawal by a *para* substituent (Figure 4).[45] The slope of the Hammett plot ($\rho = 0.26$) is indicative of a substantial effect. Density functional theory calculations and natural bond order analysis indicated that this

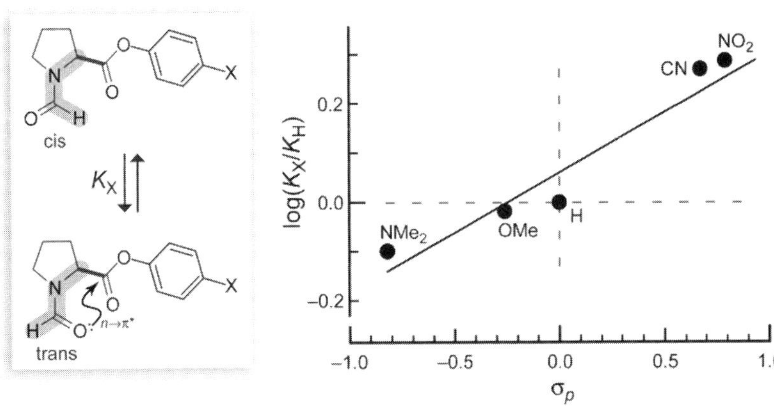

Fig. 4. Hammett plots depicting the relationship between the trans/cis ratio and electron-withdrawing ability of X in Fm-ProOC$_6$H$_4$ p-X.[45] *(A) Values of K_X were determined by ^{1}H NMR spectroscopy, and yield $\rho = 0.26$.*

effect arises from a favorable $n \rightarrow \pi^*$ interaction between the amide oxygen and ester carbon.

In textbooks, the well-known preference of the peptide bond for the *trans* conformation is attributed to steric effects. Our data indicate that this preference cannot be explained by steric effects alone. Rather, the $n \rightarrow \pi^*$ interaction, which is only extant in the trans isomer, contributes significantly to this preference. We found the $n \rightarrow \pi^*$ interaction to be worth $\Delta G° \cong -0.7$ kcal/mol in a simple model system.[39,43] This free energy is now being exploited in the *de novo* design of peptides, peptoids, and proteins by other workers.[46-48] Hence, we propose that the $n \rightarrow \pi^*$ interaction should be added to the list of noncovalent interactions that direct a polypeptide chain to assume a particular folded structure.

Longer Synthetic Collagen—Self-Assembly. One long-term goal of our work is to develop collagen-based biomaterials with tunable attributes that can be used as both collagen surrogates and templates for nanotechnological applications. A major barrier in achieving this goal is that triple helices derived from synthetic peptides are much shorter (<10 nm) than natural collagen (~300 nm). Inspired by the self-assembly of double-helical fragments of DNA, we envisioned that sticky-ended fragments of synthetic collagen could self-assemble into long triple helices. Though unlike the situation with DNA, there is no "code" for the noncovalent association of collagen strands, other than the need for one Xaa, Yaa, and Gly residue to be in each cross section of a triple helix. We have demonstrated how to access long triple helices with molecular self-assembly.

We synthesized short collagen fragments in which the three strands were held in a staggered array by disulfide bonds (Figure 5).[49,50] These fragments were synthesized directly on a solid support by using a strategy based on the orthogonal deprotection of cysteine residues. Data from circular dichroism spectroscopy, dynamic light scattering, analytical ultracentrifugation, atomic force microscopy, and transmission electron microscopy indicated that these "sticky-ended" fragments self-assembled via intermolecular triple helix formation. The resulting fibrils resembled natural collagen, and some were longer (nearly 1 μm) than any known collagen. Assemblies with Yaa=Hyp tended to be longer and more stable than those with Yaa=Pro. Additional control over the length and stability was attained by modulating the temperature and solvent. Minimalist fragments like these can be elaborated by chemical synthesis to display motifs that promote cell adhesion for engineering tissues, lateral packing for accessing two- and three-dimensional architectures, and metal coordination for producing nanowires. Hence, we anticipate that our self-assembly strategy can provide synthetic collagen-mimetic materials for a variety of applications.

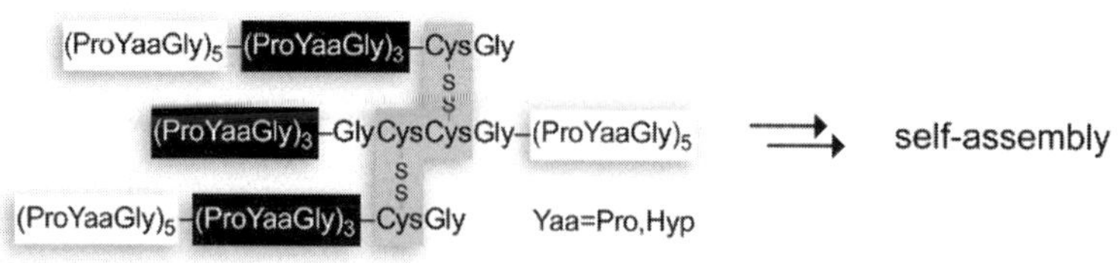

Fig. 5. Structure of a synthetic collagen fragment that self-assembles into fibrils of 1-nm width and nearly 1-μm length.[49,50]

Acknowledgments

The 2007 Makineni Lectureship is dedicated to the graduates students and postdoctorates who have worked with me on collagen, especially Dr. Frank W. Kotch, who performed the experiments in Ref. 49 and 50. This work was supported by grant AR44276 (NIH).

References

1. Myllyharju, J. and Kivirikko, K. I. *Ann. Med.* **33**, 7-21 (2001).
2. Ricard-Blum, S., Ruggiero, F. and van der Rest, M. *Top. Curr. Chem.* **247**, 35-84 (2005).
3. Schweitzer, M. H., Suo, Z., Avci, R., Asara, J. M., Allen, M. A., Arce, F. T. and Horner, J. R. *Science* **316**, 277-280 (2007).
4. Asara, J. M., Schweitzer, M. H., Freimark, L. M., Phillips, M. and Cantley, L. C. *Science* **316**, 280-285 (2007).
5. Willerslev, E., Hansen, A. J., Binladen, J., Brand, T. B., Gilbert, M. T. P., Shapiro, B., Bunce, M., Wiuf, C., Gilichinsky, D. A. and Cooper, A. *Science* **300**, 791-795 (2003).
6. Binladen, J., Gilbert, M. T. and Willerslev, E. *Biol. Lett.* **3**, 55-56 (2007).
7. Radzicka, A. and Wolfenden, R. *J. Am. Chem. Soc.* **118**, 6105-6109 (1996).
8. Smith, R. M. and Hansen, D. E. *J. Am. Chem. Soc.* **120**, 8910-8913 (1998).
9. Wolfenden, R. and Snider, M. J. *Acc. Chem. Res.* **34**, 938-945 (2001).
10. Schroeder, G. K., Lad, C., Wyman, P., Williams, N. H. and Wolfenden, R. *Proc. Natl. Acad. Sci. USA* **103**, 4052-4055 (2006).
11. Ramshaw, J. A. M., Shah, N. K. and Brodsky, B. *J. Struct. Biol.* **122**, 86-91 (1998).
12. Prockop, D. J. and Kivirikko, K. I. *Annu. Rev. Biochem.* **64**, 403-434 (1995).
13. Prockop, D. J. *Matrix Biol.* **16**, 519-528 (1998).
14. Prockop, D. J. *Biochem. Soc. Trans.* **27**, 15-31 (1999).
15. Byers, P. H. *Clin. Genet.* **58**, 270-279 (2000).
16. Werkmeister, J. A. and Ramshaw, J. A. M. (Eds.) *Collagen Biomaterials*, Elsevier Science, Barking, Essex, England, 1992.
17. Ramshaw, J. A. M., Werkmeister, J. A. and Glattauer, V. *Biotechnol. Genet. Eng. Rev.* **13**, 335-382 (1995).
18. Lynn, A. K., Yannas, I. V. and Bonfield, W. *J. Biomed. Mater. Res. B Appl Biomater.* **71**, 343-354 (2004).
19. Johnson, G., Jenkins, M., McClean, K. M., Griesser, H. J., Kwak, J., Goodman, M. and Steele, J. G. *J. Biomed. Mater. Res.* **51**, 612-624 (2000).
20. Towe, K. M., In Billingham, J. (Ed.) *Life in the Universe*, MIT Press, Cambridge, MA, 1981.
21. Guzman, N. A. (Ed.) *Prolyl Hydroxylase, Protein Disulfide Isomerase, and Other Structurally Related Proteins*, Marcel Dekker, New York, 1998.
22. Myllyharju, J. *Matrix Biol.* **22**, 15-24 (2003).
23. Bella, J., Eaton, M., Brodsky, B. and Berman, H. M. *Science* **266**, 75-81 (1994).
24. McCaldon, P. and Argos, P. *Proteins: Struct. Funct. Genet.* **4**, 99-122 (1988).
25. Bulleid, N. J., Wilson, R. and Lees, J. F. *Biochem. J.* **317**, 195-202 (1996).
26. Walmsley, A. R., Batten, M. R., Lad, U. and Bulleid, N. J. *J. Biol. Chem.* **274**, 14884-14892 (1999).
27. Snellman, A., Keranen, M.-R., Hagg, P. O., Lamberg, A., Hiltunen, J. K., Kivirikko, K. I. and Pihlajaniemi, T. *J. Biol. Chem.* **275**, 8936-8944 (2000).
28. Byers, P. H. *Philos. Trans. R. Soc. Lond. B Biol. Sci.* **356**, 151-158 (2001).
29. Friedman, L., Higgin, J. J., Moulder, G., Barstead, R., Raines, R. T. and Kimble, J. *Proc. Natl. Acad. Sci. USA* **97**, 4736-4741 (2000).
30. Fields, G. B. and Prockop, D. J. *Biopolymers* **40**, 345-357 (1996).
31. Jenkins, C. L. and Raines, R. T. *Nat. Prod. Rep.* **19**, 49-59 (2002).
32. Engel, J. and Bächinger, H. P. *Top. Curr. Chem.* **247**, 7-33 (2005).
33. Raines, R. T. *Protein Sci.* **15**, 1219-1225 (2006).

34. Berg, R. A. and Prockop, D. J. *Biochem. Biophys. Res. Comm.* **52**, 115-120 (1973).
35. Bella, J., Brodsky, B. and Berman, H. M. *Structure* **3**, 893-906 (1995).
36. Holmgren, S. K., Taylor, K. M., Bretscher, L. E. and Raines, R. T. *Nature* **392**, 666-667 (1998).
37. Holmgren, S. K., Bretscher, L. E., Taylor, K. M. and Raines, R. T. *Chem. Biol.* **6**, 63-70 (1999).
38. Bretscher, L. E., Jenkins, C. L., Taylor, K. M., DeRider, M. L. and Raines, R. T. *J. Am. Chem. Soc.* **123**, 777-778 (2001).
39. DeRider, M. L., Wilkens, S. J., Waddell, M. J., Bretscher, L. E., Weinhold, F., Raines, R. T. and Markley, J. L. *J. Am. Chem. Soc.* **124**, 2497-2505 (2002).
40. Hodges, J. A. and Raines, R. T. *J. Am. Chem. Soc.* **125**, 9262-9263 (2003).
41. Hodges, J. A. and Raines, R. T., In Chorev, M. and Sawyer, T. K. (Eds.) *Peptide Revolution: Genomics, Proteomics & Therapeutics: Proceedings of the 18th American Peptide Symposium*, Kluwer Academic, Norwell, MA, 2004, p. 472-473.
42. Shoulders, M. D., Hodges, J. A. and Raines, R. T. *J. Am. Chem. Soc.* **128**, 8112-8113 (2006).
43. Hinderaker, M. P. and Raines, R. T. *Protein Sci.* **12**, 1188-1194 (2003).
44. Inouye, K., Sakakibara, S. and Prockop, D. J. *Biochim. Biophys. Acta* **420**, 133-141 (1976).
45. Hodges, J. A. and Raines, R. T. *Org. Lett.* **8**, 4695-4697 (2006).
46. Haduthambi, D. and Zondlo, N. J. *J. Am. Chem. Soc.* **128**, 12430-12431 (2006).
47. Kümin, M., Sonntag, L. S. and Wennemers, H. *J. Am. Chem. Soc.* **129**, 466-467 (2007).
48. Gorske, B. C., Bastian, B. L., Geske, G. D. and Blackwell, H. E. *J. Am. Chem. Soc.* **129**, 8928-8929 (2007).
49. Kotch, F. W. and Raines, R. T. *Proc. Natl. Acad. Sci. USA* **103**, 3028-3033 (2006).
50. Kotch, F. W. and Raines, R. T., In Blondelle, S. E. (Ed.) *Understanding Biology Using Peptides: Proceedings of the 19th American Peptide Symposium*, Springer, New York, NY, 2006, p. 688-689.

The Vincent du Vigneaud Award*

2006— Sam Gellman, University of Wisconson- Madison
Barbara Imperiali, Massachusetts Institute of Technology
2004—Steven B. H. Kent, University of Chicago
Dieter Seebach, ETH Zurich
2002—Robert S. Hodges, University of Colorado-Denver
Horst Kessler, Technical University, Munich
2000—Charles M. Deber, University of Toronto
Richard A. Houghten, Torrey Pines Institute for Molecular Studies
1998—Peter W. Schiller, Clinical Research Institute of Montreal
James A. Wells, Genentech, Inc.
1996—Arthur M. Felix, Hoffmann-La Roche, Inc.
Richard G. Hiskey, University of North Carolina
1994—George Barany, University of Minnesota-Minneapolis
Garland R. Marshall, Washington University-St. Louis
1992—Isabella L. Karle, Naval Research Laboratory
Wylie W. Vale, The Salk Institute for Biological Studies
1990—Daniel H. Rich, University of Wisconsin-Madison
Jean E. Rivier, The Salk Institute for Biological Studies
1988—William F. De Grado, DuPont Central Research
Tomi K. Sawyer, The Upjohn Company
1986—Roger M. Freidinger, Merck Sharpe & Dohme
Michael Rosenblatt, Massachusetts General Hospital
James P. Tam, The Rockefeller University
1984—Betty Sue Eipper, The Johns Hopkins University
Lila M. Gierasch, University of Delaware
Richard E. Mains, The Johns Hopkins University

*Sponsored by BACHEM Inc.

Peptide Society Travel Grants

Name	Institution	Country
Christopher J Armishaw	University of Copenhagen	Denmark
Luigia Auriemma	University of Naples	Italy
Angeles Bonache	University of Florence	Italy
Monica Branco	University of Delaware	USA
Yann Brouillette	Université de Montpellier	France
Jason Chang	University of Oxford	UK
Stijn Claerhout	University of Leuven	Belgium
Daniel J Cline	University of North Carolina	USA
Sonya Cressman	University of British Columbia	Canada
Dattachow Bhaswati	University of Kansas	USA
Vikram Deep	National institute of Pharmaceutical Education & Research Mohali	India
Nina Deppermann	University of Giessen Intitute of Organic Chemistry	Germany
Mark J Dixon	University of Bath	UK
Ronald C Elgersma	Medical Chemistry & Chemical Biology	USA
Cinzia Exposito	University of Salerno	USA
Wei-Jie Fang	University of Kansas	USA
Steve M Fernandes	University of Arizona	USA
Lucille Fischer	Cnrs, Ibmc, Upr 9021	France
Jordan Fletcher	Monash University	Australia
Christian W Gruber	University of Queensland	Australia
Sunithi Gunasekera	University of Queensland	Australia
Lisa Ann Haines-Butterick	University of Delaware	USA
Zvi Hayouk	Hebrew University of Jerusalem	Israel
Kai Hilpert	University of British Columbia	Canada
Mohammed Akhter Hossain	University of Melbourne	Australia
David C Ireland	University of Queensland	Australia
Manoj Kurian Jacob	Rajiv Gandhi Centre for Biotechnology	India
Jatinder S Josan	University of Arizona	USA
Joseph C Kappel	UC Davis Medical Center	USA
Daniel Knappe	University of Leipzig	Germany
Dipankar Koley	India Institute of Chemical Technology	India
Joeri Kuil	University of Utrecht	USA
Hee-Kyoung Lee	University of Utah	USA
Yaniv Linde	The Hebrew University	Israel
Veronica Maes	Vrije Univesiteit Brussel	Belgium
Luciana Malavolta	Universidale Federal de Sao Paulo	Brazil

Name	Institution	Country
Marcus Marra	Federal University of Sao Paulo	Brazil
Norman Metanis	Technion - Israel Institute of Technology	USA
Byoun J Min	University of Arizona	USA
Barbara Mulinacci	University of Florence	Italy
Nadine Pannier	University of Giessen	Germany
Emanuel Perugia	Weizmann Institute of Science	Israel
Stephan Pritz	Leibniz-Institute of Molecular Pharmacology	Germany
Bettina Proneth	University of Florida	USA
Nir Qvit	The Hebrew University	Israel
Arianna Rath	The Hospital for Sick Children	Canada
Feliciana Real-Fernandez	University of Florence	Italy
Fabio Rizzolo	University of Florence	Italy
Luisa Ronga	Universita Degli Studi Di Napoli Federico ii & institut Des Biomolecules Max Mousseron	Italy
Nicolette Christa Ross	The University of Kansas	USA
Marina Sala	University of Naples	Italy
Daphne Salick	University of Delaware	USA
Mario Scrima	University of Salerno	Italy
Rohit Kumar Sharma	National Institute of Pharmaceutical Education & Research (NIPER)	India
Pavla Simerska	University of Queensland	Australia
Maciej Stawikowski	Florida Atlantic University	USA
Theresa K Tiefenbrunn	The Scripps Research Institute	USA
Larissa Yeomans	The University of Arizona	USA
Yan Zhe	University of Colorado	USA

* Thanks to generous donations from our sponsors, “Peptides for Youth” succeeded in awarding more travel grants than previous meetings, nearly doubling the amount of support for travel awards to young investigators.

Bert L. Schram Young Investigator Mini Symposium
Generously supported by the ESCOM Science Foundation

Winners:
Masa Cemazar, University of Queensland
Arianna Rath, University of Toronto

Participants:
Chris Armishaw, Copenhagen University
Monica Branco, University of Delaware
Marie-Stephanie Camus, Ecole Polytechnique Federale de Lausanne
Pawell Czechura, University of Ottawa
Zvi Hayouka, Hebrew University
Kai Hilpert, University of British Columbia
Manoj Kurian Jacob, Rajiv Gandhi Centre for Biotechnology
Kazuki Kawahara, Osaka University
Jennifer Lanning, University of Chicago
Jeffrey-Tri Nguyen, Kyoto Pharmaceutical University
Bettina Proneth, University of Florida
Nir Qvit, Hebrew University
Shai Rahimipour, Bar Ilan University
Nicole Van der Weerden, La Trobe University
Larisa Yeomans, University of Arizona

Young Investigator Poster Competition/Peptide Idol
Generously Supported by C.S. Bio and Amgen

First Place:
Meritxell Teixido, University of Barcelona

Second Place:
Swapna Bagwanth, University of Minnesota

Participants in Peptide Idol:
Jennifer Chaytor, University of Ottawa
Bhaswati Dattachowdhury, University of Kansas
Feliciana Real Fernandez, University of Florence
Norman Metanis, The Scripps Research Institute
Theresa Tiefenbrunn, The Scripps Research Institute

LIBRARY GENERATION ANALYSIS AND COMBINATORIAL CHEMISTRY

Dynamic Combinatorial Assembly of Peptide-Rhenium Coordinates: Application to the Selection of hCyp-18 inhibitors from a Library of 12 × 16 Components

C. Dugave*, C. Clavaud, J. Le Gal, R. Thai and M. Moutiez

CEA, Institute of Biology and Technology of Saclay (iBiTecS), building 152, 91191 Gif-sur-Yvette, FRANCE.

Introduction

Self-assembly of peptides and pseudopeptides through oxorhenium coordination by a composite NS_2+S motif enables the easy synthesis of libraries of oxorhenium complexes in various solvents including buffers [1]. The kinetically-controlled coordination may be tuned by introducing a reversible step, in example the complex dissociation by an exogenous thiol (Figure 1). In these conditions, the effector of highest affinity is selected by the protein from a dynamic combinatorial library (DCL) and hence, accumulates (protein protecting/enhancing effects) to the detriment of other combinations that give complexes of lower affinity [2].

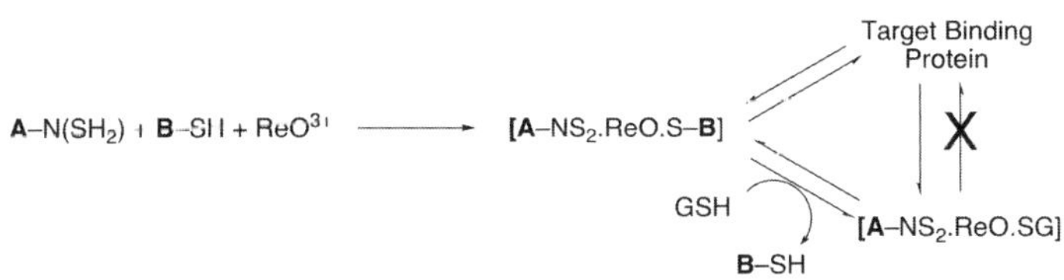

Fig. 1. Formation of Dynamic combinatorial libraries through oxorhenium coordination.

We applied this concept to the selection of modular ligands of hCyp-18, an important peptidyl-prolyl isomerase (PPIase). Each complex was composed of an **A** motif which is anticipated to bind the S1-S1' sub-site while module B is supposed to bind the S2'-S3' sub-site of the enzyme. Glutathion (GSH) was employed for substituting the **B**–S moiety of oxorhenium complexes that display a low affinity for hCyp-18. This strategy was applied to the selection of novel inhibitors of the PPiase activity of hCyp-18 from a DCL of 192 components.

Results and Discussion

Sixteen combinations of twelve modules **A**–NS_2 with a single module $(\mathbf{B}\text{–S})_2$ were readily performed after reduction of modules and trans-chelation from $[ReO.(gluconate)_2]$ in a 35 mM Hepes buffer pH 7.8 overnight at room temperature (192 complexes). The chemical nature of modules clearly influences both the yield of coordination and the ability of complexes to ionize by electrospray mass spectrometry (positive mode). Therefore, all data were standardized for accurate comparison of results. Complexes were detected by selection of the typical $^{185/187}Re^{32/34}S_3$ isotopic motif.

Integration of the ionic current corresponding to a given m/z for different concentrations in GSH and hCyp-18 showed that all 192 complexes formed in buffer except 4 of them which were not observed, though they were obtained using $[Bu_4N^+.ReOCl_4^-]$ in methanol. All oxorhenium complexes dissociate upon addition

S. Del Valle et al. (eds.), *Peptides for Youth: The Proceedings of the 20th American Peptide Symposium*, 3
DOI: 10.1007/978-0-387-73657-0_1, © Springer Science+Business Media, LLC 2009

of GSH, however, 5 complexes displayed a significant resistance to exogenous thiols. This result can be explained by the “hCyp-18 protective effect” which was assumed to reflect the tight binding of the complexes to the enzyme. The standard trypsin-coupled kinetic assay showed that these compounds are able to inhibit the PPIase activity of hCyp-18 in the micro to sub-micromolar range. Conversely, complex **III-3a**, which is highly sensitive to GSH in presence of hCyp-18, is not able to inhibit the enzyme. In the case of complexes **III-1b**, **III-5b** and **III-12n**, the ‘hCyp-18 protecting effect’ was clearly related to an interaction of the complex with the active site since these compounds inhibited the PPIase activity of the enzyme with IC_{50} in the submicromolar range (Table 1). In contrast to complex **III-1b**, S-acetamidomethylated module **Bb** (Acm-(N-Ac)Cys-*pNA*) binds to the active site and inhibits the PPIase activity with an apparent Kd of 158 ! M.

Table 1. Apparent Kd[a] and IC_{50}[b] of complexes selected from the dynamic chemical library III.

Complex #	Structure	App. Kd (! M)	IC_{50} (! M)
III-1a ([*L*], X = O, R = H_2^+)		107 ± 4	ND
III-1b ([*L*], X = O, R = Ac)	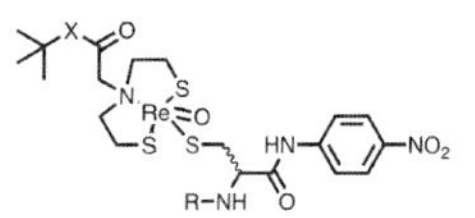	2 ± 0.1	5.3 ± 0.2
- ([*D*], X = O, R = Ac)		158 ± 9	ND
III-2g (X = NH, R = BnNHCO)		ND	3.5 ± 0.3
III-3a	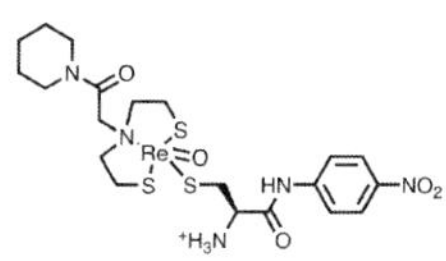	ND	> 10,000
III-5b	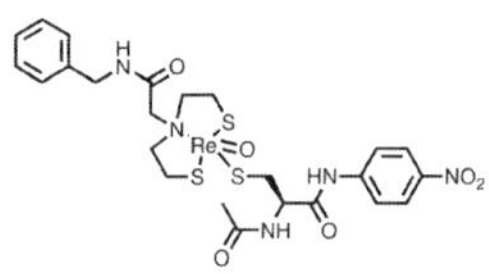	0.3 ± 0.04	0.28 ± 0.05
III-6n (X = CH)	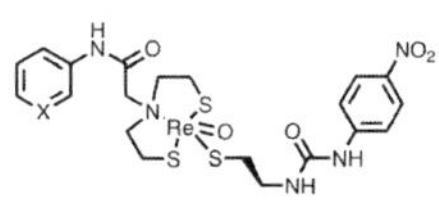	ND	1.6 ± 0.2
III-12n (X = N)		0.3 ± 0.03	0.16 ± 1
Cyclosporin A		0.3 ± 0.04	0.016 ± 0.04

[a]*Fluorescence assay, [hCyp-18] = 320 nM;* [b]*Trypsin-coupled PPIase assay, [hCyp-18] = 16 nM.*

As previsouly observed, hCyp-18 also facilitates the coordination process in absence of GSH for complexes **III-5b**, **III-6n** and **III-12n**, suggesting that the components assemble at the active site (in situ coordination chemistry). Conversely, this ‘hCyp-18 enhancing effect’ was not observed for complex **III-1b** though it displayed a micromolar affinity for cyclophilin. This seems to confirm the coexistence of two distinct pathways (assisted/not assisted) for the assembly of oxorhenium complexes through a dynamic combinatorial process.

References

1. Clavaud, C., *et al. ChemBioChem.* **7**, 1352-1355 (2006).
2. Ramström, O., Lehn, J-M. *Nature Rev. Drug Discovery*, **1**, 26-36 (2002).

Dynamic Combinatorial Assembly of Peptide-Rhenium Coordinates: Coordination Chemistry for Generating New Analogs of hCyp-18 Ligands

C. Clavaud, J. Le Gal, R. Thai, M. Moutiez and C. Dugave*

CEA, Institute of Biology and Technology of Saclay (iBiTecS), building 152, 91191 Gif-sur-Yvette, FRANCE.

Introduction

We recently described the oxorhenium-mediated combinatorial assembly of series of n modules **A**–N(SH)$_2$ with n' modules **B**–SH to provide libraries of n x n' complexes [**A**–NS$_2$.ReO.S–**B**]. Libraries of peptide-oxorhenium complexes were screened towards the human cyclophilin hCyp-18, an important peptidyl-prolyl isomerase (PPIase) which participates in protein folding as well as the regulation of the activity of many proline-containing peptides and proteins [1]. The active site of hCyp-18 includes the S1-S1' subsite which is anticipated to bind an amino acyl-prolyl analog (**A** moiety), and the S2'-S3' subsite which binds to an aminoacyl-(*para*-nitroanilide) derivative (**B** moiety) [1].

In a previous report, we described the systematic screening of libraries of 288 composite substrate analogs of hCyp-18 and the selection of the medium affinity hCyp-18 inhibitor **II-8i** ($IC_{50} = 12 \pm 2$ μM). In the present work, we investigated the cyclophilin-directed selection of compound **II-8i** from a 16 component dynamic combinatorial library of oxorhenium complexes in buffer [2].

Results and Discussion

Reduction and combination of 4 modules **A**–NS$_2$ with 4 modules (**B**–S)$_2$ through oxorhenium coordination was readily performed either in methanol using [NBu_4^+.$ReOCl_4^-$], or by trans-chelation from [ReO.(gluconate)$_2$] in a 35 mM Hepes buffer pH 7.8 overnight at room temperature, to yield a mixture of 16 complexes, most of them as a mixture of 2 diastereomers. As previously observed, addition of exogenous thiols such as glutathione (GSH) resulted in a fast **B**–S/thiol substitution which might be used for adding a reversible step to the kinetically-controlled coordination process. This should lead to the amplification of high-affinity components to the detriment of complexes that display weaker affinities for hCyp-18 (Figure 1).

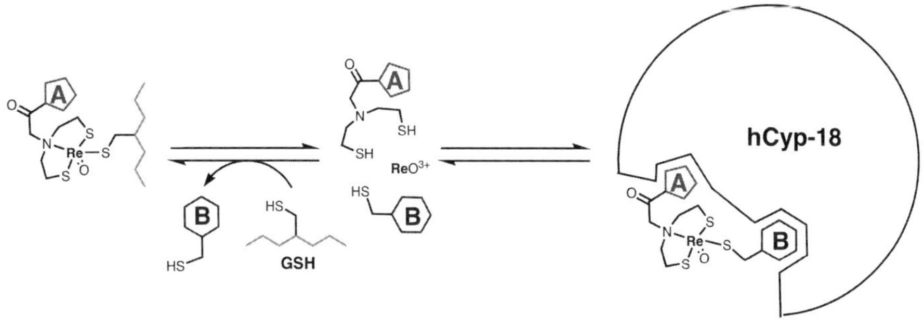

Fig. 1. hCyp-18-directed selection of a dynamic chemical library of oxorhenium complexes

S. Del Valle et al. (eds.), *Peptides for Youth: The Proceedings of the 20th American Peptide Symposium*, 5
DOI: 10.1007/978-0-387-73657-0_2, © Springer Science+Business Media, LLC 2009

After quenching of the reaction upon sudden pH decrease and dissolution in methanol, the products were identified by LC-MS analysis (positive mode) after selection of the typical $^{185/187}Re^{32/34}S_3$ isotopic motif. Addition of hCyp-18 to the incubation mixture did not affect the complexation process significantly (i.e. complex **II-3o** which does not bind hCyp-18) except in the case of complex **II-8i** whose formation was tightly related to the concentration in cyclophilin. These complexes (including **II-8i**) readily dissociate by simple thiol exchange upon addition of increasing concentrations in GSH. Conversely, addition of hCyp-18 efficiently protected some compounds against GSH substitution. In particular, complex **II-8i** partially resist to GSH substitution (up to 1.0 equiv. GSH relative to module concentration) whereas compound **II-3o** did not. This suggests that hCyp-18 is able to protect the bound complex from external reagents (cyclophiline protective effect) and therefore, indirectly confirms the affinity of **II-8i** for hCyp-18 (Figure 2).

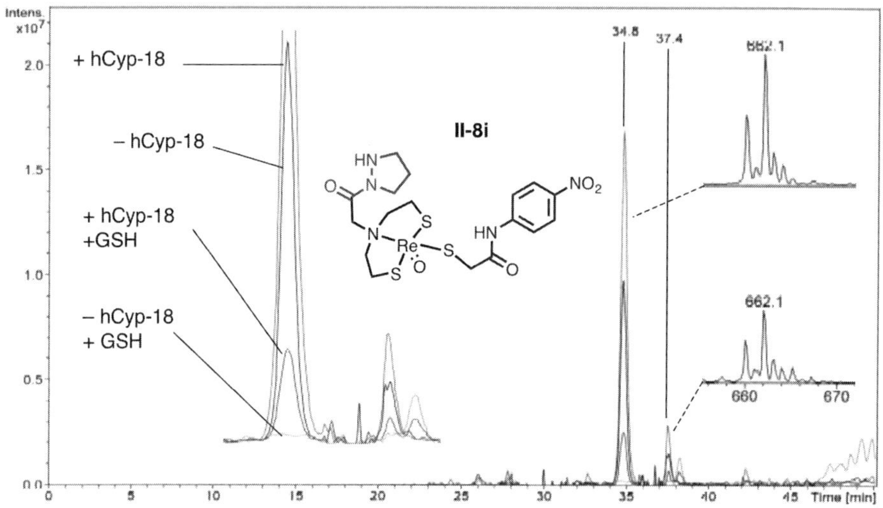

*Fig. 2. Integrated ionic current of complex **II-8i** formed under various conditions (the complex appeared as a mixture of 2 syn/anti diastereomers).*

We observed that hCyp-18 not only increases the overall yield of oxorhenium complexation by modules **A8** and **Bi**, but also improves the kinetics of formation of complex **II-8i**, suggesting a cyclophilin-directed assembly of the complex (cyclophilin enhancing effect). In this respect, the mechanism of cyclophilin-dependent selection of oxorhenium complexes might involve a site-specific assembly of modules **A8** and **Bi**, though simultaneous addition of both modules only resulted in a weak binding of compound **Bi** to hCyp-18 (Kd ≈ 500 μM). The coexistence of a dual selection process (protective effect plus enhancing effect), as well as the influence of the oxorhenium core on the binding, may be evoked to explain these results.

References

1. Clavaud, C., *et al. ChemBioChem.* **7**, 1352-1355 (2006).
2. Ramström, O., Lehn, J-M. *Nature Rev. Drug Discovery*, **1**, 26-36 (2002).

Oxorhenium-Mediated Cyclization and Self-Assembly of Peptides: A Combinatorial Approach for the Selection of New Ligands of the Human Cyclophilin hCyp-18

C. Clavaud, M. Heckenroth, C. Stricane, M-A. Lelait, A. Ménez and C. Dugave*

CEA, Institute of Biology and Technology of Saclay (iBiTecS), building 152, 91191 Gif-sur-Yvette, FRANCE.

Introduction

Rhenium belongs to the group VIIb transition metals and hence tightly binds to electron-rich elements such as N, O, P, S which are commonly found in biomolecules including peptides. In particular, coordination of the oxorhenium core (ReO^{3+}) by a $N(CH_2\text{-}CH_2S)_2$ motif and a free thiol (respectively noted NS_2 and S), unambiguously leads to a "NS_2.ReO.S" complex which may be used either to cyclize small peptides or to assemble independent peptide modules via a combinatorial approach. In this respect, reduction and oxorhenium-mediated assembly of series of n modules **A**–NS_2 and n' modules (**B**–S)$_2$ was expected to provide libraries of n × n' complexes [**A**–NS_2.ReO.S–**B**] which might be screened towards a target protein, in example the human cyclophilin hCyp-18.

HCyp-18 is an important peptidyl-prolyl isomerase (PPIase) which is involved in the protein folding as well as the regulation of the activity of many proline-containing peptides and proteins. The active site of hCyp-18 is divided into 2 subsites: the S1-S1' subsite which specifically binds to various amino acyl-prolyl motifs and analogs, and the S2'-S3' subsite which shows a marked preference for an aminoacyl-(*para*-nitroanilide) (Aa-*p*NA) derivative in model substrates and peptide ligands [1].

Firstly, we investigated the effect of the bulky oxorhenium core on the binding of a cyclic substrate analog to hCyp-18, and a series of linear metallated peptides which were anticipated to bind to the protein. Then, we screened libraries of composite substrate analogs of hCyp-18 of general formula [**A**–NS_2.ReO.S–**B**], readily assembled through oxorhenium coordination.

Results and Discussion

Peptides and peptide analogs were prepared in parallel as previously reported [2,3]. Metallation of each compound using [NBu_4^+.$ReOCl_4^-$] in methanol led to the formation of a green precipitate which was isolated after repeated centrifugation/washing cycles and was of sufficient purity for screening except the cyclic peptide **1** which was purified by RP-HPLC. The biochemical activity of these compounds was evaluated using the routine fluorescence assay and was confirmed with the trypsin-coupled PPIase assay using Suc-Ala-Ala-Pro-Arg-*p*NA as a substrate [4].

Compounds **1**, **2a and 3c** (Figure 1) bind to hCyp-18 with affinities slightly better than the corresponding S-acetamidomethylated peptide βAla-Ala-Pro-Cys (Acm)-*p*NA (Table 1). This result suggests that introduction of an oxorhenium core close to the active site has only little effect on the binding of the peptide to the S2'-S3' subsite which contains the fluorescent Trp residue. Evaluation of the IC_{50} of complexes **2a** and **3c** confirmed that both of them bind to the active site and inhibit the PPIase activity of hCyp-18.

S. Del Valle et al. (eds.), *Peptides for Youth: The Proceedings of the 20th American Peptide Symposium*, 7
DOI: 10.1007/978-0-387-73657-0_3, © Springer Science+Business Media, LLC 2009

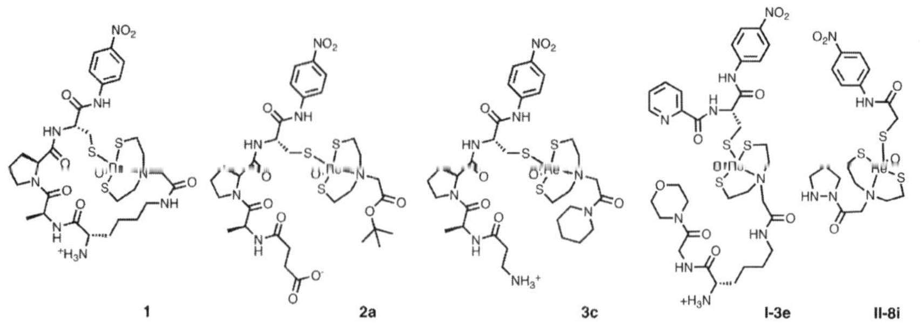

1 **2a** **3c** **I-3e** **II-8i**

Two libraries of oxorhenium complexes [**A**–NS_2.ReO.S–**B**], where **A** contains an analog of the aminoacyl proline motif and **B** is an analog of Aa-*p*NA, were synthesized as two matrixes of 7 × 16 and 11 × 16 complexes respectively. Screening of all 288 compounds revealed the preferential binding of complexes **I-3e** and **II-8i** which bind to hCyp-18 with affinities improved by more than one order of magnitude whereas separate modules that compose complexes **I-3e** and **II-8i** have lower affinities. Complex **II-8i** was shown to specifically inhibit the PPIase activity of hCyp-18 with an IC_{50} of about 12 μM (Table 1).

Table 1. Biochemical activity of cyclic and acyclic oxorhenium complexes.

Compounds	Kd_{app} (μM)[a] ([hCyp-18] = 320 nM)	IC_{50} (μM)[b] ([hCyp-18] = 16 nM)
Suc-Ala-Ala-Pro-Phe-*p*NA	135 ± 20	540 ± 70
βAla-Ala-Pro-Cys(Acm)-*p*NA	195 ± 29	288 ± 45 μM
1	108 ± 2	ND
2a	65 ± 8	51 ±24 μM
3c	75 ± 5	109 ± 26 μM
I-3e	15 ± 1	ND
II-8i	11 ± 1	12 ± 2 μM

[a]*Fluorescence quenching at 322 nm (λ_{act} = 285 nm) at 20°C;* [b]*Trypsine-coupled PPIase assay at 10°C;* [a,b]*(in Hepes buffer 35 mM pH 8.6).*

These results demonstrated that it is possible to synthesize new inhibitors of hCyp-18 by oxorhenium-mediated assembly of 2 independent modules which mimic respectively the P1-P1' and the P2'-P3' motifs of peptide ligands.

References

1. Dugave, C. in *Cis-Trans Isomerization in Biochemistry*, Wiley-VCH, 261-294 (2006).
2. Clavaud, C., *et al. Bioconjugate Chem.* **17**, 807-814 (2006).
3. Clavaud, C., *et al. ChemBioChem.* **7**, 1352-1355 (2006).
4. Dugave, C. in *Cis-Trans Isomerization in Biochemistry*, Wiley-VCH, 143-166 (2006).

Thieno[3,2-*e*]diazepinediones and 3-Thienylimidazolidinediones Syntheses: Regio-Controlled Ring Opening of Thiaisatoic Anhydride by α-Amino Acids.

Yann Brouillette,* Vincent Lisowski and Jean Martinez

Institut des Biomolécules Max-Mousseron (IBMM), CNRS UMR 5247,UMI-UMII, UFR de Pharmacie, 15 Avenue Charles Flahault, 34093 Montpellier Cedex 5, France.
yann.brouillette@univ-montp1.fr

Introduction

In previous biological studies we demonstrated that two specific p38 mitogen-activated protein (MAP) kinase inhibitors, derived from imidazole (SB202190 and SB203580), were also able to directly and selectively interact with a G-protein-coupled receptor, the cholecystokinin receptor subtype CCK1. In the same way, it was also shown that two CCK benzodiazepines antagonists (L364,718 and L365,260) were able to regulate p38 MAP kinase activity [1]. These cross-interactions between these MAP kinase inhibitors and these CCK receptor antagonists with the CCK1 receptor and the p38 MAP kinase suggested that new kinase inhibitors could be designed on the model of CCK receptor antagonists. The chemistry of diazepines being exhaustively described in the literature with the benzene series, we have focused our chemical work on the bioisosteric thieno[1,4] diazepine moiety, that has already proven its effectiveness in medicinal chemistry to design drugs such as Clotiazepam. Our first chemical goal was to develop an efficient methodology to access this scaffold from the regio-controlled ring opening of thiaisatoic anhydride.

Results and Discussion

The reactivity of such thiaisatoic anhydride **1** was investigated towards various nucleophiles such as amines [2] and alcohols [3]. In an opposite way to isatoic anhydride [4], its benzene analogue, the nucleophilic attack only proceeded on the carbonyl group of the carbamate function and not on the carboxylic carbonyl, leading respectively to ureidoacids and carbamates derivatives. However, the recent synthesis of pyrrolothienodiazepines demonstrated that in the particular cases of proline and hydroxyproline, the nucleophilic attack can favourably be oriented towards the carboxylic carbonyl of the oxazinedione **1** [5][6].

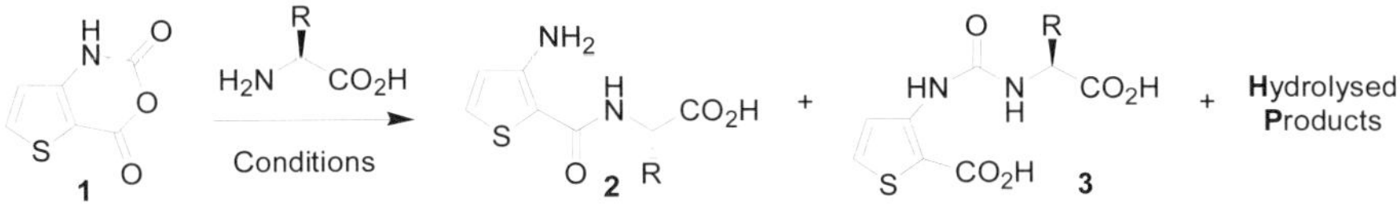

*Table 1. Nucleophilic opening conditions of thiaisatoic anhydride **1**.*

entry	solvent	T (°C)	Time (h)	Ala (mol%)	**1** (%)	**2** (%)	**3** (%)	**HP** (%)
a	Dioxane-H_2O (1:1)	100	2	150	0	21	29	50
b	Dioxane	100	14	150	59	0	41	0
c	H_2O	100	4	150	0	0	0	100
d	H_2O	40	18	1000	0	75	15	10
e	H_2O-Et_3N (1.1)	rt	0.1	150	0	0	100	0

S. Del Valle et al. (eds.), *Peptides for Youth: The Proceedings of the 20th American Peptide Symposium*, 9
DOI: 10.1007/978-0-387-73657-0_4, © Springer Science+Business Media, LLC 2009

Therefore, we investigated the access to bicyclic thienodiazepines from acyclic natural α-amino acids and thiaisatoic anhydride (Table 1).

The method describing condensation of thiaisatoic anhydride **1** with proline analogs, in a mixture of dioxane-water (1:1) at 100°C for 1h, was not suitable when acyclic α-amino acids were used. When *L*-alanine (Ala, R=Me) was reacted in these conditions (entry a, Table 1), only a mixture of amide **2** and urea **3** were recovered in equal amount along with hydrolysed products. Finally, an efficient regio-selectivity was obtained in neutral protic conditions at 40°C to afford the amide **2** (entry d, Table 1). Nevertheless, in the same protic conditions, in the presence of a base like Et_3N, we demonstrated that the latest reactivity towards the same amino acid can be totally reversed to form only the ureidoacid **3** (entry e, Table 1) [7].

We took advantage of these results to develop efficient methodologies to synthesize two libraries of optically pure 3,4-dihydro-1*H*-thieno[3,2-*e*][1,4]diazepine-2,5-diones **4** and 3-(thien-3-yl)imidazolidine-2,4-diones **5** by condensation of thiaisatoic anhydride **1** with all natural α-amino acids [7][8] (Scheme 1).

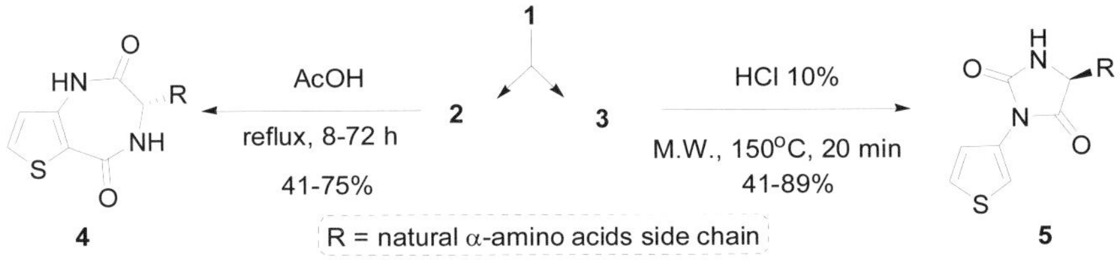

*Scheme 1. Syntheses of thienodisazepinediones and thienylimidazolidinediones from **1**.*

The further diversification of these scaffolds offers a practical means for making novel libraries of structurally diverse potential kinase inhibitors.

Acknowledgments

This work was supported by grants from *Société de Chimie Thérapeutique* (SCT) and *les Laboratoires Servier*.

References

1. Morel, C., Ibarz, G., Oiry, C., Carnazzi, E., Berge, G., Gagne, D., Galleyrand, J. C., Martinez, J. *J. Biol. Chem.* **280**, 21384-21393 (2005).
2. Le Foulon, F. X., Braud, E., Fabis, F., Lancelot, J. C., Rault, S. *J. Comb. Chem.* **7**, 253-257 (2005).
3. Fabis, F., Jolivet-Fouchet, S., Robba, M., Landelle, H., Rault, S. *Tetrahedron* **54**, 10789-10800 (1998).
4. Gates, M. *J. Org. Chem.* **45**, 1675-1681 (1980).
5. Lisowski, V., Fabis, F., Pierre, A., Caignard, D. H., Renard, P., Rault, S. *J. Enzym. Inhib. Med. Chem.* **17**, 403-407 (2002).
6. Jolivet-Fouchet, S., Fabis, F., Rault, S. *Tetrahedron Lett.* **39**, 5369-5372 (1998).
7. Brouillette, Y., Lisowski, V., Martinez, J. *J. Org. Chem.* **72**, 2662-2665 (2007).
8. Brouillette, Y., Lisowski, V., Guillon, J., Massip, S., Martinez, J. *Tetrahedron*, **63**, 7538-7544 (2007).

Solid-Phase Synthesis of Phosphinic Dipepetide Isosteres and β Amino Acids via Activated N-Terminal Acrylamides

Florian Manzenrieder and Horst Kessler

Department Chemie, Lehrstuhl II für Organische Chemie, Technische Universität München, Lichtenbergstrasse 4, D-85747 Garching, Germany.

Introduction

In recent years proteases have increasingly become important drug targets. An often used very powerful approach is the replacement of the scissile peptide bond of the substrate by a dipeptide isostere, thereby generating a protease inhibitor[1]. Usually Fmoc protected dipeptide isosteres are synthesized in solution and then used in solid-phase peptide synthesis (SPPS). However, most dipeptide isostere syntheses are time consuming and in some cases, many steps and harsh purification are needed leading to low overall yields. Phosphino dipeptide (PDP) isosteres[1] are among the possible dipeptide isosteres. The P-C bond of PDP isosteres is usually formed by Michael type addition of activated hypophosphite to an acrylate. Linkage of the acrylate on the growing peptide in SPPS provides an acrylamide that is much less active in the Michael type additions. In the literature, SPPS of PDP isosteres has required reaction temperatures of 100°C[2] which are not compatible with Fmoc SPPS protection groups. To overcome the drawbacks of solution synthesis of phosphino dipeptide isostere, we describe here a direct and mild synthesis of phosphino dipeptide isostere containing peptides, that may be suitable for SPPS.

Results and Discussion

The general strategy was to change the unreactive acrylamides on solid phase by covalently attaching an eletron-withdrawing group to the amide moiety, thereby generating a reactive Michael acceptor. A well studied motif is the o-nitrobenzenesulfonyl (oNBS) moiety which was extensively investigated in our lab for

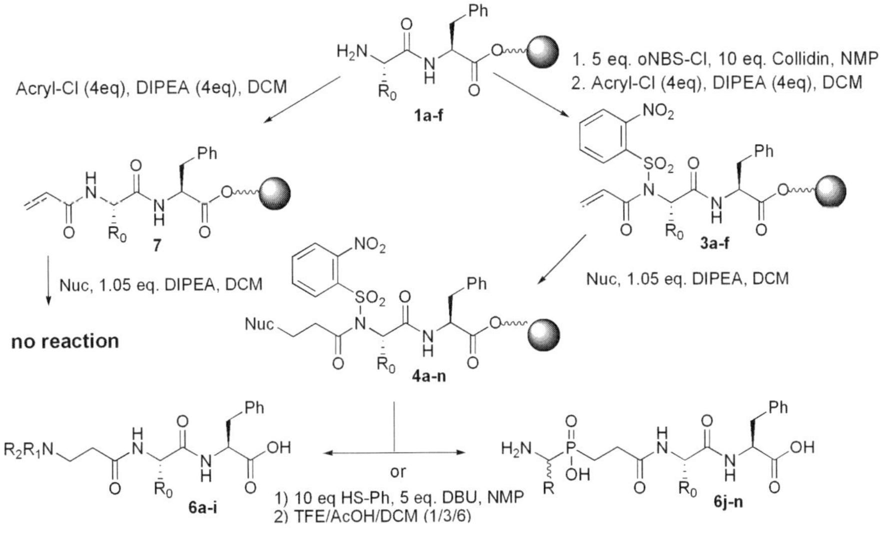

Fig. 1. SPPS of phosphino dipeptide isosteres and β amino acids.

DOI:10.1007/978-0-387-73657-0_5, © Springer Science+Business Media, LLC 2009

Table 1. Isolated 1,4 addition produtcts.

Entry	R_1	R_2	Aminoacid (R_0)	Yield[a]/ (Ratio[b])	Entry	R	R_1	R_2	Aa (R_0)	Yield[a]/ (Ratio[b])
6a	Bn	Bn	Ile	85/(94)	6h	-	Allyl	Allyl	Ile	64/(67)
6b	Bn	Bn	Ser(tBu)	79/(95)	6i	-	H	$CHPh_2$	Ile	77/(84)
6c	Bn	Bn	Lys(Boc)	59/(68)	6j	*i*butyl	-	-	Ile	60/(68)
6d	Bn	Bn	Asn(Trt)	78/(95)	6k	methyl	-	-	Ile	63/(71)
6e	Bn	Bn	Glu(tBu)	74/(85)	6l	*n*butyl	-	-	Ile	59/(65)
6f	Bn	Bn	Cys(Trt)	49/(68)	6m	*i*propyl	-	-	Ile	60/(70)
6g[c]	H	H	Ile	45/(nd)	6n	2-butyl	-	-	Ile	59/(65)

[a] *isolated yield after HPLC purification* [b] *ratio of 1,4 addtion to [(1,4)+(1,2) addition] analyzed by HPLC-HPLC-MS* [c] *the used nucleophile was HN-Trt*

N-methylation of peptide bonds.[3] It can easily and orthogonally be removed by use of thiols. The procedure involves (1) amine activation and protection by the *o*NBS group, (2) *N*-acrylation of the sulfonamide by direct *N*-acrylation with acrylchloride, (3) Michael addition with different nucleophiles, and (4) selective removal of the *o*NBS group with thiol compounds and DBU to generate the amide bond. First results were encouraging but showed a major side reaction in the 1,2 addition to the activated acrylic residue, which cleaves the acrylate from the resin. To avoid this unwanted reaction, the amount of the nucleophile as well as the reaction time was optimized. After establishing the reaction conditions for every step we used the following protocol for the Michael addition on SP. The N^{α}-acryl-N^{α}-*o*NBS peptides **3a-f** were treated with different nucleophiles (1eq, an excess should be avoided) and in case of amines with additional DIPEA (1eq) for 30 min in NMP to yield **4a-n**. The nucleophiles used to build up PDP isosteres were synthesized from the readily available Fmoc-aminoalkylphosphonic acids converted to the corresponding bis(trimethylsilyl) phosphonite using bis(trimethylsilyl)acetamide in chloroform at elevated temperatures and reacted for 2h with activated acrylamide on SP. After the Michael addition the *o*NBS group was directly removed by thiophenol (10eq) and DBU (5eq) in NMP for 30 min. In case of the PDP isostere Fmoc was also removed due to the strong base DBU. Cleavage from the resin was achieved by treatment with TFE/AcOH/DCM (1/3/6) for 3h. The crude products **6a-n** were analyzed by HPLC-MS. Compatibility with Fmoc-SPPS concerning side chain protection was investigated by treating dipeptides of the type **3b-f** with standard protection groups under the above described protocol. All reactions to the 1,4 addition products were isolated in good to moderate yields. The obtained data is shown in table 1.

Conclusion

We described a practical and efficient protocol for Michael addition to activated acrylamides on SP. This new approach for SPPS of phosphino dipeptide isostere containing peptides by Michael type addition of activated Fmoc phosphino amino acids to *o*NBS activated acrylamides may enable the fast synthesis of phosphino peptide libraries compatible with Fmoc-SPPS side chain protection.

References

1. Manzenrieder, F., Frank, A. O., Huber, T., Dorner-Ciossek, C. and Kessler, H. *Bioorg. Med. Chem.* **15**, 4136-4143 (2007).
2. Buchardt, J., Meldal, M. *J. Chem. Soc., Perkin Trans. 1* (**19**), 3306-3310 (2000).
3. Biron, E., Kessler, H. *J. Org. Chem.* **70** (13), 5183-5189 (2005).

Parallel Solid-Phase Synthesis of Macrocyclic Peptidomimetics Using Ru-Catalyzed Ring-Closing Metathesis and a New Application of Cross Metathesis

Sylvie Beaubien*, Hamid R. Hoveyda, Martin Vézina, Éric Marsault, Hugo Tremblay, Laurence Foucher and Kamel Benakli

Tranzyme Pharma Inc., 3001, 12e Avenue Nord, Sherbrooke (PQ), Canada J1H 5N4

Introduction

As part of our efforts towards the development of methodologies for the parallel synthesis of small molecule macrocyclic peptidomimetics, we report a new application of Ru-catalyzed ring-closing metathesis (RCM) cyclative-release for the solid phase synthesis of this class of molecules.[1] In addition, we achieved the synthesis of linker **C4a**, during the course of these studies, through the use of an unusually selective cross-metathesis (CM) reaction.[2]

Chemistry

The synthesis of linker **C4a** began with the CM reaction between but-4-en-1-ol and allyl cyanide, two Type I olefins.[3] The success of this reaction depends both on the Ru-complex and the concentration of the reaction.[2] The nitrile intermediate **1** thus obtained by CM was then subjected to the synthesis route shown in Scheme **1** to effect both the chemoselective reduction of the nitrile moiety and the protection of the resultant amine in a one-pot procedure.

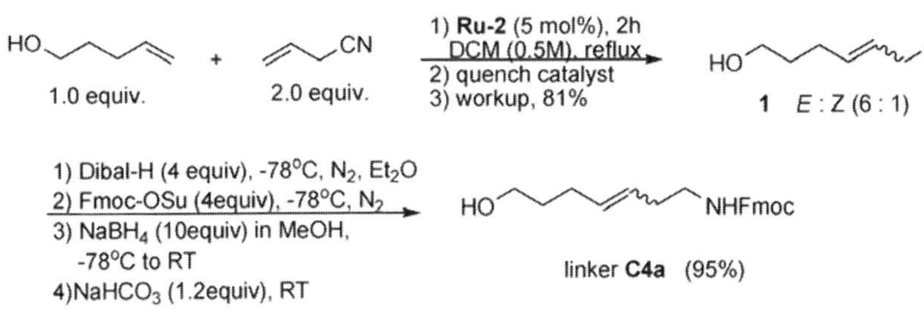

*Scheme 1. Synthesis of linker **C4a** by CM reaction.*

The general synthesis of macrocycles by parallel synthesis on solid phase utilizing RCM as a key step is represented in Scheme **2**.[1] Macrocycles were successfully synthesized with varied ring sizes (12- to 18-membered), a diversity of amino acids and different combinations of stereochemistries in the tripeptide unit. This methodology allowed the efficient introduction of aliphatic, aromatic and sugar-derived tethers. Four sequences were chosen to explore optimization of the RCM cyclative-release using different solvents and Ru catalysts. The two best RCM conditions were: Grubbs-Hoveyda[4] catalyst in 1,2-DCE and Zannan[5] catalyst in DCM at 40°C for 24 h with 10 mol% of Ru-catalyst. These conditions were applied to 25 sequences to determine the scope and limitations of this RCM methodology.

Results

This RCM approach can be used for the synthesis of cyclic peptides with different types of tethers (aliphatic, aromatic and sugar-derived) and the ease of macrocycle

S. Del Valle et al. (eds.), *Peptides for Youth: The Proceedings of the 20th American Peptide Symposium*,
DOI: 10.1007/978-0-387-73657-0_6, © Springer Science+Business Media, LLC 2009

formation follows the order: 16 > 18 > 15 = 17 > 14-membered ring when comparable tripeptide backbones are used. With pre-organized substrates, LDL configuration, NMeAla at AA_2 and Aib at AA_2, improved yields are obtained compared to non-pre-organized congeners (LLL). Seven different possibilities for stereochemistry of the tripeptide were tested. These configurations can be sorted into three classes: high yield (98% to 50%): LDL, moderate yield (50% to 20%): DLL, LDD, pre-organized LLL and low yield (20% to 1%): LLL, DDD, DDL, DLD. A wide variety of macrocycles comprised of amino acids with diverse stereochemistry linked by different tethers were obtained using this RCM approach. RCM cyclative-release is an efficient and robust solid-phase synthetic strategy enabling rapid, parallel synthesis of this new class of macrocyclic peptidomimetics.

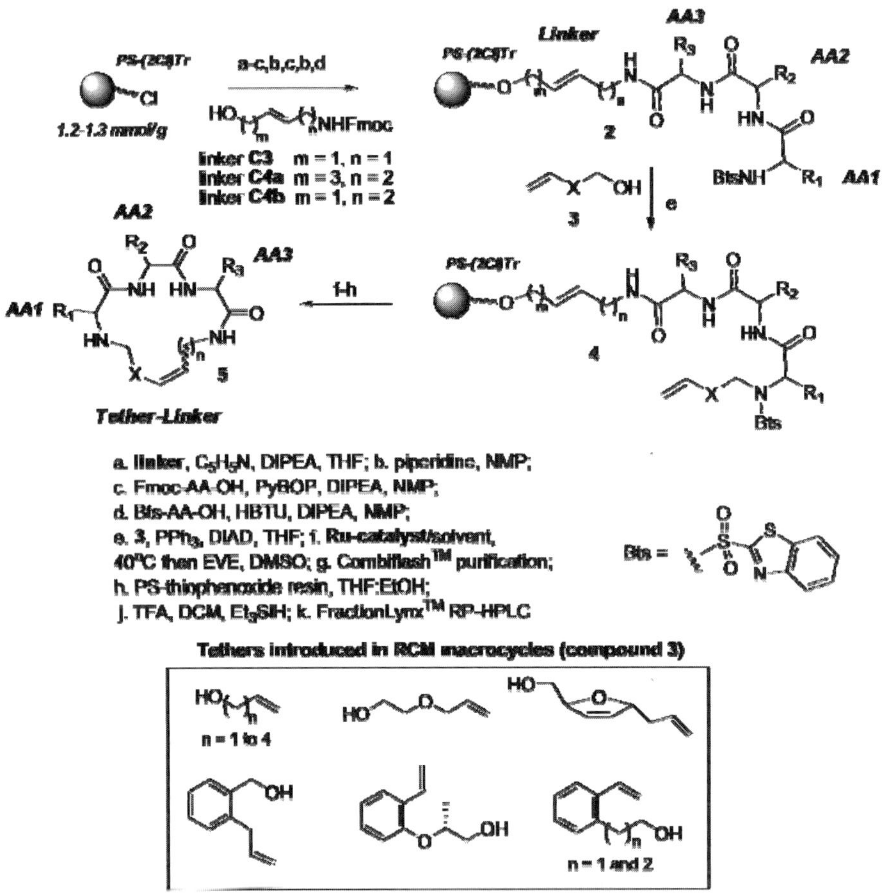

Scheme 2. General Solid-Phase Approach to Macrocyclic Peptidomimetics using RCM.

References

1. For an application see: Marsault, E. et al., *J. Med. Chem.* **2006**, 7190.
2. Hoveyda, H. R.; Vézina, M. *Org. Lett.* **2005**, 2113.
3. Grubbs, R. H. et al. *J. Am. Chem. Soc.* **2003**, 11360.
4. Hoveyda, A. H. et al. *J. Am. Chem. Soc.* **1999**, 791.
5. China Patent No. CN2005100803792.

High Throughput Solid Phase Parallel Synthesis of Macrocyclic Peptidomimetics

E. Marsault*, H. R. Hoveyda, M. L. Peterson, R. Gagnon, M. Vézina, J.F. Pinault, A. Landry, C. Saint-Louis, L. G. Ouellet, S. Beauchemin, K. Benakli, S. Beaubien, M. Brassard, Z. Wang, M. Champagne, F. Galaud, N. Fortin, D. Fortin, V. Plourde, M. Ramaseshan, S. Bhat, F. Bilodeau, D. Lonergan, R. Lan, S. Li, G. Berthiaume, L. Foucher, X. Peng, Y. Dory and P. Deslongchamps

Tranzyme Pharma Inc., 3001, 12e Avenue Nord, Sherbrooke (PQ), Canada J1H 5N4

Introduction

We hereby present a novel class of macrocyclic peptidomimetics exemplified by structure **1** (Figure 1), enabled by an efficient solid phase parallel synthesis method for producing macrocycles in large numbers.[1] Macrocycle **1** is composed of a tripeptide cyclized backbone-to-backbone by a nonpeptidic tether, and also bears a secondary amine. The synthetic method relies on solid phase parallel synthesis on a polystyrene support via a thioester linker, and makes use of the Irori™ platform.

Figure 1. Representative structure of macrocyclic peptidomimetics (L stereochemistry indicated for reference)

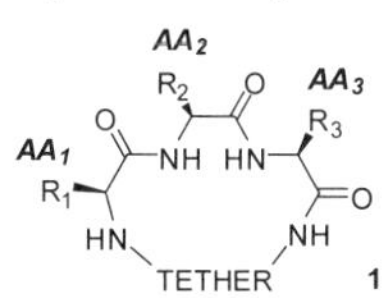

1

Results and Discussion

The use of the Ddz group allowed orthogonal deprotection in the presence of Boc, tBu esters and tBu ether protection on the side chains (Scheme 1). Using standard peptide chemistry, tripeptide **4** was elongated on a thioester linker, followed by attachment of the tether via a Fukuyama-Mitsunobu reaction allowed by the use of a Bts moiety for activation and protection on the AA_1 moiety.[2] Subsequent Ddz deprotection gave macrocyclization precursor **6**. Macrocyclization takes advantage of a cyclative release approach,[3] and was performed using DIPEA in THF, aided by the presence of a macroporous carbonate resin. Silver trifluoroacetate was found to generally enhance yields of macrocyclization.[4] Subsequent Bts deprotection and side chain protective group acidolysis gave the desired product **1**.

The method was applied to the generation of Tranzyme Pharma's HitCreate™ library of 35,000 chemically and conformationally diverse macrocycles, with the size of macrocycles varying from 14- to 22-membered. At each position, a broad diversity of amino acids was introduced including L-, D-, α-, β-, γ- amino acids, with acidic, basic, aromatic, aliphatic, polar, neutral and cyclic amino acids. Tether diversity included aromatic, aliphatic, heteroaromatic, sugar, ether and amine functionalities. The use of this technology for the discovery of potent antagonists to the motilin receptor has already been reported.[5]

S. Del Valle et al. (eds.), Peptides for Youth: The Proceedings of the 20th American Peptide Symposium,
DOI: 10.1007/978-0-387-73657-0_7, © Springer Science+Business Media, LLC 2009

Macrocyclization yield was generally increased by the use of alternating stereochemistry on the amino acids, or by the use of a turn-inducing element at position AA_2. Inversely, the use of a bulky amino acid in AA_3 reduced the yield. Silver trifluoroacetate was generally found to increase not only the yield of macrocyclization, but also accelerated the rate of the reaction. It was found to be particularly efficient for cyclization of molecules containing more sterically hindered amino acids such as Val in AA_3.

***Scheme 1**. Solid phase parallel synthesis of macrocycles*

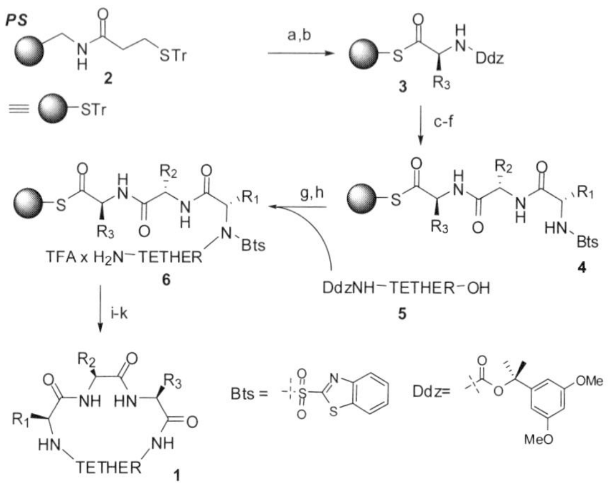

a. Reagents and conditions:
(a) 10%, TFA, Et_3SiH, DCM; (b) Ddz-AA_3-OH, PyBOP, DIPEA, NMP; (c) 2% TFA, Et_3SiH, DCM; (d) Ddz-AA_2-OH, HBTU, DIPEA, NMP; (e) 1%TFA, Et_3SiH, DCM; (f) Bts-AA_1-OH, HBTU, DIPEA, NMP; (g) **5**, PPh_3, DIAD, THF; (h) 2% FA, Et_3SiH, DCM; (i) $Ag(OCOCF_3)$, DIPEA, THF, MP-carbonate; (j) PS-thiophenol, KOTMS, THF:EtOH; (k) 50% TFA, Et_3SiH, DCM.

References

1. For parallel synthetic methods to obtain macrocyclic peptidomimetics, please refer to (a) Blackwell, H. E.; Sadowsky, J. D.; Howard, R. J.; Sampson, J. N.; Chao, J. A.; Steinmetz, W. E.; O'Leary, D. J.; Grubbs, R. H. J. Org. Chem. 2001, 66, 5291; (b) Barrett, A. G. M.; Hennessy, A. J.; Le Vézouët, R.; Procopiou, P. A.; Seale, P. W.; Stefaniak, S.; Upton, R. J.; White, A. J. P.; Williams, D. J. *J. Org. Chem.* **2004**, *69*, 1028; (c) Shi, Z. D.; Lee, K.; Wei, C. Q.; Roberts, L. R.; Worthy, K. M.; Fisher, R. J.; Burke, T. R. *J. Med. Chem.* **2004**, *47*, 788.
2. Vedejs, E.; Lin, S.; Klapars, A.; Wang, J. *J. Am. Chem. Soc.* **1996**, *118*, 9796.
3. (a) Li, W.; Burgess, K. *Tetrahedron Lett.* **1999**, *40*, 6527; (b) Nicolaou, K. C.; Pastor, J.; Winssinger, N.; Murphy, F. *J. Am. Chem. Soc.* **1998**, *120*, 5132; (c) Pernerstorfer, J.; Kramer, T. *Solid Phase Chemistry: Methods and Principles in Medicinal Chemistry*, **2000**, *9*, 99.
4. Zhang, L.; Tam, J. P. *J. Am. Chem. Soc.* **1999**, *121*, 3311.
5. Marsault, E.; Hoveyda, H. R.; Peterson, M. L.; Saint-Louis, C., Landry, A.; Vézina, M.; Ouellet, L.; Wang, Z.; Ramaseshan, M.; Beaubien, S.; Benakli, K.; Beauchemin, S.; Peeters, T.; Déziel, R.; Fraser, G. *J. Med. Chem.* **2006**, *49*, 7190.

Development of an Efficient Solid-Phase Synthetic Methodology to Construct a Combinatorial Library of a Potent HDAC Inhibitor

Salvatore Di Maro[1,2] and Jung-Mo Ahn*[1]

[1]*Department of Chemistry, University of Texas at Dallas, Richardson, TX 75080, USA;* [2]*Department of Pharmacological and Toxicological Chemistry, University of Naples "Federico II", Naples, Italy*

Introduction

Acetylation is one of the key post-translational modifications and regulates gene transcription and expression in eukaryotes. Two families of enzymes responsible for this process are histone acetyltransferases (HATs) and histone deacetylases (HDACs), and HDACs are in general classified into three, class I (HDAC1, 2, 3, and 8), class II (HDAC4, 5, 6, 7, 9, 10, and 11), and class III (SIRT1-7) [1]. Since they serve vital functions in many cellular pathways, HDACs have been intensively investigated and their link to carcinogenesis processes was well demonstrated. This correlation provides a rationale employing HDAC inhibitors as potential therapeutic agents for cancers, and a number of compounds have been developed as inhibitors against HDACs (class I and class II). They cause cell cycle arrest, differentiation and apoptosis in tumor cells, and depending on their structures HDAC inhibitors can be grouped into short-chain fatty acids (e.g., sodium butyrate), hydroxamic acids (e.g., TSA and SAHA), electrophilic ketones (e.g., Trapoxin A), benzamide (e.g., MS-275), and cyclic peptides (e.g., apicidin and FK228) [2].

Isolated from *Chromobacterium violaceum*, FK228 (also known as depsipeptide) possesses a unique chemical structure and is found to be the most potent HDAC inhibitor *in vitro* and *in vivo* [3]. Its highly constrained bicyclic structure constructed with several unnatural amino acids makes it extremely stable *in vivo* and promotes facile membrane penetration to find its cellular targets. Its first synthesis was reported by Li and coworkers with moderate yield (18% overall yield) [4], however several synthetic challenges associated with its preparation prevented structure-activity relationship studies which are generally practiced to achieve higher potency and selectivity.

Results and Discussion

Thus, we have designed a novel structural analogue of FK228 so that a large number of FK228 analogues can be produced with high efficiency. To achieve this, we have modified the most synthetically challenging moiety, (3*S*,4*E*)-3-hydroxy-7-mercapto-

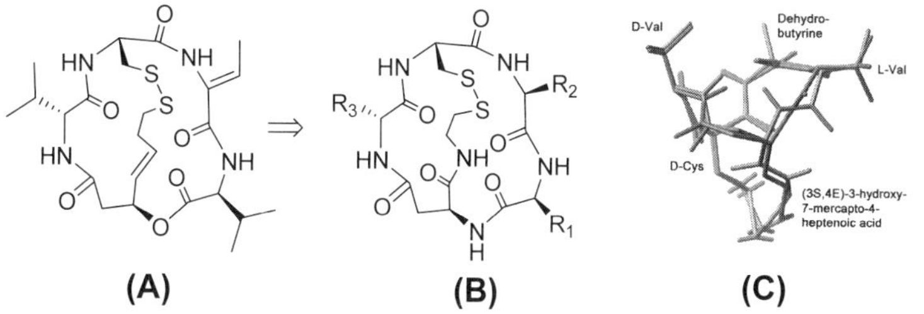

Fig. 1. Structures of (A) FK228, (B) its modified structure, and (C) their superimposition.

S. Del Valle et al. (eds.), *Peptides for Youth: The Proceedings of the 20th American Peptide Symposium*, 17
DOI: 10.1007/978-0-387-73657-0_8, © Springer Science+Business Media, LLC 2009

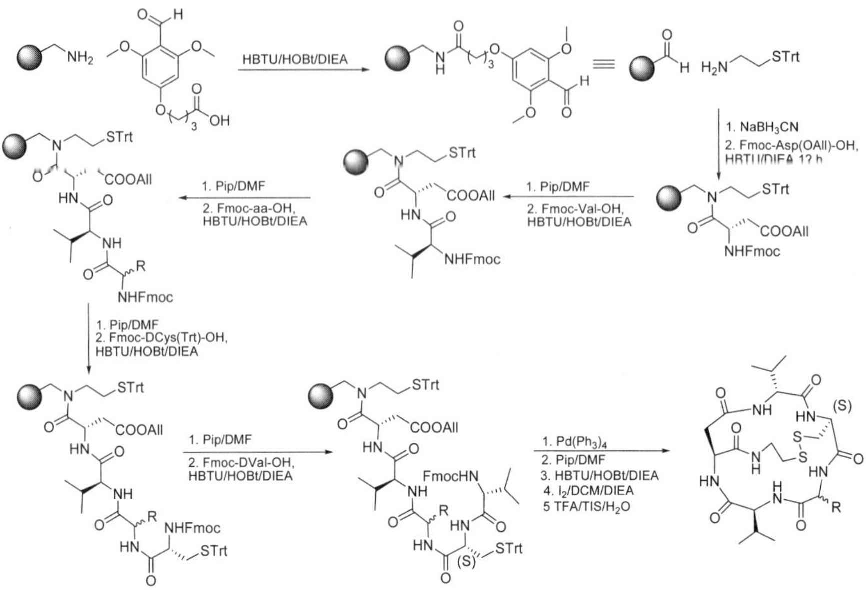

Scheme 1. Synthesis of FK228 analogues.

heptenoic acid, to a structure which can be easily assembled using readily available chemicals (Fig. 1). The conformation of the modified FK228 analogue was studied using MacroModel, and a Monte Carlo conformation search showed nearly identical structure compared to the native FK228.

The synthesis of FK228 analogues was developed using standard *N*-Fmoc/*t*-butyl solid-phase peptide synthesis (Scheme 1). Starting with a PALdheyde linker coupled to aminomethyl resin, a cysteamine was anchored by reductive amination, followed by coupling amino acids to prepare a linear and reduced intermediate. After deprotection of allyl and Fmoc groups, a macrolactam was formed, and the resulting monocyclic peptide was further oxidized using iodine to produce a bicyclic FK228 analogue. This synthetic scheme allowed us to prepare a number of FK228 analogues with high yields and purity.

Acknowledgments

The work was supported by University of Texas at Dallas and Welch Foundation (J.-M.A. AT-1595); and University of Naples "Federico II" (S.D.M).

References

1. Hassig, C. A. and Schreiber, S. L. *Curr. Opin. Chem. Biol.* **1**, 300-308 (1997).
2. Monneret, C. *Eur. J. Med. Chem.* **40**, 1-13 (2005).
3. Furumai, R., *et al. Cancer Res.* **62**, 4916-4921 (2002).
4. Li, K. W., *et al. J. Am. Chem. Soc.* **118**, 7237-7238 (1996).

A Novel Hydrogel Functionalized with Specific Peptidomimetic Ligands for 2-D and 3-D Cell Culture

Juntao Luo, Ekama Onofiok, Changying Shi, Ruiwu Liu, Kit S. Lam*

Division of Hematology and Oncology, Department of Internal Medicine, UC Davis Medical Center, 4501 X street, Sacramento, California 95817 USA

Introduction

It is believed that there is a close relationship between the three-dimensional (3-D) architecture of the cells and their differentiation. The morphologic cell integrity in the 3-D cell culture as well as the numerous cell–matrix and cell–cell interactions promotes intercellular communications, which favors the proper biochemical regulation of the tumor cell *in vivo*. Various 2-D and 3-D supports based on natural or synthetic materials have been used for the cell culture. PVA is a promising biomaterial accepted by food and drug organizations (e.g., FDA) in various countries, and widely used for biomedical applications such as artificial blood vessels, synthetic vitreous, and implantable medical materials because of its bioinertness. PVA is ready to be functionalized and PVA aqueous solution below 15% remains to be liquid at room temperature which is feasible for the conjugation of biomolecules onto PVA matrix. Agarose is a natural biomaterial which forms gel in water at room temperature and is widely used for electrophoresis as well as cell culture.

In this study, a hybrid PVA-agarose hydrogel has been explored combining the advantage of easy modification of PVA and stable gelation of agarose. The stable hybrid hydrogels have been observed at room temperature with the concentration of agarose above 0.2% together with the varying concentration of PVA. The time of gelation is decreased with the increasing concentration of agarose. LLP2A, a highly specific peptidomimetic ligand targeting to the α4β1 integrin on lymphoma cell surface [1], as well as other peptide ligands targeting to various cancer cells have been identified in our lab by OBOC combinatorial chemistry [2].

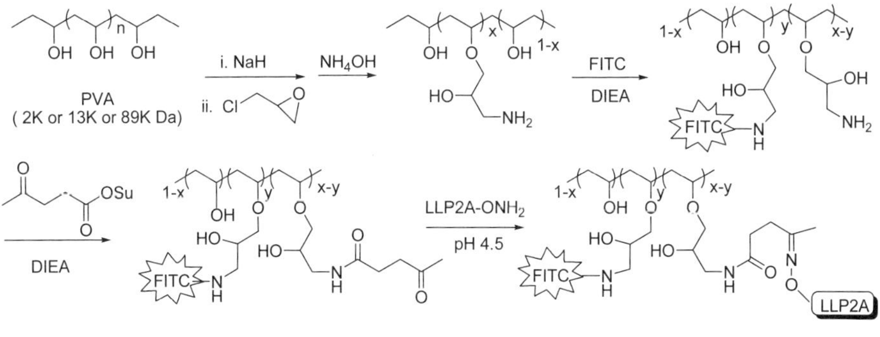

Scheme 1. Modification of PVA with amino groups and the conjugation of FITC and ligand (LLP2A) onto PVA scaffold.

Results and Discussion

Primary amino groups have been introduced onto PVA polymers with varying molecular weights via chemically inert ether bonds (Scheme 1). The substitution of

DOI: 10.1007/978-0-387-73657-0_9, © Springer Science+Business Media, LLC 2009

Table 1. Melting temperature of PVA-agarose hybrid gels with the different concentrations of PVA and agarose in water.

Melting point of the PVA/Agarose hybrid gel (°C)												
Agarose	PVA – 13 KDa (%)						PVA – 89 KDa (%)					
(%)	0	0.1	0.2	0.6	1.0	2.0	0	0.1	0.2	0.6	1.0	2.0
0.1	-	-	-	-	-	-	-	-	-	-	-	-
0.2	54	54	50	48	48	48	54	54	54	54	54	48
0.4	57	57	57	57	57	57	57	57	57	57	57	57
0.6	60	60	60	60	60	60	60	60	60	60	60	60

amino groups on PVA can be fine-tuned from 1% to 7% by the feeding amount of NaH. A conjugate of FITC and the cell adhesion ligand LLP2A onto amino-PVA, was prepared via oxime chemistry and the binding activity of this conjugate to α4β1-expressing Jurkat cells was investigated with flow cytometry using the blank FITC-PVA conjugate as a control. Incubated with 30 nM of LLP2A conjugate, 75% of Jurkat cells were shown stained by FITC-LLP2A conjugates. Physical properties of a PVA-agarose hybrid gel were studied, and the gelling and melting properties of these gels were found to depend mostly on the concentration of agarose (Table 1). PC-3N cells (prostate cancer cell) were cultured in a semi-solid hydrogel composed of a mixture of 1% agarose and a PVA conjugated targeting ligand (ikmvikw) [3] (Fig. 1, left) for 3 days, the cell grew well in an extended way. However, most of PC-3N cells had no attachment after cultured in a similar PVA-agarose hydrogel but without ikmvikw ligand (Fig. 1, right). The presence of a cell adhesion ligand conjugated to PVA in the hybrid gel facilities cell attachment in 2D culture. 3D cell cultures where cells are adequately suspended in the gel matrix are currently being further investigated.

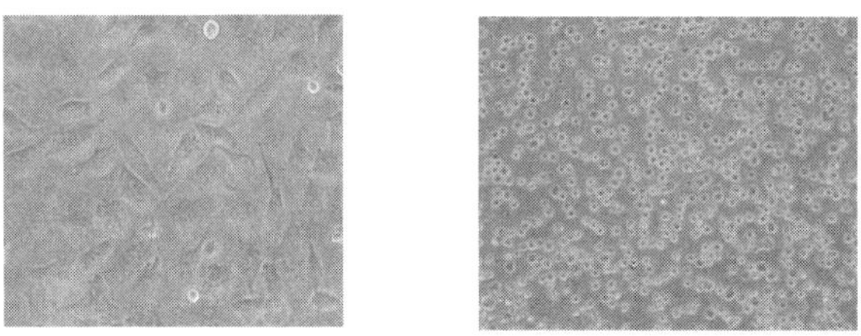

Fig. 1. PC-3N cells (prostate cancer cell) cultured in a semi-solid hydrogel composed of a mixture of 1% agarose and an alpha-6 all D-amino acid targeting ligand (ikmvikw) conjugated to PVA (left panel). PC-3N cells grown in a similar PVA-agarose hydrogel but without ikmvikw ligand (right panel).

Acknowledgments

The financial supports from NSF CH 0302122; NIH/NCI NCDDG U19 CA113298; NIH/NCI R01CA115483 are greatly acknowledged.

References

1. Li, P., Lui, R., Marik, J., Wang, X.B., Takada, Y. and Lam, K.S. *Nat. Chem. Biol.* **2**, 381-389 (2006).
2. Lam, K.S., Salmon, S.E., *et al. Nature* **354**, 82-84 (1991).
3. Sroka, T.C., Marik, J., Pennington, M.E., Lam, K.S. and Cress, A.E. *Cancer Biol. Ther.* **5**, 1556-1562 (2006).

Application of the "Libraries from Libraries" Concept to "One-bead One-compound" Combinatorial Chemistry

Joseph C. Kappel, Yi C. Fan and Kit S. Lam

Division of Hematology and Oncology, Department of Internal Medicine, UC Davis Cancer Center, 4501 X Street, Sacramento, CA 95817, U.S.A.

Introduction

The "one-bead one-compound" (OBOC) method is a convenient way to prepare large numbers of compounds rapidly and efficiently and, in conjunction with on bead screening assays, allows one to identify high affinity ligands for biological targets [1]. Beads which display positive activity must be analyzed to determine the molecular structure of the hit compounds. For molecules not amenable to Edman microsequencing this can be accomplished with chemical encoding. The on-bead screening and chemical decoding processes can be facilitated by the use of topologically segregated bilayer beads in which the testing molecule resides on the outside surface of the bead where interaction with biological targets occur and the coding molecules are segregated in the interior of the bead. Two methods, partial amine protection (PAP) [2] and partial alloc deprotection (PAD) [3], have been developed for bilayer bead preparation. The PAD method has been used in a ladder synthesis strategy to prepare libraries decodable with mass spectrometry.

A variety of small molecule and acyclic combinatorial libraries have been prepared utilizing the libraries from libraries concept [4]. In this approach, amino acids and peptides are modified via global transformation reactions (reduction, alkylation, etc.) to give molecules with different physiochemical properties and potentially enhanced therapeutic properties. In addition, new libraries can be used to make further libraries.

In an effort to expand the efficiency of OBOC library synthesis and the range of chemical space accessible via OBOC strategies, a MALDI-TOF MS-based encoding strategy for OBOC peptidomimetic libraries prepared using the libraries from libraries approach of globally modifying resin bound peptides is described.

Results and Discussion

A cleavable linker (CL) was coupled to Tentagel resin and coding tag elements Arg(Tos), Phe(Br), and a PEG-like spacer were added in turn. This resin was used to generate OBOC dipeptide library **1** using a PAP step to generate the bilayer. Resin **1** was subjected to diborane reduction conditions to generate polyamine library **2**. Twenty beads from **2** were selected for decoding. Following TFA treatment to remove all protecting groups, the twenty beads were treated in individual wells with gaseous hydrogen fluoride (HF) to cleave the ladders from the resin [5]. The cleavage products were then analyzed with MALDI-TOF MS. Nine of these beads were fully decoded. Speculating that poor cleavage yields could be responsible for this, an attempt was made to activate the beads for cleavage by acetylation. Thus, twenty single polyamine beads were individually acetylated then treated with HF. Using this method fifteen out of twenty beads were fully decoded.

S. Del Valle et al. (eds.), *Peptides for Youth: The Proceedings of the 20th American Peptide Symposium*, 21
DOI: 10.1007/978-0-387-73657-0_10, © Springer Science+Business Media, LLC 2009

Resin **2** was used in turn to prepare a poly-*N*-acetylamine sub-library **3** by treating **2** with acetic anhydride. Twenty beads of **3** were picked and subjected to the HF cleavage/ms procedures for decoding. In this case eighteen beads were successfully decoded.

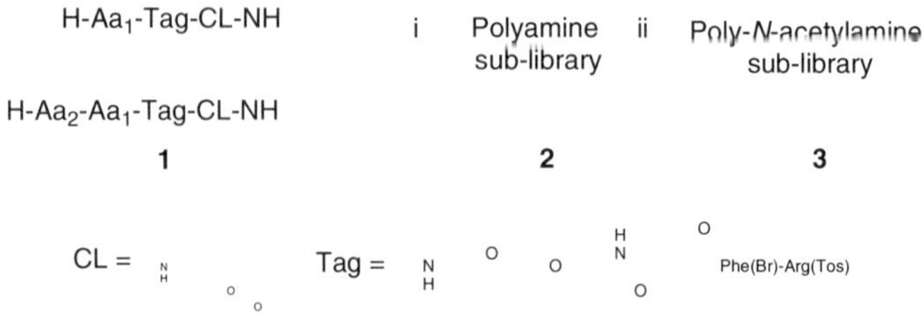

Fig. 1. Synthesis of OBOC dipeptide-based peptidomimetics. Reagents and conditions: (i) $B(OH)_3$, $B(OMe)_3$, 1 M BH_3–THF, 65 °C, 72 h then piperidine, 65 °C, 16 h; (ii) Ac O, DIEA, DMF, 20 °C, 12 h.

In conclusion, a MALDI-TOF MS based chemical encoding strategy for the synthesis of OBOC combinatorial libraries from libraries has been developed. Random beads were selected from OBOC peptide, polyamine, and poly-N-acetylamine libraries and fully decoded with the following respective success rates: 95, 45, and 90%. Single bead acetylation of polyamine library beads prior to cleavage improved the decoding success rate to 75%. Future goals include synthesizing peptidomimetic libraries incorporating higher degrees of diversity, applying this methodology to make OBOC heterocyclic libraries, and using on-bead screening assays to identify cancer-targeting ligands.

Acknowledgments

The work was funded by NSF CH 0302122, NIH/NCI NCDDG U19 CA113298, NIH/NCI R01CA115483, and NIH 1P41 GM076151.

References

1. Liu, R. W., Wang, X. B., Song, A. M., Bao, T., Lam, K. S. *Qsar & Combinatorial Science*, **24**, 1127-1140 (2005).
2. Liu, R., Marik, J., and Lam, K. S. *J. Am. Chem. Soc.*, **124**, 7678-7680 (2002).
3. Wang, X., Peng, L., Liu, R., Gill, S. S., Lam, K. S. *J. Comb. Chem.*, **7**, 197-209 (2005)
4. Nefzi, A., Ostresh, J. M., Yu, J., Houghten, R. A. *J. Org. Chem.*, **69**, 3603-3609 (2004).
5. Kerschen, A., Kanizsai, A., Botros, I., Krchňák, V. *J. Comb. Chem.* **1**, 480-484 (1999).

An Appreciation of the Scientific Life and Achievements of Bruce Merrifield

Alexander R. Mitchell

Energetic Materials Center, Lawrence Livermore National Laboratory, University of California, 7000 East Avenue, L-282, Livermore, CA 94550, USA

Introduction

Bruce Merrifield's scientific biography, "Life During a Golden Age of Peptide Chemistry: The Concept and Development of Solid-Phase Peptide Synthesis," provides a history of solid phase-peptide synthesis (SPPS) from 1959 to 1993 [1]. While many readers will be familiar with SPPS literature after 1963, the inclusion of unpublished material from Merrifield's early laboratory notebooks opens a fascinating window on the development of SPPS from the formulation of concept in 1959 (p. 56, ref. 1) to the synthesis of a tetrapeptide four years later [2]. This early period was characterized by slow progress interrupted by numerous setbacks that led Bruce to later record (p. 90, ref. 1): "*At the end of the first two years the results were so poor, I wonder what made me think that this approach would ever succeed. But from the outset I had a strong conviction that this was a good idea, and I am glad that I stayed with it long enough*". Garland Marshall, Bruce's first graduate student (1963-1966), as well as later colleagues, were essentially unaware of the many highways, byways and dead ends that Bruce had explored in the early years [3].

From Concept to Ribonuclease A Synthesis (1959 -1969)

The use of an insoluble polymer covalently linked to a growing peptide chain was without chemical precedent when Bruce began his studies on SPPS in 1959. The search for an acceptable polymer support and appropriate chemistry was especially challenging with Bruce later writing (p. 89, ref. 1):*"When I look back at my old notebooks, I am amazed at how inefficient the early developmental work was. I seemed always to choose the wrong reaction to do first and was not able to identify the most important parameters as the work was progressing."* The synthesis of Leu-Ala-Gly-Val on a polystyrene support was a watershed event that provided the proof of concept needed for SPPS [2]. A series of increasingly larger, biologically active peptides were prepared over the next three years beginning with bradykinin (1964) and culminating with bovine insulin (1967). The arrival of Bernd Gutte, Bruce's first postdoctoral fellow, from Germany (1967) provided the opportunity to push existing SPPS methodology to the limit and undertake the synthesis of the 124-residue enzyme RNase A. In early 1969 Bernd Gutte and Bruce Merrifield published the use of SPPS to achieve the total synthesis of an enzyme with RNase A activity [4]. This achievement, coupled with a similar effort by the Merck group using classical solution chemistry [5] attracted global attention in the scientific and popular press.

Critical Assessments of SPPS

Garland Marshall has recalled the early "vehement and vitriolic" critics in his discussion of SPPS as a paradigm shift [3]. Some of the most vehement tirades surfaced at meetings of the European and American Peptide Symposia in the late 1960s and early 1970s. Brenner essentially summarized early criticism of SPPS in

S. Del Valle et al. (eds.), *Peptides for Youth: The Proceedings of the 20th American Peptide Symposium*,
DOI: 10.1007/978-0-387-73657-0_11, © Springer Science+Business Media, LLC 2009

stating: *"The invention of the solid-phase method looked like an ingenious trick to overcome some of the unpleasant features of the classical methods. As we know today, the ingenuity of the trick remains, but only a large investment of heavy real effort will eventually, if ever, work it into a real progress over the classical approach* [6]." Wünsch, upon reflecting on the problems of synthetic peptide research, concluded in 1971 that SPPS exhibited "inborn defects" and was "unsuitable for the satisfactory synthesis of higher natural peptides (with more than 15 amino acid residues) [7]." Bruce Merrifield, a man modest in demeanor but strong in character, persevered. The rest, of course, is history [1].

Onward and Upward (1970-1984)

By the early 1970s it had became apparent that the solid-phase synthesis of RNase A could not be generalized. Consequently, virtually every aspect of SPPS was reexamined and improved during the decade of the 1970s (pp. 151-179, ref. 1). The sensitive detection and elimination of possible side reactions (amino acid insertion, N^{α}-trifluoroacetylation, $N^{\alpha\varepsilon}$-alkylation) was examined. An optimization of the HF cleavage reaction based on an understanding of the mechanism was developed. The quantitation of coupling efficiency in SPPS as a function of chain length was studied. A new and improved support for SPPS, the "PAM-resin," was prepared and evaluated. In addition to considerable methodological work on SPPS, parallel synthetic efforts on biologically active peptides such as glucagon, thymosin α_1, epidermal growth factor and antimicrobial peptides were undertaken (pp. 180-195, ref.1). These and many other studies from the Merrifield laboratory and elsewhere increased the general acceptance of SPPS.

The success of SPPS dramatically influenced the chemical synthesis of DNA [8]. The chemical synthesis of DNA had been extremely laborious and time-consuming prior to the development of solid-phase syntheses of DNA. For example, the preparation of a *lac* operator (a 21 base paired DNA duplex) required the equivalent of four years of highly skilled and intense effort. When the appropriate chemistry (phosphoramidite method of DNA synthesis) and support were discovered, the rapid preparation ($\leq$ 1 day) of deoxyoligonucleotides the size of a *lac* operator became possible. Use of automated DNA synthesis machines now leads to very high yields of relatively pure polynucleotides having 100 or more mononucleotides [8]. In retrospect, it seemed inevitable that Bruce would receive a call from Stockholm. The decisions of the Nobel Committee are not always obvious, however. To the great delight of friends and colleagues the call came on October 17, 1984 and we must take Bruce at his word when he states (p. 241, ref.1): "Some are dubious when I say I did not ever expect such a thing, but it is surely true. And I still do not know how it happened, but I am grateful."

Bruce Merrifield

How to best describe the man and his science? Garland Marshall has assessed the scientific impact of Bruce's work in his incisive review "Solid-Phase Synthesis: A Paradigm Shift" [3]. Solid-phase synthesis as used for the synthesis of biopolymers (peptides, proteins, nucleic acids), synthesis of natural products, chemical ligation and materials development has indeed provided a paradigm shift in the molecular biology, biotechnology and chemistry communities. The man who emerges from the pages of *Life During a Golden Age of Peptide Chemistry,* and the man his colleagues knew and respected, was tough and dedicated but also caring and modest. He deeply cared about his two families, the family at home and the family in the laboratory (pp. 208-227, ref. 1). Libby Merrifield, his wife, friend and colleague for over 55 years provided the bedrock for his career. He did not voice anger when the early critics maligned him and his work, just as he did not complain about a long-term progressively invasive skin cancer and the increasingly draconian medical treatments. It would have been out of character and a waste of energy that could be better used in the laboratory. Early in 2003, prior to the final combinations of treatments (surgery, antibiotics, and radiation), I asked Bruce if he had considered retirement as an option. He smiled and said, "sure, I think I'll retire just about 2 minutes before I drop dead." Bruce, thank you for your life, your work and your inspiration to all who knew you.

Acknowledgments

This paper was prepared under the auspices of the U.S. Department of Energy by the University of California, Lawrence Livermore National Laboratory, under Contract No. W-7405-Eng-48.

References

1. Merrifield, B. *Life During a Golden Age of Peptide Chemistry : The Concept and Development of Solid-Phase Peptide Synthesis*. American Chemical Society: Washington, DC, 1993.
2. Merrifield, R. B. *J. Am. Chem. Soc.* **85**, 2149–2154 (1963).
3. Marshall, G. R. *J. Pept. Sci.* **9**, 534-544 (2003).
4. Gutte, B. and Merrifield, R. B. *J. Am. Chem. Soc.* **91**, 501-502 (1969).
5. Hirschmann, R., Nutt, R. F., Veber, D. F., Vitali, R. A., Varga, S. L., Jacob, T. A., Holly, F. W. and Denkewalter, R. G. *J. Am. Chem. Soc.* **91**, 507-508 (1969).
6. Fankhauser, P. and Brenner, M. in "The Chemistry of Polypeptides," Katsoyannis, P. G., Ed., Plenum Press, New York, 1973, p 389-411.
7. Wünsch, E. *Angew. Chem. Int. Ed. Engl.* **10**, 786-795 (1971).
8. Caruthers, M. H. *Acc. Chem. Res.* **24**, 278-284 (1991).

FOLDING, RECOGNITION, AND CATALYSIS

The "Bip Method" for Spectroscopic Assignment of the Absolute Configuration of the Spin-Labelled, Cyclic $\beta^{2,3}$-Amino Acids β-TOAC and POAC

Karen Wright[1], Laurence Dutot[1], Anne Gaucher[1], Michel Wakselman[1], Jean-Paul Mazaleyrat[1], Cristina Peggion[2], Fernando Formaggio[2] and Claudio Toniolo[2]

[1]ILV, UMR CNRS 8180, University of Versailles, F-78035 Versailles, France; [2]Institute of Biomolecular Chemistry, CNR, Department of Chemistry, University of Padova, I-35131 Padova, Italy

Introduction

We have previously reported that in linear dipeptides Boc-Bip-αXaa*-OMe, with αXaa* [L/D-Ala, L/D-Val, L/D-Leu, L-(αMe)Val, L-(αMe)Leu] at the C-terminal position of the pro-atropoisomeric, C^{α}-tetrasubstituted α-amino acid residue Bip, the onset of an *equilibrium between two diastereoisomers* with unequal populations could be observed by CD and ^{1}H NMR [1]. The phenomenon of induced circular dichroism (ICD) represents the basis for the *"Bip method"*, an easy and fast configurational assignment of chiral α-amino acids. We have now investigated the application of the Bip method to βXaa* amino acids, L-β^3-HAla, L-β^3-HVal, L-β^3-HLeu, L-β^3-HPro, *trans (1R,2R)/(1S,2S)*-ACHC, *trans (1R,2R)/(1S,2S)*-ACPC, as well as *unique spin-labelled, cyclic, chiral $\beta^{2,3}$-amino acids* synthesized and resolved in our groups, namely *cis/trans* β-TOAC [2] and *trans* POAC [3,4] (Fig. 1).

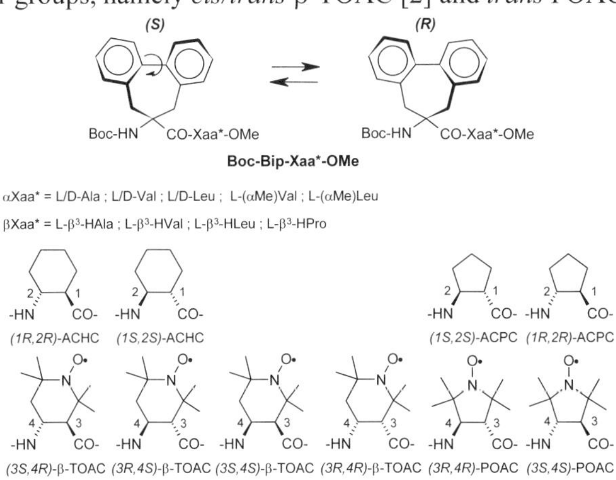

Fig. 1. Chemical structures and diastereoisomer equilibrium of the terminally protected dipeptides Boc-Bip-Xaa-OMe (Xaa* = α-, β^3- and cyclic $\beta^{2,3}$-amino acids).*

Results and Discussion

The terminally protected dipeptides Boc-Bip-βXaa*-OMe listed in Figure 1 were prepared in solution by coupling Boc-Bip-OH with the β-amino ester hydrochlorides HCl.βXaa*-OMe readily obtained from Boc- or Fmoc-βXaa*-OH. The EDC/HOAt activation method was mostly used for coupling at the sterically demanding C-terminus of Bip. Our CD analysis in MeOH solution shows that:

DOI: 10.1007/978-0-387-73657-0_12, © Springer Science+Business Media, LLC 2009

(i) The ICD resulting from the induced axial chirality in the biphenyl core of the Bip residue gives clear information on the βXaa* configuration, for both β^3- and cyclic $\beta^{2,3}$-amino acids with a negative sign of the Cotton effect of the A band (250 nm), known to be associated with a P torsion of the biphenyl axial bond of Bip, being preferentially induced by the L-β^3-Xaa* as well as by the cyclic *(1S,2S)*-$\beta^{2,3}$-Xaa* C-terminal residues (not shown). Interestingly, this sign is the same as that for the corresponding Boc-Bip-L-αXaa*-OMe dipeptides [1], although in the present case the CD signal would be weaker (but still informative).

(ii) The C-terminal enantiomers of both O-acetyl protected nitroxides *cis*- and *trans*-β-TOAC(OAc) exhibit significant Cotton effects. The signs of their A band, when compared with those of the corresponding enantiomers of the C-terminal *trans*-ACHC derivative, present a perfect correlation for the absolute configuration of their carbon atoms at positions 4 and 2, respectively, bearing the nitrogen atom (Fig. 2, left). From these results, the absolute configurations *(-)-(3S,4S)* and *(+)-(3R,4R)* for the enantiomers of *trans*-POAC may be assigned by comparison of the Cotton effects of the A band of Boc-Bip-*(-)*-POAC(OAc)-OMe and Boc-Bip-*(+)*-POAC(OAc)-OMe with those of Boc-Bip-*(1R,2R)*-ACPC-OMe and Boc-Bip-*(1S,2S)*-ACPC-OMe, respectively (Fig. 2, right).

(iii) Very similar ICD spectra as those mentioned above are obtained for the Boc/OMe Bip derivatives of the unprotected nitroxide amino acids *cis*- and *trans*-β-TOAC and *trans*-POAC.

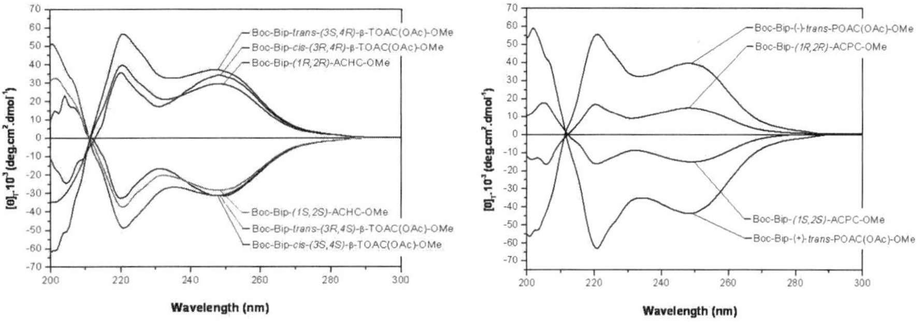

Fig. 2. CD spectra (200-300 nm range) of the terminally protected dipeptide series Boc-Bip-cis- and trans-β-TOAC(OAc)-OMe compared with Boc-Bip-(1R,2R)- and (1S,2S)-ACHC-OMe (left), and Boc-Bip-(+) and (-)-trans-POAC(OAc)-OMe compared with Boc-Bip-(1R,2R)- and (1S,2S)-ACPC-OMe (right), in MeOH solution (c 10^{-3} M).

References

1. Mazaleyrat, J.-P., Wright, K., Gaucher, A., Toulemonde, N., Dutot, L., Wakselman, M., Broxterman, Q. B., Kaptein, B., Oancea, S., Peggion, C., Crisma, M., Formaggio, F. and Toniolo, C. *Chem. Eur. J.* **11**, 6921-6929 (2005).
2. (a) Wright, K., Crisma, M., Toniolo, C., Török, R., Péter, A., Wakselman, M. and Mazaleyrat, J.-P. *Tetrahedron Lett.* **44**, 3381-3384 (2003). (b) Wright, K., De Castries, A., Sarciaux, M., Formaggio, F., Toniolo, C., Toffoletti, A., Wakselman, M. and Mazaleyrat, J.-P. *Tetrahedron Lett.* **46**, 5573-5576 (2005).
3. Tominaga, M., Barbosa, S. R., Poletti, E. F., Zukerman-Schpector, J., Marchetto, R., Schreier, S., Paiva, A. C. M. and Nakaie, C. R. *Chem. Pharm. Bull.* **49**, 1027-1029 (2001).
4. Wright, K., Crisma, M., Toniolo, C., Török, R., Péter, A., Wakselman, M. and Mazaleyrat, J.-P. *Tetrahedron Lett.* **44**, 3381-3384 (2003).

Enzyme Mechanism and Function of a Novel Plant PDI Involved in the Oxidative Folding of Cystine Knot Defense Peptides

Christian W. Gruber[1], Maša Čemažar[1], Marilyn A. Anderson[2] and David J. Craik[1]

[1]*Institute for Molecular Bioscience, University of Queensland, Brisbane, Australia*
[2]*Department of Biochemistry, LaTrobe University, Melbourne, Australia*

Introduction

Cyclotides are disulfide-rich plant peptides containing a head-to-tail cyclized backbone and a cystine knot formed by three conserved disulfide bonds, which makes them extremely stable molecules. Their native function is believed to be plant defense molecules against insect pests. Besides being explored for their various bioactivities, their use as agrochemicals and as a pharmaceutical drug-scaffold is currently under investigation. Cyclotides are synthesized in plants as larger precursor proteins and although their biosynthesis is so far unknown, they are believed to mature by a combination of oxidative folding of the three disulfide bonds, excision from the precursor and cyclisation [1].

Auxiliary proteins, such as protein disulfide isomerase (PDI), could be involved in their oxidative folding *in vivo*. PDI is an oxido-reductase enzyme whose major function is the oxidative folding of polypeptides in the endoplasmic reticulum of eukaryotic cells. The exact mechanism of action of PDI is not clear, but it is believed to bind polypeptides through hydrophobic interactions and forms ("oxidation"), breaks ("reduction") and/or shuffles ("isomerization") disulfide bonds in substrate molecules via a dithiol-disulfide exchange between its active site CXXC motif and the substrate [2]. In this study we have isolated a novel protein disulfide isomerase from the Rubiaceae plant *Oldenlandia affinis* (OaPDI) that accumulates cyclotides, characterized the function and mechanism of the enzyme and studied its effects on the oxidative folding of the model cyclotide kalata B1.

Results and Discussion

The novel plant PDI was compared to other human PDI proteins, by means of enzyme activity assays and redox potential measurements and was found to be a functional oxido-reductase. Transcriptional co-expression data of OaPDI and the cyclotide precursor Oak1 and biomolecular interaction studies using surface plasmon resonance revealed a possible interaction between the proteins and suggested a role of OaPDI in assisting the oxidative folding of cyclotides *in vivo*. We showed further that OaPDI is able to correctly fold cyclotides to yield functional insecticidal molecules, which is presumed to be their native function in the plant [3].

The mechanism of enzyme-assisted folding of cyclotides was investigated by comparing the folding of linear and cyclic kalata B1 derivatives in the presence and absence of OaPDI. OaPDI was found to dramatically enhance the correct oxidation and folding of kalata B1 at physiological p*H*. Folding without the PDI under these conditions did not yield any significant amount of correctly oxidized peptides, but rather led to the accumulation of mostly non-native 3SS species and some 2SS species [3]. In contrast, the addition of OaPDI to the folding mixture produced significant proportions of both linear and cyclic kalata B1 folded into their native disulfide connectivities. Without PDI the folding still proceeded to the initial stage

S. Del Valle et al. (eds.), *Peptides for Youth: The Proceedings of the 20th American Peptide Symposium*,
DOI: 10.1007/978-0-387-73657-0_13, © Springer Science+Business Media, LLC 2009

of non-specific disulfide bond formation by oxidation ("packing") to the point where mostly oxidized and partially oxidized (3SS and 2SS) peptides were present, but it appears that they were trapped and are not able to effectively undergo disulfide re-shuffling ("consolidation") to the native species, resulting in a lower yield of correctly oxidized peptides (Figure 1). In short, oxidation proceeds without the enzyme, but the shuffling of misfolded or partial folded intermediates into their native disulfide connectivity is significantly reduced. This reinforces the notion that isomerization is a major function of the PDI isolated from *O. affinis*.

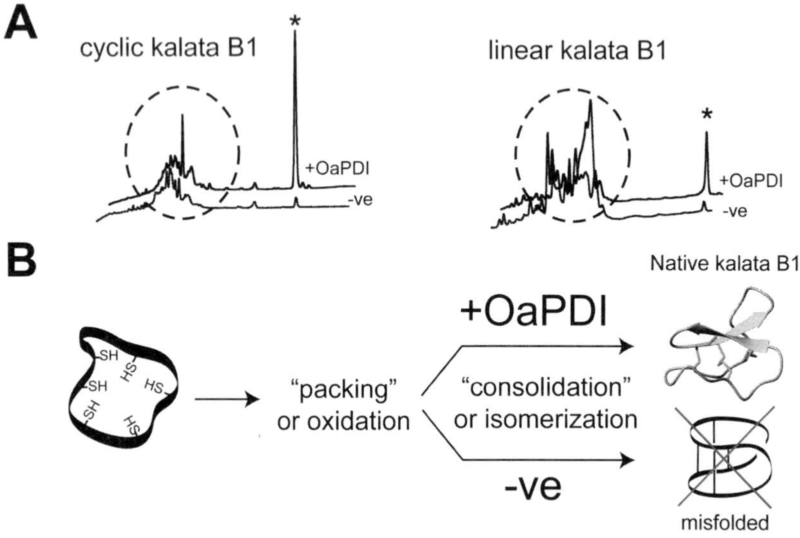

Fig. 1. ***Oxidative folding of kalata B1 using OaPDI.*** ***(A)*** *HPLC traces of linear and cyclic kalata B1 folded for 24 h at pH 7.5 with OaPDI or without the addition of OaPDI. OaPDI significantly increases the yield of correctly folded peptide (asterisk) after 24 h of folding under physiological conditions.* ***(B)*** *Analysis of folding intermediates of both linear and cyclic kalata B1 without OaPDI (dashed oval in A) showed mainly three-disulfide misfolded and two-disulfide partially folded peptide species. Regardless of the presence or absence of the enzyme, the peptides undergo thiol oxidation to form disulfide bonds, but shuffling of non-native disulfide bonds into their native connectivity is greatly compromised without OaPDI. Hence, OaPDI has an important role in the isomerization of disulfide bonds during the folding of cystine knot peptides.*

Acknowledgments

The authors thank all members of our laboratories who have contributed by providing materials or help with experiments or discussions. This work was funded by the Australian Research Council and the University of Queensland.

References

1. Gruber, C. W., Cemazar, M., Anderson, M. A., and Craik, D. J. (2007) *Toxicon* **49**, 561-575.
2. Gruber, C. W., Cemazar, M., Heras, B., Martin, J. L., and Craik, D. J. (2006) *Trends Biochem. Sci.* **31**, 455-464.
3. Gruber, C. W., Cemazar, M., Clark, R. J., Horibe, T., Renda, R. F., Anderson, M. A., and Craik, D. J. (2007) *J. Biol. Chem.* **282**, 20435-20446.

Biosynthesis of Siderophore-Peptides, A Class of Potent Antimicrobial Peptides from Enterobacteria, Requires Two Precursors

Gaëlle Vassiliadis, Jean Peduzzi, Delphine Destoumieux-Garzón, Séverine Zirah, Xavier Thomas and Sylvie Rebuffat

Chimie et Biochimie des Substances Naturelles, CNRS-Muséum National d'Histoire Naturelle, UMR 5154, CP 54, 57 rue Cuvier, 75005 Paris, France

Introduction

Microcin E492 is an 84-residue antimicrobial peptide naturally secreted by *Klebsiella pneumoniae* RYC492 [1]. Its antibacterial activity against *Escherichia coli* is dependent on the outer membrane catecholate receptors FepA, Cir and Fiu and the ManYZ inner membrane complex involved in mannose uptake [3], and involves inner membrane permeabilization [2]. Microcin E492-producing bacteria secrete both an unmodified (u-MccE492) and a post-translationally modified (MccE492) form of the microcin (Fig. 1) [4], the latter being endowed with higher activity. The modification consists of a *C*-glucosylated linear trimer of *N*-(2,3-dihydroxybenzoyl)-L-serine (DHBS), a catecholate siderophore related to salmochelins and enterobactin. This modification is reminiscent of siderophores, which are molecules involved in ferric iron uptake by bacteria. MccE492 was therefore defined as the first siderophore-peptide to be isolated. MccE492 and u-MccE492 both derive from the precursor protein MceA, by elimination of a 19-residue leader peptide. The *mceABCDEFGHIJ* gene cluster responsible for MccE492 production, export and immunity was cloned and sequenced [5]. The genes *mceC* and *mceD* encode proteins similar to IroB and IroD, which are proteins involved in the biosynthesis of salmochelin from enterobactin. Therefore, they are proposed to be involved in MccE492 maturation. Other microcins, such as microcin H47 (MccH47) display similar genetic systems, suggesting they could also be siderophore-peptides [6].

In order to characterize the molecular mechanisms governing the biosynthesis of siderophore peptides, we have investigated the potential role of MceC and MceD in MccE492 maturation, together with the potential involvement of enterobactin as a precursor for the post-translational modification.

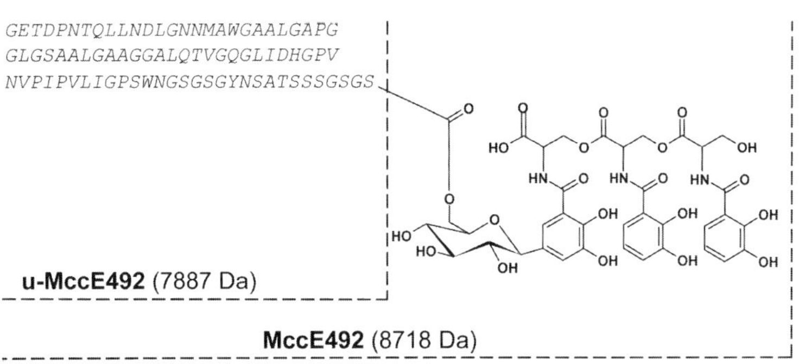

Fig. 1. Structures and molecular masses of MccE492 bearing the Glc-DHBS$_3$ siderophore post-translational modification and of the unmodified peptide u-MccE492.

S. Del Valle et al. (eds.), *Peptides for Youth: The Proceedings of the 20th American Peptide Symposium*, DOI: 10.1007/978-0-387-73657-0_14, © Springer Science+Business Media, LLC 2009

Results and Discussion

Inactivation of *mceC* or *mceD* resulted in the secretion of u-MccE492 only, as monitored by MALDI-TOF mass spectrometry. Restoration of MccE492 secretion was achieved by complementation of the strains carrying an inactivated *mceC* or *mceD* gene with the MccH47 gene cluster, which harbors *mchA* and *mchS1*, two genes homologous to *mceC* and *mceD*, respectively. Therefore, MceC and MceD are involved in the acquisition of MccE492 post-translational modification. Furthermore, the similar proteins encoded by the MccH47 gene cluster, MchA and MchS1, appear to display the same function as MceC and MceD, respectively.

Several factors impairing enterobactin biosynthesis (high iron concentration, presence of Tyr and Trp) inhibited the acquisition of MccE492 post-translational modification. Furthermore, exogenous enterobactin restored the production of MccE492 post-translational modification in a bacterial strain deficient in enterobactin synthesis. It appears therefore that enterobactin serves as a precursor for the synthesis of MccE492 post-translational modification.

Taken together, our data show that two precursors, namely enterobactin and the protein MceA, are the substrates for the MccE492 enzyme machinery, and that u-MccE492 is an incompletely processed form of the mature microcin. A sequence of events is proposed for the biosynthesis of MccE492 post-translational modification: (i) enterobactin would be *C*-glycosylated by MceC in an IroB-like manner, (ii) the *C*-glucosylated enterobactin would be hydrolyzed by MceD, an homologue of IroD, the enterobactin esterase, (iii) the resulting glucosylated linear trimer of DHBS would then be transferred onto MceA by enzymes from the gene cluster (MceI, MceJ and presumably MceC); (iv) finally, MccE492 would be generated concomitantly with its export through removal of the leader peptide.

Acknowledgments

The work was funded by MNHN (phylogenomic PPF grant).

References

1. de Lorenzo, V., and Pugsley, A. P. *Antimicrob. Agents Chemother.* **27**, 666-669 (1985).
2. Destoumieux-Garzón, D., *et al. Biometals* **19**, 181-191 (2006).
3. Bieler, S., *et al. J. Bacteriol.* **188**, 7049-7061 (2006).
4. Thomas, X., *et al. J. Biol. Chem.* **279**, 28233-28242 (2004).
5. Wilkens, M., *et al. J. Bacteriol.* **179**, 4789-4794 (1997).
6. Poey, M. E., Azpiroz, M. F., and Laviña, M. *Antimicrob. Agents. Chemother.* **50**, 1411-1418 (2006).

Post-Translational Modification and folding of A Lasso-Type Gene-Encoded Antimicrobial Peptide Require Two Enzymes only in *Escherichia coli*

Sophie Duquesne[1], Delphine Destoumieux-Garzón[1], Séverine Zirah[1], Thomas A. Knappe[2], Christophe Goulard[1], Jean Peduzzi[1], Mohamed A. Marahiel[2] and Sylvie Rebuffat[1]

[1]Chimie et Biochimie des Substances Naturelles, CNRS-Muséum National d'Histoire Naturelle, UMR 5154, CP 54, 57 rue Cuvier, 75005 Paris, France ; [2]Fachbereich Chemie-Biochemie, Philipps Universität Marburg, Hans-Meerwein-Strasse, 35032 Marburg, Germany

Introduction

Lasso peptides are 16- to 21-residue naturally occurring peptides that result from a side-chain to backbone internal linkage, which forms an *N*-terminal ring of 8 to 9 residues in which the *C*-terminal tail is threaded [1, 2]. Most of them were isolated from *Streptomyces* species and were shown to be enzyme inhibitors. Microcin J25 (MccJ25) is a gene-encoded lasso peptide secreted by *Escherichia coli* AY25 [3, 2]. It exerts a potent antibacterial activity by using the iron-siderophore receptor FhuA to enter bacteria [4] and blocking the bacterial RNA polymerase [5]. The *C*-terminal tail of MccJ25 is sterically trapped in the ring by the two bulky aromatic side-chains from Phe19 and Tyr20 on each side of the ring (Fig. 1), thus creating a β-hairpin region over the ring. The *mcjABCD* gene cluster (Fig. 1) required for MccJ25 biosynthesis has been entirely sequenced [6], making this microcin a good candidate for studying the biosynthesis of lasso peptides.

Here, we demonstrate that two enzymes, McjB and McjC, encoded by the *mcjB* and *mcjC* genes belonging to the MccJ25 gene cluster, catalyze the maturation of MccJ25 and act in concert to convert the precursor McjA into MccJ25.

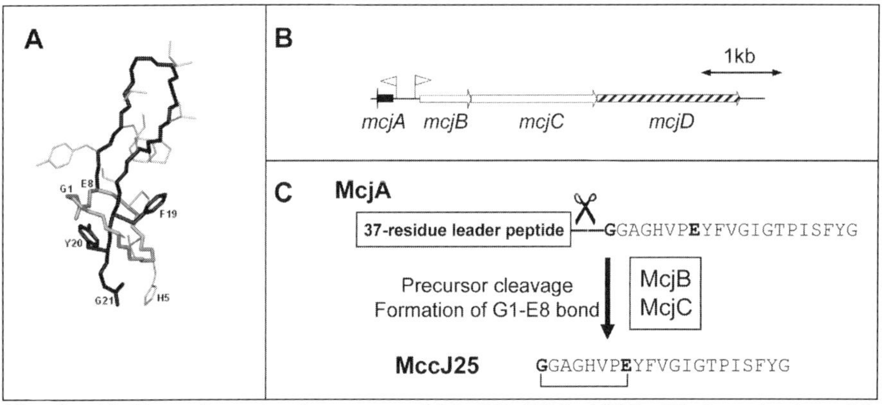

Fig. 1. (A) Three-dimensional structure of MccJ25 (PDB, 1Q71). (B) Organization of the MccJ25 gene cluster: the genes mcjA (black arrow), mcjB and mcjC (white arrows) and mcjD (hatched arrow) encode the MccJ25 precursor McjA, the putative maturation enzymes, McjB and McjC, and the immunity/export protein, respectively. The promoters are indicated by flags. (C) MccJ25 maturation: McjA is subject to proteolytic cleavage and formation of an amide bond between the amino group of Gly1 and the carboxyl side-chain of Glu8.

S. Del Valle et al. (eds.), *Peptides for Youth: The Proceedings of the 20th American Peptide Symposium*, 35
DOI: 10.1007/978-0-387-73657-0_15, © Springer Science+Business Media, LLC 2009

Results and Discussion

The involvement of both *mcjB* and *mcjC* in MccJ25 maturation was shown by gene inactivation and complementation assays. In addition, the precursor of MccJ25, McjA, and the two putative modification enzymes, McjB and McjC, were produced as *N*-His-tagged proteins by recombinant expression in *E. coli*. They were purified by immobilized-metal affinity chromatography (IMAC). The maturation of MccJ25 *in vitro* was followed by liquid chromatography coupled to mass spectrometry (LC-MS) and by antibacterial assays. We showed that the two enzymes encoded by the MccJ25 gene cluster (McjB and McjC) are sufficient to convert McjA into the lasso-structured MccJ25 endowed with antibacterial activity, through a process that requires ATP and Mg^{2+}. Similarity searches and multiple alignments revealed that McjB would exert the proteolytic activity, while McjC, which has strong similarity with β-lactam synthetases, would be the ATP/Mg^{2+}-dependent enzyme responsible for the formation of the Gly1-Glu8 amide bond.

To our knowledge, this is the first example of the *in vitro* reconstitution of a lasso peptide biosynthesis.

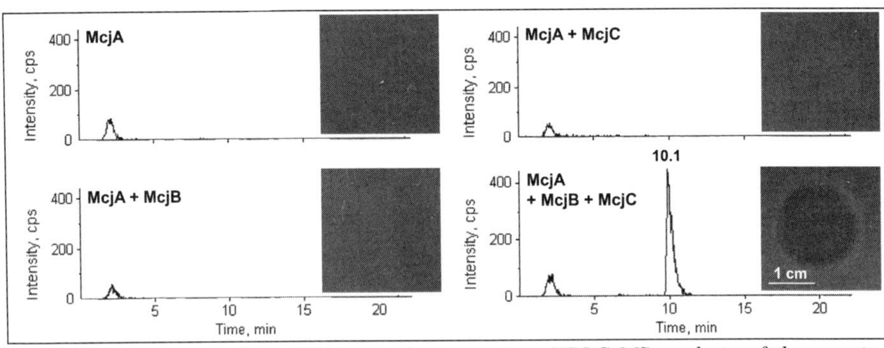

Fig. 2. Reconstitution of MccJ25 biosynthesis in vitro. HPLC-MS analysis of the reaction mixtures (0.5 nmol His_6-McjA alone or in the presence of 0.05 nmol His_6-McjB or/and His_6-McjC). The extracted-ion chromatograms for m/z 703, corresponding to the MccJ25 $[M+3H]^{3+}$ species, are presented together with the antibacterial activities against S. enterica serovar Enteritidis.

Acknowledgments

The work was funded by MNHN (phylogenomic PPF grant) and by a CNRS Procope project.

References

1. Katahira, R., *et al. Bioorg. Med. Chem.* **4**, 121-129 (1996).
2. Rebuffat, S., *et al. Curr. Protein Pept. Sci.* **5**, 383-391 (2004).
3. Rosengren, K.J., *et al. J. Am. Chem. Soc.* **125**, 12464-12474 (2003).
4. Destoumieux-Garzón, D., *et al. Biochem. J.* **389**, 869-876 (2005).
5. Mukhopadhyay, J., *et al. Mol. Cell* **14**, 739-751 (2004).
6. Solbiati, J.O., *et al. J. Bacteriol.* **181**, 2659-2662 (1999).

Synthesis and 3D-Structure of Conformationally Controlled Nucleo-Peptides

Piero Geotti-Bianchini[1,2], Marco Crisma[1], Cristina Peggion[1], Alberto Bianco[2] and Fernando Formaggio[1]

[1]ICB, CNR, Department of Chemistry, University of Padova, 35131 Padova, Italy; [2]CNRS, Institut de Biologie Moléculaire et Cellulaire, Laboratoire d'Immunologie et Chimie Thérapeutiques, 67084 Strasbourg Cedex, France

Introduction

In the last fifteen years polyamide backbones with nucleo-bases as pendants have gained much attention in view of their potential applications. Indeed, such hybrid molecules are insensitive to phosphatase hydrolytic actions, while retaining (or even improving) polynucleotide complexation properties [1]. Therefore, molecules of this class are under investigation in antisense therapy and immunotherapy studies.

By taking advantage of our long-standing experience in the synthesis and 3D-characterization of conformationally constrained α-amino acids and peptides, particularly those based on C^{α}-tetrasubstituted α-amino acids [2], we designed and synthesized short, β-turn- or helix-forming nucleo-peptides. In particular, alanyl-thymine residues (AlaT, Figure 1) were inserted into host, homo-peptide stretches of Aib (α-aminoisobutyric acid, Figure 1). The primary structure of the longest peptide, a hexamer (Figure 1), was designed in order to allow the two nucleo-bases to be aligned on the same face of the helical molecule.

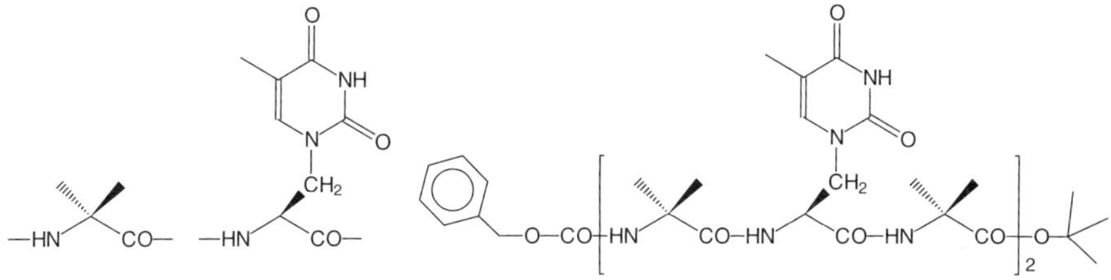

Aib **AlaT** **Z-(Aib-L-AlaT-Aib)$_2$-O*t*Bu**

Fig. 1. Chemical structure of Aib, AlaT and the longest nucleo-peptide synthesized.

Results and Discussion

The synthesis of L-AlaT was achieved according to literature procedures [3], starting from the N^{α}-protected β-lactone of L-Ser. The sterically hindered peptide bonds between L-AlaT and Aib were successfully formed by means ot the highly effective EDC/HOAt procedure. After completing the step-by-step synthesis of Z-Aib-L-AlaT-Aib-O*t*Bu, the hexapeptide Z-(Aib-L-AlaT-Aib)$_2$-O*t*Bu was prepared in satisfactory yield by condensing the 5(4*H*)-oxazolone from Z-Aib-L-AlaT-Aib-OH with H-Aib-L-AlaT-Aib-O*t*Bu in CH_3CN under refluxing conditions.

The results of our preliminary FT-IR absorption, NMR, and X-ray diffraction analyses strongly support the conclusion that these AlaT-containing oligomers are folded into β turn or helical conformations. This finding is of particular significance in view of the possible disturbing role of main-chain-to-side-chain H bonds. Indeed,

S. Del Valle et al. (eds.), *Peptides for Youth: The Proceedings of the 20th American Peptide Symposium*,
DOI: 10.1007/978-0-387-73657-0_16, © Springer Science+Business Media, LLC 2009

in addition to the H-bond stabilizing the 10-membered pseudocycle of the β-turn (type-I), in the crystal structure of Z-Aib-AlaT-Aib-O*t*Bu a H-bond between the thymine C(2)=O and the NH of AlaT was clearly observed (Figure 2).

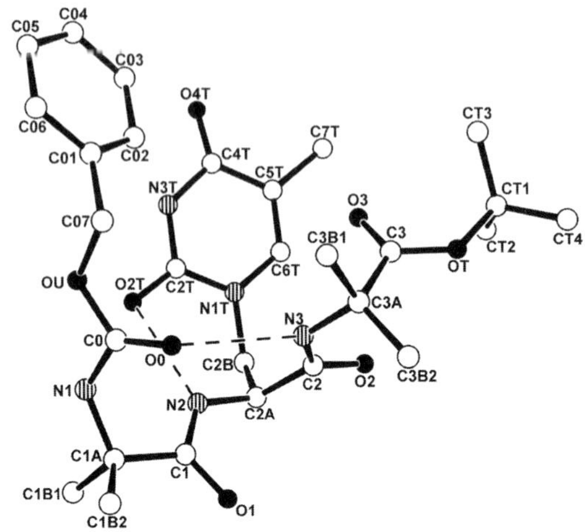

Fig. 2. X-Ray crystal structure of Z-Aib-AlaT-Aib-OtBu with atom numbering. The two intramolecular H-bonds are represented by dashed lines.

All nucleo-peptides exibit also a remarkable tendency to form dimers. The ability of the thymine moieties to form H-bonds with each other is probably the main cause of the peptide dimerization processes observed. In the crystal structure of Z-Aib-AlaT-Aib-O*t*Bu two molecules are held together thanks to the contribution of two intermolecular H-bonds involving the thymine groups. Dimers were detected in the mass spectra of all peptides. By using NMR, FT-IR absorption and CD techniques this high dimerization tendency was revealed also in chloroform solution. Such a phenomenon was particularly evident for the bifunctionalized hexapeptide.

In conclusion, innovative features of these new nucleo-peptides are (i) the markedly rigid peptide backbones, (ii) the 3_{10}-helical conformation, (iii) the high predictability of their 3D-structures and (iv) the strong tendency to form nucleo-base mediated dimers. Altogether, we expect these molecules to show new and interesting biological properties, as they might be able to force the polynucleotide chains to adopt unusual conformations.

Acknowledgments

P. Geotti-Bianchini is grateful to Università italo-francese/Université franco-italienne for financially supporting his international PhD (VINCI program 2005).

References

1. Nielsen, P. E., Egholm, M., Berg, R. H., Buchardt, O. *Science* **254**, 1497-1500 (1991).
2. Toniolo, C., Crisma, M., Formaggio, F., Peggion, C. *Biopolymers (Pept. Sci.)* **60**, 396-419 (2001).
3. Chaltin, P., Lescrinier, E., Lescrinier, T., Rozanski, J., Hendrix, C., Rosemeyer, H., Busson, R., Van Aershot, A., Herdewijn, T. *Helv. Chim. Acta* **85**, 2258-2283 (2002).

Mixing Urea and Amide Bonds: Synthesis and Self-Organization of New Hybrid Oligomers

Lucile Fischer[1], Gersande Lena[1], Jean-Paul Briand[1], Claude Didierjean[2], and Gilles Guichard[1]

[1]*CNRS;Institute of Molecular and Cellular Biology, Strasbourg, France;* [2]*CNRS UMR 7036, LCM3B Groupe Biocristallographie, Université Henri Poincaré, Vandoeuvre, France*

Introduction

We have shown previously that the urea moiety, by its capacity to form auto-complementary and bi-directional hydrogen bonds can be substituted for the amide linkage to generate linear oligomers that self-organize at the molecular level through remote intrastrand interactions to form structural patterns reminiscent of secondary protein structures. Linear *N,N'*-linked oligoureas **A** as short as heptamers adopt a stable 2.5-helical fold in solution [1]. Interestingly, the predictability of helical folding can be integrated to develop molecules with function. It is also possible to take advantage of the hydrogen-bond donor/acceptor properties of the urea linkage to generate flat ring structures **B** that self-organize at the supramolecular level to form H-bonded hollow tubular structures [2].

The possibility to combine different building blocks and generate new heterogeneous backbones as candidate foldamers has recently emerged [3]. This concept, successfully applied to oligoamides, spectacularly expands the structure space attainable with a relatively small pool of residue types. In this context, mixing amide and urea linkages is of particular interest to create folded shapes and to develop new molecules with function.

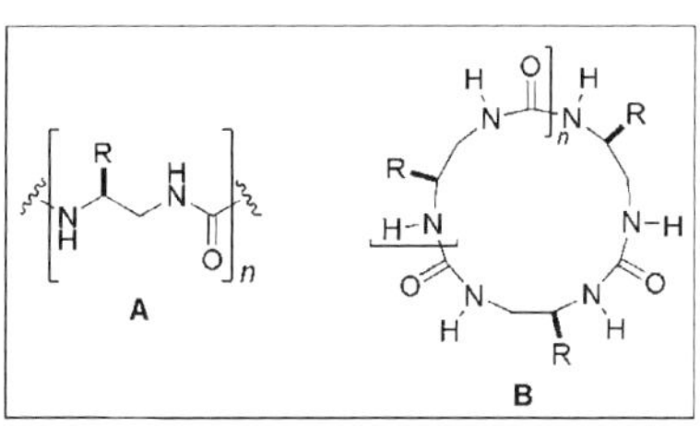

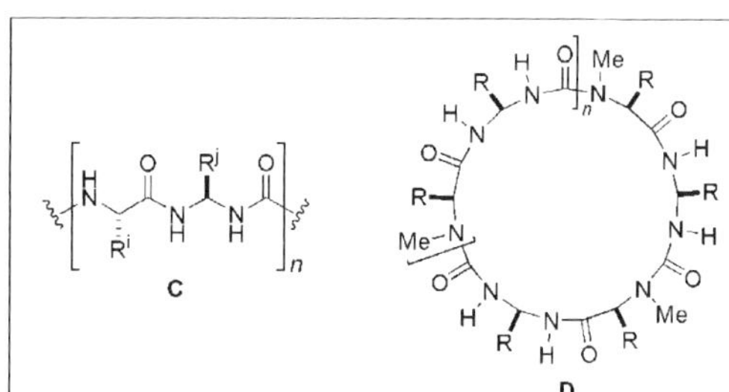

In this work, we have investigated the synthesis and conformational behaviour of short chain linear and cyclic oligo(amide/urea) hybrids **C** and **D**, respectively.

Results and Discussion

Linear oligo(amid/urea) hybrids of type **C** were obtained from succinimidyl [1-(acylamino)-1-X-methyl] carbamates derived from Z-protected dipeptides. Information about the local conformation induced by the urea fragment in such molecules was gained by examination of amide and urea NH chemical shifts upon progressive addition of DMSO-d_6 in $CDCl_3$, as well as by ROESY experiments. In the case of Z-Leu-gVal-CO-Ala-gLeu-NH*i*Pr (**1**), intense ROE connectivities were observed between N^5H and $^{\alpha}CH$ of the preceding *g*Leu residue as well as between N^3H and $^{\alpha}CH$ of the *g*Val residue (Figure 1). A short interproton distance is possible only if one assumes that the urea adopts a *cis,trans* (E,Z) geometry as shown in

S. Del Valle et al. (eds.), *Peptides for Youth: The Proceedings of the 20th American Peptide Symposium*, 39
DOI: 10.1007/978-0-387-73657-0_17, © Springer Science+Business Media, LLC 2009

Figure 1. This observation is in line with previous IR and ^{1}H-NMR spectroscopic investigations of model *N*-Boc-*N'*-carbamoyl-gem-diaminoalkyl derivatives in low polar solvents showing that the urea moiety adopts a stable *cis,trans* conformation that can be stabilized by an intramolecular H-bond (the "*urea turn*") [4]. Observed ROEs in **1** are compatible with the population of consecutive *urea turns*. It remains to be seen whether such *urea turns* can favor nucleation of overlapping γ-turns and the formation of a 7/8-helical structure.

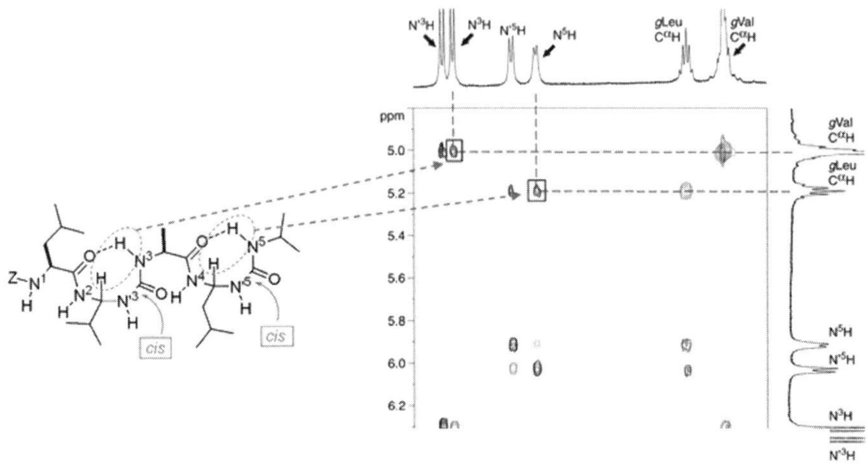

*Fig. 1. Part of the ROESY spectrum of Z-Leu-*g*Val-CO-Ala-*g*Leu-NHiPr (***1***) in DMSO-*d_6.

In parallel, macrocyclization of the TFA salts of succinimidyl [1-(amino)-1-X-methyl] carbamates derived from Boc-protected dipeptides has been evaluated. The ratio of macrocyclic species (dimers or trimers) over cyclic monomer (1,3,5-triazepan-2,6-dione) depends in part on the propensity of the central amide to populate the *cis* conformer, on the experimental conditions (tertiary base, dilution, solvent) as well as on the nature of the amino acid side chains. Preliminary X-ray diffraction analysis on single crystals revealed interesting self-assembly properties.

Acknowledgments

This research was supported in part by Centre National de la recherche Scientifique (CNRS), Agence Nationale pour la Recherche (grant number NT05-4-42848), Région Alsace and ImmuPharma France.

References

1. Violette, A., Averlant-Petit, M. C., Semetey, V., Hemmerlin, C., Casimir, R., Graff, R., Marraud, M., Briand, J. P., Rognan, D., Guichard, G. *J. Am. Chem. Soc.* ***127***, 2156-2164 (2005).
2. Semetey, V., Didierjean, C., Briand, J. -P., Aubry, A., Guichard, G. *Angew. Chem. Int. Ed.* ***41***, 1895-1898 (2002).
3. Le Grel, P. Guichard, G. In Foldamers; Huc I and Hecht S., Eds.; Wiley-VCH Verlag: Weinheim, pp. 35-74 (2007)
4. Fischer, L., Semetey, V., Lozano, J. M., Schaffner, A. P., Briand, J. P., Didierjean, C., Guichard, G. *Eur. J. Org. Chem.* 2511-2525 (2007).

Optimizing the Fold Stability of Miniprotein Sequences

D. Victoria Williams, Jasper C. Lin, Brandon L. Kier and Niels H. Andersen

Department of Chemistry, University of Washington, Seattle WA 98195

Introduction

Peptides (< 25 residues) rarely have well-defined, deep energy wells for folding with the possible exception of a reported few hairpin and miniprotein structures. However, these systems display relatively low melting temperatures and broad thermal transitions for unfolding. Our lab has recently employed several strategies that provide exceptional stability for both types of folds.

In the case of the Trp-cage motif, combining helix stabilizing mutations with one other feature provided three mutants that display T_M values greater than 75ºC, with two-state folding (Fig. 1). In the β hairpin, Trp/Trp pairings flanking a turn have been shown to greatly stabilize folds [1]. These residue pairs adopt an edge-to-face (EtF) geometry, as evidenced by circular dichroism (CD) exciton couplets and an upfield Trp Hε3 proton (-2.2 ppm). Adding a backbone-NH/indole H-bonding interaction to a turn motif stabilized by an EtF Trp/Trp interaction, produced a 10-mer hairpin that is greater than 85% folded at 25ºC (T_M > 60ºC).

WP2	`Ac-WIpGLWTGPS`	>85% folded @ 25ºC	
TC10b	`DAYAQWLKDGGPSSGRPPPS`	$T_M = 56ºC$	$\Delta G_U = 12.2$ kJ/mol
CycloTC	`(DAYAQWLADGGPSSGRPPPSGG)`	$T_M = 96ºC$	$\Delta G_U = 18.6$ kJ/mol
TC16b	`DAYAQWLADaGPASaRPPPS`	$T_M = 87ºC$	$\Delta G_U = 16.3$ kJ/mol
Trp-2 cage	`DAYAQWLADGGWASGRPPPS`	$T_M \geq 75ºC$	$\Delta G_U = \sim 15$ kJ/mol*

*Fig. 1. Peptide sequences, lower case amino acid abbreviations indicate D-AA. The melting temperatures, defined at 50% folded, were obtained from the temperature dependence of the CD. The ΔG values for unfolding (280K) were calculated from H/D exchange protection factors of backbone NHs, except *, which was estimated from T_M.*

Methods

Linear peptides were synthesized using standard Fmoc solid-state chemistry and purified using reverse-phase HPLC as previously described [1,2]. The cyclized peptide was synthesized employing a 4-sulfamylbenzoyl AM "safety catch" resin, which produces a C-terminus thioester when cleaved by a mercaptan. Under aqueous conditions that favor Trp-cage folding, the termini are spatially near resulting in the efficient attack of the N-terminal amide on the thioester producing the cyclic peptide. Backbone-amide H/D exchange, NMR and CD data collection, as well as the methods for calculating T_M and ΔG_U, have been previously described. [1-3]

Results and Discussion

To date, two strategies have been reported for stabilizing the Trp-cage fold, helix stabilization by alanine insertion [2] and substitutions at Pro12 [4-5]. Bunagan *et al.* [5] reported that a P12W mutation results in enhanced stability (ΔT_M = 15ºC, compared to TC5b) and faster folding, and also noted that an EtF Trp-Trp interaction could be observed. We now establish that the P12W mutation (ΔT_M = 19ºC) combines synergistically with helix stabilization. Trp2-cage provided the details of the EtF W/W

S. Del Valle et al. (eds.), *Peptides for Youth: The Proceedings of the 20th American Peptide Symposium*, 41
DOI: 10.1007/978-0-387-73657-0_18, © Springer Science+Business Media, LLC 2009

interaction: upfield shifts at the Hζ2 and Hη2 protons of W6 and a modest exciton couplet in the CD spectra; in addition, the P18 Hβ3 showed the largest upfield chemical shift difference (-2.7 ppm) for a Trp-cage to date. The NMR structure ensemble established that the P12W mutation does not change the backbone torsion angles of the Trp-cage motif (distant from the site of mutation).

Two residues (Y3, P19) positioned close to the termini of the Trp cage motif have been shown to interact strongly: mutating either to Ala decreases stability by 16 kJ/mol [6]. As a result, we predicted that cyclizing the Trp-cage should provide additional stabilization without disrupting the basic structure. A GG linker, with a stabilizing K8A mutation that also removes a potential competitor in the cyclization reaction, afforded cycloTC which showed a dramatic increase in melting temperature of 40°C and in ΔG_U of 6.4kJ/mol (Fig. 1) [7]. We attribute the stabilization to the decreased contact order for the stabilizing Y3/P19 hydrophobic interaction combined with a smaller entropic effect associated with the cyclization. This stabilization cannot exclusively be described by the decrease in entropy change from the unfolded to folded state as a cyclic 22-mer can adopt a large number of configurations.

Fold stability can also be enhanced by decreasing ΔS_U by mutating glycines adopting positive Φ angles to D-alanine [8]. For the Trp-cage miniprotein, G10 and G15 Φ/Ψ values are *circa* 83/22 and 86/14, respectively [6]. Combining L-alanine mutations in both the α helix and 3_{10} helix with D-alanines at positions 10 and 15 provided a gain of 31°C in T_M and 4.1 kJ/mol in ΔG_U.

It is now established that a variety of fold stabilizing strategies combine synergistically with hydrophobic core formation in the Trp-cage system and that 20-mers with truly remarkable stability result. The stability has been enhanced to the extent that even a sidechain hydroxy group (Ser14) displays a protected OH signal in water at pH 7. This feature, rare even in large proteins, has now been observed in a decapeptide, WP2 (see Fig. 1). WP2 is an optimized version of an EtF cross-strand W/W stabilized β-hairpin motif. WP2 was derived from Ac-WINGKWTG-NH_2, which displayed the diagnostics of a flipped EtF W/W interaction [3] and an upfield CSD (-2.71ppm) for G8H_N that suggested that an additional interaction involving the NH and the indole ring of W6 was present. Turn stabilization (N3 → D-Pro), improvement of the hydrophobic cluster (K5L), and a C-terminal tag (TG-NH_2 → TGPS) increased stability: the G8H_N upfield CSD increased to -3.40 ppm. In WP2, the hydroxyl group of T7 is evident in NMR spectra (presumably due to H-bonding with the acetyl carbonyl) and displays both inter and intra-residue NOEs. The lower size limit for observing protein-like fold stability has now been reduced to the 10-mer level.

Acknowledgments

The work was funded by an NIH grant (GM050658) to N. H. Andersen.

References

1. Andersen, N. H., *et al., JACS* **128**, 6101-6110 (2006).
2. Lin, J. C., Barua, B. and Andersen, N. H. *JACS* **126**, 13679-13684 (2004).
3. Kier, B. L., Eidenschink, L. A., Huggins, K. N. 20th APS poster #335.
4. Naduthambi, D. and Zondlo, N. J. *JACS* **128**, 12430-12431 (2006).
5. Bunagan, M. R., *et al., J. Phy. Chem. B* **110**, 3759-3763 (2006).
6. Barua, B. Ph.D. thesis, University of Washington, Seattle, WA (2005).
7. Lin, J. C. Ph.D. thesis, University of Washington, Seattle, WA (2007).
8. Anil, B., Song, B., Tang, Y. and Raleigh, D. P. *JACS* **126**, 13194-13195 (2004).

Antimicrobial Peptides Chelating Lanthanide Ions: the Case of Trichogin GA IV Analogues and Terbium(III)

Mariano Venanzi[1], Emanuela Gatto[1], Lorenzo Stella[1], Gianfranco Bocchinfuso[1], Antonio Palleschi[1], Fernando Formaggio[2] and Claudio Toniolo[2]

[1]*Department of Chemical Sciences and Technologies, University of Rome "Tor Vergata", 00133 Rome, Italy and* [2]*ICB, Padua Unit, CNR, Department of Chemistry, University of Padua, 35131 Padua, Italy*

Introduction

Lipopeptaibols are the members of a family of antimicrobial peptides characterized by the presence of a lipid chain at the N-terminus and a high content of α-aminoisobutyric acid (Aib). Trichogin GA IV is one of the most representative compounds of this class, being characterized by peculiar conformational properties and a high affinity to the membrane phase. X-Ray diffraction studies showed that this peptide is folded in a mixed 3_{10}/α-helix conformation, while NMR, CD and IR absorption experiments indicated that this structure is also predominantly populated in solution. Recently, we have characterized the aggregation properties and partition equilibria of trichogin GA IV fluorescent analogues in aqueous solution and in the membrane phase, showing that their membrane permeabilization capability is strictly correlated to the concentration of aggregated forms in the lipid phase [1]. We have also recently found that trichogin GA IV shows peculiar binding properties to Ca(II) and several lanthanide ions [e.g. Gd(III), Tb(III)] [2].

Lanthanide chelates are currently actively explored for their potential applications in medical diagnostics, bioimaging, and bioanalytical assays. In particular, Tb(III) has been widely used as a fluorescent probe to characterize the active site of Ca-binding proteins. We report here on the Tb(III)-binding properties of two fluorescent analogues of trichogin GA IV, the primary structure (and acronyms) of which are reported below:

F0 Fmoc-Aib-Gly-Leu-Aib-Gly-Gly-Leu-Aib-Gly-Ile-Leu-OMe

F10 nOct-Aib-Gly-Leu-Aib-Gly-Gly-Leu-Aib-Gly-Dab(Fmc)-Leu-OMe

In the two analogues the fluorescent moieties Fmoc (fluoren-9-methyloxycarbonyl) and Fmc (fluoren-9-ylmethylcarbonyl) replace, respectively, the n-octanoyl (nOct) group (**F0**) and the Ile^{10} residue (**F10**) of [Leu^{11}-OMe]trichogin GA IV (OMe, methoxy). It has been shown that these two substitutions do not alter the membrane permeabilization properties of the two analogues [3].

Results and Discussion

All spectroscopic experiments reported here (CD and steady-state fluorescence) were carried out in acetonitrile solution.

CD titration experiments with Tb(III) revealed that, upon ion binding, both **F0** and **F10** undergo a conformational transition, with the peptide chain attaining folded, non helical, structures. By adding Tb(III) to the **F0** or **F10** solution a marked change in the CD curves of both peptides can be readily observed. In particular, a strong positive dichroic absorption, peaked at 214 nm, suggesting the occurrence of a type-II β-turn structure, increases as the bound ion concentration increases. If the fractional variation of the CD intensity at 214 nm is plotted *versus* the ratio of the

S. Del Valle et al. (eds.), *Peptides for Youth: The Proceedings of the 20th American Peptide Symposium*,
DOI: 10.1007/978-0-387-73657-0_19, © Springer Science+Business Media, LLC 2009

ion/peptide concentration, a sigmoidal curve is obtained. This finding suggests the occurrence of a cooperative binding process.

Fluorescence experiments have shown that, upon ion binding, the fluorescence quantum yield of Tb(III) markedly increases. Two factors are responsible for such effect: (i) release of water molecules from the ion coordination inner shell caused by the formation of the Tb(III)/peptide complex, which depletes the Tb-water ligand non radiative relaxation pathways and (ii) sensitization of Tb(III) emission via an energy transfer process from the fluorene donor group to the Tb(III) acceptor. Excitation spectra carried out at the acceptor emission wavelength (545 nm) confirmed the transfer of excitation energy in the fluorene→Tb(III) direction. The Förster energy mechanism correlates the efficiency of the energy transfer process to the donor···acceptor distance R, according to the equation:

$$E = \frac{R_0^6}{R_0^6 + R^6}$$

where R_0 is the Förster radius, a parameter determined by the spectral overlap between the emission of the donor molecule and the absorption of the acceptor. In the case of the fluorene-Tb(III) donor-acceptor pair R_0=4.9Å. The efficiency of the energy transfer process can be determined from the excitation spectra of Tb(III) when complexed to **F0** and **F10** or free in solution.

From FRET measurements it is therefore possible to estimate the distance between the fluorene group and the Tb(III) ion. In the case of Tb(III)/**F0** we obtained E<0.01(±0.01), while in the case of Tb(III)/**F10** E=0.25 (±0.02). In the former case, the large error associated to the very low FRET efficiency allowed us to estimate only a lower limit to the Tb···Fmoc distance, i.e. R>12Å, while in the case of Tb(III)/**F10** the Tb···Fmc distance can be more accurately determined (R=5.8Å). These results suggest that the Tb(III) binding site is close to the C-terminus of the main chain of both trichogin GA IV analogues. Preliminary conformational calculations confirm this hypothesis, showing that the ion binding site involves a partial unfolding of the C-terminal segment of the peptide chain, that forms a pseudo-ring structure strongly coordinating the Tb(III) ion.

Acknowledgments

This work was funded by the grants PRIN 2006 (MIUR of Italy) and MAE.

References

1. Mazzuca, C., Stella, L., Venanzi, M., Formaggio, F., Toniolo, C. and Pispisa, B. *Biophys. J.* **88**, 3411-3421 (2005).
2. Gatto, E., Ph.D. Dissertation, 2007, University of Rome "Tor Vergata".
3. Gatto, E., Mazzuca, C., Stella, L., Venanzi, M., Toniolo, C. and Pispisa, B. *J. Phys. Chem. B* **110**, 22813-22818 (2006).

Spectroscopic Characterization of the Fully-Extended, Planar, Peptide 2.0_5-Helix Based on Chiral, C^{α}-Ethylated, α-Amino Acids

Claudio Toniolo[1], Marco Crisma[1], Fernando Formaggio[1], Alessandro Moretto[1], Cristina Peggion[1], Bernard Kaptein[2] and Quirinus B. Broxterman[2]

[1]ICB, Padova Unit, CNR, Department of Chemistry, University of Padova, 35131 Padova, Italy; [2]DSM Research, Life Sciences, Advances Synthesis and Catalysis, MD 6160 Geleen, The Netherlands

Introduction

We originally reported that homo-peptides based on the *achiral* $C^{\alpha,\alpha}$-diethylated glycine (Deg) residue strongly favor the fully-extended, planar, 2.0_5-helical conformation (multiple, consecutive, C_5 forms) in solvents of low polarity and in the crystal state as well [1]. More recently, Tanaka, Suemune, and their coworkers [2] extended these findings to homo-*chiral*, C^{α}-ethylated, homo-peptides derived from (S)-C^{α}-*n*-butyl, C^{α}-ethylglycine (Beg).

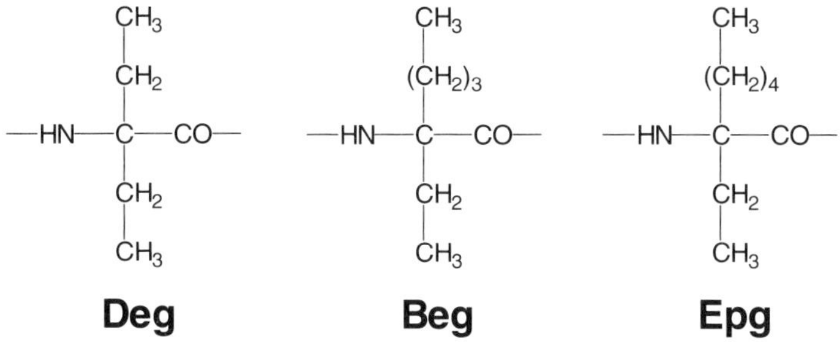

Results and Discussion

To obtain further insight into the spectroscopic properties of this relatively poorly investigated peptide secondary structure, we have prepared and performed a detailed FT-IR absorption, ^{1}H NMR, and X-ray diffraction analysis of the terminally protected Epg (C^{α}-ethyl, C^{α}-*n*-propylglycine) homo-peptide series (P)-$(Epg)_n$-O*t*Bu, where (P) is either Tfa or Ac and *n*=1-6. Peptide main-chain elongation was achieved *via* the Tfa-(S)-Epg-5(4*H*)-oxazolone method.

All homo-peptides are fully-extended, 2.0_5-helical, both in the crystal state (Fig.1) and in $CDCl_3$ solution. The FT-IR absorption maximum of the internal peptide NH stretching mode is found at 3359 cm^{-1} (Fig. 2). In the NMR spectra all NH proton chemical shifts are insensitive to the DMSO addition. All NH(i)→NH(i+1) connectivities are observed. For a full characterization of the CD spectra in an optically transparent solvent of low polarity, where the integrity of the 2.0_5-helix is presumably preserved, we are currently synthesizing an N^{α}-lipoylated (S)-Epg homopeptide series.

S. Del Valle et al. (eds.), *Peptides for Youth: The Proceedings of the 20th American Peptide Symposium*, 45
DOI: 10.1007/978-0-387-73657-0_20, © Springer Science+Business Media, LLC 2009

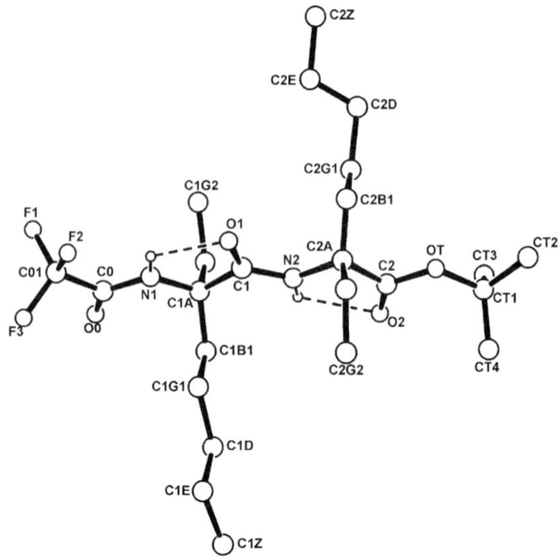

Fig. 1. X-Ray diffraction structure of Tfa-[(S)-Epg]$_2$-OtBu. The two intramolecular H-bonds are represented by dashed lines.

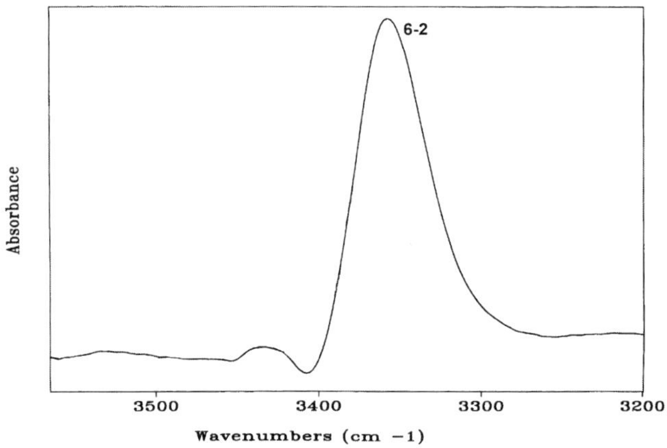

Fig. 2. FT-IR absorption difference spectra (N-H stretching region) between Ac-[(S)-Epg]$_6$-OtBu and the corresponding homo dimer in $CDCl_3$ solution.

References

1. For a recent review article see: Toniolo, C., Crisma, M., Formaggio, F., Peggion, C., Broxterman, Q.B. and Kaptein, B. *Biopolymers (Pept. Sci.)* **76**, 162-176 (2004).
2. Imawaka, N., Tanaka, M. and Suemune, H. *Helv. Chim. Acta* **84**, 2823-2835 (2000).

Monitoring Peptide Folding by Time-Resolved Spectroscopies: the Effect of a Single Gly to Aib Susbtitution

Mariano Venanzi[1], Emanuela Gatto[1], Gianfranco Bocchinfuso[1], Antonio Palleschi[1], Lorenzo Stella[1], Chiara Baldini[2], Fernando Formaggio[2], Claudio Toniolo[2] and Basilio Pispisa

[1]*Department of Chemical Sciences and Technologies, University of Roma "Tor Vergata", 00133 Rome, Italy;* [2]*ICB, Padua Unit, CNR, Department of Chemistry, University of Padua, 35131 Padua, Italy*

Introduction

We have recently reported on a new fluorescent analogue of trichogin GA IV, a natural peptide showing interesting antimicrobial activity [1]. The primary structure (and acronym) of the peptide investigated is:

Fmoc-Aib-Gly-Leu-Aib-Gly-Gly-Leu-TOAC-Gly-Ile-Leu-OMe **(F0T8)**

where Fmoc is fluorenyl-9-methyloxycarbonyl, Aib is α-aminoisobutyric acid, TOAC is 2,2,6,6-tetramethylpiperidine-1-oxyl-4-amino-4-carboxylic acid, and OMe is methoxy.

The double substitution of an energy donor (Fmoc) and an acceptor (TOAC) pair in the trichogin sequence enabled us to make use of time-resolved optical spectroscopies, spanning from the nanosecond to the microsecond time regime, to investigate the conformational propensity and the dynamical features of **F0T8**. Experimental and computational results indicated that the structural and dynamical properties of **F0T8** are characterized by a transition from an elongated helical conformation to a more compact structure mimicking a helix-turn-helix motif. To further investigate the role of the Gly^5-Gly^6 central motif we have synthesized the new [Leu^{11}-OMe]trichogin analogue having the Gly^6 residue substituted by Aib

Fmoc-Aib-Gly-Leu-Aib-Gly-Aib-Leu-TOAC-Gly-Ile-Leu-OMe **(F0A6T8)**

This replacement is expected to stiffen the peptide backbone, reducing the flexibility around the crucial -Gly^5-Gly^6- dipeptide unit.

Results and Discussion

All spectroscopic measurements (CD, steady-state and time-resolved fluorescence, and nanosecond transient absorption) were carried out in acetonitrile.

The CD spectra revealed that both the **F0T8** and **F0A6T8** analogues predominantly populate helical conformations, their CD curves being very similar to that measured for trichogin GA IV. The ratio of the molar ellipticities measured at 222 and 208 nm, i.e. $R=[\theta]_{222}/[\theta]_{208}$, is 0.40 for **F0T8** and 0.53 for **F0A6T8**, suggesting that the latter peptide has a larger α-helical content with respect to **F0T8**.

Steady-state fluorescence experiments showed that the two peptides have similar quantum yields and static quenching efficiencies. More detailed information was obtained from time-resolved fluorescence experiments, that point to a bi-exponential time decay for both peptides (Table 1). The results suggest that **F0T8** and **F0A6T8** mainly populate two different conformers, the more abundant being associated to the longer lifetime, i.e. to a larger Fmoc···TOAC distance. Interestingly, the time decay components of **F0A6T8** are invariably longer than those of **F0T8**, suggesting a higher population of the more elongated structure in the former peptide.

S. Del Valle et al. (eds.), *Peptides for Youth: The Proceedings of the 20th American Peptide Symposium*,
DOI: 10.1007/978-0-387-73657-0_21, © Springer Science+Business Media, LLC 2009

The most interesting results are provided by time-resolved transient absorption experiments in the microsecond time regime. In this case, the triplet decay rate constant (k_T) of the fluorene probe is determined by the peptide dynamics. Due to the rigidity of the peptide backbone, the conformational interconversion that brings the TOAC group at a very close distance to the fluorene moiety occurs in the microsecond time scale [2]. The relative higher rigidity of **F0A6T8** is supported by the measured k_T value, that is smaller by a factor of two with respect to the triplet decay rate constant measured for **F0T8**.

Table 1. Experimental time-resolved parameters from fluorescence and transient absorption measurements.

Peptide	Time-resolved fluorescence				Transient absorption
	α_1	$\tau_1(ns)$	α_2	$\tau_2(ns)$	$k_T(\cdot 10^{-5}s^{-1})$
F0T8	0.23	3.1	0.77	6.6	22
F0A6T8	0.22	3.8	0.78	7.6	12

Significant conformational differences between the two peptides were also determined by studying the binding properties of **F0T8** and **F0A6T8** to the Ca(II) and Gd(III) ions. Titrations of **F0T8** with Ca(II) and Gd(III) cause a conformational transition that can be easily monitored by CD. Indeed, upon ion binding, an intense dichroic absorption at 214 nm, characteristics of a type-II β-turn structure, stands out clearly. On the other hand, **F0A6T8** titrations with Ca(II) and Gd(III) give rise to a different conformational transition, probably because the steric contraints exerted by the Aib^6 residue inhibit the motions leading to β-turn like structures. Furthermore, while the **F0T8**/Gd(III) complex saturates its binding curve at an ion/peptide ratio r=1, the **F0A6T8**/Gd(III) complex saturates at r=2.5, very likely because the higher rigidity of this latter peptide makes more difficult the structural rearrangements of the peptide backbone needed to fulfill the steric requirements leading to an optimal ion binding.

Acknowledgments

This work was funded by the grant PRIN 2006 (MIUR of Italy).

References

1. Venanzi, M., Gatto, E., Bocchinfuso, G., Palleschi, A., Stella, L., Baldini, C., Formaggio, F. and Toniolo, C. *J. Phys. Chem. B* **110**, 22834-22841 (2006).
2. Venanzi, M., Gatto, E., Bocchinfuso, G., Palleschi, A., Stella, L., Formaggio, F. and Toniolo, C. *ChemBioChem* **7**, 43-45 (2006).

First Homo-Peptides Undergoing a Reversible 3_{10}-Helix to α-Helix Transition

Alessandro Moretto[1], Marco Crisma[1], Fernando Formaggio[1], Claudio Toniolo[1], Ling Wu[2], Timothy A. Keiderling[2], Bernard Kaptein[3] and Quirinus B. Broxterman[3]

[1]*ICB, Padova Unit, CNR, Department of Chemistry, University of Padova, 35131 Padova, Italy;* [2]*Departemnt of Chemistry, University of Illinois at Chicago, M/C 111, Chicago, IL, 60607-7061;* [3]*DSM Research, Life Sciences, Advances Synthesis and Catalysis, MD 6160 Geleen, The Netherlands*

Introduction

Very few examples of reversible transitions between a 3_{10}-helix (in solvents of low polarity, particularly at low temperatures) and an α-helix (in solvents of low polarity at high temperatures or, better, in polar solvents) have been already published, namely for N^{α}-acetylated Aib- and L-(αMe)Val- rich *co*-oligopeptides of different main-chain length (from an hexapeptide ester to a pentadecapeptide amide) [1-3]. X-Ray diffraction analyses of crystals of two peptides grown from solvent mixtures of low polarity confirmed the overwhelming population of 3_{10}-helical molecules in those environments.

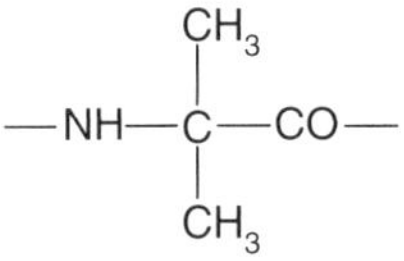

Aib

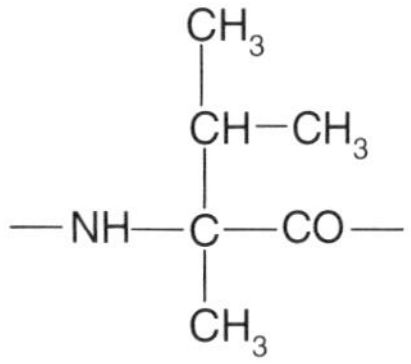

(αMe)Val

Results and Discussion

In this work we have established that amino acid sequence complexity is not a prerequisite for this reversible conformational transition. As determined by ECD measurements (Fig. 1), even the simple L-(αMe)Val *homo*-peptides listed below undergo a solvent-driven, reversible transition from 3_{10}-helix (MeOH) to α-helix (HFIP) and viceversa: Ac-[L-(αMe)Val]$_n$-OMe (n=6-8), Ac-[L-(αMe)Val]$_7$-NH*i*Pr, and Z-[L-(αMe)Val]$_7$-NH*i*Pr. A VCD analysis has also shown that the latter heptapeptide amide overwhelmingly adopts the 3_{10}-helical conformation in TFE solution (50 mg/ml).

We were also able to crystallize from a HFIP solution and solve by X-ray diffraction the 3D-structure of the N^{α}-blocked *homo*-heptapeptide alkylamide (Fig. 2). It is folded in a mostly α-helical structure (an extremely uncommon finding for a *homo*-peptide molecule), as expected from its crystallization environment. Solvent molecules are found to be H-bonded to the peptide chains. Conversely, the same peptide is folded in a regular 3_{10}-helix when its crystals are grown from a MeOH solution.

S. Del Valle et al. (eds.), *Peptides for Youth: The Proceedings of the 20th American Peptide Symposium*, DOI: 10.1007/978-0-387-73657-0_22, © Springer Science+Business Media, LLC 2009

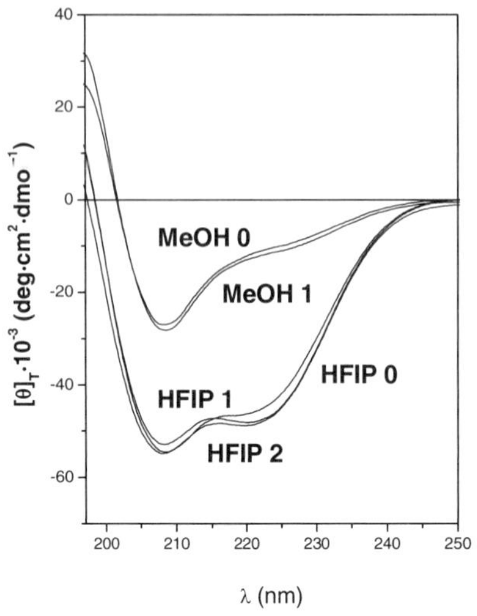

Fig. 1. ECD spectra of Ac-[L-(αMe)Val]$_7$-NHiPr in different alcohols (MeOH and HFIP). Repeated cycles of helix conversion have been performed (peptide concentration: 1 mM).

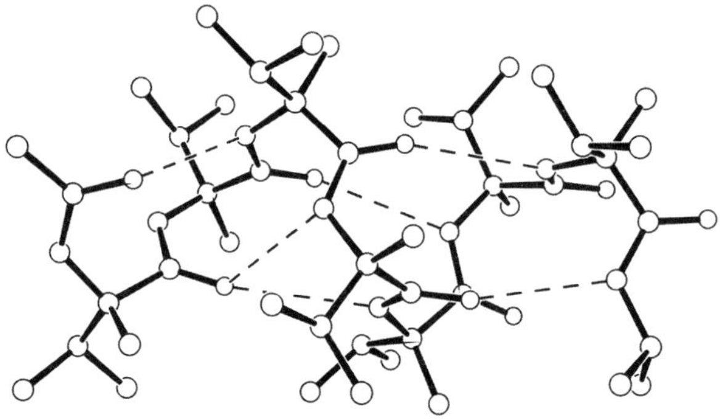

Fig. 2. X-Ray diffraction structure of one of the two independent, mostly α-helical, molecules (A) in the asymmetric unit of Ac-[L-(αMe)Val]$_7$-NHiPr (crystal grown from a HFIP solution). The intramolecular C=O···H-N H-bonds are represented by dashed lines.

References

1. Hungerford, G., Martinez-Insua, M., Birch, D.J.S. and Moore, B.D. *Angew. Chem. Int. Ed.* **35**, 326-329 (1996).
2. Pengo, P., Pasquato, L., Moro, S., Brigo, A., Fogolari, F., Broxterman, Q.B., Kaptein, B. and Scrimin, P. *Angew. Chem. Int. Ed.* **42**, 3388-3392 (2003).
3. Bellanda, M., Mammi, S., Geremia, S., Demitri, N., Randaccio, L., Broxterman, Q.B., Kaptein, B., Pengo, P., Pasquato, L. and Scrimin, P. *Chem. Eur. J.* **13**, 407-416 (2007).

Molecular Modeling Approach to Achieving Erythropoietin Mimctic Pcptides

Yanfeng Qi[1], Xuefeng Gao[2], Liping Wang[1] and Wei Li[1,*]

[1]*BioMacroMolecular Lab;* [2] *Edmond H. Fischer Cellular Signal Transduction Lab, College of Life Science, Jilin University, Changchun, 130023, China*

Introduction

Erythropoietin (EPO) is a 34 kDa glycoprotein which regulates the production of red blood cells. Deficits in EPO production result in anemia in human, the most prevalent form of anemia is associated with kidney failure. At present, the only treatment for this form of anemia is administration of recombinant EPO via subcutaneous or intravenous injection. The use of recombinant EPO has significantly improved the quality of life of these patients; however, this treatment requires repeated administration of recombinant protein, which is both inconvenient and expensive. Finding small molecules and peptides to alternative EPO, change the administration route of EPO is very popular in recent years.

The crystal structures of EPO-EPOR and small peptide agonist (EMP-1)-EPOR complexes have been solved and reveal a different configuration of the EPOR dimer in each of the complexes [1, 2]. Through(By) analysis the structure data of these complex, Wilson etc propose that there's a plasticity of recognition mechanism [3] between the receptor and the agonist, and they point some "Hot Spot" that play a key role in receptor-ligand recognition and binding. These "Hot Spot" are Phe93, Met150 and Phe205 residues (see Fig1-A). Interestingly, the interaction between "Hot Spot" and the agonist are all hydrophobic interaction.

Recently, with the development of the computer science, software technology and molecular mechanics, molecular dynamics, quantum mechanics; molecular modeling method(MM) become a very important technology to construct and analyze the three dimension structure of biological macromolecular, the interaction between receptor and ligand, drugs discover and development, and so on. Based on the known EpoR-EMPs complex structural data, we use molecular dynamics simulations methods to investigate the structure, dynamics and thermodynamics of these complex. Then, we create a molecular dynamic model to analyze the surface electrostatic potential around the binding site and the interaction with EMPs.

Results and Discussion

The simulations were based on the complex EMPs-EpoR (Protein Data Bank code: 1CN4, 1EBP etc.) using GROMACS 3.3(http://www.gromacs.org)[4], performed on a five node cluster workstation. Surface electrostatic potential energy analysis was applied when we got stable conformation of the complex with lower energy and Consistent Valence Force Field (CVFF) was used to investigate the interaction between EMPs and the receptor binding site amino acid residue which the distance to EMPs less than 1 nm (see Tab-1).

Through our simulation and analysis, we found there is a very strong negative electrostatic potential region distributed in the periphery of the binding site on the receptor. This was briefly mentioned by Wilson etc. in their plasticity of recognition theory. Through further surface electrostatic potential analysis of EPO-EpoR complex, our results investigate that the potential complementarities was a very

S. Del Valle et al. (eds.), *Peptides for Youth: The Proceedings of the 20th American Peptide Symposium*,
DOI: 10.1007/978-0-387-73657-0_23, © Springer Science+Business Media, LLC 2009

important property of ligand, which have great impact on the ability of ligand to induce receptor dimerizition.

The interaction analysis shows that some individual amino acid residue within the negative potential region (such as Glu25, Glu51, Glu236, Glu262, Ser293 and His144) has significant coulomb and hydrophobic interaction with EMPs (see Fig1-B). The total energy force of the interaction is much higher than "Hot Spot" residues. These energy forces are very important to the maintenance and stability of receptor-ligand complex. This is very crucial to the design and development of the small molecular agonist and antagonist. We consider the four "coulomb" sites and two "hydrophobic" sites as members of "Hot Spot" family.

According to these results, we designed, developed and synthesized more than 8 series mimetic binding peptide derivatives (MBPs), such as hydrophobic interaction series, coulomb series and Alanine substitution series, to calculate the binding and receptor dimerizition activities. The biological activity of the potential peptides with significant results in MD analysis will be determinate by micro-plate fluorescence technology. The results will be published separately in another paper.

Table 1. Interaction between receptor amino acid residues and EMP-1.

Residue*	*number*	*R6-R12*	*Coulomb*	*Total*
Glu	235	-3.9338	-101.1192	-105.053
Glu	51	5.5681	-106.2994	-100.7313
Glu	262	2.4715	-95.6171	-93.1456
Glu	25	-2.1696	-31.0862	-33.2558
Phe	84	-14.6288	-6.8765	-21.5053
Ser	293	-3.9396	-17.209	-21.1486
Met	141	-10.564	-9.5078	-20.0719
His	144	-1.9357	-14.7559	-16.6917
Phe	295	-11.5908	-3.0672	-14.658

* *Only shown the residue has equal or more force energy than Phe295. New "Hot Spot" was in black and italics.*

References

1. Livnah O., *et al. Science*, **273**, 464-471 (1996).
2. Wrighton NC, *et al. Science*, **273**, 458-463 (1996).
3. Ian A Wilson *et al. Current Opinion in Structural Biology* **9**, 696-704 (1999).
4. Lindahl, E., Hess, B. and van der Spoel, D. *J. Mol. Mod.*, **7**, 306-317 (2001).

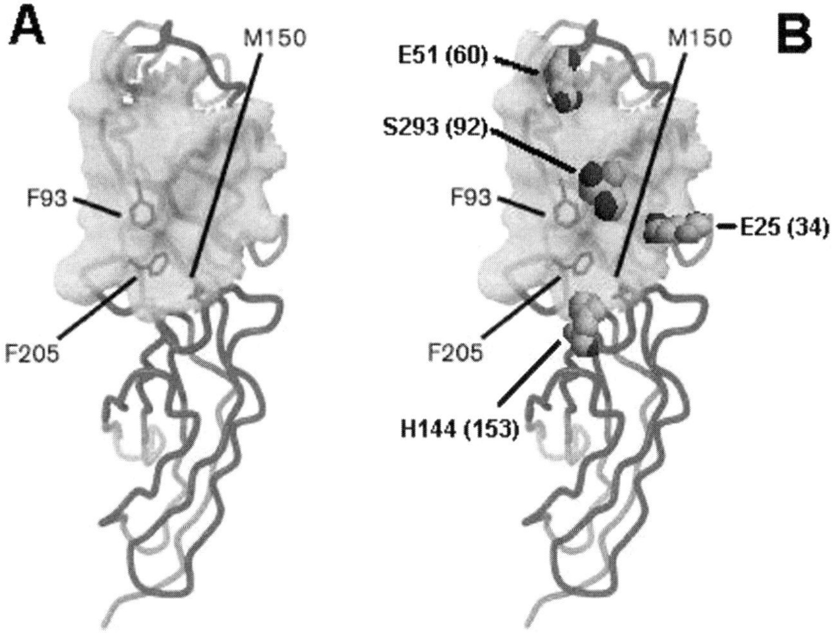

Fig-1. A comparision of the EMPs binding site of EpoR. The new hotspot in right was displayed in spacefill mode.

Conformation Analysis of Parathyroid Hormone Residues Critical for Bioactivity

Zhanna Potetinova,[1] Feng Ni,[2] Manghaiko Mayele,[2] Ping Xu[2] and Gordon E. Willick[1]

[1]*Institute for Biological Sciences, National Research Council, Ottawa, ON, Canada, K1A 0R6;* [2]*Biotechnology Research Institute, National Research Council, Montréal, QC, Canada, H4P 2R2*

Parathyroid hormone (PTH) is a major regulator of serum calcium homeostasis. The smallest fully bioactive PTH(1-31) consists of two helixes joined by a short flexible sequence. The C-terminal amphiphilic α-helix binds to the N-terminal extracellular region of the PTH receptor (P1R). Backbone methylation [1] and Ala-scan [2] studies showed that three residues, Arg20, Trp23 and Leu24 are critical for bioactivity and play a critical role in PTH-P1R interaction.

We studied 2D-NMR structures of hPTH(1-31) analogues modified in the C-terminal helical region: bioactive cyclic(Glu22-Lys26), backbone methylated at Leu24 inactive linear and its cyclic (Glu22-Lys26) analogue with restored bioactivity. Detailed conformation studies and geometrical optimization of the structures revealed similar orientations of Arg20, Trp23 and Leu24 side-chains in the bioactive analogues. In both cyclic active analogues, Arg20 aligned with Trp23 and Leu24. The inactive linear methylated analogue has the critical Arg20 and Leu24 side-chains oriented away from Trp23.

References

1 Barbier, J.R. *et al.* (2005) *J. Biol. Chem.* **280**, 23771.
2 Dean, T. *et al.* (2006) *J. Biol. Chem.* **281**, 32485.

S. Del Valle et al. (eds.), *Peptides for Youth: The Proceedings of the 20th American Peptide Symposium*,
DOI: 10.1007/978-0-387-73657-0_24, © Springer Science+Business Media, LLC 2009

Light-Switchable Folding/Unfolding of the Collagen Triple Helix with Azobenzene-Containing Model Peptides

Ulrike Kusebauch[1], Lisa Lorenz[2], Sergio A. Cadamuro[1], Hans-Jürgen Musiol[1], Martin O. Lenz[2], Christian Renner[1], Josef Wachtveitl[2] and Luis Moroder[1]

[1]*Max-Planck-Institute of Biochemistry, Martinsried;* [2]*Institute of Physical and Theoretical Chemistry, Institute of Biophysics, Johann Wolfgang Goethe-University, D-60438 Frankfurt/Main, Germany*

Introduction

Azobenzene derivatives have been used as backbone constituents or side-chain clamps to photocontrol in reversible manner conformational states of model peptides [1,2]. Because of the ultrafast isomerization (within picoseconds), this chromophore allows spectroscopic monitoring of the fast kinetics of folding/unfolding of model peptides with ordered secondary structure motifs such as β-turns, α-helices and most recently even of β-hairpins [3,4]. Aim of the present work was to gain new insights into the kinetics of assembly of tertiary structure motifs such as the collagen triple helix. For this purpose model peptides were conformationally restricted with a suitable azobenzene clamp in the *trans*-isomeric state (see Figure 1), which upon photoisomerization provokes unfolding of the triple helix [5].

Ac-$(GPO)_2$-GXO-GPO-GPY-$(GPO)_2$-GG-NH_2 (**1**)
(i) (i+7)

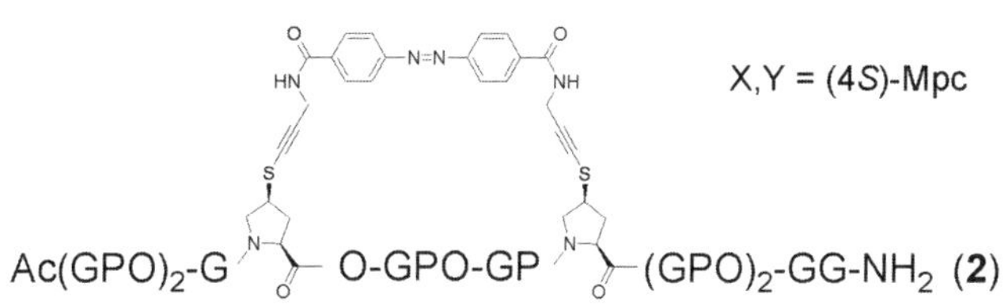

*Fig. 1. Structure of the azobenzen-collagen peptide with (4S)-Mpc in i and i+7 position (**1**), which was side-chain bridged via thioether bonds with an azobenzene derivative yielding peptide **2**.*

Results and Discussion

Replacement of a Pro and (4*R*)-Hyp residue in Ac-$(GPO)_7$-GG-NH_2 (CD: T_m = 43°C) with two (2*S*,4*S*)-mercaptopyrrolidine-2-carboxylic acids (Mpc) did not prevent a triple helical fold, but only decreased its thermal stability (CD: T_m = 34.5°C) in aqueous solution. This is due to the significantly weaker stereoelectronic effects of (4*S*)-Mpc than (4*S*)-Hyp on the geometrical parameters of the triple helix. Indeed (4*S*)-Mpc induces a preference for the *trans*-peptide bond and γ-exo pucker (unpublished results), while (4*S*)-Hyp prefers the γ-*endo* pucker [6]. Because of the insolubility in water of azobenzene-bridged **2** the thermal stability of its triple helix was analyzed in aqueous alcoholic mixtures, known to favour this fold [7]. While in MeOH/0.1 M AcOH, 4:1, the transitions could not be fully registered by CD because of the low b.p. of this alcohol, in EG/H_2O, 2:1, the T_m of **2** is 55°C *vs* 60°C of

S. Del Valle et al. (eds.), *Peptides for Youth: The Proceedings of the 20th American Peptide Symposium,* 57
DOI: 10.1007/978-0-387-73657-0_25, © Springer Science+Business Media, LLC 2009

Ac-(GPO)$_7$-GG-NH$_2$. This would suggest that the parent peptide **1** is slightly stabilized by the azobenzene clamp. Thermal denaturation of **2** in EG/H_2O, 1:1, was monitored very efficiently even by IR (see Figure 2) with a T_m of 53°C as average value from transitions at different wavelengths.

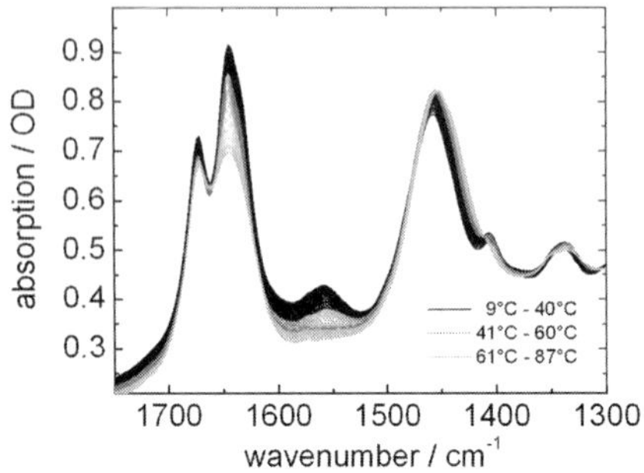

*Fig. 2. IR spectra of 10 mM peptide **2** in EG:H_2O, 1:1 as a function of temperature (9- 86°C).*

The thermal excursion leads to intensive and largely reversible variations of the characteristic absorption bands of the collagen triple helix in the amide I region [8,9] as shown in Figure 3b. Conversely, the spectral variations in the 1350-1550 cm^{-1} region are not fully reversible supporting the presence of very slow components in the refolding process. It is well known that refolding processes of non-crosslinked collagen chains are accompanied by slow readjustment processes [10].

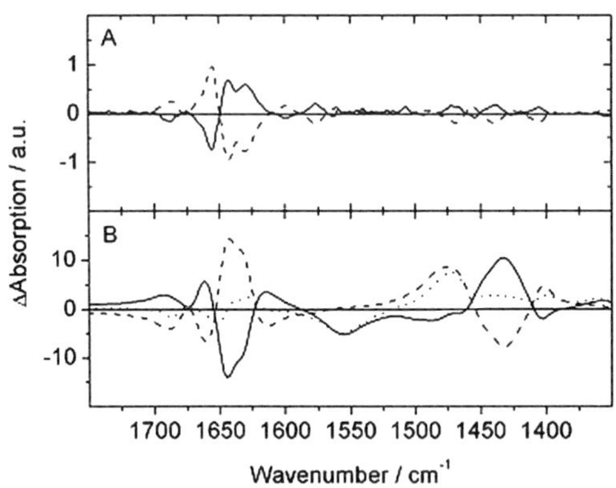

*Fig. 3. Difference IR spectra of 10 mM **2** in EG/H_2O, 1:1: A) trans-to-cis (—) and cis-to-trans (- - -) photoisomerization; B) upon heating from 9 to 86°C (—), upon cooling from 86 to 9°C (- - -) and the difference spectrum between the denatured-refolded peptide and the original IR spectrum at 9°C (·····).*

Photoisomerization of the azobenzene at rt is practically not affected by the triple helical structure as well assessed by comparative analysis of the time resolved dynamics of this process for the azobenzene-4,4'-*N*-(4-chloro-2-butenyl)-carboxamide and the peptide-bound chromophore (Figure 4). This result confirms the full functionality of the azobenzene switch when incorporated into the collagen peptide including, most importantly, the very rapid progression and completion of the isomerization process. This is making azobenzene so valuable for photo-switching as subsequent dynamics are not disturbed or mixed with the action of the switch. As desired, the photoisomerization induces almost instantly a new conformational space in which the peptide can relax and unfold.

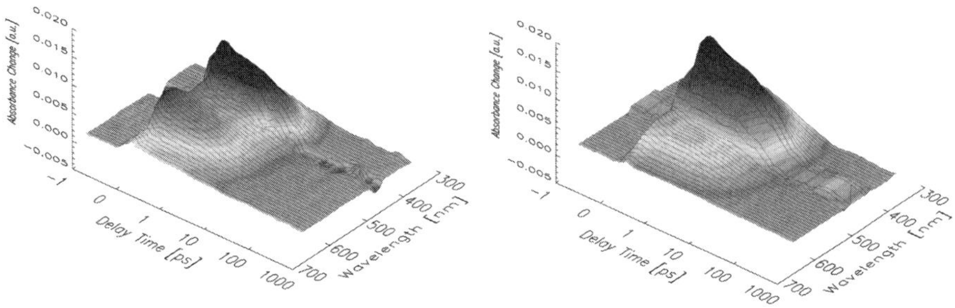

*Fig. 4. Time-resolved photodynamics of the trans-to-cis isomerization of the azobenzene chloride derivative in DMSO (left) and of **2** in EG/H_2O, 1:1 (right) at rt as monitored by UV.*

Most interestingly, the *trans*-to-*cis* and viceversa *cis*-to-*trans* photoisomerization of **2**, although of relatively weak effect on the triple helical structure at rt, generate well detectable difference spectra in the amide I region (see Figure 3A) as index of local folding/unfolding events of the triple helix. Even in MeOH/0.1M AcOH, 4:1, NMR experiments clearly revealed partial unfolding of **2** upon photoisomerization at 27°C [5]. EG/H_2O, 1:1, is stabilizing the triple helix more than aqueous MeOH, so that higher temperatures are required to destabilize the triple helix sufficiently to enable unfolding by the weak driving force of the azobenzene clamp. A search for the optimal conditions for monitoring the fast kinetics of the photocontrolled folding/unfolding of the collagen molecule in nano- to picosecond time scales by ultrafast IR spectroscopy is under way.

Acknowledgments

The work was funded by the DFG (SFB 533, grant A8).

References

1. Woolley, A. *Acc. Chem. Res.* **38**, 486-493 (2005).
2. Renner, C. and Moroder, L. *ChemBioChem* **7**, 868-878 (2006).
3. Dong, S. L. *et al. Chem-Eur. J.* **12**, 1114-1120 (2006).
4. Aemissegger, A. *et al. J. Am. Chem. Soc.* **127**, 2929-2936 (2005).
5. Kusebauch, U. *et al. Angew. Chem. Int. Ed.* **45**, 7015-7018 (2006).
6. Panasik, N. Jr. *et al. Int. J. Pept. Prot. Res.* **44**, 262-269 (1994).
7. Privalov, P. L. *Adv. Prot. Chem.* **35**, 1-104, (1982).
8. Lazarev, Y. A. *et al. Biopolymers* **24**, 1449-1478 (1985).
9. Lazarev, Y. A. *et al. Biopolymers* **32**, 189-195 (1992).
10. Engel, J. and Bächinger, H. P. *Matrix Biol.* **19**, 235-244 (2000).

Structural Studies of Parathyroid Hormone Analogue with C-Terminal Aldehyde Function

Zhanna Potetinova[1], Feng Ni[2], Ping Xu[2] and Gordon E. Willick[1]

[1]*Institute for Biological Sciences, National Research Council, Ottawa, ON K1A 0R6, Canada;* [2]*Biotechnology Research Institute, National Research Council, Montréal, QC H4P 2R2, Canada*

Introduction

The C-terminal function affects the structure (CD) and bioactivity of hPTH(1-28) analogues [1]. The hPTH with an amide C-terminus is more helical and bioactive than the analogue in a natural carboxyl form. The monomethylamide exhibited the greatest helicity and binding activity in contrast to the dimethylamide, which has helical parameters similar to the acid. Thus, removal of potential H-bond NH donors on the C-terminus destabilizes α-helix structure and decreases bioactivity. The CD profile of the aldehyde analogue is close to the bioactive monomethylamide, implying that the aldehyde function stabilizes the C-terminal helical structure of hPTH(1-28). However, the aldehyde, like the dimethylamide, demonstrated a weaker receptor binding than the monomethylamide.

The structure of hPTH(1-28) analogues with C-terminal aldehyde, mono- and dimethylamides were studied by high resolution 2D-NMR in aqueous solution at neutral pH.

Results and Discussion

The NMR spectra were recorded on a Bruker Avance-800 spectrometer at 5°C. The experiments were carried out in 25 mM sodium phosphate, 0.2 mM EDTA, 300 mM NaCl, and 10% D_2O at pH 6.8. The water signal was suppressed by the WATERGATE method. Chemical shifts of proton resonances were internally referenced to the water signal at 4.98 ppm.

The NMR data confirm two well-formed helical fragments in the N- and C-terminal regions as was found in hPTH(1-31)NH_2 [2]. The stronger negative deviations of the monomethylamide identify a greater helical C-terminal segment than in the aldehyde and dimethylamide.

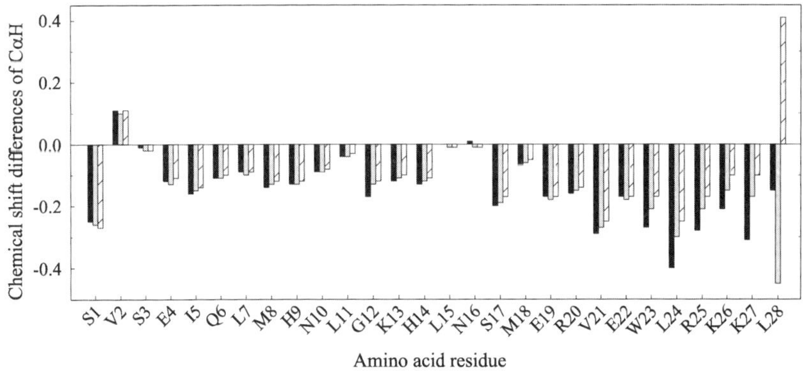

Fig. 1. Chemical shifts of Cα protons for C-terminal modified hPTH(1-28) analogues: monomethylamide (black bars), aldehyde (gray bars) and dimethylamide (open bars)

S. Del Valle et al. (eds.), *Peptides for Youth: The Proceedings of the 20th American Peptide Symposium*,
DOI: 10.1007/978-0-387-73657-0_26, © Springer Science+Business Media, LLC 2009

The dimethylamide has a positive shift at Leu28 indicating that the residue is not included in helical contacts. In the aldehyde, NOESY interactions identify the C-terminal helix between Ser17 and Lys27.

Peptide aldehydes, being very sensitive to pH and chemically very active, can exist in three covalent forms in aqueous solution: free aldehyde (I), hydrated (II) and cyclic carbinolamine (III). At neutral pH, the expected molecular mass, corresponding to Forms I or III, with the same mass, was observed in the aldehyde MS spectrum (M^+ 3361.4, expected 3362.0). A signal with M+18 in the MS spectrum confirmed the presence of Form II in acidic media.

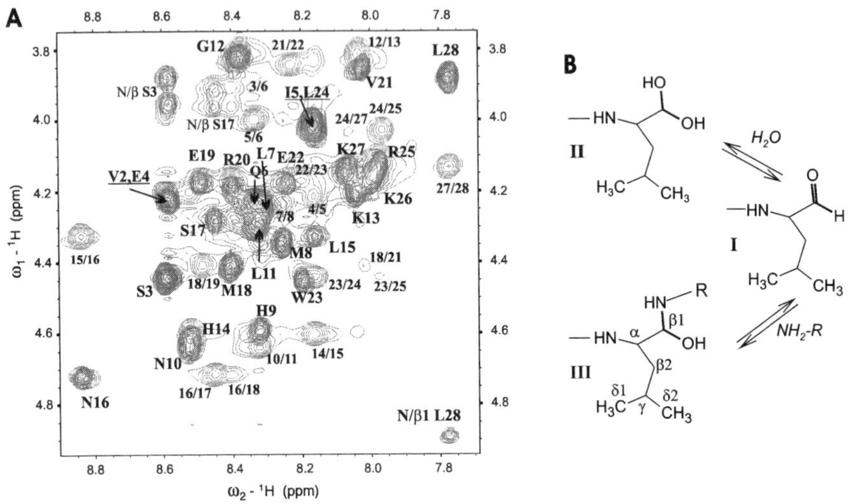

Fig. 2. (A) NH-CαH region of TOCSY and NOESY spectra of hPTH(1-28)aldehyde. (B) Equilibrium forms of C-terminal leucinal peptide: I, free aldehyde; II, gem-diol; III, carbinolamine.

Integration of Leu28 aldehyde and Trp23 Nε1 indole proton signals in the ^{1}H NMR spectrum determined that only 4% of the hPTH(1-28)aldehyde is present in the free aldehyde form under the recording conditions. The large negative shift of CαH and the presence of an additional signal at 4.90 ppm (Fig. 2A), associated with Leu28, demonstrated that the aldehyde structure is almost in a stable carbinolamine form. The C-terminal chemically-active aldehyde can easily react with an Arg guanidino (20 or 25) or Lys amino group (26 or 27), forming a cyclic carbinolamine.

The unusually strong positive shift, associated with Arg20 side-chain proton, and NOESY connectivities of Arg20 with Ser17, Trp23 and C-terminal Lys27, Leu28 residues suggest that Arg20 is not only involved in internal helical contacts but also lies in close proximity to the C-terminus. Structural calculations showed this covalent cyclic formation (Arg20-Leu28) does not exert a destabilizing effect on the C-terminal α-helix, unlike the other Arg or Lys residues. The secondary structure of the C-terminal region probably determines this predominant interaction.

References

1. Potetinova, Z., *et al. Biochemistry* **45**, 11113-21 (2006).
2. Chen, Z., *et al. Biochemistry* **39**, 12766-77 (2000).

Synthesis and Conformational Studies of Novel, Side-Chain Protected, L-(αMe)Ser Homo-Peptides

Fernando Formaggio[1], Cristina Peggion[1], Claudio Toniolo[1], Bernard Kaptein[2], Quirinus B. Broxterman[2], Minghong Ni[3], Laura Muzi[3] and Alma Dal Pozzo[3]

[1]ICB, Padova Unit, CNR, Department of Chemistry, University of Padova, 35131 Padova, Italy; [2]DSM Research, Life Sciences, Advances Synthesis and Catalysis, MD 6160 Geleen, The Netherlands; [3]"G. Ronzoni" Institute for Chemical and Biochemical Research, 20133 Milan, Italy

Introduction

In search for new, conformationally constrained, amino acid residues endowed with potentially interesting features, we started a research programme aimed at investigating the physico-chemical properties of (αMe)Ser-rich peptides. In addition to the 3D-structural peculiarities of $C^{\alpha,\alpha}$-disubstituted glycines, (αMe)Ser is expected to impart valuable solubility and biological (e.g., serine protease-like) properties when inserted into appropriately designed peptides. In the first step of this programme we synthesized and investigated the preferred conformation of the terminally- and side-chain blocked L-(αMe)Ser(Bzl) homo-peptides to the hexamer level shown below.

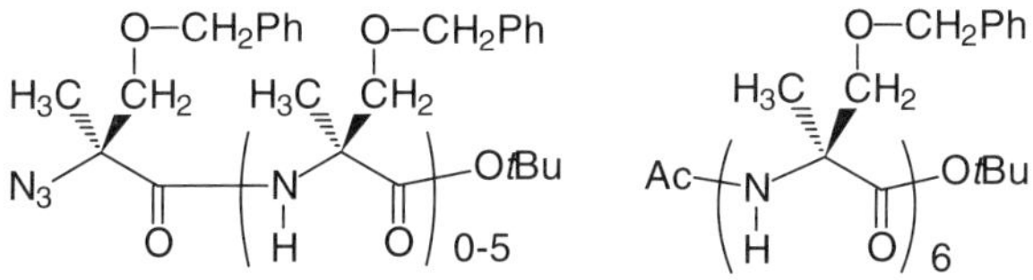

Results and Discussion

Optically pure H-L-(αMe)Ser(Bzl)-OH was prepared according to a chemo-enzymatic method developed by the DSM Research group [1]. α-Azido-peptides were prepared in excellent yields by the step-by-step, highly efficient, α-azido/acylbromide coupling procedure [2], using 2-azido-2-methyl-3-benzyloxypropionic acid bromide as the acylating agent (Fig. 1). Reduction of the hexamer α-azido group, followed by N^{α}-acetylation, afforded Ac-[L-(αMe)Ser(Bzl)]$_6$-O*t*Bu.

The results of our preliminary FT-IR absorption and 2D NMR analyses strongly support the conclusion that the longest homo-oligomers of the L-(αMe)Ser(Bzl) series are folded in a helical conformation. The FT-IR absorption spectrum in the amide A region of Ac-[L-(αMe)Ser(Bzl)]$_6$-O*t*Bu in a solvent of low polarity ($CDCl_3$) shows a very intense band centered at 3328 cm^{-1} (Fig. 2A), typical of highly populated, intramolecularly H-bonded, conformations. An analysis of the ROESY spectrum (Fig. 2B), in a solvent even more polar (MeOH), indicates that the same peptide adopts indeed a helical structure, as all NH(i)→NH(i+1) connectivities are clearly detectable.

S. Del Valle et al. (eds.), *Peptides for Youth: The Proceedings of the 20th American Peptide Symposium*,
DOI: 10.1007/978-0-387-73657-0_27, © Springer Science+Business Media, LLC 2009

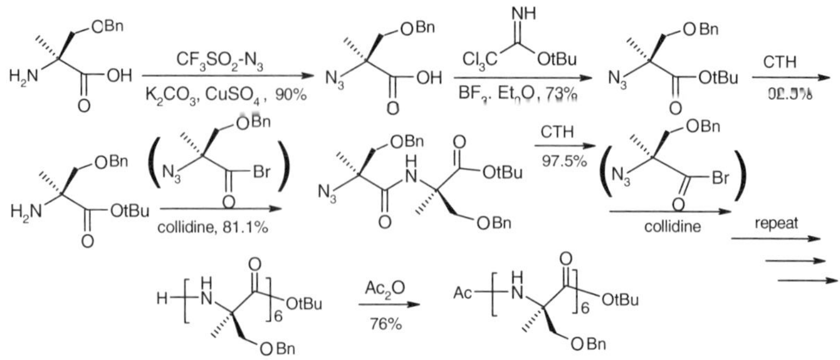

Fig. 1. Synthesis of the L-(αMe)Ser(Bzl) homo-hexamer; 1.2 equivalents of the acylbromide were used for couplings I-III and 2.0 equivalents for further couplings. Average coupling yields, after purification with flash chromatography, were more than 80% and deprotection yields more than 90%. CHT, catalytic transfer hydrogenation.

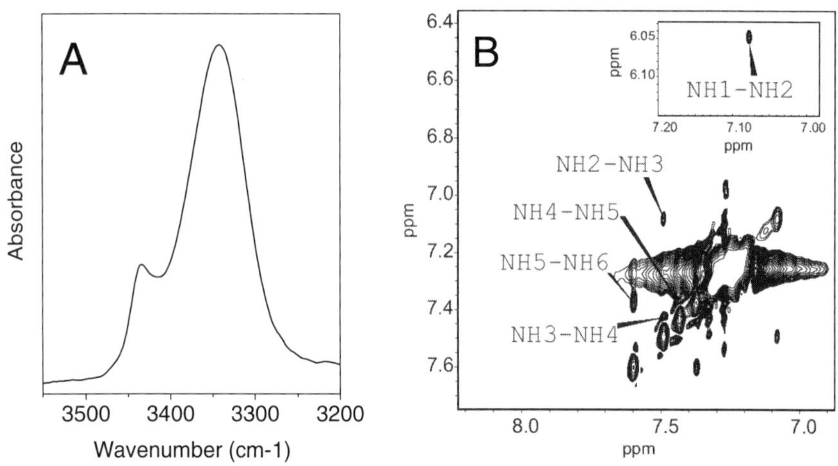

Fig. 2. A) FT-IR absorption spectrum of Ac-[L-(αMe)Ser(Bzl)]$_6$-OtBu in $CDCl_3$ solution (peptide concentration: 1 mM). B) Section of the ROESY spectrum of the same peptide in MeOH, d_3 solution. The NH(i)→NH(i+1) connectivities are marked.

References

1. Sonke, T., Kaptein, B., Boesten, W.H.J., Broxterman, Q.B., Schoemaker, H.E., Kamphuis, J., Formaggio, F., Toniolo, C. and Rutjes, F.P.J.T. in *Stereoselective Biocatalysis*, Patel, R.N., Ed., Dekker, New York, 1999, pp. 23–58.
2. Dal Pozzo, A., Ni, M., Muzi, L., Caporale, A., De Castiglione, R., Kaptein, B., Broxterman, Q.B. and Formaggio, F. *J. Org. Chem.* **67**, 6372-6375 (2002).

Introduction of N-alkyl Residues in Proline-rich Peptides: Effect on SH3 Binding Affinity and Peptide Conformation

Paolo Ruzza[1], Chiara Rubini[1], Giuliano Siligardi[2], Rohanah Hussain[2], Andrea Calderan[1], Andrea Guiotto[1], Luca Cesaro[3], Anna M. Brunati[3] and Arianna Donella-Deana[3]

[1]CNR Institute of Biomolecular Chemistry, Padua Unit, 35131 Padua, Italy; [2]Diamond Light Source, Diamond House, Chilton, Didcot, Oxfordshire, OX11 0DE, UK; [3]Department of Biological Chemistry, University of Padua, 35121 Padua, Italy

Introduction

Proline-rich peptides are involved in the interactions with SH3 domains. The analysis of these interactions emphasizes that the peculiar role of the proline residue can be explained by its direct involvement in the domain recognition, rather than by the increased structural stability of the peptide-ligand. In particular, the N-alkyl substitution of the proline residue is to be regarded as the main responsible for the ligand specificity, and the removal of this moiety has a deleterious effect on the binding affinity mainly because of the loss of van der Waals contact, rather than the destabilization of the PPII helix conformation [1].

We have studied the effect of N-alkyl amino acids substitution of proline residues of SH3 ligand HPK1 proline rich decapeptide PPPLPPKPKF (P2). Several peptide derivatives have been synthesized containing the proline homologous pipecolic acid (Pip), N-methyl amino acids sarcosine (Sar) and N-methylalanine (NMeA) respectively (Figure 1). In particular, for the peptides containing two or three substitutions, proline residues were replaced at the *i, i+3* positions of the SH3 binding sequence.

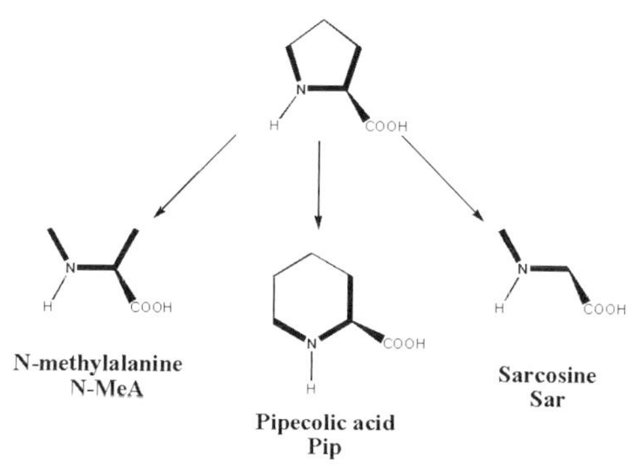

Fig. 1. N-alkyl amino acid substitutes of praline.

Results and Discussion

Here we report the results of our investigations on the interaction of these HPK1 analogue peptides with the SH3 domain of cortactin. Such interactions were analyzed by non-immobilized ligand interaction assay by circular dichroism

S. Del Valle et al. (eds.), *Peptides for Youth: The Proceedings of the 20th American Peptide Symposium*, 65
DOI: 10.1007/978-0-387-73657-0_28, © Springer Science+Business Media, LLC 2009

Table 1. The dissociation constants (K_d) of peptide-SH3 domain complexes.

Peptide	Sequence	K_d (μM)
P2	H-Pro-Pro-Pro-Leu-Pro-Pro-Lys-Pro-Lys-Phe-OH	0.8
NMeA-2	H-Pro-Pro-**NMeA**-Leu-Pro-**NMeA**-Lys-Pro-Lys-Phe-OH	27.8
NMeA-3	H-Pro-**NMeA**-Pro-Leu-**NMeA**-Pro-Lys-**NMeA**-Lys-Phe-OH	36.2
NMeA-6	H-**NMeA**-**NMeA**-**NMeA**-Leu-**NMeA**-**NMeA**-Lys-**NMeA**-Lys-Phe-OH	n.b.d.
Sar-2	H-Pro-Pro-**Sar**-Leu-Pro-**Sar**-Lys-Pro-Lys-Phe-OH	14.6
Sar-3	H-Pro-**Sar**-Pro-Leu-**Sar**-Pro-Lys-**Sar**-Lys-Phe-OH	28.3
Sar-6	H-**Sar**-**Sar**-**Sar**-Leu-**Sar**-**Sar**-Lys-**Sar**-Lys-Phe-OH	n.b.d.
Pip-2	H-Pro-Pro-**Pip**-Leu-Pro-**Pip**-Lys-Pro-Lys-Phe-OH	21.8
Pip-3	H-Pro-**Pip**-Pro-Leu-**Pip**-Pro-Lys-**Pip**-Lys-Phe-OH	5.8
Pip-6	H-**Pip**-**Pip**-**Pip**-Leu-**Pip**-**Pip**-Lys-**Pip**-Lys-Phe-OH	146

n.b.d.: no binding detectable

(NILIA-CD). Upon peptide addition, the binding was monitored by the CD changes of the Trp side-chain of the conjugate GST-SH3$_{cort}$. The dissociation constants K_d were determined (see Table 1) analyzing the CD data at 294 nm using a non-linear regression method [2].

The conformational properties of these peptides were studied by CD spectroscopy in buffer solution (Tris-HCl 25 mM, pH 7.0) and in n-propanol (95% v/v), which is a known PPI helices inducing solvent. In buffer solution, only Pip-2 and Sar-2 peptides adopt an ordered poly-proline II (PPII) helix structure at both the selected experimental temperatures (5°C and rt). As the PPII→PPI conversion is slow, samples were incubated in n-propanol for 6 days before measurements. Most peptides have CD spectra that significantly differ from those acquired in aqueous solution. Sar-2 peptide adopts a PPI helix conformation, while Sar-6 shows a PPI helix in both solvent conditions. The overall shape of Pip-2, Pip-6 and NMeA-2 spectra is similar to that of a PPI helix, even if the diagnostic band is negative in all cases. Surprisingly, the CD spectra of Sar-3 and Pip-3 in n-propanol have a shape similar to those shown in aqueous solution. The CD spectra of NMeA-3 and NMeA-6 in both the selected solvents are very similar to those reported by Goodman *et al.* [3] for polymeric N-methylalanine, and assigned to represent a helical conformation.

Our results show that the replacement of Pro residues into the P2 sequence affects in different ways the conformational properties of peptides in both aqueous and organic solvents. On the other hand, neither the presence of N-alkyl moieties nor the presence of the proline homologous Pip increase the peptide binding affinity to cortactin-SH3 domain.

References

1. Nguyen, J.T., *et al. Science* **282**, 2088-2092 (1998).
2. Siligardi, G., *et al. J. Biol. Chem.* **277**, 20151-20159 (2002).
3. Goodman, M. and Fried, M. *J. Am. Chem. Soc.* **89**, 1264-1267 (1967).

Thermodynamic Stability And Native-State Dynamics Of Porcine Pancreatic Phospholipase A_2 Studied By ^1H NMR Spectroscopic Measurements

Carmen Mrestani-Klaus, Knut Kölbel and Renate Ulbrich-Hofmann

Institute of Biochemistry and Biotechnology, Martin-Luther-University Halle-Wittenberg, Halle, 06120, Germany

Introduction

Phospholipases A_2 (EC 3.1.1.4) are ubiquitous, intra- and extracellularly occurring enzymes that catalyze the stereospecific cleavage of the *sn*-2 acylester bond in glycerophospholipids releasing the corresponding lysophospholipids and free fatty acids. Porcine pancreatic phospholipase A_2 (PLA_2) is a small secreted protein (124 amino acids) with digestive function. In industry, it is used for the production of lysophospholipids that are important emulsifiers of pharmaceutical and dietary products. PLA_2 contains seven disulfide bridges and one calcium ion, coordinated by the amino acid residues of the so-called calcium binding loop and acting as cofactor for catalysis. Although the crystal structure of PLA_2 is known [1] (Figure 1A), there is only little information on the stability of this protein.

Here, we report on differences in stability and dynamics of PLA_2 at various pH values in the absence (apoPLA_2) and presence of $CaCl_2$ (holoPLA_2).

Results and Discussion

Transition curves of guanidinium isothiocyanate-induced unfolding of PLA_2, measured at pH 4.0 and pH 8.0 by intrinsic fluorescence, revealed that PLA_2 is very resistant against denaturant action and strongly stabilized by its cofactor Ca^{2+} [2]. This effect is more pronounced at acidic pH values, where the enzyme is inactive (Figure 1B). In order to relate the results of the studies on global unfolding by spectroscopic measurements to local unfolding processes of PLA_2 under native conditions, one-dimensional hydrogen-deuterium (H/D) NMR exchange experiments were carried out at pD values in the range from 4 to 11 at 293 K and were recorded as a two-dimensional matrix. Based on the previously reported sequence-specific ^1H and ^{15}N NMR assignment of PLA_2 [3], we were able to assign the resonances of the amide protons of W3 (indole), C27, Y28, V38 and S74 of the unlabeled protein using a combination of homonuclear 1D and 2D NMR spectra measured in H_2O/D_2O (95:5) at different temperatures.

From the H/D exchange experiments, k_{ex} values were determined on the basis of the time-dependent decrease of the amide proton signal intensities of C27, Y28, V38 and S74. The data were evaluated according to Bai et al. for the EX2 case (bimolecular exchange) [4,5] and allowed the determination of the equilibrium opening constant K_{op} and of the Gibbs free energy for the opening reaction according to $\Delta G_{HX} = -RT\ln K_{op}$ (Figure 1B). The values for ΔG_{HX} of C27, Y28 and S74 and for $\Delta G^{0,\,H2O}$ of the global unfolding are in good agreement suggesting that these opening reactions and global unfolding coincide.

Since in the acidic pH range PLA_2 is not saturated at 50 mM $CaCl_2$ (K_D (Ca^{2+}) = 0.25 mM at pH 8.0 and 12 mM at pH 5.0 [6]), the increase in $CaCl_2$ concentrations at a certain pD value resulted in a decrease of the k_{ex} values (Figure 1C) reflecting the enhancement of the thermodynamic stability of PLA_2.

S. Del Valle et al. (eds.), *Peptides for Youth: The Proceedings of the 20th American Peptide Symposium*, 67
DOI: 10.1007/978-0-387-73657-0_29, © Springer Science+Business Media, LLC 2009

To elucidate the influence of pH and Ca^{2+} concentration on local structure regions of PLA_2, two-dimensional NOESY NMR measurements carried out at 313 K and 200 ms mixing time were used (spectra not shown). From the NOE intensity pattern it appears that holoPLA_2 retains its native global structure at pH 4.3 and various $CaCl_2$ concentrations apart from local conformational changes caused by Ca^{2+} binding. Furthermore, the structure of Ca^{2+} saturated PLA_2 at basic pH values strongly resembles that of the Ca^{2+} unsaturated PLA_2 species at acidic pH values as concluded from 2D NOESY spectra of PLA_2 at 50 mM $CaCl_2$ and pH values of both 4.3 and 8.0. 2D NOESY spectra of apoPLA_2 at both pH 4.3 and 8.0 revealed the enhanced flexibility of the calcium-free PLA_2. The differences in the NOESY spectra are much larger compared to holoPLA_2 indicating that local structural changes are influenced by Ca^{2+} binding.

In summary, the Gibbs free energy for the opening reaction of the local structures of PLA_2 protecting the backbone amides of C27, Y28 and S74 determined from H/D exchange in the absence of denaturant was shown to be identical to the Gibbs free energy of the global unfolding. We further demonstrated that PLA_2 retains its native global folding in the presence and absence of $CaCl_2$ at pH values in the range from 4 to 8. Depending on pH and Ca^{2+} concentration, however, the enzyme undergoes local conformational changes.

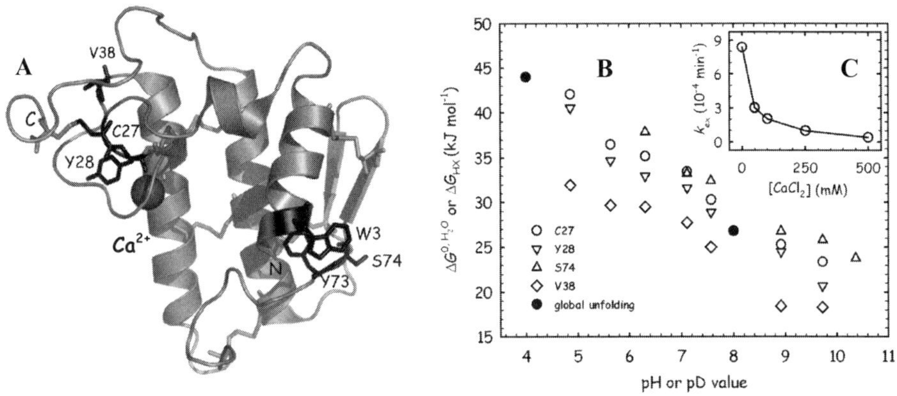

Fig. 1. A. Crystal structure of PLA_2 (pdb file 4P2P) [1]. B. $\Delta G^{0, H2O}$ or ΔG_{HX} of holoPLA_2 as a function of pH or pD. C. k_{ex} of C27 at pD 4.9 as a function of the $CaCl_2$ concentration.

Acknowledgments

The work was supported by grants of the Deutsche Forschungsgemeinschaft (DFG), GRK 1026, and the Kultusministerium of Sachsen-Anhalt to K.K. and by research funding of the DFG to C.M.-K., KON 795/2007 MR 13/3-1. The authors thank J. Balbach (Univ. Halle) and Ch. Lücke (M.-Planck Unit Halle) for helpful discussions.

References

1. Finzel, B. C., *et al. Acta Crystallogr B* **47 (Pt 4)**, 558-559 (1991).
2. Kölbel, K. *PhD thesis,* Martin-Luther-University Halle-Wittenberg, Halle, Germany (2006).
3. Dekker, N., *et al. Biochemistry* **30**, 3135-3147 (1991).
4. Bai, Y., Milne, J. S., Mayne, L. and Englander, S. W. *Proteins* **20**, 4-14 (1994).
5. Bai, Y., Milne, J. S., Mayne, L. and Englander, S. W. *Proteins* **17**, 75-86 (1993).
6. Pieterson, W. A., Volwerk, J. J. and de Haas, G. H. *Biochemistry* **13**, 1439-1445 (1974).

The amyloidogenic domains of the human serum amyloid A protein -New Insights-

Emanuel Perugia, Elena Monticelli, Sarah Rubinraut, Aviva Kapitkovsky and Mati Fridkin

Department of Organic Chemistry, Weizmann Institute of Science, 76100 Rehovot, Israel

Introduction

Serum amyloid A (SAA) is an acute phase protein associated with pathological amyloid deposits. The present study was aimed toward identification of the minimal amyloid-forming fragment of human SAA1 [1]. This information may lead to a better understanding of the pathological process of amyloidogenesis and direct to the design of amyloid inhibitors of clinical value. Employing synthetic peptides relevant to the known SAA amyloidogenic domain (i.e. SAA 1→12 RSFFSFLGEAFD) [2] we show that amyloid formation can be easily achieved with the pentapeptide SAA 2→6 (Figure 1, A). Moreover, we were able to identify a secondary amyloidogenic domain (SAA 50→59: GGAWAAEVIS) within the human SAA (Figure 1, C).

Table 1. SAA derived peptides tested in the present study for amyloid formation.

Peptide sequence	Name	MW	Amyloid formation
1. RSFFSFLGEAFDGARDMWRA	SAA 1→20	2365	Yes (TEM, ThT)
2. RSFFSFLGEAFDGA	SAA 1→14	1549	Yes (TEM, ThT, CD, IR)
3. RSFFSFLGEAFD	SAA 1→12	1421	Yes (TEM, ThT)
4. RSFFSFLGEA F	SAA 1→11	1306	Yes (TEM, ThT, CD)
5. RSFFSFLGEA	SAA 1→10	1159	Yes (TEM, ThT, FTIR)
6. **E**SFFSFLGEA	SAA 1→10 R1E	1131	Yes (TEM, ThT)
7. RSFFSFLGE	SAA 1→9	1088	Yes (TEM, ThT, FTIR)
8. **A**SFFSFLGE	SAA 1→9 R1A	1003	Yes (TEM, ThT, FTIR)
9. RSFFSFLG**A**	SAA 1→9 E9A	1031	Yes (TEM, ThT, FTIR)
10. **A**SFFSFLG**A**	SAA 1→9 R1A, E9A	946	Yes (TEM, ThT, FTIR)
11. SFFSFLGE	SAA 2→9	932	Yes (TEM, ThT, FTIR)
12. SFFSFLG**A**	SAA 2→9 E9A	875	Yes (TEM, ThT, FTIR)
13. RSFFSF	SAA 1→6	789	No (TEM, ThT)
14. **A**SFFSF	SAA 1→6 R1A	704	Yes (TEM, ThT, FTIR)
15. **K**SFFSF	SAA 1→6 R1K	761	No (TEM, ThT)
16. RSFFS	SAA 1→5	642	No (TEM, ThT)
17. SFFSF	SAA 2→6	633	Yes (TEM, ThT, FTIR)
18. GGAWAAEVIS	SAA 50→59	959	Yes (TEM, ThT, FTIR)

S. Del Valle et al. (eds.), *Peptides for Youth: The Proceedings of the 20th American Peptide Symposium*, 69
DOI: 10.1007/978-0-387-73657-0_30, © Springer Science+Business Media, LLC 2009

Results and Discussion

Synthetic peptides were prepared according to the Fmoc solid-phase strategy and purified by reverse-phase HPLC, their M.W. was confirmed by mass spectrometry. Amyloid formation was induced by dissolving the peptides at a concentration of 10 mg/ml in 10% v/v acetic acid and its characterization was performed by TEM, ThT binding, CD or FT-IR. We suggest, employing chemical point mutations, a destabilizing effect of the first N-terminal Arginine of the protein on amyloid formation. SAA derived peptides 1→5 and 1→6 are incapable of forming amyloid-like fibrils whereas the mutant SAA 1→6 R1A and the SAA 2→6 are capable (Figure 1). Interestingly, it has been documented that the Amyloid A deposits extracted from patients often lack the first arginine [3]. In addition, a desArg isoform of SAA has been detected at high concentrations in the serum of chronic patients [4]. In our model, the Arginine destabilizing effect is abolished with the increasing length of the peptide. Thus, from the ninth amino acid the fibrils appear again more similar to the classical amyloid. We intend to further investigate the implication of this finding on the progression of the amyloidosis A through generation of a truncated protein and comparison of its amyloidogenicity with the native one.

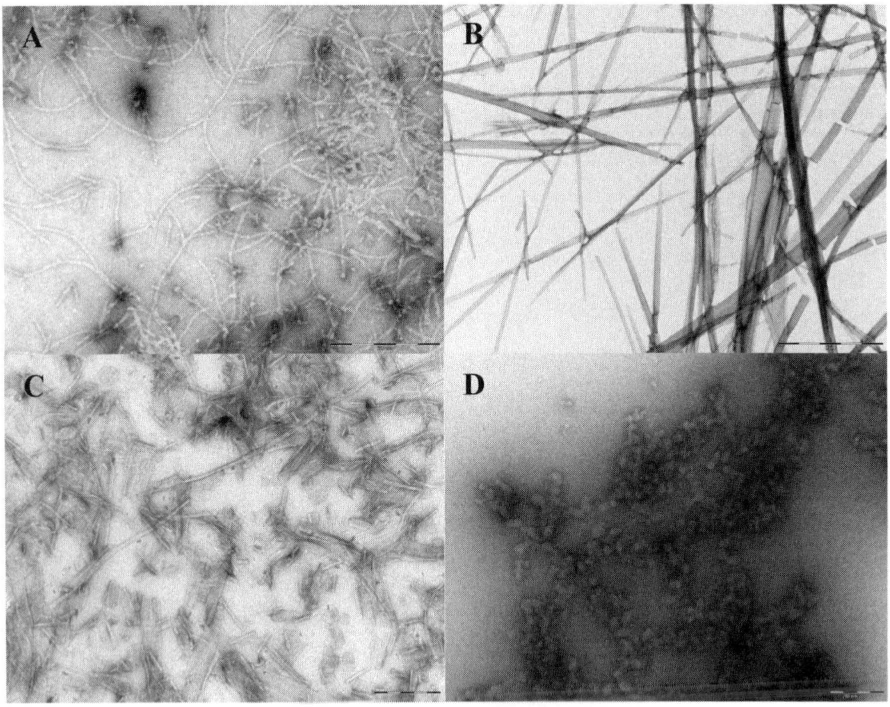

Fig. 1. TEM micrographs of amyloid-like fibrils obtained from: ***A)*** *SAA 2→6;* ***B)*** *SAA 1→6 R1A;* ***C)*** *SAA 50→59.* ***D)*** *Amorphous aggregates obtained from SAA 1→6 R1K. (Scale: A, C, D - 200nm; B - 2μm).*

References

1. Gazit E., *The FASEB Journal.* **16**, 77-83 (2002).
2 Westermark G.T., Sletten K., *Biochem Biophys Res Commun.* **82**, 27-33 (1992).
3. Röcken C., *et al. Ann Rheum Dis* **64**, 808-815 (2005).
4. Urban A., *et al. FEBS Letters* **537**, 166-1704. (2003).

Quantitating Amino Acid β-Strand Preferences, Turn Propensities and Cross-Strand Interactions in a Designed Hairpin Peptide

Kelly N. Huggins and Niels H. Andersen
Department of Chemistry, University of Washington, Seattle, WA 98195

Introduction

Beta-hairpins have gained popularity in many research groups as simple models for β sheets. The present study focuses on optimizing β-hairpin peptide sequences, quantitating residue β propensities, and determining the energetic contributions from cross-strand interactions. Studies began with the 16-residue peptide (MrH4 = KKLTVSIXGKKITVSA). In our nomenclature the turn locus residues are designated as T1 and T2 with strand positions numbered S±# based on how far residues are before or after the turn site. The sites probed in the present study are the S±2 and S±4 strand sites and the T1and T2 turn sites. These are shown below.

-K-K-L-(***S-4***)-V-(***S-2***)-I-**XG**-K-(***S+2***)-I-(***S+4***)-V-S-A-

└┘ ⇐ **T = turn sites, positions T1 and T2**

X= **Aib** (MrH4c -KKLTVSIUGKKITVSA in 20% hexafluoroisopropanol estimated to be 96%-folded and served as a basis for fold population estimation for all analogs lacking aromatic residues.

Results and Discussion

Thr/Thr -- The more stable MrH4b peptide (X = D-Pro) [2] was selected to probe T→A mutations at positions S±4, in order to retain a measurable extent of folding for the double mutant. Fold populations were determined from Hα and H_N backbone chemical shift deviations (CSD's). Threonine increased the β-hairpin population relative to alanine; the energetic contribution was 1.0 (at S+4) to 1.9 (at S-4) kJ/mol. An additional 0.84 kJ/mol of stabilization in the MrH4b peptide appears to be due to a favourable cross-strand T/T interaction.

Trp/Trp -- Mutations to Trp were made at both the S±2 and S±4 sites with X = Asn. The resulting analogues had a noticeable increase in fold population. The Trp/Trp pair at S±2, was >79% folded at 280 K and showed the edge-to-face (EtF) geometry previously seen for GB1 analogues [3] and trpzip-derived peptides [4]. For the MrH4 peptides, this geometry is characterized by a 2.1 ppm upfield shift of the Trp6 Hε3 proton which points toward the indole ring of Trp11. With Trp at positions S±4, the peptide was 66% folded at 280 K and the chemical shifts indicate that both the indole rings spend time as the "edge" residue in EtF interactions.

A double Ala mutational cycle was then carried out to estimate the β propensity of Trp. The corresponding ΔG_U's calculated from CSD's are shown in Table 1. These values suggest that Trp's at the S±2 positions are more stabilizing than at the S±4 positions. A cross-strand interaction energy at these sites could not be determined since ring current effects on neighboring residues affect the backbone CSD's and prevent accurate estimation of the 100% folded reference values. For Trp at the S±4 sites the unexpected stability of W/A (ΔG_F = –0.4 kJ/mol) also complicates an accurate determination of cross-strand interaction energy at this site. However, the data does indicate that a Trp/Trp pair is more stabilizing than Thr/Thr pair in the MrH4 peptide (X = N, $\Delta\Delta G_F^{280}$ = –4.3 kJ/mol at S+4) and that Trp has a higher β propensity than Thr at all positions examined.

S. Del Valle et al. (eds.), *Peptides for Youth: The Proceedings of the 20th American Peptide Symposium*, 71
DOI: 10.1007/978-0-387-73657-0_31, © Springer Science+Business Media, LLC 2009

Table 1. ΔG_U and $\Delta\Delta G_F$ (aq) values at 280 K for Trp at positions ±2 and ±4.

Trp at ±2	$\Delta G_{U(aq)}^{280}$ kJ/mol	$\Delta\Delta G_{F(aq)}^{280}$ kJ/mol
A/A	– 5.0	
W/A	+ 0.2	W/A→A/A (5.2)
A/W	+ 0.5	A/W→A/A (5.5)
W/W	≥ +3.0	W/W→A/A (≥8.0)
Trp at ±4		
W/W	+ 1.5	
W/A	+ 0.4	W/W→W/A (1.1)
A/W	– 1.4	W/W→A/W (2.9)

Turn forming propensity

Turn site mutations were studied in order to quantitate the residue turn preferences at T1 and T2 positions. These turn mutations were carried out on both the MrH4a peptide with a Trp/Trp and a Thr/Thr at the S±4 positions. Shown below is the order of decreasing stability as well as the $\Delta\Delta G_F$'s at 300 K in water.

Trp/Trp at S±4: pG > **p**A ≈ **p**N > NG > AG >> AN >> AA
$\Delta\Delta G_F^{300K}$ (kJ/mol) 0.3 1.0 0.7 3.1 3.6

Thr/Thr at S±4: **UG ≥ pG > NG > (GG) >> AN > (PG)**
$\Delta\Delta G_F^{300K}$ (kJ/mol) 2.4 0.5* 3.1 0.8*
p = D-Pro , *extrapolated from a related peptide series KKYTVSIXGKKITVSA (MrH3)

Fold populations are significantly reduced when Ala is present at the T1 site for most peptides. The -AG- mutation however has an unexpectedly high stability, which suggests that Gly at T2 within the Trp/Trp peptides can compensate for the poor turn propensity of Ala at the T1 position.

The differences in fold stability over these two peptide series likely reflect changes in turn geometry. While previous studies suggest that T1 = Asn or Aib results in hairpins with type I' turn (T1 ϕ/φ = 60°/30°, T2 ϕ/φ = 90°/0°) and D-Pro at T1 forms a II' turn [1] (T1 ϕ/φ = 60°/-120°, T2 ϕ/φ = -80°/0°), these are not the only feasible structures. Further NMR analysis and NOE-based structure ensembles are required to fully establish turn types for the full series of turn mutants.

Acknowledgments

The work was funded by grants from the National Science Foundation.

References

1. Keiderling, T.A., Hammer, R.P. *et al. J. Am. Chem. Soc.* **127,** 4992-4993 (2005).
2. Fesinmeyer, R.M., Hudson, F.M. *et al. J. Biol. NMR.* **33**, 213-231 (2005).
3. Andersen, N.H., Olsen, K.A. *et al. J. Am. Chem. Soc.* **128**, 6101-6110 (2006).
4. Guvench, O. and Brooks III, C.L. *et al. J. Am. Chem. Soc.* **127**, 4668-4674 (2005).

Determinants of fold stabilizing aromatic-aromatic interactions in short peptides

Lisa A. Eidenschink, Brandon L. Kier and Niels H. Andersen

Department of Chemistry, University of Washington, Seattle, WA 98195, USA

The significance of interactions involving aromatic side-chains in stabilizing protein structure is well accepted, while the geometry and specificity of these interactions are more elusive. Hydrophobic clustering plays a significant role; these interactions can be distinguished as aromatic/aliphatic and aromatic/aromatic interactions, with the aromatic/aromatic interactions displaying two dominant geometries – edge-to-face (EtF) and parallel displaced (PD) stacking [1].

Cross-strand aryl/aryl pairings occur predominantly at non-H-bonded sites in β-sheets [2]. In β-hairpin models these have been found to be stabilizing at turn flanking positions [3]. The excised N-terminal hairpin of the B1 domain of Protein G has a Tyr/Phe pair at such a position and is required for hairpin formation [4]. Trpzip4 and its analogs display two EtF Trp/Trp interactions at non-H-bonded sites, the turn flanking pair accounts for the majority of stabilization. HP6 and HP7, two peptide series more remotely related to GB1, have also shown remarkable stability attributed to a turn flanking EtF Trp/Trp pair. In all of its incarnations the interaction has been seen to maintain a specific geometry, with the edge of the N-terminal Trp abutting the face of the C-terminal Trp.

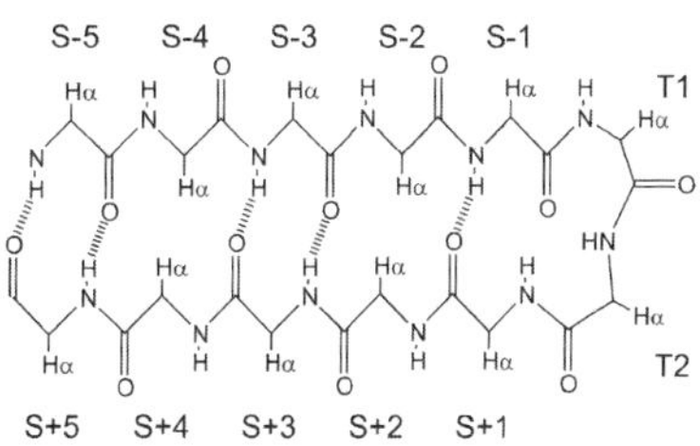

Figure 1. Nomenclature for residue postions in a β-hairpin with tight β turn.

The stability of HP7 and its truncated version, as well as chignolin [5], suggest that an EtF aromatic/aromatic interaction immediately flanking a turn sequence is particularly stabilizing in small β-hairpins. HP6V has the same stabilizing EtF interaction but a β-turn that is less favorable in systems with short β-strands, provided an excellent system to test the limit of the W/W EtF interaction. The first truncated peptide of this design, AW-SNGK-WT, displayed the usual CD exciton couplet, in fact larger than expected, and an upfield Trp Hε3 with a melting curve suggesting a Tm of *circa* 25°C, apparently more stable than Ac-WNPATGKW-NH_2, the 8-mer with the optimized reversing loop. NOESY sequencing, however, indicated that the upfield Hε3 signal was in the C-terminal, rather than N-terminal Trp. A battery of small Trp containing peptides was examined to ascertain the determinants of the EtF geometry between the Trp sidechains. The transition was not found to possess an absolute boundary; a "middle ground" with two folded states corresponding to the two EtF geometries was observed.

A 2.7 ppm upfield shift (CSD) seen for the G8-H_N of Ac-WINGKWTG-NH_2 (peptide WP) suggested a local aryl-amide interaction at the N-terminus.

Aryl-X-Gly i → i+2 interactions can act as modest structuring elements even in denatured proteins [6] and peptides, evidenced by Gly H_N CSDs up to −1.4. Only in proteins have Aryl-XG shifts as large as that in peptide WP been observed.

S. Del Valle et al. (eds.), *Peptides for Youth: The Proceedings of the 20th American Peptide Symposium*, 73
DOI: 10.1007/978-0-387-73657-0_32, © Springer Science+Business Media, LLC 2009

Further Results and Discussion

Peptide Ac-W-NPATGK-W-NH_2 maintained the Trp side-chain geometry seen in all of its longer predecessors indicating that the Trp/Trp flip reversal is, to some degree, dependent on loop length. But this doesn't explain why a four residue loop can adopt either conformation. In the short strand peptides with four residue turns, mutations at the S-1 position and T1 positions did not alter the "reversed EtF" geometry, which has been confirmed by comparing NMR structure ensembles (Fig. 2). We therefore turned to strand length as the potential culprit and a series of mutants demonstrated that indeed the interaction geometry is dependent on strand length (Table 1).

Sequence	S-2WHε3	S+2WHε3	S+4HN
Ac-W-NPATGK-W-NH_2	-0.87	-0.17	n.a.
KYVW- INGK -WTVE	-1.87	-0.56	0.55
ATW- INGK -WTG	-1.35	-0.76	0.11
Ac-TW- INGK -WTG-NH_2	-0.53	-1.62	-1.05
TW- INGK -WTG	-0.41	-1.92	-0.48
AW- SNGK -WT	-0.19	-1.56	n.a.
Ac-W- INGK -WT-NH_2	-0.39	-1.68	-2.70
Ac-W- INGK -WTG-NH_2	-0.38	-1.68	-2.71

Table 1. CSDs that map the Trp flip transition.

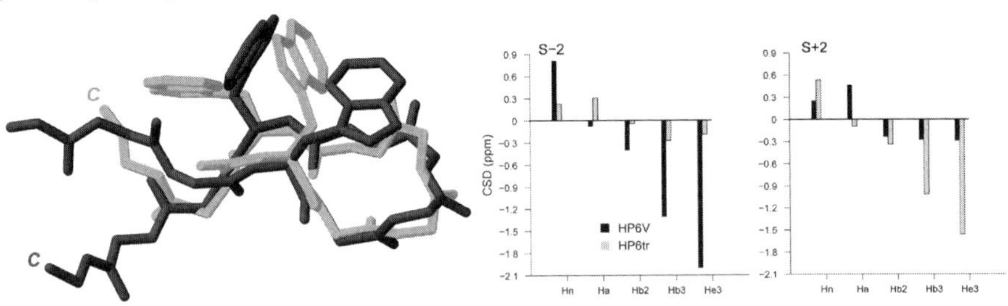

Figures 2a & b. Representative structures from the NMR ensembles of Ac-WINGKWTG-NH_2 and KYVWINGKWTVE obtained exclusively from NOE distance constraints. Figure 2b displays the CSDs of the S+/-2 Trp residues at 280K.

Synergy is observed between a W/W cross-strand interaction with flipped EtF geometry and an aryl-amide interaction at the C-terminus. The latter requires an indole ring that lies down on the C-terminal backbone. The aryl-amide interaction is disrupted by N-terminal extension of the hairpin (see Table 1), and is completely incompatible with the original EtF geometry.

Acknowledgments
The work was funded by NSF and NIH grants to Dr. Niels Andersen.

References

1. Guvench, O., and Brooks, C. *J. of Am. Chem. Soc.* **127**: 4668-4674 (2005).
2. Jackups, R., and Liang, Jie. *J. of Mol. Biol.* **354**: 979-993 (2005).
3. Andersen, N., Olsen, K., Fesinmeyer, R., Tan, X., Hudson, F., Eidenschink, L., and Farazi, S. *J. of Am. Chem. Soc.* **128**: 6101-6110 (2006).
4. Kobayashi, N., Honda, S., Yoshii, H., and Munekata, E. *Biochemistry* **39**: 6564-6571 (2000).
5. Honda, S, Yamasaki, K., Sawada, Y., and Morii, H. *Structure* **12**: 1507-1518 (2004).
6. Kemmink, J., Mierlo, C., Sheek, R., and Creighton, T. *J. of Mol. Biol.* **230**, 312-322 (1993).

Spectroscopic analysis of synthetic PrP helix2-binding peptide constructs

Luisa Ronga[1,2], Pasquale Palladino[1], Raffaele Ragone[3], Jean Martinez[2], Ettore Benedetti[1], Filomena Rossi[1] and Muriel Amblard[2]

[1]Università degli Studi di Napoli "Federico II", Dipartimento delle Scienze Biologiche C.I.R.Pe.B., Via Mezzocannone 16 (80134) Napoli, Italy; [2]Institut des Biomolécules Max Mousseron, U.M.R. CNRS 5247 Faculté de Pharmacie, 15 Avenue Charles Flahault (34093) Montpellier Cedex 2, France; [3]Seconda Università di Napoli, Dipartimento di Biochimica e Biofisica & CRISCEB, Via S. Maria di Costantinopoli 16 (80138) Napoli, Italy.

Introduction

Prion diseases are associated with the conversion of the α-helix-rich prion protein (PrP^C) into an insoluble β-structure-rich conformer (PrP^{Sc}) that is thought to be infectious. The mechanism for the $PrP^C \rightarrow PrP^{Sc}$ conversion and its relationship with the pathological effects of prion diseases are poorly understood, partly because of our limited knowledge of the structure of PrP^{Sc} [1]. The 2.5-Å-resolution crystal structures of three scrapie-susceptible ovine PrP (ovPrP) variants complexed with an antibody (Fab) that binds both PrP^C and PrP^{Sc} have been recently reported [2]. The epitope of the antibody basically consists of the last two turns of ovPrP helix 2 (perfectly conserved in the human domain) and is a structural invariant in the $PrP^C \rightarrow PrP^{Sc}$ conversion [2]. We have previously reported that the 173-195 fragment, derived from the human PrP helix 2 (hPrP[173-195]), is structurally ambivalent [3] and could be regarded as a potential target for therapeutic approaches [4] owing to its conformational flexibility [5]. It is therefore evident that the helix 2 is a suitable model to investigate both structural determinants of PrP^C misfolding and rational structure-based drug design of compounds able to block or prevent prion diseases.

We have identified the Fab[30-35] and Fab[46-53] fragments (TNYGMN and RLIYLVSR, respectively), which are able to form hydrogen bonds with the helix 2 C-terminal end in the Fab-ovPrP X-ray structures, and designed peptide constructs putatively suitable to model the ovPrP-Fab interaction. In fact, these two peptide fragments were linked by spacers of different size, rigidity and chemical nature, to obtain the peptide constructs listed in Table 1, henceforth identified as JMV. Furthermore, we have functionalized the helix 2-derived peptide using fluorescein and investigated its interaction with these JMV compounds by steady-state fluorimetry.

Results and Discussion

The Fab peptide constructs were synthesized by Fmoc chemistry standard protocol on SynPhase Rink Amide Lanterns. Subsequently, the emission spectra of fluorescein-modified hPrP[173-195] (Fl-NNFVHDC(Me)VNITIKQHTVTTTTKG-NH_2) in aqueous solution were recorded in the presence of increasing amounts of the JMV compounds. The fluorescence intensity decrease following JMV addition was appreciated by plotting modification of the fluorescein emission intensity at 518 nm versus JMV concentration. All titration curves show the hyperbolic trend typical of saturating dose-response effects. Fluorescence intensities were then used to evaluate the fractions of derivatized hPrP[173-195] bound to each JMV construct and calculate the apparent dissociation constants (K_D') of the corresponding complexes

S. Del Valle et al. (eds.), *Peptides for Youth: The Proceedings of the 20th American Peptide Symposium*, 75
DOI: 10.1007/978-0-387-73657-0_33, © Springer Science+Business Media, LLC 2009

according to 1:1 binding interaction (Table 1). As can be judged from K_D' values, these preliminary experiments suggest that all Fab peptide constructs strongly interact with the helix 2-derived fragment, particularly the JMV 3391, which shows the highest affinity (K_D' = 2.3 nM). Apparently, the affinity is poorly correlated to size, rigidity or chemical nature of the spacer, suggesting that all of the spacers could be able to correctly orientate the peptide arms of the constructs. Further studies are necessary to clarify structural details of the interactions. New experiments with more significant prion protein fragments could open interesting perspectives for the diagnostic or therapeutic use of these constructs in PrP-associated prion diseases.

Table 1. Code, structure and K_D' of JMV constructs.

Code	Structure	K_D'(nM)
JMV3302	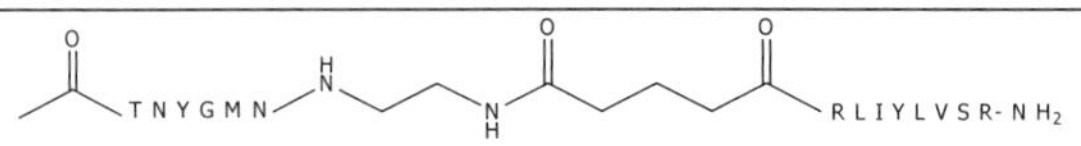	11.3
JMV3307	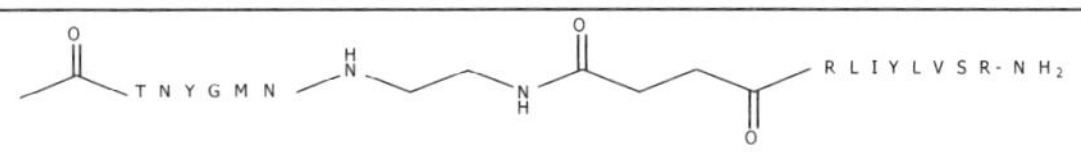	25.4
JMV3388	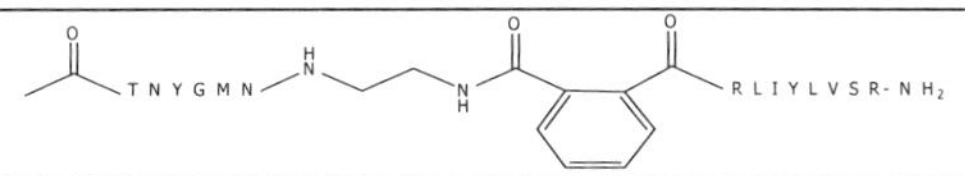	18.8
JMV3308	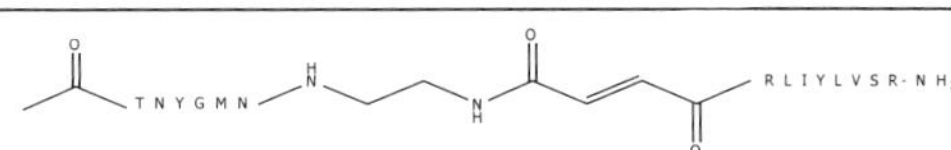	64.3
JMV3389	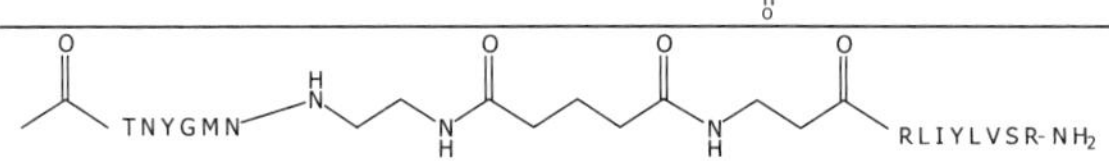	37.8
JMV3390	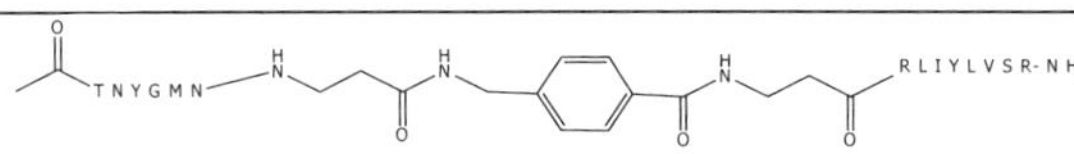	29.4
JMV3391	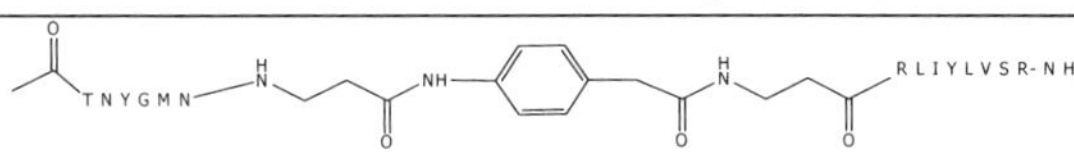	2.3

Acknowledgments

The authors acknowledge the Ministero dell'Istruzione, dell'Università e della Ricerca (MIUR) FIRB N°RBNE03PX83 (2003) and the Internazionalizzazione del Sistema Universitario 2004-2006 MIUR DM 5 agosto 2004 N°262 art. 26 for their support to this research. L.R. thanks the Università Italo-Francese (Programma VINCI 2005) for financial support.

References

1. Prusiner, S. B. Proc. Natl. Acad. Sci. USA 95, 13363-13383 (1998).
2. Eghian, F., Grosclaude, J., Lesceu, S., Debey P., Doublet, B., Tréguer, E., Rezaei, H. and Knossow, M. Proc. Natl. Acad. Sci. USA 101, 10254-10259 (2004).
3. Ronga, L., Palladino, P., Costantini, S., Facchiano, A., Ruvo, M., Benedetti, E., Ragone, R. and Rossi, F. Curr. Protein Pept. Sci. 8, 83-90 (2007).
4. Ronga, L., Langella, E., Palladino, P., Marasco, D., Tizzano, B., Saviano, M., Pedone, C., Improta, R. and Ruvo, M. Proteins 66, 707-715 (2007).
5. Tizzano, B., Palladino, P., De Capua, A., Marasco, D., Rossi, F., Benedetti, E., Pedone, C., Ragone, R. and Ruvo, M. Proteins 59, 72-79 (2005).

Unique Role of Clusters of Electrostatic Attractions in Controlling the Stability of Two-stranded α-Helical Coiled-Coils

J. Paul Kirwan[1], Stan C. Kwok[1], Susanna McReynolds[1,2], Janine Mills[1], David Osguthorpe[3] and Robert S. Hodges[1]

[1]*Department of Biochemistry & Molecular Genetics;* [2]*Department of Microbiology;* [3]*Department of Pharmacology, University of Colorado at Denver and Health Sciences Center, Aurora, CO 80045, USA*

Introduction

A major goal of protein research is to relate amino acid sequence to protein structure and function. Describing all inter-residue interactions that contribute to the threshold stability required to initiate protein folding is key to achieving this end. The two-stranded α-helical coiled-coil is a simplified protein motif well suited for studying these interactions. Our laboratory has shown that clusters of stabilizing and destabilizing residues in the hydrophobic core control coiled-coil stability [1,2]. We have also shown that the effect of chain length on coiled-coil stability depends on the composition of the added/removed heptads [3]. Here we investigate the existence of stabilizing sequence electrostatic-clusters and whether they trigger folding and confer final stability, or confer final stability without governing folding.

Results and Discussion

We prepared a series of C-terminal deletion fragments of the 284 residue coiled-coil tropomyosin (1-81, 1-92, 1-99, 1-119, 1-131, and 1-260) and examined their folding and stability. All constructs were fully folded, two-stranded α-helical coiled-coils. However, native wild-type-like stability (T_m=43°C) was not achieved until a length of 119 residues was reached (Figure 1).

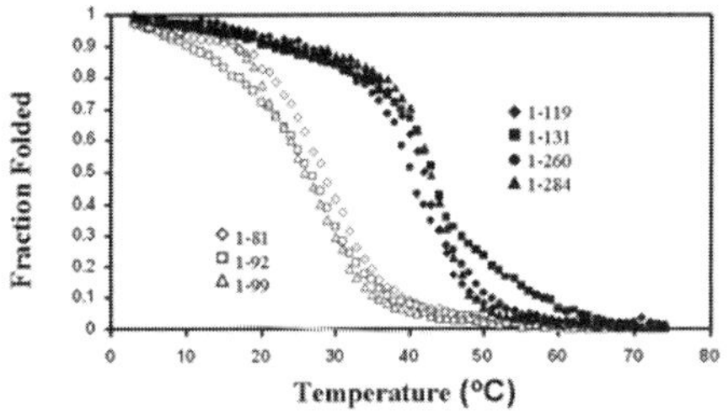

Figure 1. Melting curves for tropomyosin and C-terminal deletion fragments. A large change in stability occurs between fragments 1-99 and 1-119. The stability difference between the two groups was a ΔT_m of 15°C.

Fragments 1-81, 1-92, and 1-99 had T_m values of 26-28.5°C, but all other fragments (1-119, 1-131, and 1-260) had T_m values of 40-43°C. Inspection of the sequence between residues 96-119 suggests the presence of two unique clusters of electrostatic attractions that must control wildtype-like stability (Figure 2). Interestingly, these two clusters are highly conserved among tropomyosin sequences. To investigate the effect of electrostatic cluster 2 on protein stability, we prepared synthetic peptides based on the native tropomyosin sequence from 85-119 with the addition of an N-terminal CGG linker to form disulfide bridged parallel two-stranded α-helical coiled-coils that are independent of the monomer-dimer equilibrium. The wild type protein was prepared along with three analogues containing K112A, K118A, and K112A+K118A substitutions. The K112A analogue

S. Del Valle et al. (eds.), *Peptides for Youth: The Proceedings of the 20th American Peptide Symposium*, 77
DOI: 10.1007/978-0-387-73657-0_34, © Springer Science+Business Media, LLC 2009

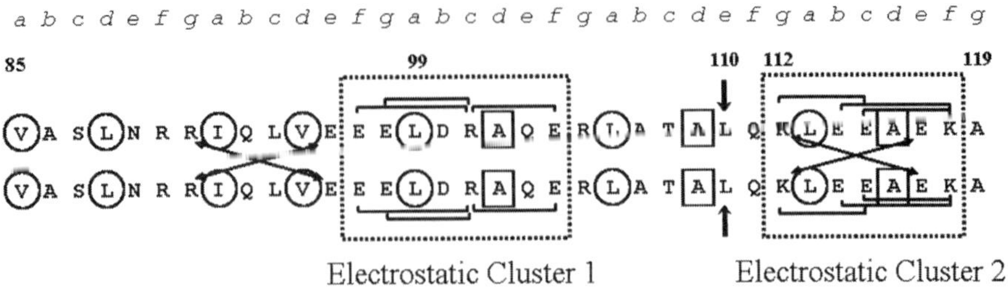

Figure 2. Tropomyosin fragment of interest. Electrostatic clusters 1 and 2 are boxed. Intra-chain electrostatic attractions are denoted by the square brackets and inter-chain electrostatic attractions are indicated by arrows.

lost the stability of one i→i+3 intra-chain and one i→i′+5 inter-chain interaction per polypeptide chain relative to the native sequence, while the K118A substitution disrupted one i→i+3 intra-chain and one i→i+4 intra-chain interaction per polypeptide chain. The resulting disruption of the K→E salt bridges is very destabilizing despite the compensatory stabilizing helical propensity contribution of Ala. The most dramatic effect on stability was observed in the K112A/K118A double substitution, where all intra- and inter-chain interactions are eliminated in electrostatic cluster 2 resulting in an 8°C reduction in thermal stability compared to the native sequence.

Table 1. Thermal denaturation data for synthesized peptides.

	T_m (C°)	ΔT_m (C°)
Native	53	---
K112A	49	- 4
K118A	48.5	- 4.5
K112A+K118A	45	- 8

Previous chain length [3] and stability studies [4] suggest that extending the polypeptide chain length in tropomyosin from 1-99 to 1-119 would be destabilizing since two alanine residues are present in the added hydrophobic core positions 109 and 116. Instead, we observed the dramatic increase in stability reflected in Figure 1. This suggests the presence of other interactions that not only overcome the destabilizing effects of the alanine residues in the hydrophobic core but also dramatically increase final stability of the protein. We examined electrostatic cluster 2 first because of its unique composition of five charged residues out of the seven present in the heptad. Interestingly, this sequence is highly conserved among orthologous tropomyosin proteins and found only once per given tropomyosin. The results reported here support the concept of previously unidentified unique stability control sites in two-stranded α-helical coiled-coils.

References

1. Kwok, S.C. and Hodges, R.S., *J Biol. Chem.* **278**, 35248-35254 (2003).
2. Kwok, S.C. and Hodges, R.S., *J Biol. Chem.* **279**, 21576-21588 (2004).
3. Kwok, S.C. and Hodges, R.S., *J. Biopolymers.* **76**, 378-390 (2004).
4. Wagschal, K., Tripet, B., and Hodges, R.S., *J. Mol. Biol.* **285**, 785-803 (1999).
5. Wagschal, K., Tripet, B., Lavigne, P., Mant, C.T. and Hodges, R.S., *Protein Sci.* **11**, 2312-2329 (1999).
6. Tripet, B., Wagschal, K., Lavigne, P., Mant, C.T. and Hodges, R.S., *J. Mol. Biol.* **300**, 377-402 (2000).

The Energetics of Weakly Polar Interactions in Model Tripeptides

József Csontos, Richard F. Murphy and Sándor Lovas

Department of Biomedical Sciences, Creighton University Medical Center, Omaha, NE 68178, USA

Introduction

Weakly polar interactions are ubiquitous in polypeptides and include aromatic-aromatic (Ar-Ar), Ar-backbone (Ar-Bb), backbone-backbone (Bb-Bb), aromatic-sulfur (Ar-S) and CH-π interactions [1]. Their interaction energies are comparable with those of hydrogen bonds and they stabilize both the secondary and tertiary structure of polypeptides [2].

The conformational potential energy surfaces (PES) of tripeptide fragments Ac-N-Phe-Gly-Gly-NH-Me (FGG), Ac-N-Trp-Gly-Gly-NH-Me (WGG) and Ac-N-Tyr-Gly-Gly-NH-Me (YGG) in aqueous phase were calculated using molecular dynamics with the OPLS-AA/L force field and density functional theory (PWPW91/6-31G*). To calculate the Ar-Bb and Bb-Bb intramolecular interaction energies, BHandHLYP/aug-cc-pVTZ computations were conducted using a combination of molecular fragmentation and a recently developed rotation method [3].

Results and Discussion

Table 1. Aromatic-backbone interaction energies (kJ/mol) in model tripeptides.

conformers	FGG			WGG			YGG		
	N[a]	*C*[b]	***NC***	*N*	*C*	***NC***	*N*	*C*	***NC***
0	0.2	0.0	**0.2**[c]	0.0	-23.6	**-23.6**	-0.5	0.0	**-0.5**
1	-2.2	-3.0	**-5.2**	-1.6	-28.5	**-30.1**	-2.8	0.0	**-2.8**
2	-1.2	0.0	**-1.2**	-0.2	-2.2	**-2.4**	0.0	-2.9	**-2.9**
3	0.0	2.8	**2.8**[c]	-1.7	-28.3	**-30.0**	0.0	-1.3	**-1.3**
4	0.0	-2.6	**-2.6**	-1.2	-19.8	**-21.0**	0.8	0.0	**0.8**[c]
5	0.9	-15.4	**-14.5**	1.0	7.0	**8.0**[c]	0.0	3.7	**3.7**[c]
6	-2.8	-2.2	**-5.0**	-0.3	-11.5	**-11.8**	-0.9	0.0	**-0.9**
7	-0.7	0.0	**-0.7**	-2.8	-2.9	**-5.7**	-1.2	-9.7	**-10.9**
8	-0.3	0.0	**-0.3**	-2.8	-9.1	**-11.9**	0.9	0.0	**0.9**[c]
9	0.7	-5.4	**-4.7**	0.0	-10.1	**-10.1**	0.0	1.9	**1.9**[c]
average	-0.5	-2.6	**-3.1**	-1.0	-12.9	**-13.9**	-0.4	-0.8	-1.2

[a]N-terminal fragment (NAc-group and NH atoms of Ar), [b]C-terminal fragment (GG-NMe-group and carboxyl-group of Ar), [c]conformer is stabilized by 2 H-bonds

The largest Ar-Bb interaction energies are -14.5 kJ/mol, -30.1 kJ/mol and -10.9 kJ/mol in FGG, WGG and YGG, respectively (Table 1). In agreement with the observation that tryptophan has the highest probability to be involved in Ar-Bb interactions [4], Ar-Bb interactions are most frequent and their average interaction

S. Del Valle et al. (eds.), *Peptides for Youth: The Proceedings of the 20th American Peptide Symposium*, 79
DOI: 10.1007/978-0-387-73657-0_35, © Springer Science+Business Media, LLC 2009

energy is greatest (-13.9 kJ/mol) in WGG. In FGG, however, the energy of the strongest Ar-Bb interaction (-14.5 kJ/mol) is significantly larger than that estimated previously by MD simulation (-9.2 kJ/mol) [2].

Table 2. The energetic contributions (kJ/mol) of weakly polar interactions to the stability of the global minimum of model tripeptides.

Interaction	FGG	WGG	YGG
Ar-Bb	0.2	-23.6	-0.5
Bb-Bb	-4.8[a] (-5.0)[b]	-1.9[b]	-3.3[b]
sum	**-4.8**	**-25.5**	**-3.8**

[a]the aromatic side-chain is included (see Figure 1b), [b]N-methylacetamide dimer calculations.

The similarity between the conformational PES of FGG, WGG and YGG is strong. In every case the global minimum is stabilized by two H-bonds and a Bb-Bb interaction which have the same geometrical arrangement (Table 2; Figure 1a, 1b). The geometries of the FGG, WGG and YGG conformers, which have strongest Ar-Bb interaction, are also similar (Figure 1c).

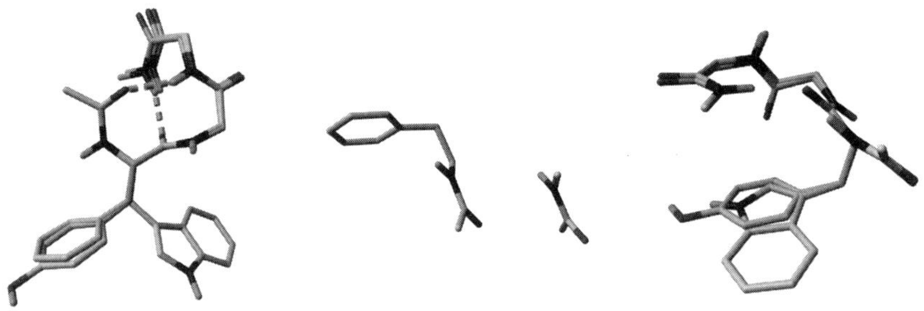

(a) (b) (c)

Figure 1. (a) the global minimum conformers; the backbone atoms are superimposed (b) Bb-Bb interaction in the fragmented FGG global minimum (c) the superimposed WGG and YGG conformers which have strongest Ar-Bb interactions.

Besides H-bonds, weakly polar interactions play a crucial role in the stabilization of the conformations of the investigated tripeptides and can clearly make an important energetic contribution to polypeptide structure and function.

Acknowledgments

This work was supported by NIH-INBRE grant (1 P20 RR16469) and the Carpenter Endowed Chair in Biochemistry, Creighton University.

References

1. Burley, S. K. and Petsko, G. A. *Adv. Protein Chem.* **39**, 125-189 (1988).
2. Toth, G., Murphy, R. F. and Lovas, S. *J. Am. Chem. Soc.* **123**, 11782-11790 (2001).
3. Palermo, N. Y., Csontos, J., Owen, M. C., Murphy, R. F. and Lovas, S. *J. Comput. Chem.* **28**, 1208-1214 (2007).
4. Toth, G, Watts, C. R., Murphy, R. F. and Lovas, S. *Proteins: Struct. Funct. Genet.* **43**, 373-381 (2001).

Peptidomimetic Ligands for the Tandem SH2 Domain of the Syk Protein Involved in Signal Transduction

Joeri Kuil, Loek T.M. van Wandelen, Nico J. de Mol and Rob M.J. Liskamp

Medicinal Chemistry and Chemical Biology, Utrecht Institute for Pharmaceutical Sciences, Department of Pharmaceutical Sciences, Faculty of Science, Utrecht University, Utrecht, The Netherlands

Introduction

The Spleen tyrosine kinase (Syk) protein takes part in a number of receptor signaling cascades [1]. It is involved in early events after activation of e.g. the IgE-, T-cell-, B-cell- and IL-15-receptor. The role of Syk in the high affinity IgE receptor (FcεRI) signaling in mast cells and basophils is shown in Fig. 1. The β- and γ-chains of FcεRI contain a specific intracellular sequence called the Immunoreceptor Tyrosine based Activation Motif (ITAM). The ITAM sequence consists of Tyr-Xxx-Xxx-(Leu/Ile)-$(Xxx)_{n=6\text{-}8}$-Tyr-Xxx-Xxx-(Leu/Ile), in which Xxx can be any amino acid. The underlined residues comprise the binding epitopes for the tandem SH2 (tSH2) domain of Syk, when tyrosine is phosphorylated. The intervening residues $(Xxx)_{n=6\text{-}8}$ (in FcεRI n = 7) are not essential for binding.

Upon IgE stimulation of the receptor, the ITAM motive is diphosphorylated (γ-dpITAM). After this, Syk is recruited to the membrane by binding of the tSH2 domain to γ-dpITAM (Fig. 1). This results in a conformational change and Syk activation, which eventually leads to cell degranulation and release of mediators. Overstimulation of this pathway leads to allergic responses and therefore Syk is an interesting target for potential anti-allergic therapy.

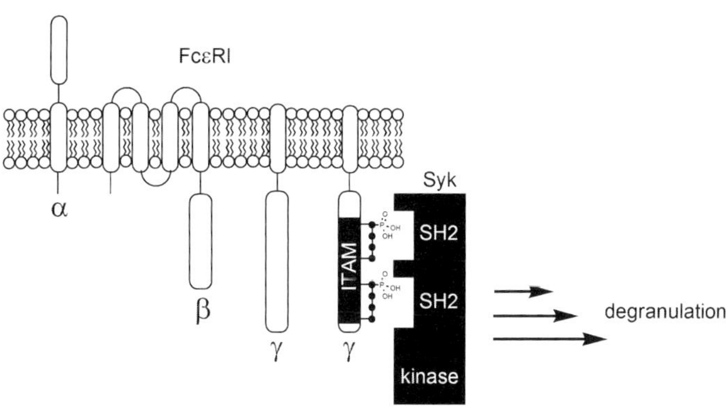

Fig. 1. Recruitment of Syk to the diphosphorylated γ-ITAM of FcεRI results in activation of its kinase domain and ultimately degranulation.

Our hypothesis is that binding of Syk to γ-dpITAM leads to a change in the inter SH2 distance of Syk and that this is necessary for Syk activation. To gain more insight into the functioning of Syk tSH2, we decided to take advantage of a photoswitchable building block and incorporate this into the ITAM sequence. Azobenzene as part of 4-aminomethyl)phenylazobenzoic acid (AMPB)) is the most widely used photoswitch because of its fast and reversible photoisomerization [2].

S. Del Valle et al. (eds.), *Peptides for Youth: The Proceedings of the 20th American Peptide Symposium*, 81
DOI: 10.1007/978-0-387-73657-0_36, © Springer Science+Business Media, LLC 2009

Binding of the cis and trans isomers of the photoswitch containing ITAM mimic was assayed using surface plasmon resonance (SPR).

Results and Discussion

With the Fmoc-protected AMBP building block a photoswitchable ITAM mimic was synthesized using standard Fmoc chemistry (Fig. 2). Integration of the signals from NMR spectra of the trans and cis isomer revealed that a maximum cis isomer percentage of 68% could be reached upon irradiation with 366 nm light, upon irradiation with visible light 100% trans isomer was obtained.

The interaction between the two isomers and the Syk tSH2 protein was assayed with SPR. Competition experiments were performed with the native ITAM peptide immobilized on the SPR chip. Prior to addition, the competing photoswitchable ligand, was irradiated with UV light (366 nm) or visible light. From these measurements the dissociation constants (K_D) for both isomers could be obtained. The K_D value of the trans isomer was 65 ± 8 nM and the K_D of the cis isomer was 860 nM. Also the effect of the cis-trans isomerization on binding could be monitored in real time by *in situ* irradiation during the SPR experiment.

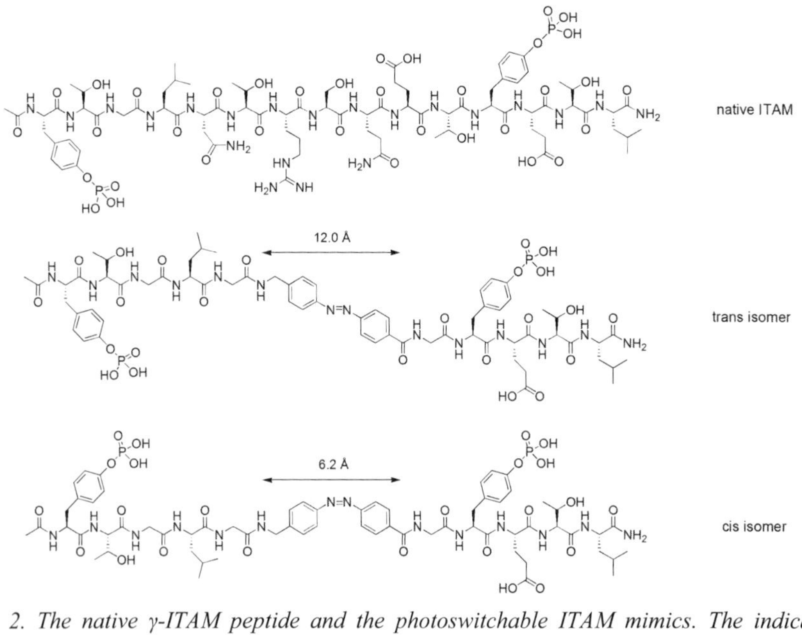

Fig. 2. The native γ-ITAM peptide and the photoswitchable ITAM mimics. The indicated distances are between the SH2 binding epitopes.

Acknowledgments

The work was funded by The Netherlands Organisation for Scientific Research (NWO).

References

1. De Mol, N.J., Catalina, M.I., Dekker, F.J., Fischer, M.J.E., Heck, A.J.R., Liskamp, R.M.J. *ChemBioChem* **6**, 2261-2270 (2005).
2. Renner, C., Moroder, L. *ChemBioChem* **7**, 868-878 (2006).

Subtilisin Kexin Isozyme-1 (SKI-1): Production, purification, inhibitor design and biochemical applications

Dayani Mohottalage[1], Natalie Goto[2] and Ajoy Basak[1]

[1]*Hormones Growth and Development Program, Regional Protein Chemistry Center, Ottawa Health Research Institute, U Ottawa, Ottawa, ON K1Y 4E9, CANADA, and* [2]*Chemistry Department, U Ottawa, D'Iorio Hall, 10 Marie-Curie, Ottawa, Ontario, K1N 6N5, CANADA.*

Introduction

Subtilisin Kexin Isozyme-1 (SKI-1) belongs to pyrolysin branch of mammalian subtilases [1, 2]. It is synthesized as a membrane bound inactive precursor protein and undergoes auto-catalytic shedding at KHQKLL^{935}⇓ to produce active 98kDa soluble form [3]. This endoprotease cleaves precursor proteins at the motif (R-X-L/I-L/T/K↓), X = any amino acid except Cys. Presence of an additional P7-Tyr greatly enhances SKI-1 recognition [3]. SKI-1 plays roles in cholesterol/lipid synthesis, cartilage development and hemorrhagic fever infections caused by Lassa, Crimean-Congo and Lymphocytic Choriomeningitis viruses [3]. For this reason attention has been focussed on the production of active soluble form of this enzyme and its inhibitors. We have previously described C-terminal modified and pseudopeptide based inhibitors as well as assay of this enzyme [4]. In this study we produced a soluble recombinant human (h) SKI-1 and designed a new miniserpin inhibitor of SKI-1 using the reactive site loop ("RSL") approach we developed for furin [5].

Results and Discussion

Soluble recombinant hSKI-1: We generated a soluble and C-terminal truncated recombinant (rec) hSKI-1 using HEK-293 cell lines and a plasmid construct of hSKI-1 cDNA lacking the transmembrane domain. The enzyme was secreted into the medium as confirmed by western blot and activity assay [4]. The content of albumin in the culture medium has been found to be crucial for SKI-1 expression and activity. Studies using 0, 0.25, 0.5, 1 and 10% albumin showed that highest SKI-1 activity was noticed with 10% albumin while poor cell growth and almost no activity was observed when there was no albumin in the medium. We therefore used 0.25% albumin containing medium for production of rec-SKI-1 to avoid excessive albumin presence. The secreted medium following 24h culture was partially purified by centricon concentration (30 x fold), DEAE ion exchange and FPLC (monoQ) columns. The eluted enzyme showed immuno-reactive band at expected 98 kDa [3] and enzyme activity against our substrates QGPC and QCMV (**Table 1**).

Miniserpin inhibitor of SKI-1: An examination of known substrates (SREBP-2, ATF-6, prodomain of SKI-1, Lassa virus glycoprotein, etc) of SKI-1 at the cleavage site revealed that in most cases they are accompanied by the tetrapeptide sequence "RRLL". In order to develop new "RSL" derived miniserpin based SKI-1 inhibitor, we substituted "AIPM^{382}" sequence of α1AT at the cleavage place of its "RSL" by SKI-1 motif "RRLL". Thus a cyclic and acyclic miniserpin peptide based on the sequence AT$^{367\text{-}394}$ (**Table 1**) was constructed.

S. Del Valle et al. (eds.), *Peptides for Youth: The Proceedings of the 20th American Peptide Symposium*,
DOI: 10.1007/978-0-387-73657-0_37, © Springer Science+Business Media, LLC 2009

Table 1. List of "RSL" of AT derived miniserpin inhibitor and fluorogenic substrates of SKI-1 (WT = wild type).

Peptide Name	Amino acid sequence	MW
$AT^{C+367-394+C}$WT	C-KGTEAAGAMFLEAIMP^{382}SIPPEVKFNKPF-C	3227
$AT^{C+367-394+C}$RRLL	C-KGTEAAGAMFLE**RRLL**^{382}SIPPEVKFNKPF-C	3354
Cyclic$AT^{C+367-394+C}$RRLL	[C-KGTEAAGAMFLE**RRLL**^{382}SIPPEVKFNKPF-C]	3352

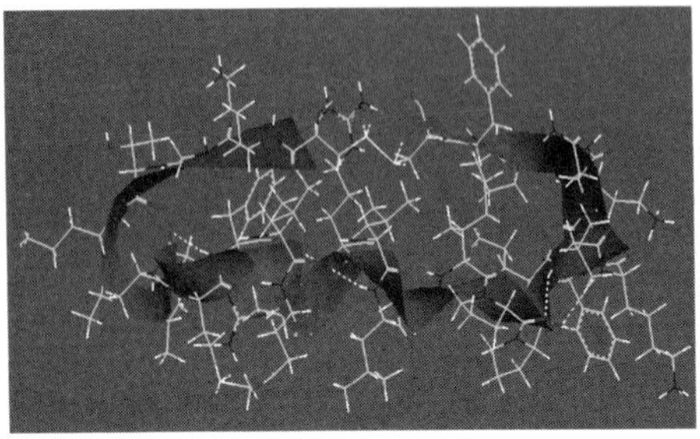

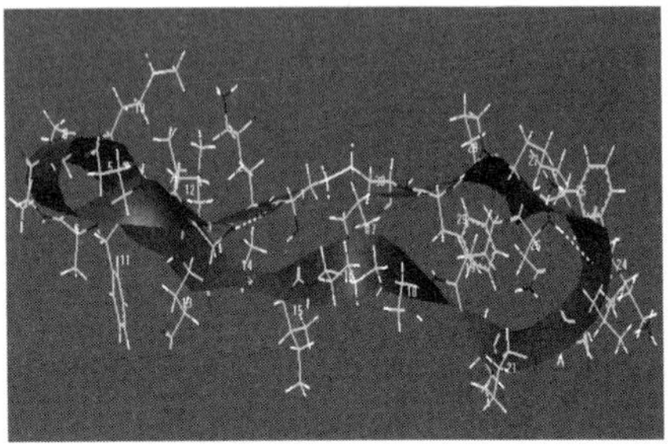

Fig. 1. 3D Model structures of cyclic $AT^{(C+367-394+C)}$RRLL (Left) and $AT^{(C+367-394+C)}$ WT (Right) peptides.

Enzyme kinetic study revealed that the cyclic and acyclic peptide inhibited SKI-1 activity in a differential manner against the fluorogenic substrate $QGPC^{251-263}$ (Abz-DIYISRRLL↓GTFTTyx-A-NH_2) (Abz = 2-amino benzoic acid, Tyx = 3-nitro Tyrosine) with inhibition constant in low μM. However, no significant inhibition of SKI-1 activity towards another substrate $Qcmv^{638-648}$ (Abz-RGVVNA↓SSRL-Tyx-A-NH_2) [4] used in this study for the first time was noticed. Cmv (Cytomegalo virus) sequence was derived from maturational (M) site of this viral protease believed to be crucial for its pathogenesis and may be mediated by SKI-1. The other site known as release (R) site is required for release of the product. We found the Qcmv substrate to be ~10 fold more potent than QGPC for SKI-1 assay. 3D model structure of cyclic $AT^{C+367-394+C}$WT and corresponding "RRLL" variant **(Fig 1)** exhibited several differences in the structure particularly the spatial orientation of all aromatic phenyl rings. Thus it is noted that the two Phe residues at positions 25 and 29 are both oriented outside of the ring in RRLL mutant compared to the wild type suggesting a significant change in structure.

Acknowledgments

Funds from NSERC, Canada.

References

1. Seidah N.G. et. al. *Proc. Nat. Acad. Sci. USA,* **96**, 1321-1326 (1999).
2. Sakai, J. et. al. *Mol. Cell.* **2**, 505-514 (1998).
3. Pasquato, A. et. al. *J. Biol. Chem.* **281**, 23471-23481 (2006).
4. Basak, S. et. al. *Pep. Prot. Lett.* **13**, 863-876 (2006).
5. Basak, A. and Lotfipour, F. *FEBS Lett*, Yao, **579**, 4813-4821 (2005).

Effects of Antifreeze Protein Fragments on the Properties of Model Membranes

Hagit Kun, Gerardo Byk and Yitzhak Mastai

Department of Chemistry, Bar-Ilan University, Ramat Gan, Zip 521900, Israel

During the past three decades a considerable amount of research has been conducted on the characterization of and structure – activity studies of antifreeze proteins. The ability of antifreeze proteins to protect cell membranes from hypothermic damage was first demonstrated by Rubinsky *et al.* [1]. In addition antifreeze proteins have been shown to prevent leakage of liposomes as they were cooled through their phase transition temperatures [2]. These results suggested interactions of antifreeze proteins with the lipid bilayer of the cell membrane. One cause of cold induced cellular damage could arise when cells are cooled through the lipid membrane thermotropic phase transition temperature. During this liquid crystalline to gel phase transition, membranes become leaky, resulting in the loss of intracellular contents and an influx of extracellular materials. Evidence consistent with this hypothesis comes from studies on human blood platelets where it was demonstrated that antifreeze proteins were able to inhibit the cold induced morphological changes that are initiated when the platelets pass through their phase transition temperature.

Liposomes are commonly used as models for chilling and freezing damage, with leakage of water-soluble contents from the aqueous interior as the most frequently used measurement of damage. In the present study liposomes were used as a model system to examine the nature of the interactions between antifreeze protein (AFP type I) and its short segments, with lipid membranes. In order to achieve an understanding of the mechanism of the interactions of antifreeze proteins with the lipid bilayer we have conducted differential scanning calorimetry (D.S.C) and isothermal titration calorimetry (I.T.C) study of the influence of Type I antifreeze protein and its short segments on the liposomes phase transitions.

Antifreeze protein type I (HPLC6 [3]) and three short segments of AFP-type I were prepared and their structures are shown in Table 1. We employ DMPC as a membrane cell model to study the interactions of our AFP's with cell membranes. The DSC measurements of the effect of AFP-type I and its short segments on the membrane phase transition temperatures are shown in Figure 1. The thermogram of the DMPC exhibits a sharp exothermic peak at T = 23.5°C, corresponding to the reported phase transition temperature of DMPC. The D.S.C scans of HK1 show a phase transition depression of 0.3°C, while for HK3 the scans show a depression of 0.4°C. The D.S.C scans for HPLC6 show a phase transition depression of 0.3°C.

Furthermore we employed isothermal titration calorimetry (ITC) method to extract the enthalpy and the stoichiometry of interaction processes of AFP's with membrane cell model. In this technique the heat produced by the stepwise injections of AFP's solution into a DMPC suspension is measured. The released heat depends on the enthalpy of the process and the equilibrium constant of the interaction. The I.T.C. thermograms displayed in figure 2 exhibits exothermic peaks for both HPLC6 and HK2. The different heat release for AFP type- I and its short segment exhibits

S. Del Valle et al. (eds.), *Peptides for Youth: The Proceedings of the 20th American Peptide Symposium*,
DOI: 10.1007/978-0-387-73657-0_38, © Springer Science+Business Media, LLC 2009

different strength of interaction as well as different tendency for interaction with DMPC.

Table 1. Native AFP type I and is short sequences.

HPLC6	DTASDAAAAAALTAANAKAAAELTAANAAAAAAATAR
HK1	TAANAAAAAAATAR
HK2	DTASDAAAAAAL
HK3	TAANAKAAAEL

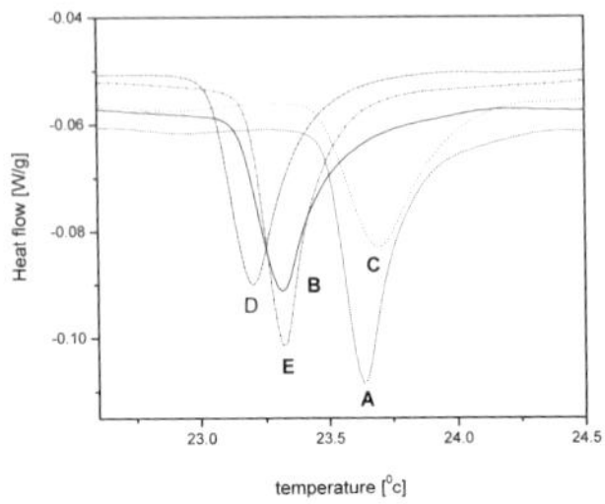

*Figure 1. Thermal histeresis measurements of the AFP-type I and its short segments using D.S.C. (**A**) Pure DMPC (**B**)HK1 (**C**)HK2 (**D**)HK3 and (**E**)HPLC6.*

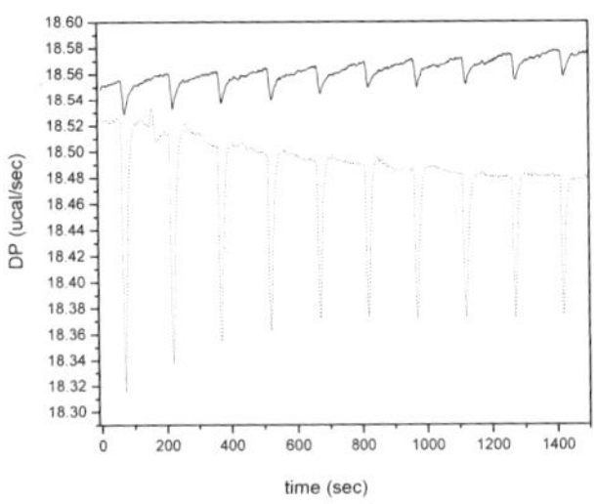

Figure 2. Heat measurements of the interaction heat produced by the stepwise injections of AFP's solution into a DMPC suspension using I.T.C. Solid line: HPLC6, Dotted line: HK2.

In conclusion, the results of our study support the hypothesis that the interaction between type I AFP's and model bilayers shift the phase transition temperature of the membrane phospholipids and leads to membrane stabilization. However, further studies are required to establish the mechanism of membrane stabilization by AFP's.

Acknowledgments

This research was supported by the Israel Science Foundation (ISF) grant No. 289.

References

1. Rubinsky, B., Arav, A., Mattioli, M. and De Vries, A.L. *Biochem.Biophys. Res Commun.* **173**, 1369-1374 (1990).
2. (a) Wu, Y., Banoub, J., Goddard S.V.,, Kao M.H. and Fletcher, G.L. *Comp. Biochem. Physiol.* **128**, 265-273 (2001). (b) Hays, L.M., Feeney, R.E., Crowe, L.M., Crowe, J.H. and Oliver, A.E. *Proc. Natl. Acad. Sci. USA* **93**, 6835-6840 (1996).
3. Harding, M.M., Ward, L.G. and Haymet, A.D.J. *Eur. J. Biochem.* **264**, 653-665 (1999).

Fluorogenic Cyclooctapeptides and Metal Ion Sensing

Maria Ngu-Schwemlein, Liliana Garcia and Kshawna Askew

Department of Chemistry, Winston-Salem State University, Winston-Salem, NC 27110

Introduction

Optical sensors for the detection of heavy metal ions have wide applications in environmental chemistry, and are directly relevant for the development of a more accurate and rapid identification of harmful metal ions in the environment. Although the design of chemical sensors for such metal ions has been reported [1,2] we are especially interested in developing cyclopeptide-based motifs for selective metal ion binding and sensing. Recent studies have shown that a model acidic cyclopeptide, c[D-Glu-Glu-D-Glu-Glu-D-Leu-Leu-D-Leu-Trp], (CP1), exhibited selective binding and specific fluorescent responses towards some divalent transition metal ions among the surveyed alkali, alkaline earth, transition and heavy metal ions [3]. The present study was undertaken to develop a better understanding of the underlying bimolecular interactions between cyclooctapeptides with acidic groups, (CP1) and c[Glu-D-Leu-Glu-D-Leu-Glu-D-Leu-D-Glu-D-Trp] (CP2), or sulfur containing residues, c[D-Leu-Leu-D-Phe-Met-D-His-Met-D-His-Lys(Dansyl)], (CP3), with Hg^{2+}, Pb^{2+} and Cd^{2+}. We report here the thermodynamic parameters for the interactions between these cyclopeptides with these metal ions as determined by Isothermal Titration Calorimetry (ITC). In order to evaluate the signal transduced upon peptide-metal ion interactions, the fluorescence emission spectral changes following titration of the cyclopeptide with these metal ions was compared. Standard Stern-Volmer formalism and the dependence of fluorescence intensity on quencher concentrations are presented.

Results and Discussion

Synthetic cyclopeptides, CP1 and CP2 were synthesized using Fmoc chemistry on Fmoc-Glu(α-*O*-Allyl)-PAL-PEG-PS (0.2 mmol/g, 1g) resins on a Pioneer peptide synthesizer at the LSU Peptide Facility as previously reported [3].

The order of the binding constant for the metal ion and peptide interaction is $Hg^{2+} > Pb^{2+} > Cd^{2+}$ (Table 1). In general, these bimolecular interactions are entropically driven. The differential binding isotherms for the interaction of these cyclopeptides with Hg^{2+} support a two binding sites model, whereas their weaker interactions with Pb^{2+} and Cd^{2+} generally involves a single site model. The fluorescence emission of the CP1 and CP2 is quenched in the presence of increasing concentration of Hg^{2+} and Pb^{2+}. In contrast, quenching of CP3 was only significant by Hg^{2+} (Fig. 1). The intensity Stern-Volmer plot for quenching by Hg^{2+} shows an upward curvature, which is indicative of both complex formation and collisional quenching (Fig. 2). Quenching by Pb^{2+} shows a downward curvature towards the x-axis, indicating that the fluorophore, Trp, is shielded from quenching.

The underlying bimolecular interactions between acidic cyclooctapeptides (CP1 and CP2) with Hg^{2+}, Pb^{2+} and Cd^{2+} are similar. In contrast, the binding affinity of CP3 (with weakly basic and methionyl residues) is selective for Hg^{2+}. Alkali and alkaline earth metal ions do not show any significant interactions with these cyclooctapeptides. CP1 and CP2 exhibit selective fluorescence quenching by Hg^{2+} and Pb^{2+}, whereas CP3 shows higher selectivity albeit lower sensitivity for Hg^{2+}.

S. Del Valle et al. (eds.), *Peptides for Youth: The Proceedings of the 20th American Peptide Symposium*, 87
doi: 10.1007/978-0-387-73657-0_39, © Springer Science+Business Media, LLC 2009

Table 1. Thermodynamic parameters of metal-ion binding to cyclooctapeptides .[a]

Metal Ion[b]	Cyclo-peptide	K_a (M^{-1})	ΔH (kJ mol^{-1})	ΔG (kJ mol^{-1})	ΔS (J K^{-1} mol^{-1})
Hg^{2+}	CP1	(2.2 ± 1.0) x 10^6	2.0 ± 0.8	-8.5 ± 1.0	21 ± 4
		(6.5 ± 1.7) x 10^3	6.5 ± 1.4	-5.2 ± 1.5	38 ± 4
	CP2	(2.8 ± 0.4) x 10^6	-11.2 ± 0.2	-8.9 ± 0.01	-7.8 ± 0.8
		(4.9 ± 0.8) x 10^4	-5.2 ± 1.2	-6.5 ± 0.1	4.2 ± 4
	CP3	(1.4 ± 0.1) x 10^6	-15.7 ± 0.3	-8.4 ± 0.8	-25.1 ±1
		(6.7 ± 0.3) x 10^4	-1.5 ± 0.4	-6.6 ± 0.5	17.2 ± 1
Pb^{2+}	CP1	(2.0 ± 0.5) x 10^5	1.4 ± 0.4	-7.3 ± 0.1	29 ± 1
	CP2	(1.5 ± 0.1) x 10^5	2.2 ± 0.4	-7.1 ± 0.01	31 ± 0.1
	CP3	(1.7 ± 0.01) x 10^4	-8.4 ± 0.1	-5.8 ± 0.01	-8.2 ± 0.4
		(2.1 ± 0.01) x 10^3	1.6 ± 0.05	-4.6 ±0.01	9.7 ± 0.2
Cd^{2+}	CP1	(1.3.0 ± 0.2) x 10^4	6.6 ± 2.2	-5.7 ± 1.1	40 ± 7
	CP2	(4.3 ± 0.01) x 10^4	4.2 ± 0.5	-6.4 ± 0.01	35.3 ± 2
	CP3	(4.05 ± 0.2) x 10^3	-0.7 ± 0.03	-4.9 ± 0.03	14.3 ± 0.1

[a]*Values correspond to the mean of three experiments and standard error.* [b]*Perchlorates.*

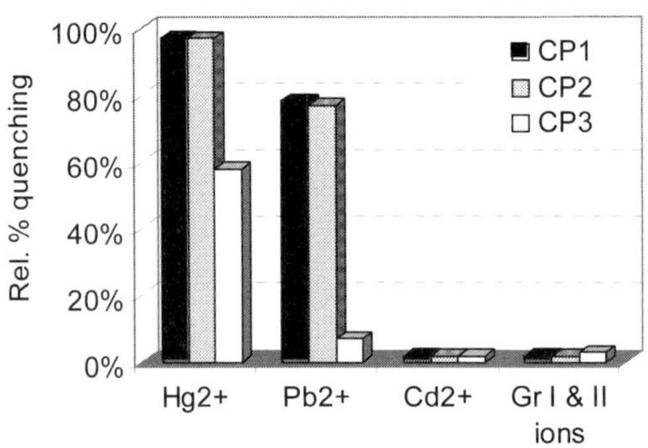

Fig. 1. Relative fluorescence quenching of 10 μM cyclooctapeptides in Hg^{2+} (80 μM) or Pb^{2+} (150 μM) and in the presence of various other metal ions (1 mM each of Cd^{2+}, Li^+, Na^+, K^+, Cs^+, Mg^{2+}, Ca^{2+}, and Ba^{2+}.

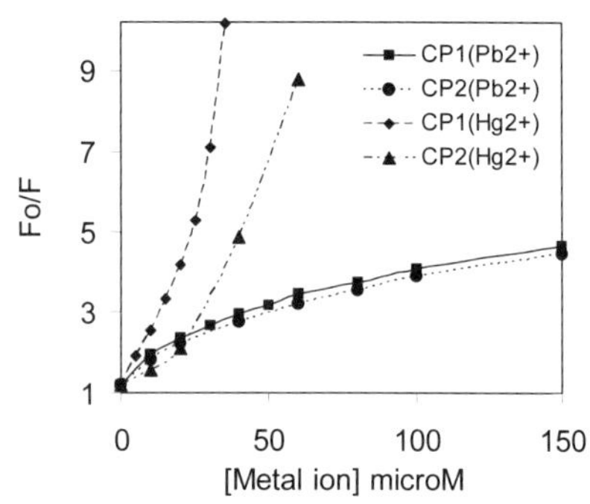

Fig. 2. Stern-Volmer plot for Hg^{2+} and Pb^{2+} quenching of 10 μM cyclooctapeptides, CP1 and CP2.

Acknowledgments

The work was funded by a NSF-RUI grant to M. N-S. We thank Martha Juban and Robert Hammer of the Peptide Facility, LSU for assistance with the SPPS of CP1&2.

References

1. Czarnik A.W., (Ed.), *Fluorescent Chemosensors for Ion and Molecular Recognition*, ACS Sym. Series 538; ACS: Washinghton, DC, 1993, Chapter 1, and references therein.
2. Aragoni, M.C., Archa, M., Demartin, F., Devillanova, F.A., Isaia, F., Garau, A., Lippolis, V., Jalali, F., Papke, U., Shamsipur, M., Tei, L., Yari, A., and Verani, G. *Inorg. Chem.* **41** (25), 6623-6632 (2002).
3. Ngu-Schwemlein, M., Butko, P., Cook, B., and Whigham, T. *J. Peptide Res.* **66** (Suppl. 1), 72-81 (2006).

The Impact of Aromatic Residues on the Tertiary Fold of Avian Pancreatic Polypeptide

Nicholas Y. Palermo[1], József Csontos[1], Richard F. Murphy[1] and Sándor Lovas[1]

[1]*Department of Biomedical Sciences, Creighton University Medical Center, Omaha, NE 68178, USA*

Introduction

Avian Pancreatic Polypeptide (aPP), is a 36 residue mini-protein with a tertiary fold formed by the packing of a polyproline-II helix (PPII) on an α-helix [1] (Figure 1A). Vibrational and electronic CD studies revealed that two fragments of aPP representing the PPII helical and α-helical structures, aPP(1-11)-NH_2 and Ac-aPP(12-36), respectively, refold into the conformation of the full structure [2]. This stabilization could be due in part to weakly polar non-bonded interactions. Aromatic (Ar) side chains , in particular, are known to participate in a variety of weakly polar interactions with the side chains and backbone (Bb) atoms of other residues. The goal of this investigation was to determine if aromatic residues have significant effects on the structural stability of aPP by calculating the strength of the pairwise interactions in which they participate, using DFT methods.

Intra-molecular interaction energies were calculated using the Jaguar 6.0 program package. The BHandHLYP level of theory was used because it closely approximates high-level results for weakly polar interactions [3]. The total interaction energy between the α-helix and the PPII helix was calculated as the sum of pairwise interactions between fragments of aPP consisting of aromatic residues and the side chains and backbone atoms of other residues (Figure 1B). The hydrogen bond between Gly9 and Ala12 and three non-canonical hydrogen bonds between the carbonyl oxygens of Pro8 and Gly9 and the aliphatic hydrogens of the Ala12 side chain, while not an aromatic interaction, were also found to participate in stabilizing the tertiary fold.

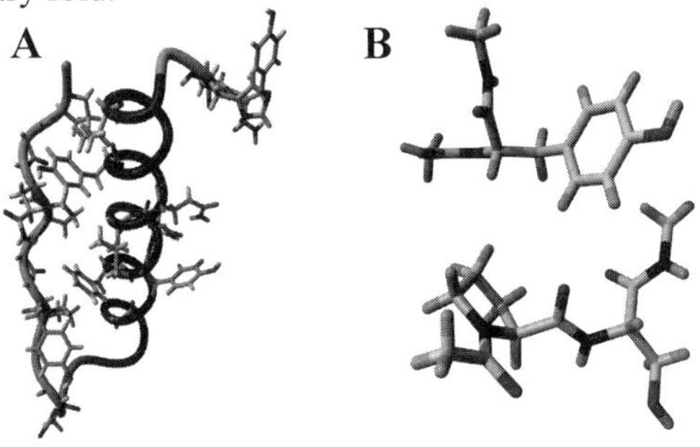

Fig. 1. A. The full structure of aPP. B. The interaction between Tyr27 and Pro2Ser3.

Results and Discussion

Ar side chains participate in several types of interactions including Ar-Pro, Ar-Ar, and Ar-Bb interactions at distances up to 5.65 Å (Table 1). Most common were interactions of aromatic rings with individual aliphatic hydrogens, also known as CH-π interactions.

S. Del Valle et al. (eds.), *Peptides for Youth: The Proceedings of the 20th American Peptide Symposium*,
DOI: 10.1007/978-0-387-73657-0_40, © Springer Science+Business Media, LLC 2009

Table 1. Distances (Å) of interactions which stabilize the tertiary structure of a PP.

Interacting Residues	Functional Group 1	Functional Group 2	Distance
Pro8Gly9-Ala12	C=O of Pro8	H^{β}	2.78
Pro8Gly9-Ala12	C=O of Gly9	H^{β}	2.98
Pro8Gly9-Ala12	C=O of Gly9	$H^{\beta2}$	2.99
Pro8Gly9-Ala12	C=O of Gly9	N-H	2.35
Phe20-Pro5Thr6	Ar	H^{β} of Pro5	4.03
Phe20-Pro5Thr6	Ar	Peptide bond 6	5.65
Phe20-Tyr7Pro8	Ar	Ar-H^{δ}	3.21
Phe20-Tyr7Pro8	Ar	H^{δ}	3.59
Tyr27-Pro2Ser3	Ar-H^{δ}	Prolyl ring	4.11
Tyr27-Pro2Ser3	Ar	C=O of Pro2	4.68
Tyr27-Pro2Ser3	Ar	Peptide bond 3	4.72
Tyr27-Gln4Pro5	Ar	H^{α} of Gln4	3.15
Tyr27-Gln4Pro5	Ar	H^{δ} of Pro5	3.71

The energies of the pairwise aromatic interactions are comparable with those of hydrogen bonds (Table 2) despite the much greater distance between the interacting functional groups. Two peptide bonds were doubly counted because of the fragmentation method. So, the peptide bonds were broken into fragments such as CH_3-NH-C(=O)-CH_3 and the interaction energy of the peptide bond was calculated as a pairwise interaction with an aromatic residue and then subtracted from the total energy (Table 2).

Table 2. Interaction energies ($kcal \cdot mol^{-1}$.)

Interactions	ΔE_{int}
Pro8Gly9-Ala12	-8.71
Phe20-Pro5Thr6	-6.05
Phe20-Tyr7Pro8	-5.25
Tyr27-Pro2Ser3	-7.43
Tyr27-Gln4Pro5	-5.53
Phe20-PB6	3.42
Tyr27-PB3	2.22
Total	-27.33

Aromatic residues Phe20 and Tyr27 participate in weakly polar interactions with seven other residues and contribute structural stability to aPP equivalent to four hydrogen bonds.

Acknowledgments

This work was supported by NIH-INBRE grant P20 RR016469.

References

1. Blundell, T. L.; Pitts, J. E.; Tickle, I. J.; Wood, S. P.; Wu, C.-W. *Proc. Natl. Acad. Acad. Sci.* **78**, 4175-4179 (1981).
2. Copps, J.; Murphy, R. F.; Lovas, S. *Biopolymers* **83**, 32-38 (2006).
3. Palermo, N. Y.; Csontos, J.; Owen, M. C.; Murphy, R. F.; Lovas, S. *J. Comput. Chem.* **28**, 1208-1214 (2007).

PEPTIDE, PROTEIN, AND PEPTIDOMIMETIC SYNTHESIS HONORING RALPH HIRSCHMANN

An Efficient Protocol for On-Resin, Vicinal Disulfide Formation: Applications to Thioredoxin Reductase

Stevenson Flemer Jr and Robert J. Hondal

Department of Biochemistry; University of Vermont College of Medicine; B415 Given Bldg; 89 Beaumont Ave; Burlington, VT; 05405; USA

Introduction

Vicinal disulfide bonds form 8-membered ring structures in proteins and occur at a low frequency in the Brookhaven Protein Data Bank. They are difficult to synthesize because **inter**molecular oxidation is preferred over *intra*molecular oxidation due to the strain on the central amide linkage possessing partial double bond character. This ring motif may have a regulatory function in several proteins where it is contained, including the nicotinic acetylcholine receptor, hepcidin, and the Janus-faced atracotoxins. Our interest in constructing these types of disulfide bonds is derived from our work on the mechanistic enzymology of thioredoxin reductase (TR) [1]. In TR, this motif is part of the catalytic cycle, and switches between reduced and oxidized states.

We wished to determine what the contributions of ring strain and entropy of this 8-membered ring were to the catalytic cycle. In order to accomplish this, we needed to isolate the ring-opening step from the rest of the catalytic cycle, which was done by constructing a truncated form of the enzyme missing this ring motif. Based upon previous work in our laboratory [2], we present here a facile method for the regioselective disulfide pairing of adjacent cysteine residues, as well as for the pairing of adjacent cysteine and selenocysteine residues to form their corresponding vicinal selenylsulfide linkage. Improving upon an existing methodology for on-resin disulfide formation [3], we demonstrate the efficient synthesis of 8-mer peptide sequences PTVTGCCG$_{(ox)}$ and PTVTGCUG$_{(ox)}$, containing the 8-membered ring motif as well as the "Open Chain" analogs of the same sequences (Figure 1). Note that U is the one letter abbreviation for selenocysteine.

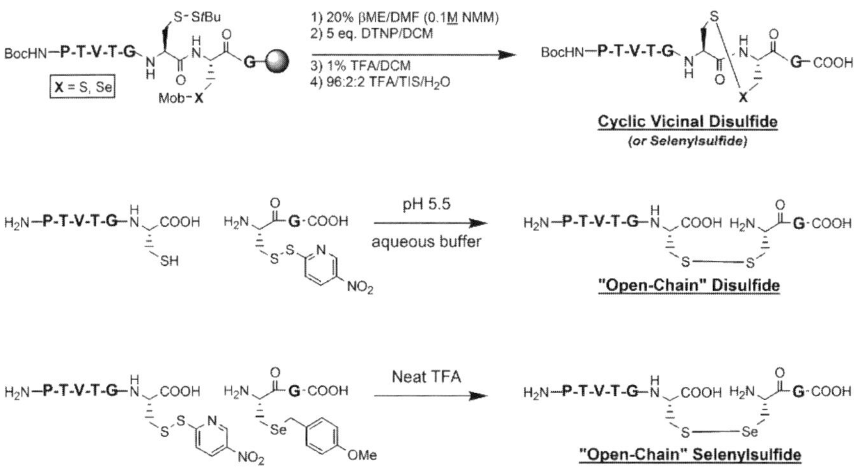

Figure 1. Synthetic protocol toward cyclic vicinal disulfide motif and "Open Chain" variants

DOI: 10.1007/978-0-387-73657-0_41, © Springer Science+Business Media, LLC 2009

Results and Discussion

Orthogonally-protected precursor peptides Boc-PTVTGC(StBu)C(Mmt)G, Boc-PTVTGC(StBu)C(Mob)G, and Boc-PTVTGC(StBu)U(Mob)G were prepared via Fmoc protocol on PAL-PEG resin. Removal of the StBu group was done by incubation of the resin with 20% βME/DMF (Galande method) or by the addition of 0.1M NMM to the βME mixture. Treatment of the resultant resin-bound thiols with excess 2,2′-dithiobis(5-nitropyridine) (DTNP) resulted in the formation of Cys (5-Npys) activated derivatives which, when treated with a solution of 1% TFA/DCM, allowed for spontaneous attack from the neighboring sulfur (or selenium) atom with concomitant removal of the Mob protecting group from Cys(Mob) or Sec(Mob). Figure 2 illustrates the marked improvement in purity of the crude cyclized disulfide product following acidolytic cleavage when using this methodology in comparison with the original protocol developed by Galande and coworkers [3]. Our method uses a thioether or selenoether as nucleophile instead of a naked sulfhydryl or selenol, suppressing intermolecular disulfide bond formation and favoring intramolecular disulfide bond formation.

We used a novel synthetic strategy in synthesizing the "Open Chain" selenylsulfide variant (Figure 1), mimicking the same approach used in the synthesis of the cyclized version [4]. This strategy uses a selenoether as nucleophile instead of a naked selenol, thus eliminating the problems associated with working with highly reactive selenols, themselves highly susceptible to diselenide formation, rendering them useless for creating selenylsulfide linkages as was done here.

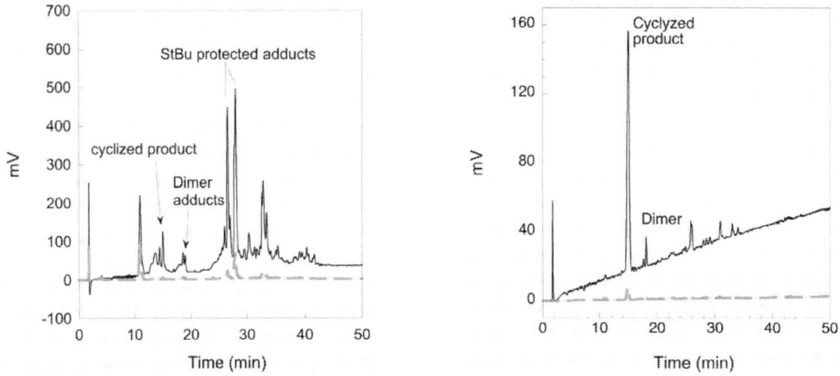

Figure 2. HPLC chromatograms of PTVTGCCG$_{(ox)}$ synthesized by the Galande method (using PTVTGC(StBu)C(Mmt)G-PAL Resin precursor) (Left) or our greatly improved method (Right).

Acknowledgments

This work was funded by National Institutes of Health Grant GM070742.

References

1. Eckenroth, B. E., *et al. Biochemistry*. **46**, 4694-4705 (2007).
2. Harris, K. M., Flemer, S. Jr., and Hondal, R. J. *J. Pep. Sci.* **13**, 81-93 (2007).
3. Galande, A. K., Weissleder, R., and Tung, C-H. *J. Comb. Chem.* **7**, 174-177 (2005).
4. Flemer, S. Jr., Lacey, B. M., and Hondal, R. J., In Press.

Potent and selective peptide agonists for human melanocortin receptor 5

Maria A. Bednarek[1], Tanya MacNeil[2], Rui Tang[2], M. Angeles Cabello[3], Marta Maroto[3] and Ana Teran[3]

[1]*Department of Medicinal Chemistry,* [2]*Department of Obesity and Metabolic Disorders, Merck Research Laboratories, Rahway, NJ, USA,* [3]*Centro de Investigacion Basica, Merck Sharp and Dohme de Espana, S.A., Madrid, Spain.*

Introduction

Melanocyte-stimulating hormones (αMSH, βMSH and γMSH) and adrenocorticotropin hormone (ACTH) are endogenous ligands for melanocortin receptors 1, 2, 3, 4 and 5 (MC1-5R). The physiological role of MC5R in mammals is not well understood despite its broad presence in various peripheral sites (skin, immune cells, exocrine glands, stomach, fat cells and others) and in the brain (cortex and cerebellum). There are indications in rodents that MC5R may affect some inflammatory and anxiolytic events, production of pheromone and lipids in the exocrine glands, and secretion of luteinizing hormone. Potent and selective hMC5R ligands could help to distinguish a role of this receptor from those of other melanocortin receptors in some skin disorders, immunomodulatory and anti-inflammatory processes and other physiological events.

The endogenous melanocortin peptides are non-selective, low-affinity agonists at hMC5R; αMSH displays the highest affinity for MC5R, and γMSH the weakest (αMSH > ACTH > βMSH >> γMSH.) In the structures of these peptides, the His-Phe-Arg-Trp segment has been recognized as critical to molecular recognition at the MC receptors. [1] This segment has subsequently become a pivotal feature of the structures of most synthetic, peptidic, melanocortin ligands. Among those ligands, the non-selective analogs NDP-αMSH, MTII, and SHU9119 have been the most extensively utilized as *in vitro* and *in vivo* research tools in various studies on the melanocortin system. [2]

Analogs of αMSH with structures similar to that of SHU9119 are reported which bind and activate the human MC5R at 1 nM concentration and which show greater than 5000-fold selectivity with respect to the other melanocortin receptors.

αMSH: Ac-Ser^{1}-Tyr^{2}-Ser^{3}-Met^{4}-Glu^{5}-His^{6}-Phe^{7}-Arg^{8}-Trp^{9}-Gly^{10}-Lys^{11}-Pro^{12}-Val^{13}-NH_2
SHU9119: Ac-Nle^{4}-cyclo[Asp^{5}-His^{6}-D-Nal(2')^{7}-Pip^{8}-Trp^{9}-Lys^{10}]-NH_2

Results and Discussion

Peptides in Table 1 were evaluated for their binding affinities to the human melanocortin receptors 1b, 3, 4 and 5 in the competitive binding assays using the radiolabeled ligand [^{125}I]-NDP-αMSH and for their agonist potency in cAMP assays employing the CHO cells expressing these receptors. The human melanocortin 1b receptor (hMC1bR) possesses pharmacological properties indistinguishable from those of its isoform, human melanocortin receptor 1a (hMC1aR) [3]. Table 1 depicts a subset of the compounds which were studied.

S. Del Valle et al. (eds.), *Peptides for Youth: The Proceedings of the 20th American Peptide Symposium,* 95
DOI: 10.1007/978-0-387-73657-0_42, © Springer Science+Business Media, LLC 2009

Analogs of SHU9119 in which Arg in position 8 was replaced with various amino acids such as Pro, Ala, Sar, Azt, Pip, Tic, Oic or D-Pro were prepared and tested. The new cyclic compounds, exemplified by Pro^8-SHU9119 in Table 1, had less affinity at the human MC receptors 1b, 3, 4 and 5 than the parent compound. However, they displayed enhanced selectivity for hMC5R with respect to hMC3,4R. Among these analogs, Pip^8-SHU9119 had the highest potency at hMC5R. Additional modification by replacement of His^6 with hydrophobic amino acids such as Ala, Val, Phg, Pro, Pip, Oic and others, resulted in peptides with higher 1/5 receptor subtype selectivity. Several analogs of SHU9119 which were modified in both position 6 (His) and 8 (Arg), displayed selectivity greater than 2000-fold with respect to the human MC 1b and 3 receptors and about 70 to 200-fold versus hMC4R (for example, Phg^6,Pip^8-SHU9119 in Table 1.)

Additional structure-function studies focused on the aforementioned compounds further substituted with various natural and unusual aromatic amino acids in position 7, such as D-Trp, D-Tyr, D-Tyr(Me), D-4Thz, D-4,4'Bip, D-Bth, D-Tic and others. Among the analogs of SHU9119 with altered residues 6, 7 and 8, Oic^6,D-Bip^7,Pip^8-SHU9119 (Table 1) was the most potent and selective hMC5R agonist with an EC_{50} ca. 1 nM at hMC5R and greater than 5000-fold selectivity with respect to the human MC 1b, 3 and 4 receptors.

Table 1. Analogs of αMSH.

	Binding Assay, IC_{50} (nM)				cAMP Assay, EC_{50} (nM)			
Peptide	hMC1bR	hMC3R	hMC4R	hMC5R	hMC1bR	hMC3R	hMC4R	hMC5R
αMSH	3.9	19	19	120	3.4	1.1	1.9	16
SHU9119	0.62	0.23	0.07	0.65	0.36	5%#	0%#	0.094
Analogs of SHU9119:								
Pro^8	880	>1000	4800	50	90	0%#	0%#	50
Pip^8	13	920	86	0.54	3.5	24%#	2%#	0.45
Phg^6,Pip^8	>1000	1090	48	0.24	23%#	3%#	24%#	0.21
Oic^6,D-4,4'Bip^7,Pip^8	>5000	>5000	>5000	0.95	18%#	5%#	49%#	0.99

percentage of cAMP accumulation, relative to [^{125}I]-NDP-αMSH, at 1 μM peptide concentration.

References

1. Hruby, V.J., Wilkes, B.C., Hadley, M.E., Al-Obeidi, F., Sawyer, T.K., Staples, D.J., Devaux, A., Dym, O., de L.Castrucci, A.M., Hintz, M.F., Riehm, J.P., Rao, K.R. *J. Med. Chem.* **30**, 2126-2130 (1987)
2. Al-Obeidi, F., de L. Castrucci, A.M., Hadley, M.E., Hruby, V.J. *J. Med. Chem* **32**, 2555-2561 (1989).
3. Tan, C.P., McKee, K.K.,Weinberg, D.H., MacNeil, T. et al. *FEBS Lett* **451**, 137-141 (1999).

Synthesis of a library of ψ-stylostatins with potential antibacterial activity

Carlos López,[1,2] Montse Cruz,[1] Javier Sánchez-Cespedes,[3] Anna Diez[1,2]*

[1]*IRB-Parc Científic de Barcelona, c/ Josep Samitier, 1-5. 08028-Barcelona, Spain.*
[2]*Laboratori de Química Orgànica. Facultat de Farmàcia. Universitat de Barcelona, 08028-Barcelona, Spain.*
[3]*Departament de Microbiologia. Hospital Clínic de Barcelona. 08036-Barcelona, Spain.*

Introduction

Using the cycloheptapeptide stylostatin 1 (*c*FNSLAIP) as our model [1,2], we have prepared a library of ψ-stylostatins with the potential to act as antibiotics directly on the procaryote cell membrane and thus circumvent mechanisms of bacterial resistance [3]. Structural and biological studies are also described.

Results and Discussion

The library consists of 15 ψ-stylostatins (Table 1) in which the original Leu has been replaced by Thr and charged aminoacids substituted for Asn. In addition, 3-aminolactam **1** (Fmoc-{S(Bn)-T(^{t}Bu)}) has been used as a Ser-Thr constrained surrogate, to modify the secondary structure.

We obtained compound **1**, in 9 steps and 16% total yield, by lactamisation of ribonolatone **3** and subsequent amination *via* the corresponding azide [4], followed by modification of compound **6** to obtain lactam **1**, suitable for solid phase synthesis.

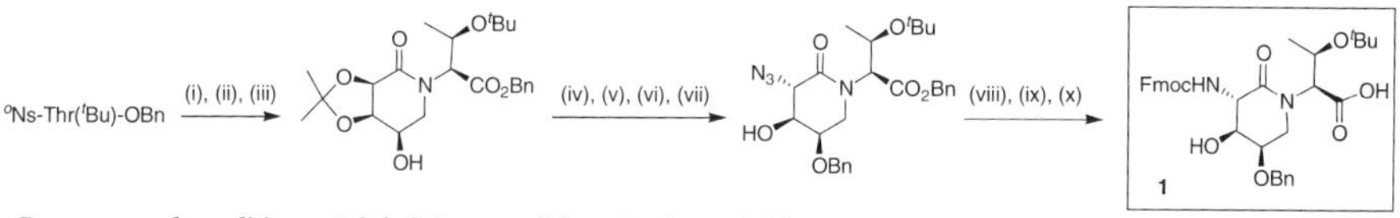

***Reagents and conditions**: i) 2,3-O-Isopropylidene-D-ribono-1,4-lactone, PPh_3, DEAD, CH_2Cl_2, 0°C, 24 h (85%); ii) K_2CO_3, PhSH, DMF, r.t., 45 min (81%); iii) 2-hydroxypyridine, toluene, 95°C, 48 h (80%); iv) BnBr, NaH, DMF, 0°C, 24 h (68%); v) PPTS, MeOH, Δ, 24h (58%); vi) Et_3N, $SOCl_2$, THF, 0°C, 1h (quant.); vii) NaN_3, DMF (81%); viii) Lindlar catalyst, H_2, EtOH, r.t., 12h (quant.); ix) FmocOSu, $NaHCO_3$, acetone/water 1:1, r.t., 24 h (90%); x) H_2, 10% Pd/C, MeOH, r.t., 3h (quant.).*

*Scheme 1. Synthesis of compound **1**.*

The linear precursors of ψ-stylostatins were prepared by Fmoc SPPS on a chlorotrityl resin (Novabiochem, 0.7 mmol/g), and recovered from the resin with AcOH:TFE:DCM (2:2:6). Cyclisation (ω Pro-Phe) was performed in solution using PyBOP (1.5 eq.) and DIEA (3 eq.) in DMF:DCM (97:3). Side-chain deprotection was achieved with TFA:H_2O (95:5), and ψ-stylostatins were purified by semi-preparative HPLC (C_{18} RP-HPLC, Waters) with a gradient of CH_3CN in aq. 0.1% TFA.

Table 1. ψ-Stylostatins obtained.

Asn:	*c*F**N**STAIP	*c*F**N**{S(Bn)T}AIP	*c*F**N**{ST}AIP	--	--
Lys:	*c*F**K**STAIP	*c*F**K**{S(Bn)T}AIP	*c*F**K**{ST}AIP	*c*(F**K**STAP)	F*c*(**K**STAIP)
Arg:	*c*F**R**STAIP	*c*F**R**{S(Bn)T}AIP	*c*F**R**{ST}AIP	*c*(F**R**STAP)	--
Asp:	*c*F**D**STAIP	--	--	*c*(F**D**STAP)	F*c*(**D**STAIP)

DOI: 10.1007/978-0-387-73657-0_43,

© Springer Science+Business Media, LLC 2009

Structural studies. Molecular modeling calculations show that the different ψ-stylostatins have different degrees of flexibility, and can show up to 3 different families of conformations (Insight II; FF: Amber; *in vacuo*).

Circular dichroism experiments show different patterns for *c*FKSTAIP and *c*FK{ST}AIP. In each case the pattern remains unchanged on temperature variation However, the characteristic curve of *c*FKSTAIP is flattened upon concentration of the sample. HPLC chromatograms of *c*FKSTAIP at room temperature consistently show a split signal that coalesces at 45°C. This observation, taken together with the CD results, indicates that *c*FKSTAIP is capable of self association.

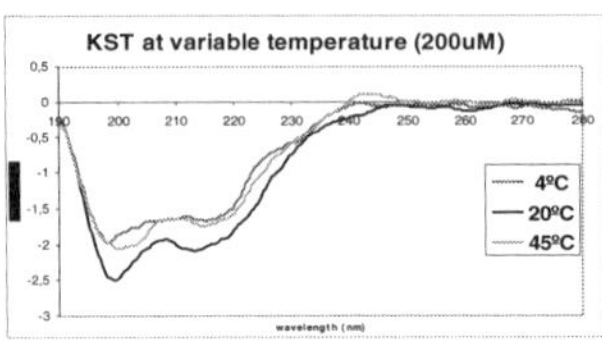

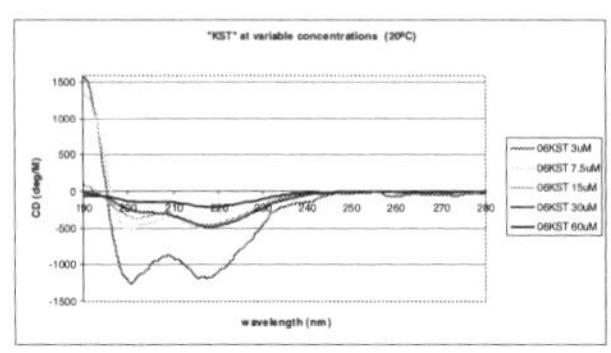

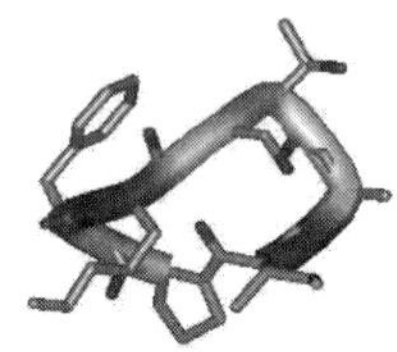

*c*FKSTAIP: ωIle-Pro *trans*

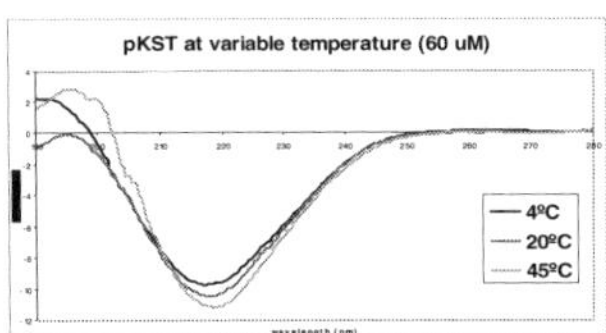

Figure 1. CD spectra of cFKSTAIP at variable temperature and at variable concentration, and a representation of its family of conformations. The CD spectrum of peptide cFK{ST}AIP at variable temperature features at the bottom-left.

Bioassays. The minimum inhibitory concentrations (MIC, μM) of ψ-stylostatins FDSTAIP, *c*FDSTAP, *c*FKSTAIP, *c*FNSTAIP and *c*FRSTAIP have been determined on two cell lines of *E. coli* (EMP5, NorE5), *A. Baumannii* (77, 77(64)), *P. aeruginosa* (wt, mexA.OprD), and *S. aureus* (1199, 1199B). No cell growth inhibition was observed at 64 μM, indicating that the tested ψ-stylostatins have no useful antibiotic activity.

Acknowledgments

This research is funded by the MEC (Spain, grant CTQ2004-01757/BQU) and the CIRIT (Generalitat de Catalunya, fellowship to C.L.). We also thank Prof. Jordi Vila (Hospital Clínic de Barcelona, Spain) for the antibiotic assays.

References

1. Pettit, G. R., Srirangam, J. K., Herald, D. L., Erickson, K. L., Doubeck, D.L., Schmidt, J. M., Tackett, L. P. and Bakus, G. J. *J. Org. Chem.* **57**, 7217-7220 (1992).
2. Forns, P., Piró, J., Cuevas, C., García, M., Rubiralta, M., Giralt, E. and Diez,A. *J. Med. Chem.* **46**, 5825-5833 (2003).
3. Ganz, T. *Nature* **412**, 392 393 (2001).
4. a. Piró, J., Rubiralta, M., Giralt, E. and Diez, A. *Tetrahedron Lett.* **40**, 4865-4868 (1999). b. Piró, J., Rubiralta, M., Giralt, E. and Diez, A. *Tetrahedron Lett.* **42**, 871-873 (2001).

Mimicking natural globular structures with rigid scaffolds based on adamantane

Nadine Pannier[1], John V. Frangioni[2] and Wolfgang Maison[1]

[1]Institute of Organic Chemistry, University Giessen, Heinrich-Buff-Ring 58, 35392 Giessen, Germany; [2]Beth Israel Deaconess Medical Center, 330 Brookline Avenue, Boston, MA 02215

Introduction

Multivalency is a common phenomenon in nature to increase affinity and specifity of receptor-ligand interactions, especially on the cell surface. Chemists have tried to make use of multivalent interactions in different context and have synthesized a number of scaffolds for the assembly of multivalent receptor ligands [1].

We have developed tetrahedral, rigid scaffolds based on adamantane for conjugation to multiple cell surface binders via amide bonds. The resulting novel multivalent conjugates **1** were designed for an efficient interaction with cell surface epitopes and are miniaturized analogs of natural systems like virus particles, resembling their globular shape and a tripodal arrangment of binding ligands. This rigid tripodal geometry is an ideal recognition motif for cell surface receptors.

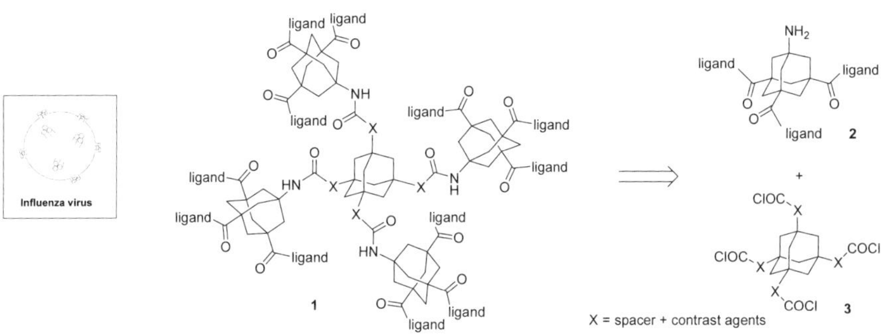

Fig. 1. Multivalent conjugates inspired by natural systems like virus particles.

Results and Discussion

We present efficient synthetic routes to symmetrical central fragments **3,** and unsymmetrically tetrasubstituted head groups **2**. The symmetrically tetrasubstituted central fragments **3** are available from 1,3,5,7-Tetracarboxymethyladamantane **6** [2]. We have developed a new practical and scalable synthesis of this well known and versatile synthetic intermediate, which is converted to the central fragment **7** in a few additional steps in good overall yield. Tetraphenyladamantane, which was prepared according to a protocol of Newman [3] **4** and tetramethoxyphenyl-adamantane **5** can be oxidized with RuO_4 to the carboxylic acid, which was directly converted to the methyl ester **6**. Tetramethoxyphenyladamantane can be synthesized from tetra-hydroxyadamantane, which was prepared acoording to a protocol of Menger and coworkers [4].

S. Del Valle et al. (eds.), *Peptides for Youth: The Proceedings of the 20th American Peptide Symposium*, 99
DOI: 10.1007/978-0-387-73657-0_44, © Springer Science+Business Media, LLC 2009

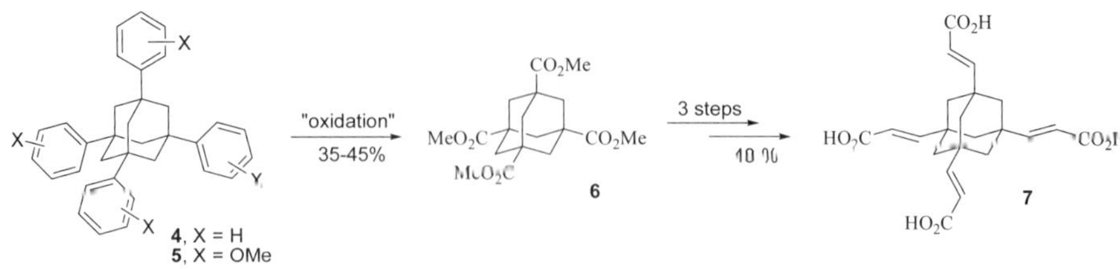

Fig. 2. Synthesis of the central fragment.

The synthesis of head groups **2** starts from adamantane **8** and gives the tetrafunctionalized derivative **9** in six steps with an overall yield of 48% [5]. An aminocaproyl-spacer was introduced and the carboxylic acids were converted to the NHS-esters **10** for coupling with specific ligands in aqueos media [6].

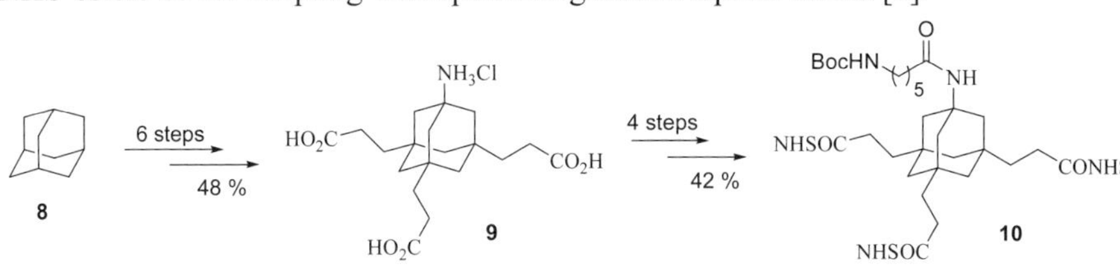

Fig. 3. Synthesis of the head group fragment.

Differentially substituted adamantane **10** is a suitable head group fragment for the assembly of large scaffolds **1**. However, it can also be used as a trivalent scaffold for multivalent receptor-ligand interactions, because it can be attached to three specific ligands and an effector molecule. It can thus be used as a general tool for affinity maturation of small molecule cell surface binders.

In this context we have coupled three prostate cancer specific ligands and a contrast agent to **10** and obtained highly specific agents for prostate tumor imaging using NIR and SPECT technology [6,7].

Acknowledgments

This work was funded by DFG MA 2519/3.

References

1. M. Mammen, S. K. Chio, G. M. Whitesides, *Angew. Chem., Int. Ed.* **1998**, *37*, 2755.
2. a) S. Landa, Z. Kamvcek, *Coll. Czech. Chem. Commun.* **1959**, *24*, 4004; b) G. R. Newkome et al., *J. Org. Chem.* **1992**, *57*, 358.
3. H. Newman, *Synthesis* **1972**, 692.
4. F. M. Menger, V. A. Migulin, *Journal of Organic Chemistry* **1999**, *64*, 8916.
5. W. Maison, J. V. Frangioni, N. Pannier, *Org. Lett.* **2004**, *6*, 4567.
6. P. Misra, V. Humblet, N. Pannier, W. Maison, J. V. Frangioni, *J. Nucl. Med.* **2007**, *48(8)*, 1379-1389.
7. V. Humblet, P. Misra, Y. Ko, N. Pannier, W. Maison, J. V. Frangioni, *Angw. Chem.* **2007**, *submitted for publication.*

Discovery and Optimization of a TRAIL R2 Agonist for Cancer Therapy

Yvonne M. Angell, Ashok Bhandari, M. Nuria De Francisco, Brian T. Frederick, Jennifer M. Green, Karen Leu, Kerstin Leuther, Reuben Sana, Peter J. Schatz, Erik A. Whitehorn, Kathy Wright, and Christopher P. Holmes

Affymax, Inc., 4001 Miranda Avenue, Palo Alto, CA 94304, USA

Introduction

TRAIL is a cytokine that induces apoptosis in a wide variety of tumor cells but rarely in normal cells. The TRAIL R2 ligand triggers tumor cell apoptosis independently of the p53 tumor-suppressor gene [1-5]; thus, peptide agonists may offer a complementary approach to conventional cancer therapy. Ligands belonging to the TNF family are anticipated to function as a homotrimer as suggested by the crystal structures of a subset of the family (TNFα, TNFβ, CD40L, and TRAIL). The TNFR family members are transmembrane proteins. The extracellular domain of the receptors is characterized by the concatenated cysteine-rich domains (CRDs) [6] that are responsible for ligand binding. TRAIL R2 is a single transmembrane receptor arranged as a homo-trimeric complex on the cell membrane, and TRAIL ligand is a Zn-coordinated trimer [7]. Formation of a complex between TRAIL and its signaling receptors, DR4 and DR5, triggers apoptosis by inducing the oligomerization of intracellular death domains.

A novel peptide sequence identified through recombinant peptide screening was found to bind the TRAIL R2 receptor with 120 nM binding affinity. The sequence of this peptide is unrelated to that of native or recombinant TRAIL yet the sequence was found to compete with TRAIL for binding the TRAIL R2 receptor. Hit to lead optimization strategies included both sequence optimization as well as architecture optimization. Truncation analysis of the hit sequence enhanced the binding affinity and identified a 15-mer as the minimal sequence necessary to maintain good binding (IC_{50} = 1 nM). Alanine scanning identified three critical residues for binding affinity including an *N*-terminal tryptophan and two residues within the disulfide loop, whereas the *C*-terminus was found to be critical for functional activity. Specific architectural modifications of the peptide structure, including dimerization, enhanced the binding affinity from 1000 to 10,000-fold depending on the architecture. In addition, the point of multimerization significantly affected functional activity in whole cell assays, resulting in peptides with potencies ranging from no functional activity to low micromolar apoptotic activity. Figure 1 illustrates how architecture influences both binding activity and functional activity. Interestingly, we observed a lack of direct correlation between binding affinity and functional activity.

Previously, we expected TRAIL mimetic peptide dimers with appropriate architecture to promote formation of the homo-trimeric receptor complex necessary for signaling apoptosis. Some synthetic dimers exhibited low micromolar apoptotic activity; however, we proposed peptide trimers may be required for full functional signaling. Thus, recently we synthesized a series of trimeric peptide architectures. Trimerization increased functional apoptotic activity 5-fold over the corresponding dimer.

S. Del Valle et al. (eds.), *Peptides for Youth: The Proceedings of the 20th American Peptide Symposium*, 101
DOI: 10.1007/978-0-387-73657-0_45, © Springer Science+Business Media, LLC 2009

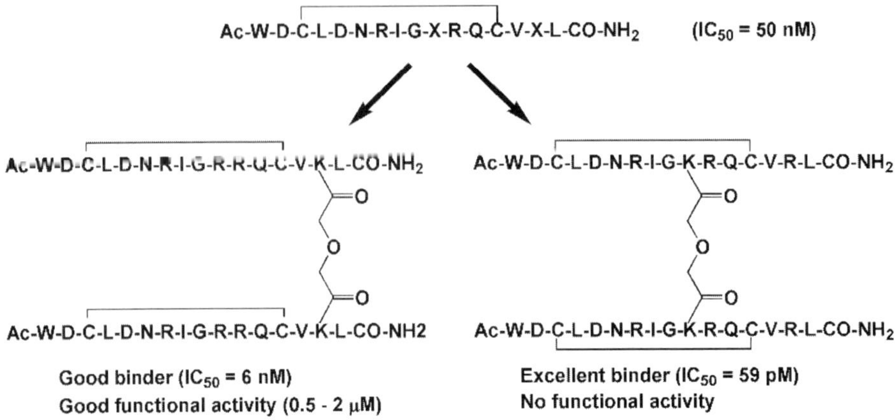

Fig. 1. Optimization of linker position.

Results and Discussion

Synthetic peptides were prepared using Fmoc chemistry on TentaGel R RAM (0.18 mmol/g, 0.4 g) resins using standard DIC/HOBt batchwise solid-phase synthesis protocols on a PTI Symphony peptide synthesizer. The *N*-terminal Fmoc-group was removed with 20% piperidine in DMF, and the *N*-terminal amine was capped with a mixture of acetic anhydride/pyridine/THF. Following resin and side-chain cleavage with 85% TFA/10% triisopropylsilane/2.5% H_2O/2.5% thioanisole, the crude peptides were precipitated with cold diethyl ether and washed twice with ether; material was solubilized in a mixture of 10% DMSO/40% acetonitrile/50%

NH_4OAc buffer (10 mM) at a peptide concentration of 1 mg/mL for oxidation of the cysteines. The oxidation was monitored by RP-HPLC and LCMS. Once the oxidation was complete (2 - 12 h depending on the sequence), the peptides were concentrated, diluted with 10% acetonitrile in water, and purified by preparative C_{18} RP-HPLC using linear gradients of acetonitrile (containing 0.1% TFA) in H_20 (containing 0.1% TFA) on either a Waters RCM Delta-Pak, 300 Å, 15 μm, 25 x 200 mm column or XTerra Prep MS, 125 Å, 5 μm, 19 x 50 mm column. Trimerization was accomplished by dissolution of peptide monomer in DMF, followed by addition of 10 eq. DIEA and portion-wise addition of 0.33 eq. Tris-succinimidyl aminotriacetate (TSAT). The reaction was monitored by HPLC and LCMS. Upon completion, the reaction mixture was diluted with water and purified by preparative C_{18} RP-HPLC using the same buffer conditions and columns used for the monomers. Final products were analyzed by analytical C_{18} RP-HPLC (Zorbax SB, 3.5 μm, 2.1 x 75 mm) with a gradient of 20-50% CH_3CN in aqueous 0.1% TFA.

Binding assays were performed as competition binding assays in AlphaQuest format. Functional assays were initially performed by evaluating the ability of synthetic peptides to induce apoptosis in Jurkat cells. Apoptosis was monitored by a MTT colorimetric viability assay. Recently, we developed a new assay to measure the activity of TRAIL agonist peptides using the colon carcinoma cell line HCT116. HCT116 cells were incubated for two days in the presence of serially diluted dimer or trimer peptides. Cell viability was measured using MTT. Trail agonist trimers

(Figure 2) exhibited 5-fold increased apoptotic activity over their corresponding dimers in this cell-based assay.

The success of this project demonstrates the feasibility of discovering and developing peptide mimetics of protein targets. Peptide-based drugs may provide superior product profiles over therapeutic proteins. Advantages of peptide-based drugs can include reduced immunogenicity, reduced dosing frequency, flexible storage and uncomplicated chemical synthesis.

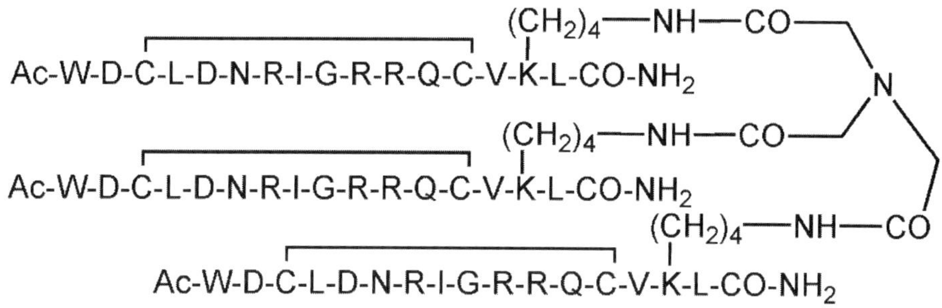

Fig. 2. TRAIL agonist peptide trimerstructure. Trimerization of this sequence enhanced apoptotic activity 5-fold over the corresponding dimer.

References

1. Wiley, S. R., Schooley, K., Smolak, P.J., Din, W.S., Huang, C.P., Nicholl, J.K., Sutherland, G.R., Smith, T.D., Rauch, C., and Smith, C.S. *Immunity*, **3**, 673-682 (1995).
2. Pan, G., O'Rourke, K., Chinnaiyan, A. M., Gentz, R., Ebner, R., Ni, J., and Dixit, V. M. *Science*, **276**, 111-113 (1997).
3. Sheridan, J. P., Marsters, S. A., Pitti, R. M., Gurner, A., Skubatch, M., Baldwin, D., Ramakrishnan, L., Gray, C.L., Baker, D., Wood, W. I., Goddard, A. D., Godowski, P., and Ashkenazi, A. *Science*, **277**, 818-821 (1997).
4. Pan, G., Ni, J., Wei, Y. F., Yu, G., Gentz, R., and Dixit, V. M. *Science*, **277**, 815-818 (1997).
5. Walczak, H., Degli-Esposti, M. A., Johnson, R. S., Smolak, P. J., Waugh, J. Y., Boiani, N., Timour, M. S., Gerhart, M. J., Schooley, K. A., Smith, C. A., Goodwin, R. G., and Rauch, C. T. *EMBO J.* **16**, 5386-5397 (1997).
6. Bazan, J. F. *Curr. Biol.* **3**, 603-606 (1993).
7. Hymowitz, S. G., Christinger, H. W., Guh, G., Ultsch, M., O'Connell, M., Kelley, R. F., Ashkenazi, A., and de Vos, A. M. *Mol. Cell*, **4**, 563-571 (1999).

Novel reactive site loop derived mini-serpin inhibitors of recombinant Proprotein Convertase 4

Ajoy Basak, Nausheen Shervani, Maria Kolajova, Shalini Cherla & Sarmistha Basak

Hormones Growth and Development Program, Regional Protein Chemistry Center, Ottawa Health Research Institute, U Ottawa, Ottawa, ON K1Y 4E9, CANADA

Introduction

The Ca^{+2}-dependent mammalian subtilase Proprotein Convertase 4 (PC4)/) is most abundant in testicular spermatogenic cells, sperm acrosome, placenta and in ovarian macrophages [1-3]. PC4 cleaves peptide bonds characterized by the motif **R/K/H-X-X/R/K-R↓**, where X = any amino acid except Cys. Our previous studies revealed that **KXKR** is one of the preferred PC4 cleavage motifs [4]. PC4 activates a variety of protein precursors found in the secretory pathway of mammalian reproduction system. Knock out animal and other studies suggest PC4's vital role in fertilization and sperm maturation [2]. PC4 substrates include (i) growth factors, proIGF-1/ proIGF-2; (ii) hormonal polypeptides, proPACAP and (iii) surface proteins of ADAM (A Disintegrin And Metalloproteinase) family, ADAM-1 (fertilin α)/ ADAM-2 (fertilin β)/ADAM-3 (procyritestin) and ADAM-5 [5]. Recently interest has grown to develop PC4-inhibitors because of therapeutic potentials as nonhormonal contraceptives. CRES (Cystatin Related Epididymal Serpin) found in epididymal fluid [5] is a candidate PC4 inhibitor. Here we describe the production of PC4 using *Leishmania tarentolae* expression system and develop new PC4-inhibitors based on CRES reactive site loop (RSL).

Results and Discussion

Recombinant PC4 production and purification: Recombinant r (rat) PC4 was generated by *Leishmania tarentolae* expression system in secreted form and purified from the culture medium. This is achieved by centricon concentration/dialysis (cut off molecular weight 10kDa) and chromatographies through diethyl amino-ethyl (DEAE)-sepharose and arginine-sepharose columns. DEAE-bound PC4 was eluted with 100mM NaCl in 25mM Tris+25mM Mes, pH 7.4 (Fraction# 28-48). The combined fraction was dialyzed and loaded onto Arginine column where the bound PC4 was eluted with the above buffer containing 20 mM free arginine (fraction #97-135) (**Fig 1**). Bands at 78, 63 and 52kDa were due to PC4 as confirmed by western blots using two antibodies (against mouse $PC4^{23-396}$ and $rPC4^{419-441}$) and proteomics analysis. The band at 37 kDa was not characterized.

Miniserpin inhibitors of PC4: Although no physiological serpin inhibitors have been reported for PCs except the drosophila dSPN4 for Proprotein Convertase furin, there may exist one serpin for each PC. CRES has been shown to inhibit PC2 which is not found in reproductive tracts [5]. However it is likely that CRES may be a natural PC4 inhibitor. To establish this notion we prepared one short and a long CRES peptide with two pairs of Cys protected as Acm for future bonding.

S. Del Valle et al. (eds.), *Peptides for Youth: The Proceedings of the 20th American Peptide Symposium*, 105
DOI: 10.1007/978-0-387-73657-0_46, © Springer Science+Business Media, LLC 2009

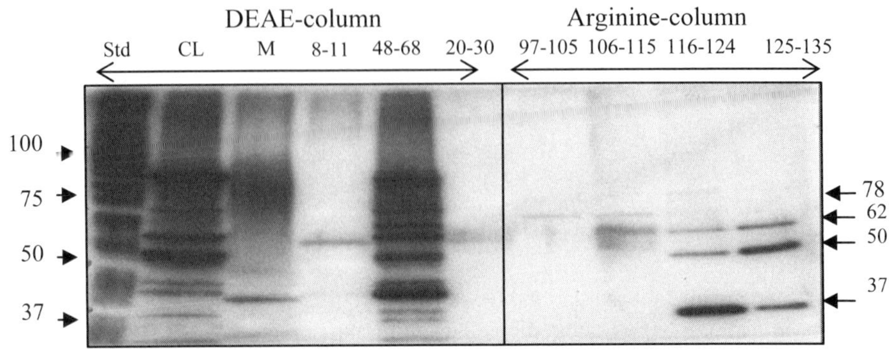

Fig 1 Coomassie stained 12%SDS-Page of various fractions obtained from DEAE (left) and Arginine (right) columns. Std=Standard, CL=cell lysate, M=cell medium.

Table 1: List of various protected CRES peptides derived from its RSL structure.

Peptide	Amino Acid Sequence
CRES-1	^{105}CAIQENSKLKRKLSC^{119}
CRES-2	^{95}CRKPLSTNEICAIQENSKLKRKLSC$_{Acm}$SFLVGALPWNGEFTVMEKKC$_{Acm}$^{139}
CRES-3	^{95}CRKPLSTNEIC$_{Acm}$AIQENSKLKRKLSCSFLVGALPWNGEFTVMEKKC$_{Acm}$^{139}
CRES-4	^{95}C$_{Acm}$RKPLSTNEICAIQENSKLKRKLSCSFLVGALPWNGEFTVMEKKC$_{Acm}$^{139}
CRES-5	^{95}C$_{Acm}$RKPLSTNEIC$_{Acm}$AIQENSKLKRKLSCSFLVGALPWNGEFTVMEKKC^{139}
CRES-6	^{95}C$_{Acm}$RKPLSTNEICAIQENSKLKRKLSC$_{Acm}$SFLVGALPWNGEFTVMEKKC^{139}

(CRES$^{95-139}$) that is crucial for enzyme inhibition (**Table 1**). Examination of PC4-inhibitory properties of these CRES peptides revealed that they all inhibit PC4 with IC_{50} in low μM ranges with the shorter CRES-1 being the most potent with IC_{50} value ~23 μM (**Fig 2**). Other protected CRES peptides inhibit PC4 with IC_{50} ranging from 35 μM (CRES-3)-352 μM (CRES-5). Preparations of various S-S bridge cyclic CRES peptides (long and short) are currently in progress.

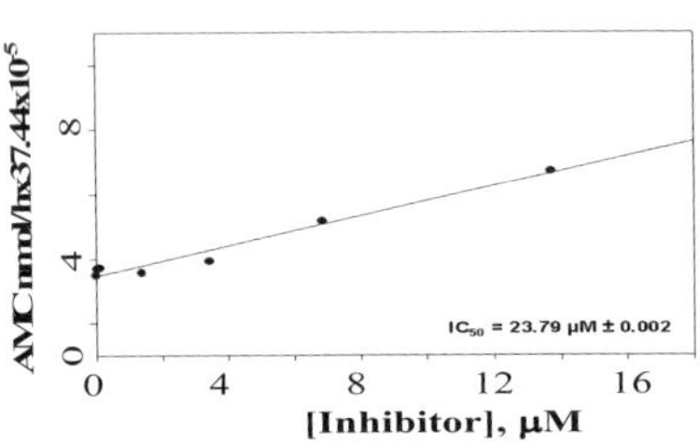

Fig 2. Inhibition by acyclic CRES-1 peptide of rPC4 activity obtained from Leishmania expression system following purification by DEAE and Arginine column chromatographies.

Acknowledgement

Funds from CIHR (AB)

References

1. Seidah, N.G. and Chrétien, M. *Brain Res* **848**, 45-62 (1999).
2. Mbikay, M. et. al. *Proc. Natl. Acad. Sci. USA*, **94**, 6842-6846 (1997).
3. Gyamera-Acheampong., C, et. al. *Biol. Reprod.* **74**, 666-673 (2006).
4. Basak, S. et. al. *Biochem. J.*, **380**, 23471-23481 (2004).
5. Cornwal, G. et. al. *Endocrinology,* **144,** 901-908 (2003).

Enzymatic ligation of peptides, peptide nucleic acids and proteins by means of sortase A

Stephan Pritz, Yvonne Wolf, Oliver Kraetke, Jana Klose, Michael Bienert, and Michael Beyermann

Leibniz-Institute of Molecular Pharmacology, Robert-Roessle-Strasse 10, 13125 Berlin, Germany

Introduction

Chemical ligation allows almost selective modification of unprotected peptides or proteins and thereby offers the opportunity for the synthesis of non-natural constructs [1]. Even more selective than chemical may be enzymatic ligations.

Sortases are transpeptidases found in Gram-positive bacteria. The *Staphylococcus aureus* sortase isoform SrtA (sortase A) cleaves proteins at a LPXTG-motif between threonine and glycine, and subsequently transfers the acyl-fragment to a N-terminal oligoglycine. *In vivo*, this reaction serves for the covalent attachment of surface proteins to the bacterial cell wall [2]. Recently, sortase-mediated ligation was introduced as a method for peptide and protein ligation [3]. Due to the low tolerance of sortase for deviation in the LPXTG-recognition motif, this enzymatic ligation is highly selective. In particular, the very limited occurrence of this motif in proteins makes the SrtA-mediated reaction interesting for protein modification.

Results and Discussion

In order to improve the yields of the transpeptidation reaction, we evaluated the impact of the residues X^n following the recognition motif (LPXTG$X^1X^2X^3X^3$) [4].

By using two small peptide libraries (Dns-LPKTGX^1RR-NH_2 and Dns-LPKTGGX^2RR-NH_2) which were subjected to the same nucleophile (H-GGGWW-NH_2) for kinetic and end-point analyses, we could show that, while residue X^2 has almost no effect, residue X^1 influences the reaction rate but not the equilibrium position. Glycine as X^1 gives the fastest reaction with the highest yield, proline in this position eliminates the transpeptidation reaction. The residues X^3 affect the equilibrium position. This could be shown by reacting Dns-LPKTGGGX^3X^3-NH_2 with different H-GGGX^4X^4-NH_2 which equilibrate at different product/educt ratios.

We utilized the ligation for the synthesis of peptide nucleic acid (PNA)–cell-penetrating peptide (CPP) conjugates like **3** (Figure 1). The 18mer PNA sequence (**1**) is targeted to the aberrant splice site of a mutated β-globin intron 2, which interrupts the coding sequence of a Luciferase reporter gene [5]. The PNA residues are flanked by three oxyethylene-spacers in order to reduce their aggregation tendency and improve solubility, and C-terminally extended by the optimized LPKTGG-motif. The peptide to be attached is the well-known CPP MAP (model amphipathic peptide) with three additional N-terminal glycine residues (**2**). However, ligation of **1** and **2** in a 1/1 ratio, gave only a yield of 38%. By using a 5-fold excess of **2** and dialyzing out the small by-product **4**, a conversion of 94% was achieved. The product was readily separable by semi-preparative HPLC and was homogeneous according to analytical HPLC and MS [4]. The biological activity of PNA-MAP **3** was investigated using a splicing-correction assay. Attachment of the CPP MAP to the PNA led to an enhanced antisense activity in a dose-dependent manner, whereas the PNA alone remained ineffective in restoring the aberrant

S. Del Valle et al. (eds.), *Peptides for Youth: The Proceedings of the 20th American Peptide Symposium*, 107
DOI: 10.1007/978-0-387-73657-0_47, © Springer Science+Business Media, LLC 2009

splicing [4,6]. Using this methodology, several conjugates with different CPPs were synthesized and their ability to enter the nucleus was determined [6].

Ac–ooocctcttacctcagttacaoooLPKTGGR–NH_2 + H–GGGKLALKLALKALKAALKLA–NH_2

1 **2**

↓ SrtA, dialysis

Ac–ooocctcttacctcagttacaoooLPKTGGGKLALKLALKALKAALKLA–NH_2 + H–GGR–NH_2

3 **4**

Fig. 1. Synthesis of CPP-PNA conjugates by sortase-mediated ligation.

As a second application we chose a protein ligation for the synthesis of an artificial receptor construct (Figure 2). The four ectodomains of the CRF_1-receptor, consisting of the N-terminus and three loops, are thought to be critical for ligand binding. Therefore, a water-soluble construct of those has been synthesized in order to study binding behavior. Attachment of the three extra-cellular loops to a soluble peptide template was accomplished by thiol-maleimide linkage. Sortase-mediated ligation was chosen for the N-terminus to not perturb its disulfide folding. By this strategy, a 23 kDa construct was obtained the identity of which was proven by HPLC and MS. The binding properties of the molecule were evaluated in a scintillation proximity assay. Here we could show that Urocortin displaces radioactively labeled Sauvagine with an EC50 of about 80 nM, in contrast to the receptor N-terminus which does not bind sauvagine at all.

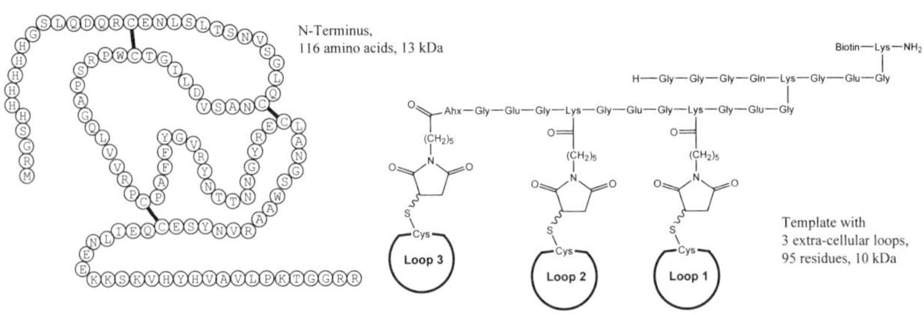

Fig. 2. The refolded receptor N-terminus with C-terminal LPKTG-motif (left) and the extra-cellular loop construct with N-terminal triglycine (right) were coupled via sortase-mediated ligation.

Acknowledgments

We thank A. Klose, G. Vogelreiter, H. Lerch, and D. Krause for excellent technical assistance. This work was supported by the European Commission (QLK3-CT-2002-01989).

References

1. Dawson, P. E. and Kent, S. B. H. *Annu. Rev. Biochem.* **69**, 923-960 (2000).
2. Mazmanian, S. K., Liu, G., Ton-That, H. and Schneewind, O. *Science* **285**, 760-763 (1999).
3. Mao, H. Y., Hart, S. A., Schink, A. and Pollok, B. A. *J. Am. Chem. Soc.* **126**, 2670-2671 (2004).
4. Pritz, S., Wolf, Y., Kraetke, O., Klose, J., Bienert, M. and Beyermann, M. *J. Org. Chem.* **72**, 3909-3912 (2007).
5. Kang, S. H., Cho, M. J. and Kole, R. *Biochemistry* **37**, 6235-6239 (1998).
6. Wolf, Y., Pritz, S., Abes, S., Bienert, M., Lebleu, B. and Oehlke, J. *Biochemistry* **45**, 14944-14954 (2006).

N-Methylated Cyclic Pentapeptides as Template Structures

Jayanta Chatterjee[1], Oded Ovadia[2], Chaim Gilon[2], Amnon Hoffman[2], Dale Mierke[3] and Horst Kessler[1]

[1]Department Chemie, Lehrstuhl II für Organische Chemie, Technische Universität München, Lichtenbergstrasse 4, Garching 85747, Germany, [2]Pharmaceutics and [2]Organic Chemistry Departments, The Hebrew University of Jerusalem, Jerusalem 91904, Israel, [3]Department of Molecular Pharmacology, Division of Biology and Medicine, Brown University, Providence, Rhode Island 02912, USA

Introduction

N-methylation of cyclic peptides can be used to modify activity and/or selectivity of biologically active peptides, and at the same time this introduces lipophilicity and drastic conformational change in the backbone of cyclic peptides, which could be employed to improve the pharmacokinetic properties of peptidic molecules. *N*-methylation introduces steric interactions and inhibits a potential hydrogen bond donor site which makes it difficult to predict the conformation of multiply *N*-methylated peptides. With the prior knowledge of the preferred conformation it would be easier to screen a large number of *N*-methylated cyclic peptides which present the pharmacophores in a desired spatial orientation and also help in refining the bioactive conformation. To understand the influence of multiple *N*-methylations on backbone conformation of cyclic pentapeptides and to search conformationally homogeneous templates, a library of 30 differentially *N*-methylated cyclic pentapeptide analogues of *cyclo*(-D-Ala-L-Ala4-) were synthesized and their conformation were studied [1]. Out of 30 *N*-methylated analogues, 15 peptides were found to strongly prefer one conformation at the NMR time scale. Based on these 15 peptides, definitive conclusions can be drawn for a correlation between the pattern of *N*-methylation and the backbone conformation.

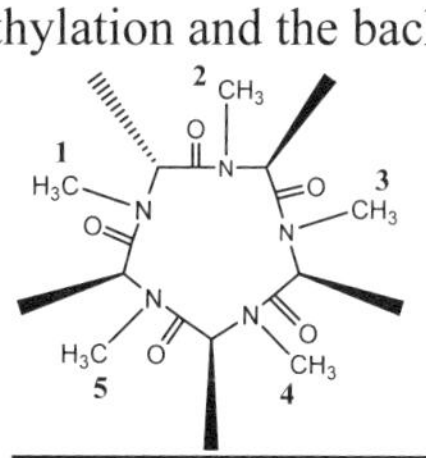

Site of *N*-	Peptide bond conformation
1	always trans-
2	always trans-
3	always cis-
4	always trans except **7** (cis)
5	trans- preferred but cis- when 3 and/or 4 *N*-

Fig. 1. The figure depicts the cyclo(-aA$_4$-) with various pattern of N-methylation and the numbers refer to the corresponding peptide bond which is N-methylated.

Results and Discussion

The 15 *N*-methylated alanine peptides, all which prefer one conformation over 80%, can nicely be analyzed [2]. As a first approach, the peptides can be grouped into five different classes (Figure 2) based on the number and position of the cis peptide bonds. It also should be noted that in these conformationally homogeneous

S. Del Valle et al. (eds.), *Peptides for Youth: The Proceedings of the 20th American Peptide Symposium*, 109
DOI: 10.1007/978-0-387-73657-0_48, © Springer Science+Business Media, LLC 2009

templates, we observe conformational dynamics (i.e., a rotation about Φ and Ψ) about peptide bonds, which are fast on the NMR shift time scale. Those processes are often found in peptides and are to be detected by non-agreement of observed and calculated distances in distinct areas.

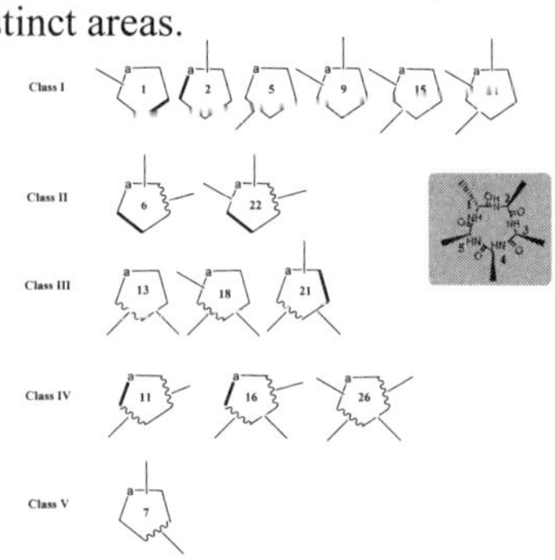

Fig. 2. The schematic representation of five different class of N-methylated cyclo(-aA4-). The wavy bonds denote the cis peptide bond and the ones in red represent the bond flipping.

As *N*-methylation was suggested to influence the permeability of peptides across the intestinal membrane, we studied the penetrating property of these peptides in Caco-2 assay. We measured the permeability rate of seven conformationally homogeneous and six conformationally inhomogeneous cyclic *N*-methylated pentaalanine peptides. Using alanine, which is devoid of functional side chain, allows us to determine the importance of conformational effects on the membrane permeability, avoiding the contribution of functionality in side chains.

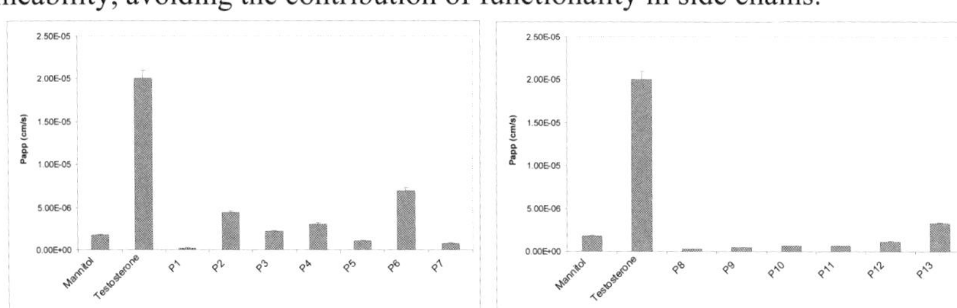

Fig. 3. The figure depicts the permeability of conformationally homogeneous peptides (left) and inhomogeneous peptides (right) in Caco-2 model.

In Figure 3 it is clearly observed that conformationally homogeneous peptides have better permeability than the inhomogeneous ones. We found out that the conformation of peptides have a strong influence on the permeability across the intestinal membrane. The results obtained suggests us that a systematic multiple *N*-methylation can enhance the membrane permeability of cyclic peptides.

Acknowledgments

We acknowledge Humboldt Foundation for the Max-Planck Forschungspreis to HK.

References

1. Chatterjee, J., Mierke, D., Kessler, H. *J. Am. Chem. Soc.* **128**, 15164-15172 (2006).
2. Chatterjee, J., Mierke, D., Kessler, H. *Chem. Eur. J.* **14**, 1508-1517 (2008).

Cyclic Dimers Of C-terminal γ_2-MSH Analogs As Selective Antagonists Of The Human Sensory Nerve-Specific Receptor (SNSR-4)

Ralf Schmidt[1], Joanne Butterworth[2], Dajan O'Donnell[3], V. Santhakumar[1] and Mirek Tomaszewski[1]

[1]Department of Medicinal Chemistry; [2]Department of Pharmacology; [3]Department of Molecular Science, AstraZeneca R&D Montreal, 7171 Frederick Banting, St.Laurent, QC, Canada H4S 1Z9

Introduction

Sensory nerve-specific G-protein coupled receptors (SNSR) have been implicated in the transmission of pain based on their unique expression and distribution in CNS and periphery. Human SNSRs are potently activated by pro-enkephalin A gene products such as BAM-22 and BAM(8-22) whereas γ_2-MSH derived sequences are the most potent activators of the rat receptor [1,2]. Both families of peptides produce a pro-nociceptive effect in various animal models of pain suggesting that a SNSR antagonist might have potential as an analgesic agent. The low homology between the human and rat receptor and widely different structure-activity relationships (SAR) of the activating ligands prompted the search for selective antagonists for either SNSR.

Starting from the most potent rat receptor agonist, the N-terminal fragment γ_2-MSH(6-12), extensive SAR studies were performed yielding linear and cyclic analogs with exclusively agonist activity at the rat receptor while being completely inactive in human SNSR based assays. The cyclization reaction on solid support to obtain c[Tyr-Arg-Trp-Asp]-Arg-Phe-Gly-OH yielded a major side product that was identified as the side chain to N-terminus cyclized dimer. Surprisingly, this compound produced the first analog activating both the rat and the human SNSR receptor with sub-micromolar potency thus providing a lead to develop antagonists.

Results and Discussion

SAR studies were initiated to convert the symmetrical dimeric agonist into an antagonist for the human SNSR receptor. Substitution of 3-iodo-Tyr in position 1/1', ring size variation and moving the basic amino acids out of the cycle finally resulted in the first chemical entity with antagonistic properties at the human SNSR (Fig. 1).

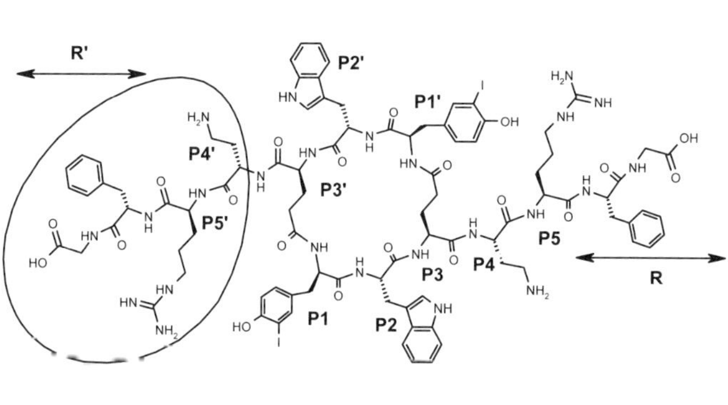

Fig. 1. Structure of γ2-MSH derived SNSR antagonists (encircled part removed in series II).

S. Del Valle et al. (eds.), *Peptides for Youth: The Proceedings of the 20th American Peptide Symposium*, 111
DOI: 10.1007/978-0-387-73657-0_49, © Springer Science+Business Media, LLC 2009

Table 1. Binding affinity and antagonist activity of ligands at the human SNSR-4 receptor.

	Peptide sequence			Binding	Antagonism
	P2/2'	P3/3'	R/R'	IC_{50} [nM]	
SI-A	Phe	Glu	NH_2/OH	12.0	9.72
SI-B	Phe	Glu	OH	7.54	22.2
SII-A	1-Nal	Glu/Glu-NH_2	Phe	253	33.4
SII-B	3-Pal/1-Nal	Glu/Glu-NH_2	Phe	475	24.9
SII-C	His/1-Nal	Glu/Glu-NH_2	Phe	351	24.0

Using the FLIPR assay with BAM8-22 as agonist, an IC_{50} value of 380nM was obtained for the original antagonist. The compound was furthermore screened for its binding selectivity versus a panel of 70 receptor assays including a variety of known pain targets revealing several interactions at a concentration of 10μM. These data made it necessary to embark on studies to improve its potency and physico-chemical properties like solubility as well as to resolve the selectivity problems.

Peptides were synthesized using Fmoc-chemistry on Rink-amide MBHA or Wang resin in a Symphony multiple peptide synthesizer. Cyclizations were performed on resin with PyBOP after side chain deprotection of Glu(Ipr). After cleavage with TFA, the crude peptides were isolated after ether precipitation and purified by preparative RP-HPLC. Homogeneity of the final products was established by analytical HPLC and identity confirmed by high-resolution MS.

Denoted as series I (SI), initial modifications of the peptide sequence were based on the symmetric core structure (Fig. 1) and included multiple positional amino acid scans, substitution of β-amino acid residues, incorporation of conformational backbone and side chain constraints and ring size variations. The guanidino group of Arg, hArg or β-hArg was absolutely essential for the interaction with the receptor regardless of their functional activity. For the antagonism, an hydrophobic core with the two Tyr(3I) and two additional aromatic residues, the D-2,4-diaminobutyric acid residues (D-Dab) and two glutamic acid residues are required. Unfortunately, it turned out that some of these pharmacophore moieties in combination with the dimeric symmetry are also responsible for the interaction with other targets.

An alternative stepwise synthetic strategy characterized by coupling of modified protected fragments starting at P3' within the sequence was developed to provide more synthetic flexibility. The application of this procedure resulted not only in symmetric dimers in high yield and purity but allowed additionally the synthesis of asymmetric dimers by substituting individual residues in the sequence and the design of asymmetrically truncated peptides as shown in series II (SII) analogs (Tab. 1). Comparison of binding and functional data showed that the removal of one exocyclic sequence lead to reduced affinity to the receptor without decreasing the antagonist potency to the same extent. More importantly the selectivity screen performed on all exemplified analogs showed high selectivity for the human SNSR receptor.

References

1. Lembo, P.M.C., *et al. Nature Neuroscience* **5**, 201-209 (2002).
2. Grazzini, E., *et al. Proc. Natl. Acad. Sci. USA* **101**, 7175-7180 (2004).

Structure-Function Relationship Study of Parathyroid Hormone (1-11) Analogues Containing D-AA

A. Caporale*[1], N. Fiori[1], E. Schievano[1], A. Wittelsberger[2], S. Mammi[1], M. Chorev[3] and E. Peggion[1]

[1]*Dept. of Chemical Sciences, Univ. of Padova, Institute of Biomolecular Chemistry, CNR, via Marzolo 1, 35131 Padova, Italy,* [2]*Dept. of Physiology, Tufts Univ. School of Medicine, 136 Harrison Avenue, Boston, MA 02111, USA;* [3]*Lab. For Translational Research, Harvard Medical School, One Kendall Square, Building 600, Cambridge, MA 02139, USA.*

Introduction

In mammals, parathyroid hormone (PTH) [1], an 84 amino acid hormone, plays a vital role in regulating the concentrations of ionized calcium and phosphate in the blood and extracellular fluids. It has been shown that the first 34 amino acid fragment of PTH is sufficient to bind to and activate the PTH type I receptor (PTH1R), a heptahelical transmembrane G protein-coupled receptor. The molecular mechanisms by which PTH(1-34) binds to and activates the PTH1R have been extensively investigated [2]. The study of miniaturized PTH agonist and antagonist analogues has been the subject of extensive research [3], for the development of safer and non-parenteral bone anabolic drugs. Recent investigations focusing on the interaction of N-terminal fragments of PTH with PTH1R showed that certain modifications can increase signalling potency in peptides as short as 11 amino acids. To understand the role of the side-chains of all amino acid residues of the most active analogue of PTH(1-11), H-Aib-Val-Aib-Glu-Ile-Gln-Leu-Nle-His-Gln-Har-NH_2 (RP), we carried out a D-amino acid scan in which every L-AA was substituted with the corresponding D-AA, obtaining a library of PTH(1-11) analogues which were tested for agonistic activity (Tab.1).

Results and Discussion

D-Homoarginine (Har) can be prepared by the guanidinylation of N-α-Fmoc-lysine with bis-Boc-triflyl guanidine, according to a reported method [4]. The fully protected homoarginine thus obtained was incorporated in the synthesis of the peptide of interest. The library was synthesized by SPPS, employing the Fmoc protocol. The conformational properties of the D-AA scan library were initially investigated by CD in (TFE)/water 20%, as in our previous experiments on potentially bioactive PTH-derived peptides [5]. The spectra in the far-UV region were collected at room temperature. The CD spectra showed a low helix-content for analogues DGlu4, DIle5, DLeu7, DNle8 and DHis9. The CD spectra of DGln10, DGln6 and DVal2 seemed to present a more ordered structure. For these analogues, the presence of some contribution of 3_{10}-helix or β-turn cannot be excluded. Finally, the spectrum of analogue DHar11 exhibits the typical shape of the α-helical conformation. NMR studies were performed on all analogues. The secondary chemical shifts of the C^{α} protons are shown in Fig.1. We observe a small effect on the structure of the peptide when a D-AA is substituted in positions 2 and 8-11. Peptide DVal2 conserves a relatively high helix-content possibly because of the influence of the $C^{\alpha,\alpha}$tetrasubstituted AAs flanking D-Val[2] (Aib in position 1 and 3) We find a less ordered structure for the analogues modified with D-Glu[4], D-Ile[5],

DOI: 10.1007/978-0-387-73657-0_50, © Springer Science+Business Media, LLC 2009

D-Leu[7], D-Nle[8] and D-His[9]. This effect is maximized for DIle5, maybe because this is the only β-branched AA in the sequence, and therefore, other then the Aib, the most hindered AA. The CD analysis results and NMR studies confirm that the D-AA substitution in the DHar11 analogue has no effect on the α-helical conformation.

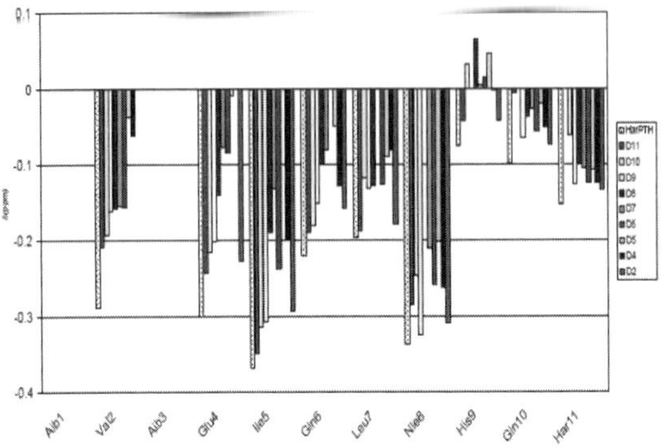

Fig.1. Secondary Chemical Shifts of PTH(1-11) analogues. $\delta\Delta = \delta$ measured - δ random coil

Table 1. results of actvity

Name	EC_{50} (M)
RP	(1.1±1.2)E-09
DVal2	(7.0±2.0)E-06
DGlu4	(2.1±1.6)E-06
DIle5	(4.1±3.0)E-07
DGln6	(4.2±1.9)E-06
DLeu7	(2.9±0.4)E-06
DNle8	(1.8±1.6)E-06
DHis9	(1.8±0.1)E-06
DGln10	(1.0±1.0)E-06
DHar11	(7.5±1.2)E-08

The results of biological tests showed that the activity of the D-Har[11] analogue is of the same order of magnitude as the most active modified PTH(1-11) analogue. This behaviour is supported by our conformational analysis, where we have observed a major α-helical structure only for DHar11. This is in agreement with previous works where a correlation between activity and helix content has been demonstrated [5, 6]. The above analysis confirms the importance of residues in the (1–9) region of PTH(1–34) for biological activity. This study also underlines the importance of the presence of the basic, guanidine group at position 11. The importance of a positively charged group in the C-terminal position is shown to be independent of the configuration of the C^{α}-carbon. The relative contribution of a positively charged residue in position 11 is amplified in the miniaturized sequences as compared to PTH(1-34) in which a much larger number of amino acid residues are involved in ligand-receptor interactions and contribute to binding responsible of different interactions with the receptor.

References

1. Kronenberg, H., Abou-Samra A, Bringhurst, F., Gardella, T., Juppner, H., and Segre, G. (1997) in *Genetics of Endocrine and Metabolic disorders* (Thakker, R., ed) pp. 389–420, Chapman & Hall, London.
2. Chorev M., Biopolymers (Pept. Sci.), 2005, **80**, 67-84.
3. Shimizu N., Guo J., Gardella T. J., J. Biol Chem., 2001, **276**, 49003-49012.
4. Feichtinger K., Zapf C., Sings H. L., Goodman, M., J Org Chem, 1998, **63**, 3804-3805.
5. Barazza, A; Wittelsberger, A; Fiori, N; Schievano, E; Mammi, S; Toniolo, C; Alexander, J. M.: Rosenblatt, M.; Peggion, E.; Chorev, M., J Pept Res 2005, **65**, 23-35.
6. Shimizu M., Carter P. C., Khatri A., Potts J. T. and Gardella T. J., Endocrinology 2001, **142**, 3068-3074.

Selective enzymatic hydrolysis of C-terminal *tert*-butyl esters of peptides

Carmen G. Boeriu[1], Ivo F. Eggen[2], Dirk-Jan van Zoelen[2] and Gilbert H. Bours[2]

[1]WUR-Agrotechnology & Food Innovations, P.O. Box 17, NL-6700 AA Wageningen, The Netherlands; [2]Organon Biosciences N.V., P.O. Box 20, NL-5340 BH, Oss, The Netherlands

Introduction

In the chemical synthesis of peptides, either a completely linear approach or a convergent approach may be followed. The latter approach, in which the final peptide is assembled from protected peptide fragments, is generally preferred for longer peptides, since overall yields are intrinsically higher and fragments may be prepared in parallel resulting in a shorter overall synthesis time. In a convergent synthesis, all fragments but the C-terminal fragment must be protected at their C-terminal with a protecting function that can be selectively cleaved. In case of peptide synthesis on a solid support, a handle on the support itself usually provides this function. However, for peptide synthesis on a manufacturing scale, solution-phase synthesis is preferred over solid-phase synthesis. Particularly preferred is a synthesis according to DioRaSSP®, a solution-phase method which combines the advantages of the solid-phase and the classical solution-phase process [1].

The C-terminal in a convergent synthesis should be stable under the conditions of the assembly of the peptide and particularly under the conditions of the deprotection of the temporary N-terminal protecting group, which is repeated each cycle of a peptide synthesis. Moreover, the N-terminal protecting group and the protecting groups of the side chains need to remain unaffected when the ester at the C-terminal is removed prior to the actual fragment coupling. On the basis of the high stability, the ease of introduction and the wide commercial availability, *tert*-butyl esters are preferred. Unfortunately, selective chemical deprotection of these esters in the presence of side-chain protecting groups is generally not applicable. Therefore, a new and mild selective method is needed for the hydrolysis of C-terminal *tert*-butyl esters of peptides.

Results and Discussion

The enzymatic deprotection of 25 synthetic model dipeptides of the general structure Z-Val-AA-OBu^t, where AA is a natural amino acid with or without side-chain protection, was studied in 50% DMF in 0.1 M phosphate buffer at pH 7.0 (Figure 1).

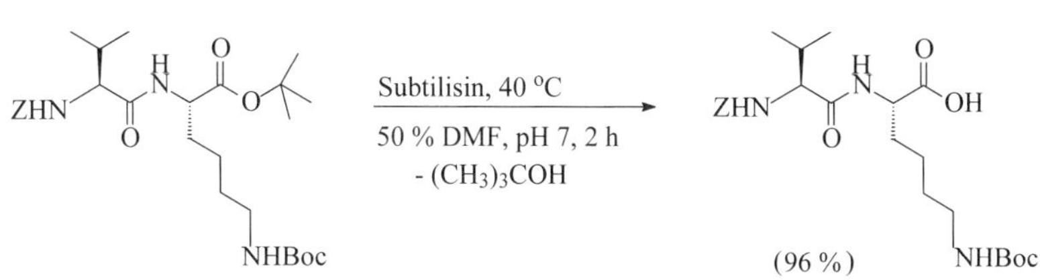

Fig. 1. Selective hydrolysis of the model dipeptide Z-Val-Lys(Boc)OBu^t.

Subtilisin from *B. licheniformis*, free or immobilized as cross-linked enzyme aggregates, was used as biocatalyst. High, up to quantitative conversion of peptide

S. Del Valle et al. (eds.), *Peptides for Youth: The Proceedings of the 20th American Peptide Symposium*, 115
DOI: 10.1007/978-0-387-73657-0_51, © Springer Science+Business Media, LLC 2009

substrates was observed for most natural amino acids (Table 1). Bulky substituents near the C^{α}-center of the C-terminal amino acid hinder the reaction.

Table 1. Reactivity of C-terminal tert-butyl protected amino acid residues.

Reactivity	Amino acid residue
Quantitative conversion (98 – 100%)	Gln, Ala, Met, Asn, His, Leu, Ser
Good conversion (50 – 98%)	Arg, Phe, Tyr, Lys(Boc), Thr
Mediocre conversion (5 – 50%)	Arg(Pbf), Glu(OBu^t), Val, Gly, Trp, Trp(Boc)
Bad conversion (1 – 5%)	Cys_2, Tyr(Bu^t), Ser(Bu^t), Asp(OBu^t)
Unresponsive (0 – 1%)	Ile, Pro, Thr(Bu^t), His(Trt)

The reaction is highly chemo- and regioselective. No hydrolysis of N-terminal and side-chain protecting groups was observed. However, some degradation of the peptide chain of longer peptides may occur due to the enzyme's endopeptidase activity. Through medium engineering, conditions were defined to enhance the ratio of esterase *vs.* endopeptidase activity. Screening a set of model tetrapeptides of the general structure Z-Val-Val-AA-Leu-OBu^t identified peptide bonds susceptible to endopeptidase scission (Chart 1).

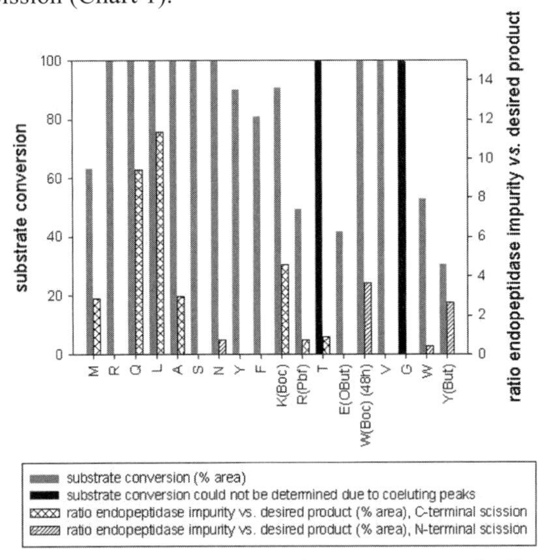

Chart 1. Esterase vs. endopeptidase activity (conditions: 50% DMF/buffer pH 7, 40 °C, 6 h).

Conclusion

A new process was developed for selective hydrolysis of C-terminal *tert*-butyl esters of peptide substrates in solution-phase synthesis, using subtilisin as biocatalyst [2]. High, up to quantitative yields were obtained for most amino acid derivatives.

References

1. Eggen, I.F. et al., *J. Pept. Sci.* ***10***, 633-641 (2005).
2. Eggen, I.F. and Boeriu, C.G. PCT/EP2007/050409 (2007).

Peptide Ligation via the In-Situ Transformation of an Amide into a Thioester at a Cysteine Residue

Toru Kawakami and Saburo Aimoto

Institute for Protein Research, Osaka University, Suita, Osaka 565-0871, Japan

Introduction

A peptide thioester is a key building block in contemporary polypeptide synthesis by ligation strategy such as the thioester method [1] and native chemical ligation [2]. During the course of our studies related to ligation methods, we found that a Cys-containing peptide was transformed into the corresponding *S*-peptide (peptide thioester) by the *N* to *S* acyl shift reaction [3]. On the other hand, in 1985, Zanotti *et al.* reported that a diketopiperazine thioester, *cyclo*(-Cys($COCH_2Ph$)-Pro-) (**1**) was formed when a dipeptide *p*-nitrophenyl (Np) ester, $PhCH_2CO$-Cys(S^tBu)-Pro-ONp (**2**), was treated under reductive aqueous conditions [4]. Thioester **1** would be formed *via* the intramolecular *N* to *S* acyl shift reaction followed by diketopiperazine formation. Based on these observations, we designed a Cys-Pro ester (CPE) autoactivating unit for peptide ligation [5]. Under neutral conditions a peptide **3** is transformed into the corresponding thioester **4**, which is ligated with a Cys-peptide **5** to give peptide **6** with diketopiperazine **7** (Fig. 1).

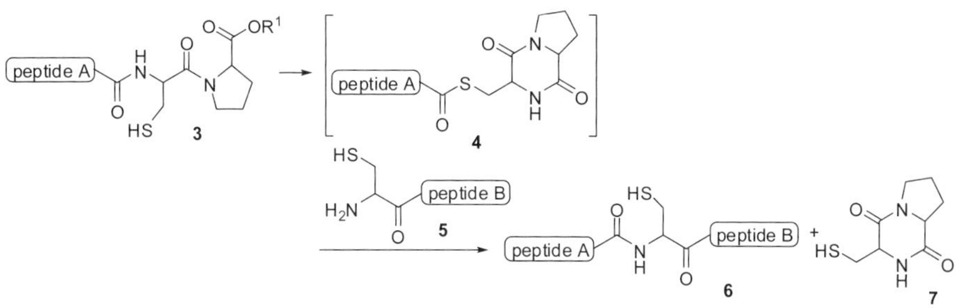

Fig. 1. CPE peptide ligation.

Results and Discussion

When Fmoc-His-Pro-Ile-Arg-Gly-Cys-Pro-OCH_2CONH_2 (**3**) was reacted with Cys-Asp-Ile-Leu-Leu-Gly-NH_2 (**5**) in a sodium bicarbonate buffer (pH 8.3) containing 10 mM tris(hydroxypropyl)phosphine (THP) and 50% acetonitrile for 24 h, Fmoc-His-Pro-Ile-Arg-Gly-Cys-Asp-Ile-Leu-Leu-Gly-NH_2 (**6**), was formed and purified by RP-HPLC in 65% yield. The Gly residue at the ligation site can be replaced by the other amino acids. For example, the reaction of Fmoc-His-Pro-Ile-Arg-Val-Cys-Pro-OCH_2CONH_2 (**3**) with **5** gave Fmoc-His-Pro-Ile-Arg-Val-Cys-Asp-Ile-Leu-Leu-Gly-NH_2 (**6**) in 68% yield.

Autoactivating function of the CPE unit is controlled by blocking the thiol [6]. Sequential ligation by the controlled CPE ligation was demonstrated by a model synthesis of Leu-Lys-Asn-Thr-Ser-Val-Leu-Gly-Ala-Ala-Cys-His-Pro-Ile-Arg-Leu-Cys-Asp-Ile-Leu-Leu-Gly-NH_2 (**13**). A blocked CPE peptide, Cys-His-Pro-Ile-Arg-Leu-Cys(4-MeOBzl)-Pro-OCH_2CONH_2 (**9**), was first ligated at the N terminus with Leu-Lys-Asn-Thr-Ser-Val-Leu-Gly-Ala-Ala-SCH_2CH_2CO-Leu-NH_2 (**8**) in sodium phosphate buffer (pH 7.2) containing 6 M guanidine and 2%

S. Del Valle et al. (eds.), *Peptides for Youth: The Proceedings of the 20th American Peptide Symposium*, 117
DOI: 10.1007/978-0-387-73657-0_52, © Springer Science+Business Media, LLC 2009

4-trimethylsilylthiophenol. Leu-Lys-Asn-Thr-Ser-Val-Leu-Gly-Ala-Ala-Cys-His-Pro-Ile-Arg-Leu-Cys(4-MeOBzl)-Pro-OCH_2CONH_2 (**11**) was isolated in 72% yield. The 4-MeOBzl group was then removed with 1 M trifluoromethanesulfonic acid in trifluoroacetic acid containing 1 M thioanisole. Peptide **12** was isolated in 90% yield. The next ligation at the C terminus was carried out with Cys-Asp-Ile-Leu-Leu-Gly-NH_2 (**10**) in a 0.1 M sodium bicarbonate buffer (pH 8.3) containing 10 mM THP and 50% acetonitrile. After 24 h, **13** was isolated by RP-HPLC in 64% yield.

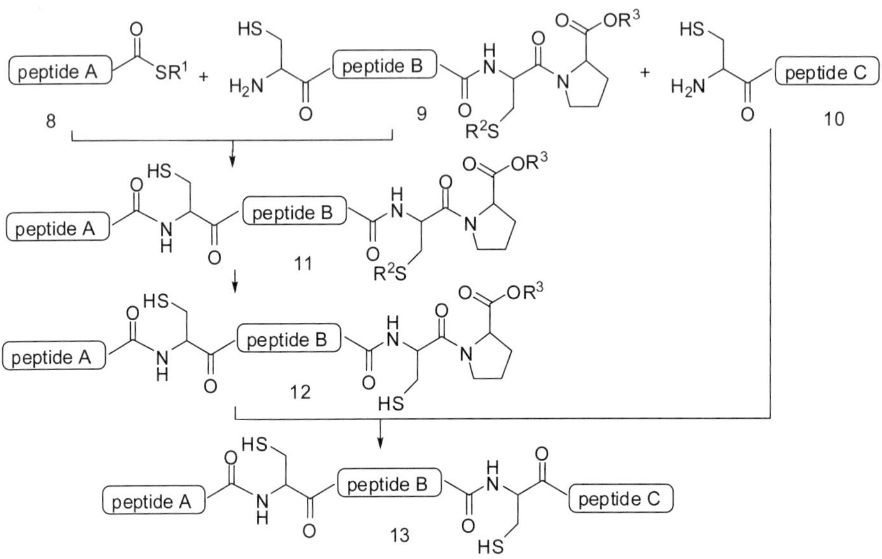

Fig. 2. Controlled CPE ligation at the N- followed by the C-terminus. R^2 = blocking group.

In conclusion, the CPE unit mediates the peptide ligation. The CPE peptide can be prepared by standard Fmoc solid phase peptide synthesis, because it does not contain a thioester moiety. The autoactivating function of the CPE unit can be quenched by introducing a blocking group to prevent inter- and intramolecular self ligation, and the direction of ligation at the N or C terminus can be controlled, thus providing a flexible ligation strategy in polypeptide synthesis using multi-component peptide building blocks.

Acknowledgments

This research was supported, in part, by Grants-in-Aid for Scientific Research from the Ministry of Education, Culture, Sports, Science and Technology, Japan.

References

1. Hojo, H., Aimoto, S. *Bull. Chem. Soc. Jpn.* **64**, 111-117 (1991).
2. Dawson, P. E., Muir, T. W., Clark-Lewis, I., Kent, S. B. H. *Science* **266**, 776-779 (1994).
3. Nakamura, K., Sumida, M., Kawakami, T., Vorherr, T., Aimoto, S., *Bull. Chem. Soc. Jpn.* **79**, 1773-1780 (2006).
4. Zanotti, G., Pinnen, F., Lucente, G. *Tetrahedron Lett.* **26**, 5481-5464 (1985).
5. Kawakami, T., Aimoto, S. *Chem. Lett.* **36**, 76-77 (2007).
6. Kawakami, T., Aimoto, S. *Tetrahedron Lett.* **48**, 1903-1905 (2007).

Facile Synthesis of Glucagon-Like Peptide-1 (GLP-1) Mimetics

Sun-Young Han[1], Martin Beinborn[2] and Jung-Mo Ahn*[1]

[1]*Department of Chemistry, University of Texas at Dallas, Richardson, TX 75080, USA;* [2]*Department of Medicine and Molecular Pharmacology Research Center, Tupper Research Institute, New England Medical Center, Boston, MA, 02111, USA*

Introduction

Small molecules have been used as a tool for inhibiting or facilitating protein-protein interactions which play crucial roles in signal transduction in living cells. Since secondary structures of proteins including an α-helix and a β-turn have been employed commonly for such interactions, they have been targeted to design suitable modulators for protein-protein interaction. While a β-turn has been studied extensively to design its mimetics due to its substantially smaller size compared to other secondary structures, only a handful non-peptide small molecules were reported to mimic the surface of an α-helix [1-5]. However, a major drawback in previously reported α-helix mimetics is inconvenient synthetic methods to prepare compounds containing various side chain functional groups found in an α-helix.

Therefore, we have focused on the development of a new scaffold for an α-helix mimetic, tris-benzamide, which can be easily produced. To demonstrate its facile synthesis and α-helix mimicry, we aimed to design peptidomimetics of a peptide hormone, glucagon-like peptide-1 (GLP-1) since small molecule agonists for its receptors are of great value to treat diabetes. As an incretin hormone, GLP-1 is known to control glucose levels by stimulating insulin secretion and inhibiting glucagon secretion through its interaction with pancreatic β-cells [6]. Structures of GLP-1 have been studied by NMR spectroscopy and found to have two α-helical regions separated by several amino acid residues called as a linker region [7]. Since the endogenous GLP-1 has extremely short half-life of approximately 1-2 min resulting from degradation by dipeptidyl peptidase IV, non-peptide small molecule GLP-1 agonists are considered as attractive therapeutic candidates for the treatment of diabetes [8,9].

GLP-1(7-36)-NH_2 HAEGTFTSDVSSYLEGQAAKEFIAWLVKGR-NH_2

Fig. 1. Sequence of glucagon-like peptide-1.

Results and Discussion

To mimic an α-helix, we have designed a tris-benzamide to assume the backbone of an α-helix that places side chain functional groups in proper orientations, and the tris-benzamide structure can hold three functionalities found at the *i*, *i+4*, and *i+7* positions in an α-helix [10].

The synthesis of GLP-1 mimetics using the tris-benzamide scaffold started from a commercially available 4-amino-3-hydroxybenzoic acid. After it was protected as an *N*-Boc methyl ester, the hydroxyl group at 3-position was alkylated to introduce a functional group to represent an amino acid side chain found in an α-helix. The *O*-alkylation reaction was carried out using NaH as a base in DMF, and the reaction with a variety of alkyl halides provided high yields. Then, an amidation reaction took place to form a bis-benzamide after hydrolysis of the methyl ester using NaOH. Among many reagents employed for the coupling reaction, we have found BOP or PyBOP most effective as well as $SOCl_2$. These steps were repeated to produce a

S. Del Valle et al. (eds.), *Peptides for Youth: The Proceedings of the 20th American Peptide Symposium*, 119
DOI: 10.1007/978-0-387-73657-0_53, © Springer Science+Business Media, LLC 2009

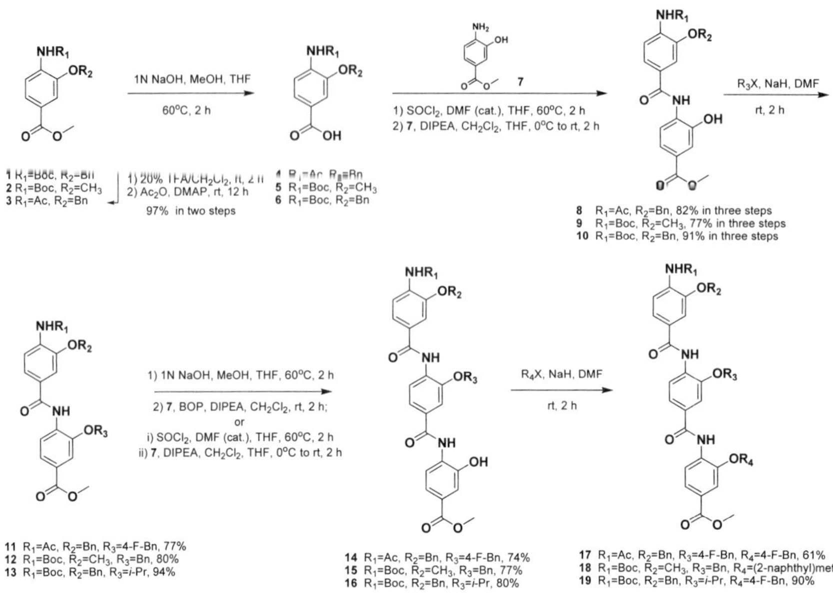

Scheme 1. Synthesis of tris-benzamides as GLP-1 mimetics.

tris-benzamide carrying three functional groups attached by ether linkages. The scheme 1 shows the synthesis of three tris-benzamides (**17-19**) as GLP-1 mimetics representing N- and C-terminal α-helical segments of GLP-1. For the synthetic convenience, 4-fluorobenzyl and 2-naphthylmethyl groups were used to represent the side chains of tyrosine and tryptophan.

Three tris-benzamides were tested with human GLP-1 receptors over-expressed on COS-7 cells and found to possess capability of stimulating cAMP production. This result clearly suggests that they are small molecule GLP-1 receptor agonists and it is noteworthy that they were discovered by *de novo* design, not from high-throughput screening of a chemical library.

Acknowledgments

The work was supported by Welch Foundation (AT-1595), American Diabetes Association (7-07-JF-02), and Texas Advanced Research Program (009741-0031-2006).

References

1. Horwell, D. C., *et al. Bioorg. Med. Chem.* **4**, 33-42 (1996).
2. Jacoby, E. *Bioorg. Med. Chem. Lett.* **12**, 891-893 (2002).
3. Orner, B. P., *et al. J. Am. Chem. Soc.* **123**, 5382-5383 (2001).
4. Ernst, J. T., *et al. Angew. Chem. Int. Ed.* **42**, 535-539 (2003).
5. Oguri, H., *et al. Tetrahedron Lett.* **46**, 2179-2183 (2005).
6. Drucker, D. J. *Diabetes Care* **26**, 2929-2940 (2003).
7. Thornton, K. and Gorenstein, D. G. *Biochemistry* **33**, 3532-3539 (1994).
8. Knudsen, L. B., *et al. Proc. Natl. Acad. Sci. USA* **104**, 937-942 (2007).
9. Chen, D., *et al. Proc. Natl. Acad. Sci. USA* **104**, 943-948 (2007).
10. Ahn, J.-M. and Han, S.-Y. *Tetrahedron Lett.* **48**, 3543-3547 (2007).

Synthetic Peptide Vaccines: The Quest to Develop Peptide Vaccines for Influenza, HIV and Alzheimer's Disease

Elisabetta Bianchi[1], Paolo Ingallinella[1], Marco Finotto[1], Joseph Joyce[2], Xiaoping Liang[2], Michael D. Miller[3], Gene G. Kinney[4], Gennaro Ciliberto[1], John W. Shiver[2] and Antonello Pessi[1]

[1]IRBM-MRL Peptide Centre of Excellence, Via Pontina Km 30.600, 00040 Pomezia (Rome) Italy; Departments of [2]Vaccine and Biologics Research, [3]Antiviral Research and [4]Alzheimer's Research, Merck Research Laboratories, Sumneytown Pike P.O. Box 4, West Point, PA, USA

Introduction

The idea that the perfect vaccine would be of total synthetic making, composed only of chemically unambiguous material, dates back to the early '80s [1,2]. However, despite considerable advances in both our understanding of the triggers of a protective immune response, and our ability to assemble increasingly complex macromolecular structures, no such vaccine has yet been approved for human use. Indeed, the initial enthusiasm has later been substituted by the prevailing belief that synthetic peptide vaccines are not possible.

In our laboratories we have been working towards this goal for a number of years, and we will highlight here some of our efforts to develop synthetic peptide vaccines for Alzheimer's disease (AD), HIV virus, and influenza virus. Our vaccines are aimed at eliciting a protective antibody response, and are based on chemically synthesized peptides corresponding to key protective B-cell epitopes, either used as such or conjugated to a protein carrier. The chosen examples cover a range of outcomes, from failure to achieve the desired result despite a convincing rationale (AD vaccine based on retro-inverso peptides), to encouraging preliminary results (HIV), to convincing preclinical data (influenza).

Results and Discussion

Retro-inverso peptides for an Alzheimer's Disease Vaccine. The key problem for a vaccine against the β-amyloid (Aβ) peptide, which is the candidate etiological agent of Alzheimer's disease (AD), is to elicit antibody formation in the absence of a T-cell response. This is because a recent clinical trial of full-length Aβ(1-42), which was showing some indication of plaque clearance and cognitive benefit, was halted when some patients experienced symptoms of meningoencephalitis, an adverse event which is widely assumed to be due to Aβ mediated T-cell activation. We reasoned that the use of retro-inverso (RI) Aβ peptides was particularly attractive in this case.

In a RI peptide, the direction of the sequence is reversed, and the chirality of each aa is inverted, resulting in inversion of all peptide bonds [3]. When used as a vaccine, a RI peptide can maintain the topology of the side chains binding to the T-cell receptor, but this requires the peptide backbone to be oriented in the opposite direction to parent peptide, with inversion of the position of all the hydrogen bond donors/acceptors to the MHC molecule: this will result in poor T-cell cross-recognition [4]. Moreover, successfully mimicry of an epitope by an RI peptide is more likely for epitopes in random coil, loop, or cyclic conformations, where the overall topology of the side-chains can be better preserved, and this is indeed the case for the N-terminal region of Aβ(1-42), which is exposed and devoid of secondary structure [5,6].

S. Del Valle et al. (eds.), *Peptides for Youth: The Proceedings of the 20th American Peptide Symposium*, 121
DOI: 10.1007/978-0-387-73657-0_54, © Springer Science+Business Media, LLC 2009

Despite this premises a systematic investigation, carried out by immunizing guinea pigs and rhesus macaques with numerous overlapping RI peptide-carrier conjugates, spanning the whole Aβ(1-40) sequence, failed to reveal cross-reactivity with the native Aβ(1-40) sequence, despite all the conjugates being able to elicit a high-titer self-reactive antibody response.

HIV vaccine. A number of HIV vaccines aimed at eliciting a vigorous cellular immunity are making their way in clinical trials. If successful, these vaccines will allow immune control of the infection, but may not prevent it, and stimulation of humoral immunity remains a key goal for an optimal vaccine. Unfortunately, despite constant progress in our understanding of how the virus escapes neutralization [7], no immunogen exists, which can elicit a broadly neutralizing antibody response.

We focused our efforts on the most conserved part of the HIV envelope glycoprotein, gp41. gp41 undergoes major conformational changes during fusion, culminating in the formation of a 6-helix bundle, where 3 α-helices of the heptad repeat region 2 (HR2) pack in antiparallel manner against a central trimeric coiled coil formed by the heptad repeat region 1 (HR1). During this process, HR1 and HR2 are transiently exposed. HR1 is highly conserved, and is the target of enfuvirtide, an HR2-derived peptide used in the treatment of therapy-experienced HIV patients. To validate HR1 as a vaccine target, we used chimeric peptides consisting of a designed trimeric coiled coil scaffold (IZ) fused to a portion of HR1, as selectors from a phage antibody library, and identified a human antibody, D5, which blocks viral infection in vivo with an enfuvirtide-like mechanism [8,9]. It followed that the mimetics used in the selection were at least partially mimicking the true structures transiently exposed by the virus.

Although these peptides were a promising starting point, they induced negligible amounts of neutralizing antibodies when used as immunogens. Improvements in the design include: (i) Covalent stabilization of the designed coiled coil, by interchain disulfide bonds. The resulting molecule, $(CCIZN17)_3$, which remaines trimeric at subnanomolar concentration, is an extremely potent inhibitor of viral infection [10]; (ii) A thioether-based synthetic route, which allows the presence in the covalent trimer of an extra thiol for conjugation to a protein carrier; (iii) Suppression of the immune response to the scaffold moiety, while leaving intact the immunogenicity of the fused HIV region, by attachment at suitable locations of small molecular weight polyethylene glycol (SMWPEG) chains.

The ability of these improved vaccines to generate neutralizing antibodies is being tested in rodents and primates.

A Universal Influenza vaccine. The efficacy of conventional influenza vaccines depends primarily on the degree of antigenic "match" between the strains used for vaccine preparation and those circulating in the population. There are three types/subtypes of influenza viruses currently circulating in the human population (influenza A, subtypes H3 and H1, and influenza B) and the vaccine must include all three components. Moreover due to continuous viral evolution in the face of immune pressure, the vaccine formulations need to be evaluated on a yearly basis and accordingly, vaccination must be performed annually. A universal influenza vaccine, effective against all circulating strains of influenza A and B, and not requiring continuous manufacturing update, would thus meet a major medical need [11].

Our approach for influenza A is based on the highly conserved, 24-amino acid extracellular domain of the M2 protein which, unlike the other two viral membrane proteins, the Hemagglutinin (HA) and the neuraminidase (NA), is highly conserved within all human influenza virus A strains [12]. When we conjugated M2 peptides to the outer membrane protein complex of *Neisseria meningitidis* (OMPC), a carrier protein widely used in human vaccines, these were highly immunogenic in mice, ferrets, and monkeys, and the immune response was protective against lethal challenge in mice [12].

The same approach however, could not be applied to influenza B, for lack of an equivalent of the extracellular domain of M2 [13], and we turned our attention to the HA. Mature HA is has two subunits, HA_1 and HA_2, which are cleaved from their precursor HA_0. Upon cleavage, the newly formed N-terminal region of HA_2, which hosts the fusion peptide, mediates fusion of the viral and cellular membranes. Therefore, cleavage of HA_0 into HA_1-HA_2 activates virus infectivity and is crucial to pathogenicity [14]. The cleavage site forms an extended, highly exposed surface loop, which is highly conserved in all influenza B viruses.

Based on the available crystal structure of the influenza A HA_0, and assuming a similar structure for the influenza B HA_0, we designed a peptide including all the cleavage site residues which are exposed to solvent, and conjugated it to OMPC [15]. The influenza B vaccine elicited a protective immune response against lethal challenge with all representative influenza B lineages [15]. The HA_0 strategy could also be applied to influenza A, although the conjugates were less efficacious against this virus.

Overall, the examples shown highlight that many obstacles still need to be overcome to realize the ultimate goal of a synthetic peptide vaccine, but also that progress is been constantly made, so that the objective may indeed be achievable.

References

1. Meloen, R.H., Langeveld, J.P.M., Schaaper, W.M.M., and Slootstra, J.W., *Biologicals* 29:233–236 (2001).
2. Zauner, W., Lingnau, K., Mattner, F., von Gabain, A:, and Buschle, M., *Biol. Chem.* 382:581–595 (2001).
3. Chorev, M., and Goodman, M., *Acc. Chem. Res*. 26:266-273 (1993).
4. Apostolopoulos, V., and Lazoura., E., Expert Rev. *Vaccines* 3:151-162 (2004).
5. Morimoto, A., Irie, K., Murakami, K., Masuda, Y., Ohigashi, H., Nagao, M., *et al*., *J. Biol. Chem*. 279:52781-52788 (2004).
6. Urbanc, B., Cruz, L., Yun, S., Buldyrev, S.V., Bitan, G., Teplow, D.B., and Stanley, H.E., *Proc. Natl. Acad. Sci. U.S.A.* 101:17345-17350 (2004).
7. Burton, D.R., Stanfield, R.L., and Wilson, I.A., *Proc. Natl. Acad. Sci. U.S.A.* 102:14943-14948 (2005).
8. Miller, M.D., Geleziunas, R., Bianchi, E. Lennard, S., Hrin, R., Zhang, H., et al., *Proc. Natl. Acad. Sci. U.S.A.*, 102:14759-14764 (2005).
9. Luftig, M., Mattu, M., Di Giovine, P., Geleziunas, R., Hrin, R., Barbato, G., et al., *Nature Struct. Mol. Biol.*, 13: 740-747 (2006).
10. Bianchi, E., Finotto, M., Ingallinella, P., Hrin, R., Carella, A. V., Hou, X.. S., et al., *Proc. Natl. Acad. Sci. U.S.A.*, 102:12903-12908 (2005).
11. Webby, R. J., and Webster, R.G., *Science* 302:1519-1522 (2003).
12. Fan, J., Liang, X., Horton, M.S., Perry, H.C., Citron, M.P., Heidecker, G.J., et al., *Vaccine* 22:2993-3003 (2004).
13. Kemble, G., and Greenberg, H., *Vaccine* 21:1789-1795 (2003).
14. Steinhauer, D. A., *Virology* 258:1-20 (1999).
15. Bianchi, E., Liang, X., Ingallinella, P., Finotto, M., Chastain, M., Fan, J., *et al*., *J. Virol.* 79: 7380-7388 (2005).

Rational Design of Peptidomimetics for Class B GPCRs: Potent Non-Peptide GLP-1 Receptor Agonists

Jung-Mo Ahn*[1], Sun-Young Han[1], Eunice Murage[1] and Martin Beinborn[2]

[1]*Department of Chemistry, University of Texas at Dallas, Richardson, TX 75080, USA;* [2]*Department of Medicine and Molecular Pharmacology Research Center, Tupper Research Institute, New England Medical Center, Boston, MA, 02111, USA*

Introduction

The family of class B G-protein coupled receptors (GPCRs) includes cognate receptors for peptide hormones like secretin, glucagon, glucagon-like peptide-1 (GLP-1), vasoactive intestinal peptide (VIP), parathyroid hormone (PTH), pituitary adenylate cyclase activating peptide (PACAP), gastric inhibitory polypeptide (GIP), corticotropin-releasing factor (CRF), and so on. Whereas class A GPCRs recognize much smaller ligands, such as dopamine, somatostatin, and enkephalins, by mostly using residues in transmembrane domains and extracellular loops proximal to membrane, all class B GPCRs possess significantly long N-terminal chains and large extracellular loops which constitute multiple binding pockets to host their large ligands [1,2].

All of these class B peptides play unique and critical functions in human physiology and are found to be attractive to treat many diseases [2]. For instance, glucagon, GLP-1 and GIP are involved in glucose homeostasis and are potential novel therapies for type 2 diabetes as well as obesity [3]. Calcitonin and PTH regulate bone turnover and calcium homeostasis and are effective to treat osteoporosis [4]. CRF is a principle mediator of the body's response to stress and have the potential to treat anxiety and depression [5].

Despite their high promises, all of them have difficulty to be used *in vivo* since they are susceptible to rapid enzymatic degradation and not orally available. Whereas ligand screening approach and *de novo* design have led to the identification of synthetic small molecules (agonists and antagonists) for many class A GPCRs, this strategy has proven much less successful when applied to class B receptors. In fact, most of non-peptide ligands identified by high-throughput screening for class B GPCRs show no functional activity, thereby classified as antagonists [2].

Structures of peptide ligands for class B GPCRs have been investigated by X-ray crystallography and NMR spectroscopy, and many of them are found to adopt highly helical structures. In addition, physiological significance of α-helices in the peptides is also demonstrated by biophysical studies undertaken in the presence of receptors [6]. Thus, choosing a different approach than conventional screening campaigns, we have developed α-helix mimetics targeting class B GPCRs, especially GLP-1 receptor.

Results and Discussion

To mimic α-helical segments found in GLP-1, we have first designed a novel α-helix mimetic using a tris-benzamide as a rigid scaffold (Figure 1B). Substitution on the tris-benzamide structure allows placement of three functional groups corresponding to the side chains of amino acids found at the *i*, *i+4*, and *i+7* positions of an α-helix, representing one helical face. To demonstrate its α-helix

S. Del Valle et al. (eds.), *Peptides for Youth: The Proceedings of the 20th American Peptide Symposium*, 125
DOI: 10.1007/978-0-387-73657-0_55, © Springer Science+Business Media, LLC 2009

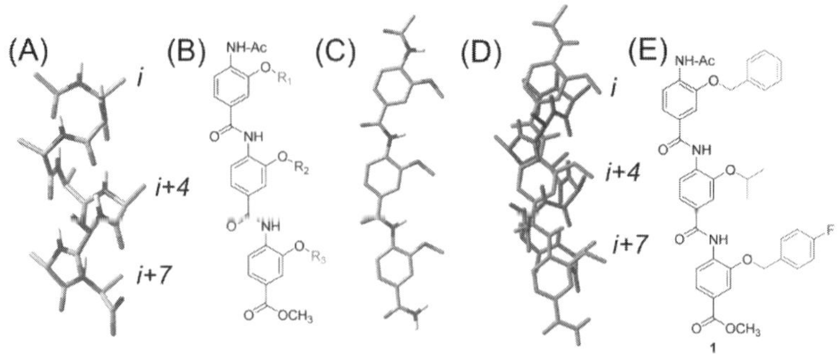

Fig. 1. Structures of α-helix mimetics. (A) ideal α-helix, (B) tris-benzamide, (C) energy-minimized structure of a tris-benzamide, (D) superimposed structure of tris-benzamide with an α-helix, (E) GLP-1 peptidomimetic.

mimicry, computer modeling was carried out using MacroModel, and a Monte Carlo conformational search was conducted using a MM3 force field implemented into the software. The energy-minimized structure of the lowest energy conformation showed that all three functional groups in the tris-benzamide mimetic are found to be well overlaid over the corresponding side chains of an ideal α-helix (Figure 1D).

Employing the tris-benzamide structure, we have designed an α-helix mimetic (**1**) representing Phe^{12}, Val^{16}, and Tyr^{19}, based on the α-helical segment identified in the NMR structure of GLP-1 [7]. To validate its α-helix mimicry, we have also substituted it for the corresponding peptide segment in GLP-1 (residues 12-21), resulting in a helix mimetic-containing GLP-1 analogue (**2**).

The synthesized compounds (α-helix mimetic (**1**), α-helix mimetic-containing GLP-1 (**2**), and Ac-GLP-1(22-36)-NH_2 as a linear control (**3**)) were assessed by receptor-binding and cAMP production assays using COS-7 cells over-expressing human GLP-1 receptors [8]. The activity shown by the α-helix mimetic (**1**) and the helix mimetic-containing GLP-1 analogue (**2**) suggests that the designed α-helix mimetic (**1**) is well recognized by the GLP-1 receptor and found to be an orthosteric agonist. In addition, attaching a GLP-1 segment (residues 22-36) to the α-helix mimetic (**1**) further enhanced its affinity and potency significantly, validating the rationale of this approach. These compounds are the first precedent of peptidomimetics rationally designed for any class B GPCRs including the GLP-1 receptor.

Acknowledgments

The work was supported by Welch Foundation (AT-1595), American Diabetes Association (7-07-JF-02), and Texas Advanced Research Program (009741-0031-2006).

References

1. George, S. R., *et al. Nature Rev. Drug Discovery* **1**, 808-820 (2002).
2. Hoare, S. R. J. *Drug Discovery Today* **10**, 417-427 (2005).
3. Drucker, D. J. *Diabetes Care* **26**, 2929-2940 (2003).
4. Neer, R. M., *et al. N. Engl. J. Med.* **344**, 1434-1441 (2001).
5. Bale, T. L. and Vale, W. W. *Annu. Rev. Pharmacol. Toxicol.* **44**, 525-557 (2004).
6. Mesleh *et al. J. Biol. Chem.* **282**, 6338-6346 (2007).
7. Thornton, K. and Gorenstein, D. G. *Biochemistry* **33**, 3532-3539 (1994).
8. Tibaduiza, E. C., *et al. J. Biol. Chem.* **276**, 37787-37793 (2001).

The depsipeptide technique for the solid phase peptide synthesis: from stepwise assembly to segment condensation

Irene Coin, Peter Schmieder, Michael Bienert and Michael Beyermann

Leibniz-Institut für Molekulare Pharmakologie, Robert-Rössle-Str. 10, 13125 Berlin, Germany

Introduction

Depsipeptide (also called O-peptide or O-acyl isopeptide) units inserted into peptides have proven to interfere very efficiently with peptide folding and association, and are therefore used to suppress aggregation during synthesis of difficult sequences, as switch elements in conformational studies and for the development of prodrugs [1] (See Figure 1).

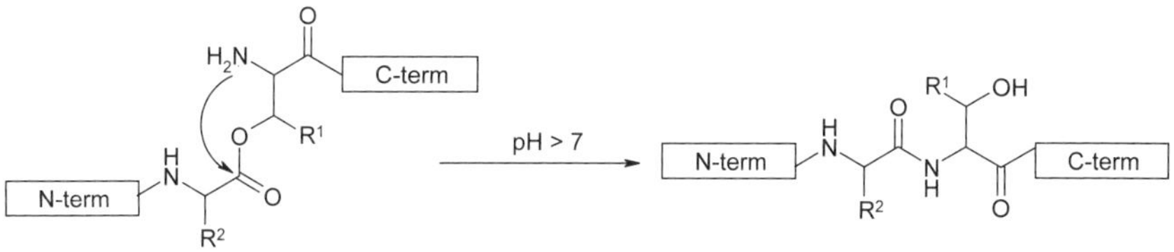

Fig. 1. Depsipeptides are converted into the all-amide form through an O,N-acyl shift, which occurs quantitatively under mildly basic conditions over a short period of time. R^1: H/CH_3 (Ser/Thr).

The most convenient way of introducing a depsipeptide unit into a sequence is the coupling of preformed depisdipeptide building blocks (Figure 2A) [2], which are easily assembled starting from commercially available amino acid derivatives using classical chemical procedures. Contrary to the case of dipeptides, depsidipeptide blocks are not likely to form optically labile oxazolones when activated for coupling, since their structure is analogous to that of urethane N-protected Ser or Thr residues (Figure 2B). Like depsidipeptide blocks, also peptide segments bearing a C-terminal depsipeptide unit (Figure 2C) are not expected to form oxazolones when activated for coupling, and should therefore provide optimal conditions for segment condensation [3].

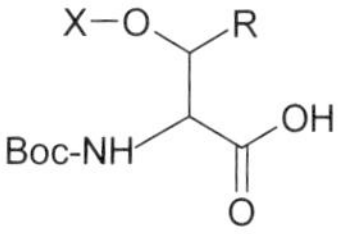

Fig. 2. A) X = protecting group; B) X = Fmoc-Xaa; C) X = polypeptide chainR = H/CH_3 (Ser/Thr).

Results and Discussion

In our work we demonstrate that coupling of depsipeptides via HBTU/DIEA activation is extremely efficient: it occurs without loss of configuration, and even much faster than the coupling of the corresponding all-amide segments.

S. Del Valle et al. (eds.), *Peptides for Youth: The Proceedings of the 20th American Peptide Symposium*, 127
DOI: 10.1007/978-0-387-73657-0_56, © Springer Science+Business Media, LLC 2009

Nevertheless, when coupling of depsipeptide segments onto solid phase bound substrates proceeds slowly, due to steric hindrance or arising of aggregation, we observed the occurrence of a serious side reaction (β-elimination) at the activated depsipeptide unit (see Figure 3).

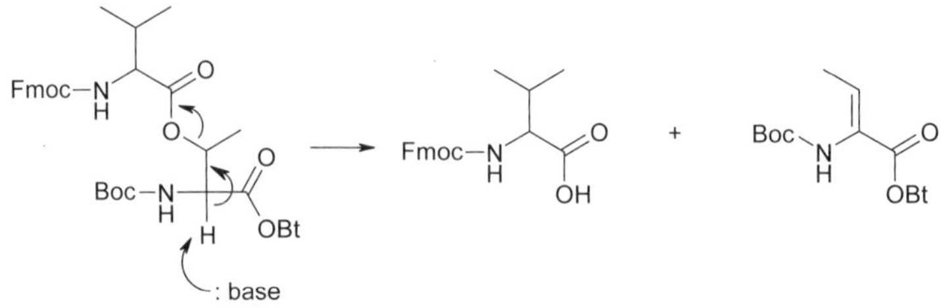

Fig. 3. β-elimination occurring at the activated depsipeptide unit.

For small peptides, β-elimination was completely suppressed by using the base-free carbodiimide/HOBt activation in non-polar solvent (DCM). Longer protected peptide segments are frequently not well soluble in DCM, so that polar media have to be used. Under these conditions, activation via carbodiimide is less efficient and the coupling occurs not fast enough to prevent the side reaction taking place. Looking for the best conditions to provide a convenient balance between efficiency of activation/coupling and β-elimination rate for depsipeptide segment condensation in polar media, we reduced either the amount and the strength of the tertiary base for HBTU activation, or the polarity of the solvent for carbodiimide activation.

For the coupling of Boc-T[Fmoc-K(Boc)LALK(Boc)LAV]-OH onto H-ALK (Boc) AALK(Boc)LA-Ram (Ram: rink amide resin) the best result was obtained by using DIC/HOBt in DMF (80% yield in 4 h coupling). Good coupling yields (~70%) and a substantial suppression of β-elimination were achieved in DMF via HBTU/collidine activation (2 h coupling) and in DMSO/toluene via DIC/HOBt (4 h coupling).

Acknowledgments

The work was funded by the Deutsche Forschungsgemeinschaft (BE 1434/5-1). We are grateful to Annerose Klose, Brigitte Schlegel and Monika Beerbaum for excellent technical assistance.

References

1. (a) Carpino, L. A. et al. *Tetrahedron Lett.* **2004**, *45*, 7519. (b) Mutter, M. et al. *Angew. Chem., Int. Ed.* **2004**, *43*, 4172. (c) Sohma, Y. et al. *Chem. Commun.* **2004**, 124.
2. (a) Coin, I. et al. *J. Org. Chem.* **2006**, *71*, 6171. (b) Sohma, Y. et al. *Tetrahedron Lett.* **2006**, *47*, 3013.
3. Yoshiya, T. et al. *Tetrahedron Lett.* **2006**, *47*, 7905.

A Novel Method for the Preparation of Peptide Thioester and its Application to Glycoprotein Synthesis

Hironobu Hojo, Yuko Onuma, Yuri Akimoto, Yuko Nakahara and Yoshiaki Nakahara

Department of Applied Biochemistry, Institute of Glycotechnology, Tokai University, Hiratsuka, Kanagawa 259-1292, Japan

Introduction

Peptide thioester has been widely used as a key intermediate for the preparation of proteins by the segment condensation methods, such as the thioester method and the native chemical ligation. The preparation of peptide thioester has been mainly carried out by the Boc method. Due to the increasing interest in the post-translational modifications, such as glycosylation, the preparation of peptide thioester by the Fmoc method, which do not use harsh acidic conditions, is desired. Recently, we have developed a novel method for peptide thioester preparation by the Fmoc strategy, in which *N*-alkyl cysteine residue at the *C*-terminus of a peptide is used as a *N* to *S* acyl transfer device as shown in Figure 1 [1]. In the presence of appropriate thiols, such as 3-mercaptopropionic acid, the peptide with *N*-alkyl cysteine at its *C*-terminus is easily converted to peptide thioester at room temperature. Thus, this method is fully compatible with conventional Fmoc method. In this paper, we applied the novel method to the preparation of a glycosylated peptide thioester, which was used for the synthesis of glycosylated CC-chemokine, CCL-27.

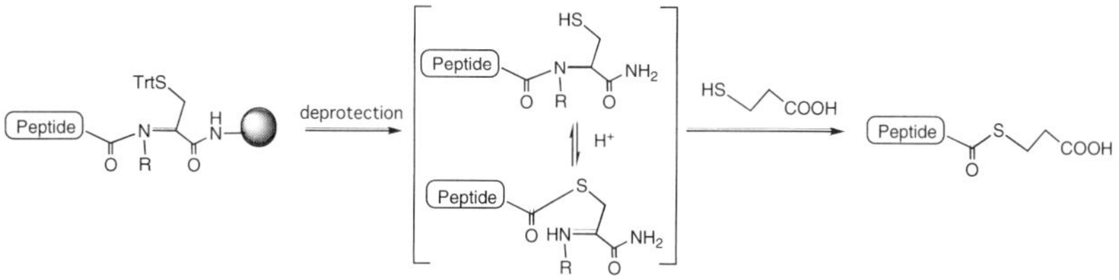

Fig. 1. Novel thioesterification reaction using N-alkyl cysteine as N- to S- acyl transfer device.

Results and Discussion

Murine CCL27 is composed of 95 AA residues: LPLPSSTSCCTQLYRQPLPSRLL RRIVHMELQEADGDCHLQAVVLHLARRSVCVHPQN*RSLARWLERQGKRL QGTVPSLNLVLQKKMYSNPQQQN (N* denotes the potential *N*-glycosylation site). The peptide chain was divided at G^{36}-D^{37} and G^{69}-K^{70}, and three segments were prepared by SPPS. The middle segment carrying *N*-linked nonasaccharide **2** was prepared as shown in Figure 2. Fmoc-*N*-Et-Cys(Trt) was coupled with CLEAR amide resin by DCC-HOBt method. Then, Fmoc-Gly was condensed by HATU in the presence of DIEA. The resin was subjected to the automated synthesis using ABI 433A peptide synthesizer by FastMoc protocol. After the sequence of (59-69) was introduced, a part of the resin was treated with 20% piperidine/NMP and acylated with Fmoc-Asn[(Gal-GlcNAc-Man)$_2$GlcNAc$_2$] by DEPBT [2, 3]. After unreacted amino group was capped by Z(2-Cl)-OSu, the remaining amino acids were introduced manually by DCC-HOBt method. After the complete assembly of CCL27(37-69), the resin was treated with Reagent K, and the crude peptide was

S. Del Valle et al. (eds.), *Peptides for Youth: The Proceedings of the 20th American Peptide Symposium*, 129
DOI: 10.1007/978-0-387-73657-0_57, © Springer Science+Business Media, LLC 2009

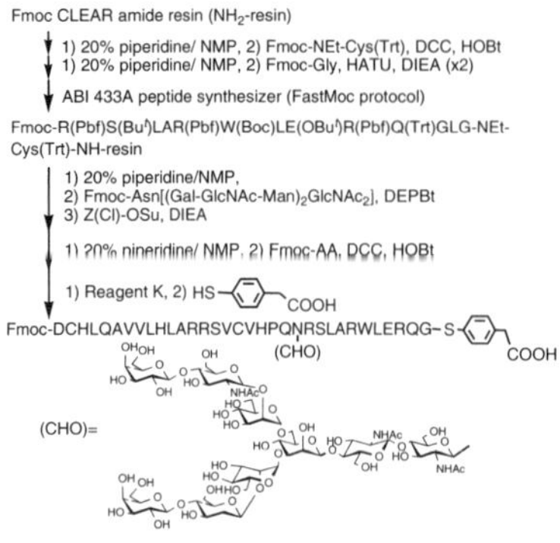

Fig. 2. Synthetic procedure for ***2****.*

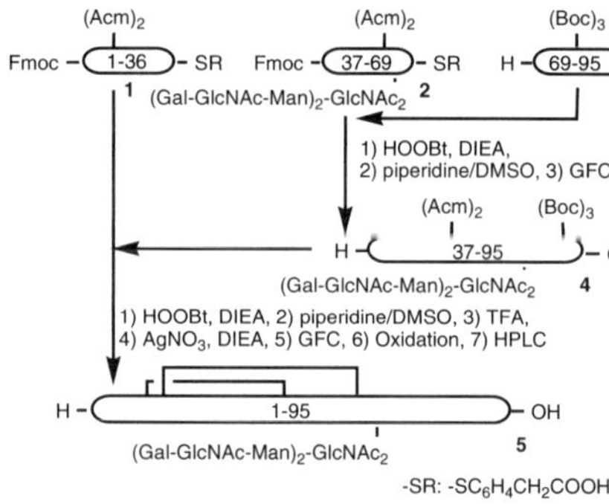

Fig. 3. Synthetic procedure for ***5****.*

dissolved in mercaptophenyl acetic acid (MPAA) solution. After overnight reaction, the peptide was converted to the thioester without significant side reactions. The product was purified by RPHPLC to obtain glycosylated peptide thioester **2**. *N*-terminal peptide thioester **1** was prepared as described above for peptide **2**. *C*-terminal peptide **3** was prepared following the previous procedure by the Fmoc strategy [4]. These segments were then condensed by the thioester method [5] as shown in Figure 3. Peptides **2** and **3** were dissolved in DMSO and the segment condensation was carried out in the presence of HOOBt and DIEA. The reaction proceeded effectively within 6 h. Piperidine was added to the reaction mixture, and the product was purified by GFC. The intermediate peptide **4** obtained was condensed with peptide thioester **1** under the same condition to give protected form of the product **5**. The peptide was successively treated with piperidine, TFA and $AgNO_3$, and purified by GFC. The reduced form of the product **5** was then oxidized in the presence of DMSO to give successfully the final product, glycosylated CCL27(1-95) **5**.

Acknowledgments

We thank Mr. Hiroaki Asai of Otsuka Chemical Company for providing Fmoc-Asn carrying nonasaccharide. We also thank Prof. Yasuhiro Kajihara of Yokohama City University for helpful discussions about the introduction of Fmoc-Asn carrying unmasked carbohydrate.

References

1. Hojo, H., Onuma, Y., Akimoto, Y., Nakahara, Y. and Nakahara *Tetrahedron Lett.* **48**, 25-28 (2007); *Tetrahedron Lett.* **48**, 1299 (2007).
2. Fan, C.-X., Hao, X.-L. and Ye, Y.-H. *Synth. Commun.* **26**, 1455-1460 (1996).
3. Kajihara, Y., Yoshihara, A., Hirano, K. and Yamamoto, N. *Carbohydr. Res.* **341**, 1333-1340 (2006).
4. Hojo, H., Haginoya, E., Matsumoto, Y., Nakahara, Y., Nabeshima, K., Toole, B. P. and Watanabe, Y. *Tetrahedron Lett.*, **44**, 2961-2964 (2003).
5. Aimoto, S. *Biopolymers (Pept. Sci.)* **51**, 247-265 (1999).

Synthesis of 4-arylmethyl-L-histidine analogues for structure-activity studies of position 10 of CGRP

D. David Smith[1], Michael Mao[2], Hayley Young[2] and Martin Hulce[2]

[1]Department of Biomedical Sciences; [2]Department of Chemistry; Creighton University, Omaha, NE 68178, USA

Introduction

Calcitonin Gene-Related Peptide (CGRP) is a potent vasodilatory, neuropeptide that is widely distributed throughout the central and peripheral nervous systems. CGRP contains 37 amino acid residues with a disulphide bridge between positions 2 and 7 that is essential for its relaxation effects[1] and a C-terminal phenylalanyl-amide that is required for binding to receptors[2].

We recently reported that N-α-benzoyl-[benzyl-His^{10}]CGRP(8-37) (**1**) is a high affinity, competitive antagonist that is selective for human CGRP receptors compared to mouse CGRP receptors.[3] In position 10 of this analogue is a derivatized histidyl residue containing a benzyl group covalently attached to the C-4 carbon of the imidazole ring. Benzylation of the imidazole ring was accomplished after treating the fully protected, resin-bound sequence of CGRP(8-37) with excess benzyl bromide under basic conditions (Figure 1). Benzylation occurs despite the presence of a benzyloxymethyl (BOM) protecting group on the π nitrogen of the imidazole ring but only in a 60% yield[4] greatly diminishing the overall synthetic yield of analogue **1**. We now report the unambiguous synthesis of Boc-His(Bn)(BOM)OH, a protected benzylated histidine that is suitable for solid-phase peptide synthesis using the Boc/benzyl protection strategy. We envisage that use of this histidyl derivative in the SPPS of analogue **1** will lead to improved overall synthetic yields and that this novel amino acid will prove useful to probe structure-activity relationships of histidyl residues in peptides.

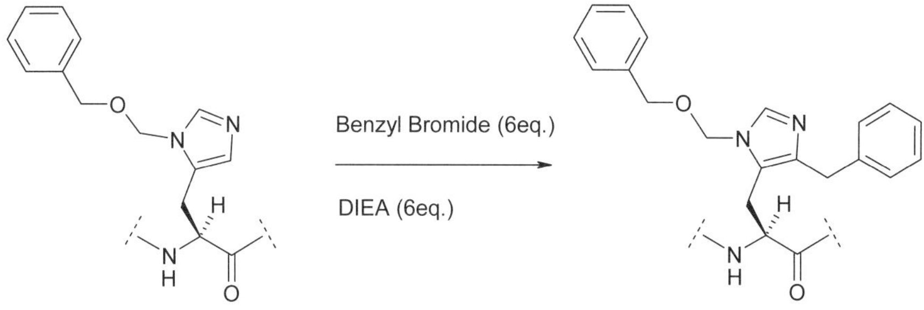

Fig. 1. Benzylation of the imidazole ring of His(BOM)^{10} of resin-bound CGRP(8-37).

Results and Discussion

4-Benzyl-L-histidine was synthesized in two steps using a previously published method[5]. L-Histidine underwent a Pictet-Spengler cyclization with benzaldehyde in alkaline aqueous methanol at reflux for 18 hours to yield benzyl carboxy spinacene. Atmospheric hydrogenolysis over palladium yielded the benzylated histidine

S. Del Valle et al. (eds.), *Peptides for Youth: The Proceedings of the 20th American Peptide Symposium*, 131
DOI: 10.1007/978-0-387-73657-0_58, © Springer Science+Business Media, LLC 2009

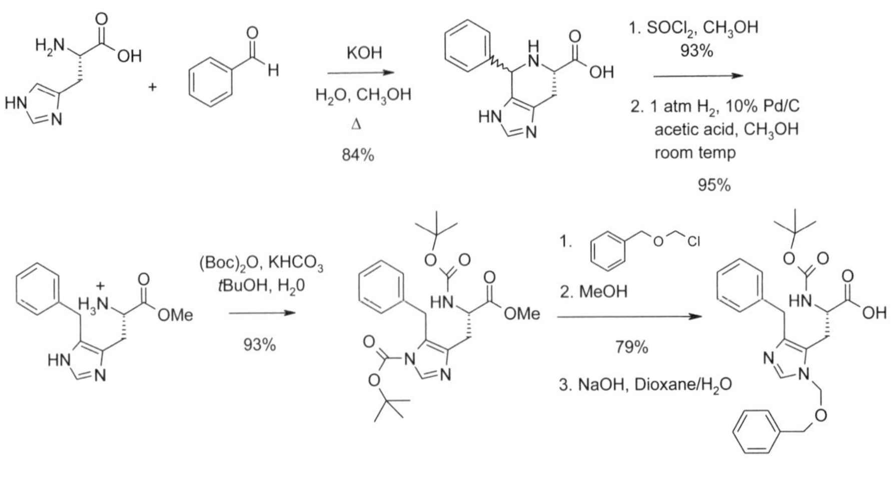

Fig. 2. Synthesis of Boc-His(Bn)(BOM)OH from L-His.

derivative in excellent yield. While conversion to the di-Boc derivative occurred without incident, no reaction was observed between this derivative and benzyloxymethyl chloride to introduce the BOM side chain protecting group.

Consequently, Boc-His(Bn)(BOM)OH was synthesized by the modified procedure outlined in Figure 2. Thus, prior to hydrogenolysis, the spinacene was converted to its methyl ester using methanolic thionyl chloride. Reaction of the resulting 4-benzyl histidine methyl ester with dibutyldicarbonate under standard conditions yielded the di-Boc derivative containing a Boc group on the τ-nitrogen of the imidazole ring. This enabled the BOM protecting group to be introduced selectively to the π-nitrogen of the imidazole ring by reaction of diBoc-4-benzyl histidine methyl ester with benzyloxymethyl chloride in dichloromethane overnight at room temperature. The imidazole bound Boc group was removed by stirring with methanol during work up to yield in an overall yield of 55% from L-histidine. Saponification[6] of the methyl ester yielded 79% of the suitably protected Boc-His(Bn)(BOM)OH for solid-phase peptide synthesis.

Acknowledgments

The work was funded by AHA-Predoctoral fellowship (CKT) and by NIH grant P20RR16469 for mass spectrometry services.

References

1. Rovero, P., Giuliani, S. and Maggi, C. A. *Peptides.* **13**, 1025-1027 (1992).
2. Smith, D. D., *et al. J. Med. Chem.* **36**, 2536-2541 (1993).
3. Taylor, C. K., *et al. J. Pharmacol. Exp. Ther.* **319**, 749-757 (2006).
4. Smith, D. D., *et al. J. Med. Chem.* **46**, 2427-2435 (2003).
5. Yutilov, Y.M., Abramyants, M.G., Smolyar, N.N. *Russ. J. Org. Chem.* **37**, 119-134 (2001).
6. Brown, T., Jones, J. H., Richards, J. D. *J. Chem Soc. Perkin Trans. I* 1553-1561 (1982).

Solid-phase synthesis of dipeptidic and pseudo-dipeptidic potential NOS inhibitors through a side-chain anchoring approach

Youness Touati-Jallabe[1], Abdallah Hamzé[1], Gilbert Bergé[1], Claudia Verna[1], Anne-Dominique Lajoix[2], Pierre Fulcrand[1], Jean Martinez[1] and Jean-François Hernandez[1]

[1]Institut des Biomolécules Max Mousseron, CNRS UMR5247; [2]Centre de Pharmacologie et Innovation dans le Diabète, CNRS UMR5232; Universités Montpellier 1 et 2; Faculté de Pharmacie 15 avenue Charles Flahault 34093 Montpellier cedex 5, France.

Introduction

Inducible NO Synthase (iNOS) is one of three isoenzymes that produce nitric oxide (NO) from arginine. Physiologically, it is mainly expressed by macrophages during an immune response and produces a cytotoxic NO. However, hyperactive iNOS has been involved in various diseases, including septic shock, chronic inflammatory diseases and the development of obesity-linked insuline resistance, which precedes the onset of type 2 diabetes. To be of high value in therapy, inhibition of iNOS has to be selective to preserve the physiological functions of the two constitutive NOS (neuronal and endothelial NOS). Search for highly selective inhibitors is hampered by the high similarity of the isoform active sites. However, Richard B. Silverman and coll. have developed dipeptides and pseudo-dipeptides containing a substrate analogue (nitro-Arg), as highly selective for neuronal NOS [1, for their last paper]. As determined by structural analyses, the basis of this selectivity is the existence of isoform-specific residues in the substrate access channel where binds the second residue [2]. Our objective is to develop selective inhibitors of iNOS by following a similar approach.

Results and Discussion

For this purpose, we combine rational design and combinatorial chemistry, which takes advantage of a solid phase strategy that we developed for the synthesis of Arg-containing peptides through a side-chain anchoring approach (Figure 1) [3] [4].

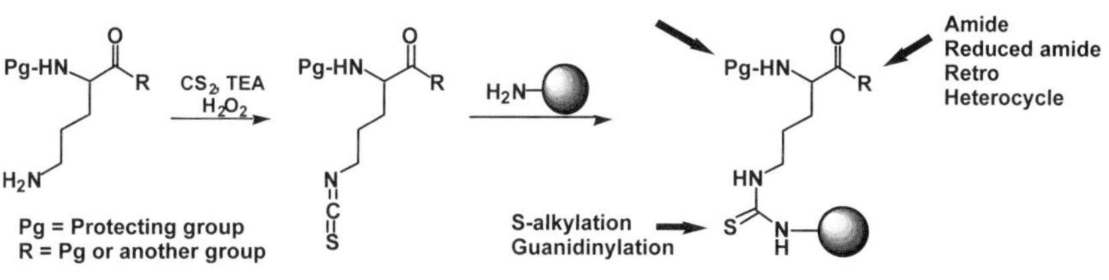

Fig. 1. Solid phase synthetic strategy through a side-chain anchoring approach. The thiocitrulline intermediate shows three points of possible variation.

Different series of compounds composed of a substrate analogue (Thiocitrulline = Tci, S-Alkyl-Isothiocitrullines, N-Alkyl-Arginines, well known but not selective NOS inhibitors) linked to a second residue through a peptide bond or a pseudo-peptide element were prepared. Figure 2 presents the synthesis of dipeptides. After saponification of the protected thiocitrulline intermediate, a protected amino acid is coupled. The obtained supported dipeptide can then be treated in different

S. Del Valle et al. (eds.), *Peptides for Youth: The Proceedings of the 20th American Peptide Symposium*, 133
DOI: 10.1007/978-0-387-73657-0_59, © Springer Science+Business Media, LLC 2009

ways: i) direct cleavage; ii) various S-alkylations and cleavage; iii) S-methylation, guanidinylation with various alkylamines and cleavage. Purity of crude products was above 85%. In this example, the protected Tci-Phe intermediate led to 8 different dipeptides.

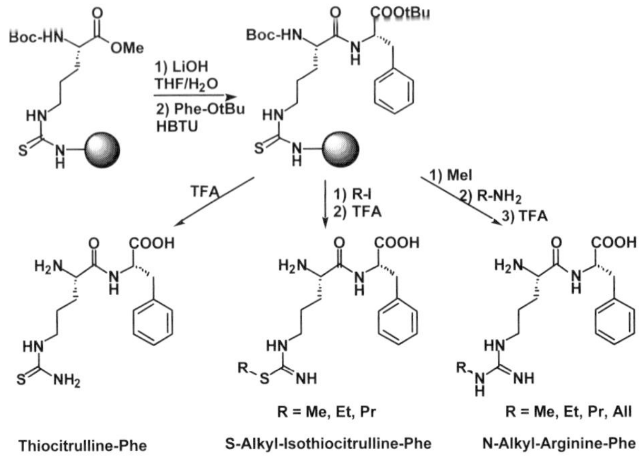

Fig. 2. Synthesis of dipeptides.

Pseudo-peptidic compounds were prepared by classical modifications of the carboxylic function of adequately protected supported Tci intermediate. In the case of reduced amide-containing molecules (Figure 3), a Weinreb amide derivative was first loaded onto the solid support. Conversion to aldehyde followed by reductive amination and cleavage gave the expected compound in satisfying yield and purity.

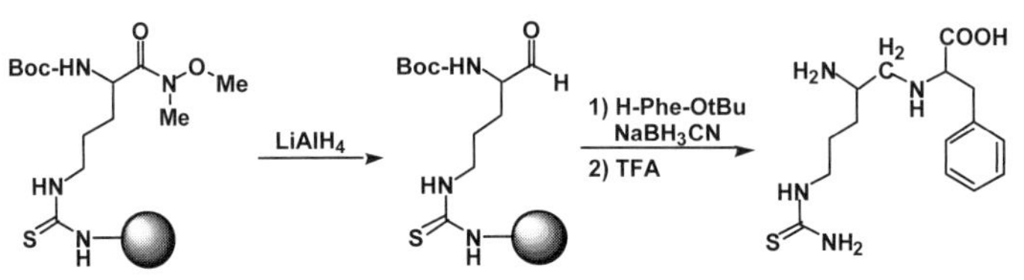

Fig. 3. Synthesis of pseudo-dipeptides with a reduced peptide bond.

The present examples show that this synthetic approach has a great potential for combinatorial chemistry of NOS inhibitors and is currently being used on Lanterns support for the preparation of libraries.

References

1. Seo, J., Igarashi, J., Li, H., Martasek, P., Roman, L.J., Poulos, T.L. and Silverman R.B. *J. Med. Chem.* **50,** 2089-2099 (2007).
2. Flinspach, M.L., Li, H., Jamal, J., Yang, W., Huang, H., Hah, J., Gomez-Vidal, J.A., Litzinger, E.A., Silverman, R.B. and Poulos, T.L. *Nature Struct. Mol. Biol.* **11,** 54-59 (2004).
3. Hernandez, J.-F., Hamze, A. and Martinez J. WO2005/068488A1.
4. Hamzé, A., Martinez, J. and Hernandez, J.-F. *J. Org. Chem.* **69,** 8394-8402 (2004).

Further Development of New Deprotection Chemistry for Cysteine and Selenocysteine Side Chain Protecting Groups

Alayne L. Schroll[1] and Robert J. Hondal[2]

[1]*Department of Chemistry and Physics, Saint Michael's College, Colchester, VT 05439, USA*
[2]*Department of Biochemistry, University of Vermont College of Medicine, B415 Given Bldg, 89 Beaumont Ave, Burlington, VT 05405, USA*

Introduction

We have previously demonstrated that electrophilic disulfides 2,2′-dithiobis (5-nitropyridine) (DTNP) and 2,2′-dithiodipyridine (DTP) could be used as deprotection reagents for the removal of *p*-methoxybenzyl (Mob) and acetamidomethyl (Acm) groups from cysteine (Cys, C) and selenocysteine (Sec, U) [1]. These protecting groups are stable to acidolysis by trifluoroacetic acid (TFA) and heretofore have required harsher removal conditions, often with sub-optimal removal of the protecting group and undesired side reactions. These electrophilic disulfides dissolve readily in TFA and provide mild removal conditions for these protecting groups. The Mob group can be easily removed from Sec using a cocktail of DTNP and TFA, while Mob and Acm can be removed from Cys with the addition of thioanisole to the cocktail [1]. Figure 1 shows the proposed mechanisms for the removal of Mob from Sec and Cys in model hexapeptides. The resulting 5-nitropyridyl derivatives can then be reduced by thiolysis. Here we report the results of using this chemistry for all commercially available Cys protecting groups as well as disulfide bond formation with these reagents.

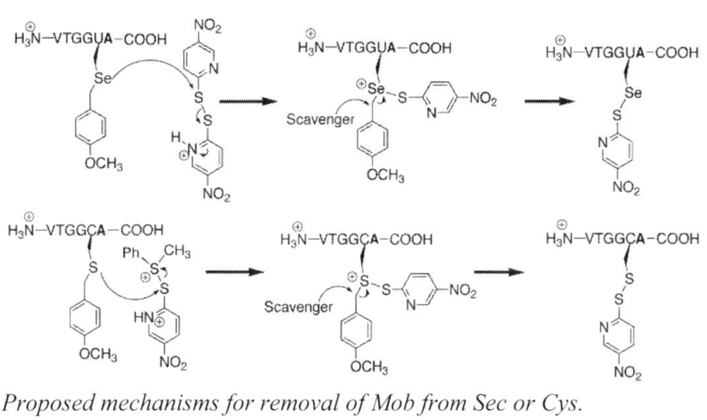

Figure 1: Proposed mechanisms for removal of Mob from Sec or Cys.

Results and Discussion

A test peptide, VTGGC(PG)A [PG=protecting group], was synthesized using standard Fmoc SPPS. The deprotection mixes contained varying amounts of DTNP dissolved in TFA with and without thioanisole (PhSMe). The reactions were carried out at 25 or 37 °C for 2 h, worked up by evaporative removal of TFA followed by extraction with water and ether. The aqueous phases were evaluated by HPLC and peak areas were used to obtain the % deprotections. Table 1 summarizes current and previous results [1]. Near quantitative removal of the *t*-butyl group (tBu) and S-*t*-butyl

S. Del Valle et al. (eds.), *Peptides for Youth: The Proceedings of the 20th American Peptide Symposium,* 135
DOI: 10.1007/978-0-387-73657-0_60, © Springer Science+Business Media, LLC 2009

group (StBu) occurred using a cocktail of 15 eq of DTNP dissolved in TFA without and with 2% thioanisole, respectively. In contrast, the benzyl (Bzl) group was not removed using this chemistry and the *p*-methylbenzyl (Meb) was removed to only a very small extent.

Table 1: % Deprotection of Cys PGs in VTGGC(PG)A

PG	%	eq DTNP	PhSMe?
Acm	>97	15	Yes
tBu	>97	15	No
StBu	>90	15	Yes
Bzl	NR	15	NA
Meb	<10	15	Yes
Mob	100	2	Yes

Table 2: % Deprotection/Cyclization of Cys/Sec in Oxytocin

Cys/Sec(PG	%	eq DTNP	PhSMe?
Cys(Acm) & Cys(Acm)	60	15	Yes
Cys(tBu) & Cys(tBu)	60	15	Yes
Cys(Mob) & Cys(Mob)	>98	6	Yes
Sec(Mob) & Cys(Mob)	>98	1	Yes
Sec(Mob) & Sec(Mob)	>98	1	Yes
Sec(Mob) & Sec(Mob)	>98	3	No

We also find this chemistry can be used for deprotection with *simultaneous* disulfide bond formation in the peptide oxytocin [CYIQNCPLG(NH_2)]. Table 2 summarizes these results and Figure 2 provides the postulated mechanism for disulfide bond formation. The disulfide bond of oxytocin was formed near quantitatively when the two Cys residues were protected with the Mob group, with less success using Cys(tBu) and Cys(Acm). Replacement of one or both Cys residues with Sec(Mob) resulted in extremely facile formation of either a selenylsulfide or diselenide oxytocin analog. These results suggest several routes to the regioselective synthesis of peptides with two or more disulfide (selenylsulfide or diselenide) bonds.

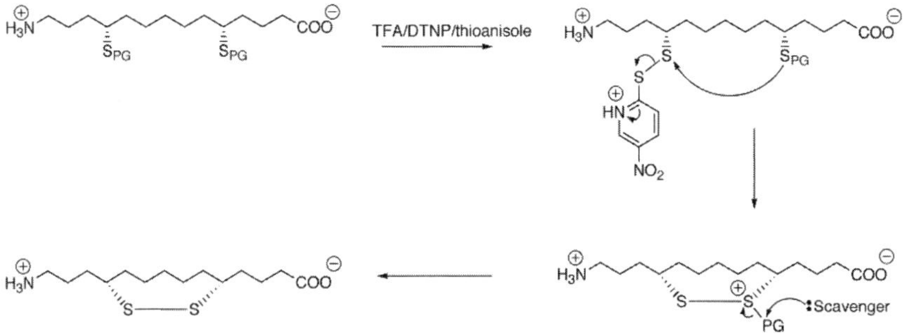

Figure 2: Proposed mechanism for disulfide bond formation.

Acknowledgments

This work was funded by NIH grant GM070742 to RJH and a NIH Vermont Genetics Network grant to ALS.

References

1. Harris, K. M., Flemer, S. Jr., and Hondal, R. J. *J. Pep. Sci.* **13**, 81-93 (2007).

A New Synthesis of (+)-Negamycin and Its Derivatives as a Potential Therapeutic Agent for Duchenne Muscular Dystrophy Treatment

Yoshio Hayashi[1,2], Shigenobu Nishiguchi[1], Magne O. Sydnes[1], Thomas Regnier[1], Jun-ya Hasegawa[3], Takahiro Katoh[3], Tetsuya Kajimoto[3], Manabu Node[3] and Yoshiaki Kiso[1]

[1]Department of Medicinal Chemistry, Kyoto Pharmaceutical University, Kyoto 607-8412, Japan; [2]Department of Medicinal Chemistry, School of Pharmacy, Tokyo University of Pharmacy and Life Sciences, Hachioji, Tokyo 192-0392, Japan; [3]Department of Pharmaceutical Manufacturing Chemistry, Kyoto Pharmaceutical University, Kyoto 607-8412, Japan

Introduction

(+)-Negamycin, a dipeptidic antibiotic containing a hydrazine peptide bond, was first isolated by Umezawa *et al.* in 1970 from culture broths of three strains closely related to *Streptomyces purpeofuscus* [1]. (+)-Negamycin, 2-[(3*R*,5*R*)-3,6-diamino-5-hydroxyhexanoyl]-1-methylhydrazinoacetic acid, exhibits low acute toxicity and strong antimicrobial activity against multiple drug resistant enteric Gram-negative bacteria including *Pseudomonas aerginosa*. Its anti-microbial activity is derived from genetic miscoding activity on bacterial ribosomal systems, thereby leading to a specific inhibition of protein biosynthesis. Because this miscoding causes read-through of termination signals in protein translation, considerable attention has been focused on negamycin as a potential therapeutic agent for genetic diseases [2]. Recently, it was reported that the aminoglycoside antibiotic, gentamicin, and the less toxic negamycin restore dystrophin expression in skeletal and cardiac muscles of *mdx* mice, an animal model of Duchenne muscular dystrophy (DMD) with a nonsense mutation in the dystrophin gene. Therefore, negamycin is regarded as a potential new therapeutic candidate for DMD and other diseases caused by nonsense mutations.

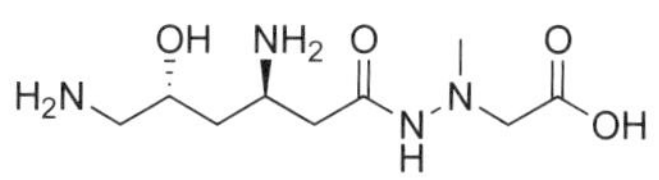

(+)-Negamycin

Although several syntheses in both racemic and optically active forms have been accomplished over three decades, medicinal chemistry has revived the synthesis of negamycin with a new synthetic route that enables molecular diversity for biological screening of the derivatives. We report here a new total synthesis of (+)-negamycin using an asymmetric Michael addition of chiral methoxybornyl-10-benzylamine as the pivotal step.

Results and Discussion

As shown in Scheme 1, commercially available (*R*)-ethyl 4-chrolo-3-hydroxybutanoate was used as a starting material and the key intermediate **8** was synthesized in seven steps in good yield (42%). With this key intermediate **8** in hands, base-catalyzed asymmetric Michael addition was performed with chiral methoxybornyl-10-benzylamine to afford desired amine adduct **9** with an excellent diastereomeric excess (>99%), followed by the regioselective cleavage of the benzyl group and chiral auxiliary with *N*-iodosuccinimide to obtain β-amino acid ester **10**.

S. Del Valle et al. (eds.), *Peptides for Youth: The Proceedings of the 20th American Peptide Symposium*, 137
DOI: 10.1007/978-0-387-73657-0_61, © Springer Science+Business Media, LLC 2009

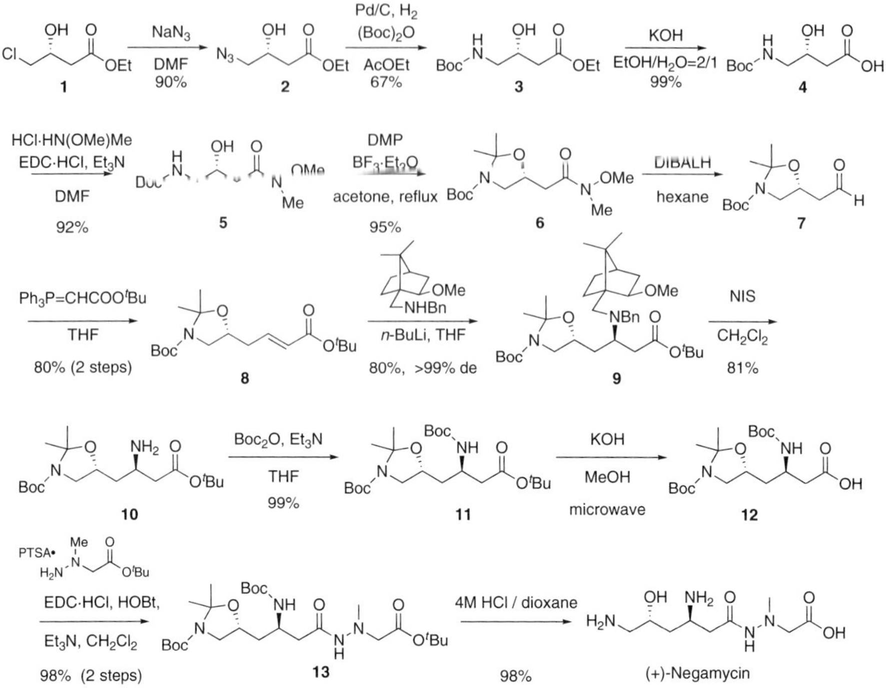

Scheme 1. Total synthesis of (+)-negamycin.

The remaining part of total synthesis consists of introducing the hydrazine unit, prior to final deprotection. After quantitative Boc-protection of **10** using standard procedure, resultant ester **11** was converted to carboxylic acid **12** by microwave-assisted saponification and this acid **12** was coupled with the hydrazine unit to obtain protected (+)-negamycin **13** in excellent yield over two steps (98%). Treatment of **13** with 4M HCl in dioxane followed by purification by ion exchange chromatography on Amberlite CG50 (NH_4^+ form) afforded the target compound with 98% yield. This compound was fully characterized and compared with the published data for the natural product to confirm the success of this new total synthesis of (+)-negamycin.

In conclusion, the synthesis of (+)-negamycin was completed in 13 steps starting from commercially available chiral ester **1** with an overall yield of 26 %. The synthesis of derivatives with potent read-through activity of termination signals in protein biosynthesis is now underway to better understand its activity against Duchenne muscular dystrophy.

References

1. Kondo, S., Shibahara, S., Takahashi, S., Maeda, K., Umezawa, H., and Ohno, M. *J. Am. Chem. Soc.*, **93**, 6305-6306 (1971).
2. Uehara, Y., Kondo, S., Umezawa, H., Suzukake, K., Hori, M. *J. Antibiot.*, **25**, 685-688 (1972).
3. Nishide, K., Ozeki, M., Kunishige, H., Shigeta, Y., Patra, P. K., Hagimoto, Y., Node M. *Angew. Chem. Int. Ed.* **42**, 4515-4517 (2003).

Design and Synthesis of Backbone Cyclic Phosphopeptides: The IκB Model

Nir Qvit[1], Ada Hatzubai[2], Deborah E. Shalev[3], Yinon Ben-Neriah[2] and Chaim Gilon[1]

[1]*Department of Organic Chemistry, The Hebrew University of Jerusalem, Jerusalem 91904, Israel.* [2]*The Lautenberg Center for Immunology, The Hebrew University-Hadassah Medical School, Jerusalem 91120.* [3]*The Wolfson Centre for Applied Structural Biology, The Hebrew University of Jerusalem, Jerusalem, 91904, Israel*

Introduction

We present the development and synthesis of a backbone cyclic (BC) phosphoserine-containing library. For our model, we used the conserved sequence derived from Inhibitor kappa B (IκB) that inhibits the interaction between IκB and the β-transduction repeat-containing protein (β-TrCp). Backbone cyclization is a method for conferring metabolic stability and enhanced bioavailability on a chosen biologically active sequence. Cycloscan is a method in which a library is designed using a single parent sequence but differing BC ring closures to allow the biologically active sequence conformational diversity. A selective and metabolically stable BC peptide may be identified from such a library [1]. Phosphopeptides are potential inhibiters of the kinome, the subset of the genome consisting of the protein kinase genes [2], but are synthetically challenging.

We applied the backbone cyclization strategy to find a peptide that inhibits nuclear factor-kappa B (NF-κB). NF-κB is a latent mammalian protein prototype of dimeric transcription factors that exists in all cell types and plays a pivotal role in activating a large number of genes such as those responsible for chronic and acute inflammatory diseases. IκB forms a stoichiometric complex with NF-κB, sequesters NF-κB away from the nucleus and interferes with DNA binding. IκB kinase phosphorylates the two serine residues in the conserved sequence of IκB, DSGXXS, allowing the β-TrCP to bind at this site and promotes the ubiquitylation and degradation of the inhibitor [3].

We show here the first case of development of a phosphopeptide-containing BC library. A BC library based on the conserved sequence was designed, synthesized and screened for IκB - β-TrCP interaction. The most active BC phosphopeptide inhibited the IκB - β-TrCP interaction at 90% at a concentration of 3 μM.

Results and Discussion

The BC peptide library was designed according to the conserved sequence of the IκB, $D^{37}S(PO3)GXXS^{32}(PO3)$. L-Phe and L-Ile were chosen from the Drosophila IκB protein, for the non-conserved “XX” residues 33 - 34 in the conserved sequence. Ser and phosphoserine were used in order to optimize the coupling and cleavage conditions. The difficult coupling of phosphoserine to the sterically hindered secondary amine of the building unit, which could not be achieved by HOBt/HBTU, was successfully performed using bis(trichloromethyl)carbonate (BTC), which has been shown to overcome difficult couplings [4].

S. Del Valle et al. (eds.), *Peptides for Youth: The Proceedings of the 20th American Peptide Symposium*, 139
DOI: 10.1007/978-0-387-73657-0_62, © Springer Science+Business Media, LLC 2009

Table 1. Structure of the BC mimetic library. A detailed description of all the peptides in the library according to ring size. n *and* m *refer to number of methylenes* Fig.1(A).

Peptide name	n	m	Ring size	MS-Cal.	MS-Obs.	Inhibition[a] (%)
1	2	2	26	965.8	965.5	62
2	2	3	27	979.8	978.2	76
3	3	2	27	979.8	979.6	90
4	3	3	28	993.8	993.7	76
5	6	2	30	1021.9	1020.7	No Data
6	6	3	31	1035.9	1034.7	No Data

[a]*Backbone cyclic phosphopeptides inhibit IκB binding to β-TrCP at a concentration of peptide 3μM.*

A 12-member backbone cyclic phosphopeptide library was designed and six peptides were synthetically feasible (Table 1 and Fig. 1 (A)). The peptides were screened for their ability to inhibit IkB binding to β-TrCP (Table 1). The most active peptide, peptide 3, showed 90% inhibition at a concentration of 3 μM. The 3D-structure of peptide 3 was solved by NMR (Fig. 1 (B)). The RMSD of the lowest energy conformations of peptide 3 is 0.7 Å (backbone) and 1.7 Å (heavy atoms).

We superimposed the backbone of peptide 3 onto the x-ray structure of the β-catenin–β-TrCP complex (PDB 1P22) [5]. Peptide 3 sterically fits into the binding pocket of β-TrCP and shows geometry conducive to hydrogen-bonding to the same residues that the β-catenin binds on its N-terminus side. The Phe residue fits into the hydrophobic pocket in the β-TrCP, occupied by Ile in the β-catenin peptide.

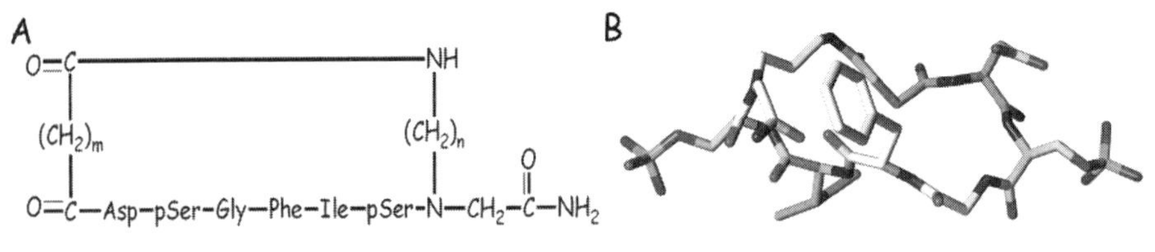

Fig. 1. (A) General structure of the phosphoserine backbone cyclic IκB mimetic library. n = *2, 3, 6;* m = *2, 3. (B) Lowest energy representative conformation of peptide 3 by NMR.*

We demonstrate the design and synthesis of a backbone cyclic library of phosphopeptides based on the IκB conserved sequence. The backbone cyclic peptides were screened for biological activity and peptide 3 was found to inhibit IkB binding to β-TrCP at a concentration of 3 μM; the 3D structure of the peptide was solved by NMR and fit the binding pocket of β-TrCP indicating that this interaction may be the basis for inhibiting β-TrCP activity.

References

1. Gilon, C. et al. (2005) PCT/IL2005.
2. Manning, G. *Science* **298**, 1912-34 (2002).
3. Ben-Neriah, Y. *Nat. Immunol.* **3**, 20-6 (2002).
4. Falb, E.; Yechezkel, T.; Salitra, Y.; Gilon, C. J. Pept. Res.**53**, 507-517 (1999).
5. Wu, G.; Xu, G.; Schulman, B. A.; Jeffrey, P. D.; Harper, J. W.; Pavletich, N. P. Mol. Cell **11**, 1445-56 (2003).

Synthesis of modified and hybrid protein derived biopolymers

Jimmy J. Weterings[1], Selina Khan[2], Jan W. Drijfhout[2], Ferry Ossendorp[2], Herman S. Overkleeft[1], Gijsbert A. van der Marel[1] and Dmitri V. Filippov[1]

[1]*Leiden Institute of Chemistry, Leiden University, P.O. Box 9502, 2300 RA Leiden, the Netherlands;* [2]*Department of Immunohematology and Bloodtransfusion, Leiden University Medical Center, P.O. Box 9600, 2300 RC Leiden, the Netherlands*

Introduction

As part of an ongoing program to synthesize well-defined synthetic polypeptide based biopolymers, natural[1] as well as modified[2], we undertook a synthesis of fluorescently labeled derivatives of the well-known[2,3] TLR-2 ligand Pam_3CysSK_4 covalently joined to a model CTL epitope (SIINFEKL). These compounds are useful tools in immunology to investigate the mechanism of action of synthetic vaccines.[2]

Results and Discussion

The synthesis is exemplified here by the preparation of a construct containing a BODIPY-FL fluorescent marker by the post-synthetic labeling of cysteine modified precursor **4a** (Scheme 1). Lipopeptide **4a** was generated via coupling of commercially available Pam_3Cys-OH (**2**, Scheme 2) to immobilized 24-mer **3a** (uneventfully prepared through standard Fmoc-SPPS using HCTU as coupling reagent). Next, treatment with the cleavage mixture TFA/TIS/H_2O/EDT gave crude **4a** which after HPLC purification over a CN-column yielded pure hydrophobic Pam_3Cys-peptide conjugate **4a** (27%). A fluorescent label was attached to the thiol function in cysteine in a 50mM phosphate buffer using an excess of BODIPY FL N-(2-aminoethyl)maleimide (Invitrogen) overnight. Purification with CN-HPLC

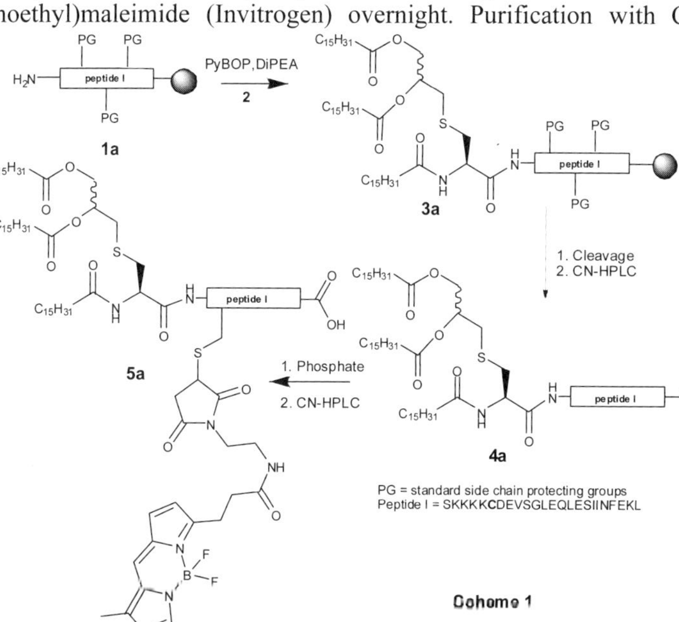

Scheme 1

S. Del Valle et al. (eds.), *Peptides for Youth: The Proceedings of the 20th American Peptide Symposium*, 141
DOI: 10.1007/978-0-387-73657-0_63, © Springer Science+Business Media, LLC 2009

finally yielded **5a** as an orange solid in 9%. CN-phase HPLC columns[4] proved to be more convenient for the purification and analysis than the standard C_{18} or C_4 based phases. The latter phases gave prohibitively long retention times for the hydrophobic lipopeptides.

Figure 1

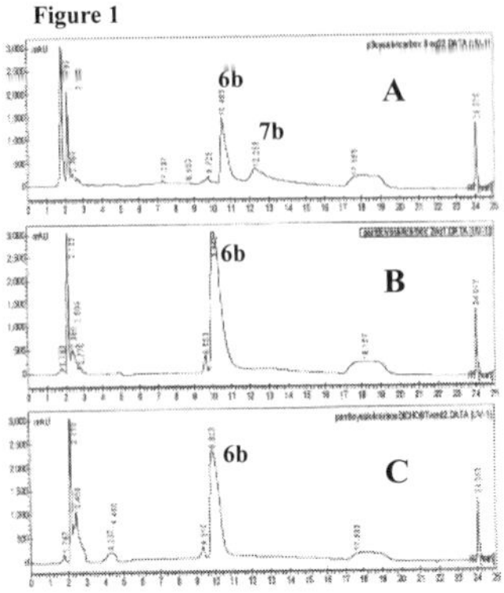

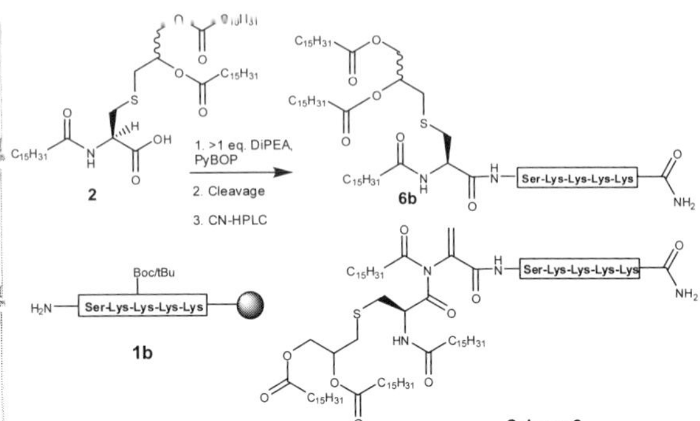

Scheme 2

In the course of our studies involving the preparation of various Pam_3Cys containing lipopeptides[2] we found that the number of equivalents of base used in the Pam_3Cys-OH (**2**) coupling step influenced the quality of the crude product. We therefore investigated the effect of the amount of base on the PyBOP assisted coupling of **2** to peptide resin using H-Ser-Lys-Lys-Lys-Lys-NH_2 as a model peptide (Scheme 2). We coupled **2** to the peptide resin **1b** under three different conditions applying decreasing amounts of base and analyzed the peptide material cleaved from the resin using LC-MS (CN-phase). When 4 eq of DiPEA (Fig. 1A), relative to **2**, were used as much as 40% of a side product (m/z 1816.8 $[M+H]^+$) was formed next to the desired **6b** (m/z 1509.4 $[M+H]^+$) (Scheme 2). This side product, with the structure **7b** tentatively assigned, was hardly detectable (Fig. 1B) if 1 eq base was applied and completely absent in base-free DIC/HOBt assisted reaction (Fig. 1C). The formation of **7b** presumably occurs via a base promoted β-elimination of the dipalmitoyloxymercaptopropane moiety from the activated cysteine building block **2** followed by acylation of the amide nitrogen by the activated Pam_3Cys-OH.

In conclusion, although commercially available Pam_3Cys-OH is a convenient building block for the preparation of tri-palmitoylated peptides, possessing interesting immunostimulatory properties, the amount of base in the coupling step should be carefully controlled.

References

1. Filippov, D. V. *et al. Tetrahedron Lett.* **47**, 9349-9352 (2006).
2. Khan, S. *et al. J. Biol. Chem.* **282**, 21145-21159 (2007)
3. Seifert, R. *et al. Biochem. J.* **267**, 795-802 (1990).
4. Phenomenex Luna 3μ CN for analysis, Phenomenex Luna 5μ CN for purification.

SPPS of the Natural Product Chitinase Inhibitor Argifin: Library Generation and Biological Evaluation

Mark J. Dixon[1], Ole A. Andersen[2], Daan M. F. van Aalten[2] and Ian M. Eggleston[1*]

[1]*Department of Pharmacy and Pharmacology, University of Bath, Claverton Down, Bath BA2 7AY, UK;* [2]*Division of Biological Chemistry and Drug Discovery, College of Life Sciences, University of Dundee, Dundee DD1 5EH, UK*

Introduction

Numerous human pathogens rely on the ability to hydrolyse chitin, a homopolymer of β(1,4)-linked *N*-acetyl-D-glucosamine (GlcNAc), at key points in their life cycles. Although chitin is absent from mammalian physiology, human chitinases have been identified and implicated in disorders such as Gaucher's disease and asthma. Consequently, chitinase inhibitors are of interest both as molecules with chemotherapeutic potential and as chemical tools to probe the role of chitinases in these diseases.[1]

The cyclic pentapeptides argifin and argadin have been identified as potent inhibitors of family 18 chitinases that offer more synthetically accessible scaffolds for rational structure-based inhibitor development than other potent inhibitors, such as the pseudotrisacharide allosamidin.[2] We have recently reported the first syntheses of both argifin[3] and argadin.[4] Our original synthesis of argifin utilised the solution phase cyclisation of a partially protected linear peptide, followed by side-chain deprotection and introduction of *N*-methyl carbamoyl (MC) modification at Arg. Although the linear assembly and cyclisation proceeded with excellent efficiency, significant side-product formation was observed during side-chain deprotection, leading to a poor yield for this step and necessitating an additional HPLC purification.

Here we report an improved, all solid-phase synthesis of argifin and demonstrate the flexibility of the synthetic strategy employed through the preparation of a library of analogues. The library was screened against a fungal chitinase from *Aspergillus fumigatus* (*Af*ChiB1) and the human acidic mammalian chitinase (*h*AMCase), which has been implicated in the onset of asthma.

Results and Discussion

Our improved argifin synthesis (Scheme 1) involves attachment to 2-chlorotrityl chloride polystyrene resin through the α-carboxylate of Asp, peptide assembly using standard Fmoc SPPS, C-terminal deprotection and cyclisation. The unusual modified Arg side-chain is then introduced through guanidination and acylation of an Orn residue before a 2-step resin cleavage/*tert*-butyl ester deprotection procedure, using a novel aqueous acidolysis. This completely eradicated the extensive aspartimide formation that was previously observed during *tert*-butyl ester removal with TFA/DCM.[3] The cyclisation step proceeded with remarkable efficiency (see inset, Scheme 1). Argifin was obtained in 21% yield following a single HPLC purification.

The design of the analogue library was based on the binding interactions observed in the *Af*ChiB1-argifin complex. Synthesis of the analogues generally proceeded smoothly, with the exception of substitution in the N-terminal position of the linear precursor, where the cyclisation completely failed. The efficiency of the

S. Del Valle et al. (eds.), *Peptides for Youth: The Proceedings of the 20th American Peptide Symposium*, 143
DOI: 10.1007/978-0-387-73657-0_64, © Springer Science+Business Media, LLC 2009

cyclisation step was found to vary significantly, and was the main reason for the observed variation in yields (3-19% overall).

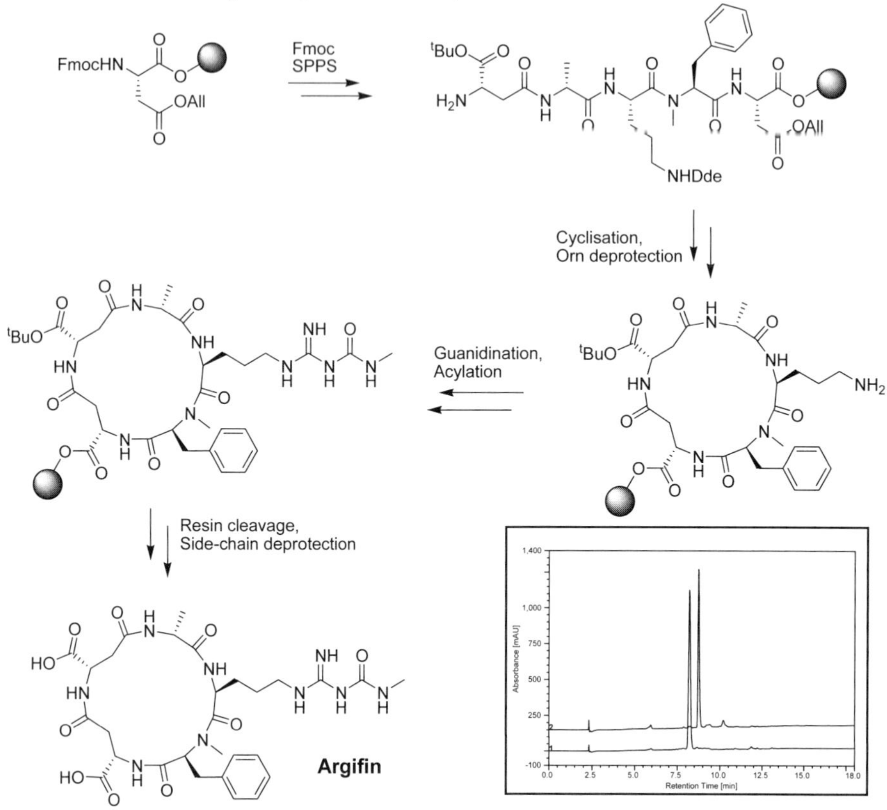

Scheme 1. Solid-phase synthesis of argifin with (inset) HPLC analysis of crude linear peptide (lower trace) and cyclised peptide (upper trace).

Screening of the library yielded extensive SAR data. The guanidino and MC moieties were found to be critical for activity, as was the *cis*-configuration of the Arg-MePhe linkage. Extension of the hydrophobic side-chain of MePhe led to a modest increase in potency, presumably through increased interaction with the hydrophobic pocket around this residue.

Acknowledgments

We thank the Wellcome Trust for funding.

References

1. Andersen, O. A., Dixon, M. J., Eggleston, I. M. and van Aalten, D. M. F. *Nat. Prod. Rep.* **22**, 563-579 (2005).
2. Berecibar, A., Grandjean, C. and Siriwardena, A. *Chem. Rev.* **99**, 779-844. (1999).
3. Dixon, M. J., Andersen, O. A., van Aalten, D. M. F. and Eggleston, I. M. *Bioorg. Med. Chem. Lett.* **15**, 4717-4721 (2005).
4. Dixon, M. J., Andersen, O. A., van Aalten, D. M. F. and Eggleston, I. M. *Eur. J. Org. Chem.* **22**, 5002-5006 (2006).

Development of Chemokine Receptor CXCR4 Antagonists Using Bio-mimetic Strategy

Hirokazu Tamamura[1], Tomohiro Tanaka[1], Hiroshi Tsutsumi[1], Nami Ohashi[1], Kenichi Hiramatsu[2], Takanobu Araki[2], Akio Ojida[3], Itaru Hamachi[3], Zixuan Wang[4], Stephen C. Peiper[4], John O. Trent[5], Satoshi Ueda[2], Shinya Oishi[2] and Nobutaka Fujii[2]

[1]*Institute of Biomaterials and Bioengineering, Tokyo Medical and Dental University, Chiyoda-ku, Tokyo 101-0062, Japan;* [2]*Graduate School of Pharmaceutical Sciences, Kyoto University, Sakyo-ku, Kyoto 606-8501, Japan;* [3]*Graduate School of Engineering, Kyoto University, Nishikyo-ku, Kyoto 615-8510, Japan;* [4]*Medical College of Georgia, Augusta, GA 30912, U.S.A.;* [5]*James Graham Brown Cancer Center, University of Louisville, Louisville, KY 40202, U.S.A.*

Introduction

G-protein-coupled receptor (GPCR) families form great drug targets in medicinal chemistry in the post-genome era. Chemokine receptors are classified into GPCR families. A chemokine receptor CXCR4 has multiple critical functions in normal and pathologic physiology. CXCR4 transduces signals of its endogenous ligand, CXCL12 (stromal cell-derived factor-1, SDF-1). The CXCL12-CXCR4 axis plays an important role in the migration of progenitors during embryologic development of the cardiovascular, hemopoietic, central nervous systems and so on. This axis has recently been proven to be involved in several problematic diseases, including HIV infection, cancer cell metastasis, leukemia cell progression, rheumatoid arthritis (RA) and pulmonary fibrosis [1]. Thus, CXCR4 is a great therapeutic target to overcome the above diseases. Fourteen-mer peptides, T140 and its analogs, were previously found to be specific CXCR4 antagonists that were identified as HIV-entry inhibitors, anti-cancer-metastatic agents, anti-chronic lymphocytic/acute lymphoblastic leukemia agents and anti-RA agents [2]. Cyclic pentapeptides, such as FC131 [*cyclo*(D-Tyr-Arg-Arg-L-3-(2-naphthyl)alanine-Gly)], were previously found as CXCR4 antagonist leads based on pharmacophores of T140 [3]. Here is described the development of low molecular weight CXCR4 antagonists based on FC131 [4].

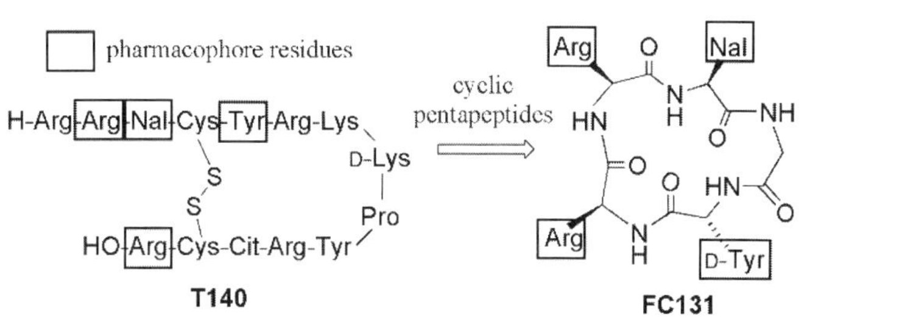

Fig. 1. Development of a cyclic pentapeptide CXCR4 antagonist FC131. Cit = L-citrulline.

Results and Discussion

Several FC131 analogues, which have substitution for Arg[2], Nal[3] and D-Tyr[5], were prepared and assessed for CXCR4-binding activity based on inhibitory activity

S. Del Valle et al. (eds.), *Peptides for Youth: The Proceedings of the 20th American Peptide Symposium,* 145
DOI: 10.1007/978-0-387-73657-0_65, © Springer Science+Business Media, LLC 2009

Table 1. Inhibitory activity of cyclic pentapeptides involving the incorporation of Phe(4-F)[1].

Peptide number	Sequence	IC_{50} (nM)[a]
1 (FC131)	*cyclo*(-Arg^1-Arg^2-Nal^3-Gly^4-D-Tyr^5-)	7.9
2	*cyclo*(-Phe(4-F)^1-Arg^2-Nal^3-Gly^4-D-Tyr^5-)	57
3	*cyclo*(-Phe(4-F)^1-Arg^2-Nal^3-Gly^4-Arg^5-)	620
4	*cyclo*(-D-Phe(4-F)^1-Arg^2-Nal^3-Gly^4-Arg^5-)	35
5	*cyclo*(-Phe(4-F)^1-Arg^2-Nal^3-Gly^4-D-Arg^5-)	88
6	*cyclo*(-D-Phe(4-F)^1-Arg^2-Nal^3-Gly^4-D-Arg^5-)	94
7	*cyclo*(-D-Tyr^1-Arg^2-Nal^3-Gly^4-Arg^5-)	300

[a] IC_{50} values are based on the inhibition of [^{125}I]-CXCL12 binding to CXCR4 transfectants of CHO cells.

against CXCL12 binding to CXCR4. SAR studies on cyclic pentapeptides having CXCR4-antagonistic activity were performed. Several analogues were synthesized to optimize side-chain functional groups, involving constrained analogues that conformationally fix the backbone and the side–chains. Taken together, Arg, Nal and D-Tyr are the most suitable at position 2, 3 and 5, respectively, than any other corresponding amino acid mimetics that were tested in the present study. Recently, a novel pharmacophore of T140-related CXCR4 antagonists, such as a 4-fluorophenyl moiety, was found in addition to the original pharmacophores of T140, Arg (x 2), Nal and Tyr. Since the phenol group of D-Tyr^5 could not be replaced by the 4-fluorophenyl group with maintenance of high activity, as seen in the D-Phe(4-F)-substituted analogue, we attempted to incorporate the 4-fluorophenyl group into the amino acid at position 1. [Phe(4-F)^1]-FC131, **2**, showed significant CXCR4-binding activity, which is less potent than that of FC131. Since another Arg residue is thought to be indispensable for high activity and an aromatic residue [L/D-Phe(4-F)] is incorporated into position 1, we tried to replace D-Tyr^5 by L/D-Arg^5. Four analogues, **3-6**, [L/D-Phe(4-F)^1, L/D-Arg^5]-FC131, were prepared and assayed. Among these compounds [D-Phe(4-F)^1, Arg^5]-FC131, **4**, showed the most potent activity, which is 10-fold more potent than that of [D-Tyr^1, Arg^5]-FC131, **7**. Thus, a novel lead involving the pharmacophores different from FC131, [D-Phe(4-F)^1, Arg^5]-FC131, **4**, was found.

Since CXCR4 is involved in several problematic diseases, low molecular weight CXCR4 antagonists such as above new leads may prove to be useful for chemotherapy of these diseases.

References

1. Tamamura, H. and Tsutsumi, H. *Chem. Biol.* **13**, 8-10 (2006).
2. Tamamura, H. and Fujii, N. *Expert Opin. Ther. Targets* **9**, 1267-1282 (2005).
3. Fujii, N., Oishi, S., Hiramatsu K., Araki, T., Ueda, S., Tamamura, H., Otaka, A., Kusano, S., Terakubo, S., Nakashima. H., Broach, J. A., Trent, J. O., Wang, Z. and Peiper, S. C. *J. Angew. Chem. Int. Ed.* **42**, 3251-3253 (2003).
4. Tamamura, H., Hiramatsu, K., Ueda, S., Wang, Z., Shuichi, K., Terakubo, S., Trent, J. O., Peiper, S. C., Yamamoto, N., Nakashima, H., Otaka, A. and Fujii, N. *J. Med. Chem.* **48**, 380-391 (2005).

Chemical Synthesis of the N-Terminal Cysteine-Rich Domain of Human OX40

Tami L. Raguse[1], Steven C Pomerantz[2], Jennifer F. Nemeth[2], Jinquan Luo[3], Audrey E. Baker[4], Jian Li[5], Diane Deriggi[6], Raymond Sweet[7] and Marian Kruszynski[4]

[1]*Protein and Peptide Synthesis;* [2]*Mass Spectrometry;* [3]*Molecular Modeling;* [4]*Biophysical Analysis;* [5]*Immunobiology;* [6]*Protein Engineering;* [7]*Molecular Discovery Technologies, Centocor Research & Development, Inc., Radnor, PA 19087, USA*

Introduction

The human OX40 receptor (hOX40) is a transmembrane protein that is part of the tumor necrosis factor receptor superfamily (TNFRSF). The extracellular region of hOX40 is composed of approximately 3.5 cysteine-rich domains (CRDs). Molecules that bind to hOX40 and modulate receptor signaling could be used to treat autoimmune diseases, inflammation, and cancer [1]. The N-terminal domain of some TNFRSF members is implicated in receptor self-association in the absence of ligand [2]. We report the synthesis of a biotinylated analogue of the 36-residue N-terminal CRD of hOX40 (**CRD-Biotin**) to evaluate the function of this domain.

Results and Discussion

Use of a glutathione redox buffer to fold and oxidize the linear, non-biotinylated, fully deprotected hOX40 N-terminal domain produced mostly a mixture of incorrect disulfide-bond isomers (data not shown). Therefore, we concluded that the isolated domain is conformationally unstable, and we chose to chemically synthesize the three disulfide bonds of **CRD-Biotin** one at a time as shown in Figure 1.

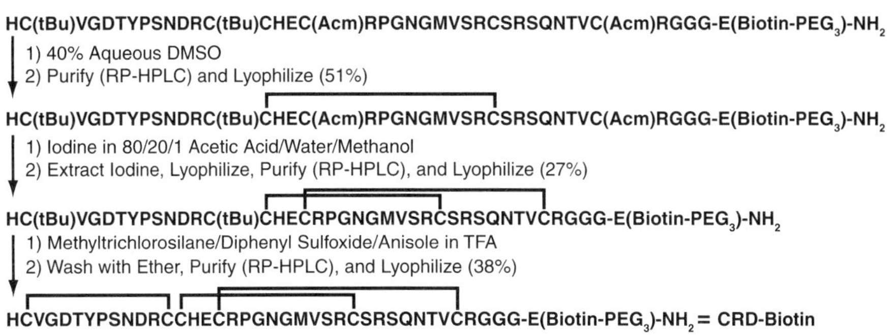

Fig. 1. Synthesis of ***CRD-Biotin****. The 36-residues of the N-terminal CRD of hOX40 are connected to E(Biotin-PEG$_3$) via a GGG spacer.*

CRD-Biotin was synthesized and cleaved using standard methods for Fmoc solid phase peptide synthesis. We used Trt, Acm, and tBu groups to protect the three pairs of cysteine side chains, because these protecting groups can be selectively removed under orthogonal conditions [3,4]. We found that proper reaction monitoring and work-up, which typically receive little attention in the reaction methodology literature, were crucial for pure product formation. Allowing reactions to proceed beyond completion leads to accumulation of side products; for example, disulfides

S. Del Valle et al. (eds.), *Peptides for Youth: The Proceedings of the 20th American Peptide Symposium*, 147
DOI: 10.1007/978-0-387-73657-0_66, © Springer Science+Business Media, LLC 2009

scramble upon extensive exposure to iodine [5] or methyltrichlorosilane/diphenyl sulfoxide [6]. Furthermore, in our hands, the rate of the methyltrichlorosilane/ diphenyl sulfoxide reaction varied inversely with scale. For these reasons, we chose to monitor the progress of each reaction using RP-HPLC and/or capillary electrophoresis. The optimal work-up after formation of the second disulfide and after formation of the final product included addition of organic and aqueous solutions and multiple organic washes.

Mass spectrometry of a trypsin/endopeptidase Glu-C digest of the purified **CRD-Biotin** final product confirmed the presence of the expected disulfide bonds (Figure 2). The P1-P2-P4, P3-P5, and P6 peptides were detected as the major species, which is consistent with formation of the correct disulfide bonds. We were able to rule out 13 of 14 incorrect disulfide-bond isomers by performing the trypsin/Glu-C double digestion (digestion with trypsin alone would have eliminated only 9 of 14 incorrect isomers).

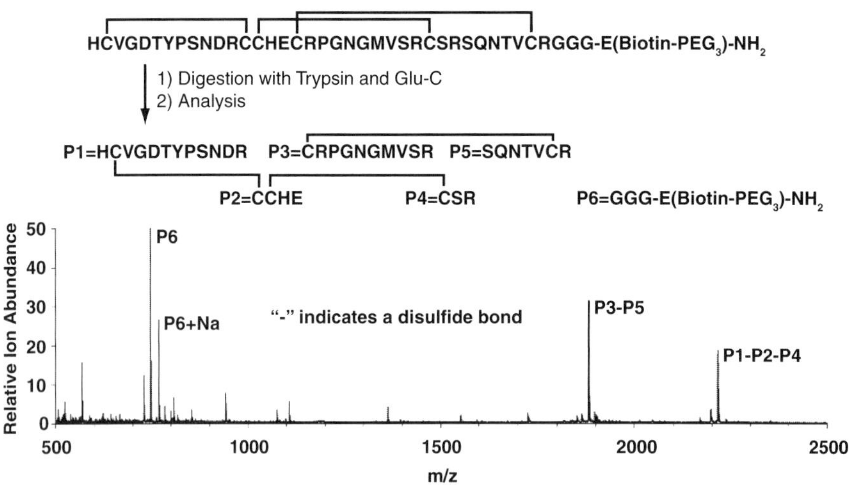

*Fig. 2. MALDI-TOF MS data for the trypsin/Glu-C double digest of **CRD-Biotin**. As expected, the P1-P2-P4, P3-P5, and P6 peptides were detected as the major species.*

We have successfully used stepwise disulfide synthesis to prepare 5 mg of the N-terminal CRD of hOX40 containing the correct disulfide bonds. **CRD-Biotin** does not self-associate at ca. 0.1 mg/mL in PBS and binds weakly and nonspecifically to hOX40-Fc, hOX40 Ligand, and a negative control antibody. The data suggest that the other domains of hOX40 may be necessary for full activity and that the exposed interdomain surface of **CRD-Biotin** leads to nonspecific binding.

References

1. Sugamura, K., Ishii, N., Weinberg, A. D. *Nat. Rev.* **4**, 420-431 (2004).
2. Bodmer, J. L., Schneider, P., Tschopp, J. *Trends Biochem. Sci.* **27**, 19-26 (2002).
3. Albericio, F., *et al.* In *Fmoc Solid Phase Peptide Synthesis: A Practical Approach* (Chan, W. C. and White, P. D., eds.) Oxford University Press, Oxford, pp. 77-114 (2000).
4. Akaji, K., Fujino, K., Tatsumi, T., Kiso, Y. *J. Am. Chem. Soc.* **115**, 11384-11392 (1993).
5. Kamber, B., *et al. Helv. Chim. Acta* **63**, 899-915 (1980).
6. Szabó, I., *et al. Biopolymers (Peptide Sci.)* **88**, 20-28 (2007).

Stereoselective Synthesis of Peptidomimetics Based on Acid-catalyzed Ring-opening of β-Aziridinyl-α,β-enoates

Hirokazu Tamamura[1], Tomohiro Tanaka[1], Hiroshi Tsutsumi[1], Koji Nemoto[2], Satoko Mizokami[2], Nami Ohashi[1], Shinya Oishi[2] and Nobutaka Fujii[2]

[1]*Institute of Biomaterials and Bioengineering, Tokyo Medical and Dental University, Chiyoda-ku, Tokyo 101-0062, Japan;* [2]*Graduate School of Pharmaceutical Sciences, Kyoto University, Sakyo-ku, Kyoto 606-8501, Japan*

Introduction

(*E*)-Alkene dipeptide isosteres (EADIs) are potential peptidomimetics, which involve backbone replacements of amide bonds in peptides. In the past several years, we have reported that ring-opening reactions of *N*-arylsulfonylaziridines bearing α,β-unsaturated esters are useful for the stereoselective synthesis of EADIs in the combinational use with organocopper (or organozinc-copper)-mediated *anti*-S_N2' reactions [1-3]. Regio- and stereo-selective ring-opening reactions of *N*-arylsulfonylaziridines bearing α,β-unsaturated esters by hard acids such as TFA, methanesulfonic acid and HCl have been reported by us. In the present study, treatment of these β-aziridinyl-α,β-enoates with alcohols, thiols or weak acids such as AcOH in the presence of Lewis acids was examined. The reactions of β-aziridinyl-α,β-enoates having no side chain group at the δ-position were also investigated. In addition, the ring-opening reactions were conducted on solid supports and also applied to the stereoselective synthesis of EADI-containing peptidomimetics.

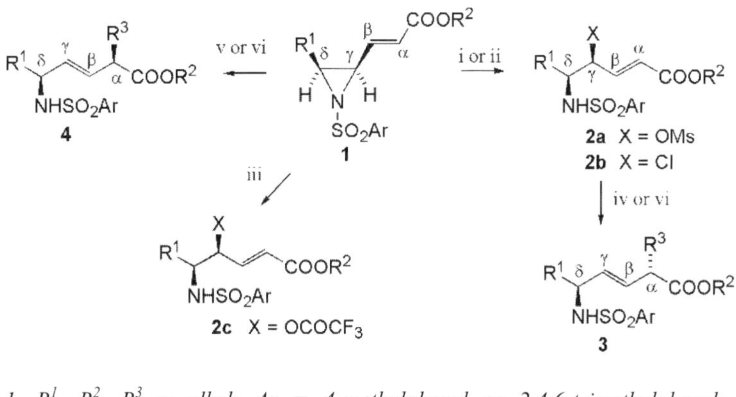

Scheme 1. R^1, R^2, R^3 = alkyl; Ar = 4-methylphenyl or 2,4,6-trimethylphenyl; Ms = methanesulfonyl; reagents: i, $MeSO_3H$ in $CHCl_3$; ii, HCl in 1,4-dioxane; iii, TFA; iv, $R^3Cu(CN)MgCl{\cdot}BF_3$; v, $R^3Cu(CN)MgCl{\cdot}2LiCl$; vi, $R^3Cu(CN)ZnI{\cdot}2LiCl$.

Results and Discussion

β-Aziridinyl-α,β-enoates, *cis*-(*E*)-enoate **5** and *trans*-(*E*)-enoate **6**, were prepared from Thr and D-*allo*threonine, respectively, as previously reported by us.[6] These β-aziridinyl-α,β-enoates **5** and **6** did not react with weak acids such as AcOH, alcohols or thiols. Thus, examined was the effect of the addition of catalytic amount of Lewis

DOI: 10.1007/978-0-387-73657-0_67, © Springer Science+Business Media, LLC 2009

acids such as TMSOTf on the ring-opening reactions with weak acids, alcohols or thiols. Treatment of **5** or **6** with AcOH, $BrCH_2COOH$, EtOH, BnSH or PhSH in the presence of catalytic amount of TMSOTf yielded the corresponding δ-aminated-γ-acyloxy (-alkoxy or -alkylthio)-α,β-enoates, **7a-e** or **8a-e**, exclusively and quantitatively, *via* the regioselective S_N2 ring-opening reaction at the γ-carbon position (Scheme 2). Regiochemical assignments for products **7a-e** and **8a-e** were readily made by ^{1}H-NMR (^{1}H-^{1}H COSY). The γ,δ-*syn* stereochemistry of **7a-e** and the γ,δ-*anti* stereochemistry of **8a-e** were based on X-ray analysis of **7a**. As a result, the addition of catalytic TMSOTf was proven to be efficient for the regio- and stereo-selective ring-opening reactions with weak acids, alcohols or thiols as nucleophiles.

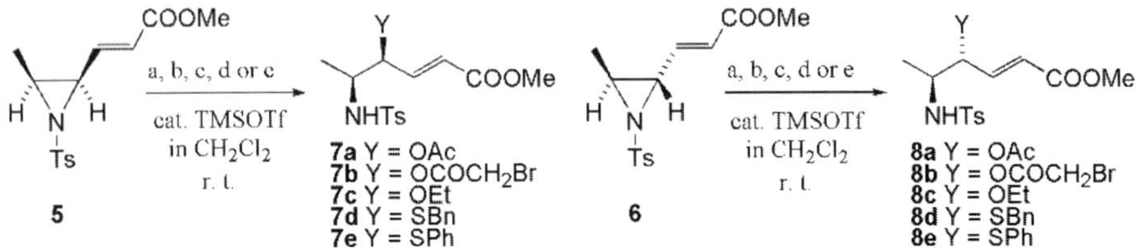

Scheme 2. Ts = 4-methylphenysulfonyl; reagents: a, AcOH; b, $BrCH_2COOH$; c, EtOH; d, BnSH; e, PhSH.

Furthermore, the regiospecific ring-opening reactions of β-aziridinyl-α,β-enoates having no side chain group at the δ-position were found. In addition, the combination of the ring-opening reactions with the Claisen rearrangement, the *O,N*-acyl transfer reaction and the organocopper-mediated *anti*-S_N2' type alkylation was efficiently applied to synthesis of EADI-containing peptidomimetics [4].

Ring-opening reactions of *N*-arylsulfonyl-γ,δ-epimino-(*E*)-α,β-enoates mediated by resin-bound sulfonic acid were applied to the synthesis of EADIs using solid-phase techniques. Treatment of β-aziridinyl-α,β-enoate **5** with toluenesulfonic acid resin (MP-TsOH, Argonaut Technologies) yielded a resin-bound γ-tosylate, which was converted into an EADI [Ts-Ala-*ψ*[(*E*)-CH=CH]-D-Leu-OMe] by organocopper reagents [4]. In this procedure, the resin-bound γ-tosylate can be purified by only washing with solvents, suggesting that the present solid-phase techniques have the advantage of manipulation.

In summary, the ring-opening reactions of β-aziridinyl-α,β-enoates with several nucleophiles involving alcohols, thiols and weak acids such as AcOH and N^{α}-protected amino acids in the presence of catalytic amount of Lewis acids such as TMSOTf have been fully investigated. The regio- and stereo-selective S_N2' ring-opening at the γ-carbon position was observed. In addition, the ring-opening reactions of β-aziridinyl-α,β-enoates were applied to the synthesis of EADIs.

References

1. Tamamura, H., Yamashita, M., Muramatsu, H., Ohno, H., Ibuka, T., Otaka A. and Fujii, N. *Chem. Commun.* 2327-2328 (1997).
2. Tamamura, H., Yamashita, M., Nakajima, Y., Sakano, K., Otaka A., Ohno, H., Ibuka, T. and Fujii, N. *J. Chem. Soc., Perkin Trans. 1* 2983-2996 (1999).
3. Oishi, S., Tamamura, H., Yamashita, M., Odagaki, Y., Hamanaka. N., Otaka A. and Fujii N. *J. Chem. Soc., Perkin Trans. 1* 2445-2451 (2001).
4. Tamamura, H., Tanaka, T., Tsutsumi, H., Nemoto, K., Mizokami S., Ohashi, N., Oishi, S. and Fujii, N. *Tetrahedron* in press.

Development of a small molecule peptidomimetic affinity Ligand for efficient purification of the large protein Factor VIII

Sebastian Knör[1], Burkhardt Laufer[1], Alexey Khrenov[2], Abdellah Benhida[3], Sabrina C. Grailly[3], Nathalie Beaufort[4], Viktor Magdolen[4], Charlotte A. E. Hauser[5], Evgueni L. Saenko[2], Jean-Marie R. Saint-Remy[3] and Horst Kessler[1]

[1]Department Chemie, TU München, Garching, Germany; [2]School of Medicine, University of Maryland, Baltimore, USA; [3]Center for Molecular & Vascular Biology, KU Leuven, Leuven, Belgium; [4]Klinische Forschergruppe der TU München, München, Germany; [5]Klinik für Frauenheilkunde & Geburtshilfe der Universität zu Lübeck, Lübeck, Germany.

Introduction

Haemophilia A, one of the most common bleeding disorders, is the result of an inherited deficiency of factor VIII (FVIII) function. [1] For medical treatment, patients are given injections of FVIII. [2] FVIII purification, however, remains a challenging task. Current procedures using immunoaffinity chromatography are expensive and suffer from the instability of the applied antibody ligands which are eluted along with the product and contaminate it. Recently, the large protein FVIII (330 kDa) was successfully purified using octapeptide ligands, [3] but as we proved, their practical use is limited due to their low protease resistance in the presence of plasma.

Results and Discussion

Starting with the octapeptidic ligand EYHSWEYC (**1**) reported by Pflegerl *et al.* [3] we have reduced the ligand to its core binding sequence WEYC (**2**). Further optimization of **2** via substitution of the Trp residue by 3-indolylacetic acid (3-IAA) and replacement of the central peptide bond (-CO-NH) by a reduced peptide bond (CH_2-NH-) resulted in the peptidomimetic (3-IAA)Eψ[CH_2NH]YC (**3**) being highly stable against proteolytic degradation and showing high FVIII binding ability. [4]

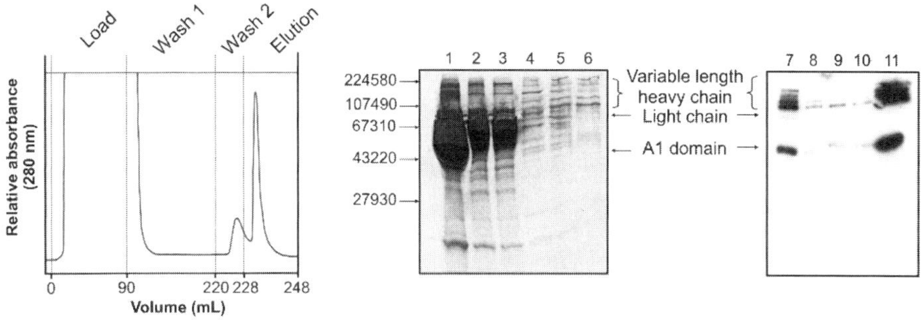

*Fig. 1. Purification of FVIII diluted in cell-conditioned FBS-containing DMEM using (**3**)-coated resin. Left hand panel: purification profile; middle panel: SDS-PAGE; right-hand panel: Western blot: Lane 1: media; Lanes 2, 7: media with FVIII; Lanes 3, 8: flow-through fraction; Lanes 4, 5, 9, 10: wash fractions (Wash 2) with 0.2 M NaCl; Lanes 6, 11: elution fraction with 0.6 M NaCl.*

S. Del Valle et al. (eds.), *Peptides for Youth: The Proceedings of the 20th American Peptide Symposium*,
DOI: 10.1007/978-0-387-73657-0_68, © Springer Science+Business Media, LLC 2009

Using **3**-coated affinity resin, we were able to efficiently purify FVIII from cell culture media presenting contaminant proteins in vast excess to FVIII (Fig. 1). To allow a cost-efficient production of the peptidomimetic (**3**) in preparative scale, a solution synthesis was developed (Fig. 2). [4]

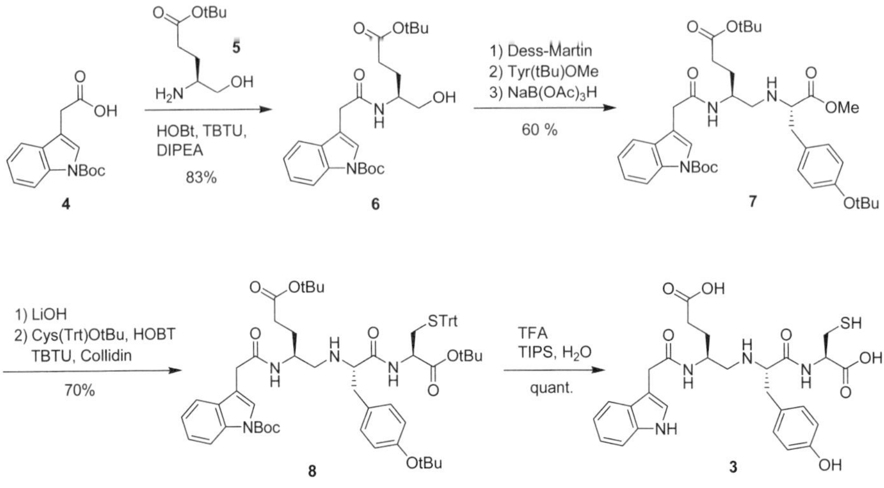

Fig. 2. Solution phase synthesis of peptidomimetic (3).

The synthesis starts with side chain protected indolylacetic acid **4** which is easily available in large scale following literature procedures. The compound was coupled with glutamol **5** to give the alcohol **6** in 83% yield. **6** was oxidized to the corresponding aldehyde in quantitative yield using Dess-Martin periodinane and further converted to the secondary amine **7** in 60% yield via reductive alkylation with Tyr(tBu)OMe as amine component and $NaB(OAc)_3H$ as reducing agent. After saponification the free acid was coupled with Cys(Trt)OtBu to achieve the protected tetrapeptide mimetic **8** in 70% yield over two steps. The compound then was quantitatively deprotected by slow acidification with TFA in presence of TIPS and water. By this procedure the peptidomimetic **3** was obtained in 35% overall yield (6 steps) and > 95% de.

In summary, the peptidomimetic (**3**) is by far the smallest ligand known to date for the separation of such a large protein. It not only binds and purifies FVIII with high efficiency but is stable, protease-resistant and cheap to be produced in preparative scale. Hence it offers a valuable alternative to current protein-based procedures in laboratory and industrial production.

References

1. Jacquemin, M., De Maeyer, M., D'Oiron, R., Lavend'Homme, R., Peerlinck, K. and Saint-Remy, J.-M. *J. Thromb. Haemost.* **1**, 456-463 (2003).
2. Ananyeva, N., Khrenov, A., Darr, F., Summers, R., Sarafanov, A. and Saenko, E. *Expert. Opin. Pharmacother.* **5**, 1061-1070 (2004).
3. Pflegerl, K., Hahn, R., Berger, E. and Jungbauer, A. *J. Pept. Res.* **59**, 174-182 (2002).
4. Knör, S., Khrenov, A. V., Laufer, B., Saenko, E. L., Hauser, C. A. E. and Kessler, H. *J. Med. Chem.* **50**, 4329-4339 (2007).

Expression and Characterization of Recombinant S2 Subunit of SARS-coronavirus S Fusion Protein

Zhe Yan[1], Kathryn V. Holmes[2] and Robert S. Hodges[1]

[1]*Department of Biochemistry and Molecular Genetics;* [2]*Department of Microbiology, University of Colorado at Denver and Health Sciences Center, Aurora, CO 80045, USA*

Introduction

The SARS-CoV Spike fusion protein subunit S2 plays an important role in viral entry by initiating fusion of the viral and cellular membranes. S2 exists in at least three different conformations within the trimeric glycoprotein complex. In the postfusion conformation (six-helix bundle, 6HB), the three helical heptad repeats of the HRC region pack in an antiparallel fashion into the hydrophobic grooves on the surface of the trimeric coiled-coil formed by the heptad repeat N-terminal (HRN) region. The HRN and HRC regions are separated by a 149 amino acid interhelical domain (IHD), which is thought to provide the flexibility required for the conformational change that occurs during the membrane fusion process. There is no structural information currently available for the interhelical domain and S2 in the prefusion state and fusion intermediate states.

In this study, we have expressed the recombinant S2 subunit of SARS-CoV S fusion protein including the interhelical domain and studied this protein using biochemical and biophysical methods. This engineered chimeric S2 fusion protein should facilitate structural studies of the prefusion state and also serve as a target to guide future attempts to develop inhibitors that prevent the formation of the postfusion 6HB.

Construct Design

We have chosen the sequence 901-1185 of the S2 subunit of SARS-CoV S protein to design the constructs. Through adding GCN4 trimeric helices and deleting 6 residues in the C-terminus of the HRC region, we proposed to stabilize the trimeric structure of HRC and prevent its binding to the HRN region, and consequently keep the S2 domain in the prefusion state. The proteins were expressed as maltose-binding protein (MBP) fusion proteins. We have designed two constructs as follows,

(1) the whole S2: HRN-IHD-HRC-GCN4 (S2);

(2) partial HRN and interhelical domain (951-1150): HRN- IHD (L4).

Results and Discussion

Both constructs were expressed in E.coli host strain BL21 cells, and purified by affinity chromatography using amylose resins and size-exclusion chromatography. The results in Fig. 1 show both the constructs have been purified as soluble proteins with high purity. Both S2 and L4 constructs have been cleaved to remove the MBP tag using protease Factor Xa.

The purified MBP fusion proteins were analyzed for their oligomerization state by size-exclusion chromatography (SEC). As shown in Fig 2a, MBP-S2 exists in an aggregated state since it is eluted in the exclusion volume. Similarly, recombinant protein MBP-L4 is also aggregated. Following treatment with 8 M urea and 10 mM

S. Del Valle et al. (eds.), *Peptides for Youth: The Proceedings of the 20th American Peptide Symposium,*
DOI: 10.1007/978-0-387-73657-0_69, © Springer Science+Business Media, LLC 2009

DTT, analysis by SEC under these denaturing conditions shows a significant concentration of monomer (Fig. 3b). After treatment of the proteins with 2% SDS and 10 mM DTT, analysis by SEC using 0.1% SDS now shows that the proteins are disrupted into the monomeric state (Fig. 3c). These results mean that the recombinant proteins were expressed and purified as soluble high molecular weight aggregates, and this aggregation may be caused by the interhelical domain (IHD).

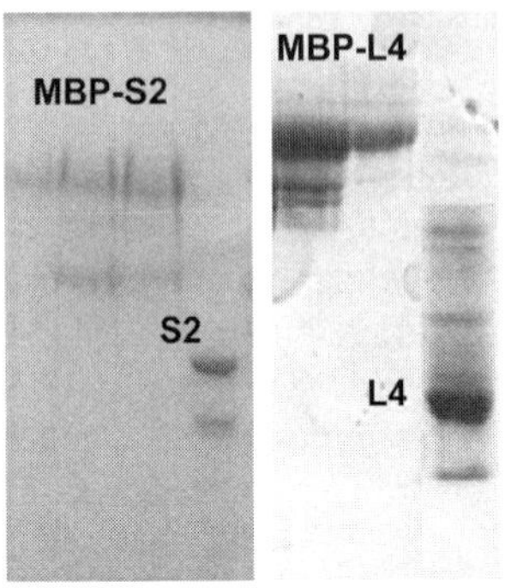

Fig. 1. SDS-PAGE analysis of MBP-S2, MPB-L4 and S2, L4

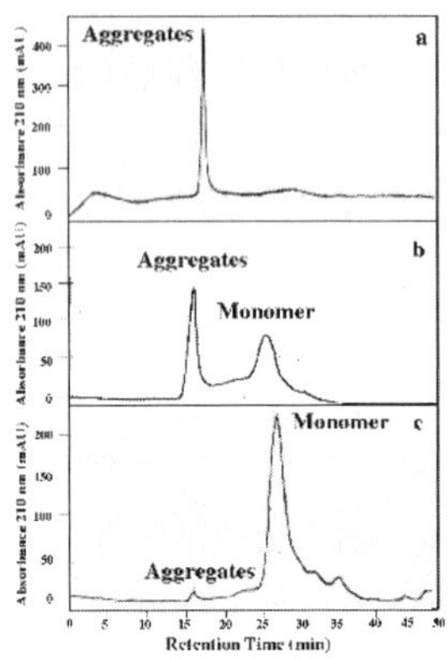

Fig. 2. SEC Analysis of MBP-S2

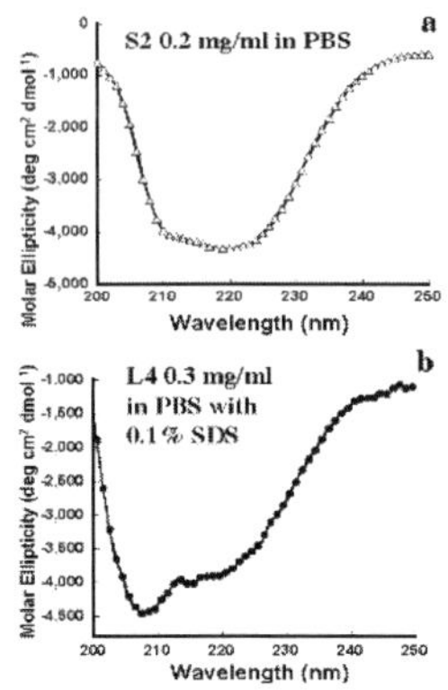

Fig. 3 CD Analysis of S2 and L4

We have used two antibodies specific for the HRC coiled-coil used to analyze the conformation of the HRC domain in the fusion protein using ELISA. HRC1 antibody could not bind to the prefusion state of HRC but the HRC2 antibody could bind to both the prefusion state and the 6HB conformation of HRC. The results showed that both HRC1 and HRC2 antibodies bind to MBP-S2, suggesting that MPB-S2 is not in a conformation containing the 6HB conformation, i.e., the postfusion state, which is consistent with our design aim.

CD spectra in Fig. 3a revealed that after cleavage from MBP, S2 in PBS showed a soluble and largely α-helical structure. After cleavage from MBP, L4 in 0.1% SDS also showed typical α-helical structure, as illustrated in Fig. 3b.

Our studies show that the recombinant proteins were expressed and purified as soluble high molecular weight aggregate proteins, and the aggregation may be caused by the interhelical domain (IHD). Thus, S2 was expressed as a soluble aggregate in the prefusion state. Further research is in progress to prevent aggregation by mutation or deletion of specific residues in the IHD and structure determination of the IHD.

Acknowledgments

This work was supported by a NIH grant (PO1AI059576) and the John Stewart Chair in Peptide Chemistry to R. S. H.

References

1. Tripet B, et al. J. Struct. Biol. 155, 176-194 (2006)
2. Xu Y, et al. J. Biol. Chem. 279, 49414-49419 (2004)
3. Wang X, et al. J. Virol. 74, 4746-4754 (2000)
4. Lay C, et al. FEBS 567, 183-188 (2004)

Synthesis of Biotinyl-TN14003, Anti-HIV Peptide Amide

H. Nedev and H.U. Saragovi

Montreal Centre for Experimental Therapeutics in Cancer,
Lady Davis Institute for Medical Research
3999, Cote St. Catherine, E-533, Montreal, Quebec, CANADA, H3T 1E2

Introduction

CXCR4 is a G-coupled major receptor for the entry of HIV. In 1998 Tamamura et al. reported the identification of the peptide T140, a specific CXCR4 inhibitor that possesses a high level of anti-HIV activity [1]. Since then, further improvements of this compound were achieved, leading to the generation of peptide-amide TN14003, which is far less toxic and more stable in serum [2].

This report describes the optimized solid phase synthesis of TN14003, its N-biotinylation on resin and the final cyclization through Cys^4-Cys^{13}. The biotinylated analogue can be used as imaging probe, specific for CXCR4 receptors.

Results and Discussion

The solid phase synthesis of the 14-peptide was carried out with the FMOC/O-tert-Bu strategy on ACT 348-Omega Synthesizer (Scheme 1). The synthesis was performed on two different resins – Rink and PAL amide type. Anchoring of the first amino acid was carried out by coupling of FMOC-Arg(Pbf)-OH to a Rink amide NovaGel Resin (subst. 0.63mmol/g), via DIC/HOBt in DMF. The same procedure was used for the preparation of the FMOC-Arg(Pbf)-OH attached to a PAL Resin (subst. 0.5mmol/g). The solid phase peptide synthesis of the linear form of the 14-mer proceeded smoothly using the DIC/HOBt as a coupling agent.

The N-biotinylation of the peptide was achieved on the resin using 8-fold excess of biotin/DIC/HOAt in DMF for 24h at RT. The peptidyl-resin was then washed repeatedly with DCM, methanol, and dried *in vacuo*.

Fmoc-R-R-Nal-C-Y-Cit-K-k-P-Y-R-Cit-C-R-PAL

↓ (i), (ii), (iii)

Biotinyl-R-R-Nal-C-Y-Cit-K-k-P-Y-R-Cit-C-R-NH_2

↓ (iv) or (v)

Biotinyl-R-R-Nal-C-Y-Cit-K-k-P-Y-R-Cit-C-R-NH_2 (Cys–Cys bridged)

Scheme 1. Synthesis of Biotinyl-TN14003. (i) 20%pip; (ii) Biotin/DIC/HOAt; (iii) EDT/thioanisole/m-cresol/TFA (1.7:10/3.3/85); (iv) 0.1M $AcONH_4$; (v) 20% DMSO

Release of the product from the solid support with concomitant side-chain protecting group removal was accomplished in two different cocktails (A: TMSBr, EDT, m-cresol and thioanisole in TFA; B: EDT, thioanisole, m-cresol in TFA) for 3 hours at 4°C. After concentration, the peptide was precipitated with diethyl ether and washed 5-fold subsequently with the same solvent. The HPLC elution profiles

S. Del Valle et al. (eds.), *Peptides for Youth: The Proceedings of the 20th American Peptide Symposium*, 155
DOI: 10.1007/978-0-387-73657-0_70, © Springer Science+Business Media, LLC 2009

showed that in this case better yield and product with higher purity was obtained with the PAL resin. The type of the scavenger was not of such importance.

Cyclization was carried out by two different approaches using respectively 20% DMSO or 0.1M $AcONH_4$ in water. The oxidation of Bt-TN14003 by $AcONH_4$ was completed within an hour. The cyclization via DMSO leads to peptide with less by-products, but takes longer time – usually overnight to complete (HPLC data not shown). The crude product was loaded onto a 250x21 Phenomenex C18 column (Varian) using a gradient 10-30% B in 30 min at RT and at a flow rate of 10ml/min. Following the purification the target product was collected and freeze-dried. The purity was confirmed by analytical HPLC. (Fig.1)

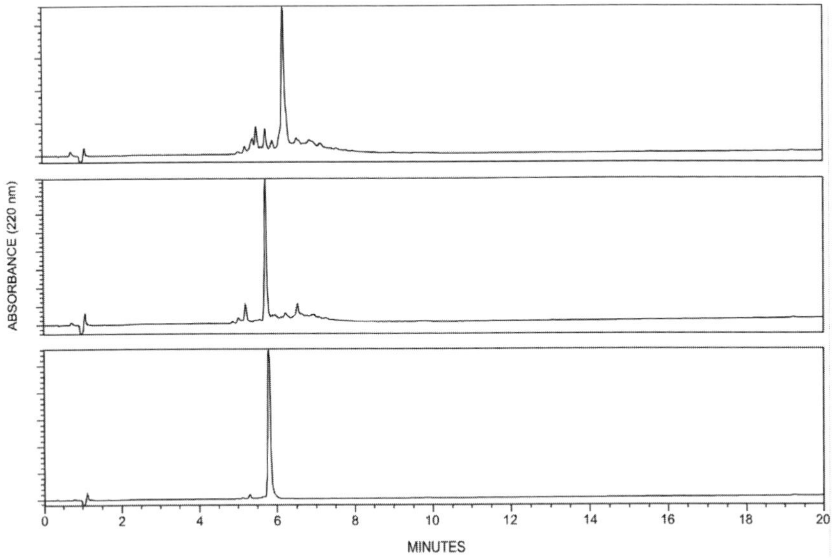

Fig. 1. HPLC elution profiles of Biotinyl-TN14003: a) Linear structure; b) Cyclic peptide – after $AcONH_4$ oxidation; c) cyclic – after purification. Analysis: 0-100% B for 20 min, A-10%ACN in water containing 0.1%TFA, B-70%ACN in water containing 0.1%TFA; FR-1ml/min

Accurate mass measurements were performed on a LC-MSD-TOF Agilent Technologies in positive electro-spray mode. The protonated molecular ions $(M+H)^+$ and fragments were used for empirical formula confirmation. The mass spectrometry data confirmed the structure of the desired cyclic biotinylated 14-peptide.

In conclusion, the strategy for preparation of Bt-TN14003 includes the use of PAL resin, DIC/HOAt for incorporation of the biotin and 20% DMSO or 0.1M $AcONH_4$ for the final cyclization of the target product followed by HPLC purification.

References

1. Tamamura, H., Xu, Y., Hattori, T., Zhang, X. *et al. Biochem. Biophys. Res. Commun.*, **253**, 877-882 (1998).
2. Tamamura, H., Omagari, A., Hiramatsu, K. *et al. Bioorganic & Medicinal Chemistry Letters* **11**, 1897-1902 (2001).

Making double-minded peptides strong-willed? The side chain impact on the backbone conformation of *N*-methylated cyclic pentapeptides

Burkhardt Laufer, Jayanta Chatterjee, Andreas O. Frank and Horst Kessler

Department Chemie, Lehrstuhl II für Organische Chemie, Technische Universität München, Garching, D-85747, Germany

Introduction

The incorporation of *N*-methylated amino acids in cyclic peptides can have a strong impact on the structure and the activity of peptides. On one hand, pharmacological features such as enzymatic stability or lipophility can be increased [1] and on the other, *N*-methylation reduces the energy difference between the *cis* and *trans* isomer. Recently, Chatterjee *et al.* showed that the *N*-methylation of cyclic pentapeptides with the sequence *cyclo*(-D-Ala-Ala_4-) can convey preferred conformation with drastic changes in its structure, as in the first approximation the conformation of cyclic peptides is determined by the sequence of D- and L-amino acids and the pattern of *N*-methylation [2]. However, in medicinal chemistry one looks for a preferred conformation which is as close as possible to the bioactive conformation, usually assumed to be identical to the bound conformation [3]. That is why we are interested in the impact of amino acids on the backbone of a conformationally inhomogeneous *N*-methylated peptide. To elucidate this effect, *cyclo*(-ala-Ala-Ala-(*N*Me)Ala-Ala-) was chosen, which exhibits two equally populated conformations in DMSO solution (Fig. 1).

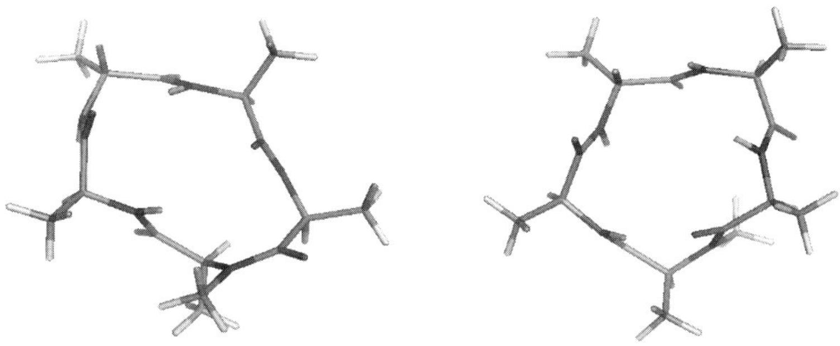

Fig. 1. The two different structures of cyclo(-ala-Ala-Ala-(NMe)Ala-Ala-).

Each alanine of the cyclic pentapeptide was substituted by hydrophobic (leucine, alanine, glycine), hydrophilic (aspartic acid, asparagine, serine), aromatic (phenylalanine, tryptophane) and *β*-branched (threonine, valine, isoleucine) amino acids generating five isomers of each amino acid class (Fig. 2) finally providing a clear picture of their specific impact on the backbone conformation.

Results and Discussion

After investigation of all the fifty peptides which were obtained by the introduction of the before mentioned amino acids at the five possible positions, it was found that in all cases, the favored conformation of the peptides was the all-*trans* conformation.

S. Del Valle et al. (eds.), *Peptides for Youth: The Proceedings of the 20th American Peptide Symposium*, 157
DOI: 10.1007/978-0-387-73657-0_71, © Springer Science+Business Media, LLC 2009

Furthermore, the position of amino acid substitution has a tremendous impact on the ratio of the different structures.

The implementation of aromatic, hydrophilic or hydrophobic amino acid residues into the cyclic alanine peptide at any position did not show significant augmentation of the *trans* to *cis* ratio. However, at the *i*-position, the position of the D-residue, a slight increase of the *trans*-conformation was noticed, especially for aspartic acid.

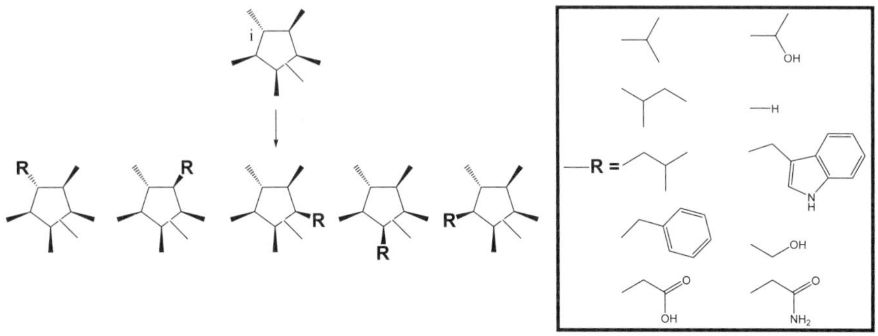

Fig. 2. Insertion of different amino acids at each position in the cyclopeptide.

As expected, dramatic changes of the 50/50 ratio could be gained by the introduction of *β*-branched amino acids threonine, valine and isoleucine into the pentapeptide. No influence of the peptide conformation was observed when the β-branched amino acids were introduced at the *i*-position; at *i+1*-position a *trans*- to *cis*-ratio of approximately 75/25 was obtained for all three amino acids. Insertion of threonine at the *i+2*-position didn't show any significant impact while valine (80/20) and isoleucine (70/30) improved the population of the *trans*-conformation. The highest impact on the ratio of the two conformations could be observed when valine (95/5) and isoleucine (100/0) were introduced at the *i+3*-position, while threonine after all showed a 75/25 ratio. Finally, at the *i+4*-position, threonine as well as the other non-*β*-branched amino acids only showed a slight impact while valine (84/16) and isoleucine (83/17) again showed their ability to increase the population of the all-*trans* conformation by having a higher bulkiness at the *β*-position compared to all other amino acids.

Here we were able to show the impact of different amino acid side chains on the backbone conformation. The non *β*-branched amino acids show nearly no impact on the ratio of the two conformations. The *β*-branched amino acids valine, isoleucine and threonine show a very strong impact on the *trans*- to *cis*-ratio when they are introduced at any position except the *i*-position.

References

1. Gilon, C., *et al.* in *Houben-Weyl, Methods of Organic Chemistry, Vol. E22C, Synthesis of Peptides and Peptidomimetics*. Georg Thieme Verlag, Stuttgart, Germany; 215-271 (2002).
2. Chatterjee, J., Mierke, D., Kessler, H. *J. Am. Chem. Soc.* **128**, 15164-15172 (2006).
3. Williams, R. J. P. *Angew. Chem. Int. Ed. Engl.* **16**, 766-777 (1977).

Deprotection of N^{in}-Formyl Group on Tryptophan by Application of New Reagent, 1,2-dimethylethylendiamine in Aqueous Solution

Takenao Odagami,[1] Yuji Kogami,[1] Hiroyuki Kouji,[1] YukoTsuda[2,3] and Yoshio Okada[2,3]

[1]*Research Institute, PRISM Bio Lab, Izunokunishi 410-2103, Japan;* [2]*Faculty of Pharmaceutical Sciences and* [3]*The Graduate School of Food and Medicinal Sciences, Kobe Gakuin University, Kobe 650-8586, Japan*

Introduction

In peptide synthesis, choosing side chain protection is one of the most important events. The Trp(N^{in}-For) derivative is widely used to overcome two main side-reactions occurring with Trp residue in the Boc chemistry. One problem is acid catalyzed oxidation or dimerization of the indole moiety during synthesis. The other is alkylation of the indole ring by t-butyl cation and t-butyl trifluoroacetate generated during cleavage.To date, a number of N^{in}-For deprotection procedures have been developed, including low-high HF treatment [1]. However, this method involves treatment of large amount of stench reagent (dimethyl sulfide) and use of special equipment. On the other hand, the N^{in}-For group can be removed in a weak alkaline medium, such as 1.0 M NH_2OH[2] or 1.0 M $NaHCO_3$[3]. Additionally, the N^{in}-For group can be cleaved by dissolving Trp(N^{in}-For) in DMF containing a 20-30 fold molar excess of hydrazine hydrate[4]. However, these procedures require application with excess amounts of alkaline reagent and formyl group migration to an α-amino group of peptide was observed [3]. In light of not only commercial manufacturing practices and but especially that of green chemistry, a new N^{in}-For deprotecting procedure with equivalent controlled reaction under aqueous conditions and without using stench reagent application would be required.

In this presentation, we deal with the examination of 1,2-dimethylethylen-diamine(DMED) as a reagent of deprotecting N^{in}-For group.

Results and Discussion

First, we investigated the conversion of H-Trp(N^{in}-For)-OH (**1**) to H-Trp-OH (**2**) under aqueous conditions with 1,2-diphenylethylendiamine (DPED) and DMED referring to the fact that DPED reacts with aldehyde to form tetrahydro-imidazole [5]. Of these compounds, the former did not cleave an N^{in}-For group due to its low solubility in water; however, the latter was active in this regard. Therefore, we directed to evaluate the reaction conditions that allow easy conversion of **1** to **2** (Table 1). A 1.0-equivalent DMED treatment did not cleave the N^{in}-For group even after 2h; however, increasing the equivalents of DMED from 1.0 to 1.5 and 2.0, the N^{in}-For group was completely removed after 1h and 20 min, respectively. 3.0-Equivalent treatment led to concomitant deprotection with addition of DMED to the reaction solution. Additionally, migration of the N^{in}-For group to α-amino group of Trp residue was not observed, while migration of the N^{in}-For group to one amino group of DMED molecule occurred to yield compound (**3**) (Table 1).

Next, we applied this reagent to the conversion of H-Phe-Trp(N^{in}-For)-Lys-Tyr-OH(**4**) to H-Phe-Trp-Lys-Tyr-OH (**5**), which was a sequence in Urotensin II. We

S. Del Valle et al. (eds.), *Peptides for Youth: The Proceedings of the 20th American Peptide Symposium,* 159
DOI: 10.1007/978-0-387-73657-0_72, © Springer Science+Business Media, LLC 2009

*Table1. Conversion of H-Trp(N^{in}-For)-OH(**1**) to H-Trp-OH(**2**) under aqueous conditions with 1,2-dimethylethylendiamine (DMED).*

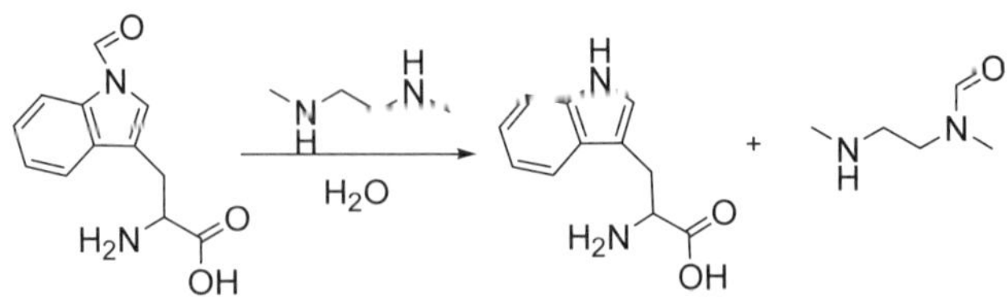

1 **2** **3**

Reaction time (min)	1.0 eq		1.5 eq		2.0 eq		3.0 eq	
	Trp%	Trp(N^{in}-For)%	Trp%	Trp(N^{in}-For)%	Trp%	Trp(N^{in}-For)%	Trp%	Trp(N^{in}-For)%
0	0.0	100.0	0.0	100.0	0.0	100.0	0.0	100.0
2	41.7	55.6	80.3	22.4	99.4	5.1	101.7	0.0
5	44.3	55.5	86.8	18.8	102.4	2.8	104.1	0.0
10	48.3	52.3	89.3	11.9	102.7	0.6	103.8	0.0
20	54.3	46.1	95.8	5.1	104.9	0.0	101.1	0.0
40	62.4	38.3	98.6	1.3	103.8	0.0	101.8	0.0
60	79.7	32.7	102.0	0.0	105.8	0.0	104.1	0.0
120	77.6	22.7	99.9	0.0	104.2	0.0	102.3	0.0

assembled a starting substrate, **4**, by using the Boc chemistry in solution method and TMSOTf-Thioanisole/TFA. The N^{in}-For group of **4** was successfully removed with treatment of 2.0-equivalents of DMED for 1h to generate **5** (90% yield) with a slight by-products. One of them was identified as a peptide, H-Phe-Trp-Lys(N^{ε}-For)-Tyr-OH by LC/MS/MS analysis.

In summary, a new method for N^{in}-For group deprotection was developed. The treatment of DMED (1.5 equivalents) in water achieved good conversion of H-Trp(N^{in}-For)-OH and H-Phe-Trp(N^{in}-For)-Lys-Tyr-OH to H-Trp-OH and H-Phe-Trp-Lys-Tyr-OH, respectively. The described procedure using DMED in aqueous solution would be practical and environmental advantageous over the previously reported methods, especially from the views of manufacturing and green chemistry.

Acknowledgments

The authors gratefully acknowledge Dr. Kiyoshi Ogawa at the Asahi Kasei Pharma Corp. for helpful analysis of RP-HPLC, LC/MS and LC/MS/MS.

References

1. Tam, P. J., Heath, F. W., and Merrifield, B. R. *J. Am. Chem. Soc.* **105**, 6442-6455 (1983).
2. Merrifield, B. R., Vizioli, D. L., and Boman, G. H. *Biochemistry* **21**, 5020-5031 (1982).
3. Yamashiro, D, and Li, H. A. *J. Org. Chem.* **38**, 2594-2597 (1973).
4. Ohno, M., Tsukamoto, S., Makisumi, S., and Izumiya N., *Bull. Chem. Soc. Japan.* **45**, 2852-2855 (1972).
5. Wanzlick, H. W. *Chem. Ber.* **86**, 1463-1466 (1953).

Racemization-free segment condensation based on the *O*-acyl isopeptide method: Toward a chemical protein synthesis on solid support

Taku Yoshiya, Youhei Sohma, Yoshio Hamada, Tooru Kimura, Yoshio Hayashi and Yoshiaki Kiso

Department of Medicinal Chemistry, Center for Frontier Research in Medicinal Science, 21st Century COE Program, Kyoto Pharmaceutical University, Japan

Introduction

The convergent synthetic method has greatly facilitated the assembly of peptides and proteins. However, a fundamental drawback of convergent synthesis is that epimerization at the C-terminal residue of an *N*-segment may occur during the condensation reaction with a *C*-segment. Particularly, in solid phase segment condensation, a large amount of epimerization is generally involved, thereby limiting the *N*-segment to contain either a C-terminal Gly or Pro residue.

We have developed a "racemization-free segment condensation" based on the "*O*-acyl isopeptide method" (Fig. 1) [1]. This method allows the use of an *N*-segment possessing a C-terminal Ser/Thr residue for segment condensation without any epimerization, as a result of the C-terminal *O*-acyl isopeptide structure with a urethane-protected Ser/Thr residue. Thus, in the synthesis of long peptides/proteins, racemization-free segment condensation becomes possible at not only the C-terminal Gly/Pro but also Ser/Thr of the *N*-segment.

Here, we report the application of the *O*-acyl isopeptide method-based racemization-free segment condensation method in the synthesis of a model pentapeptide.

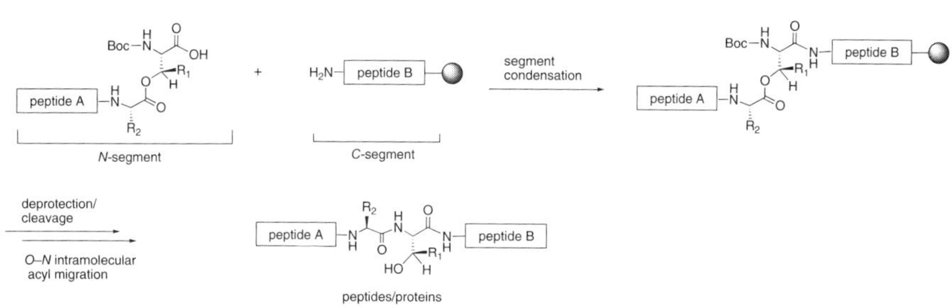

Fig. 1. Racemization-free segment condensation method.

Results and Discussion

As a model peptide, pentapeptide Ac-Val-Val-Thr-Val-Val-NH_2 (**1**) was adopted. In the condensation of Ac-Val-Val-Thr(*t*Bu)-OH with H-Val-Val-NH-resin (as a standard segment condensation), a large amount of epimerization (37.5%) at the activated Thr residue occurred in a DIPCDI (1,3-diisopropylcarbodiimide)–HOBt (1-hydroxybenzotriazole) method, which was confirmed by an independent synthesis of Ac-Val-Val-D-*allo*-Thr-Val-Val-NH_2.

On the other hand, in the *O*-acyl isopeptide method-based segment condensation (Scheme 1), *N*-segment Boc-Thr(Ac-Val-Val)-OH (**5**), which was synthesized using "*O*-acyl isodipeptide unit" [2] Boc-Thr(Fmoc-Val)-OH, was coupled to *C*-segment H-Val-Val-NH-resin (**2**) by the DIPCDI (2.5 eq)–HOBt (2.5 eq) method in DMF

S. Del Valle et al. (eds.), *Peptides for Youth: The Proceedings of the 20th American Peptide Symposium*, 161
DOI: 10.1007/978-0-387-73657-0_73, © Springer Science+Business Media, LLC 2009

(2 h). After TFA treatment, *O*-acyl isopeptide **4**·TFA was obtained with an isolated yield of 69%. HPLC analysis of crude **4** exhibited high purity of the desired product without any byproduct derived from epimerization at Thr, which was confirmed by an independent synthesis of H-D-*allo*-Thr(Ac-Val-Val)-Val-Val-NH_2 (Fig. 2). Isopeptide **4** was quantitatively converted to **1** in phosphate buffered saline (pH 7.4) [3].

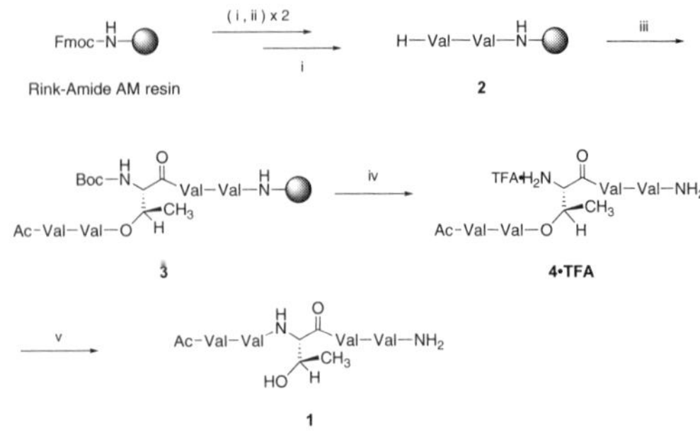

*Scheme 1. Reagents and conditions: i, 20% piperidine/DMF, 20 min; ii, Fmoc-Val-OH (2.5 eq), DIPCDI (2.5 eq), HOBt (2.5 eq), DMF, 2 h; iii, Boc-Thr(Ac-Val-Val)-OH (**5**, 2.5 eq), DIPCDI (2.5 eq), HOBt (2.5 eq), DMF, 2 h; iv, TFA–m-cresol–thioanisole–H_2O (92.5:2.5:2.5:2.5), 90 min; v, phosphate buffered saline (pH 7.4), 25 °C.*

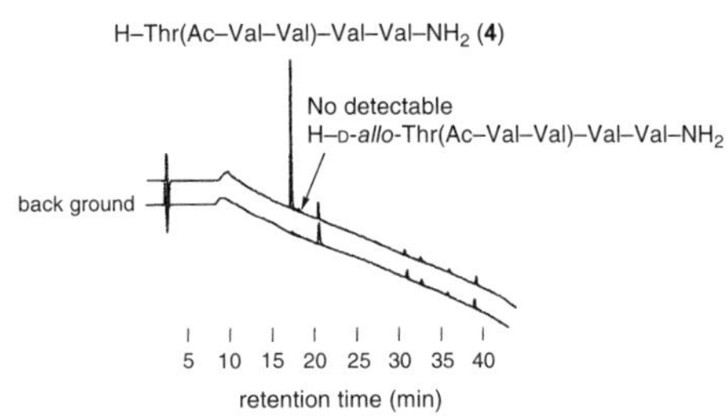

*Fig. 2. HPLC profile of crude isopeptide **4**.*

In summary, we report a racemization-free segment condensation based on the *O*-acyl isopeptide method with a successful synthesis of a model pentapeptide. This method allows the use of an *N*-segment possessing a C-terminal Ser/Thr residue for segment condensation without any epimerization, as a result of the C-terminal *O*-acyl isopeptide structure with a urethane-protected Ser/Thr residue. Thus, in the synthesis of long peptides/proteins, racemization-free segment condensation becomes possible at not only the C-terminal Gly/Pro but also Ser/Thr residue of the *N*-segment. Additionally, final deprotected peptides/proteins synthesized using the *O*-acyl isopeptide method-based segment condensation are effectively purified by HPLC, because a simple isomerization to an *O*-acyl isopeptide changed the physicochemical properties of the native peptide [3].

Acknowledgments

This research was supported in part by The "Academic Frontier" Project for Private Universities: matching fund subsidy from MEXT (Ministry of Education, Culture, Sports, Science and Technology) of the Japanese Government, and The 21st Century COE Program from MEXT.

References

1. Yoshiya, T., Sohma, Y., Kimura, T., Hayashi, Y. and Kiso, Y. *Tetrahedron Lett.* **47**, 7905-7909 (2006).
2. Yoshiya, T., Taniguchi, A., Sohma, Y., Fukao, F., Nakamura, S., Abe, N., Ito, N., Skwarczynski, M., Kimura, T., Hayashi, Y. and Kiso, Y. *Org. Biomol. Chem.* **5**, 1720-1730 (2007).
3. Sohma, Y., Yoshiya, T., Taniguchi, A., Kimura, T., Hayashi, Y. and Kiso, Y. *Biopolymers (Pept. Sci.)* **88**, 253-262 (2007).

Optimized Coupling Protocols for the Incorporation of Cys Derivatives during Fmoc-SPPS

Guenther Loidl, Fritz Dick, Monika Mergler and Ralph O. Schoenleber

Bachem AG, Bubendorf, Switzerland

Introduction

Incorporation of cysteine during Fmoc-SPPS can result in significant racemization depending on the coupling conditions [1-4]. During the preparation of cysteine containing peptides as APIs, we identified the necessity to investigate coupling conditions for Fmoc-Cys(Trt)-OH and Fmoc-Cys(Acm)-OH based on previous results of other groups [2-4] in more detail.

Using the model peptide Z-Ala-Cys(PG)-Pro-OH, we studied the influence of coupling reagent, solvent, activation time, and temperature during preactivation and coupling on the racemization of Cys. SPPS was performed on H-Pro-2-chlorotrityl resin with 3 equiv. of amino acid derivative / coupling reagent / base or additive. Peptides were cleaved from the resin with 1% TFA in DCM. After evaporation, the residue was analyzed on HPLC (C18; gradient of ACN in 0.1% TFA). The content of the LDL-isomer was determined using the relative peak areas A from HPLC as: A(LDL-isomer) / [A(LDL-isomer) + A(LLL-isomer)].

Coupling Reagent

To investigate the influence of the coupling reagent, DMF was used as solvent. As described previously for other peptides [2-4], TBTU as coupling reagent with DIPEA or collidine as base lead to significant formation of the LDL-isomer with Fmoc-Cys(Trt)-OH and Fmoc-Cys(Acm)-OH (see Table 1). With Fmoc-Cys(Acm)-OH, replacing DIPEA with the weaker base collidine resulted in a clear reduction of racemization. Couplings with the carbodiimides DCCI and DIC in the absence of base gave < 1.0% of the LDL-isomer.

Table 1.

Reagent	Base/ Additive	LDL-Isomer (%) Cys(Trt)	Cys(Acm)
TBTU	DIPEA	2.6	4.6
TBTU	collidine	2.5	0.6
DCCI	HOBt	0.6	n.d.
DIC	HOBt	0.9	0.7

Table 2.

LDL-Isomer (%)		
DMF	DMF/DCM	DMF/toluene
0.9	0.2	0.3
NMP	NMP/DCM	NMP/toluene
0.4	0.3	0.3

Solvent

The influence of the solvent on the degree of racemization during the coupling of Fmoc-Cys(Trt)-OH was examined using DIC. DMF and NMP were compared to their less polar 1 / 1 mixtures with DCM, toluene (see Table 2), and THF (data not shown). In accordance with the results of Han et al. [3], the 1/1 mixture of DMF and DCM gave less racemization than neat DMF. Compared to DMF, also NMP yielded less LDL-isomer. Replacing DCM with toluene or THF retained the positive effect of DCM in mixtures with DMF and NMP.

S. Del Valle et al. (eds.), *Peptides for Youth: The Proceedings of the 20th American Peptide Symposium*, 163
DOI: 10.1007/978-0-387-73657-0_74, © Springer Science+Business Media, LLC 2009

Preactivation Time and Temperature

For coupling of Fmoc-Cys(Trt)-OH with DIC / HOBt, the influence of preactivation time and temperature during preactivation and coupling was investigated (see Table 3). With DMF as well as with DMF / toluene = 1 / 1, best results were obtained with preactivation times of 5' and 15' at 25°C. These findings differ from previously published data [3]. Therefore, we chose a preactivation time of 5' and varied the temperature during preactivation and coupling. At 0°C and at 32°C, we found higher levels of racemization than at 15 °C or at 25 °C.

Table 3.

Preactivation Time (min)	Temp. (°C)	LDL-Isomer (%) DMF	LDL-Isomer (%) DMF/toluene
0	25	1.2	0.8
5	25	0.9	0.5
15	25	0.9	0.6
30	25	1.0	0.8
60	25	1.3	0.9
5	0	1.8	1.0
5	15	0.9	0.6
5	25	0.9	0.5
5	32	1.6	1.4

Figure 1.

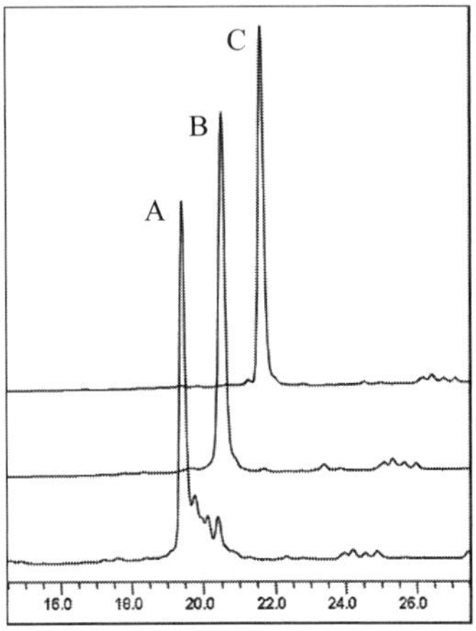

A Case-Study: SPPS of an Enterotoxin-like Peptide

Recently, we developed the SPPS for an Enterotoxin-like peptide containing 6 Cys and the following cysteine pattern: ---CCXXCCXXXCXXCX---. As a control experiment, we performed the SPPS using Fmoc-Cys(Trt)-OH and TBTU / DIPEA in DMF. HPLC analysis of the resulting crude, linear peptide showed large amounts of related, later eluting impurities (see Figure 1, line A). By changing the coupling reagent to DIC / HOBt, a clear increase in quality of the crude peptide was achieved (see Figure 1, line B). A minor additional improvement was observed with DMF / DCM = 1 / 1 as solvent and DIC / HOBt as coupling reagent (see Figure 1, line C).

Conclusion

Carbodiimides like DCCI and DIC with an additive (e.g. HOBt) in the absence of base are the reagents of choice for coupling of Fmoc-Cys(Trt)-OH and Fmoc-Cys(Acm)-OH. With a preactivation time of 5' and a reaction temperature of 15 - 25°C, the resulting amount of LDL-isomer in the model peptide Z-Ala-Cys(PG)-Pro-OH was always below 1% with DMF or NMP as solvent. The level of racemization can be further reduced by using less polar solvents, for example 1 / 1 mixtures of DMF or NMP with DCM, toluene, or THF, respectively.

References

1. Musiol, H.-J., Siedler, F., Quarzago, D. and Moroder, L. *Biopolymers* **34**, 1553 (1994).
2. Kaiser, T., Nicholson, G. J., Kohlbau, H. J. and Voelter, W. *Tetrahedron Lett.* **37**, 1187 (1996).
3. Han, Y., Albericio, F. and Barany, G. *J. Org. Chem.* **62**, 4307 (1997).
4. Angell, Y. M., Alsina, J., Albericio, F. and Barany, J. *J. Peptide Res.* **60**, 292 (2002).

Further studies on peptide synthesis at high temperature

Carina Loffredo, Nilson A. Assunção and M. Terêsa M. Miranda

Department of Biochemistry, Institute of Chemistry, University of São Paulo, Av. Prof. Lineu Prestes, 748, 05508-000, São Paulo, Brazil

Introduction

It is well known that conventional heating and microwave irradiation accelerate the reactions and may minimize aggregation in stepwise Solid-Phase Peptide Synthesis (SPPS) [1-5]. Nevertheless, SPPS-conventional heating and SPPS-microwave irradiation have not yet been employed routinely. This is credited mainly to the fear of the enhancement of secondary reactions.

In previous works we investigated the following aspects of stepwise SPPS-conventional heating: suitability of a few solvent systems, swelling properties of a few polymeric resins, efficiency of a few coupling agents and occurrence of enantiomerization. Among the temperatures and solvents studied, 60°C and 25%DMSO/toluene were the most favorable. No difficulties related to the aggregation phenomenon were encountered. In 25%DMSO/toluene at 60°C, enantiomerization was not significantly enhanced during the syntheses of Ac-Ile-Gly, **D**Tyr-Lys-Trp, Ile-Phe-Gly-Thr-NH_2 using DIC or TBTU/DIPEA [2-4].

To expand our knowledge on enantiomerization in SPPS at 60°C in 25%DMSO/toluene and aware of the increasing interest in microwave assisted SPPS, recently we studied the following new simple model peptides containing His, Ser and Cys, amino acids highly prone to enantiomerize during peptide chain assembly: *Ac-Ala-Cys-Pro-Lys* (acetylated fragment 204-207 of HIV-1 gp120), *Gly-Cys-Phe-NH₂* (model peptide earlier reported), *Ac-Pro-Ser-His-Arg* (acetylated fragment 18-21 of unsulfated human CCK-33) and *Ac-His-Gly-Ser-Ala* (acetylated fragment 50-53 of bovine α-hemoglobin). To further investigate the occurrence of other side reactions, we also studied the fragment 24-33 of unsulfated human CCK-33 (*Asp-Arg-Asp-Tyr-Met-Gly-Trp-Met-Asp-Phe-*NH_2). Therefore, the present work is focused on: *i*) the synthesis of each model peptide by traditional SPPS, SPPS at 60°C-conventional heating and SPPS at 60°C-microwave irradiation; *ii*) the examination of each resulting crude peptides; *iii*) attempts to identify and quantify their components.

Results and Discussion

Traditional SPPS, SPPS at 60°C-conventional heating and SPPS at 60°C-microwave irradiation were performed manually starting from N-urethane protected **L**-amino acids. SPPS at 60°C-microwave irradiation was carried-out using the CEM Discover™ system. Boc and Fmoc chemistries employed customized protocols [2, 4] as well as PAM and Rink amide resins. All peptide-resins, characterized by amino acid analysis, were submitted to full deprotection/peptide detachment from the polymeric resins. The crude resulting peptides were analyzed by LC/ESI-MS, RP-HPLC and/or CE, the later using (+)-(18-crown-6)-2,3,11,12- tetracarboxylic acid as chiral selector. Stereoisomers of the desired **L**-peptides were synthesized by traditional SPPS, purified by RP-HPLC and characterized by LC/ESI-MS. Identification and quantification of the stereoisomers contained in the crude peptides

S. Del Valle et al. (eds.), *Peptides for Youth: The Proceedings of the 20th American Peptide Symposium*, 165
DOI: 10.1007/978-0-387-73657-0_75, © Springer Science+Business Media, LLC 2009

were achieved by RP-HPLC and CE spiking experiments using the standard stereoisomers or by total hydrolysis of the crude peptides in 6 N DCl/D_2O followed by GC/MS analysis of the hydrolyzate.

Regarding the model peptides *Ac-Ala-Cys-Pro-Lys* and *Gly-Cys-Phe-NH*$_2$, the RP-HPLC spiking experiments revealed that: i) the crude peptides resulting from traditional SPPS and SPPS at 60°C-conventional heating were equivalent (the stereoisomers/L-peptide ratios were similar); ii) for the crude peptides resulting from SPPS at 60°C-microwave irradiation, the stereoisomers/L-peptide ratio was higher (2-fold).

The RP-HPLC and CE spiking experiments did not reveal the presence of the stereoisomers of the L-peptides in the crudes *Ac-Pro-Ser-His-Arg* and *Ac-His-Gly-Ser-Ala* resulting from the three SPPS approaches used. In fact, their hydrolyzates presented very low amounts of D-amino acids.

Concerning the fragment 24-33 of unsulfated human CCK-33, all syntheses furnished similar crude peptides. Most contaminants resulted from incomplete couplings and premature Fmoc removal. Nevertheless, SPPS at 60°C-conventional heating produced the best crude peptide in terms of desired L-peptide:byproducts ratio).

In conclusion, the novelty of the present work is that it provides new information on the SPPS at elevated temperature using conventional heating and microwave irradiation. In fact, our results: *i*) showed that, in the syntheses of the model peptides studied in 25%DMSO/toluene at 60°C using DIC/HOBt or TBTU/DIPEA, the coupling reactions occurred without significant increase of epimerization, S-alkylation, methionine oxidation and formation of γ-lactam during the carboxy-activation of arginine derivatives. Exception was done for N-urethane protected **L**-cysteine, which was clearly more susceptible to epimerization when microwave irradiation was employed; *ii*) confirmed that RP-HPLC, LC/ESI-MS and CE/(+)-(18-crown-6)-2,3,11,12-tetracarboxylic acid can serve as powerful complementary tools for the separation and identification of stereoisomers present in intact synthetic peptides. These conclusions corroborate earlier reports [4-5].

Acknowledgments

We thank *FAPESP* (financial support to MTMM), *CNPq* (postgraduate scholarship to CL), *CEM Corporation* (manual microwave peptide synthesizer), *J. Gerhardt* [total hydrolyses in 6 N DCl/D_2O and GC/MS in C.A.T.GmBH & Co. Cromatographie und Analysentechnik KG (Tübingen, Germany)], *C.W. Liria* (amino acids analyses) and *F.M. Prado* (technical support in the LC/ESI-MS analyses).

References

1. Rabinovich, A. K. and Rivier, J. E. *Am. Biotechnol. Lab.* June, 48-51 (1994).
2. Varanda, L. M. and Miranda, M. T. M. *J. Pept. Res.* **50**, 102-108 (1997).
3. Rivier, J. E. and Miranda, M. T. M. Solid-phase peptide synthesis at elevated temperature: optimized condition. In *Synthesis of Peptides and Peptidomimetics*; Goodman, M., Felix, A., Moroder, L., Toniolo, C., Eds.; Georg Thieme (Houben-Weyl): Stuttgart, vol. E22, pp. 806-813 (2002, 2004).
4. Souza, M. P., Tavares, M. F. M. and Miranda, M. T. M. *Tetrahedron* **60**, 4671-4681 (2004).
5. Palasek, S. A.; Cox, Z. J. and Collins, J. M. *J. Pept. Sci.* **13**, 143-148 (2007).

Preparation of C-terminal modified peptides through alcoholysis and thiolysis mediated by metal ions

Patrícia B. Proti and M. Terêsa M. Miranda

Department of Biochemistry, Institute of Chemistry, University of São Paulo, Av. Prof. Lineu Prestes, 748, 05508-900, São Paulo, Brazil

Introduction

C-terminal modified peptides can be used as enzyme substrates or inhibitors, act as peptide analogs with altered biological activity and serve as acyl donors in chemical or enzymatic convergent synthesis [1]. These compounds can be obtained by the following routine: *1*) stepwise peptide synthesis on polymeric resin; *2*) appropriate nucleophilic attack on the peptide-resin linkage to generate the C-terminal modified protected peptide; *3*) removal of the side-chain blockers; *4*) purification of the crude product; *5*) final chemical characterization. Since the use of conventional procedures to perform step *2* may not always furnish products of good quality with satisfactory yields, we have studied a mild procedure to produce C-terminal modified protected peptides based on mediation by metal ions of the nucleophilic attack on the peptide-resin linkage [2]. The Kaiser Oxime Resin (KOR) has shown to be the most suitable polymeric resin and, among Ca^{2+}, Zn^{2+}, Co^{2+} and Cu^{2+}, the first one has revealed to be the most effective mediator [3].

Recognizing that many peptide chemists still employ Boc strategy and that Nα-acyl-peptide esters and thioesters are essential for convergent peptide synthesis, in the present work we attempted to verify whether our alternative procedure, first successfully applied for preparing Ac-Ile-Ser(Bzl)-Asp(OcHx)-OMe from the corresponding peptide-KOR, was suitable for preparing larger protected peptide methyl esters and protected peptide thioesters. For such purpose, we chose fragment 1-7 of the antimicrobial vespid chemotactic peptide M and fragment 22-24 of cholecystokinin-33 as models. Once deprotected, Ac-Phe-Leu-Pro-Ile-Ile-Gly-Lys-OMe, Ac-Ile-Ser-Asp-OMe and Ac-Ile-Ser-Asp-S$(CH_2)_2$COOEt could be used as acyl donors in peptide segment condensation [1,4]. In this context, we also examined the stability of such thioester against HF treatment and incubated Ac-Phe-Leu-Pro-Ile-Ile-Gly-Lys-OMe and Ac-Ile-Ser-Asp-OMe with Gly-NH_2 in the presence of the crude porcine pancreatic lipase (cPPL).

As Z-Ala-OMe and Z-Ala-OBzl can be used as acyl donors in esterase-catalyzed synthesis of the bitter dipeptide Z-Ala-Phe-OH, we also performed Ca^{2+} mediated-alcoholysis of the oxime ester bond of Z-Ala-KOR aiming to produce such reagents.

Results and Discussion

The amino acid and peptide-KORs were manually synthesized using Boc/Bzl chemistry. Their initial substitution levels (SL) were determined through total hydrolysis followed by amino acid analysis of the hydrolysate. The alcoholysis and thiolysis reactions were monitored by RP-HPLC and the crude products were characterized by LC/ESI-MS. ^{1}H NMR and amino acid analysis of purified esters complemented the chemical characterization. The final SLs of the remaining amino acid- and peptide-resins were determined in the same way that initial SLs. The difference between the initial and final SLs allowed for calculating the yields of peptide detachment from the resins.

S. Del Valle et al. (eds.), *Peptides for Youth: The Proceedings of the 20th American Peptide Symposium*, 167
DOI: 10.1007/978-0-387-73657-0_76, © Springer Science+Business Media, LLC 2009

Ac-Ile-Ser(Bzl)-Asp(OcHx)-OMe was obtained through Ca^{2+} assisted-methanolysis of the corresponding peptide-KOR using the best reaction condition previously determined [50% MeOH/DMF, 50°C, peptide and $Ca(OAc)_2$ molar ratio of 1:2] with yield of 88% in 2 h. The removal of side chain protecting groups was attained through HF treatment. The resulting Ac-Ile-Ser-Asp-OMe was successfully used as acyl donor in a cPPL-catalyzed coupling to Gly-NH_2 in 20% Tris buffer, pH 8, 0.5 M/*n*-hexane [5]. The yield obtained in 1 h was 65%.

Under the experimental condition cited above, Ac-Phe-Leu-Pro-Ile-Ile-Gly-Lys(2ClZ)-OMe was effectively obtained with yield of 95% in 4 h; the side chain deprotection of Lys was attained through HF treatment. Preliminary incubations of Ac-Phe-Leu-Pro-Ile-Ile-Gly-Lys-OMe with Gly-NH_2 and cPPL were performed in: *A*) 80% *n*-hexane/0.5 M Tris buffer, pH 8; *B*) 80% DCM/0.5 M Tris buffer, pH 8; *C*) 60% DMF/0.1 M phosphate buffer, pH 7.5; *D*) 80% 1-butyl-3-methylimidazolium tetrafluoroborate/0.5 M Tris buffer, pH 8; *E*) 80% 1-butyl-3-methylimidazolium hexafluorophosphate (BP6)/0.5 M Tris buffer, pH 8; *F*) BP6. The RP-HPLC monitoring of these reactions indicated that the coupling efficiencies followed the order: $D > C > B > E > F > A$.

Regarding the Ca^{2+} assisted-thiolysis of the oxime ester bond of Ac-Ile-Ser(Bzl)-Asp(OcHx)-KOR, the experimental condition employed was 20% ethyl-3-mercaptopropionate/DMF, 60°C, peptide: $Ca(OAc)_2$ molar ratio of 1:1, a variation of that used in the methanolyses described above able to dissolve the calcium salt. While the reaction carried out in the presence of the metal ion provided Ac-Ile-Ser(Bzl)-Asp(OcHx)-$S(CH_2)_2COOEt$ with yield of 93% in 2 h, the thiolysis performed in the absence of Ca^{2+} furnished the product with yield of only 28% in the same period of time. Ac-Ile-Ser-Asp-$S(CH_2)_2COOEt$ was successfully attained through HF treatment of the fully protected peptide thioester.

Last but not least, our attempt to prepare Z-Ala-ORs from Z-Ala-KOR employing the optimized condition of Ca^{2+} mediated-methanolysis of the oxime ester bond was well succeeded. Actually, Z-Ala-OMe and Z-Ala-OBzl were efficiently produced with yields of 99% in 4 h and 65% in 6 h, respectively.

The present results confirmed that our alternative procedure is suitable for preparing not only N^{α}-acylated peptide esters, but also N^{α}-acylated peptide thioesters and N^{α}-acylated amino acid esters. As expected, such compounds can be efficiently employed as acyl donors in esterase-catalyzed peptide bond formation.

Acknowledgments

We thank FAPESP for the grant to MTMM and the postgraduate scholarship to PBP, C.W. Liria for the amino acids analyses and F.M. Prado for the technical support in the mass spectrometry analyses.

References

1. Songster, M.F.; Barany, G. *Method. Enzymol.* **289**, 126 (1997).
2. Moraes, C. M.; Bemquerer, M.P.; Miranda, M.T.M. *J. Pept. Res.* **55**, 279 (2000).
3. Proti, P.B.; Oliveira, P.V.; Miranda, M.T.M. In *Understanding Biology Using Peptides* (Blondelle S.E., ed.), Springer: New York, 82 (2006).
4. Kitagawa, K.; Adachi, H.; Sekigawa, Y.; Yagami, T.; Futaki, S.; Gu, Y.J.; Inoue, K. *Tetrahedron* **60**, 907 (2004).
5. Liria, C.W.; Miranda, M.T.M In *Peptides 2000 - Proceedings of the 26th European Peptide Symposium* (Martinez, J.; Fehrentz, J.-A., eds.), EDK: Paris, 331 (2001).

Synthesis and Biological Activity of Amino Acid Esters of Acyclovir

Ivanka Stankova[1], Tatyana Dzimbova[2] and Stoyan Shishkov[3]

[1]Department of Chemistry, South-West University "Neofit Rilski", 2700 Blagoevgrad, Bulgaria, [2]Institute of Molecular Biology, Bulgarian Academy of Sciences, 1113 Sofia, Bulgaria, [3]St. Kl. Ohridski Sofia University, Faculty of Biology, Laboratory of Virology, Sofia 1164, Bulgaria

Introduction

The discovery of acyclovir (ACV) as a selective antiherpetic agent heralded a new era in antiviral chemotherapy, that of a selective approach to attack virus infections. ACV would, later on, become the gold standard for the treatment of herpesvirus (HSV-1, HSV-2 and VZV) infections [1]. ACV has limited oral bioavailability (15-20%) and also limited solubility in water (~0.2%, 25°C). A possible way to increase the bioavailability is by modifying the known antiviral drugs with various amino acids. The L-valyl ester of aciclovir (valaciclovir) with bioavailability of 60% is obtained in this manner [2-5].

Modification of anti-herpes agents like aciclovir by peptidomimetics, whose chemical structure is different from the natural peptides but have the same ability to interact with specific receptors, is of definite interest [6]. Modification of acyclovir with thiazole containing dipeptide mimetics derived from glycine, L-valine, L-alanine and L-leucine can enhance the oral bioavailability of the parent drugs.

Moreover, such prodrugs may have longer half-live time, for the identification of the peptidomimetics (wich are pro-moiety) by the cellular enzymes may be hindered. The slower release of the prodrug could result in reduced toxicity.

The aim of this study was to design and to synthesize of new thiazole containing amino acids ester prodrugs of acyclovir and to explore their activity on the HSV-1.

Results and Discussion

Boc-2-aminomethyl-thiazole-4-carboxylic acid, Boc-2-Val-thiazole-4-carboxylic acid, Boc-2-Ala-thiazole-4-carboxylic acid and Boc-2-Leu-thiazole-4-carboxylic acid were prepared according to ref. [7-8].

Synthesis of prodrugs

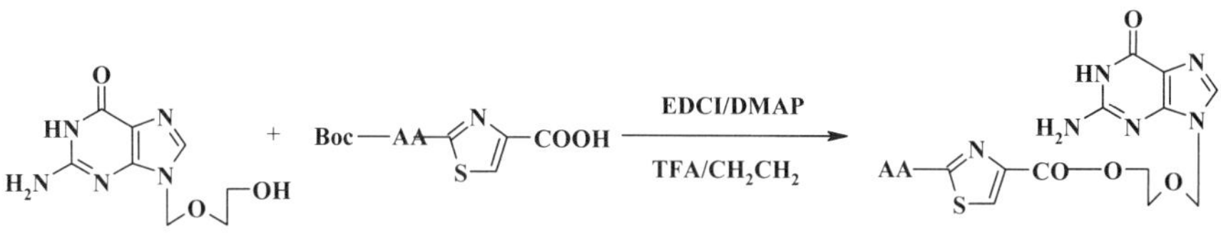

1 2 3

AA=NHC(iPr)H-, -NHCH2-, NHC(CH3)H-, NHC(iBu)H-

Scheme 1

S. Del Valle et al. (eds.), *Peptides for Youth: The Proceedings of the 20th American Peptide Symposium*, 169
DOI: 10.1007/978-0-387-73657-0_77, © Springer Science+Business Media, LLC 2009

The amino acid esters were prepared in two steps (Scheme 1). A solution of acyclovir (**1**) in dimethylformamide (DMF) was treated with the N-protected amino acid (**2**) using the coupling agent 1-[3-(di-methylamino)propyl]-3-ethyl carbodiimide hydrochloride (EDCI) and 4-(dimethylamino)-pyridine (DMAP) as a catalyst, to produce the N-Boc blocked ester derivates and deprotection was done with trifluoroacetic acid.

Antiviral activity

Virus:

Herpes simplex virus type 1 (HSV-1), strain DA.

The guanosine analogues did not affect the cell morphology in investigated concentrations. Acyclovir was not cytotoxic according to data in references. Results from the application on their two derivatives were not unexpected - 20 μg/ml no effected cells.

The four examined derivatives (all compounds were converted into HCl salts) and acyclovir as reference drug were applied in concentrations 10, 5, 1 and 0.5μg/ml. The 2-aminomethyl-thiazole-4-yl-ACV, HCl.Val-thiazole-4-yl-ACV, HCl.Ala-thiazole-4-yl-ACV and HCl.Leu-thiazole-4-yl-ACV shown insignificant effects on the herpesvirus replication – 20, 10, 20 and 8% inhibition respectively. Whereas the reference drug inhibited the viral replication completely in same dose (10μg/ml).

These results suggest that 2-aminomethyl-thiazole-4-yl-ACV, HCl.Val-thiazole-4-yl-ACV, HCl.Ala-thiazole-4-yl-ACV and HCl.Leu-thiazole-4-yl-ACV may be attractive in higher concentrations for antiviral chemotherapy obligatory with lower cytotoxicity effect in comparison with the effective nucleoside analogs.

Compound Ala-thiazole-4-yl-ACV is active against resistant HSV-1 strain.

Conclusion

Design of amino acid prodrugs seems to be an attractive strategy to enhance the solubility of the otherwise poorly aqueous soluble compounds and also to afford a targeted and possibly enhanced delivery of the active drug.

An implicit proof of this assumption lies in the fact that L-valyl ester of acyclovir (valacyclovir) shows bioavailability of 60%.

References

1. De Clercq E, Field HJ, *British J. Pharmac.* **147**, 1-11 (2006)
2. Beauchamp, L. M., Orr, G. F., de Miranda, P., Burnette, T. and Krenitsky T. A., *Antiviral Chemistry & Chemotherapy,* **3**, 157-164 (1992).
3. Beauchamp LM and Krenitsky TA, *Drugs Future* **18**: 619-628 (1993).
4. Anand BS, Katragadda S and Mitra AK, *J. Pharmacol. Exp. Ther.* **311**: 659–667 (2004).
5. Anand BS and Mitra AK, *Pharm Res.,* 1194–1202 (2002).
6. Moss, N., Beaulieu, P., Duceppe, J., Ferland, J., Garneau, M., Gauthier, J., Ghiro, E., Goulet, S., Guse, I., Jaramillo, J., Llinas-Brunet, M., Malenfant, E., Plante, R., Poirier, M., Soucy, F., Wernic, D., Yoakim, C. and Deziel, R., *J. Med. Chem.*, **39**, 4173-4180 (1996).
7. G. Videnov, D. Kaiser, C. Kempter and G.Jung , *Angew. Chem*, **108**, 1604-1607 (1996).
8. Stankova I. G., Videnov G. I., Golovinsky E.V., and Jung G., *J. Peptide Sci*, **5**, 392-398 (1999).

Poly-(Proψ[CH(OAc)CH_2]Aib)$_3$ a Protected Trimer of Pro-Aib Hydroxyethylene Isosteres

T. Lama, G. Fridkin and W. D. Lubell

Departement de Chimie, Université de Montréal, C.P. 6128, Succursale Centre Ville, Montréal, Québec, Canada H3C 3J7

Introduction

The β-bend ribbon secondary structure has been characterized in channel forming peptaibol antibiotics, such as zermavicin [1]. The alternation of L-Pro and Aib residues leads to a right-handed β-bend ribbon helix [2] as characterized by X-Ray analyses of *p*-BrBz-Aib-(L-Pro-Aib)*n*-OMe (*n* = 3 and 4) [3]. Exploring the importance of the amides in the β-bend ribbon, we have begun replacing Pro-Aib units with the corresponding hydroxyethylene isosteres. Syntheses of *p*BrBz-(Pro-[ψ-CH-(OAc)-CH_2]-Aib)$_3$-OMe, possessing (*R*)- and (*S*)-acetate stereochemistry, have now been achieved and the latter analog has been characterized by X-Ray analysis.

Results and Discussion

Both alcohol stereoisomers of the hydroxyethylene isostere of Pro-Aib were made in enantiopure form by routes commencing with Cu-catalyzed cascade addition of 2-methyl-1-propenylmagnesium bromide to Boc-Pro-OMe [4]. Homoallylic acetates (*S*,*R*)- and (*S*,*S*)-**2** were respectively obtained from alcohols (*S*,*R*)- and (*S*,*S*)-**1** [4] in 90% and 93% yield by acetylation using Ac_2O and pyridine in DMF. Ozonolysis of olefin **2** and oxidation of crude aldehyde with an aqueous solution of NaH_2PO_4 and $NaClO_2$ gave δ-amino γ-acetoxy acids (*S*,*R*)- and (*S*,*S*)-**3** in 93% and 95% overall yields, respectively.

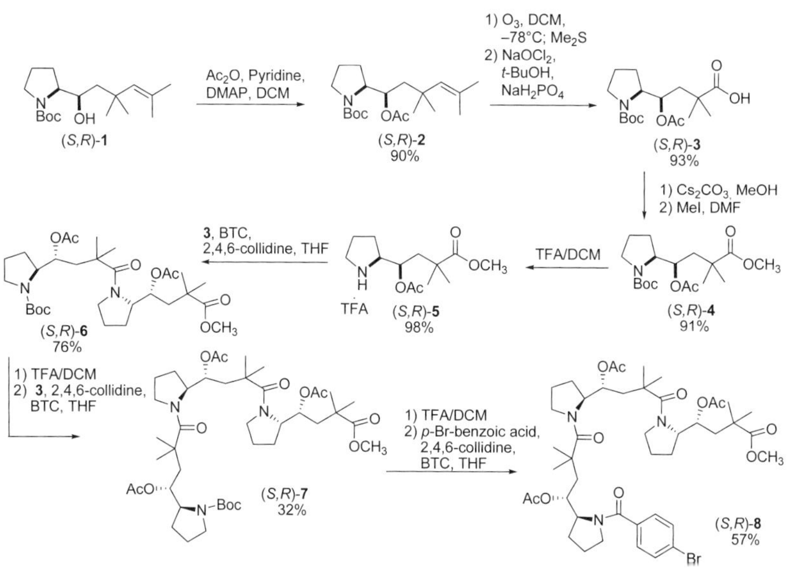

*Scheme: Synthesis of pBrBz (Pro-[ψ-CH-(OAc)-CH2]-Aib)$_3$-OMe (S,R)-**8***

S. Del Valle et al. (eds.), *Peptides for Youth: The Proceedings of the 20th American Peptide Symposium*, 171
DOI: 10.1007/978-0-387-73657-0_78, © Springer Science+Business Media, LLC 2009

Methyl esters (*S*,*R*)- and (*S*,*S*)-**4** were made by alkylation of the cesium salt of acid **3** using methyl iodide in DMF in 91% and 95% yields, respectively. Removal of the Boc protection was achieved quantitatively using a 1:1 TFA/DCM solution to afford δ-amino γ-acetoxy methyl esters (*S*,*R*)- and (*S*,*S*)-**5** as TFA salts.

Oligomers **8** were respectively assembled by sequential couplings of acid **3** to amino ester **5**, as described for the (*S*,*R*)-isomer in the Scheme. Activation of acids (*S*,*R*)- and (*S*,*S*)-**3** using bis-(trichloromethyl)carbonate (BTC) in THF in the presence of 2,4,6-collidine and respective coupling to methyl esters (*S*,*R*)- and (*S*,*S*)-**5** gave the desired amides (*S*,*R*)- and (*S*,*S*)-**6** in 76% and 89% yield, respectively, after chromatographic purification on silica gel. Employing similar conditions for removal of the Boc group from **6** and for coupling of acid **3**, as described above, oligomers (*S*,*R*)- and (*S*,*S*)-**7** were obtained in 32% and 59% overall yields, respectively. Similarly, removal of the Boc group from **7** and acylation with *p*-bromobenzoic acid gave *p*BrBz-(Pro-[ψ-CH-(OAc)-CH_2]-Aib)$_3$-OMe (*S*,*R*)- and (*S*,*S*)-**8** in 57% and 62% yields, respectively.

Oligomer (*S*,*S*)-**8** crystallized from EtOH and was analyzed by X-Ray diffraction. Unlike the parent *p*-BrBz-Aib-(L-Pro-Aib)$_3$-OMe, acetoxyethylene isostere oligomer (*S*,*S*)-**8** adopted a relatively extended structure (Figure). This conformational difference may be ascribed to the absence of hydrogen-bond donors preventing intra-molecular hydrogen bonds. Attempts to crystallize (*S*,*R*)-**8** have yet to be successful.

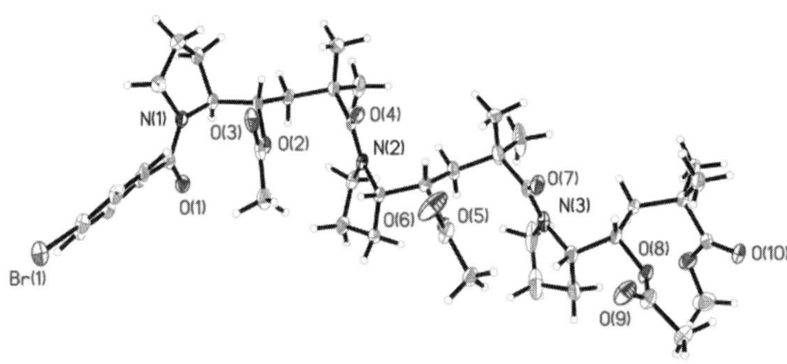

*Figure: X-ray structure of pBrBz-(Pro-[ψ-CH-(OAc)-CH$_2$]-Aib)$_3$-OMe (S,S)-**11***

Acknowledgments

This research was supported by the Natural Sciences and Engineering Research Council (NSERC) and the Ministére de l'Éducation du Québec. The authors would like to thank Françine Bélanger-Gariépy from the Regional Laboratory of X-ray Analysis at the Université de Montréal for X-ray structure determination.

References

1. Pandey, R.C., Cook, J.C., Rinehart, K.L. *J. Am. Chem. Soc.* **99**, 8469-8483, (1977).
2. Benedetti, E. *Biopolymers* **40**, 3-44, (1996).
3. Di Blasio, B., Pavone, V., Saviano, M., Lombardi, A., Nastri, F., Pedone, C., Benedetti, E., Crisma, M., Anzolin, M., Toniolo, C. *J. Am. Chem. Soc.* **114**, 6273-6277, (1992).
4. Lama, T., Del Valle, S.E., Genest, N., Lubell, W.D. *Int. J. Pep. Res. Ther.* **13**, 355-366, (2007).

Fast Conventional Synthesis of Human ß-Amyloid (1-42) on the Symphony® and Prelude™

German Fuentes, Christina Hood, Jae H. Park, Hirendra Patel, Karen Page and Mahendra Menakuru

Protein Technologies, Inc. Tucson, AZ 85714, USA.

Introduction

Human ß-amyloid (1-42) (Sequence: H_2N-DAEFRHDSGYEVHHQKLVFFAEDV GSNKGAIIGLMVGGVVIA-COOH) is a major component of the plaque deposits found in the brains of Alzheimer's disease patients [1]. Its synthesis has been reported to be difficult due to the high hydrophobicity of the C-terminal segment and subsequent on-resin aggregation [2]. During the years, different ways of chemically speeding up solid-phase peptide synthesis reactions have been proposed [3]. Several investigators have contributed to these studies [4,5,6,7] and the conclusion is that by choosing the most appropriate resin, deprotectant, and coupling agent, it is possible to reduce the total synthesis time of some peptides [2b] without compromising the quality of the crude material obtained. In this study the goal was to synthesize the human ß-amyloid (1-42) peptide at conservative (54 hrs) and reduced (12 hrs) synthesis times and observe the effects of different resins and activators (HCTU/6Cl-HOBT and HATU) on the purity of the crude peptide. 2% DBU was added to the 20% piperidine/DMF deprotectant solution to speed up the deprotection reaction.

Results and Discussion

The peptide was synthesized in a Symphony® (25 μmol scale) and Prelude™ (20 μmol scale) peptide synthesizer from Protein Technologies, Inc. on Fmoc-Ala-Wang ChemMatrix resin (0.56 mmol/g), Fmoc-Ala-HMPB-ChemMatrix resin (0.54 mmol/g) or Fmoc-Ala-Wang-PS-LL resin (0.25 mmol/g) using various protocols (Table 1). Crude peptides were precipitated with ether, dissolved in hexafluro 2-propanol and analyzed by C_{18} RP-HPLC (Varian ProStar) heated at 60°C with a gradient of 5-95% acetonitrile in aq. 0.1% TFA. Shorter synthesis times produced cleaner crude peptide (Figures 1-3). The mixture of 2% DBU/20% piperidine/DMF provided a fast and efficient deprotection agent. Peptides coupled with HCTU/6-Cl-

Table 1. Reaction, washing, cycle and synthesis times for the different synthesis protocols.

	PROTOCOL 1 [a]	PROTOCOL 2 [b]	PROTOCOL 3
Deprotection Time:	1 x 15 min	2 x 2.5 min	2 x 1 min
Coupling Time:	1 x 35 min	2 x 10 min	1 x 5 min
Capping Time:	1 x 5 min	None	None
Washing Time:	6 x 30 sec	6 x 30 sec	2 x 30 sec
Cycle Time:	1.3 hours	44 min	18 min
Total Synthesis Time:	54 hours	30 hours	12 hours

[a] *Modified protocol from Garcia-Martin [8] for the synthesis of β-amyloid (1-42)*
[b] *PTI's standard quality control protocol*

S. Del Valle et al. (eds.), *Peptides for Youth: The Proceedings of the 20th American Peptide Symposium*,
DOI: 10.1007/978-0-387-73657-0_79, © Springer Science+Business Media, LLC 2009

HOBt produced peptides of comparable or slightly lower purity than those coupled with HATU, but with significant cost savings. HMPB-ChemMatrix resin with HATU produced the highest purity crude peptide, although Wang-PS-LL resin with HCTU produced crude peptide of comparable purity at a lower cost, as well as higher crude peptide yields. Isolation of the major product peak by HPLC and subsequent analysis by MALDI-TOF mass spectroscopy (4513.70 m/z) identified it as the full-length β-amyloid peptide (data not shown)

References

30 Hours	**12 Hours**	**12 Hours**
HCTU/6-Cl/HOBt	HCTU/6-Cl/HOBt	HATU

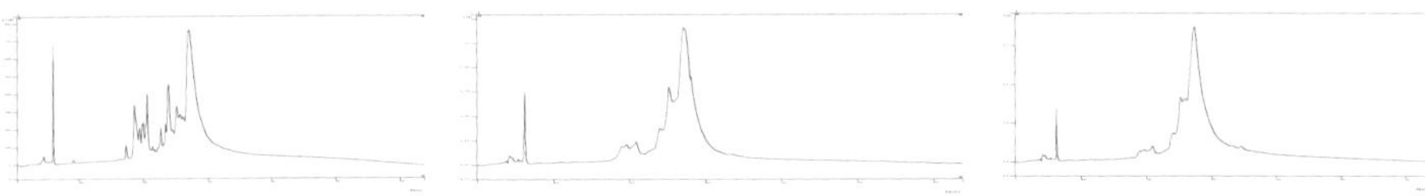

Fig. 1. β-amyloid (1-42) synthesized on Fmoc-Ala-Wang-ChemMatrix resin (0.56 mmol/g).

30 Hours	**12 Hours**	**12 Hours**
HCTU/6-Cl/HOBt	HCTU/6-Cl/HOBt	HATU

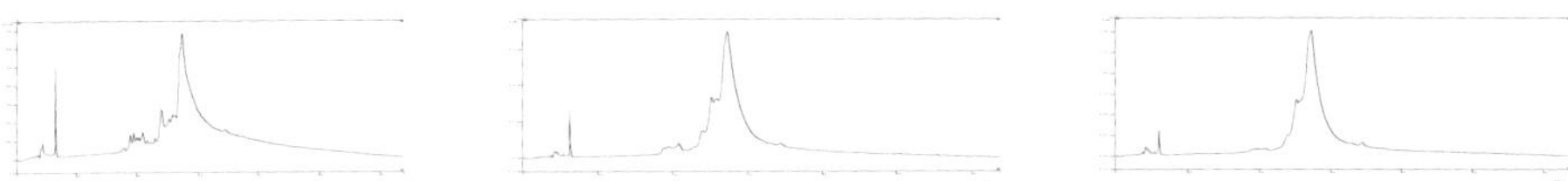

Fig. 2. β-amyloid (1-42) synthesized on Fmoc-Ala-HMPB-ChemMatrix resin (0.54 mmol/g).

30 Hours	**12 Hours**	**12 Hours**
HCTU/6-Cl/HOBt	HCTU/6-Cl/HOBt	HATU

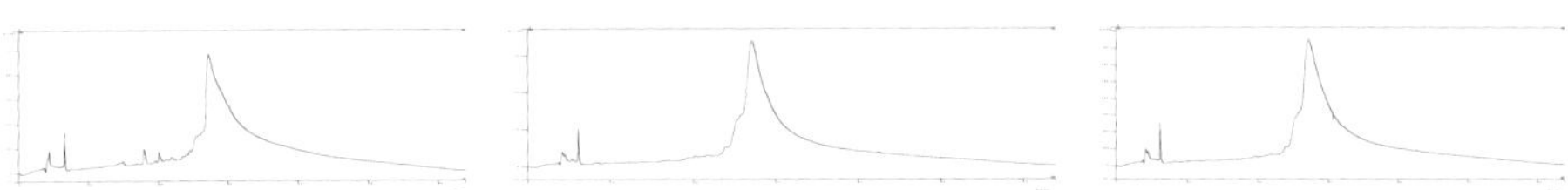

Fig. 3. β-amyloid (1-42) synthesized on Fmoc-Ala-Wang-PS-LL resin (0.25 mmol/g).

1. Burdick, D., Soreghan, B., Kwon, M., Kosmoski, J., Knauer, M., Henschen, A., Yates, J., Cotman, C., and Glabe, C. *J. Biol. Chem.* **267**, 546-554 (1992).
2. (a) Quibell, M., Turnell, W. G., and Johnson, T. *J. Org. Chem.* **59**, 1745-1750 (1994). (b) Tickler, A. K., Clippingdale, A. B., and Wade, J. D. *Prot. Pep. Lett.* **11**, 377-384 (2004).
3. Alewood, P., Alewood, D., Miranda, L., Love, S., Meutermans, W., and Wilson, D. *Methods in Enzymology* **289**, 14-29 (1997).
4. Miranda, L. P. and Alewood, P. F. *Proc. Natl. Acad. Sci USA* **96**, 1181-1186 (1999).
5. Hetnarski, B. and Merrifield, R. B. *Peptides, Chemistry and Biology*, Marshall, G.R., Ed. (ESCOM, Leiden, the Netherlands, 1988) pp. 220-224.
6. Atherton, E. and Sheppard, R. C. *Solid Phase Peptide Synthesis: A Practical Approach* (IRL Press at Oxford University Press, 1989), p. 116.
7. Wade, J. D., Bedford, J., Sheppard, R. C. and Tregear, G. W. *Pep. Res.* **4**, 194-199 (1991).
8. Garcia-Martin, F., Quintanar-Audelo, M., Garcia-Ramos, Y., Cruz, L. J., Gravel, C., Furic, R., Cote, S., Tulla-Puche, J., and Albericio, F. *J. Comb. Chem.* **8**, 213-220 (2006).

Side chain-to-Side chain Cyclization by Intramolecular Click Reaction - Building Blocks, Solid Phase Synthesis and Conformational Characterization

Sonia Cantel[1,2], Jose A. Halperin[1,2], Michael Chorev[1,2], Mario Scrima[3], Anna M. D'Ursi[3], Jay J. Levy[4], Richard D. DiMarchi[4], Alexandra Le Chevalier[5], Paolo Rovero[5] and Anna M. Papini[5]

[1]Laboratory for Translational Research, Harvard Medical School, One Kendall Square, Building 600, Cambridge, MA 02139, USA; [2]Department of Medicine, Brigham and Women's Hospital, 75 Francis Street, Boston, MA 02115, USA; [3]Dipartimento di Scienze Farmaceutiche, Via Ponte Don Melillo 11C, Salerno 84084, Italy; [4]Department of Chemistry, Indiana University, 800 E. Kirkwood, Bloomington, IN 47405, USA; [5]Laboratorio di Chimica dei Peptidi, Dipartimento di Chimica Organica "Ugo Schiff", Via della Lastruccia 13, 50019 Sesto F.no (Firenze), Italy.

Introduction

Side chain-to-side chain cyclization is frequently employed to stabilize and select specific conformations and to reduce susceptibility toward proteolytic degradation. Among the numerous modes of cyclization bioisosteric modifications that can be accomplished without elaborate orthogonal protection schemes are of great interest. The recently introduced Cu^I-catalyzed azide–alkyne 1,3-dipolar Huisgen's cycloaddition [1] as a prototypic "Click reaction" [2] presented a promising opportunity to develop a new paradigm for intramolecular side chain-to-side chain cyclization. The inherent high energetics, chemical selectivity, and the demonstrated bioorthogonal characteristics of the azide and alkynyl functions, as well as the proteolytic stability, and the peptidomimetic nature of the cycloaddition product, the 1,4-disubstituted [1,2,3]triazolyl moiety, suggest that 1,4-bridged [1,2,3]triazolyl-mediated side chain-to side chain cyclization could offer an interesting mode of structural constraint.

To this end, we have developed the building blocks, explored the methodologies and synthesized a model peptide in which side chains of residues in positions i and i+4 are cyclized through a 1,4-disubstituted [1,2,3]triazole moiety formed by an intramolecular click reaction (Fig. 1.). The parent cyclic i-to-i+4 side chain to side chain peptide [Lys[13],Asp[17],Tyr[34]]hPTHrP(7-34)NH_2 was previously reported to be a potent antagonist of the parathyroid receptor 1 (PTHR1). This peptide contains a lactam ring between Lysine and Aspartic acid residues at positions *i* and *i+4*, respectively, that stabilizes an extended α-helical conformation [3,4].

hPTHrP (7-34) H-AVSEHQLHDKGKSIQDLRRRFFLHHLIAEIHTA-NH_2

Lactam model (I) Ac-KGKSIQDLR-NH_2

┌1,4-[1,2,3]triazolyl┐

Triazole model (II) Ac-KG-Nle----S-----I----Q----A-LR-NH_2

Fig. 1. Sequences of hPTHrP (7-34)NH_2, lactam- and1,4-[1,2,3] triazolyl-cyclopeptides.

S. Del Valle et al. (eds.), *Peptides for Youth: The Proceedings of the 20th American Peptide Symposium*, 175
DOI: 10.1007/978-0-387-73657-0_80, © Springer Science+Business Media, LLC 2009

Results and Discussion

To introduce the L-2-amino-6-azido-hexanoic acid residue [Nle(ε-N_3)] into the resin-bound peptide, we followed two different strategies. While in the first pathway, the Nle(ε-N_3) residue is formed via a diazo transfer reaction on the selectively deprotected ε-amino moiety of the otherwise fully protected resin-bound peptide. In the second pathway, Fmoc-Nle(ε-N_3)-OH was incorporated as part of the standard stepwise on-resin peptide assembly involving repetitive cycles of coupling and deprotection steps.

All our attempts to carry out on-resin cyclization via intra-chain Cu^I-catalyzed side chain-to-side chain azide-alkyne 1,3-dipolar cycloaddition employing a wide range of reported conditions failed. Failure to carry out on-resin intramolecular Cu^I-catalyzed click reaction led to the undertaking of the solution phase approach. Simultaneous deprotection and cleavage from resin generated the free peptide that was subjected to effective intramolecular click reaction-mediated cyclization. The model lactam peptide was generated on-solid support as previously reported [4].

The comparison of the most representative NMR structures relative to lactam- and [1,2,3]triazolyl-containing cyclopeptide **I** and **II** shows that both peptides assume an α-helical structure in the cyclic part of the molecules (Fig. 2.). The structures of two peptides differ regarding the location of the turn-helical segment, which in **I** involves non-cyclized residues while in **II** it overlaps with residues involved in the cyclic structure. A careful inspection of the cyclic portions shows that, despite a slight difference of the backbone arrangements, the side chains share the same spatial orientation. In particular those of Ser, Gln, and Ile that are located at the i+1-to-i+3 of the ring forming sequence (Fig. 2). This common spatial orientation of the side chains in both cyclic peptides suggests that they will share a common interaction with the same macromolecular target, which may result in similar biological response if these interaction are critical for biological activity.

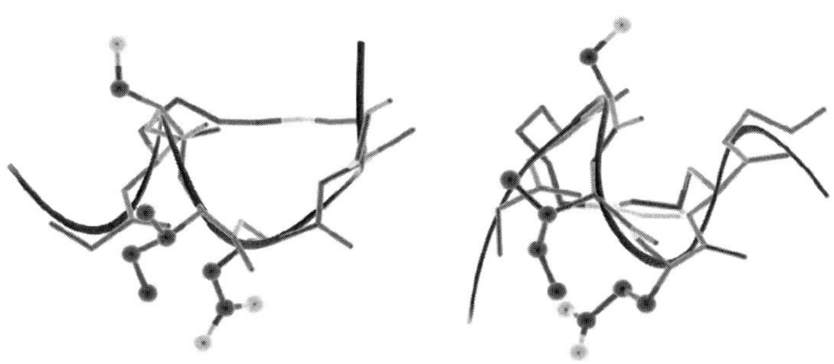

Fig. 2. Comparison of representative NMR structures of lactam- and [1,2,3]triazolyl-containing cyclopeptides (left and right structures, respectively).

Acknowledgments

The work was funded by FIRB internazionalizzazione 2005 RBIN04TWKN.

References

1. Rostovtsev, *et al. Angew Chem Int Ed Engl* **41**, (14), 2596-9, 2002.
2. Kolb, *et al. Angew Chem Int Ed Engl* **40**, (11), 2004-21, 2001.
3. Chorev, *et al. Biochemistry* **30**, (24), 5968-74, 1991.
4. Maretto, *et al. Biochemistry* **36**, (11), 3300-7, 1997.

An Efficient Approach to Synthesize Multiple Peptides with Carboxyl C-termini

Koichiro Hayashi, Sylvia Chen, Paul Opare-Addo and Ashok Khatri

Protein/Peptide Core, Endocrine Unit, Massachusetts General Hospital, Boston, MA 02114, USA

Introduction

There is continuing search for strategies that can be used in the easy preparation of small scale multiple peptides with free carboxyl C-termini in a 96-well reaction block. Automated Fmoc solid phase chemistry is very convenient for the synthesis of these small scale multiple peptides that can serve as biologically active molecules for immunological and physiologic studies. Several types of pre-loaded resins such as Wang, HMP, Tentagel, and 2-Chlorotrityl resins are commercially available for peptide synthesis [1]. However, transferring of resins with different C-terminal amino acids into designated wells of a 96-well reaction block is a cumbersome and time-consuming process. This study presents an easier method for resin transfer into wells and the advantage of using unloaded 2-Chlorotrityl chloride resins over the pre-loaded resins. In this study, four peptides (Table 1) were chosen because of their easy synthesis and successfully documented results. One of them was Acyl Carrier Protein, ACP (65-74) that is used as a standard for testing peptide synthesizers as well as the efficacy of different synthesis protocols. Three Parathyroid hormone (PTH) analogs were also included in this study. Unloaded 2-Chlorotrityl chloride resin, pre-loaded Wang resin, and pre-loaded Tentagel resins were used to synthesize the above four peptides.

Table 1. Sequences of ACP and Three Parathyroid hormone (PTH) analogs.

ACP (65-74)	VQAAIDYING-OH
h-PTH (1-14)	SVSEIQLMHNLGKH-OH
h-PTH (1-20)	SVSEIQLMHNLGKHLNSMER-OH
h-PTH(1-34)analog	AVAEIQLMHQRAKWLNSMEAVERLRKKLQDVHNF-OH

This paper compares the automated loading of 2-Chlorotrityl chloride resins vs. the use of pre-loaded resins for the synthesis of these peptides.

Results and Discussion

The peptides were assembled on the loaded resins using 5-micromoles scale, 5-fold excess of Fmoc-amino acids using Fmoc/tBu chemistry and HBTU/HoBt/DIEA activation [2] on a MultiPep (Intavis AG, Germany) Instrument. Resins were swollen in DMF and DCM mixture (1:1, v/v), and the mixture was well shaken for equal distribution of the resin beads by pipetting aliquots into individual wells of a 96 well-reaction block. In the case of the 2-Chlorotrityl chloride resins, the first amino acids were coupled to the unloaded resin using a five-fold molar excess of Fmoc-amino acid, 10 fold molar excess of DIPEA for 90 minutes double coupling [3]. Methanol capping was done with DCM/MeOH/DIPEA (80:15:5 v/v) for 60 minutes, and peptides were synthesized immediately after capping was complete. Peptides were cleaved from resins and deprotected using Reagent K [4] for 3 hours at ambient

S. Del Valle et al. (eds.), *Peptides for Youth: The Proceedings of the 20th American Peptide Symposium*, 177
DOI: 10.1007/978-0-387-73657-0_81, © Springer Science+Business Media, LLC 2009

temperature. Crude peptides were precipitated using cold MTBE. Unpurified peptides were evaluated on Waters HPLC using a Vydac, C-18, 4.6x150mm column. Buffer A was 0.1%TFA/Water and Buffer B was 0.1%TFA/Acetonitrile. A linear (5% to 65% B) gradient with a flow rate of 1.0ml/min was employed and the spectra were recorded at 214nm wavelength. For molecular weight verification of crude peptides, MALDI-TOF/MS using Voyager DE-Pro MALDI mass spectrometer, Applied Biosystems (Foster City, CA) was used.

HPLC profiles of the unpurified peptides indicate that all resins-types eventually had nearly similar results (Figure 1) for the peptides up to 20 amino acids in length even though this synthesizer does not have any shaking of resins during the synthesis process. However, the quality of the 34mer peptide was superior with the tentagel resin due to higher diffusion properties and low amino acid substitution. Wang resin has higher loading capacity than other resins used in this study but it does not swell as well as the 2-Chlorotrityl chloride resin. In addition, the Wang and Tentagel resins required longer cleavage times and higher volumes of cleavage mixture. Cleavage had to be repeated in order to get good results for tentagel and wang resins.

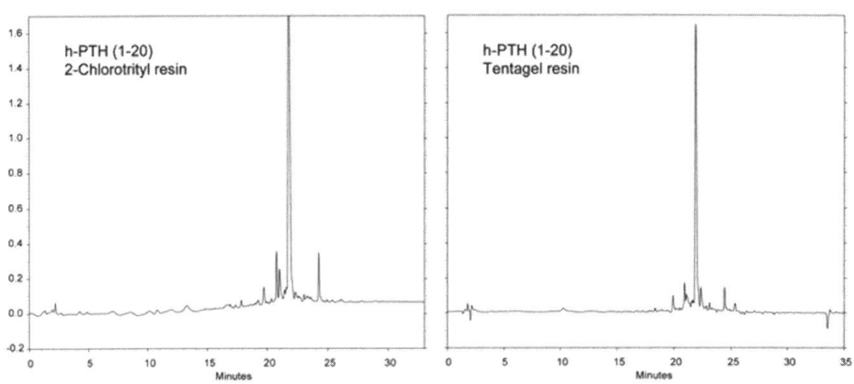

Fig. 1. HPLC profiles for h-PTH (1-20) on different types of resins.

In our experience, loading of the 2-Chlorotrityl chloride resin is simple, fast, convenient and economical. Moreover, as the cleavage of the peptide from this resin is more efficient, higher yields of peptides were achieved as compared to other resins.

Acknowledgments

We would like to thank Tufts University Core Facility for providing assistance in Mass Spectrometry analysis of peptides.

References

1. Chan, W. C., and White, P. D. in *Fmoc Solid Phase Peptide Synthesis*, Oxford University Press 45, (2000)
2. Fields, C. G., Lloyd, D. H., Macdonald, R. L., Otteson, K. M., and Noble, R. L. *Peptide Research.* **4**, 95-101 (1991)
3. Barlos, K., Chatzi, O., Gatos, D., and Stavropoulos, G. *Int. J. Pept Protein Res* **37**, 513-520 (1991)
4. King, D., Fields, C., and Fields, G. *Int. J. Pept. Protein Res.* **36**, 255-266 (1990)

Synthesis and Biological Activity of Novel α-Conotoxin Analogues Incorporating Substituted Proline Derivatives

Christopher J. Armishaw[1], Anders A. Jensen[1], Christian Skonberg[2], Tommy Liljefors[1] and Kristian Strømgaard[1]

[1]Department Medicinal Chemistry and [2]Department of Pharmaceutics and Analytical Chemistry; Faculty of Pharmaceutical Sciences, University of Copenhagen, Universtitetsparken 2, DK-2100, Copenhagen Ø, Denmark

Introduction

α-Conotoxins are peptide neurotoxins, which are potent and selective inhibitors of neuronal nicotinic acetylcholine receptors (nAChRs)[1]. They are characterised by the presence of two disulfide bonds and a highly conserved proline residue (Pro6) that contributes to their well defined 3D structures (Figure 1). Acetylcholine binding protein from *Aplysia californica* (*Ac*-AChBP) shares sequence homology with the ligand binding domain of nAChRs, thus provides insight for the rational design of new analogues with improved pharmacology. Recent crystal structures of α-conotoxins ImI and PnIA[A10L,D14K] complexed with *Ac*-AChBP provide further insight of ligand binding at the molecular level [2, 3]. These show that upon binding, the side chain of Pro6 is oriented towards the hydrophobic binding pocket in AChBP (Figure 2). We propose that additional interactions with the hydrophobic binding pocket could be gained through introduction of substituents on the proline side-chain without perturbing the overall fold of the peptide. Further experimental data into the role of specific amino acids in the nAChR will provide further insight into α-conotoxin / α7 nAChR binding.

ImI	GCCSD**P**RCAWR----C
PnIA[A10L]	GCCSL**P**PCALNNPDYC

Fig. 1. Sequences of α-Conotoxins ImI and PnIA[A10L]. The conserved Pro6 highlighted. The disulfide bond connectivity is 1-3, 2-4.

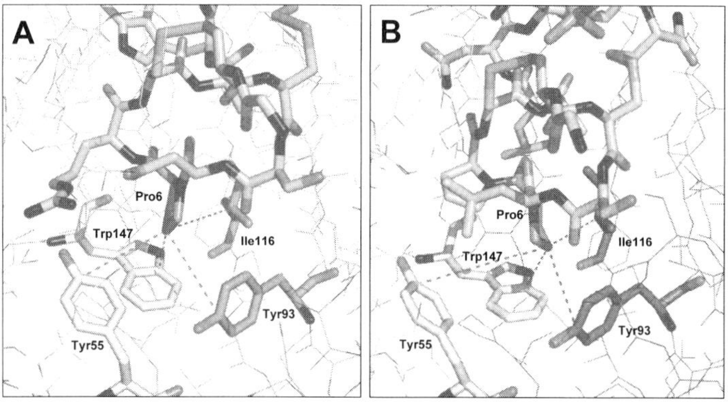

Fig. 2. X-ray crystal structures of (A) α-Ctx ImI and (B) PnIA[A10L,D14K] bound to Ac-*AChBP, showing the interactions with the hydrophobic binding pocket.*

DOI: 10.1007/978-0-387-73657-0_82, © Springer Science+Business Media, LLC 2009

Results and Discussion

Proline derivatives were selected based on expected interactions with hydrophobic binding pocket, including π-cation and π-stacking. The derivatives selected included 4-(*R/S*)-hydroxy, 4-(*R/S*)-amino, 4-(*R/S*)-guanidino and 4-(*R/S*)-betainamidyl proline, 4-(*R/S*)-phenyl and 3-(*S*)-phenyl. Analogues were assembled in parallel using Fmoc-SPPS, with the exception of 4-(*R/S*)-betainamidyl, which required a Boc-SPPS strategy. Oxidation and folding was achieved by dissolving the reduced peptide in 0.1M NH_4HCO_3 using 30% isopropanol. Each analogue folded correctly as confirmed by CD-spectroscopy, however difficulties were encountered folding 4-(*S*)-hydroxy, 4-(*S*)-guanidino and 4-(*S*)-betainamidyl analogues, thus no recoverable quantity was obtained. This was presumably a result of *trans-cis* isomerism of Pro6.

Pharmacological testing at the α7 nAChR was performed in two assays, including [^{3}H]MLA binding experiments with the α7/5-HT_3 chimera and a Ca^{2+}/Fluo-4 functional assay at human α7 in GH3 cells using a NOVOstar system with EC_{80}-EC_{90} values of the agonist (30 μM ACh) (Table 1). These data show that polar and charged substituents on Pro6 have an IC_{50} > 300μM. However, aromatic substituents demonstrated moderate activity, possibly a result of π-stacking of the phenyl substituents with aromatic residues inside the ligand binding pocket. This stacking is also dependent on the orientation and position of the aromatic ring on the proline side chain, with the 3-(*S*) and 4-(*S*) positions in ImI, and 4-(*R/S*) in PnIA[A10L] demonstrating the most potent activity. Future work using different positions on side chain ring and different aromatic functional groups will lead to a greater understanding of the interactions of α-conotoxin Pro6 with nAChRs.

Table 1. Pharmacological assay results.

Analogue	Assay	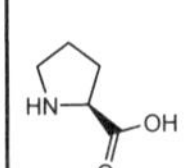	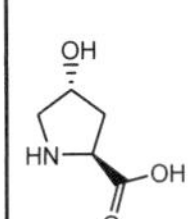	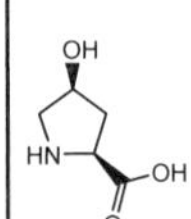	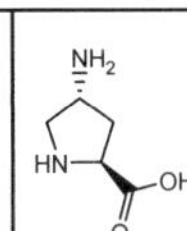	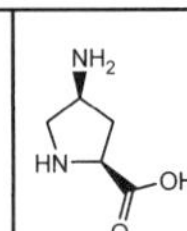	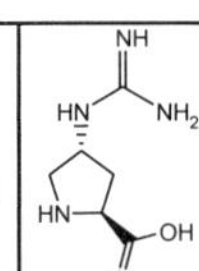	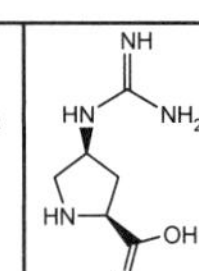
ImI	α7 / 5HT$_3$ Human α7	1.2 μM 2.6 μM	≈100 μM ≈100 μM	ND ND	>100 μM >300 μM	≈100 μM ≈300 μM	≈100 μM >300 μM	ND ND
PnIA[A10L]	α7/5HT$_3$ Human α7	0.26 μM 0.51 μM	≈30 μM >300 μM	ND ND	≈100 μM >300 μM	>100 μM ≈300 μM	>300 μM >300 μM	ND ND

Analogue	Assay	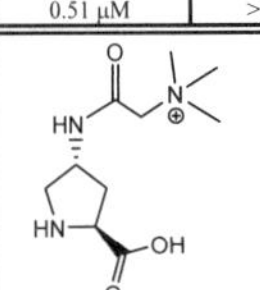	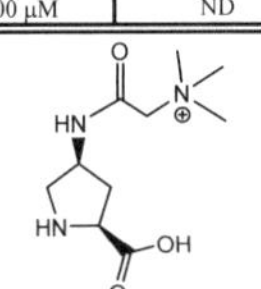	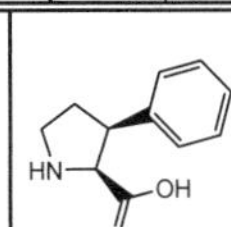	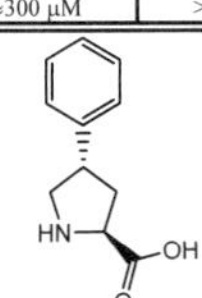	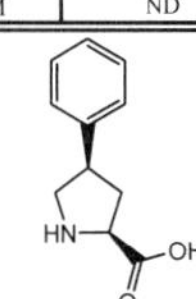
ImI	α7 / 5HT$_3$ Human α7	>100 μM >300 μM	ND ND	6.5 μM 11 μM	>100 μM >300 μM	7.0 μM 14 μM
PnIA[A10L]	α7 / 5HT$_3$ Human α7	≈100 μM ≈300 μM	ND ND	≈100 μM >300 μM	7.6 μM 33 μM	1.7 μM 26 μM

Acknowledgements

CJA is funded by a Drug Research Academy post-doctoral fellowship.

References

1. Armishaw, C. J. and Alewood, P. F., *Curr. Protein Pept. Science*, **6**, 221-240 (2005).
2. Celie, P. H. N., *et al.*, *Nature Struc. Mol. Biol.*, **12**, 582-588 (2005).
3. Hansen, S. B., *et al.*, *EMBO J.*, **24**, 3635-3646 (2005).

Application of Triethylsilane and Palladium-Charcoal-Induced Reductions in the Synthesis of Fmoc-Glutamic Acid Analogues

Pijus K. Mandal and John S. McMurray

The University of Texas M. D. Anderson Cancer Center, Department of Experimental Therapeutics, Houston, TX 77030, U.S.A

Introduction

Glutamine is an important recognition element in protein-protein interactions. Natural amino acid peptides containing glutamine are susceptible to proteolytic cleavage and deamidation by glutaminases. Therefore, to impart stability into peptidomimetics, new glutamine analogues are needed. We report here a facile method for the preparation of Fmoc-protected glutamic analogues in which the α-carboxyl group is replaced by other functionalities. Attachment to Rink resin and peptide elaboration result in peptide mimetics possessing modified glutamine at the C-terminus.

Fukiyama and colleagues, in 1990, reported the use of triethylsilane (TES) and palladium-charcoal catalyst for the reductive conversion of thioesters to aldehydes [1]. Following this lead we found that TES/Pd-C is a widely applicable, very convenient catalytic transfer hydrogenation (CTH) reagent. We observed *in situ* generation of molecular hydrogen by addition of triethylsilane to palladium-charcoal catalyst. These conditions produce rapid and efficient reduction of multiple bonds, azides, imines, and nitro groups as well as benzyl group and allyl group deprotection under mild, neutral conditions [2].

The reactions are carried out at room temperature and are often complete in 10 min or less using excess TES and 10-20% Pd-C (by weight) in MeOH. The conditions are neutral, thus acid or base sensitive substrates can be reduced without harm. Cleavage of benzyl ethers required longer reaction time than benzyl esters or the Z group, thus to some extent a selectivity was observed. The Fmoc group is stable under these conditions. This protecting group, highly used in peptide synthesis, is subject to hydrogenolysis using H_2 and Pd-C [3].

Results and Discussion

Since Fmoc group is stable under TES/Pd-C mediated hydrogenation, selective deprotection of benzyl-based protecting groups is possible. This observation formed the basis of our strategy for the synthesis of glutamic acid analogues. A modification of the procedure of Loukas et al. [4] was employed (Scheme 1).

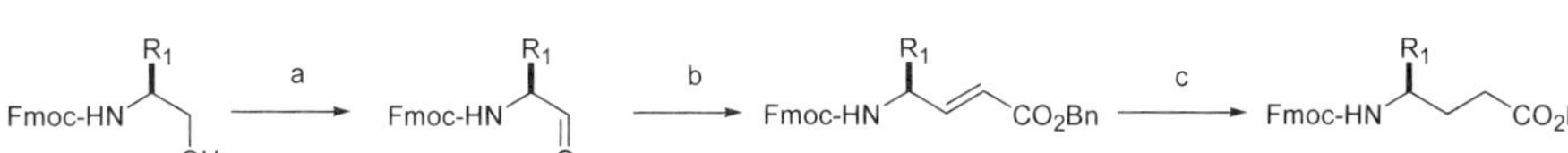

Scheme 1. Synthesis of glutmaine analogues: a) DMSO, Et_3N, ClCOCOCl, CH_2Cl_2, -78°C-room temp. b) Ph_3PCHCO_2Bn, CH_2Cl_2, room temp. c) TES, Pd/C, MeOH, room temp.

Fmoc-amino acids were first reduced to alcohols (**1**) via formation of an active ester followed by sodium borohydride teatment. The alcohols were then oxidized to aldehydes (**2**), which were elongated by Wittig coupling with Ph_3PCHCO_2Bn (**3**)

S. Del Valle et al. (eds.), *Peptides for Youth: The Proceedings of the 20th American Peptide Symposium*, 181
DOI: 10.1007/978-0-387-73657-0_83, © Springer Science+Business Media, LLC 2009

(Scheme 1). Concurrent reduction of the double bond and hydrogenolysis of the benzyl group with TES and Pd/C [2] gave Fmoc-protected glutamic acid analogues (**4**) in which the α-carboxyl group was replaced with various groups (Table 1). D-amino aldehydes result in L-Glu analogues.

The Fmoc-Glu analogues were well behaved in peptide synthesis

Table 1:

Substrate	Product	Yield (%)
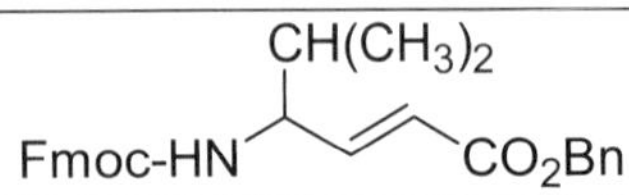	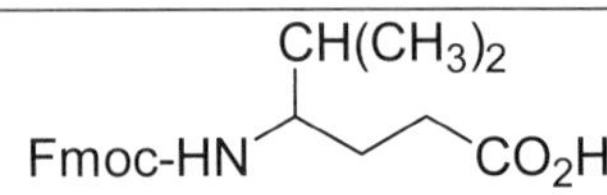	90
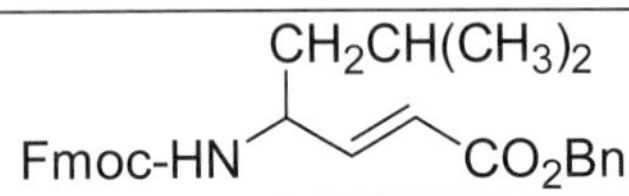	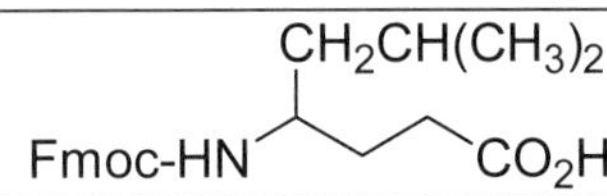	91
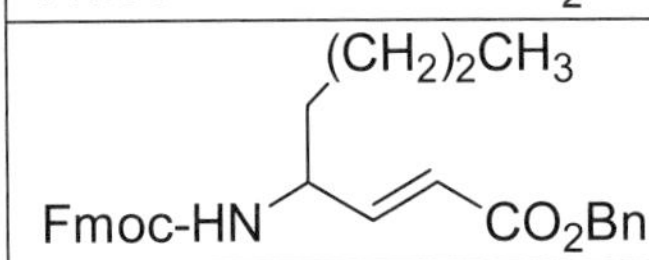	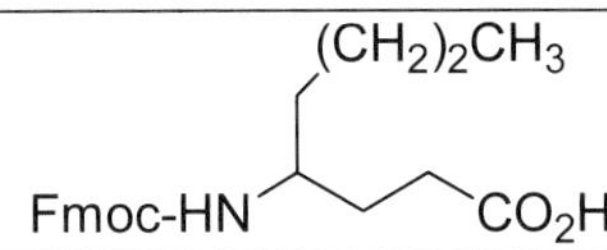	90
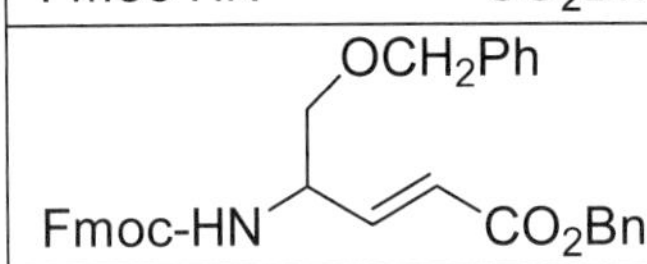	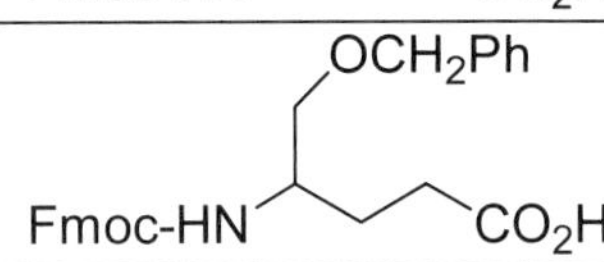	74
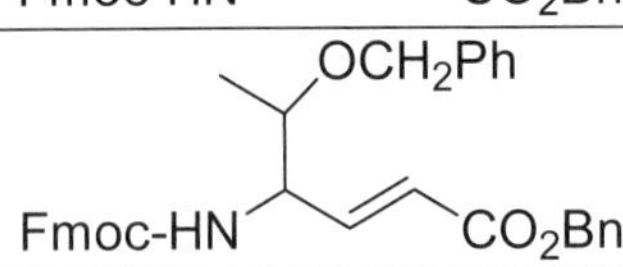	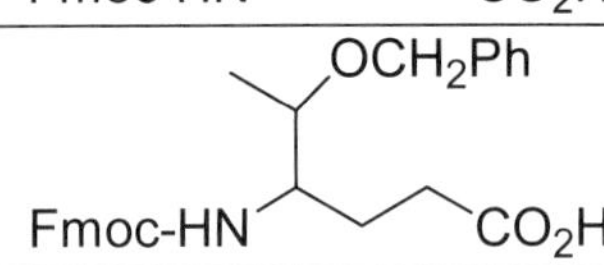	82

Acknowledgments

- MDACC CTDP
- MDACC UCF
- NCI RO1 CA096652
- NCI P30 CA016672

References

1. Fukiyama et al. *JACS*, **1990**, *112*, 7050-7051.
2. Mandal, P.K. et al. *J.Org.Chem,* **2007**, in press.
3. Martinez, J. et al. *J.Org. Chem.* **1979**, *44*, 3596-3598.
4. Loukas et al., *J. Peptide Sci.* **2003**, *9*, 312.

Synthesis of pyrrolo[3,2-*e*][1,4]diazepin-2-ones as potential γ-turn mimetics

Philippe Deaudelin and William D. Lubell

Département de chimie, Université de Montréal, C.P. 6128, Succursale Centre-Ville, Montréal, Québec, Canada, H3C 3J7
lubell@chimie.umontreal.ca

Introduction

Conformationaly rigid mimics of peptides secondary structure have been pursued to overcome issues such as poor bioavailability and rapid metabolism that are associated with peptides.[1] For example, 1,4-diazepin-2-ones have been shown by X-ray analysis to mimic γ-turns.[2] Privileged structures that have shown therapeutic potential against inflammation and autoimmune diseases, 1,4-diazepinones have served as receptor ligands and enzyme inhibitors.[3] For example, the pyrrolobenzodiazepine structure is contained in the anthramycin family of antitumor antibiotics and non-nucleoside reverse transcriptase inhibitors.[4]

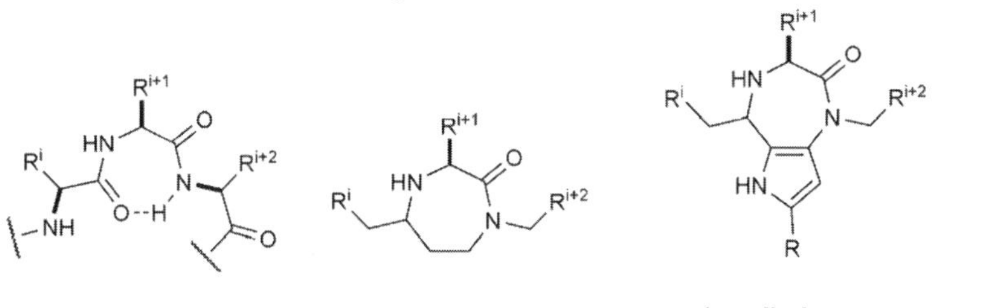

Fig. 1: Representative γ-turn and diazepinone mimics.

Results and Discussion

Pyrrolo[3,2-*e*][1,4]diazepin-2-one **3** was synthesized from 4-hydroxyproline by a route featuring an intermolecular Pictet-Spengler cyclization using an aminoacyl-aminopyrrole intermediate. Methyl 4-*N*-benzylaminopyrrole carboxylate **1** was synthesized in four steps from 4-hydroxyproline using the same protocol reported for the synthesis of its benzyl ester counterpart.[5] Methyl 4-[*N*-(Fmoc) isoleucinyl]aminopyrrole-2-carboxylate was prepared by acylation of aminopyrrole **1** using Fmoc-Ile-OH, triphosgene and 2,4,6-collidine in THF in 75% yield.[6] Removal of the Fmoc group with 5% piperidine in DMF yielded the free amine which was converted to the corresponding HCl salt **2** using a 0.1N HCl solution in THF and isolated after freeze-drying in 85% yield.

Pyrrolo[3,2-*e*][1,4]diazepin-2-one **3** was isolated as a single diastereoisomer in 43% yield after chromatography of the Pictet-Spengler reaction mixture from treating amine hydrochloride **2** and benzaldehyde (0.95 mol%) in toluene with TFA (1.1 mol%). In addition, air oxidation of **3** occurred during its synthesis to afford 3,6-dihydropyrrolo[3,2-*e*][1,4]diazepin-2-one **4** in 10% yield. The configuration of the newly formed chiral center in diazepinone **3** was assigned a *cis* relative stereochemistry with respect to the side-chain derived from Ile by a NOESY experiment. Transfer of magnetization was observed between the α-proton of the Ile residue and the proton at the position 5 of the diazepinone ring.

S. Del Valle et al. (eds.), *Peptides for Youth: The Proceedings of the 20th American Peptide Symposium*, 183
DOI: 10.1007/978-0-387-73657-0_84, © Springer Science+Business Media, LLC 2009

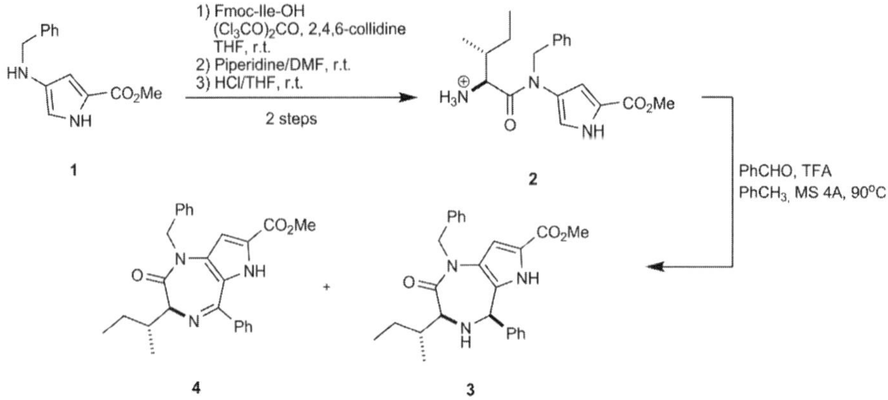

*Fig. 2: Synthesis of pyrrolo[3,2-e][1,4]diazepin-2-ones **3** and **4**.*

The proof-of-concept Pictet-Spengler cyclization has now afforded diazepinone **3**. Work is now in progress to develop this route for making diazepinone libraries by employing different amines, amino acids and aldehydes in the sequence.

Acknowledgments
This work was funded by the Natural Sciences and Engineering Research Council of Canada and the Fonds de Recherches sur la Nature et les Technologies du Québec.

References

1. a) Cluzeau, J.; Lubell, W. D. *Biopolymers Pept. Sci. Rev.* **2005**, *80*, 98.
   b) Glenn, M. P.; Fairlie, D. P. *Mini Rev. Med. Chem.* **2002**, *2*, 433.
2. Iden, H. S.; Lubell, W. D. *Org. Lett.* **2006**, *8*, 3425.
3. Wattanasin, S.; Kallen, J.; Myers, S.; Guo, Q.; Sabio, M.; Ehrhardt, C.; Albert, R.; Hommel, U.; Weckbecker, G.; Welzenbach, K.; Weith-Schmidt, W. *Bioorg. Med. Chem. Lett.* **2005**, *15*, 1217.
4. a) Antonow, D: Cooper, N.; Howard, P. W.; Thurston, D. E. *J. Comb. Chem.* **2007**, *9*, 473.
   b) De Lucca, G. V.; Otto, M. J. *Bioorg. Med. Chem. Lett.* **1992**, *2*, 1639.
5. a) Sharma, R.; Lubell, W. D. *J. Org. Chem.* **1996**, *61*, 202.
   b) Marcotte, F.A.; Lubell, W. D. *Org. Lett.* **2002**, *4*, 2601.
6. Falb, E.; Yechzelkel, T.; Salitra, Y.; Gilon, C. *J. Pept. Res.* **1999**, *53*, 507.

Improvements in the Chemical Synthesis of Insulin-Like Peptides

Panagiota Tsetseni, Aggeliki Karkantzou, Spyros Markos, Kostas Barlos, Dimitrios Gatos and Kleomenis Barlos

Department of Chemistry, University of Patras, GR 26500 Patras, Greece

Introduction

The synthesis of insulin and insulin-like peptides is performed by the selective combination of the independently synthesized A- and B-chains. The main strategy followed for the chain combination is to protect the side-chains of the cysteine residues by three different protecting groups of gradually increased acid and oxidative sensitivity. Before starting the chain combination, the intramolecular disulfide bond formation in A-chain is performed by treatment the thiol-free peptide with dipyridyldisulfide. The interchain combination is then performed again by using selective disulfide bond formation by activating one thiol function of the A-chain with pyridine sulfide and combining with the thiol free B-chain [1]. Although the method is theoretically straightforward, the reported yields during the chain synthesis and the interchain combination are very poor. In this study we report on the results of the synthesis of the A- and B-chains of human insulin and relaxin and the analysis of the proceeding of the interchain formation.

Results and Discussion

Insulin A-chain was obtained by the condensation of the resin-bound 14-21 *C*-terminal fragment, with the already oxidized 6-13 protected fragment. This was obtained by iodine oxidation of the *S*-Trt protected segment, selectively in the presence of Cys^7(Acm). The completion of the synthesis was performed by the step-by-step manner. Human relaxin A-chain was prepared by a similar procedure. The deprotection of the A-chains was performed either in the presence or absence of 2,2'-dithiobis(5-nitropyridine) (DTNP). The finally obtained insulin *[Cys^7(Acm)]-A-chain,* relaxin *[Cys^{11}(Acm)]-A-chain,* insulin *[Cys^7(Acm), Cys^{20}(Npyr)]-A-chain* and relaxin *[Cys^{11}(Acm), Cys^{24}(Npyr)]-A-chain* were purified and analyzed by RP-HPLC and RP-UPLC and their correct mass and parts of the sequences was determined by MS and MS-MS.

For the preparation of the insulin B-chain we condensed protected fragments prepared on 2-chlorotrityl resin in solution. In a not optimized procedure we treated the resulting protected B-chain with TFA/H_2O/DTT (85:10:5) for 3 h at RT and obtained the crude insulin *[Cys^7(Acm)]-B-chain* in approx. 40% purity. The purification was performed by RP-HPLC and the correct mass and parts of the sequence were determined by MS and MS-MS.

We tested several polystyrene resins for the step by step synthesis of the entire B-chain of relaxin. Best results were obtained with the application of the 4-methylbenzhydryl (MBH) resin. After chain assembly and cleavage from the resin and simultaneous deprotection with TFA/H_2O/DTT (85:10:5) for 3 h at RT, we obtained the crude B-chain in approx. 50 % purity. Purification and identification was performed as above.

The combination of the insulin *[Cys^7(Acm), Cys^{20}(Npyr)]-A-chain* with the insulin *[Cys^7(Acm)]-B-chain* and of relaxin *[Cys^{11}(Acm), Cys^{24}(Npyr)]-A-chain* with

S. Del Valle et al. (eds.), *Peptides for Youth: The Proceedings of the 20th American Peptide Symposium*, 185
DOI: 10.1007/978-0-387-73657-0_85, © Springer Science+Business Media, LLC 2009

the relaxin *[Cys[11](Acm)]-B-chain* was performed in DMSO or mixtures of DMSO with fluorinated alcohols, such as trifluoroethanol. Within 10 min at RT the reaction was completed. Surprisingly, the reaction mixture besides the expected insulin *[Cys[7](Acm)]-A-chain/[Cys[7](Acm)]-B-chain* and the starting materials, contained also the insulin *[Cys[7](Acm), Cys[23](Npyr)]-B-chain*. Characteristic was also the absence of the expected 5-nitro-2-pyridylthiol (Npyr-SH) in the mixture. Similar results were obtained in the combination of the relaxin chains. In a characteristic UPLC-Q-Tof analysis of the reaction mixture during the combination of the insulin chains, we obtained the components A-D in 36, 25, 7 and 32% respectively (Fig. 1). This ratio remained unchanged independently of the reaction time. So, we concluded that the equilibrium is established very fast.

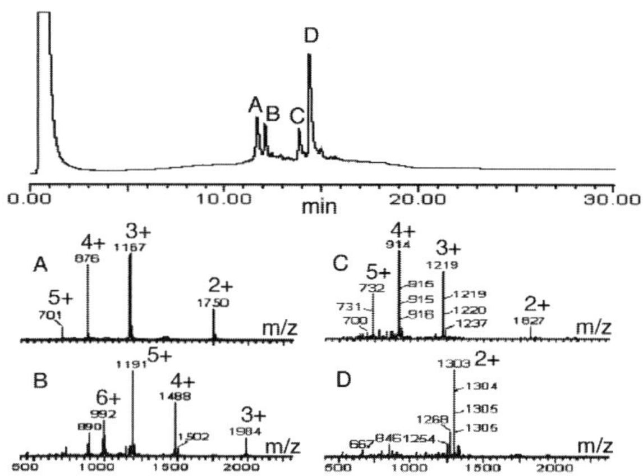

Fig. 1. UPLC-Q-Tof analysis of the reaction mixture during the combination of the insulin chains in DMSO; components: A=[Cys[7](Acm)]-B-chain, B=[Cys[7](Acm)-A-chain/[Cys[7](Acm)-B-chain], C=[Cys[7](Acm), Cys[23](Npyr)]-B-chain, D= [Cys[7](Acm), Cys[20](Npyr)]-A-chain.

Acknowledgments

The work was funded by CBL-Patras S.A.

References

1. Samuel, C. *et al. Biochem.* **46**, 5374-5381 (2007).

Design And Parallel Synthesis Of New Bicyclic Small Molecules For Targeting The Melanocortin Receptors

Byoung J. Min, Alexander V. Mayorov, Minying Cai, Erin Palmer and Victor J. Hruby

Department of Chemistry, University of Arizona, Tucson, AZ 85721, USA

Introduction

The melanocortin receptors (MCRs) belong to the G-protein coupled receptor family, which are seven transmembrane α-helical bundles involved in signal transduction. The melanocortin receptors regulate a large number of physiological functions including pigmentation, sexual function, steroidogenesis, energy homeostasis, exocrine secretion, analgesia, inflammation, immunomodulation, temperature control, cardiovascular regulation, and neuromuscular regeneration. There are two types of peptide ligands targeting MCRs, natural linear peptide ligands- α, β, and γ melanocyte stimulating hormones (MSHs) as agonists, and natural melanocortin antagonists such as agouti, and agouti-related protein (AGRP) [1]. With regard to peptide ligands for MCRs, the importance of rational design and synthesis toward conformationally restricted ligand developments is briefly summarized in *Fig. 1* [2]. Among the three endogenous MSHs, α-MSH is particularly interesting, and numerous analogues of α-MSH have been and are being synthesized in many research groups including ours.

α-MSH Ac–Ser–Tyr–Ser–Met–Glu–His–Phe–Arg–Trp–Gly–Lys–Pro–Val–NH_2

NDP-α-MSH Ac–Ser–Tyr–Ser–Nle–Glu–His–*D*-Phe-Arg-Trp–Gly–Lys–Pro–Val–NH_2

MT-II Ac–Nle–Asp-His–*D*-Phe-Arg-Trp-Gly–Lys–NH_2

SHU-9119 Ac–Nle–Asp-His–*D*-Nal(2')-Arg-Trp–Gly–Lys–NH_2

Fig. 1. Development of MSH peptide ligands.

It should be emphasized that all naturally occurring MSHs contain a common tetrapeptide unit (underlined in *Fig. 1.*), His-Phe-Arg-Trp. We earlier found that this sequence was the significant core structure for biological activities. Also, it has been found that there is a β-turn structure which resides in this core sequence, and we recently performed comprehensive NMR and MacroModel simulation structures of MT-II and SHU-9119 [3]. In order to test the importance of the β-turn structure found in the core sequence of MSHs and the feasibility to develop small molecules having a minimal pharmacophore, we and others report on the synthesis of novel bicyclic templates to mimic a β-turn structure [4]. Using this novel template, several β-turn mimetics were synthesized in parallel fashion shown in *Fig. 3*.

Results and Discussion

The design of the bicyclic template was achieved by the calculation of the superimposed global minimum of the desired bicyclic structure using the large scale low frequency mode in Monte Carlo simulation with the NMR structure of MT-II shown in *Fig. 2*.

S. Del Valle et al. (eds.), *Peptides for Youth: The Proceedings of the 20th American Peptide Symposium*,
DOI: 10.1007/978-0-387-73657-0_86, © Springer Science+Business Media, LLC 2009

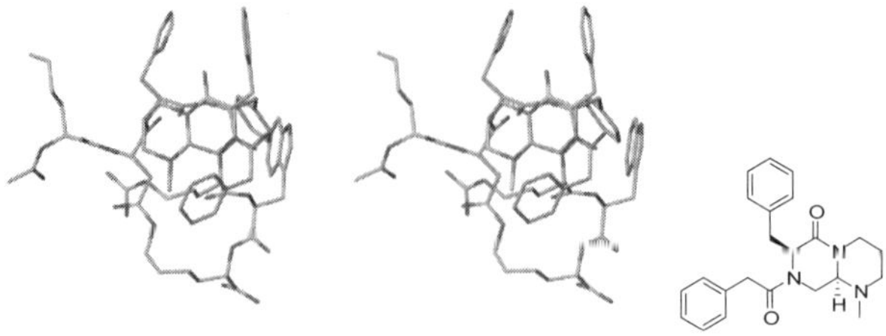

Fig. 2. Large scale low frequency mode in Monte Carlo simulation of the bicyclic structure (brown) overlapped with NMR structure of MT-II (blue).

Boc-phenylalanine **1** was coupled to mono-Cbz protected diamine **3**. The corresponding amine **4** was subjected to reductive alkylation followed by acylation with various aromatic carboxylic acids. The synthesized acetals **7** were subjected to the acid catalyzed *N*-acyliminium ion cyclization to yield the desired bicyclic compounds **9**. The other β-turn mimetics were synthesized using similar procedures.

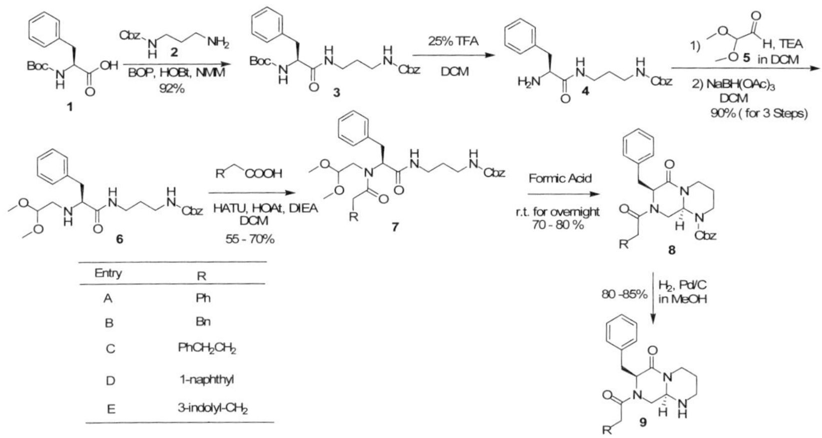

Fig. 3. Synthesis of bicyclic β-turn mimectics.

Preliminary binding data showed that all of synthesized compounds do not have substantial binding to any of MCR subtypes except compound **9D**, which showed moderate potency and selectivity. Based on these results bicyclic derivatives with additional mimetics for MCR interactions are being synthesized and their biological evaluations are also being pursued.

Acknowledgments

The work was funded by grants from US Public Health Service and NIDA.

References

1. Hruby, V. J., Cai, M., *et al. Ann. N. Y. Acad. Sci.* **2003**, *994*, 12-20.
2. Zimanyi, I. A., Pelleymounter, M. A. *Current Pharm. Design*, **2003**, *9*, 627-641.
3. Ying, J., *et al. Biopolymers* **2003**, *71*, 696-716.
4. Cain, J. P., *et al. Bioorg. Med. Chem. Lett.* **2006**, *16*, 5462-5467.

Fully Automated Fast Fmoc Synthesis and On-Resin Disulfide Bridge Formation with Pseudoproline Dipeptides on the *Prelude*™ [1]

Karen Page, Jae H. Park, Christina A. Hood, Hirendra Patel, German Fuentes and Mahendra Menakuru

Protein Technologies, Inc, Tucson, AZ 85714, USA

Introduction

Human islet amyloid polypeptide or amylin ($hAmylin_{1-37}$: H-KCNTATCATQRLAN FLVHSSNNFGAILSSTNVGSNTY-NH_2, C^2-C^7), is a 37-residue major protein contributor to the amyloid deposits found in the pancreases of type-II diabetes patients [2]. In 2005, Abedini and Raleigh reported [3] that they used Mutter's dimethyloxazolidine dipeptide (pseudoproline dipeptide) derivatives to obtain crude linear $hAmylin_{1-37}$ successfully. They cyclized the peptide between Cys-2 and Cys-7 with air oxidation after the peptide was cleaved from the resin. However, it took about 24 hr to fully react and form the disulfide bridge. Our internal studies and Yajima *et. al.* [4] show that $Tl(tfa)_3$, a mild oxidant, sometimes gives better yields and purities of the desired disulfide products, with respect to methods using I_2 or air oxidation. This report illustrates the synthesis of long, difficult peptides incorporating pseudoproline dipeptides followed by on-resin disulfide bridge formation and the effect of pseudoproline dipeptides on on-resin disulfide formation.

Results and Discussion

To test the effects of pseudoproline dipeptides on on-resin disulfide bridge formation, $hAmylin_{1-13}$ was synthesized with and without a pseudoproline dipeptide, then cyclized with $Tl(tfa)_3$. The first peptide was synthesized with Fmoc-Ala-Thr-$\Psi^{Me,Me}$pro at A^8T. The second peptide was synthesized with Fmoc-Ala-OH and Fmoc-Thr(Boc)-OH at A^8 and T^9, respectively. Each linear protected peptide was treated on-resin with $Tl(tfa)_3$ at room temperature, and the cyclization was monitored (Figure 1). The $hAmylin_{1-13}$ containing the pseudoproline dipeptide cyclized to completion in 45 min, while the $hAmylin_{1-13}$ without the pseudoproline dipeptide did not cyclize to completion even after 120 min.

After determining that the presence of a pseudoproline dipeptide actually promoted the cyclization reaction, $hAmylin_{1-37}$ was synthesized with Fmoc-Ala-Thr-$\Psi^{Me,Me}$pro (A^8T), Fmoc-Ser-Ser-$\Psi^{Me,Me}$pro ($S^{19}S$) and Fmoc-Leu-Ser-$\Psi^{Me,Me}$pro ($L^{27}S$).[3] A second $hAmylin_{1-37}$ was synthesized without pseudoproline dipeptides for comparison purposes. The purity of the $hAmylin_{1-37}$ without pseudoproline dipeptides was so poor it could not be cyclized. Cyclization of the peptide containing pseudoproline dipeptides was easily accomplished on-resin with $Tl(tfa)_3$, and went to completion within 10 minutes. Crude HPLCs (Figure 2a & b) show the retention time shifts from 25.7 minutes to 26.8 minutes upon cyclization, and the mass spectrometry data show a change in mass from 4047 m/z to 3905 m/z upon cyclization, thus verifying the cyclization occurred.

The initial synthesis of $hAmylin_{1-37}$ used coupling times of 30 min x 2 and deprotection times of 3 min followed by 20 min, giving a total synthesis time of 58 hours. Based on previous studies with ACP_{65-74} [5], linear $hAmylin_{1-37}$ was

S. Del Valle et al. (eds.), *Peptides for Youth: The Proceedings of the 20th American Peptide Symposium*, 189
DOI: 10.1007/978-0-387-73657-0_87, © Springer Science+Business Media, LLC 2009

synthesized with reduced deprotection and coupling times (Figure 2c). It was found that pseudoproline dipeptides did not require additional reaction times, and the total synthesis time was reduced from 58 hours to 8.5 hours without reducing the crude product purity (Table 1).

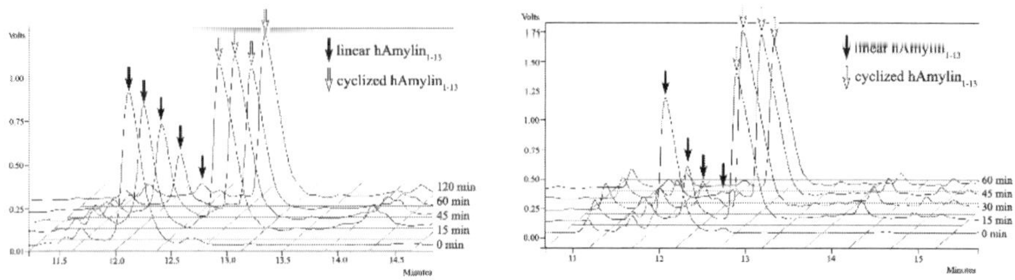

(a) (b)

Fig. 1. Disulfide bridge formation progress (a) without and (b) with the pseudoproline dipeptide.

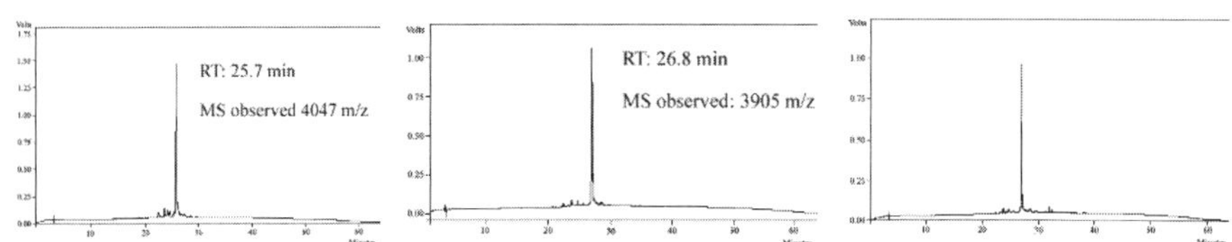

(a) (b) (c)

Fig. 2. HPLC of crude (a) linear hAmylin$_{1\text{-}37}$ and (b) cyclized hAmylin$_{1\text{-}37}$ using standard synthesis times and (c) cyclized hAmylin$_{1\text{-}37}$ using reduced synthesis times.

Table 1. Summary of hAmylin$_{1\text{-}37}$ syntheses to reduce total synthesis time from 58 hr to 8.5 hr.

Synthesis	Deprotection Time	Coupling Time	Number of Washes[a]	Cycle Time[a]	Total Synthesis Time
Standard Time	3 & 20 min	30 min x 2	6	94 min	58 hr
Reduced Time	1 min x 2	2.5 in x 2	3	10 min	8.5 hr

[a]Peptides were synthesized on a Prelude[TM] peptide synthesizer to determine cycle times and total synthesis times.

References

1. Page, K., Hood, C. A., Patel, H., Fuentes, G., Menakuru, M., and Park, J. H. *J. Pept. Sci.* accepted for publication (2007).
2. (a) Abedini, A. and Raleigh, D.P. *Biochemistry* **44**, 16284-16291 (2005); (b) Clark, A., Wells, C. A., Buley, I. D., Cruickshank, J. K., Vanhegan, R. I., Matthews, D. R., Cooper, G. J., Holman, R. R. and Turner, R. C. *Diabetes Res.* **9**, 151-159 (1988).
3. Abedini, A. and Raleigh, D. P. *Org. Lett.* **7**, 693-696 (2005).
4. Fujii, N., Otaka, A., Funakoshi, S., Bessho, K., Watanabe, T., Akaji, K. and Yajima, H. *Chem. Pharm. Bull.* **35**, 2339-2347 (1987).
5. Fuentes, G., Hood, C. A., Page, K., Patel, H., Park, J. H. and Menakuru, M. Poster Presented at the European Peptide Symposium (2006). http://www.pti-instruments.com/images/29EPS_P559.pdf.

Efficient Synthesis of Protected *L*-Phosphonophenylalanine (Ppa) Derivatives Suitable for Solid Phase Synthesis

Satendra S. Chauhan, Varshney, A., Verma, B. and Michael W. Pennington

BACHEM Bioscience Inc., 3700 Horizon Drive, King of Prussia, PA 19406, USA

Introduction

Phosphotyrosine (*p*Tyr) containing peptides are useful as pharmacological tools as well as therapeutics [1]. However, due to hydrolytic lability of the phosphate ester bond of *p*Tyr under physiological conditions, a number of nonhydrolyzable mimetics of *p*Tyr have been designed. Figure 1 shows the structures of *p*Tyr mimicking moieties [2]. In this communication, we describe an efficient synthesis of Fmoc-*L*-Phosphonophenylalanine (Fmoc-Ppa-OH).

1; X = O; Phosphotyrosine (*p*Tyr)
2; X = CH_2; Phosphonomethyl phenylalanine (Pmp)
3; X = CHOH; Hydroxyphosphonomethyl phenylalanine (HPmp)
4; X = CHF; Fluorophosphonomethyl phenylalanine (FPmp)
5; X = CF_2; Difluorophosphonomethyl phenylalanine (F_2Pmp)

6; Phosphonophenylalanine (Ppa)

Figure 1. Structures of Phosphotyrosine mimetics.

Results and Discussion

Scheme 1 shows two plausible strategies for the synthesis of Fmoc-Ppa-OH. Strategy 1 utilizes *L*-tyrosine as the starting material and is reported in the literature [3]. However, Ppa can also be synthesized from *L*-4-iodophenylalanine by employing Michaelis-Arbuzov reaction to form the aryl carbon phosphorus bond (Strategy 2) [4].

Scheme 1

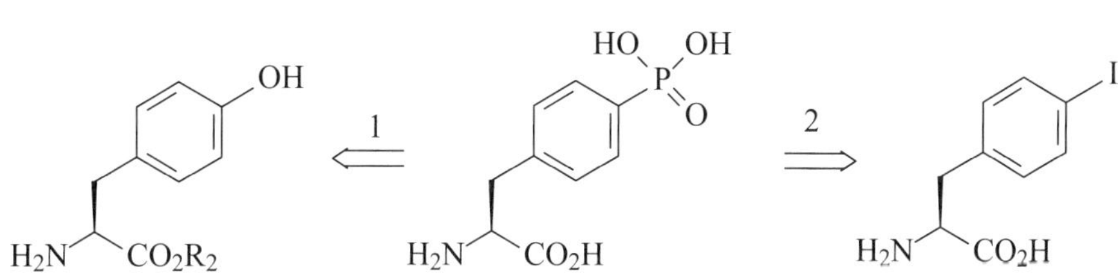

DOI: 10.1007/978-0-387-73657-0_88, © Springer Science+Business Media, LLC 2009

In our first attempt, we tried the Michaelis-Arbuzov reaction on Fmoc-Phe(I)-OEt with diethylphosphite in the presence of Pd(0). However, Fmoc-Phe(I)-OH could not be successfully phosphonylated.

Next, we tried Michaelis-Arbuzov reaction on Boc-Phe(I)-OEt. As shown in Scheme 2, H-Phe(I)-OH was esterified with EtOH saturated with $HCl_{(g)}$. H-Phe(I)-OEt thus obtained, was reacted with $(Boc)_2O$ to afford Boc-Phe(I)-OEt. The Boc-Phe(I)-OEt was treated with diethylphosphite in the presence of Pd(0) and triethylamine in acetonitrile at 72 +2°C for 20 h to afford Boc-Ppa$(Et)_2$-OEt in 87% yield. All the protecting groups were simultaneously removed by refluxing Boc-Ppa$(Et)_2$-OEt in 6 N aqueous HCl at 110°C for 6 h. Concentration of the reaction mixture followed by trituration with ether afforded phosphonophenylalanine (Ppa) in quantitative yield. Ppa thus obtained, was treated with Fmoc-OSu in the presence of sodium carbonate to afford Fmoc-Ppa in excellent yield [5].

Scheme 2

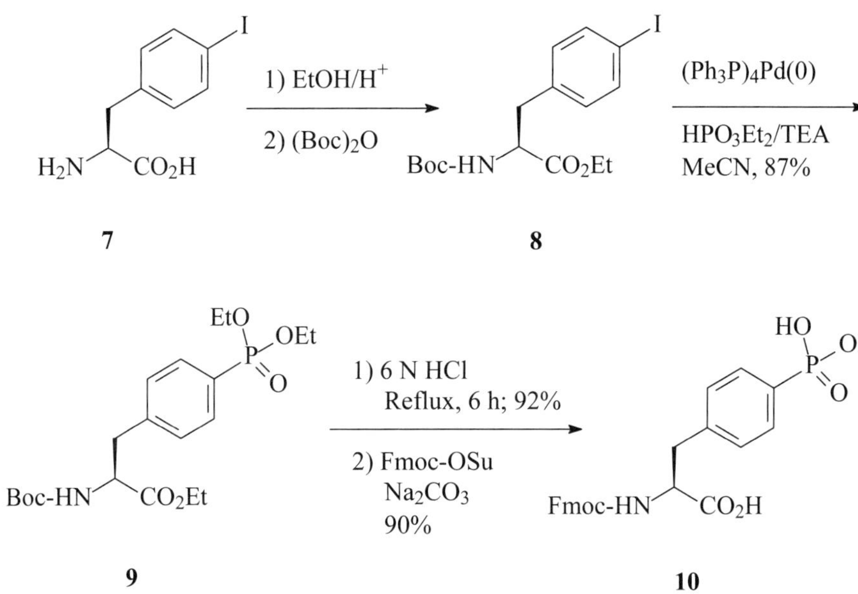

References

1. Beeton, C., Pennington, M. W., Wulff, H., Singh, S., Nugent, D., Crossley, G., Khaytin, I., Calabresi, P. A., Chen, C.-Y., Gutman, G. A., Chandy, K. G. *Mol. Pharmacol.* **67**, 1369-1381 (2005).
2. Burke, T. R. Jr., Smyth, M. S., Nomizu, M., Otaka, A., Roller, P. *J. Org. Chem.* **1993**, *58*, 1336-1340.
3. Thurieau, C., Simonet, S., Paladino, J., Prost, J.-F., Verbeuren, T., Fauchere, J.-L. *J. Med. Chem.* **37**, 625-629 (1994).
4. Hirao, T., Masunaga, T., Ohshiro, Y., Agawa, T. *Synthesis,* 56-57 (1981).
5. Chauhan, S. S., Varshney, A., Verma, B., Pennington, M. W. *Tetrahedron Lett.* **48**, 4051-4054 (2007).

ALLESSAN CAP -
A Highly Versatile And Affordable Coupling Agent

Richard Amitrano[1], Peter Koch[2], Marita Neuber[2], Doris Neumann[2], Peter Petrovic[2], Michael Roos[2] and Christoph Vedder[2]

[1]AC Chem Partners LLC, 10800 Sikes Place, Suite 130, Charlotte, NC 28277, USA;
[2]AllessaChemie GmbH, Alt Fechenheim 34, 60386 Frankfurt, Germany.

Introduction

Peptides are continuously gaining more and more attention in the chemical and pharmaceutical industry [1]. In many peptide syntheses, enantiomerically pure, chemically protected, highly valuable amino acids are used as starting materials. Therefore, a loss in yield during coupling reactions of amino acids either by direct loss of the product caused by complex work up procedures or by racemization is an expensive side effect that needs to be minimized.

Enabling mild reaction conditions as well as simple work up by aqueous removal of the coupling reagent, Allessan CAP (propylphosphonic acid anhydride, PPA) allows access to very effective peptide syntheses, combined with highly suppressed racemization compared to other reagents (see Table 1) [2].

Applications

The main application of PPA is in the field of coupling reactions of amino acids and / or smaller peptides to form new, more complex peptide molecules. Coupling reactions using PPA deliver up to quantitative yields (see Figure 1) [3].

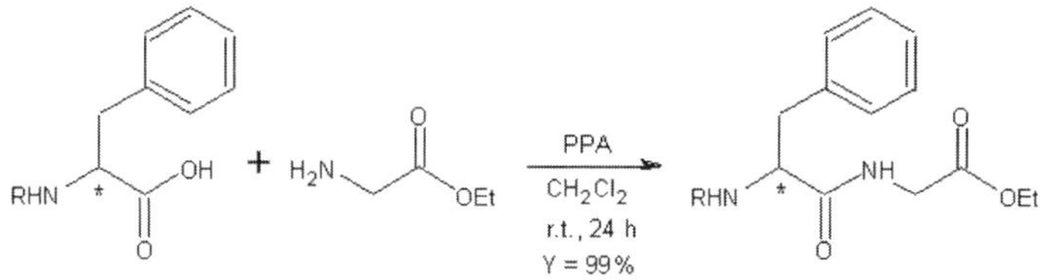

Fig. 1. High-yield peptide synthesis using PPA as a coupling reagent [3].

Table 1. Comparison of epimerization levels with different coupling reagents [2].

Coupling reagent	Epimerization[a]
PPA	1.76
HOBt / DCC	5.91
HOAt / DCC	11.60
PyBOP	14.22
HBTU	16.14

[a]ratio of side product to desired product

S. Del Valle et al. (eds.), *Peptides for Youth: The Proceedings of the 20th American Peptide Symposium*, 193
DOI: 10.1007/978-0-387-73657-0_89, © Springer Science+Business Media, LLC 2009

Furthermore, these very effective peptide syntheses are accompanied with highly suppressed racemization, compared to other reagents. Table 1 shows the degree of racemization when different coupling reagents were used in the same peptide coupling reaction under identical reaction conditions [2].

As a highly efficient reagent for condensation reactions in general, other applications of PPA lie in amide formations similar to peptide syntheses [4], and esterifications of highly valuable acids as substrates (see Figure 2) [5].

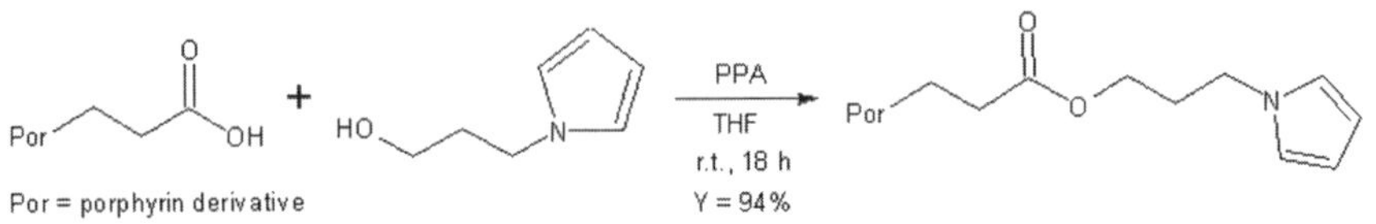

Fig. 2. PPA supporting esterification of highly valuable acids [5].

PPA's property as excellent water removing reagent is furthermore applied for heterocycle syntheses via intramolecular dehydration [6], alkene formations starting from alcohols [7], and the preparation of nitriles by dehydration of oximes [8].

Allessan CAP offers advantages in handling, compared to several alternative reagents. Dosing of the liquid Allessan CAP formulations is technically much easier and more precise than the addition of solid coupling reagents. Further facilitation in handling is provided by its very low toxicity and very low sensitization potential. All in all Allessan CAP is a highly affordable reagent with much better economics than many of its alternatives.

Delivery Form

Allessan CAP is typically offered as a solution in an organic solvent with an assay of 50 to 60 weight-%. The broad range of possible solvents includes ethyl acetate, butyl acetate, acetonitrile, tetrahydrofuran (THF), *N*-methyl pyrrolidinone (NMP), and *N,N*-dimethylformamide (DMF).

References

1. No. of hits for "peptide" in SciFinder database: 1970 3.114, 1980 7.654, 1990 16.515, 2000 23.589, 2005 29.413.
2. J. Hiebl, D.P. Alberts, A.F. Banyard, K. Baresch, H. Baumgartner, I. Bernwieser, P.K. Bhatnagar, M. Blanka, M. Bodenteich, T. Chen, P.M. Esch, H. Kollmann, I. Lantos, K. Leitner, G. Mayrhofer, R. Patel, A. Rio, F. Rovenszky, D. Stevenson, K.D. Tubman, K. Undheim, H. Weihtrager, W. Welz, K. Winkler, *J. Peptide Res.* **1999**, *54*, 54-65.
3. H. Wissmann, H.-J. Kleiner, *Angew. Chem. Int. Ed. Engl.* **1980**, *19*, 133-134.
4. F. Burkhart, M. Hoffmann, H. Kessler, *Angew. Chem. Int. Ed. Engl.* **1997**, 36, 1191-1192.
5. M. Wedel, A. Walter, F.-P. Montforts, *Eur. J. Org. Chem.* **2001**, 1681-1687.
6. B. Napierski, H.-P. Rebenstock, E. Holla, US 6,407,258 B1, **2002**.
7. S. Scherer, C. Böhm, WO 2005/123632 A1, **2005**.
8. A. Meudt, S. Scherer, C. Böhm, WO 2005/123661 A1, **2005**.

Novel Analogues of Bifunctional Ligands for Opioid and Melanocortin 4 Receptor

Vinod V. Kulkarni[1], Yeon Sun Lee[1], Christine Salibay[1], Peg Davis[2], Shou-wu Ma[2], Frank Porreca[2], and Victor J. Hruby[1]

[1]Department of Chemistry; [2]Department of Pharmacology; University of Arizona, Tucson, AZ 85721, USA

Introduction

Neuropathic pain and its cure is the most sought and challenging research in the area of pain. Targeting individually opioid, cholecystokinin, neurokinin and recently melanocortin receptors has been a developing approach in the treatment of neuropathic pain. Targeting multiple receptors with a single ligand is a new paradigm that has been established in our group [1]. Biologically many disease states lead to changes in expressed proteins, therefore, "system changes" that occur must be considered in any treatment for the disease. This new approach to drug design and discovery will be particularly applicable to the diseases that involve adaptive changes in the central nervous system and spinal cord, such as neuropathic pain.

The multifaceted properties of chimeric ligands incorporating bifunctional pharmacophores is targeted towards complex diseases such as neuropathic pain. Recent research has shown that an antagonist at the melanocortin-4 receptor (MC4R) can produce analgesic, anti-allodynic, anti-nociception and reversal of opiate tolerance effects [2]. The post-translated opioid β-endorphin and melanocortin peptides are co-synthesized and released from the proopiomelanocortin protein. These peptides exert opposite physiological actions towards spinal cord mediated pain. Developing chimeric peptide ligands takes advantage of our new paradigm. For treatment of neuropathic pain we seek to design novel ligands which incorporate opioid agonist and melanocortin antagonist pharmacophores. The opioid pharmacophore sequence Tyr-*D*Ala-Gly-Phe is thus linked with the linear sequence of SHU9119, a known super antagonist towards the MC3 and MC4 receptors.

The following ligands were found to interact with both opioid and melanocortin-4 receptors.

VVK052	H-Tyr-*D*Ala-Gly-Phe-Nle-His-*D*Nal(2')-Arg-Trp-Gly-Lys-Pro-NH_2
VVK054	H-Tyr-*D*Ala-Gly-Phe-Nle-His-*D*Nal(2')-Arg-Trp-Gly-NH_2
VVK055	H-Tyr-*D*Ala-Gly-Phe-Nle-His-*D*Nal(2')-Arg-Trp- NH_2

Fig. 1. Analogues of bifunctional/chimeric ligands for Opioid and Melanocortin 4 receptors.

Results and Discussion

Ligands VVK052 and VVK054 show excellent affinity towards the human melanocortin 4 receptor (Table 1). The drop in binding of VVK055 may be indicative of the loss of the β-turn in the structure due to the C-terminal truncation which otherwise is present in VVK052. The binding affinities of the above ligands toward δ and μ opioid receptors are in the nM range. The ligands are potent in both MVD and GPI assays and are moderately selective for the δ opioid receptor. The designed bifunctional novel analogues thus exhibit and indicate unaltered potencies

S. Del Valle et al. (eds.), *Peptides for Youth: The Proceedings of the 20th American Peptide Symposium*, 195
DOI: 10.1007/978-0-387-73657-0_90, © Springer Science+Business Media, LLC 2009

Table 1. Binding affinities.

Peptide	Binding affinities			Functional assay	
	K_i(nM)			IC_{50} (nM) (agonist)	
	hDOR	rMOR	hMC4R	MVD	GPI
VVK052	0.48	0.9	15.9	2.7	54
VVK054	0.24	1.0	11.6	1.5	52
VVK055	0.18	0.8	43.5	2.0	83

towards their respective receptors despite an additionally incorporated pharmacophore region. Such an observation, invites us to explore further design and synthesis of bifunctional peptides. Cyclic ligands based on SHU9119 are currently being analyzed, and tethered with various known opioid peptides like DAMGO, Enkephalin and Endomorphin moieties at the N-terminal.

Bifunctional ligands of opioid and MSH are thus a valuable and novel design for the treatment of neuropathic pain and definitely holds great promising future.

Acknowledgments

Grants supported by USPHS and NIDA

References

1. Hruby, V. J., et al. In Lebl, M. and Houghten, R. A. (Eds) *Peptides: The Wave of the Future,* American Peptide Society, San Diego, 2001, 969-970.
2. Starowicz, K., et al. *International Jounal of Neuroprotection and Neuroregeneration*, **2**, 17-26, (2005)

Fmoc as a Probe in Peptide Purification

Jiangli Song, Zhenchao Gao, Lanlan You, Niansheng Li, XianGang Yang, Wentao Zhang*

Scilight Biotechnology, Beijing 100094, China

Introduction

In general, RP-HPLC separation method is sufficient to obtain high purity peptides synthesized by SPPS[1-2]. However, for the peptides with difficult sequences, typically with high percentage of hydrophobic residues, homogeneous impurities such as deletion sequences that lack a few residues can be generated and will be chromatographically indistinguishable from the target sequence. This presents the major problems for the removal of the impurities, or even results in miss identification of the product. In this study, we kept the last amine protecting Fmoc group[3-8] as a probe to move the targeted product away from the closely related impurities. Due to the hydrophobicity of Fmoc group, the labeled target product will have longer retention time on the HPLC profile that will make it more recognizable and easier to be purified.

Result and Discussion

We have studied twelve peptides with minimum length of 18 residues. The degree of the synthesis difficulty is the main criteria for the sequence selection. The most representative data of 4 peptides are shown here in Table 1. Peptide 1-3 represent those that are hard to synthesize while Peptide 4 represents those with intermediate degree of difficulties.

Table 1: Comparison of the purities and overall yields of the products in Group A & B.

Peptide sequence	# DS	Purity		Overall Yield	
		A	**B**	**A**	**B**
Peptide1: ALNRGLKKKTILKKAGIG	7	81%	92%	28%	20%
Peptide2:RKKSAFATYKVKAAASAH	8	29%	84%	6.1%	7.5%
Peptide3:GIGMCVKVSSIFFINKQKP	7	0%	70%	0%	4%
Peptide4:DKKRATFLLALWECSLPQ	1	92%	65%	26%	8%

DS: Difficult Sequence (in Red) predicted by the Peptide Companion software.

The results for Peptide1 are shown in Figure 1. The synthesis was quite successive. The Fmoc probe shifted the peptide 7 min away from most of the impurities. The purity of the final product from the probe method is much higher than the standard method.

S. Del Valle et al. (eds.), *Peptides for Youth: The Proceedings of the 20th American Peptide Symposium*, 197
DOI: 10.1007/978-0-387-73657-0_91, © Springer Science+Business Media, LLC 2009

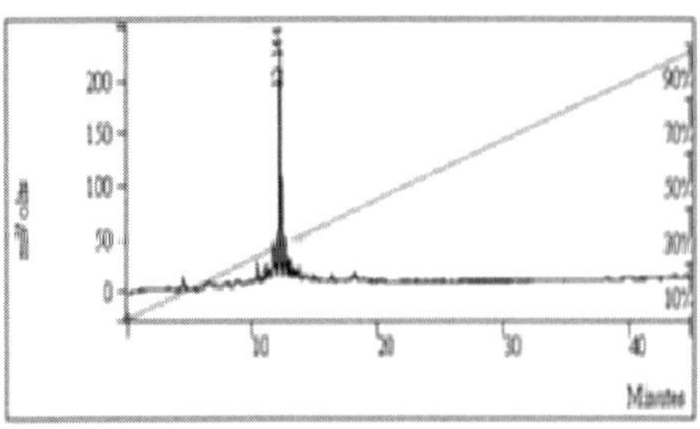
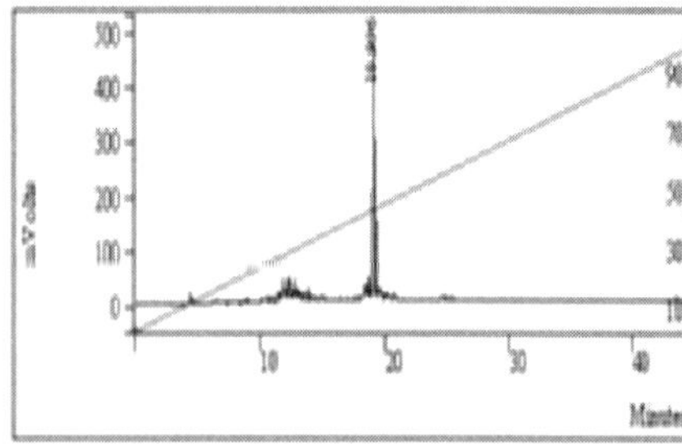

Crude Peptide 1-A (Rt=12.1min) Crude Peptide 1-B with Fmoc probe (Rt=19.0)

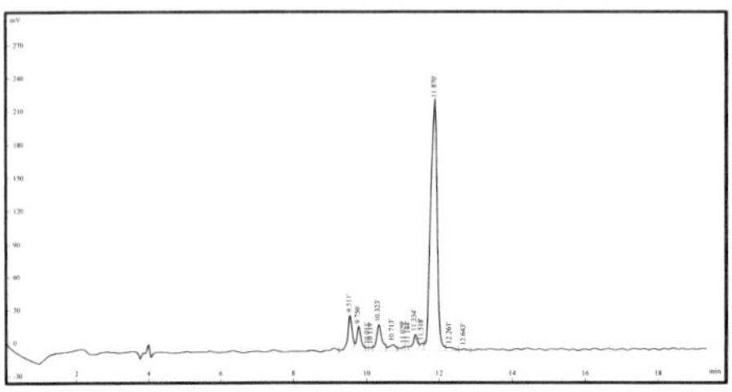
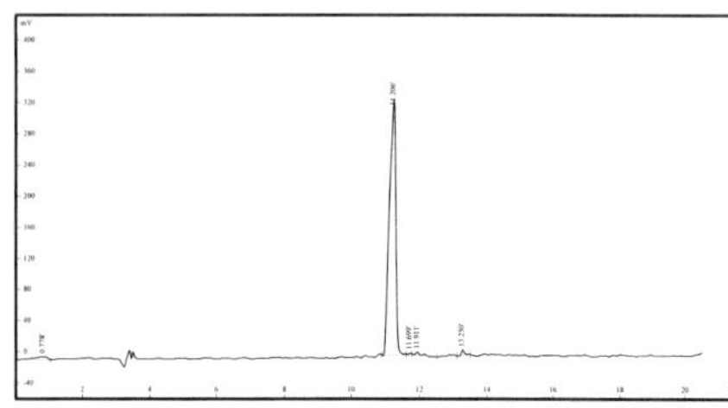

Purified Peptide1-A (83%) **Purified Peptide 1-B (92%)**

Figure1: HPLC Profiles of Peptide 1

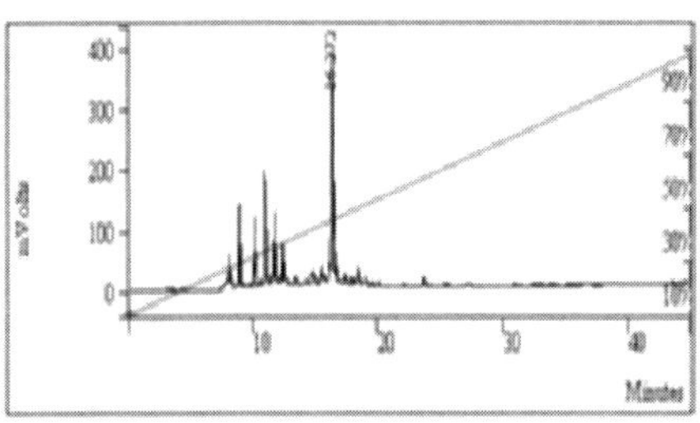
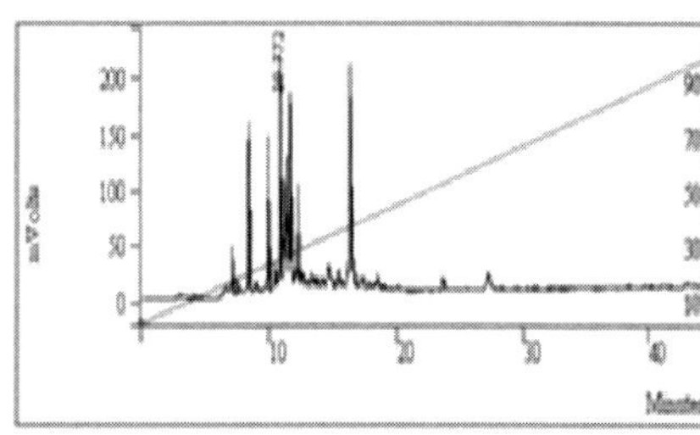

Crude Peptide 2-A (Rt=10.8min) Crude Peptide 2-B with Fmoc probe (Rt=16.3)

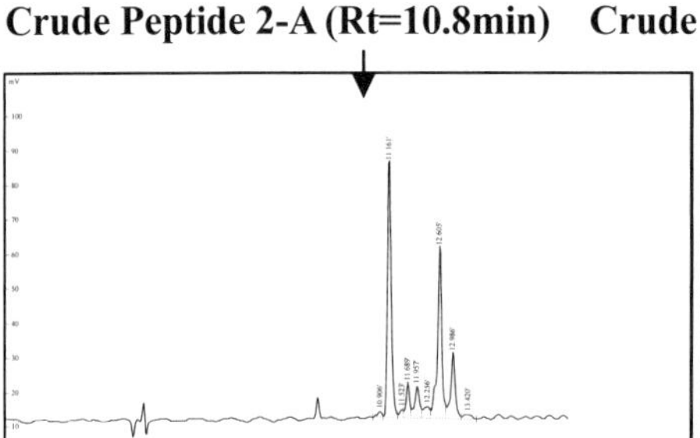
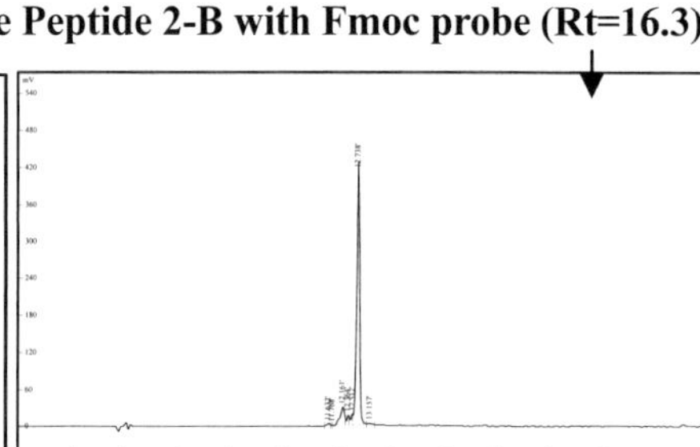

Purified Peptide 2-A (29%) **Purified Peptide 2-B (83%)**

Figure 2: HPLC Profiles of Peptide 2

The results for Peptide 2 are shown in Figure 2. It is very hard to identify the targeted product in Group A and it's more difficult to remove the closely related impurities. In Group B, Fmoc probe has made the target molecule stand out and moved 5 min away from the major side products. The advantage of the probe method is clearly shown in the final purified products. (84% purity in B vs 29% in A)

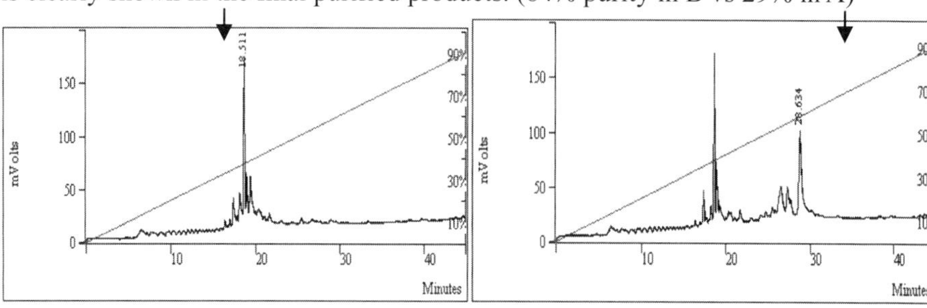

Crude Peptide 3-A **Crude Peptide 3-B with Fmoc probe (Rt=28.6)**

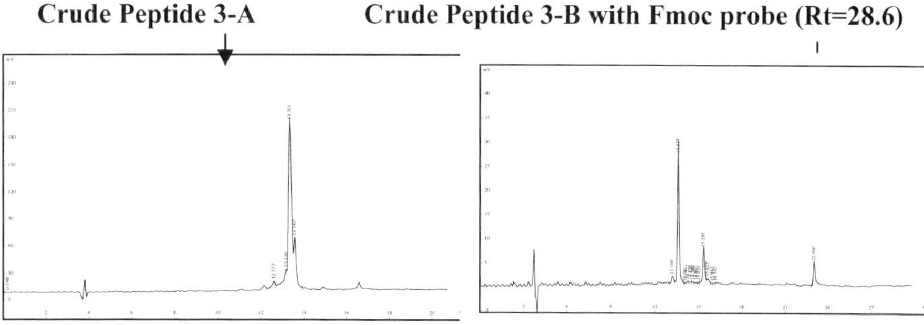

Purified peptide with severe deletion **Purified Peptide 3-B.**

Figure 3: HPLC Profiles of Peptide 3.

The results for Peptide 3 are shown in Figure 3.The only major product in Group A was the deletion sequence: Ac-VKVSSIFFINKQKP judging by the MS data (Theoretical: 1676; Actual:1675). In Group B, there were 2 major peaks. The later one (t=28.6) turned out to be the correct product with Fmoc probe. Again, Fmoc probe method has demonstrated it's power in the purification of peptide with difficult sequence.

Peptide 4 was the easiest one to make among these 4 based on the computational prediction and confirmed by the actual synthesis. It is clear that Fmoc probe method is not very useful in this case.

Conclusions

1. Fmoc-probe method can aid the purification of the targeted peptide when the synthesis is difficult.
2. The desired peptide with the Fmoc probe has longer retention time than most of the homogeneous impurities resulted from the synthesis. This has not only made the identification and purification easier, but also enabled us to obtain higher purity product.
3. The probe method may not be necessary when the synthesis is quite easy.

Reference

1. Fields, G.B and Noble, R.L. (1990). Int. J. Peptide
2. Grant, G.(ed.)(1992). Synthetic peptides, W.H. Freeman & Co., New York

3. Carpino, L.A. and Han, G.Y.(1973). J.Am.Chem.Soc., 92, 5748
4. Mascagni, P., Ball, H.L., and Bertolini, G. (1997). Anal. Chim. Acta, 352, 375
5. Merrifield, R.B. and Bach, A.E. (1978). J.Org.Chem., 43, 25
6. Ramage, R. and Raphy, G (1992).Tetrahedron Lett., 33, 385
7. Brown, A.R., Irving, S.L., Ramage, R., and Raphy, G (1995). Tetrahedron,51, 11815
8. Mutter, M and Bellot, D (1984). Helv.Chim.Acta, 67, 2009

1,3,5-Triazepan-2,6-diones as Conformationally Constrained Dipeptide Mimetics. In Silico Guided Identification of sPLA2 Inhibitors

Gilles Guichard[1], Gersande Léna[1], Joel Boeglin[1], Pascal Muller[2], Eric Boilard[3], Gérard Lambeau[3], Didier Rognan[2], Claude Didierjean[4]

[1]*CNRS;Institut de Biologie Moléculaire et Cellulaire, Strasbourg, France* [2]*CNRS, Bioinformatique du médicament, Faculté de Pharmacie, Illkirch, France* [3]*CNRS, Institut de Pharmacologie Moléculaire et cellulaire, Valbonne, France* [4]*Université Henri Poincaré, CNRS, LCM3B groupe Biocristallographie, Vandoeuvre, France*

Introduction

The design and synthesis by combinatorial chemistry techniques of cyclic/polycyclic molecular frameworks that can efficiently distribute selected pharmacophores in the 3D space is an important method to identify small-molecules capable of modulating biological processes. In this context, heterocyclic systems derived from peptides (e.g. DKP, **A**) are of particular interest owing to the facile access, the chemical and stereochemical diversity of peptide derivatives, as well as enhanced diversity resulting from post-cyclization appending operations.

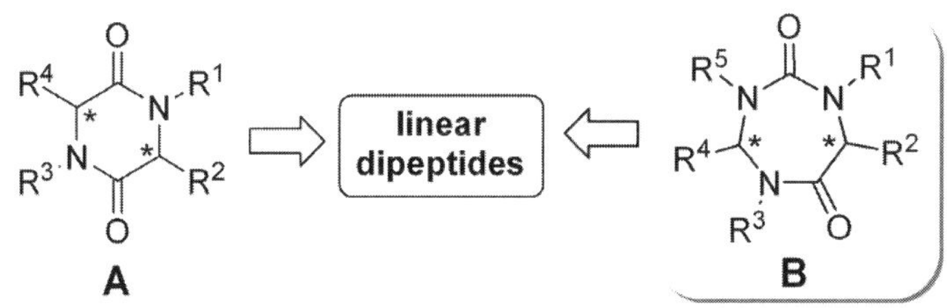

We recently developped the 1,3,5-triazepan-2,6-dione system **B** as a novel, conformationally restricted, and readily accessible class of dipeptidomimetics [1-3]. The synthesis of this densely functionalized 1,3,5-triazepan-2,6-dione skeleton is achieved in only 4 steps from a variety of simple linear dipeptide precursors. Molecular and structural diversity can be increased further through post-cyclization appending operations at urea nitrogens (e.g. mono and di-alkylation, acylation).

Because they are structurally diverse and rapidly accessible in a library format from dipeptides, 1,3,5-triazepan-2,6-diones have a strong potential for use in biological screens. Herein, we describe the application of inverse docking to the identification of the most likely targets of a small panel of triazepandiones **B** [4].

Results and Discussion

A collection of 2,150 druggable active sites from the Protein Data Bank [5] was screened by high-throughput docking to identify putative targets for five representative molecules of a combinatorial library sharing a 1,3,5-triazepan-2,6-dione scaffold. Five targets were prioritized for experimental evaluation by computing enrichment in individual protein entries among the top 2% scoring targets.

Out of the five proposed proteins, secreted phospholipase A2 (sPLA2) was shown to be a true target for a panel of 1,3,5-triazepan-2,6-diones (e.g. compounds **1-4**) which exhibited micromolar affinities toward two human sPLA2 members, i.e.

S. Del Valle et al. (eds.), *Peptides for Youth: The Proceedings of the 20th American Peptide Symposium*, 201
DOI: 10.1007/978-0-387-73657-0_92, © Springer Science+Business Media, LLC 2009

human group V and X sPLA2s (Fig. 1a). Interestingly, these two sPLA2s and other members of this emerging family of enzymes have become attractive therapeutic targets in a number of inflammatory diseases including asthma, rheumatoid arthritis, atherosclerosis, septic shock, as well as in different types of cancer.

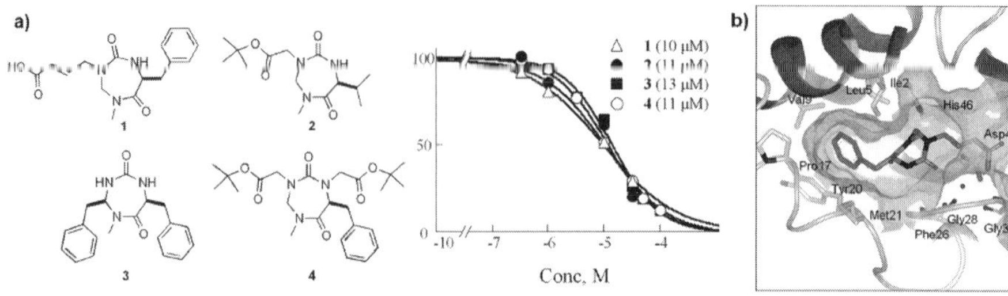

Fig. 1. a) Formulae of identified sPLA2 inhibitors and IC_{50} values for compounds **1-4** *to hGX sPLA2; b) Predicted binding mode of compound* **1** *to hGX sPLA2 (modeled from the pdb entry 1le6)*

The predicted binding mode for compound **1** to the catalytic site of hGX sPLA2 is shown in Fig. 1b. The triazepandione core is embedded in the catalytic site of the enzyme, similarly to a known series of indole derivatives exemplified by indoxam. The long chain carboxylic acid can directly coordinate the catalytic calcium ion at the vicinity of His46 and Asp47 and place the triazepandione in the center of the binding site, the urea carbonyl moiety completing the prototypical heptacoordination of calcium.

To the best of our knowledge, this is the first report of an in silico-guided target identification for drug-like compounds. The simplicity of the approach makes it particularly attractive for prioritizing a few targets for experimental validation and is therefore a good complement to experimental target identification strategies. Although our current triazepandione compounds have modest affinities for sPLA2s, they appear to be selective for group V and X sPLA2s, and thus are interesting lead compounds to develop in order to design more potent yet selective inhibitors of these particular sPLA2 subtypes.

Acknowledgments

This research was supported in part by Centre National de la recherche Scientifique (CNRS), Région Alsace and ImmuPharma France. CIFRE fellowships from ImmuPharma France and the Association Nationale pour la Recherche Technique (ANRT) for G.L. and J.B. are gratefully acknowledged.

References

1. Lena, G., Lallemand, E., Gruner, A. C., Boeglin, J., Roussel, S., Schaffner, A. P., Aubry, A., Franetich, J. F., Mazier, D., Landau, I., Briand, J. P., Didierjean, C., Rénia, L. & Guichard, G. *Chem. Eur. J.* ***12***, 8498-8512 (2006).
2. Schaffner, A. P., Lena, G., Roussel, S., Wawrezinieck, A., Aubry, A., Briand, J. P., Didierjean, C., Guichard, G. *Chem. Commun.* ***39***, 4069-4071 (2006).
3. Aubert, E., Lena, G., Gellenoncourt, M., Durain, E., Guichard, G., Didierjean,C. *Acta Cryst.E* ***63***, o2306-o2308 (2007).
4. Muller, P., Lena, G., Boilard, E., Bezzine, S., Lambeau, G., Guichard, G. Rognan, D. *J. Med. Chem.* ***49***, 6768-6778 (2006).
5. Paul, N., Kellenberger, E., Bret, G., Muller, P., Rognan, D. *Proteins* ***54***, 671-680 (2004).

Proline-based diketopiperazine-scaffolds

Nina Deppermann and Wolfgang Maison*

Institute for Organic Chemistry; Justus-Liebig-University; Giessen; Germany

Introduction

Proline-based diketopiperazine structures for the core of several interesting natural product classes such as marcfortines, paraherquamides, brevianamides and tryprostatines which show promising activities as potent cytotoxic compounds and have a known specifity for certain cancer cell lines[1]. They are thus evaluated scaffolds for targeted cancer therapy[2] and have been the targets for different synthetic approaches in the last years[3].

We have developed a diversity oriented, efficient and stereoselective synthetic strategy to highly functionalized proline-based diketopiperazines **4**.

Our precursors are azabicycloalkenes **1**, which can be obtained from a diastereoselective hetero-Diels-Alder-reaction in defined but variable stereochemistry and ring sizes[4].

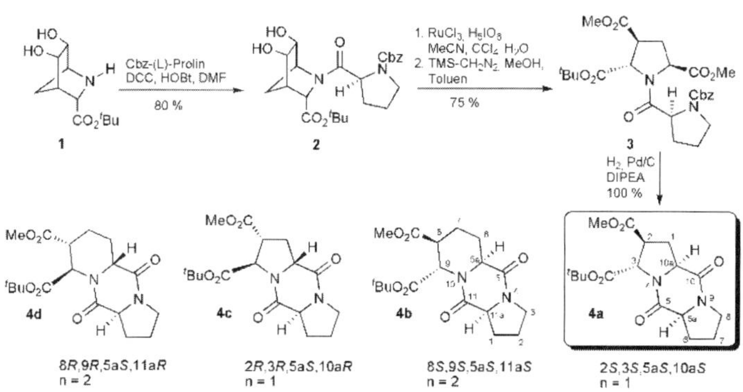

*Fig. 1. Synthetic route to proline-based diketopiperazines **4**.*

The first step is a peptide coupling of enantiomerically pure azabicycloalkane **1** with (L)-proline. Oxidative cleavage of the diol **2** and subsequent esterification gives triester **3** in very good yield[5]. Lactam cyclization to the desired diketopiperazine **4b** occurs quantitatively under alkaline hydrogenolytic conditions. Following this route, we have already prepared a family of enantiomerically pure diketopiperazines **4**. A selection is depicted above.

Results and Discussion

Our scaffolds 1 and 4 have been designed according to the following design criteria: 1. they are easily diversified to priviledged natural product structures and analogues thereof, 2. the chemistry is adaptable to the solid phase and 3. they provide and anchor group for immobilization and conjugation of pharmaceutical effector molecules. All three points are highlighted in the schematically drawing of a diketopiperazine library above.

S. Del Valle et al. (eds.), *Peptides for Youth: The Proceedings of the 20th American Peptide Symposium*, 203
DOI: 10.1007/978-0-387-73657-0_93, © Springer Science+Business Media, LLC 2009

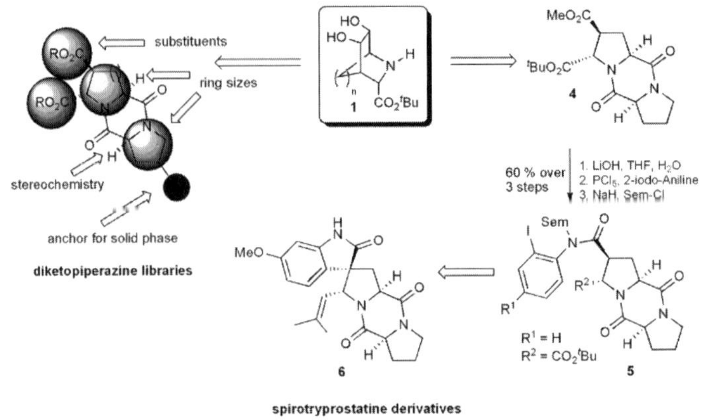

Fig. 2. Diversification of diketopiperazine scaffolds.

We chose a primary amine as a suitable anchor group that is easily installed by using hydroxyproline **7** instead of proline as an educt.

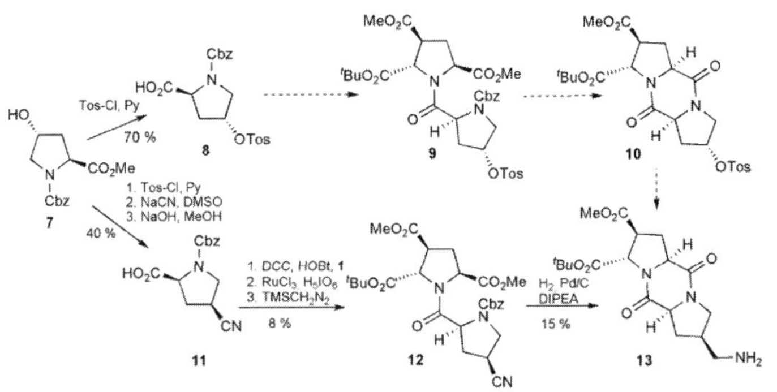

Fig. 3. Introduction of an amine as a solid phase anchor.

Starting from hydroxyproline **7**, the desired scaffold **13** was successfully obtained. However, a synthetic strategy *via* cyanide **11** and dipeptide **12** turned out to be low yielding due to side reactions during oxidation on the way to **12** and also reduction and cyclization to the diketopiperazine **13**. In consequence, we are now focusing on the installation of the amine group at a later stage *via* intermediates **8**, **9** and **10**.

Acknowledgments

Financial support comes from the German Science Foundation DFG (MA 2529).

References

1. *For a review see:* Williams, R. M., Cox, R. J. *Acc. Chem. Res.* **36**, 127 – 139 (2003).
2. Maison, W., Frangioni, J. V. *Angew. Chem.* **115**, 4874 – 4876 (2003); *Angew. Chem. Int. Ed. Engl.* **42**, 4726 – 4728 (2003).
3. *For a review see:* (a) Marti, C., Carreira, E. M. *Eur. J. Org. Chem.* 2209 – 2219 (2003); (b) Maison, W. *Targets in Heterocyclic Systems* **9**, 87 – 113 (2005).
4. Stella, L., Abraham, H. *Tetrahedron* **48**, 9707 – 9718 (1992).
5. Maison, W., Grohs, D. C., Prenzel, A. H. G. P. *Eur. J. Org. Chem.* 1527 – 1543 (2004).

Concept and Synthetic Approach for a Kilogram Scale Synthesis of Octa-D-Arginine Amide Nonahydrochloride Salt

Califano*, J. C., Uribe, A., Chang, J. L., Becker, C. L., Napier, J.J., Kishore, V., Zhou, D., Love, G., Gernhardt, K. and Tolle, J. C.

Abbott Laboratories, GPRD Process Research & Development, Dept R-450, BLDG R13/4, 1401 Sheridan Road, North-Chicago, IL, 60064-6290, USA.

Introduction: Timing and development of scalable processes are critical to the successful production of a peptide-based drug. Although solid phase synthesis is a common approach used to prepare oligopeptides, this method can have limitations, such as the high cost of excess reagents and protected amino acids, the high volume of solvents used, potential limited scalability, and the usual requirement for a chromatographic purification once the peptide is cleaved from the resin. A solution phase synthesis appeared to be a more attractive approach for the preparation of our target molecule, octa-D-arginine amide nonahydrochloride salt **(6)** (Scheme 2) in terms of robustness, scalability, and the potential to minimize labor-intensive purification. The limited availability and cost of protected D-arginine derivatives led us to explore the development of a synthesis in which D-ornithine is substituted for the D-arginine residues. A per-guaniginylation step would then be used for conversion of the 8 ornithine residues to the corresponding arginine residues.

Results and Discussion: The synthetic process, as described in Scheme 2, resulted in the preparation of 0.7 kg of the desired compound, isolated as a hydrochloride salt **(6)**. The main features of the process were the synthesis of a key crystalline tetrapeptide ester intermediate **(1)** with only one isolation by applying a Minimal Isolation Peptide Synthesis (MIPS), followed by the preparation of two crystalline tetrapeptide intermediates **(2)** & **(3)** from the common tetrapeptide ester **(1)**.

The general procedure for conducting the MIPS process is described in Scheme 1 below. Excess N-benzyloxycarbonyl-protected amino acid N-hydroxysuccinimide ester (Z-AA-OSu) is used to drive the acylation to completion. Residual Z-AA-OSu is captured/scavenged by the addition of amine-derivatized insoluble resin or silicagel, which is then removed from the reaction mixture by filtration. Subsequent N-deprotection of the soluble peptide by hydrogenolysis completes one MIPS cycle. The process can be repeated (n-1) times for a peptide of (n) residues in length.

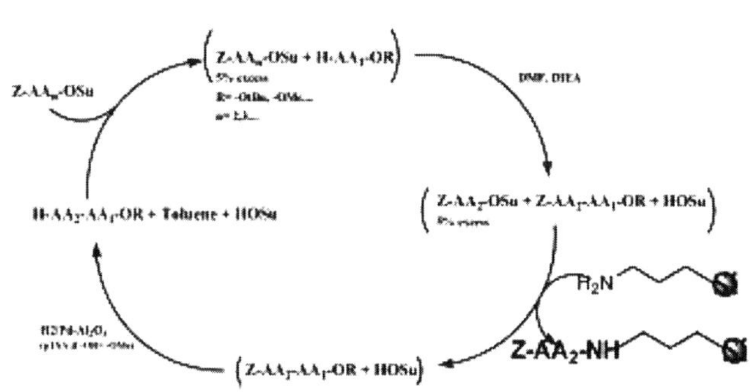

S. Del Valle et al. (eds.), *Peptides for Youth: The Proceedings of the 20th American Peptide Symposium*, 205
DOI: 10.1007/978-0-387-73657-0_94, © Springer Science+Business Media, LLC 2009

Coupling of the N-terminal deprotected **(2)** with C-terminal deprotected **(3)** gave the fully protected octa-D-ornithine intermediate **(4)**, which upon global removal of 8 amine-protecting groups followed by global guanidinylation of the free amines in one step gave the desired D-arginine residues. No chromatographic purification was used at any step of the process. At the final step a simple activated carbon treatment of the desired octa-D-arginine amide hydrochloride salt **(6)** in water was sufficient for purification.

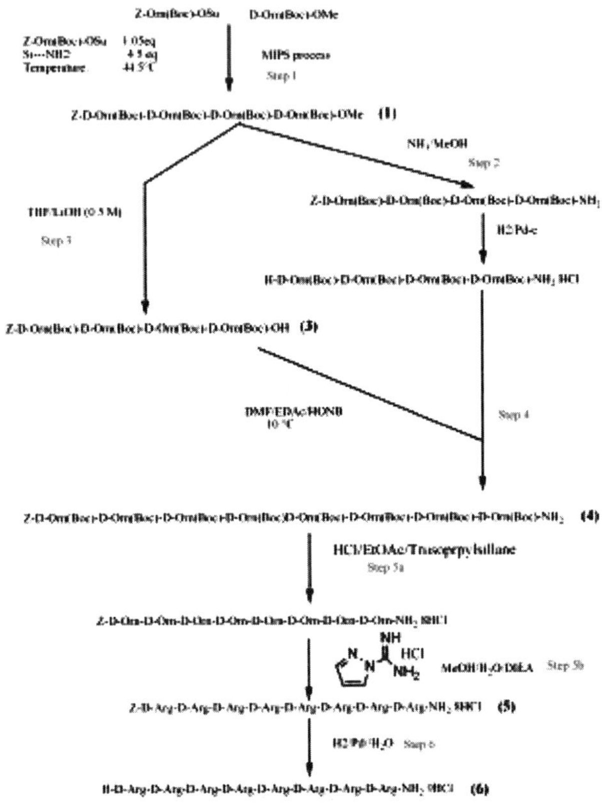

Scheme 2: Octa-D-arginine amide - Overall Process

Conclusion:

A solution phase synthesis of H-(D-Arg)$_8$-NH_2.9HCl (6) was developed and demonstrated at the kilogram scale. The minimum isolation process for the preparation of tetrapeptide (1) was an efficient approach to this key intermediate. Conversion of the fully protected octa-D-ornithine intermediate (4) to the protected octa-D-arginine intermediate (5) avoided the need to develop what would be a more costly and complicated process if protected D-arginine was the starting material. Given the need and opportunity, the process as described herein could be used for the preparation of larger quantities of octa-D-arginine amide hydrochloride.

Reference:

Kirshberg, T. A.; VanDeusen, C. L.; Rothbard, J. B.; Yang, M.; A.; Wender, P. A. *Organic Letters*, **2003**, 5(19), 3459-3462.

Califano J.C.; Uribe A.; Chang J. L.; Becker C. L.; Napier J. J.; Kishore V.; Zhou D.; Love G.; Gernhardt K. and Tolle J. C. *Tetrahedron*, 61 (**2005**), 8821-8829.

Stereoselective Synthesis of (3*S*)- and (3*R*)-3-Hydroxy-3-(2,6-Dimethyl-4-Hydroxyphenyl)Propanoic Acid and its Incorporation into a Cyclic Enkephalin Analogue

Grazyna Weltrowska, Carole Lemieux, Nga N. Chung, Brian C. Wilkes and Peter W. Schiller

Laboratory of Chemical Biology and Peptide Research, Clinical Research Institute of Montreal, Montreal, Quebec, Canada H2W 1R7

Introduction

2',6'-dimethyl substitution of the Tyr[1] residue of opioid agonist peptides and deletion of the N-terminal amino group have been shown to represent a general structural modification to convert opioid peptide agonists into antagonists [1]. This conversion required the syntheses of opioid peptide analogues containing 3-(2,6-dimethyl-4-hydroxyphenyl)propanoic acid (Dhp) in place of Tyr[1]. The cyclic enkephalin analogue Dhp-c[D-Cys-Gly-Phe($p$$NO_2$)-D-Cys]$NH_2$ is a potent μ opioid antagonist and a less potent δ and κ antagonist [2]. An analogue of this peptide with β-methylated Dhp, (3*S*)-3-methyl-3-(2,6-dimethyl-4-hydroxyphenyl)propanoic acid [(3*S*)-Mdp], in place of Dhp showed increased antagonist activity at all three receptors, whereas an analogue containing β-isopropyl-substituted Dhp [(3*R*)-Idp] turned out to be a mixed κ agonist/μ antagonist [3]. To further investigate the SAR at the β position of Dhp in the cyclic analogue, we examined the effect of substituting β-hydroxylated Dhp on the opioid activity profile. This required the development of stereoselective syntheses of (3*S*)- and (3*R*)-3-hydroxy-3-(2,6-dimethyl-4-hydroxyphenyl)propanoic acid [(3*S*)-and (3*R*)-Hdp].

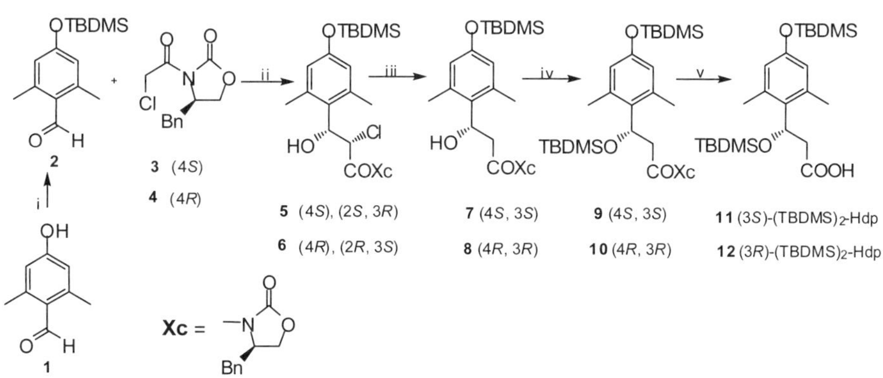

Scheme 1: (i) TBDMSCl, imidazole, DMF, rt., 95%; (ii) Cl-CH_2-COXc, TEA, Bu_2BOTF, CH_2Cl_2, -78°C to rt., 1,5h, phosphate buffer pH7, 30% H_2O_2, ether, 0°C, 1h; (iii) Zn, AcOH, THF, rt., 15min, 50%; (iv) TBDMSCl, imidazole, DMF, 80%; (v) LiOH, 30% H_2O_2, THF/H_2O, 0°C to rt., 90%.

S. Del Valle et al. (eds.), *Peptides for Youth: The Proceedings of the 20th American Peptide Symposium*, 207
DOI: 10.1007/978-0-387-73657-0_95, © Springer Science+Business Media, LLC 2009

Results and Discussion

(3*S*)- and (3*R*)-Hdp were synthesized using Evans chemistry [4] (Scheme 1). Aldol reaction between silyl-protected aldehyde **2** and the dibutylboron enolate derived from chloroacetyloxazolidinone **3** or **4** afforded the desired intermediate **5** or **6** in 70% and 75% yield, respectively. The diastereoselectivity ratio for **5** (78:22) and for **6** (82:18) was determined by ^{1}H NMR spectroscopy. Similar to this moderate diastereomeric excess, limited stereoselectivity had also been observed upon introduction of alkyl substituents in the β-position of Dhp [2]. However, the diastereomers were easily separated by silica gel chromatography, yielding the pure diastereomer **5** or **6**. Dechlorination of **5** or **6** was accomplished by treatment with zinc dust and glacial acetic acid to afford **7** or **8** in 50% yield in both cases. Silylation of the free hydroxyl group in 80% yield for **9** and 85% yield for **10** was followed by removal of the chiral auxiliary in standard manner [5] to give **11** or **12**. Structures were confirmed by ^{1}H NMR spectroscopy and EI mass spectrometry.

The target peptides (3*S*)- and (3*R*)-Hdp-c[D-Cys-Gly-Phe(*p*NO_2)-D-Cys)]NH_2 were prepared by a combination of manual solid-phase and solution techniques. The linear precursor peptide of H-c[D-Cys-Gly-Phe(*p*NO_2)-D-Cys]NH_2 was assembled on a *p*-methylbenzhydrylamine resin, using Boc-protection of the α-amino function and Acm protection of the Cys side chain. After cleavage from the resin by HF/anisole treatment, disulfide bond formation was carried out in MeOH/H_2O using iodine as oxidation agent and the peptide was purified by preparative reversed-phase HPLC. (3*S*)- or (3*R*)-Hdp(TBDMS)$_2$ (**11**,**12**) was attached to the N-terminal amino group of H-c[D-Cys-Gly-Phe(*p*NO_2)-D-Cys]NH_2 using HBTU as coupling agent. After removal of the silyl protection by 90% TFA/anisole treatment, the final peptide products were purified by reversed-phase HPLC, and their purity and structural identity were established by TLC, analytical HPLC and FAB-MS.

(3*S*)- and (3*R*)-Hdp-c[D-Cys-Gly-Phe(*p*NO_2)-D-Cys-NH_2 both showed neither agonist nor antagonist activity in the guinea pig ileum and mouse vas deferens assays at concentrations up to 10 μM. In the opioid receptor binding assays, both the (3*S*)-Hdp[1]- and the (3*R*)-Hdp[1]-analogues showed no μ and κ receptor binding affinity at concentrations up to 10 μM and very weak δ receptor binding affinity with K_i values of 7.65 and 26.3 μM, respectively. This result indicates that introduction of the hydrophilic hydroxyl group at the β-position of Dhp is not tolerated at all three receptors.

Acknowledgments

This work was supported by grants from the CIHR and the NIH.

References

1. Lu, Y., Nguyen, T.M.-D., Weltrowska, G., Berezowska, I., Wilkes, B.C., Lemieux, C., Chung, N.N. and Schiller, P.W. *J. Med. Chem.* **44**, 3048-3053 (2001).
2. Weltrowska, G., Lu, Y., Lemieux, C., Chung, N.N. and Schiller, P.W. *Bioorg. Med. Chem. Lett.* **14**, 4731-4733 (2004).
3. Schiller, P.W., Weltrowska, G., Berezowska, I., Lemieux, C., Chung, N.N. and Wilkes, B.C. In Blondelle, S.E., Ed. *Understanding Biology Using Peptides (Proceedings of the 19th American Peptide Symposium)*, American Peptide Society, San Diego, 2005, pp. 31-35.
4. Evans, D.A., Fitch, D.M., Smith, T.S., Cee, V.J. *J. Am. Chem. Soc.* **122**, 10033-10046 (2000).
5. Evans, D.A., Britton, D.C. and Ellman, J.A. *Tetrahedron Lett.* **28,** 6141-6144 (1987).

Highly Selective Cyclic Hexapeptides Antagonist of GPIIb-IIIa by Multiple *N*-Methylation

Jayanta Chatterjee[1], Oded Ovadia[2], Grit Zahn[3], Luciana Marinelli[4], Amnon Hoffman[2], Chaim Gilon[2], and Horst Kessler[1]

[1]*Department Chemie, Lehrstuhl II für Organische Chemie, Technische Universität München, Lichtenbergstrasse 4, Garching 85747, Germany,* [2]*Pharmaceutics and Organic Chemistry Departments, The Hebrew University of Jerusalem, Jerusalem 91904, Israel,* [3]*Jerini A.G., Invalidenstrasse 130, 10115 Berlin, Germany,* [4]*Farmaceutica e Tossicologica, Università di Napoli 'Federico II', Via D. Montesano, 49-80131 Napoli, Italy*

Introduction

N-methylation is a precious tool to modify lipophilicity, proteolytic stability, bioavailability and induce conformational rigidity to peptide backbone. Mono *N*-methylation of peptide ligands has been employed over the years to enhance potency, new receptor subtype selectivity and tuning of an agonist to antagonist. However, multiple *N*-methylation to our knowledge has never been reported, probably owing to difficult coupling, availability of *N*-methylated amino acid and unpredictable conformational change. We show here that a systematic multiple *N*-methyl scan can be employed to achieve surprisingly enhanced receptor selectivity. For that purpose, we envisioned a 'Design Approach' instead of the commonly used 'Library Approach'. In this method, the prerequisite is the knowledge of the bioactive conformation of the stem peptide (lead structure). Then a systematic *N*-methylation of the externally oriented (solvent exposed) peptide bonds are carried out. Here, we employed cyclic -RGD- peptides as a model system to obtain a highly selective αIIbβ3 integrin antagonist. Thus compounds that compete with αIIbβ3 in binding to fibrinogen can act as potent antithrombotic agents. In order to improve the oral bioavailability of the cyclic peptide *cyclo*(-G¹R²G³D⁴f⁵L⁶-) (**1**), which was reported by Pfaff et al. [1] to be selective towards αIIbβ3 compared to αvβ3, we targeted a design approach by *N*-methylation. The conformation of the stem cyclic hexapeptide **1** (Figure 1) reveals a βII' turn about D-Phe-Leu and a βII turn about Arg-Gly [2] which is the recognition motif; with two internal hydrogen bonds between AspCO--HNGly and GlyCO--HNAsp. We describe a biased small library in which all externally oriented amide bonds except Gly³, which is involved in the receptor binding, were *N*-methylated (Table 1).

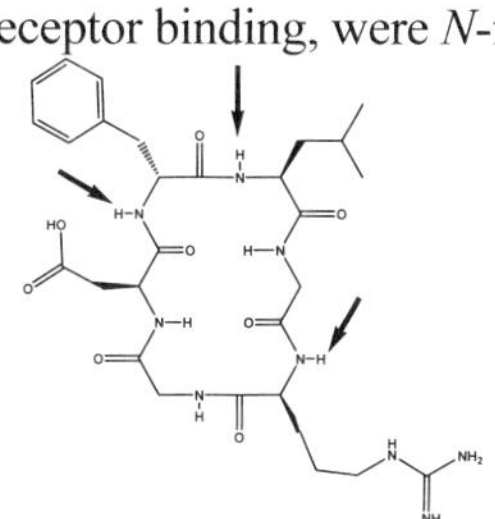

Fig. 1. cyclo(-G¹R²G³D⁴f⁵L⁶-). The -RGD motif is shown in red. The sites targeted for N-methylation are indicated by the arrows.

Results and Discussion

It is worth noting that in spite of our previous report, we found no receptor selectivity for **1**, and there is an inclination towards αvβ3 selectivity with

S. Del Valle et al. (eds.), *Peptides for Youth: The Proceedings of the 20th American Peptide Symposium*, 209
DOI: 10.1007/978-0-387-73657-0_96, © Springer Science+Business Media, LLC 2009

N- methylated leucine (**2**) or D-phenylalanine (**3**). Significant selectivity is first obtained in **4** with *N*-methylated arginine residue, which corroborates with previous results. Extending further the *N*-methylation of **4** to leucine, there is almost no loss in the activity in **6** but a further gain in selectivity. Exchanging the site of *N*-methylation from leucine to phenylalanine in **7**, there is a sudden loss in the activity. However, interestingly the activity is gained back with an additional *N*-methylation of **7** giving rise to **8**, with a tremendous enhancement in the selectivity. It is really surprising that a single *N*-methyl group when present at phenylalanine is responsible for the loss of activity in **7** and gaining back the activity and enormous selectivity when present at leucine in **8**.

Table 1. Different N-methylated analogues of cyclo(-$G^1R^2G^3D^4f^5L^6$-) and their IC50 values (nM) towards different integrins. The bold letters indicate N-methylated residue.

No.	Analogue	$\alpha_5\beta_1$	$\alpha_v\beta_3$	$\alpha_{IIb}\beta_3$	$\alpha_v\beta_3/\alpha_{IIb}\beta_3$
1	C(-GRGDfL-)	740	100	195	0.5
2	c(-GRGDf**L**-)	3900	103	560	0.2
3	c(-GRGD**f**L-)	4300	490	2000	0.2
4	c(-G**R**GDfL-)	1200	770	12	64
5	c(-GRGD**fL**-)	>20,000	1200	620	2
6	c(-G**R**GDf**L**-)	~20,000	1300	15	86
7	c(-G**R**GD**f**L-)	>20,000	2730	165	16
8	c(-G**R**GD**fL**-)	>20,000	12,200	30	406

In the stem peptide **1** or when leucine (**2**) and/or phenylalanine (**3**) are *N*-methylated, there is considerable flexibility in the βII turn resulting in unspecific binding. In all the peptides lacking the *N*-methylated arginine residue, the NMR and the MD show the possibility of forming a γ turn about the glycine[3], which ultimately brings the side chains of the aspartic acid and arginine close to each other, resulting in a comparatively better fit into the αvβ3 binding pocket. In addition to the γ turn, we observe a 'kink' in the backbone conformations of **1** and **2**, giving rise to a 'folded' structure [3], resulting in two further γ turns about Asp and Gly[1]. This kinked conformation is favored for αvβ3. This kink is lost by *N*-methylation of D-Phe and/ or Arg which blocks the γ turns and presents the peptide in a flattened conformation. In case of **8**, there is no indication of any γ turn about glycine[3] and the peptide is in a flat conformation, which eventually results in holding the aspartic acid and arginine side chains farther apart, fitting well into the αIIbβ3 pocket. In conclusion, we demonstrate that a systematic multiple *N*-methylation by knowing the bioactive conformation of the stem peptide, can be employed for enhancing receptor selectivity and activity of a moderately active ligand.

Acknowledgments

We acknowledge Humboldt Foundation for the Max-Planck Forschungspreis to HK.

References

1. Pfaff, M. et al. *J. Biol. Chem.* **269**, 20233-2023872 (1994).
2. Gurrath, M. et al. *Eur. J. Biohem.* **210**, 911-921 (1992).
3. Huang, Z. W. et al. *J. Am. Chem. Soc.* **114**, 9390-9401 (1992).

Synthesis of *N*-Boc Protected Hydrazine Diacids as Key Structural Units for the Formation of α-Helix Mimics

Laura Anderson[1,2], Melissa Topper[1,2], Priyesh Jain[1,2], Emily McIntosh[2] and Mark L. McLaughlin[1,2,*]

[1]Department of Chemistry, University of South Florida, 4202 E. Fowler Ave, Tampa, FL 33620, USA; [2]Drug Discovery Program, H. Lee Moffitt Cancer Center & Research Institute, 12902 Magnolia Drive, Tampa, FL 33612, USA

Introduction

Many diseases, such as autoimmunity, cancer, inflammatory, and neurodegenerative disorders are the result of disregulation of the apoptotic process. Within the B-cell lymphoma-2 (Bcl-2) family, the overall interaction of anti and pro-apoptotic proteins play a major role in regulating apoptosis. Several studies have demonstrated that overexpression of anti-apoptotic Bcl-2 and Bcl-x_L proteins is associated with tumor progression and drug resistance [1]. Previously reported peptidomimetics that inhibit the interaction between pro-apoptotic BH3 domains and Bcl-2 family proteins contain hydrophobic scaffolds that would diminish their use as potential drugs [2]. We report the synthesis of key monomer building blocks for the formation of more hydrophilic compounds based on hydrazine linked piperazine-2,6-dione scaffolds (Figure 1).

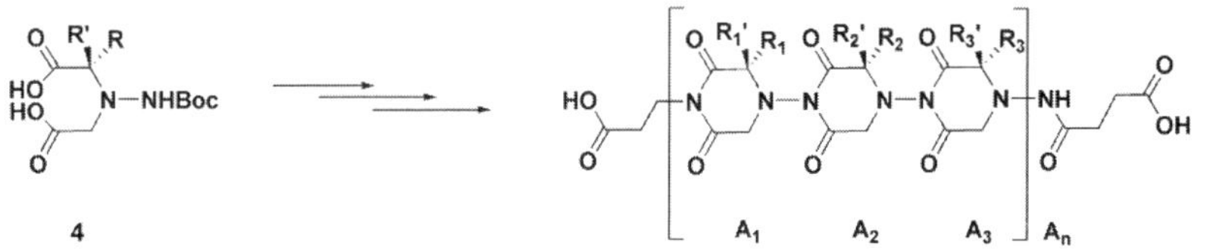

Fig. 1. Hydrazine linked piperazine-2,6-dione peptidomimetics.

Results and Discussion

N-Boc protected hydrazine diacid **4** was synthesized in three steps (Figure 2). In the first step, the methyl esters of selected natural and unnatural amino acids were *N*-alkylated with ethyl bromoacetate in the presence of diisopropylethylamine. Although dialkylation side reactions can occur, it has been found that using the α-halo-alkyl acetate is an efficient method for the preparation of diester **2** in good yields. The alternative reductive amination route with glyoxalic acid or ethyl glyoxalate gave poorer yields.

The key step of the overall synthesis is the formation of the *N*-Boc protected hydrazine **3**. Several procedures for the formation of substituted hydrazines have been reported; some of these include: reaction of alkyl ureas with hypochlorite under basic conditions, direct substitution of hydrazine with triphenylbismuthane in the presence of copper acetate [3], nitrosation of a secondary amine followed by its selective reduction and Boc protection [4], and electrophilic amination with oxaziridines [5]. Although the formation of nitrosamines is widely used and it has been previously explored in our lab, we adopted the latter method due to its efficiency and reduced exposure to potential carcinogenic compounds, such as

DOI: 10.1007/978-0-387-73657-0_97, © Springer Science+Business Media, LLC 2009

nitrosamines. Preparation of the oxaziridine reagent was achieved following the procedure described by Vidal [5]. One advantage of this method is the use of the *N*-alkoxycarbonyloxaziridine to cleanly transfer the desired *N*-Boc protected group to the secondary amine. Oxaziridines are also common reagents for oxygen transfer. In fact, hydroxylamine by-products were observed, but this side reaction was reduced by lowering the reaction temperature. The best results were obtained when the reaction was performed at temperatures not exceeding 5°C. Compound **2** was treated with *tert*-butyl 3-(trichloromethyl)-1,2-oxaziridine-2-carboxylate in methanol at -78°C to yield Boc protected hydrazine ester **3** in good yields.

In order to retain the Boc group, hydrolysis of compound **3** was performed under basic conditions. Less hindered amino acid derivatives, such as phenylalanine and leucine, were completely hydrolyzed to the diacid with 1N NaOH in methanol. However, treatment of more hindered amino acids, such as isoleucine and valine, with 1N NaOH resulted in a partial hydrolysis to the monoacid. Although further attempts to hydrolyze with NaOH were finally successful without racemization, the use of LiOH formed the desired compound with less harsh conditions that might lead to the racemization of the α-carbon with some substrates. Compounds **4a-d** were utilized as building blocks for the synthesis of peptidomimetics whose preparation and biological activity are described elsewhere herein. Syntheses of **4e-g** and other diacids derived from unnatural amino acids are currently being pursued.

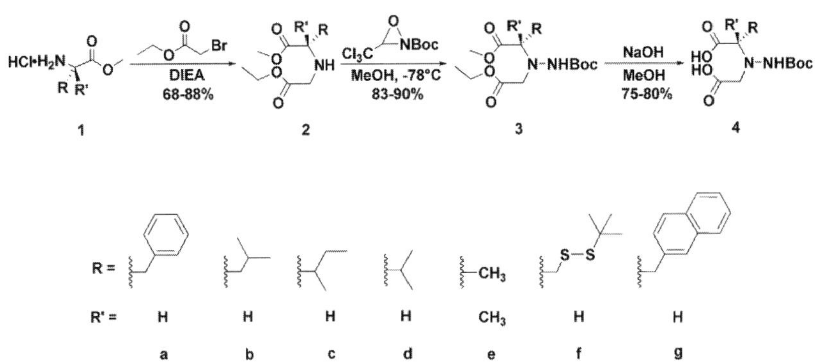

Fig. 2. Synthesis of N-Boc protected hydrazine diacids.

Acknowledgments

We thank Dr. Ted Gauthier and Dr. Umut Oguz for technical assistance, the NIH for financial support through grants R21GM072772 and P01CA118210, and the American Cancer Society and Moffitt Cancer Center for the SPARK program. We also thank the FASEB MARC Program for financial support through a travel award.

References

1. Wendt. M. D., *et al. J. Med. Chem.* **49**, 1165-1181 (2006).
2. Yin, H., *et al. J. Am. Chem. Soc.* **127**, 5463-5468 (2005).
3. Ragnarsson, U. *Chem. Soc. Rev.* **30**, 205-213 (2001).
4. Oguz, U., Guilbeau, G. G. and McLaughlin, M. L. *Tetrahedron Lett.* 43, 2873-2875 (2002).
5. Vidal, J., *et al. J. Org. Chem.* **58**, 4791-4793 (1993).

Solid-Phase Synthesis of 1,3,5-Trisubstituted 1,4-Diazepin-2-one Peptide Mimic

Hassan S. Iden and William D. Lubell*

Département de chimie Université de Montréal, C.P. 6128, Succursale Centre Ville, Montréal, Québec, Canada H3C 3J7

lubell@chimie.umontreal.ca

Introduction

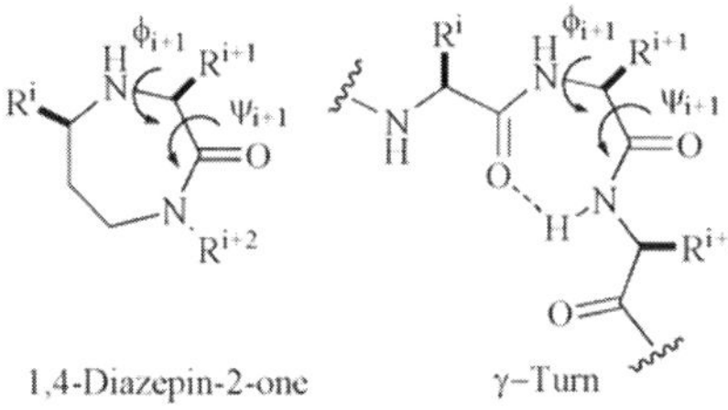

1,4-Diazepin-2-ones display a wide range of pharmacological properties likely due to their potential to mimic peptide γ and β-turn secondary structures [1]. The seven-membered 1,4-diazepin-2-one ring adopts a γ-turn conformation, as supported in our study by X-ray diffraction [2]. The 1,4-diazepin-2-ones have been reported to posses anticonvulsant activity and antitumor properties as well as lymphocyte function-associated antigen-1/ immunoglobulin superfamily ICAM-1 receptor antagonism. We recently reported a solution-phase synthesis route that allowed access to 1,4-diazepin-2-one. While numerous literature reports concerning the synthesis of 1,4-benzodiazepin-2-one using solid-phase synthesis, less attention has been devoted for the synthesis of 1,4-diazepin-2-one [3]. The development of methodology for the synthesis of this class of pharmaceutically important molecule is thus well deserve with potential for creating new biologically active peptide mimics.

Results and Discussion

Resin-bound aldehyde **1** was prepared as described [4]. Conversion of the aldehyde **1** to supported secondary amine **2** was achieved by a reductive amination of the imine produced by treating the resin **1** with β-alanine methyl ester using sodium triacetoxyborohydride in DMF containing a catalytic amount of acetic acid [5]. The reductive amination reaction was monitored by a dinitrophenyl hydrazine test and IR spectroscopy, by following the disappearance of the aldehyde stretch at 1681 cm^{-1}. An amine **2** loading of 1.0 mmol/g was determined by measurement of nitrogen content using elemental analysis. The *N*-Boc-phenylalanine was then coupled to secondary amine **2** using HATU and DIEA in DMF. The progress of the reaction was monitored by chloroanil test. Resin γ,δ-unsaturated ketone **2** was synthesized by treating tertiary amide **3** with freshly prepared vinyl magnesium bromide (2000 mol %) in the presence of copper cyanide (500 mol%) in THF at -45 °C [6]. γ,δ-Unsaturated ketone formation was indicated in the FT-IR spectrum of resin **4** compressed in KBr tablets by the strong carbonyl C=O streching band at 1706 cm^{-1}.

S. Del Valle et al. (eds.), *Peptides for Youth: The Proceedings of the 20th American Peptide Symposium*, 213
DOI: 10.1007/978-0-387-73657-0_98, © Springer Science+Business Media, LLC 2009

Supported diazepinone **5** was prepared by sequence of reactions initiated by Boc group removal using TFA:DCM (1:1), followed by intramolecular imine formation

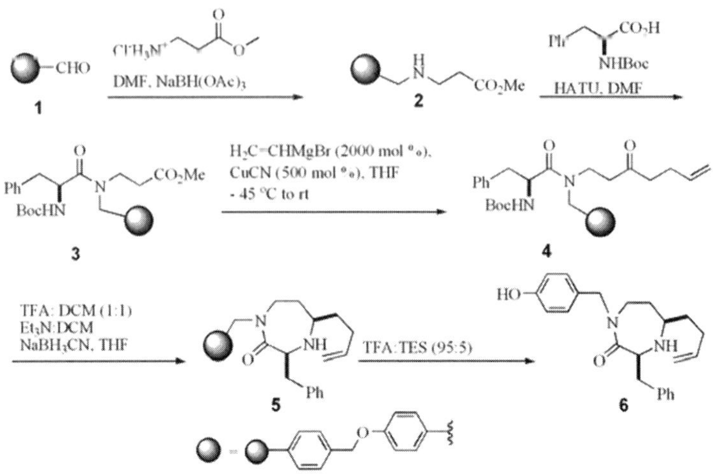

Fig. 1. Diazepinone synthesis using a 4-alkoxybenzaldehyde resin

and neutralization of the TFA salt using triethylamine; imine reduction using sodium cyanoborohydride in THF buffered with 1% acetic acid. The disappearance of the primary amine was monitored by the Kaiser test. Diazepinone **6** was cleaved from the solid support using TFA:TES (95:5). The crude product was obtained in 80% purity as esteemed by HPLC analysis. Purification of the crude product by chromatography on silica gel afforded the diazepinone **6** in 18% overall yield. In accordance with results in solution, the newly formed chiral center in diazepinone **6** was assigned a *cis*-relative stereochemistry [2]. In summary, an effective acid labile linker strategy has been developed for the synthesis of enantiopure 1,4-diazepin-2-one from inexpensive amino acid building blocks. Efforts are underway to exploit this proof-of-concept solid-phase synthesis to generate 1,4-diazepin-2-one libraries.

Acknowledgments

The authors acknowledge gratefully financial support from the Natural Sciences and Engineering Research Council of Canada (NSERC) and the Fonds Québécois de la Recherche sur la Nature et les Technologies (FQRNT).

References

1. Weitz, I. S., Pellegrini, M., Mierke, D. F., Chorev, M. *J. Org. Chem.* ***62***, 2527-2534 (1997).
2. Iden, H., Lubell, W. D. *Org. Lett.* ***8***, 3425-3428 (2006).
3. Horton, D. A., Bourne, G. T., Smythe, M. L. *Chem. Rev.* ***103***, 893-930 (2003).
4. Kobayashi, S., Akiyama, R. *Tetrahedron Lett.* ***39***, 9211-9214 (1998).
5. Boojamra, C. G., Burow, K. M.; Thompson, L. A., Ellman, J. *J. Org. Chem.* ***62***, 1240-1256 (1997).
6. Hansford, K. A., Dettwiler, J. E., Lubell, W. D. *Org. Lett.* *5*. 4887-4890 (2003).

General Solid Phase Phosphopeptide Proteomis with Affinity Pull-Down

Kim Larsen[1], Kasper K. Sørensen[2], Steen Gammeltoft[1] and Knud J. Jensen[2]

[1]*Department of Clinical Biochemistry, Glostrup Hospital, 2600 Glostrup, Denmark;*
[2]*Department of Natural Sciences, Faculty of Life Sciences, University of Copenhagen, 1871 Frederiksberg, Denmark*

Introduction

Phosphorylation and glycosylation are the most common posttranslational modifications (PTMs). Protein phosphorylation is regulated by the interplay between protein kinases and protein phosphatases in response to extracellular signals [1]. Serine/threonine phosphorylation plays a central role in cellular regulation, either by altering protein activity directly or by inducing specific protein-protein interactions. Protein kinases and phosphatases act in dynamic interaction to make and break phosphoester bonds, while phosphopeptide binding proteins control function. Human Protein Phosphatase 2Cδ, PP2Cδ (Wip1, PPM1D) belongs to the PPM phosphatase family. PP2Cδ is thought to be involved in the regulation of a wide range of physiological functions including cellular stress responses, apoptosis and cell cycle regulation. PP2Cδ is regulated by divalent metal ions, typically Mg or Mn and binds to pSer, pThr and pThrXxxpTyr motifs, however, its substrate specificity is poorly understood. New methods for chemical phosphopeptide proteomics and their use in identification and characterization of phosphopeptide binding proteins are desirable.

Here we describe the automated, parallel synthesis of immobilized phosphopeptides directly on a solid support and their use in affinity pull-down studies, especially for characterization of substrate specificity. The central element in this strategy is the on-resin synthesis and display of phosphopeptides on a support, poly(ethylene glycol) polyacrylamide (PEGA), which is compatible with both chemical synthesis and affinity pull-down of proteins from cell lysates in aq. buffers [2] [3]. Using microwave heating, reaction times were significantly reduced, while maintaining high purities of the crude products. Here relative slow steps benefit most from this approach [4].

Results and Discussion

Immobilized peptides were synthesized on a PEGA support through a TFA-stable BAL type linker. Parallel synthesized phosphopeptidyl-resins were used as baits in affinity pull-down (fishing) experiments to determine PP2Cδ substrate specificity. We used tagged PP2Cδ transfected into COS-7 cells. More generally, this approach was used for screening of substrate specificity of kinases, phosphatases, and phosphopeptide binding proteins, and well as for identification of proteins from cell lysates interacting with phosphopeptide sequences. Proteins were identified by SDS PAGE gel separation, followed by tryptic digestion of cut-out protein bands, and then nanoHPLC-MS/MS.

A range of phosphopeptide sequences from the literature reported to be substrates for PP2Cδ together with the corresponding non-phosphorylated peptides were selected for chemical synthesis followed by affinity pull-down, while the peptides remained BAL anchored to the PEGA support:

S. Del Valle et al. (eds.), *Peptides for Youth: The Proceedings of the 20th American Peptide Symposium*, 215
DOI: 10.1007/978-0-387-73657-0_99, © Springer Science+Business Media, LLC 2009

(1) Ac-TDDEMpTGpYVATGG-NH2
(2) Ac-TDDEMTGYVATGG-NH2
(3) Ac-TDDEMpTGYVATGG-NH2
(4) Ac-TDDEMTGpYVATGG-NH2
(5) Ac-TDDEMpSGpYVATGG-NH2
(6) Ac-AFEEGpSQSTTIGG-NH2
(7) Ac-AFEEGSQSTTIGG-NH2
(8) Ac-WEPPLpSQETFSGG-NH2
(9) Ac-WEPPLSQETFSGG-NH2
(10) Ac-QGISFpSQPTCPDGG-NH2
(11) Ac-QGISFSQPTCPDGG-NH2
(12) Ac-LETVSpTQELYSIPGG-NH2
(13) Ac-LETVSTQELYSIPGG-NH2
(14) Ac-RTLCGpTPTYLAPEGG-NH2
(15) Ac-QVLAQPpSTSRKRPGG-NH2
(16) Ac-PAVQGpTGVAGVPGG-NH2
(17) Ac-HQYFMpTEpYVATRGG-NH2
(18) Ac-TSFMMpTPpYVVTRGG-NH2

Solid-phase bound phosphopeptides **1**, **6**, **10** and **12** bound to PP2Cδ whereas the corresponding nonphosphorylated peptides (**2**, **7**, **11**, and **13**) were not able to bind PP2Cδ. Sequences **1** and **3-5** illustrate the importance of the motive pThrXxxpTyr for PP2Cδ recognition. Peptides **14-17** confirm the literature by not being binding substrates for PP2Cδ whereas **18** contradicts literature by being a good binder of PP2Cδ

Acknowledgments

We thank Birte Kofoed, Glostrup Hospital, for excellent technical assistance and Dr. Andrew Coffey, Varian-Polymer Laboratories, for generous supply of resins.

References

1. Jensen, O.N. *Nature Rev. Mol. Cell Biol.* **2006** 7, 391.
2. Larsen, K. *et al*, *Chimica Oggi* **2007** 25, 34.
3. Brandt M. *et al*, *Chembiochem* **2006** 7, 623.
4. Brandt M. *et al* *Int. J. Pep. Res. Therap.* **2006** 12, 349.

Synthesis and Peptide Coupling of Protected Pyrrolylalanine

Aurélie A. Dörr, and William D. Lubell

Département de Chimie, Université de Montréal, C.P. 6128, Succursale Centre Ville Montéal, QC H3C 3J7, Canada

Introduction

Arylalanines possessing varying degrees of aromatic-ring electron density may serve as tools for studying the importance of π-interactions for peptide folding and biological activity [1]. For example, by varying the electronic density of arylalanine residues *N*-terminal to prolyl residues, the prolyl *cis*-amide isomer population has been controlled [2,3]. The π-deficient Phe derivative, (2*S*)-*N*-Boc-3-(6-methylpyridazinyl) alanine was synthesized and incorpored into Ac-Xaa-Pro-NHMe in which relative to Phe, the pyridazinyl analog exhibited significantly reduced prolyl *cis*-amide isomer populations [3]. The synthesis of the π-enriched Phe analog, (2*S*)-pyrrolylalanine **7** has now been pursued as a means for augmenting the prolyl *cis*-amide isomer population.

Results and Discussion

β-Amino ester **1** was synthesized in 4 steps and 47% yield from L-Asp as described [3]. The Cu-catalyzed cascade addition of vinyl Grignard reagent to ester **1** gave homoallylic ketone **2** in 63% yield and subsequent olefin oxidation by using $NaIO_4$ and 2,6-lutidine in dioxane/H_2O gave 4-ketoaldehyde **3** in 95% yield. Pyrrole **4** was prepared by treatment of 4-ketoaldehyde **3** with ammonium formate in the presence of a mix of NaOAc/AcOH under Paal-Knorr conditions and isolated in 72% yield. Pyrrole protection with *t*-BuOK, 18-c-6 and $PhSO_2Cl$ in THF furnished phenylsulfonylpyrrole **5** in 80% yield [4]. Oxazolidine **5** was then ring opened using 80% aqueous AcOH at 50°C overnight to afford *N*-(Boc)-amino alcohol **6**. Removal of the Boc group with 25% TFA in CH_2Cl_2, followed by *N*-sulfonation with $PhSO_2Cl$ provided *N*-(phenylsulfonyl)amino alcohol **7** in 64% yield. (2*S*)-*N,N'-Bis*-(phenylsulfonyl)-3-(2-pyrrolyl)alanine **8** was subsequently made by oxidation of alcohol **7** using TEMPO, sodium chlorite, and sodium hypochlorite in a sodium-phosphate-buffered aqueous acetonitrile solution in 74% yield [5]. Pyrrolylalanine **8** was coupled to proline *N'*-methylamide hydrochloride using TBTU, HOBT and DIEA in acetonitrile to give dipeptide **9** in 77% yield. Conformational analysis of protected dipeptide **9** showed a 13:87 ratio of amide *cis*- and *trans*-isomers *N*-terminal to proline. Deprotection of the sulfonamides and *N*-acetylation of dipeptide **9** is being pursued to obtain the model Ac-Xaa-Pro-NHMe dipeptide for gaging the influence of pyrrolylalanine on the prolyl amide geometry.

S. Del Valle et al. (eds.), *Peptides for Youth: The Proceedings of the 20th American Peptide Symposium*, 217
DOI: 10.1007/978-0-387-73657-0_100, © Springer Science+Business Media, LLC 2009

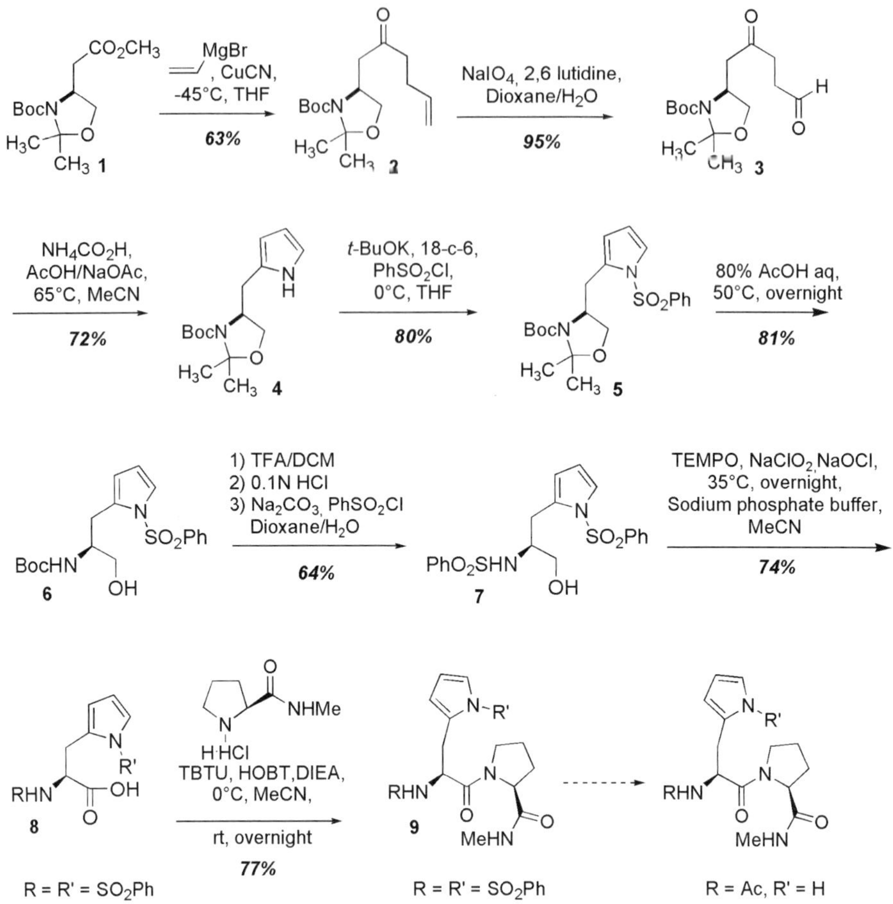

Scheme 1. Synthesis and peptide coupling of pyrrolylalanine ***8***

Acknowledgments

The authors thank NSERC (Canada), and FQRNT (Québec) for financial support.

References

1. Schmidt, G. *Top. Curr. Chem.* **136**, 109-159 (1986).
2. Thomas, K. M., Naduthambi, D. and Zondlo, N. J. *J. Am. Chem. Soc.* **128**, 2216-2217 2006).
3. Dörr, A. A. and Lubell, W. D. *Biopolymers* **88**, 290-299 (2007).
4. Jolicoeur, B.; Chapman, E. E.; Thompson, A.; Lubell, W. D. *Tetrahedron.* **62**, 11531-11563 (2006).
5. Dettwiller, J. E. and Lubell, W. D. *J. Org. Chem.* **68**, 177-179 (2003).

Synthesis of aryl sulfamate and phenol small peptide derivatives using a multidetachable sulfamate linker strategy

Diane Fournier, Pierre-Alain Breuil and Donald Poirier

Oncology and Molecular Endocrinology Laboratory, CHUL Research Center (CHUQ), Quebec, Qc, G1V 4G2, Canada

Introduction

Breast cancer is the most frequent cancer among women in the U.S., representing 31% of all newly diagnosed cancers in 2006[1]. It is estimated that 75% of breast tumors express the estrogen receptor (ER) in a significant amount of cells. In these cases, it is possible to use hormone therapy as a treatment, alone or as an adjuvant to classical radio- or chemotherapy. Hormone therapy can consist of either an ER antagonist (antiestrogen) or an inhibitor of the biosynthesis of estradiol (E_2), the most potent endogenous estrogen. This work focuses on the second strategy, with steroid sulfatase (STS) and type 1 17β-hydroxysteroid dehydrogenase (17β-HSD1) as target enzymes. STS, a membrane enzyme widely distributed in the body and often overexpressed in breast tumors[2], catalyzes the transformation of estrone sulfate, inactive on ER, into estrone (E_1), which is active. 17β-HSD1, expressed in aggressive breast tumors[3], transforms E_1 into E_2. Arylsulfamates are known irreversible inhibitors of STS[4,5] and phenols are inhibitors of 17β-HSD1[6] in addition to being reversible inhibitors of STS[7]. To address the need to rapidly synthesize arylsulfamate and phenols as inhibitors for these enzymes, our laboratory has developed the multidetachable sulfamate linker[8] (see Figure 1) for solid phase synthesis. When treated with an acidic solution, this linker yields arylsulfamates whereas when submitted to a nucleophilic treatment, phenol is obtained. Nine small model libraries, 5 of sulfamates and 4 of phenols, were synthesized using this linker and a sulfamate precursor obtained in two steps from commercially available tyramine.

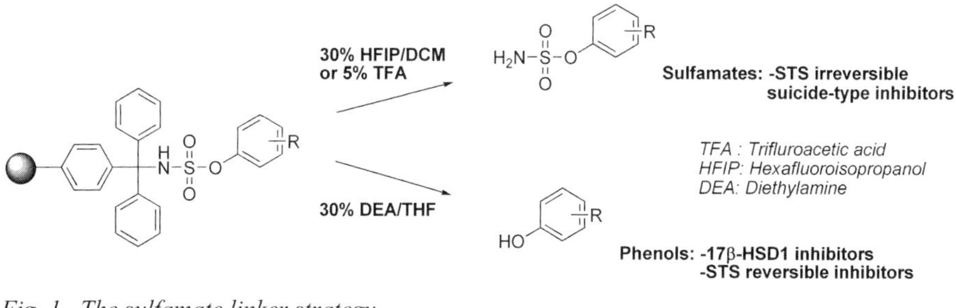

Fig. 1. The sulfamate linker strategy

Results and Discussion

The sulfamate precursor was quantitatively loaded onto polystyrene trityl resin using DIPEA in DCM overnight under argon. The libraries (see Figure 2) were then synthesized using Fmoc chemistry and PyBrOP/HOBt/DIPEA in DMF peptide coupling conditions. For libraries 3, 4 and 5, the isopropyl group(s) were introduced using a one pot reductive amination reaction, and peptide coupling on the secondary amines was made with *in situ* formed acyl chlorides using triphosgene and 2,4,6-collidine in THF. The acetate deprotection steps needed in libraries 8 and 9 were

S. Del Valle et al. (eds.), *Peptides for Youth: The Proceedings of the 20th American Peptide Symposium*, 219
DOI: 10.1007/978-0-387-73657-0_101, © Springer Science+Business Media, LLC 2009

done on resin, using sodium methoxide in methanol and THF. The reduction step needed in library 8 was also done on resin using sodium borohydride in a methanol/DCM mix. For the acetal deprotection steps needed in both libraries 8 and 9, the products were first cleaved from resin and then submitted to 10% aqueous HCl in methanol/DCM or ethanethiol and boron tribromide in DCM

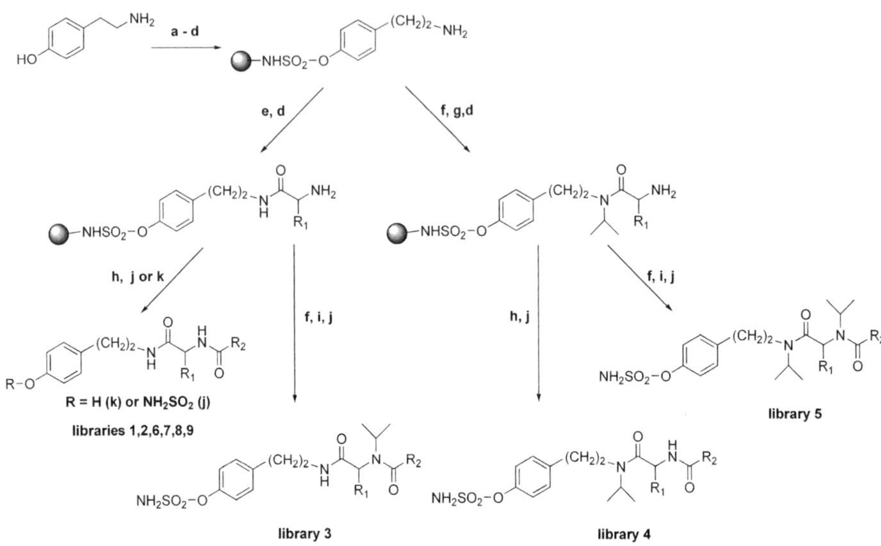

a) FmocOSu, $NaHCO_3$, H_2O/THF, 1 h, R.T.; b) NH_2SO_2Cl, DMA, 3 h R.T.; c) Trityl resin, DIPEA, CH_2Cl_2, 16 h, R.T.; d) 20% piperidine/CH_2Cl_2 1 h, R.T.; e) Amino acid ($FmocNHCHR_1COOH$), HOBt, DIPEA, PyBOP, DMF, Argon, 3 h, R.T.; f) Acetone, $NaBH(OAc)_3$, DCE, 2 h, R.T. g) Amino acid, triphosgene, 2,4,6-collidine, THF, 1 h, 50°C; h) Carboxylic acid (R_2COOH), HOBt, DIPEA, PyBrOP, Argon, 3 h, R.T.; i) Carboxylic acid, triphosgene, 2,4,6-collidine, THF, 1 h, 50°C; j) 5% TFA/CH_2Cl_2, 3h, R.T.; k) 30% DEA/THF, 48 h, R.T.

Fig. 2. Synthesis of libraries 1-9

Synthesis of libraries 1, 2, 4, 6, 7, 8 and 9 generally proceeded in good yields (average 90% crude) and purities (from 80% to 50%). The products of library 3 were not obtained. This is probably due to the incompatibility of the *in situ* formation of chloride peptide coupling method with a free amide NH. Products of library 5 were isolated in poor yields (10 – 15%) and purities. The sulfamate linker allowed us to synthesize 8 of the 9 model libraries. In addition to Fmoc chemistry yielding secondary amides, the sulfamate linker is also compatible with the use of a one-pot reductive amination and subsequent one-pot acyl chloride peptide coupling reaction to synthesize tertiary amides.

Acknowledgements
The work was funded by a CIHR grant to D. Poirier.

References

1. American Cancer Society, 2006 statistics (http://www.cancer.org).
2. Nussbaumer, P., Billich, A., *Curr. Med. Chem. - Anti-cancer Agents* **5**, 507-528 (2005).
3. Gunnarson, C. et al., *Br. J. Cancer* **92**, 547-552 (2005).
4. Poirier, D. et al., *Exp. Opin. Ther. Pat.* **9**, 1083-1099 (1999).
5. Reed, M.J. et al., *Endocr. Rev.* **26,** 171-202 (2005).
6. Poirier, D., *Curr. Med. Chem.* **10**, 453-477 (2003).
7. Boivin, R. et al., *J. Med. Chem.* **43**, 4465-4478 (2000).
8. Ciobanu, L.C., Poirier, D., *J. Comb. Chem.* **5**, 429-440 (2003).

Phe-Aib Hydroxyethylene Dipeptide Isostere Synthesis

Nicolas Genest[2], Susan Del Valle[1], Teresa Lama[2] and Willam D. Lubell[2]

[1] Department of Chemistry, New Mexico State University,PO Box 30001, MSC 3C, Las Cruces, New Mexico, 88003; [2]Departement de Chimie, Université de Montréal, C.P. 6128, Succursale Centre-Ville, Montréal, Québec, Canada, H3C 3J7

Hydroxyethylene dipeptide isosteres have been used as transition-state mimic scaffolds to make potent inhibitors of aspartic, serine and cysteine proteases (Fig. 1).[1] For example, Darunavir© is a hydroxyethylene dipeptide isostere marketed as an HIV-1 protease inhibitor.[2] Amino *iso*-butyric acid (Aib) is known to induce turn conformations in peptides.[3] For example, Boc-Phe-Aib-Leu-OMe is known to adopt a type II β-turn centered on the Phe-Aib residues. To explore the importance of the central amide bond in such turns, we have been studying the synthesis and conformation of hydroxyethylene variants of Xaa-Aib.[4] The Boc-Phe-ψ-[CH(OAc)CH_2]-Aib-OH **1**(Fig. 1) has now been made by a strategy featuring a Cu-catalyzed conjugate addition of 2-methylpropenyl magnesium bromide to *N*-protected phenylalanine methyl ester,[5] reduction of the resulting ketone and oxidative cleavage of the olefin to provide the protected δ-amino γ-hydroxy acid.

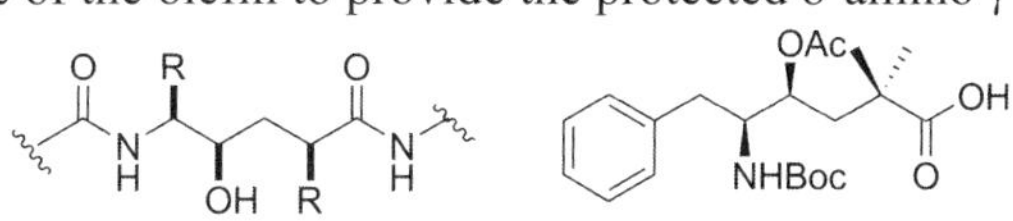

(2*s*, 3*s*)-**1**

Figure 1. Hydroxyethylene Isosetere Derivatives

Results and Discussion

Boc-Phe-ψ-[CH(OAc)CH_2]-Aib-OH **1** was synthesized starting from the diphenyl ketimine Schiff base of Phe-OMe **2** (Scheme 2).[6] Ester **2** was treated with 2-methylpropenyl magnesium bromide in the presence of CuCN to give γ, δ-unsaturated ketone **3** in 84% yield after chromatography on silica gel. Ketone **3** was reduced with $LiAlH_4$ in toluene at –78°C to provide alcohol **4** as a 1:2 mixture of diastereomers in 98 % yield.

The major diastereoisomer, (2*S*, 3*S*)-**4** was isolated in 65% yield by triturating the crude crystals with heptane. The minor (2*S*, 3*R*)-diastereoisomer was purified by evaporation of the washings and chromatography with EtOAc in hexane as eluant to afford alcohol (2*S*, 3*R*)-**4** in 33% yield. Ketimine (2*S*, 3*S*)-**4** was solvolyzed in quantitative yield with pyridinium *p*-toluenesulfonate salt in toluene at reflux to provide the amine salt which was subsequently protected with Boc anhydride to give carbamate **5**. The alcohol was acetylated with acetic anhydride to give acetate **6** in good yield. Oxidative cleavage of olefin **6** was performed with ozone and the resulting aldehyde was converted to the acid without purification on treatment with sodium chlorite to give Boc-Phe-ψ-[CH(OAc)CH_2]-Aib-OH (2*S*, 3*S*)-**1**.

S. Del Valle et al. (eds.), *Peptides for Youth: The Proceedings of the 20th American Peptide Symposium*, 221
DOI: 10.1007/978-0-387-73657-0_102, © Springer Science+Business Media, LLC 2009

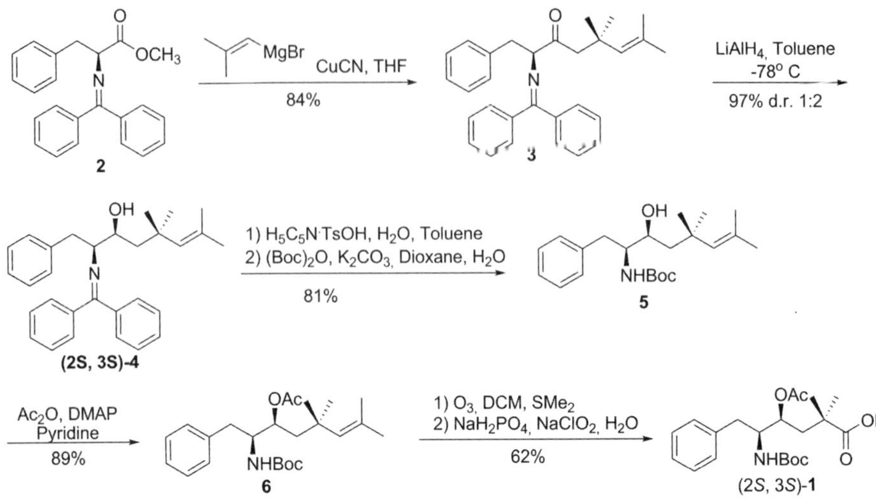

Scheme 1. Synthesis of δ-amino-γ-hydroxy acid dipeptide isostere (2S, 3S)-1

The major diastereoisomer (2*S*, 3*S*)-**5** was assigned the syn stereochemistry after formation of the corresponding oxazolidinone (2*S*, 3*S*)-**7** by treatment with NaH in DMF and subsequent measurement of the vicinal coupling constant for the ring protons (J = 3.6 Hz).[7] The syn stereochemistry would result from hydride addition to the ketone by way of a Felkin-Anh transition state.[7]

*Scheme 2. Assignment of relative stereochemistry of **5***

In conclusion, Boc-Phe-ψ-[CH(OAc)CH_2]-Aib-OH ((2*S*, 3*S*)-**1**) was prepared in 6 steps and 32% overall yield. Incorporation of (2*S*, 3*S*)-**1** into peptides is now being performed.

Acknowledgments

The work was funded by the GRUM and NSERC.

References

1 Leung, D., Abbenante, G. and Fairlie, D. P. *J. Med. Chem.* **43**, 305-341 (2000).
2. Ghosh, A. K. et *al. J. Med. Chem.* **49**, 5252-5261 (2006).
3. Banerjee, A., Drew, M. G. B. and Haldar, D. *Tetrahedron* **63**, 5561-5566 (2007).
4. Lama, T., Del Valle, S. E., Genest, N. and Lubell, W. D. *Int. J. Pept. Res. Ther.* **13**, 355-366 (2007).
5. Hansford, K. A., Dettwiler, J. E. and Lubell, W. D. *Org. Lett.* **5**, 4887-4890 (2003).
6. O'Donnell, M. J. and Polt, R. L. *J. Org. Chem.* **47**, 2663-2666 (1982).
7. Cervantes-Lee, F., Maslouh, N. and Hoffman, R. V. *J. Org. Chem.* **67**, 1045-1056 (2002).

Poly (vinyl alcohol)-*graft*-Poly (ethylene glycol) Supported Hydroxyproline: Synthesis and Application in the Enantioselective Aldol Condensation

Tarek Kassem, X. X. Zhu and William D. Lubell

Département de Chimie, Université de Montréal, C.P.6128, Succursale Centre Ville, Montréal, Québec H3C 3J7, Canada

Introduction

L-Proline has served as catalyst in Michael additions, Robinson annulations and aldol condensations among other reactions [1]. Although the chemistry of this amino acid in solution is being intensively investigated, only a limited number of heterogeneous proline-based catalysts have thus far been reported [2-4]. Hydroxyproline is an attractive building block for proline-like catalysts, because immobilization *via* the hydroxy group leaves the amine and carboxylate functions available for catalysis. Poly (vinyl alcohol)-*graft*-poly (ethylene glycol) (PVA-*g*-PEG) resins have recently been introduced by our laboratory in solid-phase synthesis [5], supported TEMPO catalysis and in on bead HR-MAS ^{1}H NMR spectral analysis [6]. Herein we report the preparation of a hydroxyl-proline-based catalyst on PVA-*g*-PEG resin and its exploration in the asymmetric aldol reaction.

Results and Discussion

Initially, the N^{α}-amino group from the commercially available and inexpensive (2S,4R)-hydroxy-L-proline was Boc protected (93% yield) [7] and then esterified using *tert*-butyl trichloroacetimidate [8]. *O*-Propargyl-*N*-Boc-hydroxyproline *tert*-butyl ester **3** was prepared by *O*-alkylation of the N-Boc-hydroxyproline *tert*-butyl ester **2** with propargyl bromide in DMF in the presence of sodium hydride (75% isolated yield) [3]. Supported hydroxyproline **6** was prepared by reacting *O*-propargyl hydroxyprolinate **3** and azide resin PVA-PEG_{20}-N_3 **4** using a Cu-(I)-catalysed [2+3]-dipolar cycloaddition reaction in 1/1 THF/DMF at 35°C followed by removal of the Boc and *tert*-butyl ester groups using 50 % TFA in CH_2Cl_2.

Characterization of the resin **6** was realized by FT-IR spectroscopy and elemental analysis. The progress of the cycloaddition reaction was monitored by withdrawing resin samples from the reaction vessel at different times and following the disappearance of the azide function by IR spectroscopy (ca 2100 cm^{-1}). For the deprotection step the IR-signals of the Boc carbonyl group and the ester had completely disappeared after 1h.

PVA-PEG supported hydroxyproline resin **6** had a loading of 1.06 mmol/g on the basis of nitrogen elemental analysis. The application of PVA-PEG-supported hydroxyproline **6** as catalyst for the asymmetric aldol reaction, was tested in the reaction of acetone and *p*-bromobenzaldehyde at room temperature using 0.5M of resin in DMF. After 3 days under these conditions, a conversion of ca 70% was observed in the ^{1}H NMR spectrum (measuring the C<u>H</u>OH peak of the aldol product at 5.12 ppm *vs* the C<u>H</u>O at 9.9 ppm of the p-bromobenzaldehyde) of the crude product from filtration of the resin and solvent evaporation. Moreover, examination of the aldol product by chiral HPLC using a Chiralpak AS Daicel (i-PrOH/hexane 15/85), UV 280 nm, flow rate 1.0 mL/min, $t_R$minor 8.950 min and $t_R$major 10.143 min) indicated an enantiomeric excess value of 70%. These results indicate that

S. Del Valle et al. (eds.), *Peptides for Youth: The Proceedings of the 20th American Peptide Symposium*, 223
DOI: 10.1007/978-0-387-73657-0_103, © Springer Science+Business Media, LLC 2009

PVA-PEG proline resin **6** can effectively catalyze the enantioselective aldol condensation reaction. The generality of the use of our resins in other enantioselective transformations is currently being investigated in our laboratories.

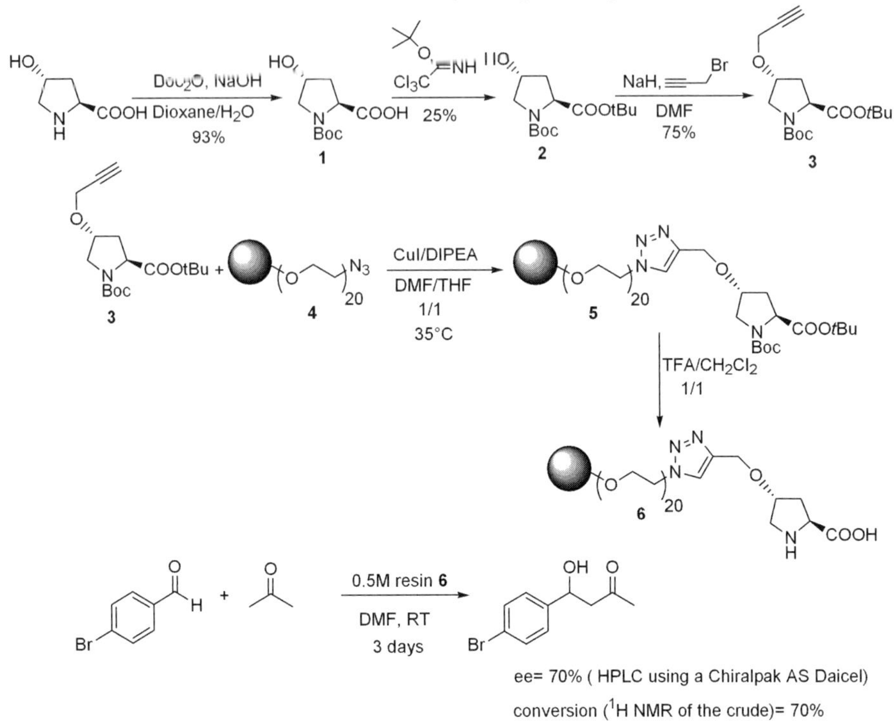

*Fig. 1. Synthesis of PVA-PEG alkoxyproline **6** and its application in an aldol reaction.*

Acknowledgments

This research was supported by NSERC (Canada) and VRQ (Québec). We thank Jad Tannous for chiral HPLC analysis.

References

1. List, B. *Tetrahedron* **58**, 5573-5590 (2002).
2. Calderón, F., Fernández, R., Sánchez, F. and Fernández-Mayoralas, A. *Adv. Synth. Catal.* **347**, 1395-1403 (2005).
3. Font, D., Jimeno, C. and Pericàs, M. A. *Org. Lett.* **8** , 4653-4655 (2006).
4. Giacalone, F., Gruttadauria, M., Marculescu, A. M. and Noto, R. *Tetrahedron. Lett.***48**, 255-259 (2007).
5. Luo, J., Pardin, C., Zhu, X. X. and Lubell, W.D. *J. Comb. Chem.* **9**, 582-591 (2007).
6. Luo, J., Pardin, C., Lubell, W. D. and Zhu, X. X. *Chem. Commun.*, 2136-2138 (2007).
7. Biel, M., *et al. Chem. Eur. J.* **12**, 4121-4143 (2006).
8. Surprenant, S. and Lubell, W. D. *J. Org. Chem.* **71**, 848-851 (2006).

Microwave-Assisted Synthesis of Rhodamine Fluorescent Tags

Wajih B. Ben tahar, Jeffrey W. Keillor, and William D. Lubell*

Département de chimie Université de Montréal, C.P. 6128, Succursale Centre Ville, Montréal, Québec, Canada H3C 3J7, lubell@chimie.umontreal.ca

Introduction

Fluorescent tags have become widely used as tools for studying cellular biology, because of their high detection sensitivity and non-destructive properties [1]. Peptides and proteins possessing fluorescent dyes have been employed to study localization and conformation in living cells [2] [3]. Rhodamine (**1**) is an inexpensive dye that posses a relatively high quantum yield, aqueous solubility, as well as relative stability to acid and neutral pH. Rhodamine (**1**) is, however, sensitive to basic pH, in which the corresponding non-fluorescent lactone **2** is formed (Fig 1).

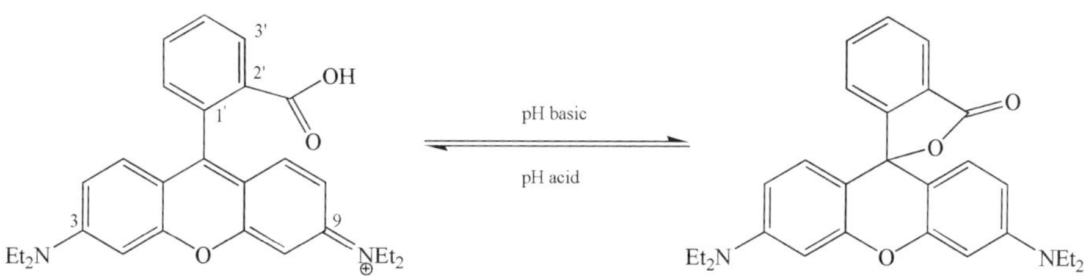

Fig. 1. Rhodamine acid-lactone equilibrium.

To avoid lactone formation, conjugated rhodamines have been described possessing modifications at various positions [4] [5]. Herein we describe a mild, and rapid microwave-assisted technique for coupling rhodamine ethyl ester **3** to different amines to form amides suitable for conjugation chemistry.

Results and Discussion

Microwave-assisted amidation was examined using ethyl ester **3** obtained from esterification of rhodamine **1** using acetyl chloride in ethanol [6]. Microwave irradiation of ester **3** (100 mol%) in the presence of different amines (300 mol%) and amino ester salts (300 mol%) was examined at 150°C in DMF from 20 min to 2h (Fig 2). Amides **4a-d** were isolated in 51-95% yields.

S. Del Valle et al. (eds.), *Peptides for Youth: The Proceedings of the 20th American Peptide Symposium*, 225
DOI: 10.1007/978-0-387-73657-0_104, © Springer Science+Business Media, LLC 2009

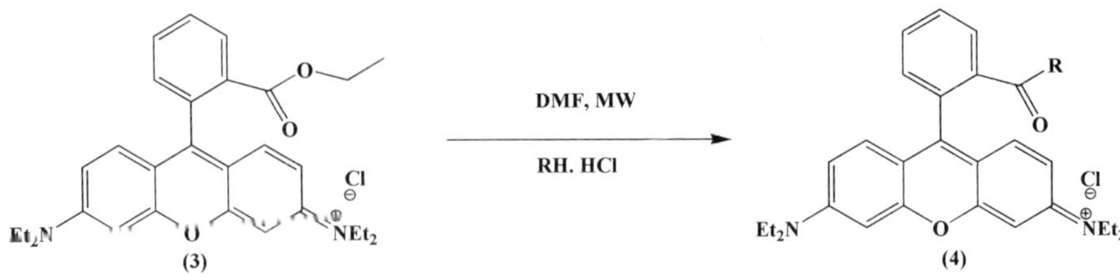

R	T (°C)	t (min)	yield (%)
(a) $NH(CH_2)_2NHBoc$	150	20	95
(b) $NH(CH_2)_2CO_2Me$	150	120	53
(c) $NH(CH_2)_5CO_2Me$	150	90	61
(d) $NHCH_2CO_2Me$	150	120	51

*Fig. 2. Microwave assisted reaction of rhodamine **(3)** with amino esters.*

For example, *N*-(Boc)ethylenediamine reacted with ester **3** in 20 min to yield amide **4a** in 95% yield, after column chromatography (5% MeOH in EtOAc). Under similar conditions, amino ester hydrochloride salts of 2-, 3- and 6-carbons provided amides **4b-d** in 51-61% yields after 90-120 min of irradiation and purification by column chromatography (5% MeOH:EtOAc) for **4c** and **4d** and by preparative HPLC for **4b**. Minor impurities in these reactions such as the unreacted ester **3** and its corresponding lactone, were removed by chromatography. Attempts failed when employing more sterically demanding amino ester hydrochloride salts, such as Phe-OtBu•HCl, Ser-OBz•HCl and Leu-OMe•HCl, as well as with methyl isonipicotate hydrochloride.

A set of rhodamine tags **4** was thus effectively synthesized using microwave-assisted heating. The chemistry of tags **4** and further development of the approach for their synthesis are currently under study in our laboratory.

Acknowledgments

The work was funded by FQRNT Quebec and NSERC Canada.

References

1. Tang, B., Xing, Y., Li, P., Zhang, N., Yu, F., Yang, G. *J. Am. Chem. Soc.* **129**, 11666-11667 (2007).
2. Berezowska, I., Chung, N. N., Lemieux, C., Zelent, B., Szeto, H. H., Schiller, P. W. *Peptides.* **24**, 1195-1200 (2003).
3. Buku, A., Masur, S., Eggena, P. *Am. J. Physiol.* **257**, E804- E808 (1989).
4. Nguyen,T., Francis, M. B. *Org. Lett.* **5**, 3245-3248 (2003).
5. Adamczyk, M., Grote, J. *Bioorg. Med. Chem. Lett.* **10**, 1539-1541 (2000).
6. Ross, J. A., Ross, B. P., Rubinsztein-Dunlop, H., McGeary, R. P. *Synth. Commun.* **36**, 1745-1750 (2006).

A novel family of diketopiperazines as a tool for the study of transport across the blood-brain barrier (BBB) and their potential use as BBB-shuttles.

Meritxell Teixidó[1], Esther Zurita[1], Roger Prades[1], Teresa Tarrago[1] and Ernest Giralt[1,2]

[1]*Institute for Research in Biomedicine, Barcelona Science Park, Universitat de Barcelona;* [2]*Dept. Organic Chemistry, University of Barcelona, Barcelona, E-08028, Spain*

Introduction

Neurodegenerative diseases such as Parkinson and Alzheimer disease, schizophrenia, epilepsy, brain tumors, HIV are pharmaceutical targets located inside the brain. In many cases there are promising compounds for their treatment, however owing to their Blood-Brain Barrier (BBB) transport problems > 98% of these potential drugs do not go to drug development stage. The BBB is a natural defense mechanism designed to keep harmful substances out of the brain. The anatomical bases of the BBB are primarily the tight junctions at the endothelial cells of the brain capillaries. Therefore, possible transport mechanisms are all transcellular. While different transport mechanisms occur at the BBB, this work has focused on the passive diffusion mechanism. We design, synthesize and evaluate the potential use of DKPs as BBB-shuttles for the delivery of cargoes that can not cross the BBB unaided.

Results and Discussion

On the basis of two initial libraries of mono- and di-*N*-methylated diketopiperazines (DKPs), the best side-chains for our BBB-shuttles were chosen. A first library of 15 DKPs with distinct side-chains was prepared using a solid-phase methodology. The DKP compounds of the first library were evaluated as a mixture, which has the advantage of speeding up the high-throughput of the PAMPA assay (Parallel Artificial membrane Permeability Assay) and facilitating comparison between the compounds and establishment of rules. We also evaluated control molecules with known high brain penetration capacities such as propranolol (a well known beta-adrenergic receptor blocker) and carbamazepine (an anticonvulsant and mood stabilizing drug, used primarily in the treatment of epilepsy and bipolar disorder).

Not all the DKPs of the first library showed efficient transport across the BBB PAMPA model. After choosing those with the best capacity, we designed and prepared a second library, in this case di-*N*-methylated DKPs, which were also synthesized in solid-phase and evaluated by PAMPA. We wanted to check whether this second library showed improved transport capacity or whether they became too hydrophobic and were retained in the membrane. A comparison of the DKPs that had both amide bonds *N*-methylated with those that only had one *N*-methylated amide bond indicated that the presence of the second *N*-methyl group in the amide bond improves BBB transport capacity (Phe, HomoPhe) because it reduces the number of potential hydrogen bonds that the DKP can form. However, in the case of highly hydrophobic DKPs (2Nal, Cha, Oct, 1 PyrenylAla), the addition of a second

S. Del Valle et al. (eds.), *Peptides for Youth: The Proceedings of the 20th American Peptide Symposium*,
DOI: 10.1007/978-0-387-73657-0_105, © Springer Science+Business Media, LLC 2009

N-methyl group can be a disadvantage because the corresponding DKP is more retained in the membrane. Comparison of the DKP scaffold (*N*-MePhe-*N*-MePhe) to the corresponding linear peptide (Ac-*N*-MePhe-*N*-MePhe-$CONH_2$) highlights that the DKP is a privileged scaffold in terms of BBB penetration potential.

The DKPs, which have the ability to cross the BBB, have been modified in order to attach drugs such as dopamine, which cannot cross the BBB unaided. In this way the DKPs will work as BBB-shuttles.

Table 1. *Percentage of transport after 4 h and effective permeability (P_e and Log P_e) in the PAMPA for the DKP-Dopamine constructs and also Dopamine (Cargo) and L-Dopa (Current drug used in the market). * These compounds were not be detected in the acceptor well of the PAMPA after 4 h.*

Compound	% Transport 4h	P_e (cm/s)	Log P_e
Dopamine	--*	--*	--*
L-Dopa	--*	--*	--*
DKP Phe(*p*-NH-CH_2-CO-NH-Dopamine)-*N*-MePhe	0.08	3.61 10^{-8}	-7.4
DKP Phe(*p*-NH-CH_2-CO-NH-Dopamine)-*N*-MeHomoPhe	0.02	0.76 10^{-8}	-8.1
DKP Phe(*p*-NH-CH_2-CO-NH-Dopamine)-*N*-MeCha	0.05	2.28 10^{-8}	-7.6
DKP Phe(*p*-NH-CH_2-CO-NH-Dopamine)-*N*-Me2Nal	0.24	1.15 10^{-7}	-6.9
DKP Phe(*p*-NH-CH_2-CO-NH-Dopamine)-*N*-MeOct	0.67	3.21 10^{-7}	-6.5
DKP Phe(*p*-NH-CH_2-CO-NH-Dopamine)-*N*-Me-1PyrenylAla	0.04	1.90 10^{-8}	-7.7

Although our DKP-Dopamine constructs enabled only limited transport of dopamine, they nevertheless greatly increased the transport of this challenging compound across the artificial model BBB membrane.

This BBB-shuttle approach could be useful to transport promising antiretroviral agents and chemotherapy drugs into the brain, thereby reducing the severe side-effects caused by the high doses currently used to overcome low permeability. Our approach implies that the search for novel drugs does not become limited only to compounds that have the capacity to cross the BBB.

Acknowledgments

This work was supported by Ministerio Educación y Ciencia (BIO 2005-00295) and the Generalitat de Catalunya [Xarxa de Referència en Biotecnologia and Grups Consolidats (2005SGR-00663)].

References

Kansy, M. et al. Physicochemical High Throughput Screening: Parallel Artificial Membrane Permeation Assay in the Description of Passive Absorption Processes. *J. Med. Chem.* **1998**, 41, 1007-1010.

New Perspective in Peptide Chemistry by *N*-Alkylation

Horst Kessler[1], Jayanta Chatterjee[1], Lucas Doedens[1], Florian Operrer[1], Eric Biron[1], Daniel Hoyer[2], Herbert Schmid[2], Chaim Gilon[3], Victor J. Hruby[4] and Dale F. Mierke[5]

[1]Department Chemie, Lehrstuhl II für Organische Chemie, Technische Universität München, Lichtenbergstrasse 4, Garching 85747, Germany, Novartis Institutes of Biomedical Research, Novartis Pharma, Basel, Switzerland, [3]Organic Chemistry Department, The Hebrew University of Jerusalem, Israel, Jerusalem 91904, [4]Department of Chemistry, The University of Arizona, Tucson, Arizona 85721, USA, [5]Department of Molecular Pharmacology, Brown University, Providence, RI-02912, USA

Introduction

Oral bioavailability is a major concern in the development of peptide based drugs. Peptides usually possess high selectivity and potency but the major drawback in their development as a drug is the extremely low oral availability. Insufficient bioavailability of peptides is caused by low membrane permeation, low uptake via tight junctions (paracellular transport), and active export into the gut and/or low resistance against enzymatic degradation. To circumvent these undesired properties, we envisioned the approach of multiple *N*-methylation. We were inspired by the cyclic peptide drug Cyclosporin which is administered orally and has seven of its eleven amide bonds *N*-methylated, and it violates all of the Lipinski's rules [1], for governing oral availability. Mono *N*-methylation has been employed over the years to improve lipophilicity, bioavailability, proteolytic stability and activity of peptides [2]. However, to our knowledge multiple *N*-methylation has never been reported, which could be owing to the availability of *N*-methylated amino acids, difficult coupling sequences [3], and unpredictable conformational change. Thus, our lab focused in understanding the biophysical characteristics like: conformational behavior and membrane permeability conferred by multiple *N*-methylation to cyclic peptides. At the same time we were also interested in developing potent analogues of Somatostatin, MTII and αIIbβ3 integrin by multiple *N*-methylation which would posses oral bioavailability.

Results and Discussion

Impact of multiple N-methylation on conformation and membrane permeability of cyclic pentapeptides of type cyclo(-DXaa-LXaa$_4$-)

We concentrated in the beginning to understand the behavior of successive multiple *N*-methylation on the backbone of cyclic pentapeptides of type cyclo(-aA$_4$-), where 'a' represents D-Alanine and 'A' as L-Alanine. With the successive multiple *N*-methylation we ended with a library of 30 peptides excluding the stem peptide and the per-methylated peptide. Out of the 30 peptides, some were conformationally homogeneous and some were inhomogeneous at the NMR time scale[4]. Studying in detail the conformation of these peptides by distance geometry and MD simulations, we could figure out that, the *N*-methylation systematically modulates the conformation of these cyclic peptides. There is a fixed pattern in which they dictate the conformation of these peptides. By knowing the conformation of these peptides, one can vary the amino acids with the fixed pattern of chirality and *N*-methylation to spatially orient the pharmacophores in order to rationally design bioactive peptides.

S. Del Valle et al. (eds.), *Peptides for Youth: The Proceedings of the 20th American Peptide Symposium*, 229
DOI: 10.1007/978-0-387-73657-0_106, © Springer Science+Business Media, LLC 2009

Once we identified the conformationally homogeneous peptides, the influence of *N*-methylation on the membrane permeability was tested, as *N*-methylation is known to influence the intestinal permeability of peptides. We tested the permeability of these peptides in the Caco-2 and PAMPA assay. The results obtained suggest that, conformationally homogeneous peptides permeate the membrane better than the inhomogeneous peptides. The *N*-methylation has a positive effect on the membrane permeability of the cyclic peptides, however, *N*-methylation failed to convert the paracellular transport route to the transcellular route (see the proceeding *N*-methylated cyclic pentaalanine peptides as template structures).

Multiply N-methylated Somatostatin analogue (Veber peptide)
In order to improve the oral bioavailability of *cyclo*(-PFwKTF-), discovered by Veber et al.[5], we synthesized a library of 30 *N*-methylated analogues. All the analogues were tested for their binding towards the five different receptor subtypes (sst1-sst5). Out of the 30 analogues, only 7 were found to bind in significant concentration to the receptors sst-2 and sst-5. The membrane permeability of all these 7 analogues was tested in Caco-2 and Ussing assay. Interestingly only one of these analogues (Figure 1) showed significant permeability in both the assays. A drug to be orally bioavailable has to be metabolically stable, so we also tested the metabolic stability of the analogue. To our expectation the analogue was highly stable towards enzymatic degradation as compared to the stem peptide. The peptide was found to permeate the intestinal membrane via the paracellular pathway inspite of tri *N*-methylation. Though, the oral bioavailability of the analogue was found to be 1% owing to metabolic stability, it stays longer in the blood stream.

As it was known that sst2 is over expressed in neuroglioblastoma, the cytotoxicity of the analogue was tested. The compound was found to be more cytotoxic than PTR-3173 [6]and Octreotide [7] inspite of a higher IC50 value of the latter.

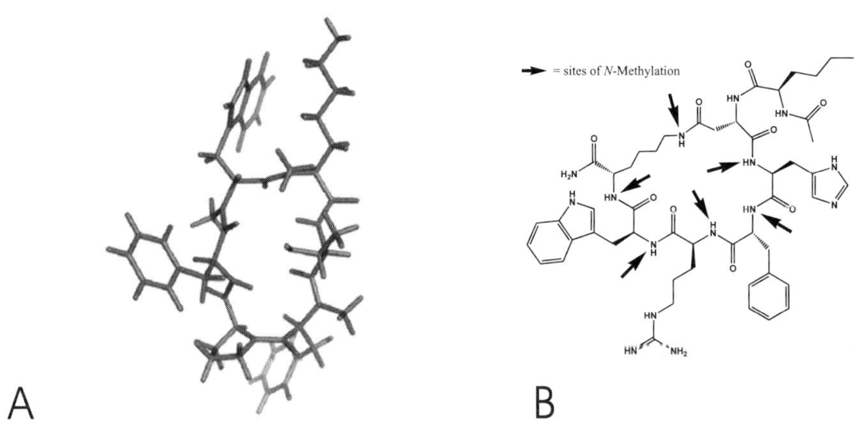

Figure 1. A) N-methylated somatostatin analogue. B) N-methylated MTII analogues.

Multiply N-methylated MTII analogues

MT-II is a super agonist for four receptor subtypes (MCR1, MCR3-5) of the melanocortin system. It is derived from the natural ligand α-MSH by truncation and substitution of several amino acids and lactam bridging by side chain to side chain cyclization [8]. This agonist binds unselectively to these receptors, therefore, we hoped to gain selectivity by synthesizing *N*-methylated derivatives of MT-II (Figure 1B). Within our library containing 32 *N*-methylated MT-II derivatives we have several compounds that act as agonists and antagonists and that in deed show selectivity for certain receptors. Final results will be published soon.

These candidates were also tested in CaCo-2 assays. The results suggest a paracellular transport of these compounds. Tests for metabolic stability and for blood brain barrier will follow.

Multiply N-methylated αIIbβ3 antagonists by a designed approach

After the results obtained from the *N*-methylated libraries of somatostatin analogue and MTII, we envisioned that a 'Designed Approach' can be employed to obtain the selective analogues, instead of the commonly employed 'Library Approach' [9]. In this approach, the preliminary requirement is the knowledge of the bioactive conformation of the lead peptide. Then systematically all the externally oriented peptide bonds are *N*-methylated. This approach reduces the number of *N*-methylated analogues to be synthesized to screen a library. This approach was employed to synthesize only 7 differentially *N*-methylated analogues of a cyclic hexapeptide cyclo(-GRGDfL-), which is partially selective for αIIbβ3 vs. αvβ3. We found that *N*-methylation of some residues like D-Phe and Leu increased selectivity towards αvβ3 but with reduced activity. On the contrary *N*-methylation of Arg led to highly active and selective ligand for αIIbβ3.

In conclusion, we show that multiple *N*-methylation can be a useful approach for achieving oral bioavailability, enhancing membrane permeability and obtaining highly active and receptor-subtype selective ligands.

Acknowledgment

We gratefully acknowledge the Humboldt Foundation for the support via the Max-Planck Forschungspreis. This work was also in part supported by the German Israel Foundation (GIF) and the Center of Integrated Protein Science Munich.

References

1. Lipinski, C. A., Lombardo, F., Dominy, B. W., Feeney, P. J. *Adv. Drug Delivery Rev.* **1997**, *23*, 3-25.
2. Fairlie, D. P.; Abbenante, G.; March, D. R. *Curr. Med. Chem.* **1995**, *2*, 654-686.
3. Teixido, M.; Albericio, F.; Giralt, E. *J. Pep. Res.* **2005**, *65*, 153-166.
4. Chatterjee, J.; Mierke, D.; Kessler, H. *J. Am. Chem. Soc.* **2006**, *128*, 15164-15172.
5. Veber, D. F. et al. Nature **1981**, *292*, 55-58.
6. Afargan, M. et al. Endocrinology **2001**, *142(1)*, 477-486.
7. Bauer, W.; Briner, U.; Doepfner, W., Haller, R.; Huguenin, R.; Marbach, P.; Petcher, T. J.; Pless, J. *Life Sci.* **1982**, *31*, 1133-1140
8. Al-Obeidi, F.; Hadley, M. E.; Pettitt, B. M.; Hruby, V. J. *J. Am Chem. Soc.* **1989**, *111*, 3413-3416.
9. Chatterjee, J.; Ovadia, O.; Zahn, G.; Marinelli, L.; Hoffman, A.; Gilon, C.; Kessler, H. (manuscript in preparation).

PEPTIDE MATERIALS SCIENCE

Interactions of peptides with single-walled carbon nanotubes

Zhengding Su, Ken Mui, Elisabeth Daub, Tong Leung and John Honek

Department of Chemistry, University of Waterloo, Waterloo, ON N2L 3G1, Canada

Introduction

Single-walled carbon nanotubes (SWNTs) have promising applications in the fields of biotechnology and medicine due to their unique electrical, metallic and structural characteristics[1, 2]. Recently the design and utilization of polypeptides specifically binding to carbon nanotubes (CNTs) has been the focus of much attention due to their functionality in biological systems[3]. Therefore, mechanism(s) of interaction between biomolecules and SWNTs is a critical focus of investigation because the extent of nanotube functionality and reactivity in biological systems remains relatively unknown and this knowledge will be important in understanding their environmental and biological activity as well as their potential for application to nanostructure fabrication.

Results and Discussion

As an initial step, three classes of phage libraries defined by length and constrained conformation were screened against SWNTs (**Figure 1a**). In these constructed peptide libraries, peptides (12-mers, 7-mers or disulfide-constrained 7-mers) are displayed on the surface of M13 phage by fusion to the N-terminal of pIII, a coat protein of this bacteriophage. The binding affinity and specificity of identified phage and peptides were quantified and compared with previously reported peptide sequences having affinity to other types of nanotubes[4,5]. Through phage displayed peptide library screening, we have identified a SWNT-binding motif, [6] *X*TH*XX*PWT*X*, where *X* is any amino acid (**Figure 1b**). One peptide, with the amino acid sequence of LLADTTHHRPWT (named as UW-1), exhibited the highest affinity to SWNTs and formed a β-turn structure which was induced by SWNT-binding[6] (**Figure 1c**), suggesting that hydrophobic interactions as well as π-π interactions between the side-chain of tryptophan at position 11 and the side-walls of SWNT could play an important role in high-affinity binding.

Thus, we have investigated the nature of the π-π interaction between the tryptophan residue and

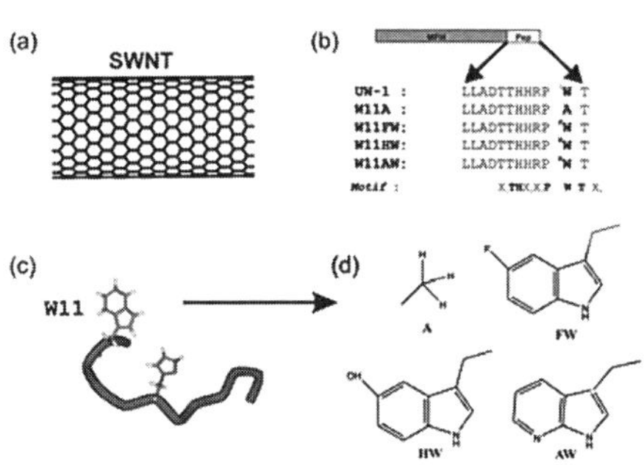

Fig. 1. Probing tryptophan's importance in SWNT-binding peptides by unnatural amino acid substitution. (a). SWNTs used for phage panning. (b). The UW-1 peptide was recombinantly expressed in a fusion protein in pMFH vector[6]. Incorporation of tryptophan analogs (i.e. 5-fluorotryptophan (^{F}W), 5-hydroxytryptophan (^{H}W) and 7-azatryptophan (^{A}W) was done with an auxotrophic strain (i.e. E. coli W3110). (c). The conformation of UW-1 peptide exhibits a β-turn like structure forming a unique surface for SWNT binding. (d). Side-chain structures of alanine and three tryptophan analogs.

S. Del Valle et al. (eds.), *Peptides for Youth: The Proceedings of the 20th American Peptide Symposium*, 235
DOI: 10.1007/978-0-387-73657-0_107 © Springer Science+Business Media, LLC 2009

SWNTs, using a single-point mutation (i.e. W11A) as well as the incorporation of unnatural tryptophan analogs i.e. (5-fluorotryptophan (FW), 5-hydroxytryptophan (HW) and 7-azatryptophan (AW)) to alter the side chain properties of the peptide (**Figure 1d**). Their binding affinities to SWNTs were experimentally determined and further investigated utilizing computational chemistry. **Table 1** summarizes the binding affinities of the UW-1 mutants to the SWNTs, which are expressed as percentages to that of the WT protein (i.e. the UW-1 peptide). These values have been calibrated with their incorporation efficiency of tryptophan analogs. The order of binding affinity for the UW-1 peptide and its analogs follows the order: W11HW > WT > W11FW >W11AW > W11AW$^+$, where AW$^+$ represents the protonated form of 7-azatryptophan.

Table 1. Binding affinities of the wild-type protein and its mutant and tryptophan analog substituents.

Fusion protein	Incorporation efficiency (%)	Binding affinity (%)	pH
WT	-	100	7.0
W11A	-	38 ± 8	7.0
W11FW	96	77 ± 12	7.0
W11HW	97	253 ± 32	7.0
W11AW	93	45 ± 8	7.0
W11AW$^+$	93	27 ± 6	5.0

Electronic structure calculations were undertaken to analyze the energy levels of each tryptophan analog side chain with respect to their lowest unoccupied molecular orbital (LUMO) and their highest occupied molecular orbital (HOMO) using several electronic calculation methods (AM1, RHF/3-21G, RHF/6-31G*, B3LYP/6-31G* and MP2/6-31G*) as incorporated in SPARTAN'06 (Wavefunction Inc, Irvine, CA) to determine if there is a correlation between either of these energies and the trends observed for interaction. The results indicated that in all the considered methods except MP2/6-31G*, the energy levels of HOMO energies are ranked in the order of HW > W > FW > AW > AW$^+$, which is consistent with that of the observed peptide binding affinities (**Table 1**). As a result, it is likely that the orbital mixing in the π-π interaction takes place between the HOMO of the tryptophan and the LUMO of the sidewall of SWNT, implying that the side wall of SWNTs acts in this instance as an electron-acceptor and the indole ring as an electron donor.

Acknowledgement

We thank NSERC Nano Innovation Platform for financial support of this research.

References

1. Bianco, A.; Kostarelos, K.; Prato, M. *Curr. Opin. Chem. Biol.* **9**, 674-679 (2005).
2. Contarino, M. R.; Sergi, M.; Harrington, A. E.; Lazareck, A.; Xu, J.; Chaiken, I. *J. Mol. Recognit.* **19**, 363-371 (2006).
3. Ortiz-Acevedo, A.; Xie, H.; Zorbas, V.; Sampson, W. M.; Dalton, A. B.; Baughman, R. H.; Draper, R. K.; Musselman, I. H.; Dieckmann, G. R. *J. Am. Chem. Soc.* **127**, 9512-9517 (2005).
4. Wang, S.; Humphreys, E. S.; Chung, S. Y.; Delduco, D. F.; Lustig, S. R.; Wang, H.; Parker, K. N.; Rizzo, N. W.; Subramoney, S.; Chiang, Y. M.; Jagota, A. *Nat. Mater.* **2**, 196-200 (2003).
5. Kase, D.; Kulp, J. L., III; Yudasaka, M.; Evans, J. S.; Iijima, S.; Shiba, K. *Langmuir* **20**, 8939-8941 (2004).
6. uSZ.; Leung, T.; Honek, J. F. *J. Phys. Chem. B*, **110**, 23623-23627 (2006).

Stimuli-responsive Peptide Self-assembly into Well-organized Nanofibers

Tomoyuki Koga, Ken-ichi Kitamura and Nobuyuki Higashi

Department of Molecular Science & Technology, Faculty of Engineering, Doshisha University, Kyotanabe, Kyoto, 610-0321, JAPAN

Introduction

Advances in peptide self-assembly provide strong incentives to fabricate novel nanostructured biomaterials. Recently, we have reported that fine tuning of peptide sequence, stereochemistry, and solution pH enables the construction of 3D-nanoarchitectures, including supramolecular nanofibers, by controlled self-assembly of 16-mer peptide building blocks [1, 2]. Here, we report a novel stimuli-responsive self-assembling system by using artificial peptides having enzyme-active site in order to control spontaneous assembling process. Amphiphilic block peptides **1** and **2**, in which triblock-type β-sheet forming peptide ($L_4K_8L_4$) was conjugated with relatively hydrophilic poly(ethylene glycol) (*PEG*) unit [3] or D,L-alternating oligolysine unit through thrombin-cleavable site (VPRGS), were designed and employed as building blocks for self-assembly (Fig. 1). The conformational and self-assembling properties of **1** and **2** in response to enzyme triggering were investigated, especially in consideration of the effect of molecular structures of hydrophilic units. Establishing a facile methodology to fabricate 3D nano-objects at controlled timing is attractive in order to advance a bottom-up nanotechnology.

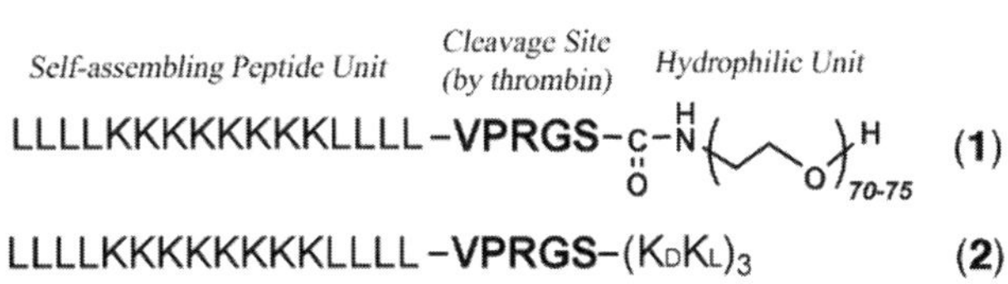

*Fig. 1. Molecular structures of amphiphilic block peptides having enzyme-active site, **1** and **2**, used as building blocks for self-assembly.*

Results and Discussion

The peptides **1** and **2** were prepared by standard solid phase synthesis using Fmoc-chemistry on PAP Tenta-Gel (Rapp Polymere) and CLEAR-acid (Peptide Institute, Inc.) resins, respectively, and characterized by ^{1}H-NMR and MALDI-TOFMS analyses. To test the biological activity, the peptides **1** and **2** were first digested by thrombin (0.8 NIH unit) at pH 9.0 and room temperature. MALDI-TOFMS analyses clearly showed the site-specific cleavage took place smoothly at the R-G bond within 30 min in both cases of **1** and **2**. In other words, the hydrophilic units (*PEG* and $(K_DK_L)_3$) can be easily separated from $L_4K_8L_4$-based self-assembling unit by the addition of thrombin.

In previous studies [1, 2], we have demonstrated that amphiphilic triblock-type $L_4K_8L_4$ forms β-sheet structure and self-assembles into nanofibers at around pH 9. In cases of peptides **1** and **2**, however, such β-sheet formation and self-assembly were obviously prevented even at pH 9.0. It seems that introductions of the hydrophilic units into the C terminal of $L_4K_8L_4$ increase the solubility in water and

S. Del Valle et al. (eds.), *Peptides for Youth: The Proceedings of the 20th American Peptide Symposium*, 237
DOI: 10.1007/978-0-387-73657-0_108, © Springer Science+Business Media, LLC 2009

inhibit the self-assembly. In particular, the D,L-alternating oligolysine unit was found to act to repress the β-sheet formation effectively rather than the *PEG*-based hydrophilic unit. Fig. 2(A) shows the CD spectra of **2** with and without thrombin-digestion in Tris/HCl buffer at pH 9.0. With freshly prepared sample solution, the peptide **2** took mainly random coil structure containing a small amount of α-helix structure with two negative maxima at 202 and 220 nm. Such random conformation of **2** was kept stably even after 24 h-incubation. AFM image also demonstrated that the peptide **2** did not self-assemble into specific nanostructure and only existed as small globules at this condition (Fig. 2(B)). On the other hand, interestingly, the addition of thrombin, which generates the $L_4K_8L_4$-VPR rapidly by detaching the hydrophilic $(K_DK_L)_3$-GS unit from **2**, was found to cause the drastic CD spectral change from random coil to β-sheet-like structure with negative maximum at 218 nm (Fig. 2(A)). In parallel with such enzyme-triggered conformational transition, well-organized nanofiber structures with nearly uniform heights of 5-6 nm were clearly observed in AFM image (Fig. 2(C)). The *PEG*-attached **1** also showed the similar self-assembling behavior in response to the addition of thrombin [3], although amorphous aggregates of **1** were slightly observed at 24 h even without digestion. Thus, OFF-to-ON switching of peptide self-assembly into nanofibers was successfully accomplished by external stimuli; specifically, by enzymatic reaction. These approaches should be useful in establishing a programmable molecular organization system, as well as for our understanding of the peptide self-assembly.

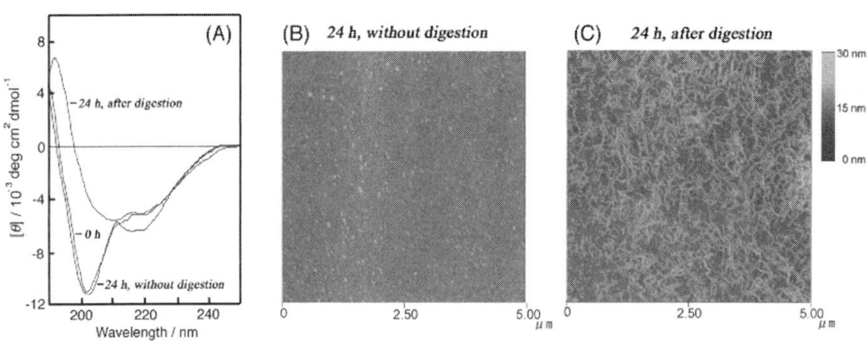

*Fig. 2. (A) CD spectra of **2** obtained just after sample preparation (0 h) and after 24 h-incubation with and without thrombin-digestion in Tris/HCl buffer at pH 9.0. [2]=40 μM. AFM images of **2** (B) and thrombin digested **2** (C) at 24 h, pH 9.0. Note that the digestion of **2** by thrombin (0.8 NIH unit) was carried out after 2.5 h-incubation.*

Acknowledgments

This work was supported in part by a grant-in-aid for scientific research (No. 19710099), and by the Academic Frontier Research Project on "New Frontier of Biomedical Engineering Research" from the Ministry of Education, Culture, Sports, Science and Technology, Japan.

References

1. Koga, T., Matsuoka, M. and Higashi, N. *J. Am. Chem. Soc.* **217**, 17596-17597 (2005).
2. Koga, T., Higuchi, M., Kinoshita, T. and Higashi, N. *Chem. Eur. J.* **12**, 1360-1367 (2006).
3. Koga, T., Kitamura, K. and Higashi, N. *Chem. Commun.* 4897-4899 (2006).

pH Controlled Aggregation Morphology of Aβ(16-22): Formation of Peptide Nanotubes, Helical Tapes and Amyloid Fibrils

Ronald C. Elgersma, Dirk T. S. Rijkers and Rob M. J. Liskamp

Medicinal Chemistry & Chemical Biology, Utrecht Institute for Pharmaceutical Sciences, Department of Pharmaceutical Sciences, Faculty of Science, Utrecht University, Utrecht, The Netherlands

Introduction

Amyloid fibrils are characterized by the (anti)parallel organization of β-sheets which leads to a reduced solubility of the protein and to the formation of amyloid plaques. Amyloid formation in living organisms is a highly undesirable process and is a (co)causative factor in several diseases for example in Alzheimer's disease, Parkinson's disease and type II diabetes. Otherwise, amyloid formation can also be exploited in the design of self-assembled bionanomaterials.

In recent studies we have shown that small modifications in the peptide backbone of amyloidogenic amylin(20-29) results in complete loss of fibril formation; instead, supramolecular assemblies were observed such as peptide nanotubes, lamellar sheets and fibrous assemblies [1,2]. These results triggered us to investigate the scope of amyloidogenic peptides to serve as a templates for the design and synthesis of a new class of self-assembled bionanomaterials.

For this purpose, a series of Aβ(16-22) peptides was synthesized with an *N*-terminal modification as a capping moiety to introduce a conformational constraint and/or a pH-sensitive moiety to control the overall charge of the peptide (Figure 1). The influence of both modifications on the formation of specific nanostructures by self-assembly was investigated as a function of pH.

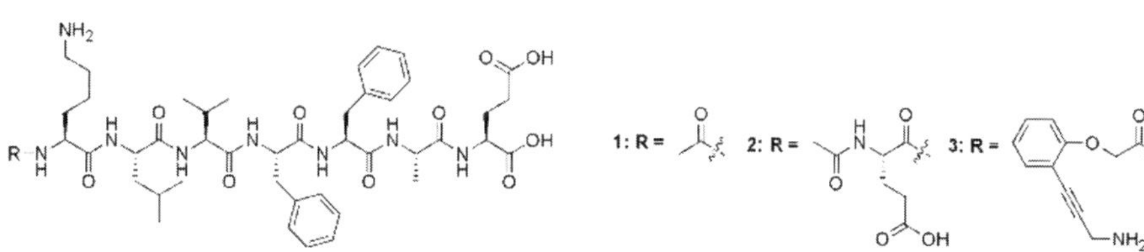

Fig. 1. Derivatives of the amyloidogenic peptide sequence Aβ(16-22).

Results and Discussion

Among others, peptides **1** – **3** were synthesized. These Aβ(16-22) derivatives were dissolved in buffer with an appropriate pH (1, 7.4 or 11). It was found that the solubility of peptides **1** – **3** was strongly dependent on the pH and the lowest solubility was generally found at pH 7.4. As a result of this, gelation behavior was also strongly influenced by pH.

After dissolution of the peptides in buffer with pH 1, clear solutions were obtained which formed upon standing a glassy gel. However, at pH 7.4 a turbid solution/suspension was obtained and after aging for three weeks, small aggregates were visible by eye but a gel was not formed. Transmission electron microscopy (TEM) of the gel formed at pH 1, showed the presence of tubules (in some cases with a length of up to 12 μm) and helical ribbons as is shown in Figure 2, while at pH 7.4, typical amyloid fibrils were observed. The presence of amyloid fibrils was

S. Del Valle et al. (eds.), *Peptides for Youth: The Proceedings of the 20th American Peptide Symposium*, 239
DOI: 10.1007/978-0-387-73657-0_109, © Springer Science+Business Media, LLC 2009

also supported by Fourier transform infrared spectroscopy (FTIR) in which the characteristic 1628 cm^{-1} amide I absorption was observed, indicative for antiparallel β-sheets.

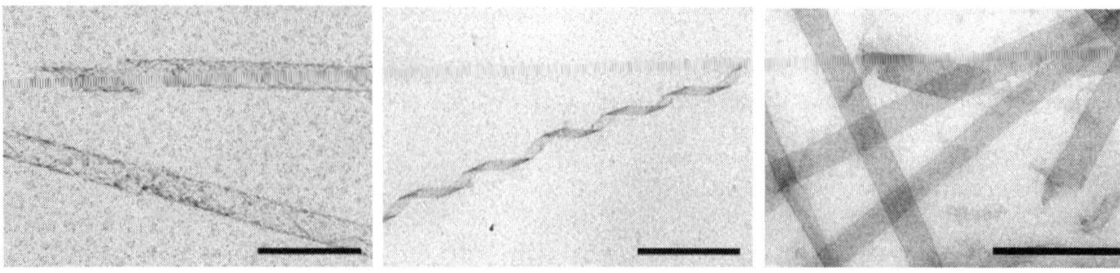

*Fig. 2. TEM images of the gels formed by peptide **1** (left), **2** (middle) and **3** (right) at pH 1; the scale bar represents 500 nm.*

Interestingly, FTIR analysis of the supramolecular assemblies as shown in Figure 2 also indicated the presence of a characteristic cross β-sheet secondary structure, since the absorptions at 1628 cm^{-1} and 1670 – 1679 cm^{-1} were visible. Despite the observed differences in supramolecular morphology, we found indications from our own experiments and in the literature [3,4] that the driving forces which form these structures rely on β-sheet forming propensities of the peptides. Nevertheless, circular dichroism spectroscopy (CD) revealed significant differences of the secondary structure in solution as a function of the pH. Apparently, variations in electrostatic interactions and bulkiness of the *N*-terminal capping moiety influence the interaction mode of the individual β-sheet peptides resulting in differences of the morphology of the supramolecular assemblies.

In conclusion, amyloidogenic peptides can be used as templates and the formation of specific supramolecular structures can be tuned, which is important for the ultimate design of peptide-based nanomaterials.

Acknowledgments

These investigations were supported by the council for Chemical Sciences of the Netherlands-Organization for Scientific Research (CW-NWO).

References

1. Elgersma, R. C.; Meijneke, T.; Posthuma, G.; Rijkers, D. T. S.; Liskamp, R. M. J. *Chem. Eur. J.* **12**, 3714-3725 (2006).
2. Elgersma, R. C.; Meijneke, T.; de Jong, R.; Brouwer, A. J.; Posthuma, G.; Rijkers, D. T. S.; Liskamp, R. M. J. *Org. Biomol. Chem.* **4**, 3587-3597 (2006).
3. Lu, K.; Jacob, J.; Thiyagarajan, P.; Conticello, V. P.; Lynn, D. G. *J. Am. Chem. Soc.* **125**, 6391-6393 (2003).
4. Reches, M.; Gazit, E. *Science* **300**, 625-627 (2003).

Immobilization of Antimicrobial Peptides on Reverse Osmosis Polyamide Membranes: Potential Biofilm Inhibitors?

Karin Berliner[1], Eyal Hershkovitz[1], Zeev Ronen[2] and Roni Kasher[1]

[1]Department of Desalination & Water Treatment; [2]Department of Hydrology & Environmental Microbiology, Zuckerberg Institute for Water Research, The Blaustein Institutes for Desert Research, Ben-Gurion University of the Negev, Sede-Boqer 84990, Isreal

Introduction

Thin-film composite (TFC) membranes for reverse osmosis (RO) and nanofiltration (NF) are widely used in water desalination, ultra-pure water production and waste water treatment, and are presently the most common membranes used in technologies for drinking water production[1]. Prevention and treatment of biofilm formation on RO and NF membranes is a major obstacle in these processes, acting as a barrier towards their large-scale utilization. Effective prevention of microbial growth on membranes is achieved only when a continuous, high chlorine concentration is maintained. However, chlorine generates harmful byproducts upon reaction with organic matter. In addition, modern TFC membranes, such as RO aromatic polyamide membrane **(1)**, are sensitive to oxidizing agents, such as chlorine and ozone.

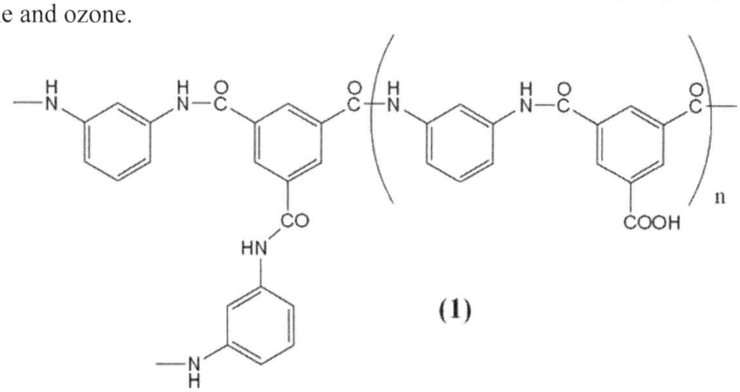

(1)

Here we propose a novel approach for prevention of biofilm formation on RO membranes that relies on antimicrobial peptides immobilized to solid polymeric membranes in a way that preserves the conformation, and hence the biological activity, of the peptides. The study is focused on the development of a synthetic method for covalent binding of antimicrobial peptides to RO polyamide membrane surfaces. Subsequent studies would be the characterization of the conformation of the bound peptides, followed by evaluation of antimicrobial activity of the attached peptides. To date, the use of surface-attached antimicrobial peptides as antibacterial agents for the prevention of biofilm growth on RO membranes has not been investigated.

Results and Discussion

We study thin-film composite RO membranes, since these are susceptible to biological fouling by virtue of the continuous supply nutrients in the feed solution, which, in turn, accelerates bacterial growth on the membrane surface. We modify

S. Del Valle et al. (eds.), *Peptides for Youth: The Proceedings of the 20th American Peptide Symposium*, 241
DOI: 10.1007/978-0-387-73657-0_110, © Springer Science+Business Media, LLC 2009

the thin-film layer (aromatic polyamide (**1**)) since this layer comprises the surface in direct contact with the feed solution and therefore it is liable to bacterial growth. Synthetic antimicrobial peptides were used, due to their short amino acid sequence.[2-3] The peptides were prepared by SPPS, using Fmoc-chemistry, or were given generous gift from Prof. Mati Fridkin (Weizmann Institute of Science, Rehovot). While widely used in solution, much less is known about conformation and activity of antimicrobial peptides when attached to a solid support. The peptides would have to be away from the membrane surface in order to allow proximity to the bacterium cell envelope. Therefore, we synthesized polyethylene glycol-based tethers to connect the peptides to the membrane surface (see Fig. 1).

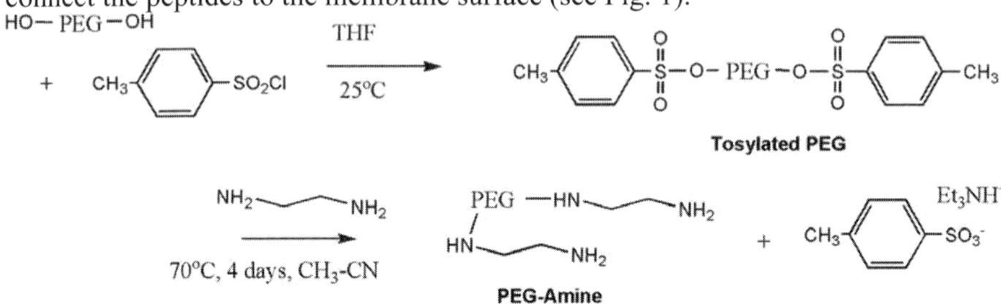

Fig. 1: Amine-functionalization of polyethylene glycol (PEG; MW=500 g/mole) by using tosyl-chloride for immobilization on thin-film polymeric membrane.

The modifications of the thin layer – the aromatic polyamide - in TFC membranes is done by coupling of an amine to a free carboxyl group on the membrane surface (see Fig. 2). The attachment of peptides may be performed by linking Cys-thiol groups to maleimide groups on the membrane surface (Fig. 2).

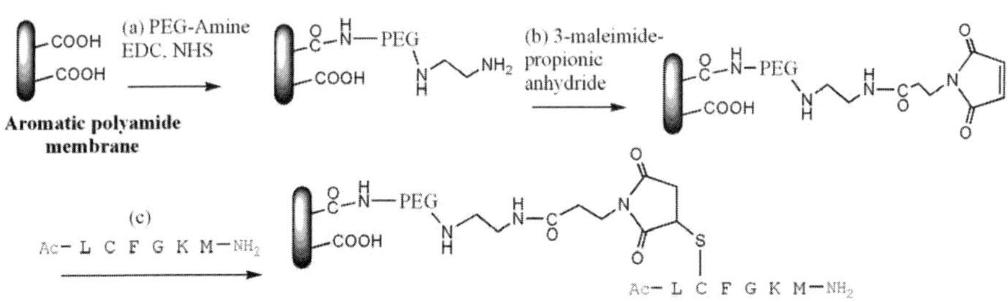

Fig. 2: Immobilization of a Cys-containing antimicrobial peptide onto aromatic polyamide membrane through PEG-Amine linker. Stages a-c were done in aqueous solutions, to keep the polymeric TFC membrane insoluble.

Acknowledgments

We wish to thank Prof. Mati Fridkin and Mrs. Aviva Kapitkovsky (Weizmann Institute of Science, Israel) for their generous gift of synthetic antimicrobial peptides

References

1. Petersen, R. J. *Journal of Membrane Science* ***83***, 81-150 (1993).
2. Blondelle, S. E., and Houghten, R. A. *Trends in Biotechnology* ***14***, 60-65 (1996).
3. Bessalle, R., Gorea, A., Shalit, I., Metzger, J. W., Dass, C., Desiderio, D. M., and Fridkin, M. *J. Med. Chem.* ***36***, 1203-1209 (1993).

Synthesis of Polylactides with Side Chain Functionality: Ring-Opening Polymerimerization of a Homobislactone Prepared from Lysine

Hiromi Kodera[1], Naoki Nishishita[2], Yoshiaki Hirano[2], Yoshiaki Kiso[1] and Yoshio Hayashi [1,3]

[1]Department of Medicinal Chemistry, Center for Frontier Research in Medicinal Science, Kyoto Pharmaceutical University, Kyoto 607-8412, Japan; [2]Department of Applied Chemistry, Faculty of Engineering, and Nanomaterials Microdevices Center, Osaka Institute of Technology, Osaka 535-8585, Japan; [3]Department of Medicinal Chemistry, Tokyo University of Pharmacy and Life Sciences, Hachioji, Tokyo 192-0392, Japan

Introduction

Poly-L-lactide (PLLA) and poly(lactide-co-glycolide) (PLGA) are known as bioabsorbable macromolecules, due to their low immunogenicity. In recent years, they have been widely used in medical and pharmaceutical fields such as surgical sutures, orthopedic implants, scaffolds for cells in tissue engineering, and sustained-release formulations in drug delivery systems. However, because of their simple chemical structure, it seems to be difficult to introduce other functionalities on the polymer. Therefore, introduction of a functional group in the side chain structure would enhance its functionality, thereby expanding the use of polylactides in medical and pharmaceutical fields.

Applying peptide chemistry, we designed and synthesized polylactides possessing a cationic group on the side chain. This moiety was further modified by attaching other functional groups, for example, thiol reactive groups such as bromoacetyl and maleimide groups. These groups can anchor bioactive peptides and proteins on the surface of particles that are derived from functionalized polylactides. Previously, during an amide bond formation with an α-hydroxy-β-amino acid derivative as an acyl component, we observed byproduct formation of a homobislactone during activation of the carboxyl group of α-hydroxy-β-amino acid via an active ester formation [1]. Because homobislactones are usually used as starting materials in a Lewis acid-catalysed polylactide synthesis [2], we applied this coupling method to the synthesis of a new homobislactone **3** from H-Lys(Z)-OH.

Result and discussion

α-Hydroxy acid **2** was derived from H-Lys(Z)-OH **1** by converting an amino group into a hydroxyl group with H_2SO_4/$NaNO_2$ under 0–10°C. This α-hydroxy acid **2** was converted to homobislactone **3** using a BOP–HOBt method [3]. The ring-opening polymerization of homobislactone **3** was performed with an initiator and a catalyst, MEE (2-(2-methoxyethoxy)ethanol) and $SnOct_2$, respectively. Resultant polymer **4** was then deprotected by a TMS-Br–thioanisole–*m*-cresol/TFA system (Scheme 1). The molecular weight of the polymer could be adjusted by varying the homobislactone/MEE ratio and measured with ^{1}H NMR, GPC and MALDI-MS (TOF) [4]. The *M*n (number average molecular weight) value in ^{1}H NMR was estimated from the integral ratio of proton signals between methoxy and ε-methylene groups. This procedure was also applied to copolymerize homobislactone **3** and the

S. Del Valle et al. (eds.), *Peptides for Youth: The Proceedings of the 20th American Peptide Symposium*, 243
DOI: 10.1007/978-0-387-73657-0_111, © Springer Science+Business Media, LLC 2009

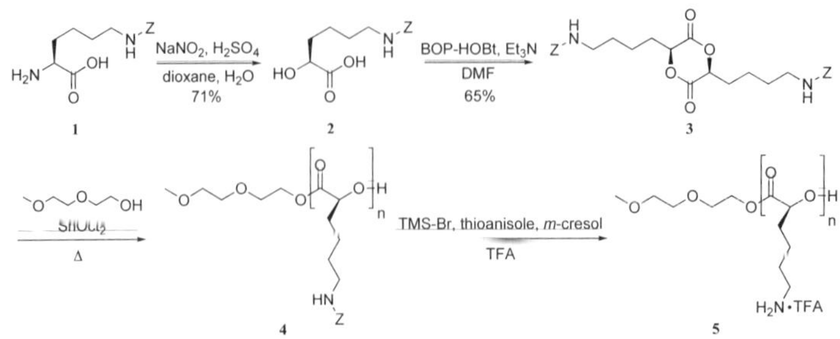

Scheme 1. Synthesis of the lactate-type cationic polymer.

lactide. The amino groups of the side chains in the deprotected copolymer were further modified with the maleimide group, and then micro- or nano-particles were formed by a solvent evaporation method using CH_2Cl_2 and DMSO as the organic solvents. The average diameter of the resultant particles from different solvent ratios was evaluated by DLS (Dynamic Light Scattering) (Table 1). The formation of particles was observed in the presence of CH_2Cl_2 (Entry: A, B, C). No particle was formed using only DMSO (Entry: D). The particles were examined by light microscope and SEM (Scanning Electron Microscope) as shown in Figure 1. These results indicated that the copolymer with maleimide was successfully formed based on an oil-in-water emulsion. Particles having various encapsulation efficiency and size could be controlled by changing the ratio of co-solvents with different water-miscibility.

Table 1. Formation of particles under different solvent conditions.

Entry	Maleimide Unit (mol%)	Solvent DMSO : CH_2Cl_2	Average Diameter[a] (μm)
A	5	0 : 6	1.67
B	5	1 : 1	2.17
C	5	5 : 1	3.89
D	5	6 : 0	-

[a] Average diameter was measured by DLS.

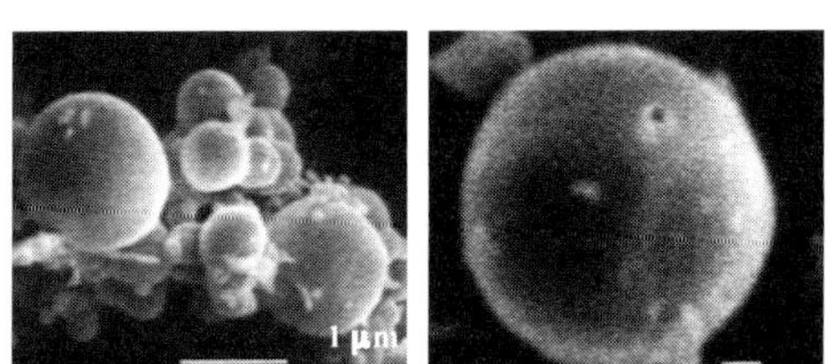

Figure 1. SEM images of maleimide-containing particles.

References

1. Hayashi, Y., Kinoshita, Y., Hidaka, K., Kiso, A., Uchibori, H., Kimura, T. and Kiso, Y. *J. Org. Chem.* **66**, 5537-5544 (2001).
2. Hyon, S., Jamshidi, K. and Ikada, Y. *Biomaterials* **18**, 1503-1508 (1997).
3. Nguyen, D. L., Seyer, R., Heitz, A., Castro, B. *J. Chem. Soc., Perkin Trans. 1* 1025 (1985).
4. de Jong, S. J., van Dijk-Wolthuis, W. N. E., Kettenes-van den Bosch, J. J., Schuyl, P. J. W. and Hennink, W. E. *Macromolecules* **31**, 6397-6402 (1998).

New Degradable Cationic Peptides for Modulated Gene Delivery

Fiana Mirkin[1], Hadar Eini[2], Amos Douvdevani[2] and Gerardo Byk[1]

[1]*Dept. of Chemistry, Bar Ilan University Ramat Gan, 52900 Israel;* [2]*Dept. of Clinical Biochemistry, Soroka Medical Center & Ben-Gurion University, Beer-Sheva, 84101 Israel*

Introduction

Synthetic DNA delivery agents are an importante alternative to viral vectors, since they display less immunogenicity and propagation risks [1]. However, non-viral DNA delivery agents display in many cases high toxicity. This is especially significant in the case of polymeric cations such as polylysine and dendrimers. The integration of a biodegradable entity into non-viral gene delivery systems promises decreased toxicity and improved gene expression [2]. We have designed and synthesized biodegradable polymers derived from Lys, Arg and His of various lengths containing disulfide bridges along their backbone. Additionally, a biodegradable dendrimer using the same approach was synthesized.

Results and Discussion

SSa was synthesized and introduced along the peptide chains of various polypeptides using conventional SPPS (see figure 1).

SSa	Fmoc-$NHCH_2CH_2$-S-S-$CH_2CH_2NHCOCH_2CH_2COOH$
Lys_n n=1,2,3	H-(K_{10}**SSa**)$_n K_{10}NH_2$
His_n n=1,2	H-(H_{10}**SSa**)$_n H_{10}NH_2$

Fig. 1. Cationic polypeptides synthesized using building block ***SSa.***

In another approach, we have developed a new method for solid phase intersite disulfide bond formation which resulted in dimers containing cystine. These dimers were generated after conventional SPPS of the polypeptides containing (S-trityl)cysteine at different positions. The cysteines were bridged on the solid support using iodine in methanol (see figure 2).

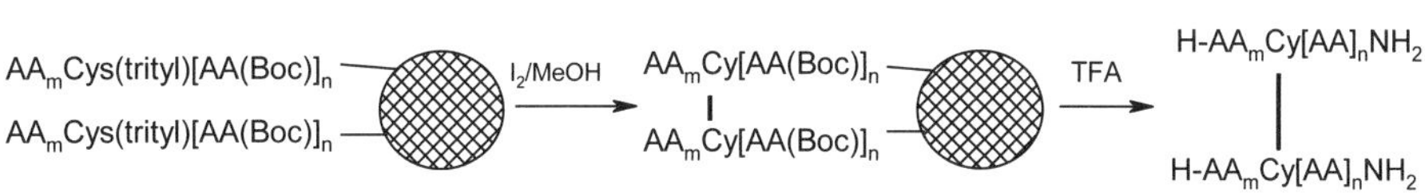

*Fig. 2. Cationic polypeptides obtained by solid phase intersite disulfide bridge formation: AA= K (**CyK**, m=0, n=10; **KcyK**, m=8, n=2); H (**CyH**, m=0, n=10; **HCyH**, m=8, n=2); R (**CyR**, m=0, n=10).*

Additionally, we have combined the use of SSa and the intersite disulfide bridge formation to obtain the long cationic polypeptide containing 40 histidines, FIM-2256: (H-H_{10}-**SSa**–**Cy**-$H_{10}NH_2$)$_2$. Finally, we have synthesized the dendrimer FIM-2126 which contains a disulfide bridge (see figure 3).

The products were complexed with DNA at different charge ratios and ran on gel electrophoresis in the presence or absence of dithiothreitol (DTT). DNA was retained at positive charge ratios similarly to non-reducible polycations. Incubation of the complexes with DTT disrupted completely the complexes and DNA ran freely

S. Del Valle et al. (eds.), *Peptides for Youth: The Proceedings of the 20th American Peptide Symposium*, 245
DOI: 10.1007/978-0-387-73657-0_112, © Springer Science+Business Media, LLC 2009

independently of the charge ratios. These results proved *in vitro* that the reducible complexes form normal complexes with DNA but when treated with reducing conditions the complexes disrupt as expected.

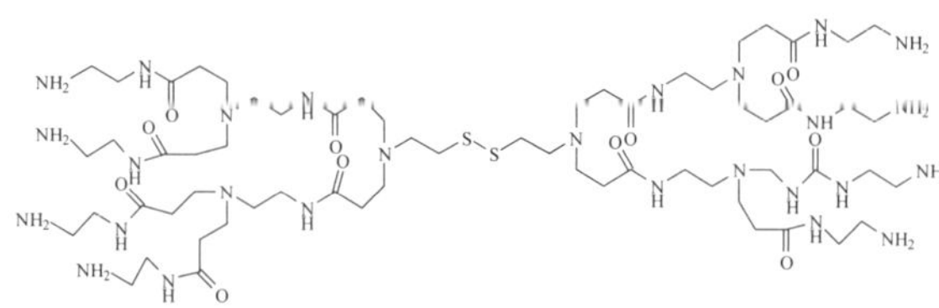

Fig. 3. Structure of biodegradable dendrimer FIM-2126.

The complexes were tested for expression of SEAP protein after incubation with 3T3-BALB/c fibroblasts accompanied by neutral red viability assays as compared to the known standard RPR-120535 [1]. All the polypeptides displayed low level of transfections (10% of the activity of the standard RPR-12535, data not shown), however dendrimer FIM-2126 displayed a significant transfection efficiency as compared to the standard RPR-120535 (see figure 4).

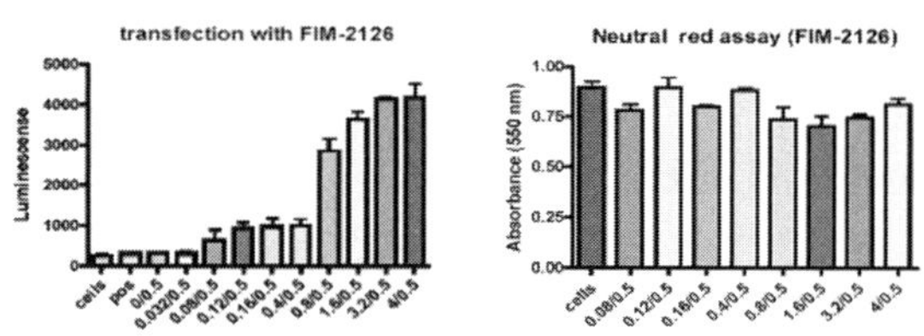

Fig. 4. Left panel: transfection levels in 3T3-BALB/c fibroblasts for dendrimer FIM-2126/SEAP-DNA complexes at different charge ratios (+/-). Right panel: XTT viability assay in 3T3-BALB/c fibroblasts of FIM-2126 at different charge ratios (+/-).

No significant toxicity was observed at any charge ratio as compared to non treated cells (levels were similar to those shown for FIM-2126 I fig.4).

We have measured the size of some of the complexes used for transfection and interestingly, the complex FIM-2126/DNA disclosed a size of about 30-40 nm, while the polycationic peptides displayed sizes of about 200-800 nm. We conclude that the introduction of the degradable disulfides along polycationic peptide chains resulted in a substantial reduction of toxicity at any charge ratio, however the complexes displayed low transfection efficiency probably due to early release of the DNA from the complexes upon cell entry, or their large size. FIM-2126 displayed significant transfection efficiency with particle size of 30-40 nm.

Acknowledgments

The work was supported by grant N⁰ 274/02-4 from the Israel; Science Foundation, Jerusalem, Israel and by the Marcus Center for Medicinal Chemistry.

References

1. Byk, G., Dubertret, C., Escriou, V. et al. *J. Med. Chem.* **41**, 224-235 (1998).
2. Byk, G., Wetzer, B., Frederic, M. et al. *J. Med. Chem.* **43**, 4377-4387 (2000).

The Dcpm protecting group is a useful alternative to the Hmb residue

Petra Henklein[1], René Röder[1,2], Hardy Weißhoff [3], Peter Henklein[1] and Louis A. Carpino[4]

[1]Institute of Biochemistry, Charité Universitätsmedizin-Berlin, Monbijoustrasse 2, D-10117 Berlin, Germany; [2]Institute of Clinical and Molecular Virology, University of Erlangen-Nürnberg, Schlossgarten 4, D-91054 Erlangen, Germany; [3]Institute of Chemistry, Humboldt-University-Berlin, Brook-Taylor-Strasse 2, D-12489 Berlin, Germany; [4]Department of Chemistry, University of Massachusetts, Amherst, MA 01003 USA

Introduction

The Hmb (2-hydroxy-4-methoxy-benzyl) protecting group is often used to prevent the aspartimide side reaction of sensitive Asp-Gly units and to avoid interchain interactions (i.e. β-sheet formation) during peptide synthesis [1]. For our studies, we synthesized the same peptide (Biotinyl-RQYRLIVHNGYCDG (X) RSERNL-COOH; X=H, Hmb, Dcpm) in three different ways. The glycine residue in the Asp-Gly unit was introduced (a) without backbone protection, (b) using the Hmb-protecting group, and (c) employing the Dcpm (dicyclopropylmethyl) nitrogen protecting group. All peptides were labeled with biotin at the N-terminus. Analysis of the deprotection process for the Dcpm group was carried out by NMR-spectroscopy.

Results and Discussion

From the synthesis of the biotin labeled Hmb peptide (Mon1_b), we observed only a by-product with an additional mass of 362 Da. By using NMR spectroscopy we demonstrated that the hydroxyl group of the Hmb residue also reacts with the biotin molecule. The signal of the protons from the β-methylene group of biotin is 2.14 ppm if the biotin is bound to the N-terminus. On the other hand if the biotin is binding to the Hmb group the signal is shifted to lower field (2.54 ppm) because of the generated ester. Furthermore in the NOESY spectra we observed a crosspeak between these protons and the proton in *meta*-position of the Hmb group.

To avoid this side reaction we used the Dcpm protection group, recently introduced by Carpino et al. [2]. The Dcpm group is applicable to Fmoc-chemistry and could easily be removed by TFA. With the Dcpm group we obtained the peptide Mon1_c in a good yield. Surprisingly the Dcpm group is removable even at very low concentrations of TFA e.g. 5% TFA in chloroform. The products of the deblocking of the Dcpm group were easy to remove.

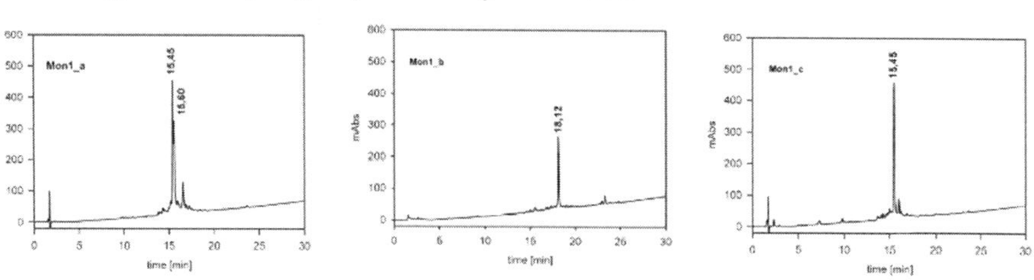

Fig. 1. HPLC chromatogram of the different Mon_1 peptides, a: without any protecting group at Gly[14], b: with the Hmb protecting group at Gly[14], c: with the Dcpm protecting group at Gly[14].

S. Del Valle et al. (eds.), *Peptides for Youth: The Proceedings of the 20th American Peptide Symposium*, 247
DOI: 10.1007/978-0-387-73657-0_113, © Springer Science+Business Media, LLC 2009

All peptides were synthesized on a 433A peptide synthesizer from Applied Biosystems®. The Hmb group was introduced via the commercially available Fmoc-Asp(tBu)-Gly(Hmb)-OH from Merck. The Dcpm group was introduced via Fmoc-Gly(Dcpm)-OH which was synthesized according to the protocol of Carpino et al. [3]. Coupling of the dipeptide and of the Dcpm protected glycine was carried out by manual synthesis according to the Fmoc strategy. i.e. 4 eq. of amino acid or dipeptide, 4 eq. of N-HATU and 8 eq. of DIEA, all relative to the substitution of the resin [4]. The crude peptides were analyzed by C_{18} RP-HPLC (Nucleosil, 125×4.6mm) with a gradient of 10-100% CH_3CN in aq. 0.1% TFA.

For further NMR studies we used Fmoc-Gly(Dcpm)-OH as a model for deprotection of the Dcpm group. Under standard conditions we established that four main by-products were derived from the Dcpm unit, either under normal (95%TFA) or mild conditions (5% TFA in DCM). Three of the four deblocking by-products were derived from opening of one of the cyclopropyl rings (Fig. 2). All were identified by NMR spectroscopy (COSY, TOCSY, HSQC, HMBC and NOESY).

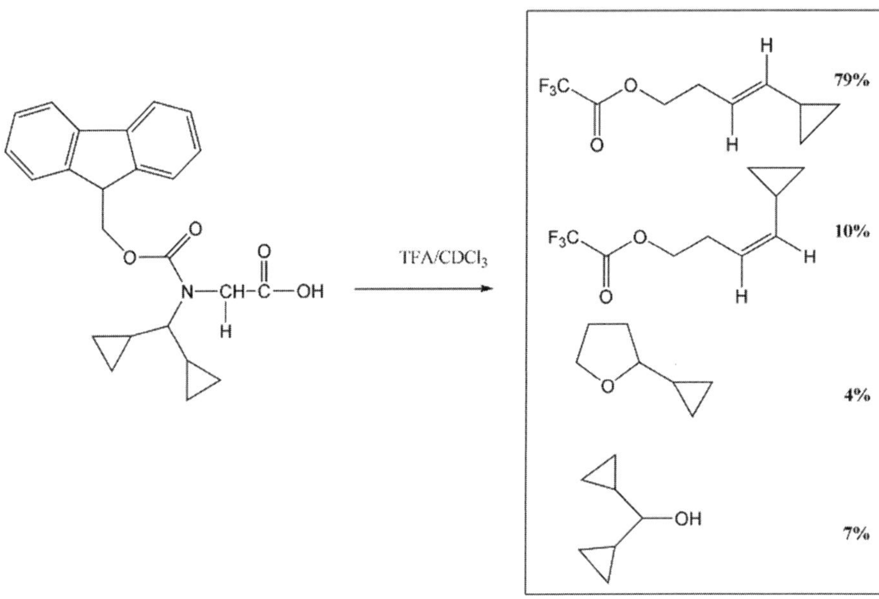

Fig. 2. Percentages of the deblocking by-products derived from the Dcpm group upon treatment with TFA as determined by ^{1}H-NMR analysis.

References

1. Johnson T., Quibell M., Sheppard R.C. *J. Pept. Sci.* **1**, 11-25 (1995).
2. Carpino L.A. et al. *J. Org. Chem.* **60**, 7718-7719 (1995).
3. Carpino L.A. et al. submitted
4. Fields C. G., Lloyd D.H., MacDonald R.L., Otteson K.M., Noble R.L. *Pept. Res.* **4**, 95-101 (1991).

Development of a RGDS-peptide modified polyurethane for tissue regeneration

A. Rechichi[1], S. Sartori[2], A. Caporale[3], G. Ciardelli[1], G. Vozzi[4], L. Mazzucco[5], E. Peggion[3] and P. Giusti [4]

[1]*Politecnico di Torino, Dep. of Mechanics Corso Duca degli Abruzzi, 24, Torino, Italy*
[2] *Univ. di Pisa, Dep. of Chemistry and Ind. Chemistry, via Risorgimento 56126 Pisa, Italy*
[3] *Università di Padova, Dep. of Chemical Sciences, Via Marzolo, 1 - 35131 Padova*
[4] *Università di Pisa, Dep. of Chemical Engineering via Diotisalvi 2 56126 Pisa, Italy*
[5] *Blood Transfusion Centre, Department of patologia clinica, immunoemat.logia e medicina trasfusionale, Ospedale Santi Antonio e Biagio, Alessandria.*

Introduction

The development of biomaterials for tissue engineering applications has focused on the design of biomaterials that can elicit specific cellular responses [1]. In multicellular organisms contacts of cells with neighboring cells and the surrounding Extra Cellular Matrix (ECM) are mediated by cell adhesion receptors (usually an integrin). A promising approach in the synthesis of biomaterials, is the modification of the polymer with peptides containing the adhesion domains of the ECM proteins. Since the identification of the RGD peptide as a minimal sequence required to mediate cell adhesion [2], researchers have been depositing RGD-containing peptides on biomaterials to promote cell attachment [3]. Polyurethanes represent a main class of synthetic elastomers used for long term medical implants [4]. They exhibit tunable chemical properties, excellent mechanical properties, good biocompatibility and can be designed to degrade in biological environment [5].

Functionalisation of biocompatible polyurethanes (PU) with cell adhesive peptides to promote cell-matrix specific interactions is still a strategy with limited diffusion. A few approaches are known [6] that can be applied for the introduction of peptide sequences into the chain back-bone, moreover these synthetic procedures reduce conformational mobility of the peptide making it less effective for the interaction with the cell receptors. In this work a strategy for functionalisation of PU was proposed by a novel approach based on the insertion of the RGDS cell binding peptide as a side chain. The presence of a spacer was provided to allow the peptide to stand out from the artificial surface and to reach the binding site of the cell receptors.

Results and Discussion

The structural unit containing the RGDS peptide (K-Amc-GRGDSG-OEt) was synthesized by SPPS (with Fmoc protocol) and used as chain extender in the polymerisation through the reactive amino groups of the Lys residue. The protecting groups were removed by acid treatment at the end of the polymerisation. UV analysis insured the completed cleavage of sidechains protective groups of peptide residue. Mechanical and physico-chemical characterisation of the functionalised polymer (PU-Amc-GRGDSG-OEt) did not highlight remarkable differences with respect to the original ones, indicating the manipulation of biological interactions without changing the bulk properties of PU materials. DSC (Differential Scanning Calorimetry) analysis evidenced an endotherm peak (T_m= 45°C) corresponding to

S. Del Valle et al. (eds.), *Peptides for Youth: The Proceedings of the 20th American Peptide Symposium*, 249
DOI: 10.1007/978-0-387-73657-0_114, © Springer Science+Business Media, LLC 2009

the melting of the PCL crystals in PU and functionalised PU curves. The related ΔH values were almost the same for both polymers, indicating that ,the introduction of the peptide had not remarkable effects on the crystallinity. Uniaxial stress/strain curves were acquired using an EnduraTEC tester. The linear-region moduli values before and after the functionalisation were similar, ranging from 50 MPa of the PU-RGDS samples to 60 MPa of the original PU.

Human Mesenchymal cells extracted by bone marrow and human fibroblasts derived by epidermis were seeded and cultured for 6 hours on PU -RGDS and PU films.

Before cell seeding the polymeric films were sterilised and incubated overnight with complete free media culture. Cell seeding density was 40,000 cells/ml. The analysis under optical microscope showed that cell adhesion is better on the RGS functionalised films.

This result was confirmed also by long time cell experiments performed on PU films not treated with serum. In fact, after 72h cells grown and proliferated on functionalised PU-RGDS, while on the PU film the cells activated apoptosis phenomena.

The original idea to use a suitable peptide as chain extender in the synthesis of polyurethanes was demostrated a viable strategy for the preparation of biomimetic materials for tissue engineering applications. Following the same strategy peptide sequences more selective than RGDS might be used to direct the growth of specific cell lines.

Acknowledgments

Co-financing by the Italian Ministry of research through the PRIN project 2005091572_002, year 2005 is acknowledged.

References

1. Temenoff J.S., Steinbis E.S., Mikos A.G., J Biomater Sci Polym Ed. 2003, 14(9), 989-1004.
2. Wermuth J., Goodman S. L., Jonczyk A., Kessler H., J. Am. Chem. Soc., 119(6), 1328-1335 (1997) .
3. Hersel U., Dahmen C., Kessler H., Biomaterials 24, 4385-4415 (2003).
4. Lambda N. M. K., Woodhouse K. A., Cooper S. L., Polyurethanes in Biomedical Applications. CRC Press: New York, 205-241 (1998).
5. Zdrahala R. J.,. Zdrahala I. J., J Biomater Appl 14: 67-90 (1999).
6. Jun H.W., West J., J Biomater Sci Polym Ed. 2004, 15(1), 73-94.

Modulating Collagen Triple-Helix Stability with 4-Chloro, 4-Fluoro, and 4-Methylprolines

Matthew D. Shoulders[1] and Ronald T. Raines[1,2]

[1]*Department of Chemistry;* [2]*Department of Biochemistry; University of Wisconsin-Madison, Madison, WI 53706, USA*

Introduction

Collagen is a fibrous protein comprising a right-handed, triple-helical bundle of three parallel, left-handed polyproline II-type helices. Each strand consists of approximately 300 repeats of the trimer (Xaa–Yaa–Gly), where Xaa is often (2*S*)-proline (Pro) and Yaa is often (2*S*,4*R*)-4-hydroxyproline (Hyp) [1]. The most abundant protein in vertebrates, collagen is of fundamental importance to the three-dimensional architecture of such animals. Understanding the chemical determinants of the structure and stability of collagen is essential for both curing collagen-related diseases and creating collagen-based biomaterials.

We have previously demonstrated that pyrrolidine ring pucker and thus triple-helical stability can be manipulated by functionalizing the γ-position of the proline ring with moieties that control the ring pucker via either steric or stereoelectronic effects (Figure 1). The presence of Hyp in the Yaa position of natural collagen greatly increases the stability of the triple helix [2]. This enhanced stability is presumably due to a stereoelectronic effect that defines the pyrrolidine ring pucker of Hyp residues as the C^{γ}-exo pucker. The C^{γ}-exo ring pucker fixes the ϕ and ψ angles of Hyp residues to those required for triple-helical assembly [3]. Substitution of non-natural proline derivatives that prefer the C^{γ}-exo ring pucker, such as (2*S*,4*R*)-4-fluoroproline (Flp) [4] or (2*S*,4*S*)-4-methylproline (Mep) [5], for proline in the Yaa position also stabilizes triple helices. In contrast, the C^{γ}-endo ring pucker is favorable for triple-helix stability in the Xaa position. Consequently, substitution of (2*S*,4*S*)-4-fluoroproline (flp) and (2*S*,4*R*)-4-methylproline (mep) for proline in the Xaa position stabilizes triple helices, because flp and mep prefer the C^{γ}-endo ring pucker [5,6].

Here, we introduce additional proline derivatives that can be utilized to endow conformational stability and unique properties on collagen triple helices, namely the 4-chloroprolines. Additionally, we summarize the ability to modulate the thermal stability of collagen over a wide range of temperatures by modifying the γ-carbon of proline residues.

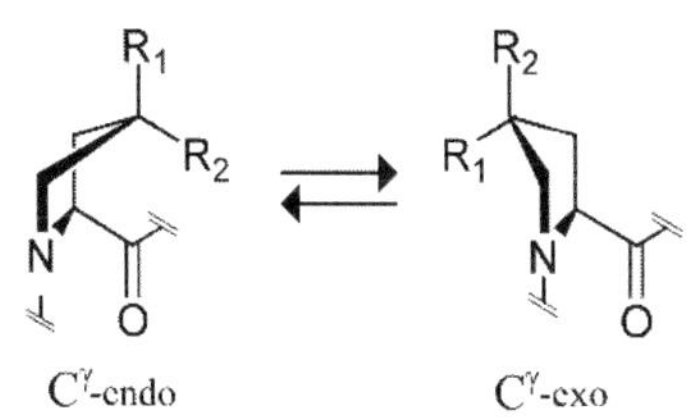

Figure 1. Ring conformations of 4-substituted prolines. The C^{γ}-endo conformation is favored strongly by stereoelectronic effects when R_1 = H, R_2 = F (flp) or Cl (clp) and by steric effects when R_1 = Me (mep), R_2 = H. The C^{γ}-exo conformation is favored strongly by stereoelectronic effects when R_1 = OH (Hyp), F (Flp) or Cl (Clp), R_2 = H and by steric effects when R_1 = H, R_2 = Me (Mep).

Results and Discussion

We hypothesized that (2*S*,4*R*)-4-chloroproline (Clp) would prefer the C^{γ}-exo pyrrolidine ring pucker and that (2*S*,4*S*)-4-chloroproline (clp) would prefer the C^{γ}-endo ring pucker (Figure 1). Thus, Clp should

S. Del Valle et al. (eds.), *Peptides for Youth: The Proceedings of the 20th American Peptide Symposium*, 251
DOI: 10.1007/978-0-387-73657-0_115, © Springer Science+Business Media, LLC 2009

Table 1. Effect of clp and Clp on collagen triple-helix stability.

Peptide	T_m
(Pro–Clp–Gly)$_{10}$	52
(Pro–Pro–Gly)$_{10}$	41
(clp–Pro–Gly)$_{10}$	33
(Pro–Clp–Gly)$_7$	23
(clp–Pro–Gly)$_7$	<10
(Pro–Pro–Gly)$_7$	<10
(clp–Clp–Gly)$_{10}$	no helix
(Clp–Pro–Gly)$_{10}$	no helix

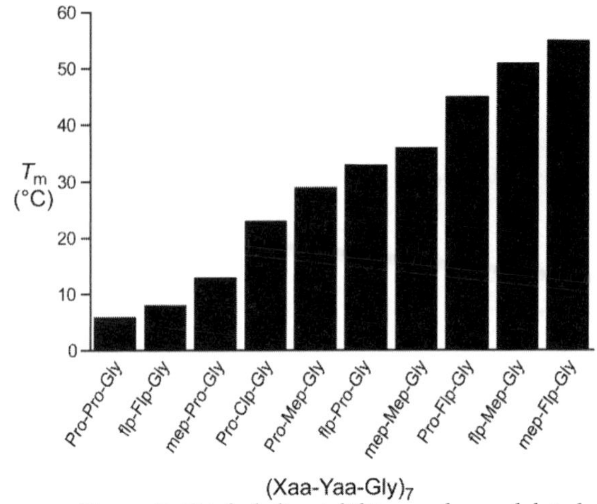

Figure 2. Triple-helix stability can be modulated over a wide range of temperatures by controlling the identity and configuration of functional groups at the γ-position of proline residues [3,5,6,8].

stabilize triple helices in the Yaa position, while clp should stabilize triple helices in the Xaa position. To confirm this hypothesis, collagenous peptides containing clp and Clp synthesized by a route similar to that of Berger and co-workers [7] were prepared by segment condensation on the solid phase. We tested the triple-helical propensity of clp and Clp and found that they enable triple helix formation when placed in the Xaa or Yaa position, respectively (Table 1), confirming our hypothesis. (Clp–Pro–Gly)$_{10}$ and (clp–Clp–Gly)$_{10}$ do not form triple helices, as expected from prior results for flp and Flp in collagen mimics [6,8].

Figure 2 illustrates our ability to fine-tune the thermal stability of triple helices of the same length using the results detailed herein and in previous results from our laboratory [3,5,6,8]. Utilizing 4-chloroprolines, 4-methylprolines, 4-fluoroprolines, and combinations of these residues, we have synthesized 21-residue, blunt-ended collagen triple helices with T_m values ranging from 0–53°C. This ability to modulate triple-helical stability by simply modifying the γ-position of proline rings portends the development of synthetic collagen mimetics for applications requiring triple-helical stability over a wide range of temperatures.

Acknowledgments

This work was supported by grant AR44276 (NIH). M.D.S. was supported by a graduate fellowship from the Department of Homeland Security.

References

1. Jenkins, C. L., Raines, R. T. *Nat. Prod. Rep.* **19**, 49-59 (2002).
2. Sakakibara, S., *et al. Biochim. et Biophys. Acta* **303**, 198-202 (1973).
3. DeRider, M. L., *et al. J. Am. Chem. Soc.* **123**, 777-778 (2001).
4. Holmgren, S. K., *et al. Nature* **392**, 666-667 (1998).
5. Shoulders, M. D., Hodges, J. A., Raines, R. T. *J. Am. Chem. Soc.* **128**, 8112-8113 (2006).
6. Hodges, J. A., Raines, R. T. *J. Am. Chem. Soc.* **125**, 9262-9263 (2003).
7. Berger, Y., *et al. J. Med. Chem.* **48**, 483-498 (2005).
8. Hodges, J. A., Raines, R. T. *J. Am. Chem. Soc.* **127,** 15923-15932 (2005).

Peptide-Chitosan Matrix: a New Multifunctional Biomaterial

Motoyoshi Nomizu[1], Natsumi Yamagata[1], Mayumi Mochizuki[1], Yamato Kikkawa[1] and Yuichi Kadoya[2]

[1] *Toyko University of Pharmacy and Life science, Hachioji, Tokyo 192-0392, Japan;* [2] *Kitasato University School of Medicine, Sagamihara 228-8555, Japan*

Introduction

Tissue engineering requires biomaterials for delivering transplanted cells to organ sites needing repair/regeneration. Extracellular matrix (ECM) plays a critical role in tissue regeneration by promoting cell adhesion, migration, proliferation, and differentiation. ECM mimetics are of importance for tissue engineering because of their functions as scaffolds for cells. Previously, we developed bioactive laminin-derived peptide-conjugated chitosan membranes and demonstrated their cell- and peptide-type specific functions [1]. We also demonstrated that a most active laminin peptide (AG73: RKRLQVQLSIRT)-conjugated chitosan membrane can deliver cells and is applicable for keratinocyte transferring to wound bed [2]. Here, we conjugated four integrin-binding peptides derived from ECM proteins onto chitosan membranes and examined biological activity. FIB1 (YAVTGRGDSPAS; from fibronectin), EF1zz (ATLQLQEGRLHFXFDLGKGR; from laminin $\alpha1$ chain), A99 (AGTFALRGDNPQG; from laminin $\alpha1$ chain) [3], and 531 (GEFYFDLRLKGDKY; from collagen $\alpha1$ (IV) chain) [4] conjugated chitosan membranes promoted integrin-dependent cell adhesion. Various integrins, including αv, $\beta1$, and $\beta3$, were involved in the cell adhesion to the peptide-chitosan membranes. Further, only the FIB1- and A99-chitosan membranes promoted neurite outgrowth with PC12 rat pheochromocytoma cells. These data demonstrate that integrin binding peptide-chitosan membranes can regulate specific integrin-mediated cell responses.

Results and Discussion

Four biologically active peptides derived from laminin, fibronectin, and type IV collagen previously shown to interact with integrins were synthesized with a Cys-Gly-Gly (CGG) sequence on the N-termini (Table. 1) [1]. As a control, AG73, which is derived from laminin and interacts with syndecans, was also prepared similarly [1]. Chitosan was reacted with N-(maleimidobenzoyloxy)-succimide (MBS). The obtained MB-chitosan was coated on tissue culture plates and prepared MB-chitosan membrane. The Cys residue was utilized for conjugation of the MB-chitosan membrane and two glycine residues were used as a spacer between the active

Table 1. Synthetic peptides derived from ECM components and biological activities of peptide-chitosan membranes

Peptide	Sequence	Protein (Residues)	Fibroblast Attachment	Neurite Outgrowth
A99	AGTFALRGDNPQG	Murine laminin a1 chain (1141-1153)	++S	+
EF1zz	ATLQLQEGRLHFXFDLGKGR (X=Nle)	Murine laminin a1 chain (2749-2768)	++S	-
531	GEFYFDLRLKGDKY	Human collagen a1, type IV (531-543)	++S	-
FIB1	YAVTGRGDSPAS	Human fibronectin (1519-1530)	++S	+
AG73	RKRLQVQLSIRT	Murine laminin a1 chain (2719-2730)	++	+

S. Del Valle et al. (eds.), *Peptides for Youth: The Proceedings of the 20th American Peptide Symposium*, 253
DOI: 10.1007/978-0-387-73657-0_116, © Springer Science+Business Media, LLC 2009

peptides and chitosan. The CGG-peptides were added into the plates and covalently coupled to the MB-chitosan membrane [1]. Human foreskin fibroblasts (HFFs), which bind to ECM via integrins, were used to evaluate the cell attachment activity on the peptide-chitosan membranes. The AG73-chitosan membrane promoted strong cell attachment with filopodia formation. When FIB1-, A99-, EF1zz-, and 531 peptides were conjugated to chitosan, the peptide-chitosan membranes promoted HFF cell attachment and spreading (Fig. 1) [5]. The morphological appearance of HFF differed on each peptide-conjugated chitosan membrane [5]. Cells strongly attached and spread on the EF1zz- and 531-chitosan membranes. FIB1- and A99-chitosan membranes showed cell spreading, where the cells migrated toward each other and formed small colonies with cell-cell interactions. Previously, we demonstrate that integrins enhanced the formation of HFF actin stress fibers on peptide-chitosan membranes.

Next, we evaluated the effect of integrin-binding peptides on rat PC12 pheochromocytoma cell differentiation (Fig. 2). PC12 cells were used to analyze cell differentiation because they have been previously shown to exhibit neurite outgrowth when attached to certain synthetic peptides [6,7]. PC12 cells were cultured on the peptide-chitosan membranes and laminin-1-coated plate as a control. After 24 h, cell adhesion and neurite extension were examined. Previously, we showed that the A99-chitosan membrane promoted neurite outgrowth with PC12 cells, and that the EF1zz peptide coated onto plastic did not promote neurite outgrowth [1, 3]. As expected, A99-chitosan membranes promoted PC12 cell neurite outgrowth [5]. Additionally, the FIB1-chitosan membrane was active for neurite outgrowth, while the EF1zz- and 531-chitosan membranes were inactive [5]. These data demonstrate that the PC12 neurite outgrowth activity is integrin type specific.

We also conjugated syndecan and integrin binding peptides (AG73 and A99) with a various ratios on a chitosan membrane. The mixed-peptide-chitosan membranes strongly promoted cell attachment and neurite outgrowth. The cell attachment activity and cell morphology were dependent on the ratios of the syndecan- and integrin-binding peptides on a chitosan membrane. The mixed-peptide-chitosan membranes are multifunctional and can control several cell functions. The multifunctional mixed-peptide-chitosan membranes have a potential to be used as a tailor-made and tissue-specific biomedical material.

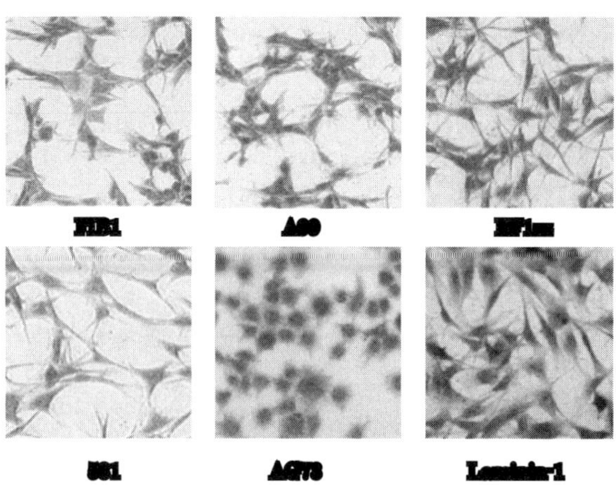

Fig. 1. Morphology of human foreskin fibroblasts adhered to peptide-chitosan membranes.

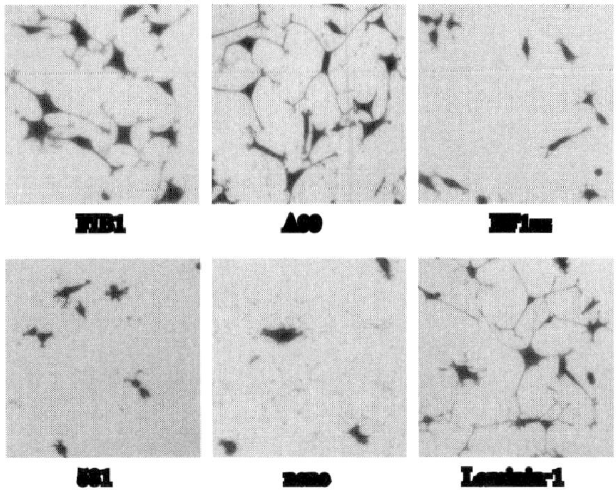

Fig. 2. Neurite outgrowth on the peptide-chitosan membranes.

Recently, we developed easy-handling peptide-chitosan membranes and demonstrated the potent for clinical applications using in vivo assays. We proposed the AG73-chitosan membrane for keratinocyte delivery [2]. The keratinocyte were incubated with the peptide-chitosan membrane for 2 h and the obtained cell-attached peptide-chitosan membrane was applied to the wounds. We found that the transplant cells grow and establish a stratified epithelium even from single cell, and concluded that the peptide-chitosan membrane system is a powerful tool for cell transplantation.

Taken together, we demonstrated that peptide-chitosan membranes can promote biological activities via various integrin interactions. The peptide-chitosan membrane approach has various advantages and has a potential to use as a biological material for cell engineering in vitro and in vivo. Our studies are a first step in developing new biomaterials that mimic in vivo ECM-integrin interactions.

References

1. Mochizuki, M., Kadoya, Y., Wakabayashi, Y., Kato, K., Okazaki, I., Yamada, M., Sato, T., Sakairi, N., Nishi, N. and Nomizu, M. *FASEB J.* **17,** 876-877 (2003)
2. Ikemoto, S., Mochizuki, M., Yamada, M., Takeda, A., Uchinuma, E., Yamashina, S., Nomizu, M. and Kadoya, Y. *J. Biomed. Mater. Res. A* **79,** 716-722 (2006)
3. Suzuki, N., Nakahara, H., Mochizuki, M., Nishi, N., Kadoya, Y., Utani, A., Oishi, S., Fujii, N., Kleinman, H. K. and Nomizu, M. *J. Biol. Chem.* **278,** 45697-45705 (2003)
4. Miles, A. J., Knutson, J. R., Skubitz, A. P. N., Furcht, L. T., McCarthy, J. B. and Fields, G.B. *J. Biol. Chem.* **270,** 29047-29052 (1995)
5. Mochizuki, M., Yamagata, N., Philp, D., Hozumi, K., Watanabe, T., Kikkawa, Y., Kadoya, Y., Kleinman, H. K. and Nomizu M. *biopolymers* **88,** 122-130 (2007)
6. Weeks, B. S., Nomizu, M., Ramchandran, R. S., Yamada, Y. and Kleinman, H. K. *Exp. Cell. Res.* **243,** 375-382 (1998)
7. Richard, B. L., Nomizu, M., Yamada, Y. and Kleinman, H. K. *Exp. Cell. Res.* **228,** 98-105 (1996)

PEPTIDES FOR YOUTH, UNMET MEDICAL NEEDS OF TOMORROW

Effect of phosphorylation on tau aggregation using model peptides and Circular Dichroism studies

Heather Palmer and Ajoy Basak

Hormones Growth and Development Program, Regional Protein Chemistry Center, Ottawa Health Research Institute, U Ottawa, Ottawa, ON K1Y 4E9, CANADA

Introduction

Human (h) tau is a microtubule (MT) associated protein that has been implicated in Alzheimer's Dementia (AD). AD leads to progressive and irreversible loss of brain activities such as memory, learning ability and cognitive functions. MTs help maintain cell shape, division, axonal transport, secretion and receptor activity. Tau through its binding with MT maintains those functions [1]. It includes 6 isoforms with 352 to 441 amino acids (aa) of which the 441aa form is the most abundant [2]. AD is characterized by the formation of (i) neurofibrillary tangles containing paired helical filaments (PHFs) composed of tau aggregates, and (ii) insoluble plaque containing aggregated amyloid beta (Aβ) peptide. Like Aβ, tau-aggregation has also become a major focus of AD research. Although the mechanism of tau aggregation is not well understood, it is linked to hyper and/or abnormal phosphorylation of Ser and/or Thr residues located at the N-terminal region. This aggregation possibly occurs via one or more of its four MT-binding domains (MTB-1, 2, 3, and 4). In tau there are at least 25 Ser/Thr phosphorylation sites some of which but not all are implicated in tau aggregation [3]. Most of the Ser/Thr sites found *in vivo* are phosphorylated by proline dependent kinases (PDKs) [4]. Upon tau-phosphorylation, the MTs dissociate leading to tau aggregation and formation of PHFs, which are highly toxic to neurons. Most likely enhanced and/or selected phosphorylation of tau may lead to increased fibril formation and aggregation resulting rapid loss of brain activity. Studies revealed that secondary structure plays an important role in this event [1, 2]. Herein our object is to study *in vitro* the effect of phosphorylation in tau aggregation using synthetic peptides from specific Tau-domains.

Results and Discussion

Studies revealed that hTau contains 4 microtubule binding (MTB) domains near the C-terminal and 3 phospo-Ser/Thr rich domains (PRD) with the middle-domain (201-240) having the most effect on Tau aggregation. In this domain abnormal and/or hyperphosphorylation leads to Tau-aggregation that likely mediates via MTB-2 or 3 domains. There are 10 potential Ser/Thr phosphorylation sites within this domain. Among these, Thr^{231} and Ser^{235} are most likely to initiate the aggregation although other sites may be possible. To study this notion *in vitro* we synthesized $Tau^{201\text{-}240}$ peptide, a fully phosphorylated and partly (T^{231} and S^{235}) phosphorylated forms as well as the MTB-2 peptide (**Table 1**). Analyses of 3D model structures suggested that upon phosphorylation at 10 possible sites ($ppTau^{201\text{-}240}$) there is a significant increase in H-bonds involving Arg, Pro and Gly residues leading to conformational change compared to $hpTau^{201\text{-}240}$ and $Tau^{201\text{-}240}$. Circular Dichroism (CD) study of these peptides in water, TFE:water (1:1) and TFE revealed differences in their

S. Del Valle et al. (eds.), *Peptides for Youth: The Proceedings of the 20th American Peptide Symposium*, 259
DOI: 10.1007/978-0-387-73657-0_117, © Springer Science+Business Media, LLC 2009

Table 1. List of various h-Tau peptides used in the present study.

Peptide	Amino acid sequence	MW
$Tau^{201-240}$	GSPGTPGSRS RTPSLPTPPT REPKKVAVVR TPPKSPSSAK	4095
$pTau^{201-240}$	GSPGTPGSRS RTPSLPTPPT REPKKVAVVR **Tp**PPK**Sp**PSSAK	4257
$ppTau^{201-240}$	G**Sp**PGTPG**Sp**R**Sp** R**Tp**P**Sp**LP**Tp**PPT REPKKVAVVR **TpPPKSpPSpSpAK**	4898
$MTB2^{275-305}$	VQIINKKLDL S-NVQSKCGS KDNIKHVPGG GS	3264
$Fl\text{-}MTB2^{275-305}$	VQIINKKLDL S-NVQSK**C(Fl)**GS KDNIKHVPGG GS	3992

MW = molecular weight, MTB = micro tubule binding, Sp = Phospho Ser, Tp = Phospho Thr, ***C(Fl)*** *= Fluorescent labeled Cysteine residue via its side chain SH group. Fl= Texas-red C_2 maleimide*

Fl =

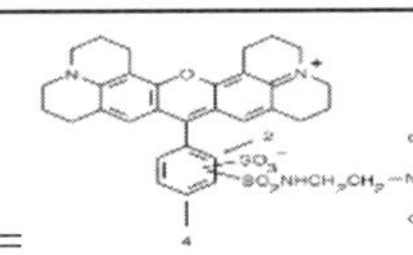

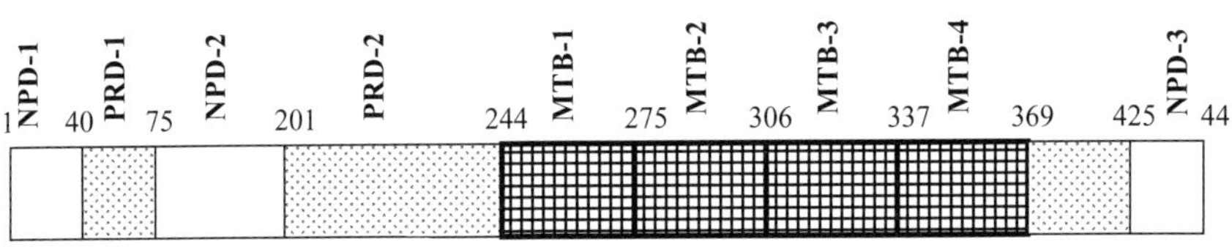

Fig. 1. Schematic presentation of hTau protein showing 3 phospho rich domains (PRD), 4 micro tubule binding domains (MTB) and 3 non-phosphorylated domains.

secondary structure profiles. Thus while presence of TFE led to more beta-sheet structures in $ppTau^{201-240}$ and $Tau^{201-240}$, the secondary structure of $pTau^{201-240}$ remains largely unaffected. In order to examine *in vitro* the effects of various $Tau^{201-240}$ forms on MTB2 peptide secondary structure, CD, mass spectroscopy and gel-electrophoresis were used. CD revealed that incubation of p and pp-$Tau^{201-240}$ increases the beta-sheet structure of MTB-2 peptide compared to the nonphosphorylated $Tau^{201-240}$ peptide. This may suggest more aggregation for MTB-2 peptide in presence of either phosphorylated $Tau^{201-240}$ peptides compared to the non-phosphorylated one.

Above finding was supported by non-denatured tricine-glycine gel electrophoresis (**Fig 2**) which showed the formation of higher oligomers (upto 22kDa heptamer) for MTB2 peptide when incubated with $ppTau^{201-240}$ suggesting that hyper-phosphorylation of $Tau^{201-240}$ promotes aggregation. While some low oligomeric forms (trimer) have been detected with $pTau^{201-240}$, there was little change in oligomeric forms of MTB-2 upon incubation with $Tau^{201-240}$. We also noticed self association of both p and $ppTau^{201-240}$ peptides as well. To provide additional evidence for above observation, we synthesized an N-terminal Cys-linked fluorescent MTB-2 peptide with Texas red C_2 maleimide and studied its fluorescence properties in absence and presence of various $Tau^{201-240}$ forms. There was a significant increase in fluorescence intensity upon 4h incubation with ppTau peptide (**Fig 3**). The result was also supported by light microscopy study.

The finding was further confirmed by SELDI-tof (Surface enhanced laser desorption time of flight) mass spectrometry. As shown in **Fig 4**, higher oligomeric MTB-2 forms (upto tetramer) have been detected only when the latter was incubated for 7 days with ppTau$^{201\text{-}240}$ or pTau$^{201\text{-}240}$ while with non-phosphorylated Tau$^{201\text{-}240}$, no forms higher than dimeric could be noticeable.

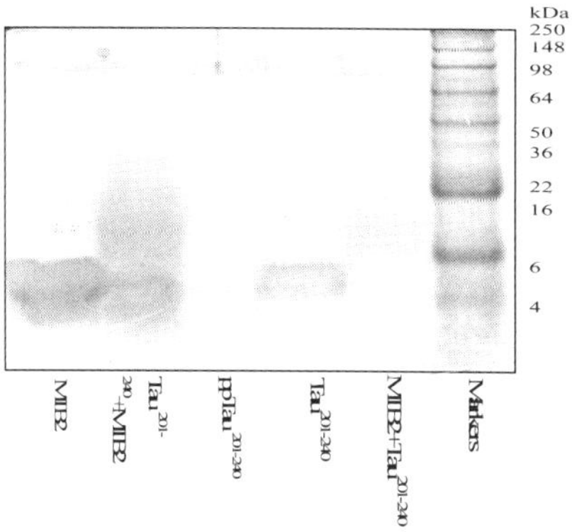

Fig. 2. SELDI tof MS of MTB2 incubation with non and phospho Tau $^{201\text{-}240}$.

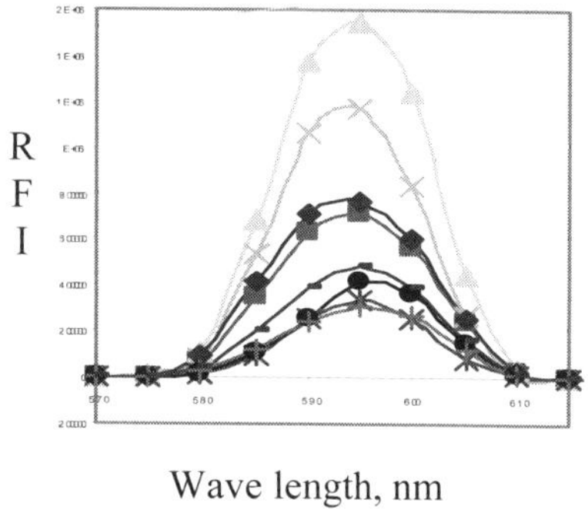

Fig. 3. Overlay of emission fluorescence spectroscopy of Fl-MTB2, from top: +pTau-0day, +ppTau-0day, alone-0day, +tau-0day, +Tau-7day, +pTau-7day, +ppTau-7day. RFI = raw fluorescence intensity.

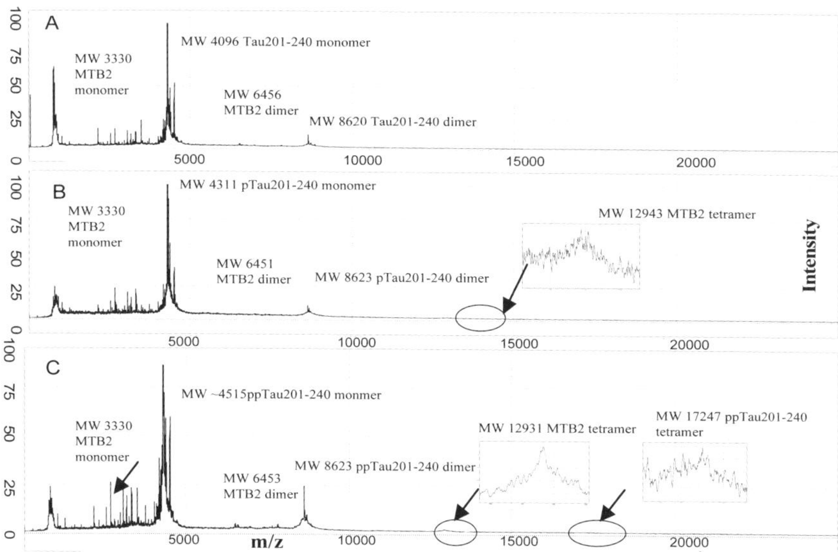

Fig. 4. SELDI MS of MTB2 after incubation with A) Tau$^{201\text{-}240}$, B) pTau$^{201\text{-}240}$ and C) ppTau$^{201\text{-}240}$

We conclude that in presence of both pTau $^{201\text{-}240}$ and ppTau$^{201\text{-}240}$, MTB2 peptide exhibited an increase in beta sheet structure and higher aggregate forms. Funds: CCRI-U Ottawa.

References

1 Goedert, M, In Microtubules (Hyams & Lloyd, eds) 183-200, Wiley-Liss, NY, USA (1994).
2. Tomoo, K, et. al. *J. Biochem. Biol. Chem.* **138**, 413-423 (2005).
3. Ko, L, et. al. *Biochim. Biophys. Acta* **1739**, 125-139 (2005).
4. Yao, T.M. et. al. *J. Biochem,* **134**, 91-99 (2003).

Peptides for prolonging youth

Marc Heidl

Pentapharm Ltd., Engelgasse 109, 4002 Basel

Introduction

Proteins and peptides play an important role in the skin health as cutaneous structural components, inhibitors, stimulating agents or regulators of biochemical processes in the skin. Small peptides are often responsible for a biochemical effect: they are either released from larger units by proteolysis or exposed as a partial sequence due to a conformational change.

Thrombospondin-1 (TSP), a high molecular weight glycoprotein, contains a tripeptide sequence -Arg-Phe-Lys- (RFK) which activates the transforming growth factor beta (TGF-beta),[1,2] inducing the synthesis of extracellular matrix proteins.[3]

Results and Discussion Palm-Lys-Val-Lys-OH

Only a few experiments were necessary to show that the original sequence pattern basic, hydrophobic and basic amino acid is required for the efficacy of such substances. By maintaining this pattern, the most differing variations have been tried on the three individual amino acids as well as on the *N*- and *C*-terminus. Part of the results obtained from the different test series has been presented at the 23rd IFSCC in Orlando.[4] The best compound regarding efficacy, safety and preparation costs, Palm-Lys-Val-Lys-OH, has been developed as a cosmetic anti-aging ingredient.

The Fig.2 illustrates in a simplified scheme how TSP binds to the LAP protein of the TGF-beta LAP complex, leading to a conformational change: the LAP-TSP complex cleaves off and TGF-beta is released in the mature form.[2] Further studies have shown that Palm-Lys-Val-Lys-OH has the ability to bind in the same way.[5]

The compound shows a dose-dependent effect of the peptide of type I collagen of *in vitro* cultured skin fibroblasts detected with an ELISA. In view of the excellent *in vitro* results, this sequence was developed as a cosmetic ingredient.

Four groups of 15 volunteers each applied a cream - on one half of the face twice a day during 84 days - containing 10 ppm of Palmitoyl Tripeptide-5, 25 ppm of Palmitoyl Tripeptide-5, 10 ppm competitor's peptide or no active ingredient (placebo).

It could be demonstrated that deep wrinkles are significantly reduced within 3 months and the overall skin appearance is improved. The colour-coded 3D-profile after the treatment shows no more area coded in black (corresponds to wrinkles deeper than 220 micrometers). The areas coded in yellow (corresponds to hills in the profile) are clearly reduced, confirming the observation that the overall profile height is reduced after treatment with the peptide.

H-beta-Ala-Pro-Dab-NH-benzyl diacetate

Mimic wrinkles (wrinkles on the forehead or around the mouth, crow's feet) could up to now only be reduced from a medical point of view by the subcutaneous injection of Botox®, a highly efficient neurotoxin and metabolite of *Clostridium botulinum*. Its effect is based on the relaxing of wrinkle producing muscles by inhibiting the acetylcholine release.

Some snake venoms cause muscle relaxation as well. Waglerin-1 is known to selectively block the epsilon-subunit of the muscular nicotinic acetylcholine receptor.[6] It is a 22-mer peptide of the *Tropidolaemus wagleri* venom with a frequent

S. Del Valle et al. (eds.), *Peptides for Youth: The Proceedings of the 20th American Peptide Symposium*, 263
DOI: 10.1007/978-0-387-73657-0_118, © Springer Science+Business Media, LLC 2009

apparition of a proline next to a basic amino acid in the sequence. The synthesized dipeptide derivatives did not show any muscle-relaxing property, but the effect was found with the tripeptide sequence of the type “unblocked amino acid-proline-basic amino acid”. The *in vitro* and *in vivo* effects are illustrated by using H-β-Ala-Pro-Dab-NH-benzyl diacetate which imitates the typical structural attribute of the Waglerin-1 peptide maintaining excellent muscle-relaxing efficacy and which was finally developed to an anti-wrinkle product.

The test model was established as a coculture of human muscle cells and neurons derived from the spinal cord of rat embryos. Active compounds should block the neuromuscular contraction signal.

The contraction frequency was measured 1 min, 2 hours, 2 days and 4 days after addition of the peptides. H-beta-Ala-Pro-Dab-NH-benzyl diacetate reduces muscle cell contraction just after 1 min by 36%; the maximal contraction decrease (82%) was measured after 2 hours which means that the compound is fast-acting. With a contraction decrease of 67% after two days it could be demonstrated that the effect is long-lasting. After 4 days total reversibility was determined.

The anti-wrinkle efficacy was evaluated by a double blind and intra-individual study on 45 volunteers using the Primos 3D® technique. Three groups of 15 volunteers each applied a cream with 100 ppm (~0.2 mMol) H-β-Ala-Pro-Dab-NH-benzyl diacetate, a cream with 100ppm of a competitor’s product or a placebo on the face (forehead and crow’s feet) twice a day during 28 days. The following cutaneous parameters have been analyzed: Ra: average roughness: A decrease of the Ra parameter expresses a smoothing effect. Rt: maximum relief amplitude and Rz: average relief on five regions of the profile. An anti-wrinkle effect is observed by a decrease of the Rt and Rz parameter.

H-beta-Ala-Pro-Dab-NH-benzyl diacetate exhibits a significant anti-wrinkle effect measured on the forehead (decrease of the Ra, Rt and Rz parameters: -21%, -20% and -15%), and an improvement of the cutaneous relief of the crow’s feet was observed for a majority of volunteers (68% for Ra, 77% for Rz and 69% for Rt).

In view of its originality and high degree of innovation, the product has been rewarded by the Swiss Technology Award 2006.

Conclusions

Knowing the influence of biochemical processes using natural active ingredients led to the development of two biomimetics that can be used to prevent wrinkle formation and reduce wrinkles. Regarding the cosmetic applicability of these natural actives they have been synthetically optimized by simplifying the structural complexity and introducing substituents to increase the stability and improve the dermal and cellular transport. Combining Palm-Lys-Val-Lys-OH, that stimulates the collagen synthesis, with H-β-Ala-Pro-Dab-NH-benzyl, that reduces mimic wrinkles by muscle relaxing, will allow to completely “rejuvenate” the skin image.

References

1. J Lawler, *J Cell Mol Med* 6 1-12 (2002)
2. JE Murphy-Ullrich and M Poczatek, *Cytokine Growth Factor Rev* **11** 59-69 (2000)
3. L Yin, A Morita and T Tsuji, *J Invest Dermatol* **120** 703-705 (2003)
4. M Heidl, D Imfeld, M Stöckli, H Ziegler, *Poster at 23rd IFSCC*, Oct 24-28, 2004
5. D Imfeld, M Stöckli, M Heidl, H Ziegler, T Schreier, *Poster at 24rd IFSCC*, October 16-19, 2006.
6. JJ McArdle et al., *J Pharmacol Exp Ther* 289, 543-550 (1999)

The Synthesis of DAMGO-Based Potential Affinity Labels with High Mu Opioid Receptor Affinity and the Formation of Cyclic O-Alkyl Thiocarbamates

Bhaswati Dattachowdhury[1], Thomas F. Murray[2] and Jane V. Aldrich[1]

[1]*Department of Medicinal Chemistry, University of Kansas, Lawrence, KS 66045, USA;* [2]*Department of Pharmacology, Creighton University School of Medicine, Omaha, NE 68178, USA*

Introduction

The objective of the present research was to synthesize peptide-based electrophilic affinity labels for μ opioid receptors (MOP). Narcotic analgesics produce pain relief through activation of the μ opioid receptors.[1] Electrophilic affinity labels, which interact with the receptors in a non-equilibrium manner, can provide detailed information on receptor-ligand interactions.[2] This information can then assist in the design of new opioid drugs, potentially with fewer side effects. In the present project DAMGO ([D-Ala^2,N-MePhe^2,glyol]enkephalin), a synthetic analog of enkephalin with high MOP affinity and selectivity,[3] was chosen as the parent ligand for further modification to incorporate electrophilic affinity labels. The synthesis and pharmacological results for a series of DAMGO derivatives containing an affinity label in position 2 are described. In the course of the synthesis and purification of DAMGO derivatives containing an isothiocyanate as the affinity label, the formation of a cyclic O-alkyl thiocarbamate side product was observed and characterized by various analytical methods. This side reaction was then successfully avoided by replacing the C-terminal glyol with glycylamide functionality. All of the affinity labels show subnanomolar affinity (IC_{50}).

Results and Discussion

Peptides were synthesized using an Fmoc solid phase peptide synthetic strategy developed in our laboratory.[4] DAMGO derivatives were synthesized on a DHP-HM (3,4-dihydro-2-H-pyran-2-yl methoxymethyl polystyrene) resin. Fmoc-glyol was loaded onto the resin in the presence of pyridinium *p*-toluene sulfonate,[5] followed by assembly of the protected peptide containing D-AA(Aloc) (D-AA = D-Orn or D-Lys) in position 2 (Figure 1). After selective Aloc removal an electrophilic affinity label (either bromoacetamide or isothiocyanate) was incorporated into the peptides, with subsequent cleavage from the resin using 95% TFA and 5% H_2O as the carbonium ion scavenger.

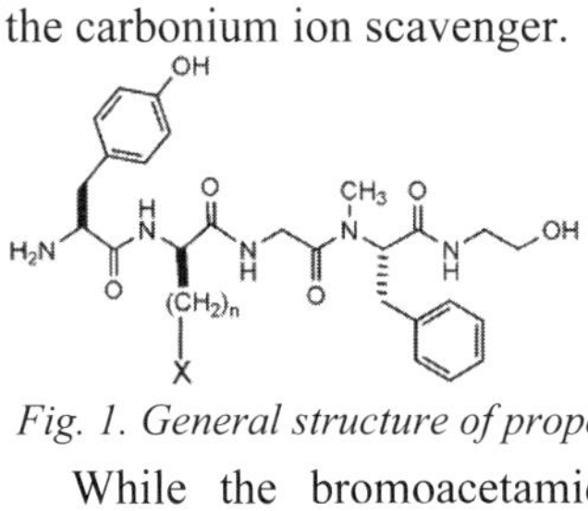

n = 3 (D-Orn) or 4 (D-Lys)
X = $NHCOCH_2Br$, -N=C=S, $NHCOCH_3$ (control)

Fig. 1. General structure of proposed DAMGO analogs.

While the bromoacetamide derivatives were successfully synthesized, two products (fractions A and B) were obtained during the synthesis of the isothiocyanate derivatives, both with the same molecular weight as the desired linear peptide (MW: 613.2) but with a difference in retention time of 7 min by HPLC (Table 1). Examination of the second product by RP-HPLC, IR, and NMR suggested

DOI: 10.1007/978-0-387-73657-0_119, © Springer Science+Business Media, LLC 2009

that it was a cyclic O-alkyl thiocarbamate derivative resulting from intramolecular attack of the C-terminal alcohol on the isothiocyanate (Scheme 1). The identity of the side product was based on the disappearance of the isothiocyanate stretch in the IR of fraction A (Table 1) and the disappearance of the broad peak corresponding to C-terminal glyol OH in the proton NMR.

To avoid this side reaction, a new series of analogs were synthesized in which the glyol functionality was replaced by a glycylamide; the affinity label group was attached to the side chain amine of D-Orn or D-Lys as before. As expected, no side reaction was observed for the isothiocyanate analogs, and the desired derivatives in the glycylamide series were successfully obtained.

Table 1: Analytical results for fractions A and B for [D-Orn(=C=S)^2]DAMGO.

Fractions	HPLC t_R (min)	MS $[M+H]^+$	IR -N=C=S stretching frequency (2117-2189 cm^{-1})
A	8.7	613.2	Absent
B	15.3	613.2	Present

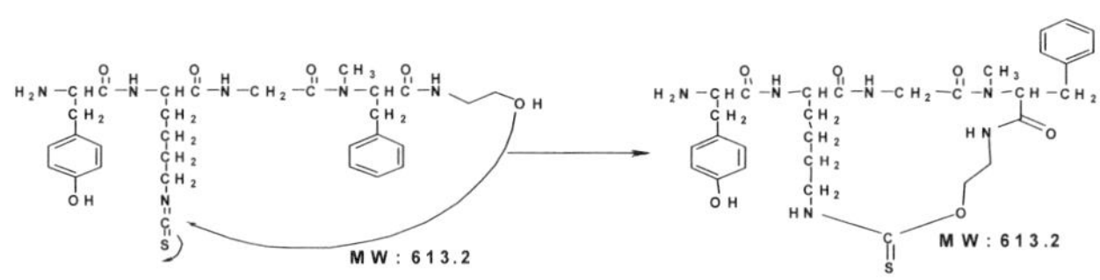

Scheme 1. Proposed reaction for the formation of the cyclic O-alkyl thiocarbamate.

All of the affinity label derivatives of DAMGO and DAMGO glycylamide exhibit subnanomolar affinity for MOP (IC_{50} = 0.38-0.93 nM). Thus these compounds have comparable MOP affinity to that of the parent ligand DAMGO (IC_{50} = 0.50 nM) and higher MOP affinity than the most potent MOP peptide affinity label reported previously.[6-8] Additional pharmacological experiments for these compounds are underway.

Acknowledgments

This research was supported by the National Institute on Drug Abuse (R01 DA 010035)

References

1. Aldrich, J.V., Vigil-Cruz, S.C. "Narcotic Analgesics" in *Burger's Medicinal Chemistry and Drug Discovery*, 6th Ed., Abraham, D.J., Ed., John Wiley, Inc: New York, 2003, vol. 6, pp 329-481.
2. Chen, C., *et al. J. Biol. Chem.* **270,** 17866-17870 (1995).
3. Handa, B. K., *et al. Eur. J. Pharmacol.* **70,** 531-540 (1981).
4. Leelasvatanakij, L., Aldrich, J.V., *J. Pept. Res.* **56,** 80-87 (2000).
5. Thompson , A. L., Ellman, J. A., *Tetrahedron Lett.* **35,** 9333-9336 (1994).
6. Garbay-Jaureguiberry, C., *et al. Proc. Natl. Acad. Sci. U.S.A.* **81,** 7718-7722 (1984).
7. Landis, G., *et al. J. Med. Chem.* **32,** 638-643 (1989).
8. Benyhe, S., *et al. Neuropeptides.* **9**, 225-235 (1987).

Amino acid Derivatives of Aporphinic Alkaloid Glaucine and their antioxidant activity

Maya Spasova[1], Stefan Philipov [2] and Tsenka Milkova [2,3]

[1]*Institute of Molecular Biology,* [2]*Institute of Organic Chemistry with Centre of Phytochemistry, Bulgarian Academy of Sciences, Sofia 1113, Bulgaria;* [3]*South-West University "Neofit Rilski" Blagoevgrad 2700, Bulgaria*

Introduction

The increasing interest of the role of reactive oxygen species and free radical reactions in a broad range of pathological processes has attracted our attention for searching for new molecules with antioxidant properties. Among the many naturally occurring types of antioxidants - the benzylisoquinoline alkaloids have recently emerged as a novel group with such kind of activity [1]. The antioxidant activity of glaucine [(S)-1,2,9,10-tetramethoxyaporthine] (one of the main representatives of benzylisoquinoline alkaloids), proposed by *Martinez L. A., et al.* [2] might be related to the presence of the biphenyl system

The facts that heterocyclic amino acid are biogenetic precursors of alkaloids and that some amino acids possess antioxidant activity and are rapidly destroyed by singlet oxygen [3] have provoked us to conjugate glaucine with different protected N^{α}-amino acids and to elucidate the antioxidant potential of the synthesized compounds against DPPH* test.

Results and Discussion

Herein we report the synthesis of a series of glaucine amides of aminoacids. The absence of the appropriate functional group in the glaucine nucleus has enforced us to modify it by introduction of amino group at C-3 in its molecule [4]. Our attempts to prepare amides of 3-aminoglaucine with different protected N^{α}-amino acids by the methods used in the peptide chemistry failed (obviously owing to the low reactivity of the amino group as a result of conjugation of its nitrogen lone pair with π- electrons of the benzene ring). To solve this problem we obtained **3-aminomethylglaucine** by known procedure [5].

The new amino acid derivatives of 3-aminomethylglaucine have been synthesized using peptide method (EDC/HOBt) (Table 1) and characterized by UV, 1H-NMR and ESI/MS.

The amides **1, 2, 3** and **4,** glaucine and 3-aminomethylglaucine were tested for their antioxidant activity against DPPH* test. Radical scavenging activity values (RSA) are presented for 10 min, as proposed by *Pekkarinen et al.* [6]. Sinapic acid was used as standard antioxidant. Most of the tested amides of 3-aminomethylglaucine have shown radical scavenging activity higher than both 3-aminomethylglaucine and glaucine. The results obtained demonstrated lower activity of the amides than the sinapic acids.

S. Del Valle et al. (eds.), *Peptides for Youth: The Proceedings of the 20th American Peptide Symposium*, 267
DOI: 10.1007/978-0-387-73657-0_120, © Springer Science+Business Media, LLC 2009

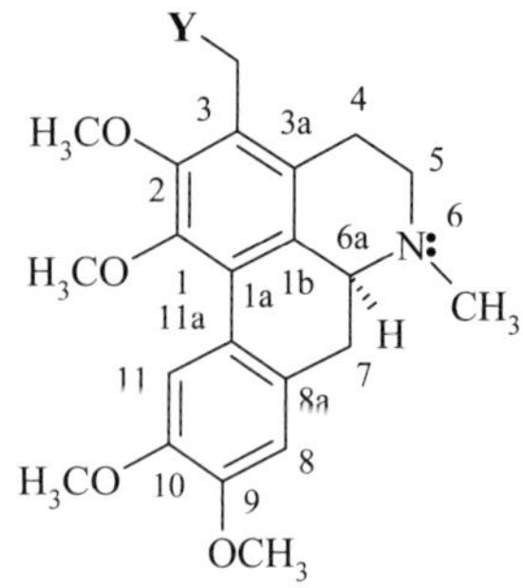

Table 1. Amino acids derivatives of 3-aminomethylglaucine.

№	Y-	Molecular weight	Yields, %
1	Boc-Phe(3-F)-CO-	649.4 $C_{36}H_{44}N_3O_7F$	63.2
2	Boc-His-CO-	621.4 $C_{33}H_{43}N_5O_7$	43.0
3	Boc-Phe-CO-	825.987 $C_{50}H_{55}N_3O_8$	81.3
4	H-Phe-CO-	531 $C_{31}H_{37}N_3O_5$	71.7

Acknowledgments

For the support of this work we are grateful to the National Found for Scientific Research of Bulgaria (Contracts VUH-07/05), South-West University "Neofit Rilski" Bulgaria and to the CEEPUS mobility program.

References

1 Cassels, B., Asencio, M., Conget, P., Speisky, H., Videla, L. and Lissi, E. Pharmac. Research 31, 103–107 (1995).
2. Martinez, L. A., Rios, J. L., Paya, M. and Alcaraz, M., J. Free Rad. Biol. Med. 12, 287 (1992).
3. Foote C. S. in Oxygen and Oxyradicals in Chemistry and Biology (Rodgers, M. A. J. and Powers, E. L., eds) p.425. Academic Press, New York.
4. Philipov, S. Dissertation "Synthesis of new derivatives of alkaloid glaucine (1985).
5. Mollov, N., Philipov, S., Ivanovska, N. and Dutschewska, H., Chem. Ber. 111, 554- (1978).
6. Pekkarinen, S., Stockmann, H., Schwarz, K., Heinonen, I. M., Hopia, A. I., J. Agric. Food. Chem. 47, 3036-3043 (1999).

Synthesis of Cyclic Tetrapeptide CJ 15,208: A Novel Kappa Opioid Receptor Antagonist

Santosh S. Kulkarni,[1] Nicolette C. Ross,[1] Jay P. McLaughlin,[2] and Jane V. Aldrich*[1]

[1]*Department of Medicinal Chemistry, The University of Kansas, Lawrence, KS 66045, USA*
[2]*Department of Psychology, Northeastern University, Boston, MA 02115, USA*

Introduction

Kappa (κ) opioid receptor (KOP) ligands are of considerable interests for the development of potential therapeutic agents [1]. The KOP antagonists have demonstrated promising activity in stress-induced cocaine seeking behavior [2,3] and a functional KOP antagonist has shown activity for the treatment of opiate abuse [4]. In addition KOP antagonists exhibit anti-depressant activity in animal models [5]. Thus KOP antagonists are potentially useful as therapeutic agents in the treatment of drug abuse and depression.

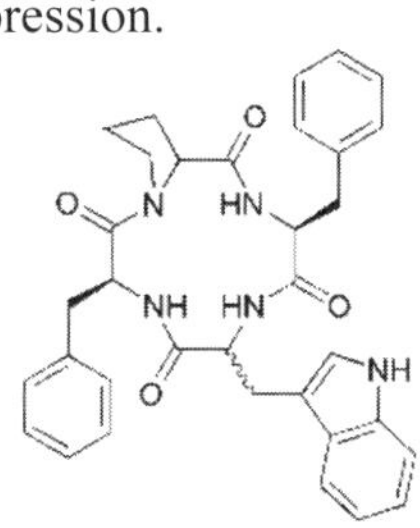

CJ 15,208 (1)

Figure 1: KOP antagonist, CJ 15,208.

Recently CJ 15,208 (**1**), a head-to-tail cyclic peptide obtained from the fermentation broth of the fungus *Ctenomyces serratus,* was reported to show antagonist activity at KOP [6]. The structural elucidation of the natural product assigned three of the amino acids as a D-proline and two L-phenylalanines; however, the stereochemistry of the tryptophan residue was not assigned. Thus here we present the synthesis and initial analytical results for characterization of the two isomers of this cyclic tetrapeptide in order to assign its structure.

The synthesis of cyclic tetrapeptides has been challenging, as the small ring size has often led to the formation of predominantly oligomeric products [7]. We used molecular modeling to preselect the linear peptide precursors which have preorganized conformations favoring cyclization. Using this strategy the compounds containing both stereoisomers of tryptophan were synthesized and characterized in an effort to assign the structure of the natural product.

Results and Discussion

All possible linear sequences for the syntheses of [L-Trp]- and [D-Trp]-CJ 15,208 were modeled using the Biopolymer module in Sybyl 7.0. The peptide conformations were analyzed for intrachain hydrogen bonds (HB) and the distances between the N- and C-termini (Table 1).

S. Del Valle et al. (eds.), *Peptides for Youth: The Proceedings of the 20th American Peptide Symposium,* 269
DOI: 10.1007/978-0-387-73657-0_121, © Springer Science+Business Media, LLC 2009

Table 1. Modeling results for L- and D-Trp-containing linear peptides.

Linear Seq.[1]	HB/Conf.[2]	N-C (Å)	Linear Seq.	HB/Conf.	N-C (Å)
FWFp	--/Extended	11.5	**FwFp**	--/Extended	10.4
WFpF	2 HB/γ-turn	6.5	**wFpF**	2 HB/γ-turn	2.9
FpFW	1 HB/β-turn	3.4	**FpFw**	2 HB/γ-turn	3.1
pFWF	--/Extended	11.9	**pFwF**	--/Extended	11.6

[1]D-amino acids are indicated by small letters. [2]HB = hydrogen bonds, Conf. = conformation

The linear peptides with proline or other turn-inducing residues at the C_{i+1} or C_{i+2} positions induce a folded conformation, which is further stabilized by intrachain hydrogen bonds. The linear sequences were synthesized on solid phase and cyclized in solution to obtain the cyclic tetrapeptides as single products (Scheme **1**).

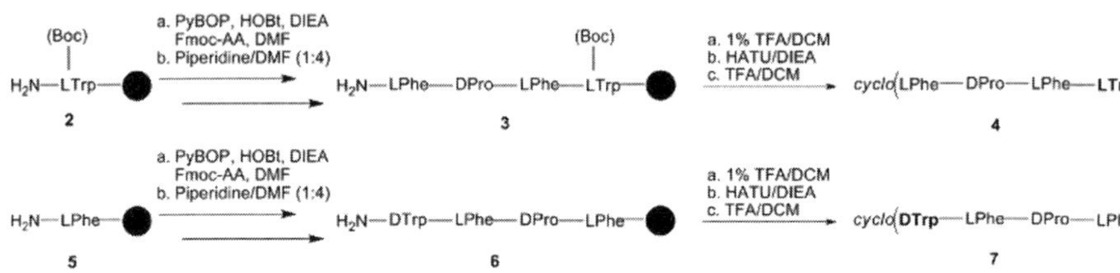

Scheme 1: Synthesis of cyclic tetrapeptides.

The analytical HPLC and HR-FABMS of the products confirmed the formation of cyclic tetrapeptides. Preliminary biological evaluation showed only a 2-fold difference in the binding affinities at KOP for the two stereoisomeric cyclic tetrapeptides. Further evaluation by NMR and pharmacological evaluations are currently being conducted to determine the structure of the natural product.

Acknowledgments

This research is supported by NIDA grant R01 DA018832 (JVA). We wish to thank Mr. Robert Drake, Dr. Todd Williams and Dr. David VanderVelde for analytical support.

References

1. Aldrich J. V.; Vigil-Cruz, S. C. Narcotic Analgesics In *Burger's Medicinal Chemistry & Drug Discovery; 6th ed.;* D. J. Abraham, Ed.; John Wiley & Sons, Inc.; New York, 2003; pp 329-481.
2. Beardsley, P. M. *et al., Psychopharmacology*, **2005**, *183*, 118-126.
3. Carey, A. N.; Borozny, K. Aldrich, J. V.; McLaughlin, J. P. *Eur. J. Pharmacol.*, **2007**, in press.
4. Rothman, R. B. *et al., J. Subst. Abuse Treat.*, **2000**, *18*, 277-281.
5. Mague, S. D. *et al., J. Pharmacol. Exp. Ther.*, **2003**, *305*, 323-330.
6. Saito, T. *et al.*, *J. Antibiotics*, **2002**, *10*, 847-854.
7. Humphrey, J. M.; Chamberlin, R. A. *Chem. Rev.*, **1997**, *97*, 2243-2266.

Exploring the relationship between turn geometry and allosteric antagonism of peptide mimic ligands for the prostaglandin F2α receptor

C.B. Bourguet[1], X. Hou[2], S. Chemtob[2], W. D. Lubell[1]

[1]*Département de Chimie, Université de Montréal, Montréal, Québec H3C 3J7, Canada;* [2]*Departments of Pediatrics, Ophthalmology and Pharmacology, Research Center of Hôpital Sainte-Justine, Montréal, Québec H3T 1C5, Canada.*

Introduction

Prematurity is an unmet medical need, with the highest per patient cost per year in the USA [1]. Term and preterm labor are largely triggered by the prostaglandin F2α (PGF2α) receptor in uterine muscle [2]; the same applies to uterine contractile spasms associated with dysmenorrhea. Peptides and mimics have been developed as allosteric antagonists [2,3] of the PGF2α receptor. For example, peptide **1** and mimic **2** (Fig. 1) were synthesized and exhibited good inhibition activity against PGF2α (IC50 340 nM and 2.5 nM respectively) [4,5]. Recently [6], the importance of the 3*S*,6*S*,9*S*-stereochemistry of the indolizidin-2-one (I2aa) type II' beta-turn mimic in ligand **2** was demonstrated by its replacement with its 3*R*,6*R*,9*R*-counterpart in ligand **3**, which was found to be inactive. Extending this study of the influence of configuration on activity, we have now synthesized the enantiomer of **2**, mimic **4** and examined its biological activity.

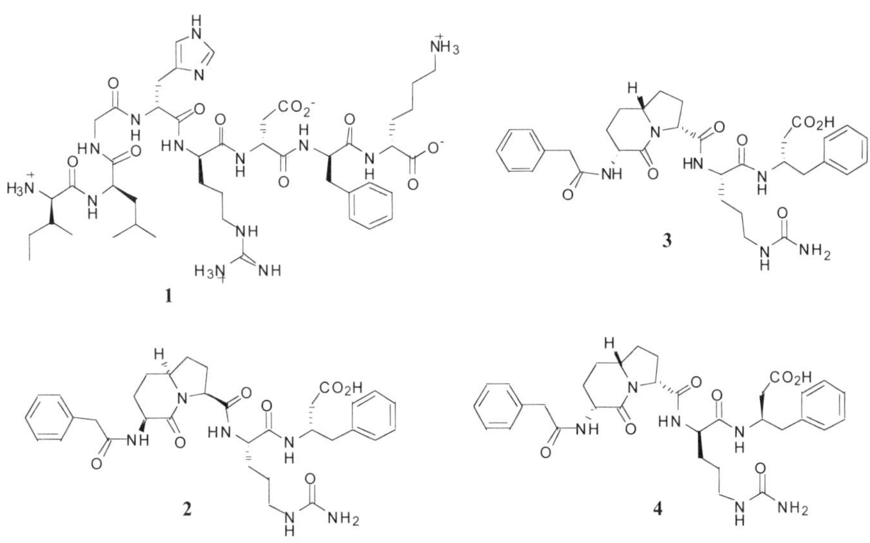

Fig. 1: Octapeptide 1 and mimics, ***2-4****.*

Results and Discussion

Enantiomer **4** was synthesized by using a Boc-protection strategy on oxime resin similar to that reported earlier [6] (Figure 2). D-Boc-Cit-OH was first coupled to the resin using DCC and ethyl 2-(hydroxyimino)-2-cyanoacetate (EACNox) in DCM.

S. Del Valle et al. (eds.), *Peptides for Youth: The Proceedings of the 20th American Peptide Symposium*, 271
DOI: 10.1007/978-0-387-73657-0_122, © Springer Science+Business Media, LLC 2009

Removal of the Boc group was performed with 20% TFA in DCM. To the resin bound peptide, (3*R*,6*R*,9*R*)-Boc-I2aa-OH [7] and phenylacetic acid were sequentially coupled using TBTU, HOBt and DIEA in DMF. Couplings were monitored by the Kaiser test, and LC/MS analysis of product from cleavage of a resin aliquot with methoxyethylamine in chloroform. The resin was cleaved by displacement with D-H-βPhe-OBn (100 mol%) in the presence of DIEA (100 mol%) and AcOH (100 mol%) in DCM. After purification of the benzyl ester by preparative HPLC, acid **4** was liberated by hydrogenolysis in EtOH and isolated by preparative HPLC for biological testing.

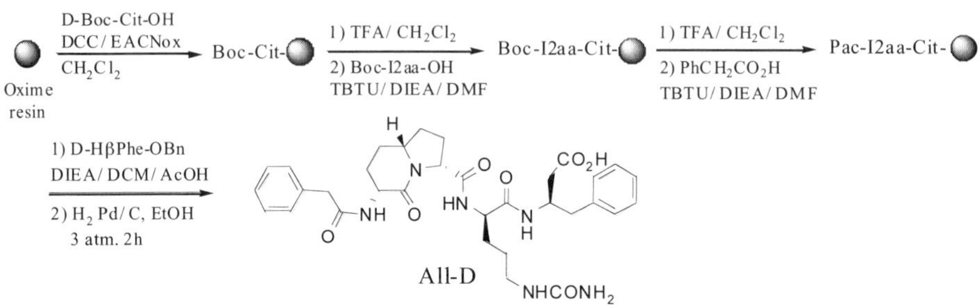

*Fig. 2: Synthesis of mimic **4**.*

The effect of peptide mimics on uterine contractility in the process of labor was tested against PGF2α responses on myometrial strips from CD-1 mice immediately after delivery in organ baths (containing Krebs buffer equilibrated with 21% oxygen at 37°C)[3]. After 1 h of equilibration, changes in mean basal tension, as well as peak, duration and frequency of spontaneous contractions in response to added agents were recorded with a Kent digital polygraph system (Fig. 3). As already mentioned [6], **2** dissipated robust PGF2α-induced contractions of mouse myometrium. In contrast, its enantiomer **4** showed no inhibition on the contractile activity of PGF2α on uterine smooth muscle.

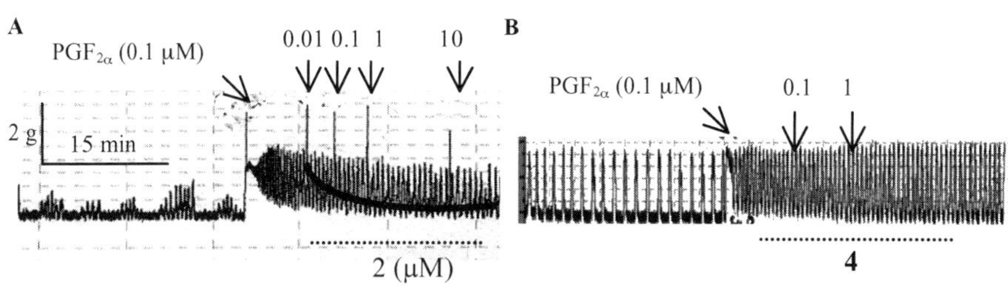

*Fig. 3: Tracings of spontaneous and PGF2α-induced myometrial contraction of pregnant mice tissue treated with A. **2** ; B. **4**.*

In conclusion, continued examination of the relationship between stereochemistry and activity in allosteric antagonists of the PGF2α receptor was performed by the synthesis and analysis of peptide mimic **4**. The lack of activity of this enantiomer of the potent antagonist **2** supports the importance of residue stereochemistry and the type II' beta-turn mimicked by the (3*S*,6*S*,9*S*)-I2aa residue as requirements for inhibition of uterine contractions induced by PGF2α. The development of bioavailable PGF2α receptor antagonist would be a significant breakthrough for managing preterm labor.

Acknowledgments

This research was supported by the Natural Sciences and Engineering Research Council (NSERC) of Canada and the Ministère de l'Éducation du Québec.

References

1. Morrison, J., Bergauer, N.K., Jacques, D., Coleman, S.K. and Stanziano, G.J. *Manag. Care* **10**, 42-6, 48-9 (2001).
2. Pathe-Neushäfer-Rube, A., Neushäfer-Rube, F., Püschel, G. P. *Biochem. J.* **388**, 317 (2005).
3. Jeremy Presland *Cur. Op. Drug Disc. Dev.* **8**, 567 (2005).
4. Peri, K.G., Quiniou, C., Hou, X., Abran, D., Varma, D.R., Lubell, W. D. and Chemtob, S. *Seminars in Perinatology* **67**, 389 (2002).
5. Peri, K.G., Polyak, F., Thouin, E., Lubell, W.D. and Chemtob, S. PCT Int. Appl. Ser. No. 60/387, 424, Jun 11, (2002).
6. Bourguet, C.B., Hou, X., Chemtob, S., Lubell, W.D. *Proceeding of the 29th European Peptide Symposium*, 215, (2006).
7. Lombart, H.G., Lubell, W.D. *J. Org. Chem.* **61**, 9437 (1996).

Optimization of the bioconvertion of the Angiotensin I Converting Enzyme inhibitors IPP and VPP

Jos Schalk

Unilever Food and Health Research Institute, Olivier van Noortlaan 120, NL-3133 AT, Vlaardingen, The Netherlands

Introduction

Hypertension is considered to be one of the main risk factors for Cardio Vascular Diseases (CVD). One of the mechanisms that regulate blood pressure is the renin-angiotensin system. This is a cascade of reactions leading to the formation of angiotensin II, which has a strong vasoconstrictive effect and hence a blood pressure-increasing effect. Inhibition of one of the key enzymes in this cascade, Angiotensin I Converting Enzyme (ACE), reduces formation of angiotensin II and thus has a blood pressure lowering effect. Since the discoveries of the first ACE inhibiting peptides, isolated from venom of the *Bothrops jararaca*, research has focussed on the identification of natural occurring peptides that can ameliorate hypertension. Especially milk appears to be a good source for a wide range of ACE inhibiting peptides. For instance, the *in vitro* ACE inhibiting effect of Ile-Pro-Pro (IPP) and Val-Pro-Pro (VPP), has been demonstrated. *Lactobacillus helveticus* strains grown in milk are able to liberate reasonable amounts of IPP and VPP from these caseins. In 1997 Calpis (Japan) launched a fermented milk product, containing these bioactive peptides, called Ameal-S, which claims to be 'suitable for those with mild hypertension'. However, the disadvantage of using lactobacillus in general is the long fermentation time, low yield in bioactive peptides and the production of lactic acid during the fermentation, which has great implications for product development. All of these issues results in a quite high production price for bioactive peptides. To overcome this problem research within Unilever has focused on the formation of IPP and VPP by using partly purified enzymes or mixtures of enzymes. This manuscript presents the results on the generation of IPP and VPP of a screening with several commercial enzyme mixtures and on the optimization to obtain the highest peptide concentration.

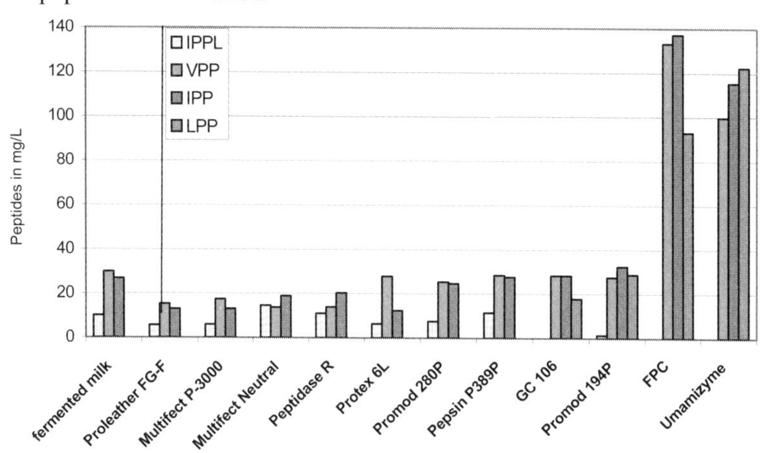

Fig. 1. A screening of comm available enzymemixtures on milk fermented with a Lb. helveticus.

S. Del Valle et al. (eds.), *Peptides for Youth: The Proceedings of the 20th American Peptide Symposium*, 275
DOI: 10.1007/978-0-387-73657-0_123, © Springer Science+Business Media, LLC 2009

Results and Discussion

Umamizyme and Fungal Protein Concentrate (FPC), both derived from *Aspergillus oryzae* showed the highest increase in XPP concentration when milk fermented with *Lb. helveticus* was used as substrate (figure 1). Both mixtures are a blend of proteases and peptidases. The concentrations for VPP and IPP were 133 mg/L (Yield: 98%) and 137 mg/L (Yield: 67%), respectively for FPC and 100 mg/L (Yield: 73%) and 115 mg/L (Yield: 56%) for Umamizyme. Much higher compared to the fermented milk with 30 and 27 mg/L, respectively. Other enzyme mixtures showed no or only a small effect. Since further increase in SMP concentration was not possible due to solubility problems, pure casienate was used as an XPP source. At 10% caseinate solution the highest XPP concentrations were observed, although enzyme activity became rate limiting. The highest IPP and VPP concentrations obtained were 374 and 378 mg/L, respectively when a 10% caseinate solution was used .

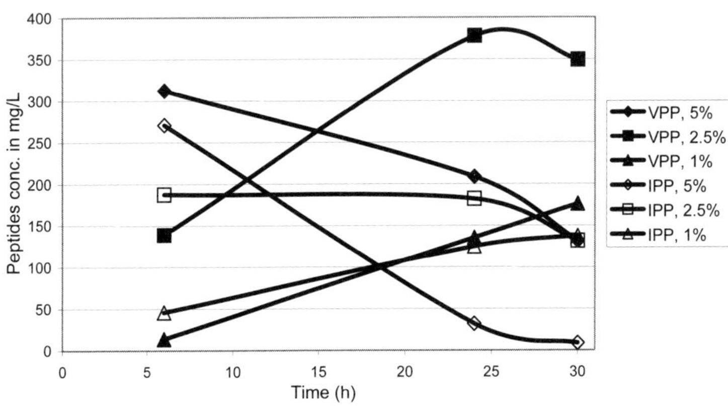

Fig. 2. Effect of enzyme/substrate ratio. VPP and IPP concentrations analysed after 6, 24 and 30 hours of 10% sodium-caseinate hydrolysis with Umamizyme at different enzyme/substate ratios of 5%, 2.5% and 1%.

The enzyme/substrate ratio is an important parameter that has a great influence on the XPP formation (Figure 2). A ratio of 5% showed the highest XPP concentrations after 6 hours. However, prolongation of the hydrolysis resulted in a large decrease, which was about 20% after 30 hours. With an enzyme/substrate ratio of 2.5% (w/w) the highest XPP concentration was measured after 24 hours. Degradation of the XPP was also noticed, although not as severe as for the 5% ratio.

Conclusions

- Umamizyme and FPC, both enzyme mixtures derived from *Aspergillus oryzae* showed to be the best mixtures to generate the most bio-active peptides IPP and VPP from milk fermented with *Lb. helveticus*.
- Further increase in concentration was achieved by using caseinate as substrate The highest IPP and VPP concentrations obtained were 374 and 378 mg/L, respectively.
- A ratio of 5% showed the highest XPP concentrations after 6 hours. However, prolongation of the hydrolysis resulted in a large decrease.
- With an enzyme/substrate ratio of 2.5% (w/w) the highest XPP concentration was measured after 24 hours.

Tailor made Pharmacokinetics of Peptides by Transient PEGylation

Ulrich Hersel, Martina Kreuzer, Harald Rau, Dirk Vetter and Thomas Wegge

Complex Biosystems GmbH, 69120 Heidelberg, Germany

Introduction

Conjugation to polyethylene glycol (PEGylation) is widely used in order to prolong the pharmacodynamic effect of proteins. For peptides, PEGylation requires extensive engineering to retain sufficient pharmacological activity (receptor affinity, tissue penetration) of the conjugate.

Transient PEGylation is based on the release of unmodified peptide from a PEG carrier thus obviating the need for engineering. Peptide liberation proceeds via slow cleavage of a traceless linker under physiological conditions.

Linker cleavage should ideally be governed primarily by pH and temperature in order to avoid enzyme dependent intrapatient variability as it is known e.g. for ester prodrugs. Several linkers have been proposed for transient PEGylation so far, mostly depending on enyzmatic activation and with half life times less than 24 h [1].

We developed a family of linker structures that cleave after initial hydrolysis of a carbamate bond (Fig. 1). The first hydrolytic step is rate determining and depends on the structure of the internal trigger and linker scaffold.

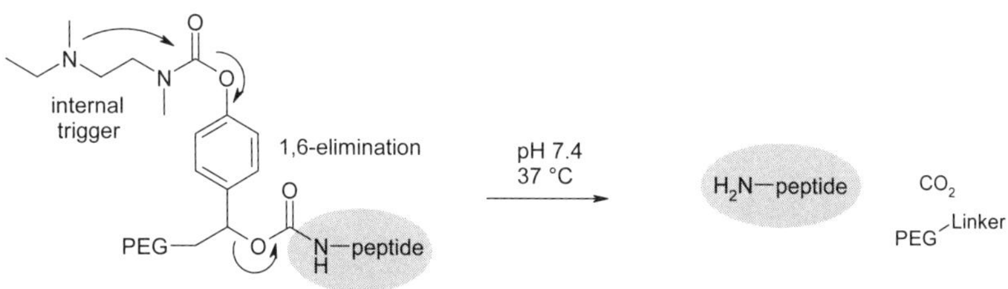

Fig. 1. Unmodified peptide is regenerated from PEG conjugate by cleavage of traceless linker.

By variation of trigger and linker scaffold we obtained cleavage half life times from hours to weeks at physiological conditions *in vitro*. *In vitro - in vivo* correlation of cleavage kinetics of a PEG-GLP-1 conjugate was assessed using a double fluorescence assay. In vivo pharmacodynamics of a slowly releasing PEG-exendin-4 conjugate was tested in diabetic db/db mice.

Results and Discussion

Activated linkers containing a protected thiol group were coupled to insulin (ε-amino group of LysB29) according to lit. [2]. After removal of thiol protecting group, insulin-linker-thiols were reacted with PEG-maleimide to obtain PEG-linker-insulin conjugates. Linker cleavage was effected by incubating conjugates at pH 7.4 and 37°C. Linker cleavage kinetics were determined by analyzing samples for released insulin by RP-HPLC. Cleavage kinetics were found to follow first order

S. Del Valle et al. (eds.), *Peptides for Youth: The Proceedings of the 20th American Peptide Symposium*, 277
DOI: 10.1007/978-0-387-73657-0_124, © Springer Science+Business Media, LLC 2009

kinetics. Linker half life times of 13 h, 45 h, 98 h, 10 d, and 29 d were obtained. Integrity of released insulin was confirmed by ESI-MS.

In vivo linker cleavage kinetics were assessed by a double fluorescence assay as described previously [3]. GLP-1(7-36 amide) was labeled with fluorescein (Em 538 nm) close to the C-terminus and permanently linked to PEG40K via the N-terminus. Compound was coinjected with permanent PEG40K-Bodipy (Em 620 nm) as an internal reference s.c. into rats. Plasma sample fluorescence revealed a constant ratio of labeled compounds thus demonstrating equal clearance rate and the absence of proteolytic breakdown of the peptide.

In a second experiment GLP-1(7-36 amide) was labeled with fluorescein close to the C-terminus and linked to PEG40K via a slowly hydrolyzing linker at the N-terminus. Compound was coinjected with PEG40K-Bodipy as internal reference s.c. into rats. Plasma sample fluorescence of linker conjugate decreased with first order kinetics vs. reference. This reflects the cleavage kinetics of the hydrolyzing linker *in vivo*. *In vivo* and *in vitro* cleavage kinetics correlated tightly, thus allowing for predictability of *in vivo* kinetics by *in vitro* results.

In order to develop a long acting conjugate of the antidiabetic peptide exendin-4, we prepared PEG40K-linker-exendin-4 conjugate with a slow releasing linker. Conjugate and exendin-4 reference were given s.c. to diabetic db/db mice (6.0 nmol/ mouse). PEG-linker-exendin-4 conjugate was able to extend exendin-4's plasma glucose lowering potency from 24 h (reference) to more than 72 h (Fig. 2). This holds promise for a once-weekly slowly releasing exendin-4 depot in man.

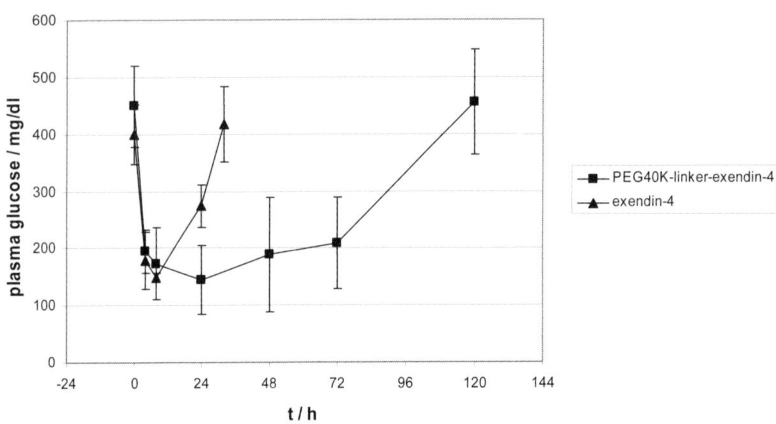

Fig. 2. Prolonged pharmacodynamic effect of slowly releasing PEG40K-linker-exendin-4 conjugate (squares) as compared to exendin-4 reference (triangles).

References

1. Greenwald R. B., Yang K., Zhao H., *et al. Bioconjugate Chem.* **14**, 395-403 (2003). Greenwald R. B., Choe Y. H., Conover C. D., *et al. J. Med. Chem.* **43**, 475-487 (2000). Tsubery H., Mironchik M., Fridkin M., *et al. J. Biol. Chem.* **279**, 38118-38124 (2004).
2. Hersel U., Rau H., Schnepf R., *et al. WO2005099768.*
3. Wegge T., Fricker G., Hersel U., *et al. 8th German Peptide Symposium*, P026 (2007).

Design and Synthesis of Cyclic Arodyn Analogues by Ring-Closing Metathesis (RCM) for Kappa Opioid Receptor (KOP) Antagonists

Wei-Jie Fang[1], Santosh S. Kulkarni[1], Thomas F. Murray[2] and Jane V. Aldrich[1]

[1]*Department of Medicinal Chemistry, University of Kansas, Lawrence, KS 66045, U.S.A.;* [2]*Department of Pharmacology, Creighton University, Omaha, NE 68178, U.S.A.*

Introduction

Our laboratory is interested in the design and synthesis of antagonists for kappa opioid receptor (KOP). In addition to their use as pharmacological tools, KOP antagonists have the potential to be used therapeutically in the treatment of cocaine[1,2] and opioid dependence[3] and also have antidepressant activity.[4] Arodyn (Ac[Phe[1,2,3],Arg[4],D-Ala[8]]Dyn A(1-11)NH_2, Fig. 1) is an acetylated dynorphin A (Dyn A) analogue identified in our laboratory that is a potent and highly selective KOP antagonist.[5,6] As a linear peptide, arodyn is very flexible and can adopt numerous conformations and is susceptible to metabolic degradation by peptidases. Therefore we are interested in the design and synthesis of cyclic arodyn analogues to restrict its conformation in order to study its possible bioactive conformation at KOP and to enhance the metabolic stability of the peptide.

Ring-closing metathesis (RCM) has emerged as a very useful method of making cyclic organic compounds as well as cyclic peptides.[7-9] Compared with the traditional approach of preparing cyclic peptides by disulfide or amide bond formation between amino acid side chains, cyclization by RCM has some advantages. The carbon-carbon bond is more stable than disulfide or amide bonds, and it is possible to maintain side chain functionalities when using RCM. Here we report the design and synthesis of cyclic arodyn analogues using RCM.

Ac-Phe-Phe-Phe-Arg-Leu-Arg-Arg-D-Ala-Arg-Pro-Lys-NH_2

Fig. 1. Sequence of arodyn.

Results and Discussion

Different strategies were applied for the design of cyclic arodyn analogues using D/L-allylglycine (AllGly) and/or Tyr(All) incorporated in different positions in the peptide. The linear peptide precursors were synthesized by standard Fmoc-solid phase synthetic methods and cyclized on resin using second generation Grubbs catalyst. Following cleavage from the resin and purification by reversed phase HPLC, the peptides were characterized by ESI-MS. The configuration of the olefinic geometric isomers was assigned based on NMR coupling constants of the vinyl protons (*cis*: $J \approx 10$ Hz; *trans*: $J \approx 15$ Hz).

The cyclization between the two side chains of D/L- AllGly generally gave high yields (76-90%) of the desired cyclic arodyn analogues. The ratio of *cis* to *trans* isomers, as determined by NMR, varied from approximately 1/4 to 1/1, depending on the position and stereochemistry of the AllGly. Surprisingly, preliminary pharmacological evaluation in the adenylyl cyclase assay indicated that several of

DOI: 10.1007/978-0-387-73657-0_125, © Springer Science+Business Media, LLC 2009

these analogues are partial to full agonists at KOP (40-110% in adenylyl cyclase at 10 μM compared to Dyn A(1-13)NH_2).

The yields for cyclizations involving Tyr(All), however, were highly variable and depended on the sequence of the linear precursor peptide. Molecular modeling suggested differences in the conformations of the linear precursors that may explain the differences in reactivity. When the two alkenes are close to each other and the cyclic peptide doesn't have a high energy structure, the cyclization gave modest yields (38-64%). Only *trans* isomers were observed. In addition, a side reaction involving the deletion of the allyl group from Tyr was observed for the first time. Model studies with Fmoc-Tyr(All)-OH were consistent with a mechanism for this side reaction involving double bond migration under the conditions of the metathesis reaction, followed by acidic cleavage of the resulting vinyl ether (Scheme 1). This side reaction dominates when the two alkenes are not in close proximity and/or the cyclic peptide could have a high-energy structure.

The successfully synthesized peptides are currently undergoing further pharmacological evaluation.

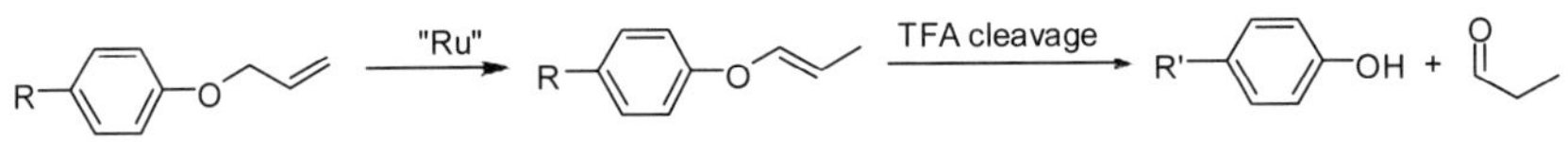

Scheme 1. Proposed mechanism for the deletion of allyl group.

Acknowledgments

The authors thank Dr. Jun Xie in Dr. Thomas Murray's laboratory for performing the pharmacological assays. This research was supported by grant R01 DA018832 from the National Institute on Drug Abuse.

References

1. Beardsley, P. M. et al. *Psychopharmacology* **183**, 118-126 (2005).
2. Carey, A. N. et al. *Eur. J. Pharmacol.* in press.
3. Rothman, R. B. et al. *J. Subst. Abuse Treat.* **18**, 277-281 (2000).
4. Mague, S. D. et al. *J. Pharmacol. Exp. Ther.* **305**, 323-330 (2003).
5. Bennett, M. A. et al. *J. Med. Chem.* **45**, 5617-5619 (2002).
6. Bennett, M. A. et al. *J. Pept. Res.* **65**, 322-332 (2005).
7. Fu, G. C. et al. *J. Am. Chem. Soc.* **114**, 5426-5427 (1992).
8. Miller, S. J. et al. *J. Am. Chem. Soc.* **118**, 9606-9614 (1996).
9. Berezowska, I. et al. *J. Med. Chem.* **50**, 1414-1417 (2007).

Switch-Peptides as Folding Precursors in Self-Assembling Peptides and Amyloid Fibrillogenesis

M.-S. Camus[1], R. Mimna[1], A. Schmid[2], A. Chandravarkar[1], G. Tuchscherer[1], H. A. Lashuel[2] and M. Mutter[1]

[1]Institute of Chemical Sciences and Engineering (ISIC); [2]Brain Mind Institute (BMI); Ecole Polytechnique Fédérale de Lausanne (EPFL), 1015 Lausanne, Switzerland

Introduction

The study of conformational transitions of peptides has obtained considerable attention recently because of their importance as a molecular key event in a variety of degenerative diseases. However, the study of peptide self-assembly into β-sheets and amyloid β fibrils is strongly hampered by their difficult synthetic access and low solubility. We have recently developed a new concept termed "switch-peptides" that allows the controlled onset of polypeptide folding and misfolding at physiological conditions[1, 2, 3]. As a major feature, the folding process is initiated by chemically or enzyme triggered O,N-acyl migration in flexible and soluble folding precursors containing Ser- or Thr-derived switch (S)-elements. This in situ intramolecular migration allows for the controlled induction or reversal of secondary structural transitions and self-assembly of small peptide chains[4]. Here, we describe a switch peptide that is designed to disrupt amyloid-like β-sheet assemblies through the controlled induction of a transition from a β-sheet to an α-helix structure (Figure 1).

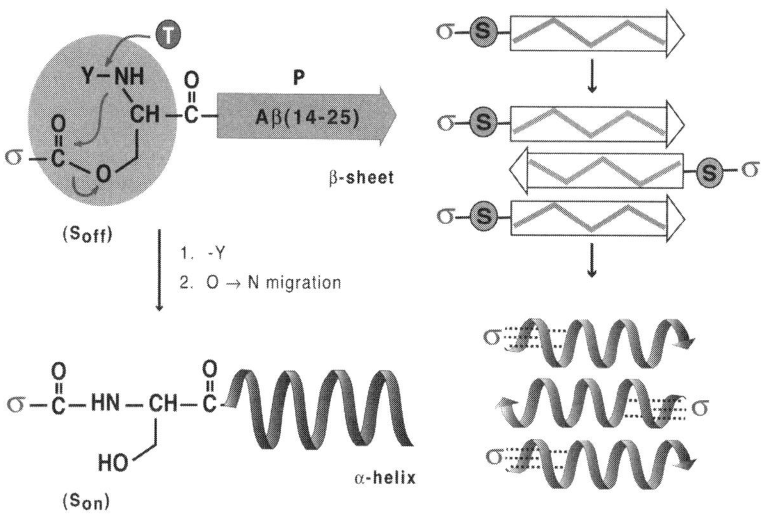

*Figure 1: Switch-peptide **I**: Aβ(14-25) is linked to a helix-inducing template **σ** (Ncap = Ac-[cyclo-1-5]-KARADA) via a Ser-derived switch-element **S** (O-acyl isopeptide unit, S_{off}-state). Upon removal of the N-protecting group Y, spontaneous O,N-acyl migration occurs (S_{on}-state) resulting in the activation of **σ** and induction of a helix structure in peptide **I** (left). Under the right conditions, the helix-inducing effect of **σ** (right) is strong enough to induce β-sheet (step 1) to an α-helix (step 2) transformation and disruption of preformed Aβ(14-25)-derived β-sheet rich assemblies within amyloid fibrils.*

S. Del Valle et al. (eds.), *Peptides for Youth: The Proceedings of the 20th American Peptide Symposium*, 281
DOI: 10.1007/978-0-387-73657-0_126, © Springer Science+Business Media, LLC 2009

Results and Discussion

To examine the properties of peptide **I** in the S_{off} and S_{on} states (Figure 1), the secondary structure and aggregation states were probed by circular dichroism (CD) and electron microscopy (EM). In the S_{off} state, the CD spectrum of peptide **I** displays the Cotton effect, which is typical of a β-sheet structure (curve 1, Figure 2A). Activation of the helix-inducing Ncap[5] (S_{on}) through a pH-induced O,N acyl-migration in presence of 25% TFE reveals an unprecedented transition from a predominantly β-sheet structure in the S_{off} state (curve 1 & 2) to a predominantly α-helix structure in the S_{on} state (curve 3). Electron microscopy revealed that the transition from β-sheet to α-helix structure is accompagnied by a dramatic change in the morphology and subsequent dissociation of the fibrils (Figure 2 B, C & D).

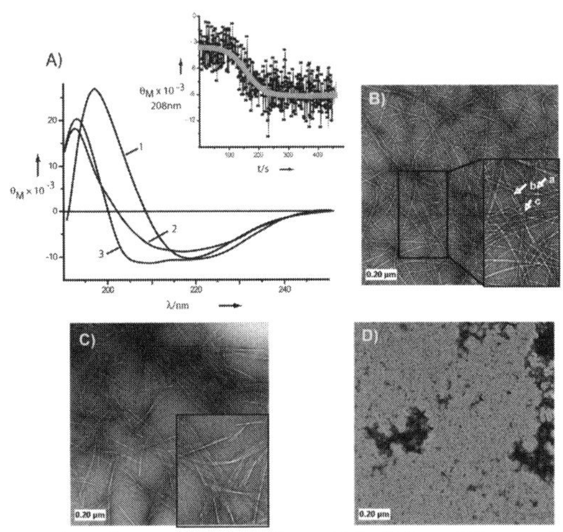

*Figure 2: Conformational transitions of peptide **I** (see Figure 1) in the CD spectrum and electron microscope (c = 5 x 10^{-5} M, T = 25°C):A) curve **1**: pH 4.5 buffer (S_{off}); curve 2: pH 4.5 buffer/TFE (75:25) (S_{on}); Curve 3: pH 7.0 buffer/TFE (75:25) (S_{on}). B) negative electron micrograph in the Soff state; C) S_{on} state 5 min and D) 12h after activation of the intramolecular O to N acyl migration.*

Using peptide **I**, we were able to investigate the feasibility of disrupting amyloid formation through the controlled induced transformation from β-sheet to α-helix structures of self-assembled peptides within the amyloid structure. The experimental data offer new insights into the stability and structural properties of amyloid fibrils.

Acknowledgments

This work was supported by the Swiss National Science Foundation.

References

1. M. Mutter et al., *Angew. Chem. Int. Ed.*, **2004**, 43, 4172
2. S. Dos Santos et al., *J. Am. Chem. Soc.*, **2005**, 127, 11888.
3. G. Tuchscherer et al., *Biopolymers (Peptide Science)*, **2007**, 88, 2, 239
4. R. Mimna et al., *Angew. Chem. Int. Ed.*, **2007**, 46, 2681.
5. R. Mimna, et al., *Int. J. Pept. Res. & Therapeutics*, **2007**, 13, 237.

PEPTIDES AND MEMBRANE PROTEINS: A SESSION DEDICATED TO MIKLOS BODANSKY

BioShuttle mediated Plasmid Transfer – a Way to new genetic Intervention

Rüdiger Pipkorn[1], Waldemar Waldeck[2], Bernd Didinger[3] and Klaus Braun[4]

[1]*Department of Central Peptide Synthesis;* [2]*Department of Biophysics of Macromolecules;* [4]*Department of Molecular Toxicology; German Cancer Research Center, Im Neuenheimer Feld 280; D-69129 Heidelberg, Germany;* [3] *University of Heidelberg*

Introduction

As a therapeutic approach genetic interventions achieved an exciting development to treat diseases. The choice of the appropriate gene transporter shaped up as the key for a safe and efficient gene transfer. Up till now the transfer based on viruses, polymers or ultrasound waves was hampered by constitutional questions like limited efficiency and poor entry of plasmid DNA from the cytoplasm into the cell nuclei and by the risk of nocuous immunological response. BioShuttle a modular transporter system seemed to be a step to face these obstacles. The modules use cell immanent transport mechanisms and enable the delivery of agents and diagnostics into specific subcellular compartments like cell nucleus, mitochondria, peroxisoms etc. After delivery across the cellular membrane into the cytoplasm facilitated by the first module, and its enzymatic cut off, the second module harbouring the nuclear localization address sequence (NLS) acts as the natural substrate for the RAN-mediated nuclear importin system. The third module, the nuclease and protease-resistant Clamp-PNA (Peptide-Nucleic-Acid) harbours a complementary sequence a part of the pUC ORI-sequence in the plasmid, hybridized before application. Here we demonstrate both the gentle BioShuttle-mediated transfer of the recombined phNIS-IRES2-EGFP with two reporter genes into DU-145 prostate cancer cells: the Sodium Iodide-Symporter activity was measured by the Iodide-125-uptake and, the transcription of EGFP was estimated integration of fluorescence intensities using the Confocal Laser Scanning Microscopy.

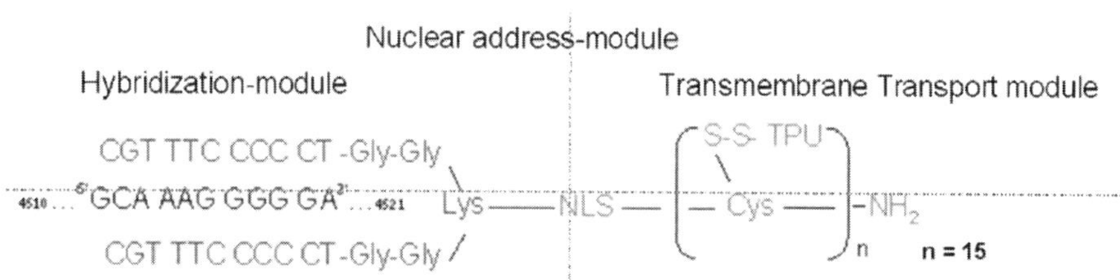

Fig. 1. Schematized modules of the Clamp-BioShuttle: the left part of the figure displays the sequence and the complementary ORI sequences of the plasmid. The middle part of the figure the nuclear address module is shown NLS; the module responsible for the transport across the cellular membrane is shown on the right part of the figure). The 15 TPU peptides correspond to the amphiphilic peptides derived from the drosophila origin and which are connected with the NLS by an intracellular enzymatic-cleavable disulfide bridge

S. Del Valle et al. (eds.), *Peptides for Youth: The Proceedings of the 20th American Peptide Symposium,* 285
DOI: 10.1007/978-0-387-73657-0_127, © Springer Science+Business Media, LLC 2009

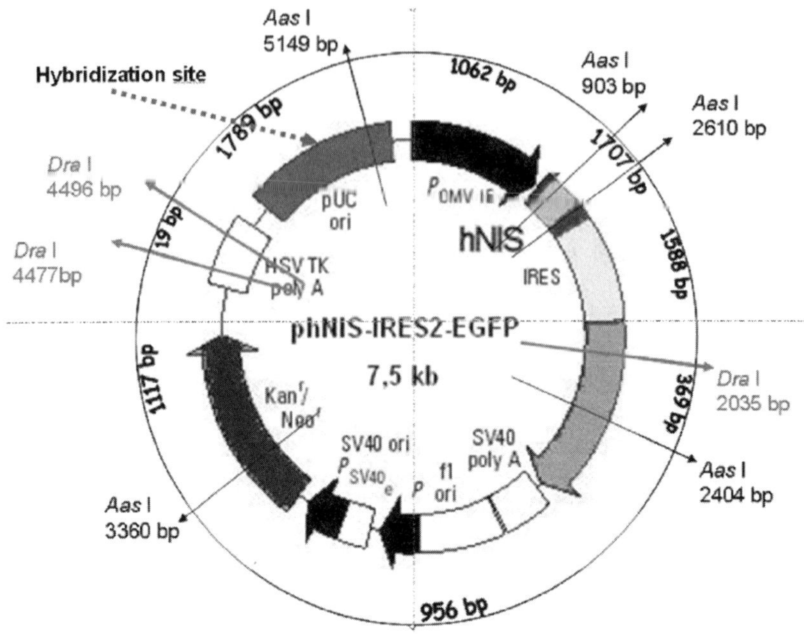

Fig. 2. The figure shows the plasmid phNIS-IRES2-EGFP construct which was used after hybridization for the gene transfer and gene activation study. The positions bp represent the restrictions sites of the enzymes Aas 1 and Dra 1 respectively. Inside of the circle circularly aligned: the size of the restriction fragments (bp); the arrow demonstrates the hybridization site of the clamp-BioShuttle at the pUC ORI section.

Results and Discussion

After hybridization [1] and activation we observed a rapid (60 min incub. time) and a near 100% cellular uptake of phNIS-IRES2-EGFP into DU-145 human prostate cancer cells [2]. By means of EGFP-fluorescence, a period of 30 s was determined to be sufficient for gene activation using the focussed ultrasound (HIFU) [3]. After 24 h the EGFP expression was measured. Figure 1 displays the functional modules of the simplified Clamp-BioShuttle construct and the pUC ORI hybridization side. The Quick plasmid is shown in figure 2. Non-activated control cells do not show any fluorescence. '*BioShuttle'* carriers [4] could be a breakthrough technology for gene transfer and gene therapeutic approaches. The intent for use in genetic interventions, the BioShuttle carrier permits a homogenous, evenly-distributed gene expression, that delivered a low rate of mortality of the transfected cells.

Acknowledgments

The work was partially supported by the Deutsche Krebshilfe.

References

1. Britten, R. J. and Kohne, D. E., *Science* ***161*** 529-540 (1968).
2. von Brasch, L. Thesis, University of Heidelberg, (2007).
3. Divkovic, G. W. et al. *Ultrasound Med.Biol.* ***33*** 981-986 (2007).
4. Braun, K. et al. *J. Mol. Biol.* ***318*** 237-243 (2002).

Incorporation of a Novel, Fluorescent and Helicogenic A-Amino Acid Into Peptaibols: Trichogin Ga IV

C. Baldini[1], F. Formaggio[1], C. Toniolo[1], A. Toffoletti[1], C. Corvaja[1], J.-P. Mazaleyrat[2], K. Wright[2], J.-F. Lohier[2], M. Wakselman[2], B. Orioni[3], L. Stella[3] and M. Venanzi[3]

[1]ICB, Padova Unit, CNR, Department of Chemistry, University of Padova, 35131 Padova, Italy; [2]ILV, UMR CNRS 8180, University of Versailles, 78035 Versailles, France; [3]Department of Chemical Sciences and Technologies, University of Rome Tor Vergata, 00133 Rome, Italy

We have exploited a novel fluorescent α-amino acid, antAib, based on a planar anthracene core and belonging to the class of cyclic $C^{\alpha,\alpha}$-disubstituted glycines (β-turn and helix inducers in peptides) to investigate the preferred conformation and membrane activity of the peptaibol antibiotic trichogin GA IV.

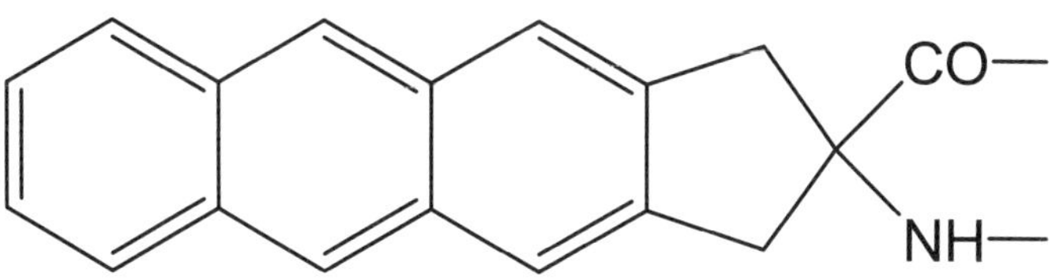

antAib

We incorporated antAib at position 1 of two trichogin analogues, one of which bears a free-radical TOAC residue at position 4. Both $C^{\alpha,\alpha}$-disubstituted glycines (antAib and TOAC) replace Aib residues in the natural sequence.

Peptide synthesis was performed by solution methods. Conformational analysis was carried out by use of a large combination of spectroscopic techniques (FT-IR absorption, CD, NMR and EPR (both cw and TR). Self-aggregation and membrane activity properties were analyzed by use of steady-state and time-resolved fluorescence spectroscopy.

S. Del Valle et al. (eds.), *Peptides for Youth: The Proceedings of the 20th American Peptide Symposium*, 287
DOI: 10.1007/978-0-387-73657-0_128, © Springer Science+Business Media, LLC 2009

Seeking for α-Helical Propensity in a Receptor-Bound Conformation of Glucagon-Like Peptide-1

Eunice Murage[1], Martin Beinborn[2] and Jung-Mo Ahn*[1]

[1]Department of Chemistry, University of Texas at Dallas, Richardson, TX 75080, USA; [2]Department of Medicine and Molecular Pharmacology Research Center, Tupper Research Institute, New England Medical Center, Boston, MA, 02111, USA

Introduction

Glucagon-like peptide-1 (GLP-1) is a 30 amino acid-containing gastrointestinal hormone released from L-cells in small intestine [1]. It enhances glucose-dependent insulin secretion; stimulates pancreatic β-cell growth; suppresses glucagon secretion; and lowers gastric emptying. Due to its highly favorable physiological functions for treating diabetes, it has been considered as a potent therapeutic agent for diabetes.

To improve potency of a peptide hormone, structure-activity relationship studies are typically conducted, and structural information of how the peptide interacts with its receptor often provides useful insights. A 2D-NMR study of GLP-1 bound to a dodecylphosphocholine micelle revealed two α-helical segments between residues 13-20 and 24-35 which are connected by a linker region (residues 22-23) [2]. However, the membrane-like environment provided by micelles cannot properly mimic complex interactions between GLP-1 and its cognate receptor.

Thus, to explore the presence and location of α-helices in GLP-1 when it binds to its receptor, we have designed and synthesized a series of conformationally constrained GLP-1 analogues. The conformational restriction used in this study is the formation of a lactam bridge between Lys at the i position and Glu at the i+4 position, which in general induces and stabilizes an α-helical structure [3]. We have positioned lactam bridges at various regions of GLP-1 and questioned the presence and location of α-helices in the receptor-bound conformation of GLP-1 by making the cyclic peptides interact with the receptor [4]. For instance, a peptide containing a correctly folded structure at a correct location will be awarded with high receptor binding affinity compared to a misfolded one or one at a misplaced position. For this study, residues that are known to be important for receptor binding and signal transduction like His7, Phe12, Ile29, and Trp31, were not substituted.

Table 1. Binding affinity of cyclic GLP-1 analogues.

		Sequence	IC_{50} (nM)	Relative Binding (%)
	GLP-1	HAEGTFTSDVSSYLEGQAAKEFIAWLVKGR	2.8	100
1C	c[Lys^{11},Glu^{15}]	----K---E---------------------	108	2.6
2C	c[Lys^{14}, Glu^{18}]	-------K---E------------------	214	1.3
3C	c[Lys^{16}, Glu^{20}]	---------K---E----------------	4.5	62
4C	c[Lys^{18}, Glu^{22}]	-----------K---E--------------	15.1	19
5C	c[Lys^{20}, Glu^{24}]	-------------K---E------------	3,020	0.09
6C	c[Lys^{21}, Glu^{25}]	--------------K---E-----------	9,180	0.03
7C	c[Lys^{23}, Glu^{27}]	----------------K---E---------	3.3	85
8C	c[Lys^{27}, Glu^{30}]	--------------------K--E------	7.8	36
9C	c[Lys^{30}, Glu^{34}]	-----------------------K---E-	1.5	190

S. Del Valle et al. (eds.), *Peptides for Youth: The Proceedings of the 20th American Peptide Symposium*, 289
DOI: 10.1007/978-0-387-73657-0_129, © Springer Science+Business Media, LLC 2009

Results and Discussion

Cyclic GLP-1 analogues used in this study were synthesized using standard *N*-Fmoc/*t*-butyl peptide chemistry in solid-phase, and the Lys and Glu which form a lactam bridge were orthogonally protected with allyl protecting groups. Before the cyclization was carried out on resin, the allyl groups were selectively removed by using $Pd(PPh_3)_4$ and N,N'-dimethylbarbituric acid used as an allyl scavenger.

To demonstrate the induction and stabilization of α-helices by the formation of lactam bridges, we have examined all cyclic GLP-1 analogues by CD spectroscopy. The cyclic peptides at 20 μM were dissolved in a TFE-containing aqueous solution with increasing content of TFE, and showed enhanced or comparable α-helix contents when compared to that of the native GLP-1.

The cyclic peptides were also examined for their receptor-binding affinities which were determined by using ^{125}I-exendin(9-39) on human GLP-1 receptors over-expressed on COS-7 cells. Among 8 positions at which the lactam bridges were introduced, high binding affinities were observed for the cyclic peptides containing lactam bridges at the N- and C-terminal regions. This suggests that GLP-1 adopts α-helical conformations between residues 16-20, 23-27 and 30-34 which were well recognized by the GLP-1 receptor. However, α-helical structure appears not to be preferred at the very N-terminus and in the linker region, showing poor binding affinities. The receptor-bound conformation determined by this study is consistent with the solution structure of GLP-1 determined by NMR studies.

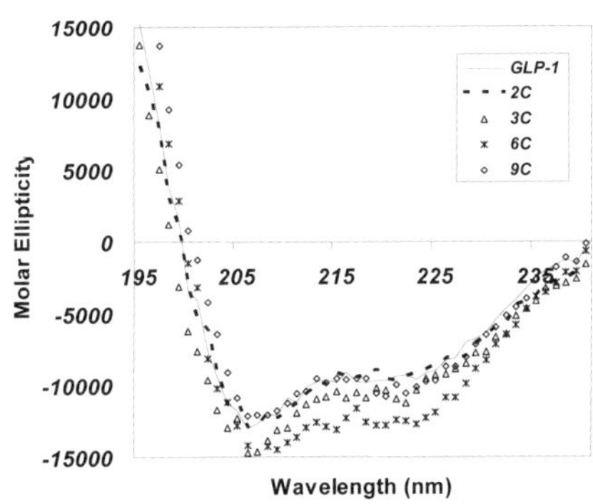

Fig. 1. CD spectra of the cyclic GLP-1 analogues.

Acknowledgments

The work was supported by Welch Foundation (AT-1595), American Diabetes Association (7-07-JF-02), and Texas Advanced Research Program (009741-0031-2006).

References

1. Drucker, D. J. *Diabetes* **47**, 159-169 (1998).
2. Thorton, K. and Gorenstein, D. G. *Biochemistry* **33**, 3532-3539 (1994).
3. Houston, M. E., *et al. J. Pept. Sci.* **1**, 274-282 (1995).
4. Ahn, J.-M., *et al. J. Med. Chem.* **44**, 3109-3116 (2001).

Conformational Analysis and Folding of Transmembrane and Matrix Peptide Segments of the Mitochondrial Uncoupling Proteins: A Comparative Study

Masoud Jelokhani-Niaraki[1], Marina V. Ivanova[1], Cheryl L. Newman[1], Elizabeth K. Young[1], Laura Wheaton[1], Bonnie L. McIntyre[2] and Matthew D. Smith[2]

[1]*Department of Chemistry;* [2]*Department of Biology, Wilfrid Laurier University, Waterloo, Ontario, N2L 3C5, Canada*

Introduction

The archetypal *Uncoupling Protein 1* (UCP-1) is a mitochondrial regulated protein carrier/ion channel, and by far the best studied isoform of the mammalian UCP subfamily of transporters located in the inner membrane of mitochondria. Thermogenic proton leak resulting in the decoupling of ATP synthesis from cellular respiration is considered to be the main physiological function of UCP-1, which is mainly found in brown adipose tissues. UCP-1 also transports anions and fatty acids. The other four mammalian UCP isoforms are located in several tissues, including brain (UCPs 4 and 5), which in addition to proton transport possess diverse physiological roles such as reduction of oxidative stress and protection against cardiac ischemia. Despite an increasing number of publications about the physiological functions of UCPs, the molecular mechanisms of the biological function as well as the structure of UCPs are not yet fully understood. To address these questions, we have synthesized the six transmembrane (TM) and the three matrix loop (ML) peptide segments of golden hamster (*Mesocricetus auratus*) UCP-1 (90% of the sequence of the native protein, Table 1) to explore their conformation, folding and ion transport properties in comparison to those of full-length, recombinant UCP-1. In addition, we have designed and synthesized the TM2 peptide fragments of the human UCPs 1 to 5 to compare their conformations and electrophysiological properties.

Table 1. Sequences of TM and ML peptide segments of the golden hamster UCP-1 [a,b].

Peptide (# of AA)	Sequence
TM1 (34)	T^{4}TSEVHPTMGVKIFSAGVAAS*LADIITFPLDTAK^{37}
ML12 (29)	V^{38}RLQIQGEGQISSTIRYKGVLGTITTLAK^{66}
TM2 (35)	T^{67}EGLPKLYSGLPAGIQRQISFASLRIGLYDTVQEY^{101}
TM3 (27)	T^{111}LGNRISAGLMTGGVAVFIGQPTEVVK^{137}
ML34 (35)	V^{138}RLQAQSHLHGIKPRYTGTYNAYRIIATTESFSTL^{172}
TM4 (28)	W^{173}KGTTPNLLRNVIINS*VELVTYDLMKGA^{200}
TM5 (30)	L^{207}ADDVPS*HLLSAFVAGFS*TTFLASPADVVK^{236}
ML56 (23)	T^{237}RFINSLPGQYPSVPSS*AMTMLT^{259}
TM6 (35)	K^{260}EGPTAFFKGFVPSFLRLASWNVIMFVS*FEQLKKE^{294}

[a] *Underlined sequences represent putative intramembrane segments.*
[b] *All Cys residues are replaced by Ser, with their positions marked with "*". Biological activity of UCP-1 is not affected by replacing Cys residues with Ser.*

S. Del Valle et al. (eds.), *Peptides for Youth: The Proceedings of the 20th American Peptide Symposium*,
DOI: 10.1007/978-0-387-73657-0_130, © Springer Science+Business Media, LLC 2009

Results and Discussion

Conformations of the peptide fragments of hamster UCP-1 were examined individually and as mixtures in aqueous, organic solvent (50% TFE), and phospholipid vesicles as models of mitochondrial inner membranes. Conformations of the individual peptide fragments in the aqueous environment were non-ordered, but all peptides and their mixtures exhibited different degrees of helical conformation in 50% TFE. Peptides in two unilamellar vesicle systems containing PC, PE and PG lipids in the absence and presence of cardiolipin had diverse conformations ranging from non-ordered to helical and β conformations. In order to compare the peptide mixtures with the native UCP-1 protein, we over-expressed mouse UCP-1 (92% homology to the hamster UCP-1) in *E. coli*. UCP-1 in digitonin, a membrane-mimicking detergent, exhibited a helical conformation with a maximum at 193 nm, a shoulder at ~208 nm and a broad minimum at ~220 nm (Fig. 1A). The typical ratio of the minima in the CD spectrum of an α-helix is close to unity. The CD spectrum of UCP-1 in Fig. 1A implies helix-helix interaction since the $\theta_{220}/\theta_{208}$ is larger than one (~1.75). This interaction between TM helices is consistent with the tight packing of helices in the structure of other mitochondrial transporters such as the ADP/ATP carrier protein. Interestingly, the CD spectra of the mixtures of six TM and nine TM and ML peptide segments of UCP-1 indicated common conformational features with the spectrum of native UCP-1 ($\theta_{220}/\theta_{208}$=1.75 for 6 TM peptides spectrum) (Fig. 1B), implying a tight folding of the peptides in membranes.

We have already shown that assemblies of the TM2 segments of UCP-2 and UCP-1 in phospholipid bilayers form stable multi-state anion channels, and inferred that these domains can be essential in the ion transport activity of UCPs [1]. To examine the common features between the conformation and ion-conducting properties of TM2 domains, TM2 peptides of the five human UCPs were synthesized. Conformations of TM2 peptides were dissimilar in model membranes. Further investigation of the ion-conducting properties of these peptides in our laboratory can lead to a better understanding of the functions of uncoupling proteins.

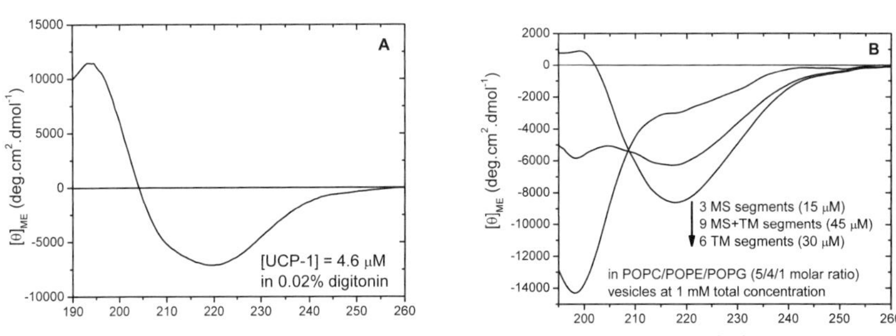

Fig. 1. CD spectra of UCP-1 (A) and the peptide segments (B) in the model membranes.

Acknowledgments

This work was funded by CFI and NSERC grants to M.J.-N. and M.D.S.

References

1. Yamaguchi, H., Jelokhani-Niaraki, M. and Kodama, H. *FEBS Letts.* **577**, 299-304 (2004).

Dynamics study on single and multiple β-sheets

Inta Liepina[1], Salvador Ventura[2], Cezary Czaplewski[3] and Adam Liwo[3]

[1]*Latvian Institute of Organic Synthesis, Riga, LV1006, Latvia;* [2]*Institut de Biotecnologia i de Biomedicina, Universitat Autonoma de Barcelona, E-08193 Bellaterra, Spain;* [3]*Faculty of Chemistry, University of Gdansk, 80-952 Gdansk, Poland*

Introduction

Short protein sequence stretches drive the protein amyloid fibril formations [1-2]. By a homology search, an aggregation-prone region in the Abl-SH3 domain of Drosophila was identified with sequence DLSFMKGE (MK), along with a less amyloidogenic human homologous region with sequence DLSFKKGE (KK).

The antiparallel flat β-sheets consisting of ten strands of MK and KK were constructed [3]. We created two multi-sheet systems: (1) six 10-strand β-sheets placed in parallel (10x6) of MK and KK (Fig.*1a* and *2a*). Each of these β-sheets systems was surrounded by a 10 Å layer of water molecules over the solute and subjected to molecular dynamics (MD) simulations with the Amber 8.0 force field in the NPT (constant number of molecules, pressure, and temperature) scheme. The MD simulations were started from the temperature of 10 K and the temperature was gradually risen by the step of the ten degrees till 300 K, then the simulations were run at the constant temperature of 300 K. The 10x6xMK system was simulated by 90 ns of MD, the 10x6xKK system was simulated by 40 ns of MD.

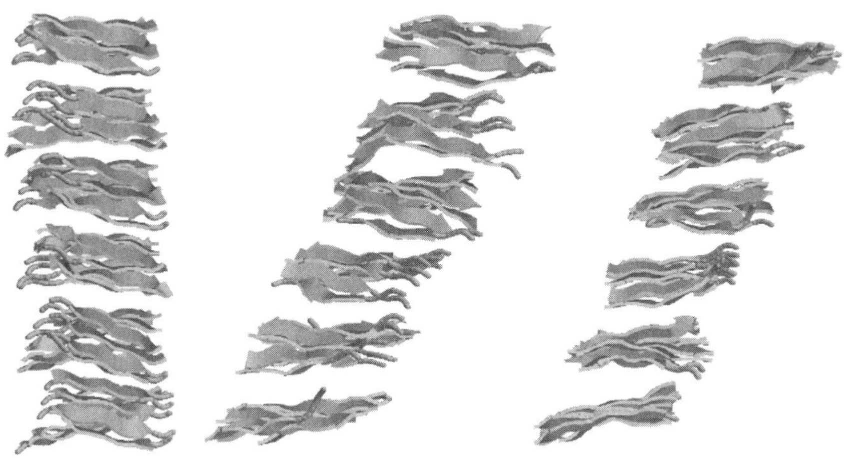

Fig. 1. Multisheet system of DLSFMKGE peptides (10x6xMK) keeps together all the time of MD simulation. The 10x6xMK system at a) 351 ps, b) 21705 ps, c) 74499 ps of MD.

Results and Discussion

The MD simulation of multisheet systems revealed that: a) The 10x6xMK β-sheet stack is stable, but the 10x6xKK β-sheet stack is not. b) The 10x6xMK β-sheet is stable because of hydrophobic interactions of methionine and phenylalanine side

S. Del Valle et al. (eds.), *Peptides for Youth: The Proceedings of the 20th American Peptide Symposium,*
DOI: 10.1007/978-0-387-73657-0_131, © Springer Science+Business Media, LLC 2009

chains and the leucine side chain of the neighboring sheets. Met, Phe, Leu make a hydrophobic core for the stack of β-sheets. c) During the MD run, the Met, Phe, and Leu residues of neighbouring sheets act as a conformational switch moving the β-sheets by two amino acid step towards each other. d) Replacement of Met by Lys destroys the hydrophobic core, which is the stability factor of the β-sheets stack. The 10x6xKK system maintains β-sheets, but loses interactions between β-sheets. e) The calclulacions of six β-sheets confirm the conclusion drawn for single sheet systems[3]: parallelly placed β-sheets stabilize each other.

Fig. 2. Multisheet system of DLSFKKGE peptides (10x6xKK) does not remain together during MD simulation. The 10x6xKK system at a) 382 ps, b) 11408 ps, c) 36771 ps of MD.

Acknowledgments

This work was supported by NATO Collaborative Linkage Grant LST.CLG.979807, by stipend for I.L. from the "Kasa im. Jozefa Mianowskiego" Fund, Poland, by Latvian Science Council Grant 05.1768, calculations are carried out at the Gdansk Academic Computer Center TASK.

References

1. Ventura, S., Zurdo, J., Narayanan, S., Parreno, M., Mangues, R., Reif, B., Chiti, F., Giannoni, E., Dobson, C.M., Aviles, F.X. and Serrano, L. Proc. Natl. Acad. Sci. U S A **101**, 7258-7263 (2004).
2. Ventura S., Lacroix E. and Serrano L. Journal of Molecular Biology **332**, 1147-1158 (2002).
3. Liepina I., Ventura S., Czaplewski C., Liwo A. Journal of Peptide Science **12**, 780-789 (2006).

Homotrimeric Collagen Peptides As Model Systems For Cell Adhesion Studies

Leslie Barth, Eva K. Sinner, Sergio A. Cadamuro, Christian Renner, Dieter Oesterhelt and Luis Moroder

Max-Planck-Institute of Biochemistry, D-82152 Martinsried, Germany

Introduction

Collagen model peptides with high triple-helix propensity can self-associate into the suprahelical structure, but folding/unfolding of such systems is strongly concentration-dependent. Cross-bridging of the three collagenous chains with synthetic templates as well as with native collagen-type or with artificial cystine knots has been successfully applied to overcome the shortcomings of self-assembled triple helices [1]. Aim of the present study was to analyze the conformational properties of a disulfide-crosslinked trimeric collagen model peptide (**2** in Figure 1) containing the specific collagen type I motif recognized by the α1β1 and α2β1 integrins, i.e. the hexapeptide sequence portion GFOGER of the α1 chain [2]. Embedding this adhesion epitope into the triple helical collagenous peptide **3** (Figure 1) allowed X-ray structural analysis of the complex with the α2-I domain, which provides within the α2β1 integrin the principal binding site for collagen [3].

$(H_2N\text{-S-GPOGPO-GFOGER-(GPO)}_3\text{-GP-CC-GGG-NH}_2)_3$ **(1)**

↓ O_2, 1 mM at 7-8 °C; yield: 100 %

$(H_2N\text{-S-GPOGPO-GFOGER-(GPO)}_3\text{-GP-CC-GGG-NH}_2)_3$ **(2)**

$Ac\text{-(GPO)}_2\text{-GFOGER-(GPO)}_3\text{-NH}_2$ **(3)**

*Fig. 1. Synthesis of the disulfide-crosslinked collagen-like homotrimer **2**; ^{15}N-labelled Gly residues are underlined; O = (2S,4R)-hydroxyproline.*

Results and Discussion

The adhesion motif was flanked by (GPO) triplets known to induce and stabilize most efficiently a triple helical fold and C-terminally by the cystine knot sequence of type III collagen (Figure 1). As a correct fold of the bis-cysteinyl precursor into a triple helical homotrimer is required for an efficient oxidative cystine knot formation [4], oxidation of the self-associated trimer **1** (CD: T_m = 17°C) was performed at 7-8°C. An almost quantitative conversion of **1** into the disulfide-bridged homotrimer **2** was achieved without detectable amounts of oxidized monomer and dimer as observed in preceding studies [4]. This unusual behaviour can possibly be attributed to interchain electrostatic interactions between the Glu and Arg residues, which would optimally align the chains in the triple helix for their correct rastering and cystine knot formation. As expected, the crosslinked trimer **2** shows a significantly increased T_m of 51°C (CD) and 54°C (DSC) in agreement with previous results [4,5]. The symmetrical endotherms would exclude a subdomain structuring effect of the imino acid-poor adhesion motif as observed for a collagenous trimer containing the type IV collagen sequences 457-468 as the non-contiguous integrin adhesion motif of this basal membrane collagen [6].

The X-ray structure of the α2-I domain complexed by the self-associated homotrimeric non-crosslinked peptide **3** [3] revealed strong deviations in the

S. Del Valle et al. (eds.), *Peptides for Youth: The Proceedings of the 20th American Peptide Symposium*, 295
DOI: 10.1007/978-0-387-73657-0_132, © Springer Science+Business Media, LLC 2009

main-chain geometrical parameters of the hexapeptide portion from the ideal triple helix with kinks and bends at the junctions between the three zones of the collagenous molecule. A similar non-ideal triple helix was found in the crystalline non-bonded peptide **3** despite the expectedly strong crystal packing effects [7]. The NOESY spectrum of **1** revealed intense signals in the ppm range indicative of hydrogen bonded amides (data not shown), however the ^{1}H-^{15}N-FHSQC spectra of both the self-associated peptide **1** and the crosslinked peptide **2** clearly indicated that Gly-2 (at the N-terminus of the molecule), Gly-11 and Gly-14 (in the adhesion motif) are not hydrogen bonded (Figure 2), a fact that is further supported by the large temperature shifts of these protons (-7.5 to -11.3 ppb/K). These findings would indicate a non-uniform helical fold with a rather loose structure of the central part of the molecule. Such effect of imino acid-poor native sequences was already observed for a disulfide-linked heterotrimeric peptide containing the sequences 772-785 (α1) and 772-784 (α2) of the collagenase cleavage site of typ I collagen [8] as well as for peptides containing the cell adhesion motif of collagen type IV [6].

Despite this loose structure of the adhesion motif, peptide **2** binds to the α1-I domain of α1β1 integrin with high affinity ($K_d = 1.45 \cdot 10^{-6}$ M). These results support the notion that binding of interacting proteins to the rod-shaped collagens are mediated by strongly breathing subdomains that adapt as ligands to the binding sites.

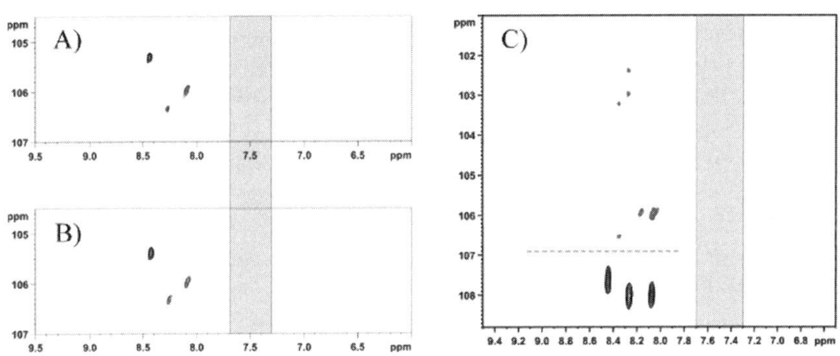

*Fig. 2. ^{1}H-^{15}N FHSQC spectra with the signals of ^{15}N-Gly-2, -Gly-11 and -Gly-14 of 1 mM **1** (A) and 30 μM **2** (B) in water at 4°C and (C) at 27°C; in (C) the Gly signals of **2** are above and of **1** below the line. The ppm range of hydrogen-bonded Gly are highlighted.*

Acknowledgments

The work was funded by the SFB 563 (grant C4) of the DFG.

References

1. Moroder, L. *et al. Biopolymers* **80**, 85-97 (2005).
2. Knight, C. G. *et al. J. Biol. Chem.* **275**, 35-40 (2000).
3. Emsley, J. *et al. Cell* **101**, 47-56 (2000).
4. Barth, D. *et al. Chem. Eur. J.* **9**, 37033-3714 (2003).
5. Boudko, S. P. and Engel, J. *J. Mol. Biol.* **335**, 1289-1297 (2005).
6. Saccà, B., Renner, C. and Moroder, L. *J. Mol. Biol.* **324**, 309-318 (2002).
7. Emsley, J. *et al. J. Mol. Biol.* **335**, 1019-1028 (2004).
8. Fiori, S., Saccà, B. and Moroder, L. *J. Mol. Biol.* **319**, 1235-1242 (2002).

Cyclic Peptide Analogs of Laminin Active Sequences Enhance the Biological Activity

Shu Takaki[1], Kozue Kato-Takagaki[1], Nobuharu Suzuki[1], Shinya Oishi[2], Yamato Kikkawa[1] and Motoyoshi Nomizu[1]

[1]*School of Pharmacy, Tokyo University of Pharmacy and Life Sciences;*
[2]*Graduate School of Pharmaceutical Sciences, Kyoto University*

Introduction

The laminin α3 chain is mainly expressed in the skin and in the floor plate of the developing neural tube, and contains a unique C-terminal globular domain, called G-domain. The G-domain consists of five tandem laminin G-like modules (LG1-5), which play crucial roles in the cell-laminin α3 chain interactions [1]. Previously, the LG4 module was found to be a major active module for cell adhesion, and heparin binding in the laminin α3 chain G domain [2-4]. We screened the cell and heparin binding sites on the LG4 module using 22 synthetic peptides. A3G756 (KNSFMALYLSKGRLVFALG, human laminin α3 chain 1411-1429) was found to be an active sequence on the loop region of the LG4 module [5, 6].

In this study, we examined the structural requirement of the loop region, A3G756 sequence, for the biological activity.

Results and Discussion

We designed and synthesized three overlapping linear peptides covering the A3G756 sequence (hEF3A: LYLSKGRLVFA, hEF3B: NSFXALYLSKGR (X:Nle), and hEF3C: KGRLVFALGTDG). The three peptides contained the KGR sequence, previously identified as an active core, in the middle or at the N- or C- terminus. Additionally, we synthesized their cyclic peptides using a similar protocol [7, 8] with an S-S bridge of two cysteins at the N- and C-termini (cyclo-hEF3A: CLYLSKGRLVFAC, cyclo-hEF3B: CNSFXALYLSKGRC, and cyclo-hEF3C: CKGRLVFALGTDGC), respectively [7, 8]. We also prepared a recombinant human laminin α3LG4 protein (rec-hα3LG4), which showed heparin binding and cell attachment activities, and promoted neurite outgrowth as shown previously [2-4].

First, we examined the inhibitory effect of the linear and cyclic A3G756 peptides on the interaction between rec-hα3LG4 and heparin-Sepharose beads. The inhibitory effect of *cyclo*-hEF3A was stronger than that of the linear hEF3A peptide. In contrast, *cyclo*-hEF3B and *cyclo*-hEF3C weakly inhibited rec-hα3LG4 heparin binding, and the inhibitory effect was weaker than that of the corresponding linear peptides. These results suggested that *cyclo*-hEF3A possess a loop structure similar to that of the connecting loop regions between the E and F strands in the α3 chain LG4 modules.

We focused on the *cyclo*-hEF3A peptide and evaluated further biological activities, including cell attachment and neurite outgrowth. Next, we evaluated the inhibitory effect of hEF3A and *cyclo*-hEF3A on cell attachment to rec-hα3LG4 using human fibroblast and PC12 rat pheochromaocytoma cells. The inhibitory effect of the *cyclo*-hEF3A peptide on fibroblast and PC12 cell attachment to rec-hα3LG4 was stronger than that of the linear peptide.

Further, we tested the neurite outgrowth activity of the hEF3A and cyclo-hEF3A. The *cyclo*-hEF3A promoted a stronger neurite outgrowth activity compared with

S. Del Valle et al. (eds.), *Peptides for Youth: The Proceedings of the 20th American Peptide Symposium*, 297
DOI: 10.1007/978-0-387-73657-0_133, © Springer Science+Business Media, LLC 2009

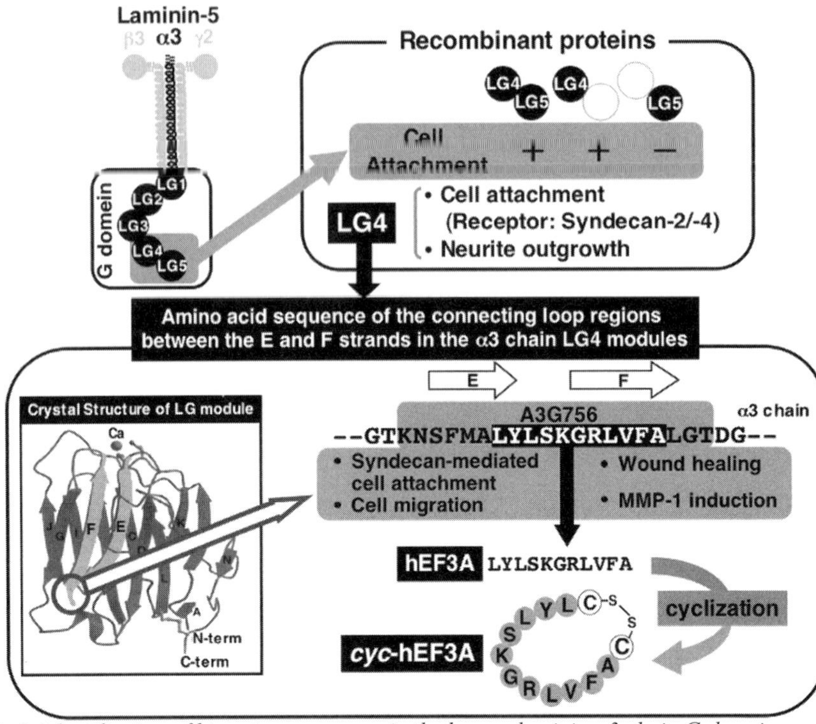

Fig 1. Identification of bioactive sequences in the human laminin α3 chain G domein.

hEF3A. These results suggest that the activity of the A3G756 sequence is conformation-dependent and the loop structure of hEF3A is important for the activity.

Here, we demonstrated that the loop structure of the A3G756 sequence is important for the biological activity of the laminin α3 chain LG4 module. A3G756 was previously found to promote syndecan mediated cell attachment and migration, and wound healing. Taken together, the *cyclo*-hEF3A peptide has potential to use as a therapeutic application for wound healing with nerve regeneration.

References

1. Suzuki, N., Yokoyama, F., and Nomizu, M. (2005) *Connect. Tissue Res.*, **46**, 142-152.
2. Utani, A., Nomizu, M., Matsuura, H., Kato, K., Kobayashi, T., Takeda, U., Aota, S., Nielsen, P. K., and Shinkai, H. (2001) *J. Biol. Chem.*, **276**, 28779-28788.
3. Kato, K., Utani, A., Suzuki, N., Mochizuki, M., Yamada, M., Nishi, N., Matsuura, H., Shinkai, H., and Nomizu, M. (2002) *Biochemistry*, **41**, 10747-10753.
4. Utani, A., Momota, Y., Endo, H., Kasuya, Y., Beck, K., Suzuki, N., Nomizu, M., and Shinkai, H. (2003) *J. Biol. Chem.*, **278**, 34483-34490.
5. Hohenester, E., Tisi, D., Talts, J. F., and Timpl, R. (1999) *Mol. Cell.*, **4**, 783-792.
6. Timpl, R., Tisi, D., Talts, J. F., Andac, Z., Sasaki, T., and Hohenester, E. (2000) *Matrix Biol.*, **19**, 309-317.
7. Suzuki, N., Nakatsuka, H., Mochizuki, M., Nishi, N., Kadoya, Y., Utani, A., Oishi, S., Fujii, N., Kleinman, H. K., and Nomizu, M. (2003) *J. Biol. Chem.*, **278**, 45697-45705.
8. Yokoyama, F., Suzuki, N., Haruki, M., Nishi, N., Oishi, S., Fujii, N., Utani, A., Kleinman, H. K., and Nomizu, M. (2004) *Biochemistry*, **43**, 13590-13597.

Hydrophobic Peptide Segments in Soluble Proteins Competent For Membrane Insertion: Role in Amyloidogenesis

Fiona Cunningham[1,2], Arianna Rath[1] and Charles M Deber [1,2]

[1] *Division of Molecular Structure & Function, Research Institute, Hospital for Sick Children, Toronto M5G 1X8; and* [2] *Department of Biochemistry, University of Toronto, Toronto M5S 1A8, Ontario, Canada.*

Introduction

Amyloid diseases such as Alzheimer's have a significant impact on human health. While the proteins implicated in these diseases vary in structure, function and cellular location, they share a common pathological mechanism characterized by the formation of amyloid fibrils with a similar morphology [1]. Fibril formation is initiated when a soluble protein misfolds from its native conformation; usually from an α-helix to a β-strand. As well, it has been suggested that regions in soluble proteins prone to amyloidogenesis may have unstable native secondary structure due to a large number of hydrophobic residues with aqueous β-strand propensity [2, 3].

It has been shown that amyloidogenic segments such as the Aβ(1-40) peptide can exhibit characteristics similar to TM segments such as insertion and adoption of α-helical structures in membrane mimetic environments [4]. Because fibrils can be formed from almost any protein, recognition of the factors influencing protein conformation changes and misfolding is essential. Our laboratory has noted that amyloidogenic segments are similar to transmembrane (TM) segments of membrane proteins in that they are enriched in hydrophobic amino acids with aqueous β-strand propensities, but which nevertheless form α-helical structures in an apolar environment [5]. This observation implies that regions of local structural instability in soluble proteins may be responsible for amyloid formation and that programs designed to predict potential TM segments may be useful in identifying such amyloidogenic regions. Application of the TM predicting program *TM Finder* [6] to a database of soluble helices revealed that 22% of soluble segments met requirements for membrane insertion [5]. We hypothesize that these segments – which we have termed "δ-helices" - may correspond to amyloidogenic regions of soluble and/or TM proteins, and that δ-helices and amyloidogenic sequences might share similar biophysical properties.

In order to elucidate the relationship between amyloidogenic and TM segments, δ-helix peptides were synthesized (Table 1). A combination of fluorescence spectroscopy and circular dichroism was used to assess their similarity to TM segments and amyloidogenic sequences in membrane mimetic environments.

Table 1. Peptides Synthesized from Soluble Protein Sequences with δ-helix Properties.

PDB ID	Sequence	Protein
1ECA	**K**-FAGAEAAWGATLDTFFGMIF-**KK**	Erythrocruorin
1MBA	**K**-ADAAWTKLFGLIIDALKAA-**K**	Myoglobin
1HCI	**KK**-ELFFWVHHQLTARFDFERL-**K**	α-actinin, Rod domain
2MNR	**K**-GLIRMAAAGIDMAAWDALGKV-**K**	Mandelate racemase
5LDH	**KK**-GYTNWAIGLSVADLIESMLK	L-lactate dehydrogenase

S. Del Valle et al. (eds.), *Peptides for Youth: The Proceedings of the 20th American Peptide Symposium*, 299
DOI: 10.1007/978-0-387-73657-0_134, © Springer Science+Business Media, LLC 2009

Results and Discussion

Synthetic peptides were synthesized via Fmoc chemistry on a PS3 Protein Technologies peptide synthesizer with PAL-PEG-PS resin to produce an amidated C-terminus. Lyophilized peptides were dissolved in ultrapure water, and purified via C_4 RP-HPLC on an acetonitrile gradient, 0.1% TFA. Lys tags were added to each δ-helix segment to facilitate peptide purification, and render peptides water-soluble.

Although these segments are helical inside their corresponding native proteins, circular dichroism spectroscopy of the δ-helix peptides in aqueous buffer indicated little or irregular secondary structure (Fig. 1). In a membrane mimetic environment such as SDS, these peptides adopt α-helical structure (Fig. 1). Secondary structure is likely due to insertion of the peptides into the SDS micelles. Inclusion into the membrane mimetic environment was confirmed via Trp fluorescence (Fig. 1 – insets) where the observed blue shift in the fluorescence of Trp is indicative of insertion into an apolar environment.

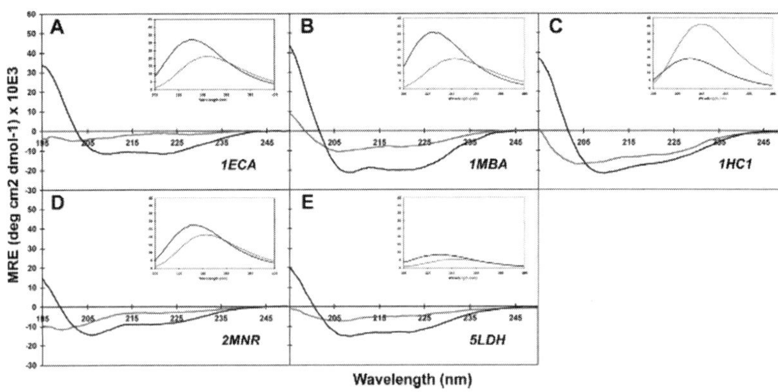

Fig. 1. A-E: Circular dichroism spectra and tryptophan emission fluorescence spectra (inset) of each indicated δ-helix peptide in aqueous buffer (grey curves) and in buffer with SDS (black curves). Values on the y-axis of the inset graphs are arbitrary fluorescence units.

The observation that sequences in soluble proteins resemble TM segments in their hydrophobicity and structural preferences may be useful in delineating regions prone to amyloidogenesis *in vivo*, with δ-helix segments representing sequences on the edge of aggregation.

Acknowledgments

Funded, in part, by a grant to CMD from the Canadian Institutes of Health Research (CIHR). FC and AR hold awards from the CIHR Strategic Training Programs.

References

1. Dobson, C. M. *Nature.* **426**, 884-90 (2003)
2. Kallberg, Y., *et al. J Biol Chem.* **276**, 12945-50 (2001)
3. Yoon, S. and Welsh, W. J. *Protein Sci.* **13**, 2149-60 (2004)
4. Bokvist, M., *et al. J Mol Biol.* **335**, 1039-49 (2004)
5. Wang, C., Liu, L. P., Deber, C.M. *Proc. Am. Pept. Symp.* **19**, 367-369 (2000)
6. Deber, C. M., *et al. Protein Sci.* **10**, 212-9 (2001)

Construction and Characterization of Melanocortin-2 and -4 Receptor Chimeras

Krista R. Wilson[1], Bettina Proneth[1], Hendrik Luesch[1], Sally A. Litherland[2] and Carrie Haskell-Luevano[1]

[1]*Department of Medicinal Chemistry;* [2]*Department of Pathology, University of Florida, Gainesville, FL 32610, USA*

Introduction

The melanocortin system consists of five receptors, MC1-5, which are members of the superfamily of G protein coupled receptors (GPCRs). The MC1R and the MC3-5R are stimulated by the endogenous agonists α-MSH (Melanocyte Stimulating Hormone), β-MSH, γ_1-MSH, γ_2-MSH, and ACTH (Adrenocorticotropic Hormone), and activate the cAMP second messenger signaling pathway. The MC2R is unique in that it is only stimulated by ACTH [1]. ACTH and α-MSH only differ in that α-MSH represents the first thirteen amino acids of ACTH. This study will focus on the MC2R and the molecular mechanisms involved in its selective binding to ACTH over α-MSH. As the hMC2R plays a role in the regulation of cortisol production, this study may lead to a better understanding of the relationship between the MC2R, ACTH, and hypertension leading to heart disease and stroke.

It has been shown that cell lines of adrenal origin are the only cell types able to express a functional MC2R. Cell lines of non-adrenal origin (such as HEK293 cell lines) can be stably transfected with the MC2 receptor, but there is no functional response when stimulated with ACTH [2, 3]. Until this time, it was assumed that the reason the MC2R did not produce a functional response when expressed in HEK293 cells was due to the fact that the receptor was not expressed on the cell surface. This study demonstrates that this may not be the case and the hMC2R is expressed on the surface of HEK293 cells as determined by FACS.

Amino acid alignment of the hMC2R with the hMC4R has identified key amino acid changes in the transmembrane (TM) regions which may be involved in ligand binding and/or signal transduction. To test the hypothesis that these amino acids are involved in the production of a functional hMC2 receptor in HEK293 cells, chimeric receptors were made to identify the role each transmembrane domain plays in ligand binding and signal transduction in the hMC2R (Figure 1).

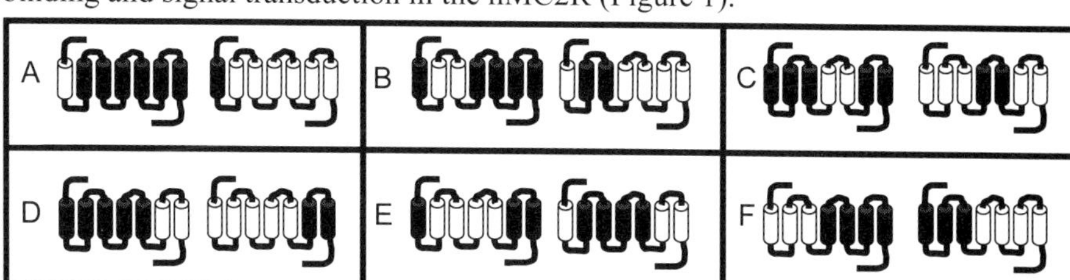

Fig. 1. Chimeras of the hMC2R and hMC4R constructed in this study. White = from hMC2R; Black = from hMC4R.

Results and Discussion

FACS analysis of the melanocortin chimeras has revealed that both the hMC2R and hMC4R wild type receptor as well as all chimeras, are expressed on the cell surface

S. Del Valle et al. (eds.), *Peptides for Youth: The Proceedings of the 20th American Peptide Symposium*, 301
DOI: 10.1007/978-0-387-73657-0_135, © Springer Science+Business Media, LLC 2009

of both HEK and OS3 cells (Figure 2). Some chimeras, such as 2A, 2C, 4C, and 4F, have significantly more surface expression of the receptor than the wild type receptor in HEK cells. Chimera 2A is also more highly expressed on the cell surface of OS3 cells than the wild type hMC2R. Functional analysis of the chimeras expressed in HEK cells by luciferase assay (Figure 3) shows that most chimeras do not exhibit any functional activity. Chimera 4A is stimulated by both ligands, but may only have 50% maximal activity when compared to the forskolin control, and chimera 4C is 10-fold less potent when stimulated with both ACTH and α-MSH than the wild type hMC2R. The relative ratios of EC_{50} values of ACTH:α-MSH remains the same (2:1), however, indicating that these TM domains may not be involved in selectivity between the ligands, and may instead be involved in ligand binding or signaling. Functional analysis of the chimeras expressed in OS3 cells and binding studies in both cell types will help reveal more information about the importance of each TM region in ligand selectivity. These studies represent a major shift in the understanding of MC2R expression, and may lead to a better understanding of the role of ACTH and the hMC2R in human heart disease and stroke.

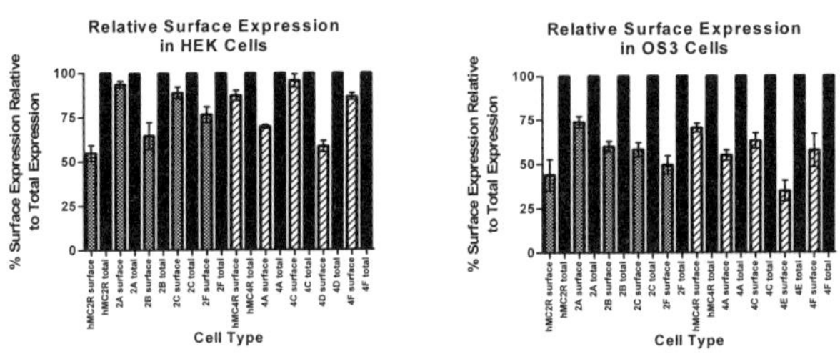

Fig. 2. FACS data showing relative surface expression of each chimera in both HEK and OS3 cell lines.

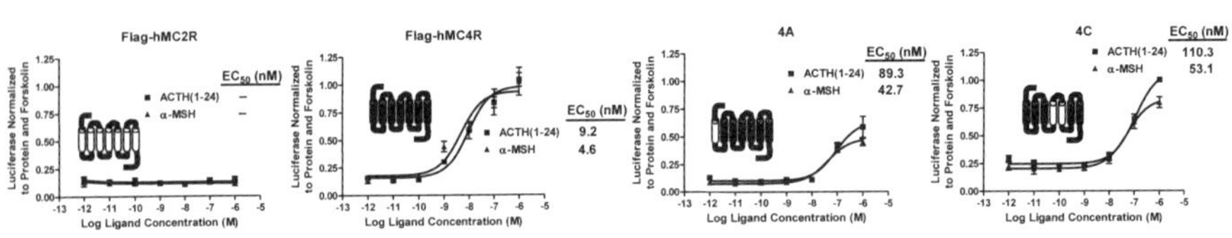

Fig. 3. Functional activity of chimeras in HEK cells. All hMC2R chimeras as well as the hMC4R chimeras 4D and 4F did not show any functional activity.

Acknowledgments

Krista R. Wilson is the recipient of a University of Florida Alumni Fellowship. This work was supported by NIH R01DK057080.

References

1. Schioth, H. B., Chhajlani, V., Muceniece, R., Klusa, V. and Wikberg, J. E. *Life Sci.* **59**, 797-801 (1996).
2. Noon, L. A., Franklin, J. M., King, P. J., Goulding, N. J., Hunyady, L. and Clark, A. J. *J Endocrinol* **174**, 17-25 (2002).
3. Penhoat, A., Naville, D., El Mourabit, H., Buronfosse, A., Durand, P. and Begeot, M. *Endocr Res* **26**, 549-557 (2000).

Expression of Double Transmembrane Domain GPCR Fragments for Biophysical Analysis

Leah S. Cohen[1], Boris Arshava[1], Racha Estephan[1], Jacqueline Englander[1], Melinda Hauser[2], Jeffrey M. Becker[2] and Fred Naider[1]

[1]Department of Chemistry and the Macromolecular Assemblies Instítiute, The College of Staten Island (City University of New York), Staten Island, NY 10314; [2]Department of Microbiology, University of Tennessee, Knoxville, TN 37996

Introduction

There are thousands of protein structures in the databases that have been determined using X-ray crystallography and nuclear magnetic resonance spectroscopy. Despite the prevalence of integral membrane proteins, or IMPs (20-30% of the human genome encodes IMPs) few structures of these proteins have been determined, and only one G protein-coupled receptors (GPCRs) structure, that of rhodopsin, has been solved by X-ray crystallography [1]. Based on biochemical and biophysical analyses, GPCRs have been shown to consist of an N-terminal extracellular region, seven transmembrane (TM) helices separated by intra- and extra-cellular loops, and a C-terminal cytoplasmic tail. One of the factors that limits the number of high resolution structures is the low level of expression for a full-length GPCR. A few groups have started to study fragments of GPCRs to help to elucidate their structures [2-3].

We have recently published high resolution structures of a 73-residue peptide corresponding to the third extracellular loop (EL3), the seventh transmembrane (TM) domain and 40 residues of the C-terminal tail (CT40) of the yeast α-factor mating receptor Ste2p in organic aqueous media [4], and completed an NMR analysis of the peptide in DPC micelles [5]. In an attempt to study larger fragments of Ste2p, we have cloned and expressed two double TM domain fragments, TM1-TM2 [Ste2p(G31-T110, M54L, C59S, M69V, M71I)] and TM6-TM7-CT40 [Ste2p(R231-S339, M250A, C252S, M294A)], consisting of 80- and 109-residues, respectively. These proteins were expressed as fusion proteins (FP) attached to an N-terminal TrpΔLE carrier protein, cleaved with CNBr and purified to >95% purity as observed by RP-HPLC and MS. [^{15}N,^{1}H]-HSQC analysis was performed to determine the probability of successful structural determination [6].

Results and Discussion

Plasmids for the expression of TM1-TM2, pLC01, and TM6-TM7-CT40, pREJ04, were generated by PCR amplification, ligation and mutagenesis to remove all Met residues and to facilitate CNBr cleavage. The amplified regions were cloned into a vector containing a TrpΔLE sequence just upstream of the insert. After expression in *E. coli* strains BL21(DE3)pLysS, BL21-AI and BL21Star(DE3)pLysS it was determined to use the BL21-AI strain due to its higher levels of expression and its ease of use. Growth optimization was performed in both rich (Luria Broth), and minimal (M9) media. Good expression was observed for TM1-TM2-FP in BL21-AI (Figure 1A, inset), whereas the expression of TM6-TM7-CT40 was lower, but still improved over that of other two strains (data not shown). Both fusion proteins were biosynthesized in labeled M9 such that uniformly [^{15}N]- and/or [^{15}N,^{13}C]-labeled or selectively [^{15}N-amino acid]-labeled fusion proteins were generated.

S. Del Valle et al. (eds.), *Peptides for Youth: The Proceedings of the 20th American Peptide Symposium*, 303
DOI: 10.1007/978-0-387-73657-0_136, © Springer Science+Business Media, LLC 2009

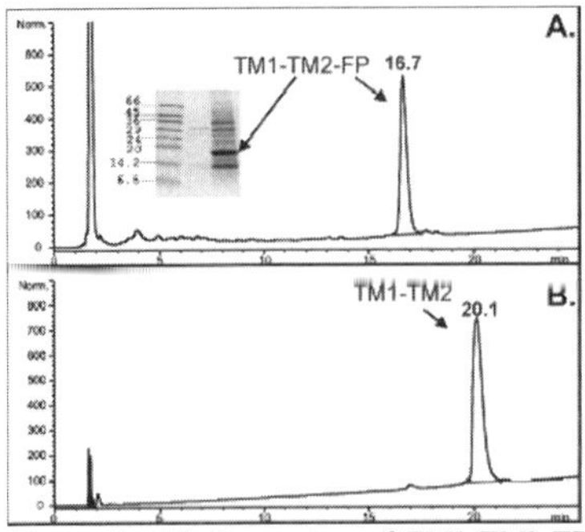

Figure 1. (A) Expression of TM1-TM2-FP observed by RP-HPLC and SDS-PAGE. (B) Purified TM1-TM2 observed by RP-HPLC.

After expression, the cells were lysed and the insoluble fraction (inclusion bodies) was separated. These inclusion bodies were solubilized in 70% TFA with sonication and treated with CNBr to release the desired peptide. Purification was accomplished on an RP-HPLC Zorbax 300SB-C3 column using an acetonitrile gradient with 10% isopropanol at 60°C (Fig. 1). TM1-TM2 was purified with a 50-90% gradient and the peptide yield was 8-10 mg for both unlabeled and isotopically labeled peptide. TM6-TM7-CT40 was purified with a 40-80% gradient and the yield was approximately 3-6 mg/L. Molecular weight of the purified peptide and percent incorporation of the isotopic labels (93-97%) was determined by ESI-MS.

CD analysis was used to determine the helical content of the peptides. The spectra showed a maximum at 192 nm and two minima at 208 nm and 222 nm. For TM1-TM2 in organic aqueous media the molar ellipticity of the 222 nm band was -28000 deg-cm^{2}decimole^{-1} indicative of a highly helical peptide. High helicity was maintained in SDS and the lysolipid 1-palmitoyl-2-hydroxy-sn-glycero-3-[phospho-RAC-(1-glycerol)] (LPPG). Due to the random structure of the 40 residue C-terminal tail, the overall helicity of TM6-TM7-CT40 was less in SDS and LPPG as seen by a decrease in the molar ellipticity at 222 nm, though the two characteristic minima were still observed. In contrast, in DPC micelles both TM1-TM2 and TM6-TM7-CT40 had very little helical tendency. [^{15}N,^{1}H]-HSQC of these peptides in SDS and LPPG showed well-dispersed chemical shifts and >90% of the expected peaks were observed in all cases. These characteristics indicate that TM1-TM2 and TM6-TM7-CT40 are probably forming unique structures in each of the micellar environments examined. Our results indicate that TM1-TM2 and TM6-TM7-CT40 of Ste2p are strong candidates for further structural analysis that could lead to the step-by-step determination of a GPCR structure.

Acknowledgments

This work was supported by grants NIH GM-22086 and NIH GM-22087 to FN and JMB.

References

1. Palczewski, K., Kumasaka, T., Hori, T., Behnke, C. A., Motoshima, H., Fox, B. A.; Le Trong, I., Teller, D. C., Okada, T., Stenkamp, R. E., Yamamoto, M., and Miyano, M. *Science* **289**, 739-745 (2000).
2. Zheng, H., Zhao, J., Sheng, W., Xie, X. Q. *Biopolymers* **83**, 46-61 (2006).
3. Yeagle, P. L. and Albert, A. D. *Biochimica et Biophysica Acta* **1768**, 808-824 (2007).
4. Estephan, R., Englander, J., Arshava, B., Samples, K. L., Becker, J. M., and Naider, F. *Biochemistry* **44**, 11795-11810 (2005).
5. Neumoin, A., Arshava, B., Becker, J., Zerbe, O., and Naider, F. *Biophys J* (2007).
6. Page, R. C., Moore, J. D., Nguyen, H. B., Sharma, M., Chase, R., Gao, F. P., Mobley, C. K., Sanders, C. R., Ma, L., Sonnichsen, F. D., Lee, S., Howell, S. C., Opella, S. J., and Cross, T. A. *J Struct Funct Genomics* **7**, 51-64 (2006).

Peptide and Non-Peptide Mimetics Utilize Different Pathways for Signal Transduction

Victor J. Hruby[1,2], Minying Cai [1], Matt Dedek [2], Hongchang Qu [1], Erin Palmer [1], Alexander Mayorov [1], Dev. Trivedi [1], George Tsaprailis[3] and Yingqui Yang [4]

[1]Department of Chemistry, [2]Department of Pharmacology, University of Arizona, Tucson, Arizona 85721; [3]Center for Toxicology, University of Arizona, Tucson, Arizona 85724; [4]University of Alabama, Birmingham, Alabama 35294.

Introduction

Despite the central importance of peptides as drugs such as insulin, oxytocin, calcitonin, ace inhibitors, etc., there has been a prejudice against developing peptide drugs. Instead large efforts are made to develop non-peptide small molecules ("peptide mimetics") to replace peptides. But are they true mimetics (same SAR, same bioactivity, same toxicity profile, etc.)? This issue generally has not been addressed directly. This is especially the case when the targets for the ligands are GPCRs. When the ligands are antagonists (for receptors, enzymes, etc.) this issue is not as critical as when the ligand is an agonist for a GPCR (or growth factors, or cytokines, etc.). The issue is highly relevant because biological function will depend on activation of a particular signaling pathway, and to obtain the same bioactivity will **require** the same SARs. To examine this conundrum we have begun to carefully evaluate whether peptide and non-peptide mimetics utilize the same signaling pathways. We find that often they do not, which raises questions regarding drug design. Here we examine the signaling of two agonist (a peptide and a non-peptide) ligands for the human melanocortin 4 receptor (hMC4R) utilizing molecular biological, pharmacological, biophysical, and biochemical methods and demonstrate that though both are agonists they utilize quite different signaling pathways. The implications for drug design are briefly discussed.

Results and Discussion

Utilizing a new biophysical method, plasmon waveguide resonance (PWR) spectroscopy we had shown that when a peptide and a non-peptide agonist for the human delta opioid receptor binds to the receptor a different structure of the complex was obtained [1], and furthermore that a different $G_{\alpha i}$ protein was preferred [2], suggesting that different signaling pathways would result. In our current approach we have sought direct structural evidence for differences in signaling pathways utilizing the hMC4R. For this purpose we have utilized five major mutants at the C-terminal intracellular domain (Figure 1) and four major mutants at the 3rd intracellular domain of the hMC4R [3], all introduced into HEK 293 cells, as was the wild type receptor, to study their ability to act as a receptor for agonists, and to examine post translational modifications (phosphorylation) on ligand interaction. As shown in Figure 1, when the receptor was C-terminal truncated by 13 amino acids to leave only the remaining 12 amino acid residues (C12), full binding of the peptide agonist was observed which diminished as further truncation occurred [3]. However, when the same experiment was run with the non-peptide hMC4R agonist THIQ

S. Del Valle et al. (eds.), *Peptides for Youth: The Proceedings of the 20th American Peptide Symposium*, 305
DOI: 10.1007/978-0-387-73657-0_137, © Springer Science+Business Media, LLC 2009

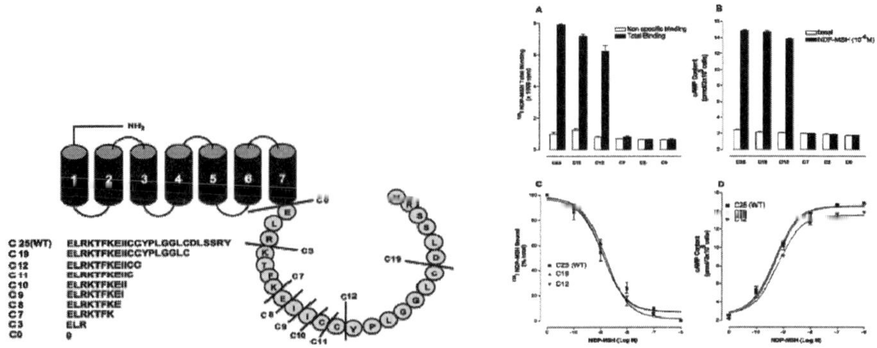

Fig. 1. Binding (A, B) and cAMP (C, D) assays of MTII at hMC4R C-terminal truncation and substitutions (3).

(*N*-[(3*R*)-1,2,3,4-tetrahydroisoquinolinium-3-ylcarbonyl]-(1*R*)-1-(4-chlorobenzyl)-2-[4-cyclohexyl-4-(1*H*-1,2,4-triazol-1-ylmethyl)piperidin-1-yl]-2-oxoethylamine) [4] no significant change in binding affinity occurs even for the completely C-terminal truncated receptor (data not shown). These results show that the peptide and non-peptide ligands do not interact with receptor in the same way and suggest that post translation modification of the receptor will be different. The 3rd intracellular loop of the hMC4R which contains 28 amino acids was modified by replacing heptapeptide units with polyalanine sequences in the loop, and examining the binding affinity and second message of cAMP activity of the peptide and small molecule agonists. These changes do not strongly affect the peptide agonist. On the other hand, all such modifications of the 3rd intracellular loop of the hMC4R led to changes in the binding affinity of the non-peptide agonist THIQ (data not shown).

To further examine the differences in signaling of peptide and non-peptide agonists for the hMC4R we have utilized biophysical methods of cell imaging. Here we used two further tools we previously developed: 1) rhodamine labeled MT-II [5] and THIQ; and 2) green fluorescent protein labeled β-arrestin now present in cells containing the C-terminal and 3rd loop modified hMC4Rs. When the C-terminal modified hMC4R-containing cells were treated with MT-II, only the C12 or longer C-terminal containing receptors led to signaling. On the other hand, when THIQ was used as the agonist there was no change in the effect on β-arrestin in any case.

Finally using state of the art isolation and purification methods to obtain quantities of the hMC4R, we have examined the sites of phosphorylation of the hMC4R following treatment with the peptide and non-peptide ligands. The method chosen for analysis was MALDI-TOF mass spectrometry on hMC4R-derived peptides from chymotrypsin digests of the receptor. The phosphorylate peptides obtained were determined by matching the observed mass spectra with the known fragment sequences of the hMC4R. In Figure 2 we show the phosphorylation sites for the wild type hMC4R treated respectively with MT-II (dark) and THIQ (white). As can be seen, the MT-II treated receptor is phosphorylated primarily at the C-terminal whereas the non-peptide THIQ-treated receptor is phosphorylated primarily at the 2nd and 3rd intracellular loops. Clearly peptide and non-peptide agonist ligands for the hMC4R lead to different ligand-receptor structure and signaling pathways.

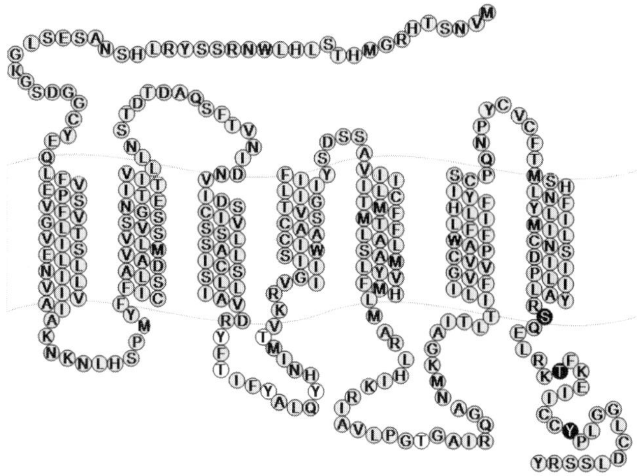

Fig. 2. Structure of the human melanocortin 4 receptor showing the sites of phosphorylation following agonist peptide MTII (dark) and agonist non-peptide THIQ (white) treatment of wild type receptor.

Conclusions

As demonstrated in these experiments, peptide ligands for GPCRs where the natural neurotransmitter is a peptide, interact with the receptor differently than a non-peptide (peptide mimetic) leading to a different ligand-receptor complex. This in turn leads to different receptor signaling. The precise effects of these differences in down stream signaling remain to be determined. Nonetheless, a central dogma of chemical biology is that a change in structure means a change in function, so it is reasonable to suggest that though the non-peptide THIQ interacts with the hMC4R to transduce a message, the message it transduces is different than that of the peptide. These observations have important implications for drug design. Since the target for most drug designs is a peptide or a peptide substructure of a protein, careful attention needs to be paid as to whether so-called peptide mimetics are actual mimetics. If not, one can do a lot more SAR, etc. on non-peptides, or one could develop a peptide which at least for hormones, growth factors, neurotransmitters, cytokins, etc. may be a wiser choice.

Acknowledgements

Supported by grants from the U.S. Public Health Service, NIDDK and NIDA. We thank Prof. Robert Lefkowitz for the β-arrestin cDNA.

References

1. Alves, I. D., Hruby, V. J., et al. *Mol. Pharmacol.* **65**, 1248-1257 (2004).
2. Alves, I. D., Tollin, G., et al. *J. Biol. Chem.* **279**, 44673-44682 (2004).
3. Yang, Y., Chen, M., Harmon, C.M, et al. *Biochemistry.* **44**, 6971-6979 (2005).
4. Sebhat, I.K., Martin, M.I., Nargund, R.P., et al. *J. Med. Chem.* **45**, 4589-4596 (2002).
5. Cai, M., Hruby, V. J., et al. *J. Am. Chem. Soc.* **126**, 7160-7161 (2004).

Structural Studies on Large Fragments of G Protein Coupled Receptors

Fred Naider[1], Alexey Neumoin[2], Boris Arshava[1], Leah S. Cohen[1], Jeffrey M. Becker[3] and Oliver Zerbe[2]

[1]*Department of Chemistry and the Macromolecular Assemblies Institute, The College of Staten Island of the City University of New York (CUNY), Staten Island, NY 10314,* [2]*Institute of Organic Chemistry, University of Zurich, Switzerland,* [3]*Department of Microbiology, University of Tennessee, Knoxville, TN 37996*

Introduction

The use of peptide fragments to study the biophysical and structural properties of regions of polytopic membrane proteins is widespread [1,2]. In most cases these studies have been limited to synthetic peptides containing 20-40 amino acid residues. Moreover, although some high resolution structures of fragments corresponding to single transmembrane regions of membrane proteins in organic-aqueous membrane mimetic solvents have appeared, there are few such structures from studies in detergent micelles. Although important information is obtained from studies on membrane peptides in media such as trifluoroethanol-water and dimethylsulfoxide, it is highly desirable to compare such structures to those obtained in micelles and bicelles. Furthermore, since long range interhelical interactions can influence the structures of individual transmembrane domains, investigations on peptides containing multiple transmembrane regions of integral membrane proteins are necessary.

We have been studying structural characteristics of regions of the α-factor receptor, Ste2p a G protein coupled receptor from the yeast *Saccharomyces cerevisiae* [3]. Previously, we presented structures of peptides with 30 to 73 residues corresponding to each of the seven putative transmembrane domains of Ste2p [4,5]. Recently, we succeeded in expressing isotopically labeled peptides of Ste2p containing two transmembrane domains connected by the contiguous intracellular loop. One of these peptides TM1-TM2 Ste2p(G31-T130) was found to form stable solutions at 0.2 mM concentration in a variety of lipids including SDS and 1-palmitoyl-2-hydroxy-sn-glycero-3-[phospho-RAC-(1-glycerol)] (LPPG) Here we report initial results from a detailed NMR analysis of TM1-TM2 in LPPG.

Results and Discussion

High-resolution NMR experiments on TM1-TM2 were performed using 0.2-0.4mM and 200mM solutions of ^{15}N-, ^{15}N/^{13}C- or ^{15}N/^{2}H-labeled TM1TM2 peptides and LPPG, respectively, in 20mM Na-phosphate buffer (pH = 6.4). The NMR sample was prepared by mixing the peptide and detergent in buffer and sonicating the resulting solution at 40°C. Using this procedure, we were able to obtain reproducible [^{15}N, ^{1}H]-HSQC spectra of good quality. Samples prepared this way were sufficiently stable for measurement of NMR spectra at 320K over a period of 1-2 weeks.

A set of the most sensitive triple-resonance spectra, HNCA, HNCOCA, HNCO and HNCACO were recorded on uniformly labelled [^{15}N, ^{13}C]TM1-TM2. These experiments sequentially correlate Cα and CO resonances with the HN resonance. In the [^{15}N, ^{1}H]-HSQC spectra, we were able to observe 75 cross-peaks leading to

S. Del Valle et al. (eds.), *Peptides for Youth: The Proceedings of the 20th American Peptide Symposium*, 309
DOI: 10.1007/978-0-387-73657-0_138, © Springer Science+Business Media, LLC 2009

well-separated correlations in both the HNCO and HNCA spectra. About 30 of these crosspeaks corresponded to intra- and interresidue Cα and Cβ cross-peaks in the [^{13}C, ^{15}NH] – strips of the HNCACB and CBCA(CO)NH experiments. Due to enhanced relaxation phenomena in the large micellar peptide-detergent complex, 20 residues did not exhibit cross-peaks in the HNCACB and CBCA(CO)NH experiments. The unique chemical shifts observed for Gly, Ala and Val residues provided starting points in the assignment process. Data from the HNCACB and CBCA(CO)NH spectrum were used to assign and confirm the sequential assignments in the regions where it was possible. Assignments were confirmed by [^{15}N, ^{1}H]-HSQC spectra from peptides that were selectively ^{15}N-labeled by Ile, Leu, Phe and Val. Due to the large number of Ile (9 residues) and Leu (11 residues) residues in the transmembrane portion of TM1TM2 significant difficulties with the backbone assignment for residues 33 to 38 and 57 to 64 sequence were experienced. Intraresidual and interresidual Hα resonances were distinguished on the basis of ^{15}N-resolved NOESY, HCCH-TOCSY and ^{13}C-resolved NOESY data. Overall the percentage of assigned resonances was above 90% for both backbone and sidechain resonances.

Chemical shift indexing according to methods developed by Wishart and coworkers [6], analysis of heteronuclear ^{15}N-^{1}H NOEs and saturatution transfer experiments allowed placement of helices in several regions of TM1-TM2. Both putative transmembrane parts of the peptide were found to be highly helical, quite rigid and protected from the solvent. In contrast the predicted first intracellular loop was flexible and exposed to the solvent as were the ends of the peptide. Interestingly, a helical domain was found in the N-terminal extension of TM1 and residues G56-G60 were relatively flexible and exposed to solvent. This structural information appears to correlate with results from affinity crosslinking studies that indicate that F55-R58 may crosslink to the pheromone [7]. We are presently attempting to determine medium and long range NOEs for TM1-TM2 in LPPG as well as RDC measurements to assess the relative orientation of the TM1 and TM2 helices. Once these experiments are completed, we will calculate a high-resolution for the TM1-TM2 peptide in LPPG.

Acknowledgments

This work was supported by grants NIH GM-22086 and NIH GM-22087 to FN and JMB.

References

1. Popot, J. L., and Engelman, D. M. *Annu Rev Biochem* **69,** 881-922 (2000).
2. White, S. H., and Wimley, W. C. *Annu Rev Biophys Biomol Struct* **28**, 319-365 (1999).
3. Naider, F. *et al. Biopolymers* **60**, 334-350 (2001).
4. Arshava, B. *et al. Biopolymers* **64**, 161-176 (2002).
5. Neumoin, A. *et al. Biophys J* **Epub ahead of print (**2007).
6. Wishart, D. S., Sykes, B. D., and Richards, F. M. *Biochemistry* **31**, 1647-1651 (1992).
7. Son, C. D. *et al. Biochemistry* **43**,13193-13203 (2004).

Isolation of the Influenza A HA2 C-terminal segment by combination of nonionic detergents

Yulia A. Smirnova[1], Natalya V. Fedorova[1], Aleksander L. Ksenofontov[1], Larisa V. Kordyukova[1], Marina V. Serebryakova[2], Ludmila A. Baratova[1] and Boris V. Vaskovsky[3]

[1]Department of Bioengineering and Bioinformatics and Belozersky Institute of Physico-Chemical Biology, Moscow State University; [2]Institute of Physico-Chemical Medicine, Moscow 119992, Russia; [3]Shemyakin-Ovchinnikov Institute of Bioorganic Chemistry Russian Academy of Sciences, Moscow 117997, Russia

Introduction

Influenza A virus hemagglutinin (HA) is a transmembrane glycoprotein of the viral envelope which plays a central role in the delivery of the viral nucleocapsid into the host cell. HA ectodomain cleaved by enzyme bromelain was solved by X-ray crystallography. On the other hand, for the C-terminal anchoring segment which includes a transmembrane domain (TM) and a cytoplasmic tail (Fig. 1a) there is no 3D-structural information available, and data about its primary structure is insufficient. Another transmembrane glycoprotein of the envelope, neuraminidase (NA) is anchored by its N-terminal domain (Fig. 1b).

(a) *NH2*-VKLESMGIY**QILAIYSTVASSLVLLVSLGAISFWMC**SNGSLQCRICI
176 222

(b) MNPNQK**IITIGSICLVVGLISLILQIGNIISIWI**SHSIQT-*COOH*
1 40

Fig. 1. Amino acid structures of the HA2 C-terminal (a) and NA N-terminal (b) segments.

Results and Discussion

Earlier a method was supposed for the HA2 C-terminal anchoring segment chloroform/methanol extraction from the bromelain-digested virions (the subviral particles) [1]. Now we have developed a novel protocol of isolation of these TM-containing segments using solubilization by combination of nonionic detergents (octylglucopyranoside and Igepal) at room temperature. There were HA2 C-terminal (major peaks) and NA N-terminal (minor peaks) segments revealed by MALDI-TOF MS analysis, similar to those detected in the chloroform extracts (Fig. 2).

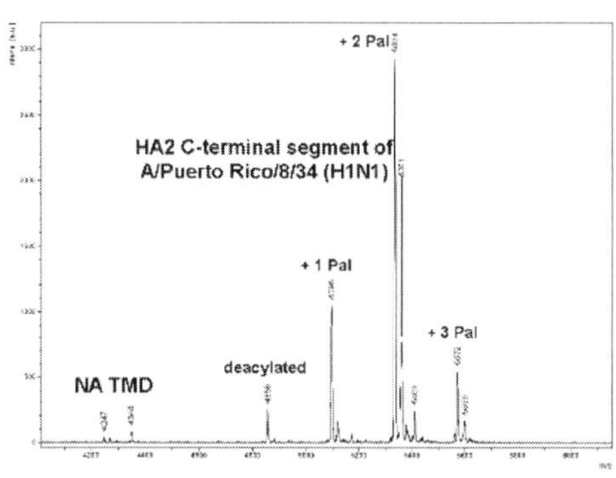

Fig. 2. MALDI-TOF MS analysis of the HA2 C-terminal anchoring segments.

S. Del Valle et al. (eds.), *Peptides for Youth: The Proceedings of the 20th American Peptide Symposium*, 311
DOI: 10.1007/978-0-387-73657-0_139, © Springer Science+Business Media, LLC 2009

Acid hydrolysis of the preparation followed by removing of lipid molecules from the hydrolyzate and subsequent amino acid analysis has demonstrated the good coincidence of the experimentally determined and announced in the database amino acid composition of the Influenza virus A/Puerto Rico/8/34 HA2 C-terminal segment (Table 1). The determined output of the HA2 C-terminal segment from the subviral particles indicated that our approach allowed us to isolate the TM-peptide almost quantitatively which had not achieved by the chloroform/methanol method. The obtained result gives us an opportunity to study mechanisms of interaction of the HA2 C-terminus with membrane-associated matrix M1 protein *in situ*.

Table 1. Amino acid composition of the A/PR/8/34 virus HA_2 C-terminal segment measured on Hitachi L-8800 Amino Acid Analyzer (ninhydrin identification at 570 nm)

Name of Amino Acid	*Number of residues of the 176-222 HA_2 C-terminal segment*		
	Experiment (Mean ± S.D., n=3)		*Literature*
Asx	2,4	± 0,1	1
Thr	1,7	± 0,1	1
Ser	8,8	± 0,4	8
Glx	2,8	± 0,1	3
Pro	0,7	± 0,1	0
Gly	3,5	± 0,2	3
Ala	3,3	± 0,0	3
Val	3,1	± 0,2	4
Met*	1,4	± 0,0	2
Ile	4,2	± 0,2	6
Leu	5,4	± 0,1	7
Tyr*	1,1	± 0,1	2
Phe	1,3	± 0,0	1
Lys	1,3	± 0,2	1
His	0,4	± 0,0	0
Arg	1,7	± 0,1	1
Cys**	0		3
Trp**	0		1

**partially and **fully destroyed during amino acid hydrolysis*

Acknowledgments

This work was funded by an ISTC/BTEP grant #2816p/82 and RFBR grant #06-04-48728.

References

1. Serebryakova, M. V., Kordyukova, L. V., Baratova, L. A. and Markushin, S. G. *Eur. J. Mass Spectrom.* **12**, 51-62 (2006).

Molecular Imaging and Orientational Changes of Antimicrobial Peptides in Membranes

Tzong-Hsien Lee[1], Kristopher Hall[1], Adam Mechler[2], Lisandra Martin[2], Jonathan Popplewell[3], Gerry Ronan[3] and Marie-Isabel Aguilar[1]

[1]*Department of Biochemistry & Molecular Biology, Monash University, Clayton Vic 3800, Australia,* [2]*School of Chemistry, Monash University, Clayton Vic 3800, Australia,* [3]*Farfield Scientific Ltd, Crewe CW1 6GU, UK.*

Introduction

Resistance to conventional antibiotics has placed antimicrobial peptides under the spotlight as alternative therapeutics for microbial infections. However, the design of specific non-toxic peptides has been elusive due largely to our poor understanding of the precise mechanism of cell lysis. Specifically, the difficulty in defining peptide conformation and location in membranes needs to be addressed. We have used a combination of dual polarisation interferometry (DPI), surface plasmon resonance spectroscopy (SPR) and atomic force microscopy (AFM) to gain an unprecedented detailed molecular picture of membrane lysis by antimicrobial peptides.

Results and Discussion

Dual Polarisation Interferometry

DPI is a new optical biosensor that provides a measure of the thickness and density of the membrane upon peptide binding, which together with the kinetic data from SPR and the high resolution images from AFM, provides an integrated approach to delineating specific molecular details of membrane lysis by antimicrobial peptides [1]. DPI allows the measurement of the changes in thickness, refractive index and mass of materials 100 nm thick or less on the surface of a glass waveguide. The dual slab waveguide is illuminated with laser light of alternating polarisations at one end, and as the light exits the 2 waveguides at the other end they interfere to produce an interference pattern [1]. The waveguide structure is integrated with a fluidic system permitting the continuous flow of material over the top waveguide, and as material is added to or removed from the waveguide, the interference pattern moves, and these changes in the position of the interference pattern can be "resolved" into changes in thickness, RI and mass of the material on the waveguide.

We have applied this technique to study the binding of melittin to phospholipid membranes. DPI allows a very detailed analysis of the deposited membrane bilayers and Table 1 lists the different parameters obtained for the deposition of two different phospholipid mixtures. For example, the thickness (21.2nm) showed that dimyristylphosphatidylcholine (DMPC) liposomes (50nm) adsorbed onto the chip surface as compressed liposomes while the DMPC: dimyristyl-phosphatidylglycerol (DMPG) (4:1) liposomes formed single bilayers (3.89nm). The subsequent binding of melittin resulted in increases in density for both types of membrane structures which is indicative of a vertical orientation relative to the planar membrane surface.

S. Del Valle et al. (eds.), *Peptides for Youth: The Proceedings of the 20th American Peptide Symposium*, 313
DOI: 10.1007/978-0-387-73657-0_140, © Springer Science+Business Media, LLC 2009

Table 1. Geometrical parameters determined for liposome deposition and peptide binding.

Liposome Mixture	Thickness (nm)	Refractive Index	Mass (ng/mm2)	Density (g/cm3)
DMPC	21.20	1.3530	4.90	0.23
DMPC + melittin	11.56	1.3744	5.70	0.49
DMPC/DMPG	3.89	1.3751	1.96	0.50
DMPC/DMPG + melittin	3.32	1.4220	3.71	1.12

Surface Plasmon Resonance

Antimicrobial peptide action is mediated by a direct interaction with cell membranes. A common feature of these interactions is the induction of cationic amphipathic secondary structure following binding of the peptides to the membrane surface. Since selective binding to different phospholipids is central to the design of non-hemolytic antimicrobial peptides, the affinity of the peptide for the membrane surface is a critical factor in the cell-lytic process. We have developed a sensitive method based on SPR, which allows the real-time measurement of peptide binding to phospholipid membranes [2,3]. We have used this SPR method to measure the membrane affinity of melittin and Fig 1 shows a series of sensorgrams for the interaction of melittin with DMPC. The data demonstrates strong concentration-dependent binding of melittin to this liposome.

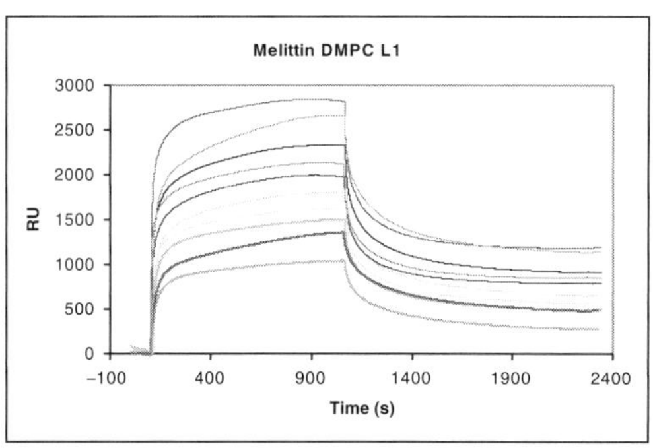

Fig. 1. Surface Plasmon resonance analysis of the interaction of melittin (2-50μM) with DMPC liposomes.

Atomic Force Microscopy

A liposome comprised of a 4:1 molar ratio mixture of DMPC and DMPG was deposited on atomic flat mica and imaged by tapping mode in solution using a Nanoscope IV Multimode scanning probe microscope (Veeco Corp., Santa Barbara, USA) with a vertical-engage "E" scanner. Silicon probes (NSC36/No Al 'B' - MikroMasch, Tallinn, Estonia) with a typical working resonance frequency of 30 kHz. Height, phase and amplitude images were simultaneously obtained using scan frequencies between 1 and 2 Hz and scan sizes of 1 to 5μm. The images were processed using WSxM software (Nanotech Electronica Madrid, Spain.). Melittin was then injected onto the surface and imaged over a period of 2 hours. Fig 2 shows the morphology of the deposited liposome 0, 15 and 30 minutes after the injection of melittin. The images demonstrate the presence of different domains which differed in height of approximately 0.5 nm. Moreover, it can be seen that melittin causes the rupture of the membrane causing the disintegration of the bilayer structure.

Fig. 2. Atomic force microscopic image of the interaction of melittin (10μM) with DMPC/DMPG (4:1)

In summary, AFM provides atomic images of the changes membrane morphology during lysis by melittin in real time that, together with the DPI and SPR studies, suggests a vertical insertion which is not pore-forming but leads ultimately to membrane destruction. These complementary techniques provide new insight into the mechanism of action of antimicrobial peptides.

Acknowledgments

The support of the Australian Research Council is gratefully acknowledged.

References

1. Popplewell, M.J. et al *Biochim Biophys Acta.* **1768**, 13-20 (2007).
2. Mozsolits, H. et al., *Biochim Biophys Acta*, **1512**, 64-76 (2001).
3. Mozsolits, H. and Aguilar, M. I., *Biopolymers – Peptide Science*, **66**, 3-18 (2002).

Glycated-CD59 antigen: exploration of synthetic approaches

S. Cantel[1,2], A. Kavishwar[1], M. Schlimme[1], A. Khatri[3], J.A. Halperin[1] and M. Chorev[1]

[1]Laboratory for Translational Research, Harvard Medical School, One Kendall Square, Building 600, Cambridge MA 02391 and Brigham and Women's Hospital, Department of Medicine, 75 Francis Street, Boston, MA 02115, USA; [2]IBMM, Max Mousseron Institute of Biomolecules, CNRS UMR 5247, Place Eugène Bataillon, CC1703, Université Montpellier 2, 34095 Montpellier Cedex 5, France; [3]Endocrine Unit, Massachusetts General Hospital and Harvard Medical School, Boston, Massachusetts 02114, USA.

Introduction

One of the most abundant non-enzymatic post-translational modifications involves reducing sugars-mediated glycation of lysine side chains in proteins leading initially to Schiff base formation. Subsequent spontaneous rearrangement to the Amadori products followed by oxidation and cross-linking reactions result in the formation of advanced glycation end products (AGEs)[1].

Hyperglycemic conditions associated with diabetes accelerate glycation of numerous proteins including human CD59 (Fig. 1), an inhibitor of the membrane attack complex (MAC) of the complement system. Increased deposition of MAC is found in blood vessels and kidneys of diabetic patients, which may induce lysis of MAC-targeted endothelium cells and the release of cytokinins and growth factors such as bFGF and PDGF. The release of these agents leads to increased proliferation of cells such as fibroblasts, smooth muscle and meseangial cells and increase synthesis of extracellular matrix proteins. These may stimulate the development of vascular proliferative disease leading to what is known micro- and macrovascular complications of diabetes.

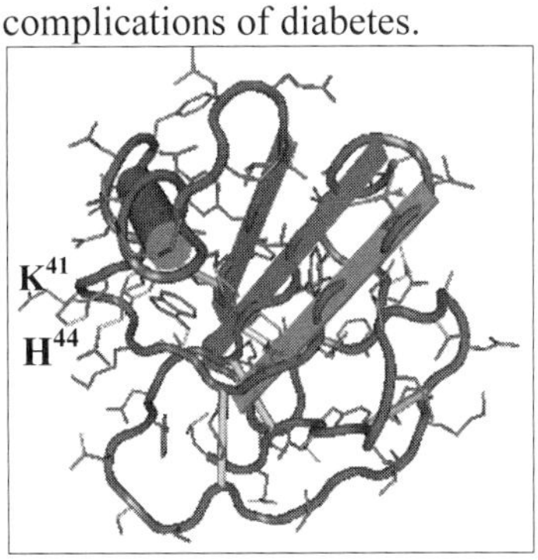

Therefore, we postulate that glycation inactivation of hCD59 on K^{41}, which is part of a glycation motif ($W^{40}KFEH^{44}$), will result in highly prevalent, mortality and morbidity causing diabetic micro- and macrovascular complications leading to nephropathy, neuropathy, retinopathy, ischemic heart disease, stroke and peripheral vascular disease[2]. As such, glycated-CD59 is a direct surrogate of these complications.

Fig. 1. Three-dimensional solution structure of soluble CD59[3].

In an effort to enable the measurement of glycated hCD59 we undertook the development of a glycated hCD59 antigen that will be instrumental in raising specific anti-glycated hCD59 antibodies. We report the design of different synthetic approaches that led to the synthesis of CD59-derived peptide glycated on K^{41}.

Results and Discussion

We report the development of several strategies for generating *K^{41}-glycated-CD59*. The development of an orthogonally protected building block which was used in the stepwise solid phase peptide synthesis was an essential step toward the successful synthesis of the antigen (Fig. 2, pathway A). We also attempted specific glycation of

S. Del Valle et al. (eds.), *Peptides for Youth: The Proceedings of the 20th American Peptide Symposium*, 317
DOI: 10.1007/978-0-387-73657-0_141, © Springer Science+Business Media, LLC 2009

K^{41} following the completion of peptide assembly both on the protected resin-bound peptide (Fig. 2, pathway B) and in solution on the partially protected peptide (Fig. 2, pathway C). The obvious advantages of the direct glycation of a fully assembled peptide is the potential to use the same peptide for generating a variety of glycated peptides that differ in the glycating moiety and in eliminating the need for generating fully protected glycated-Lys moieties as building blocks. We were able to obtain the desired antigen through all three pathways. However, we found the use of the building block in a stepwise assembly of the peptide as the method of choice (see HPLC tracings of the crude reaction mixtures in Fig. 2).

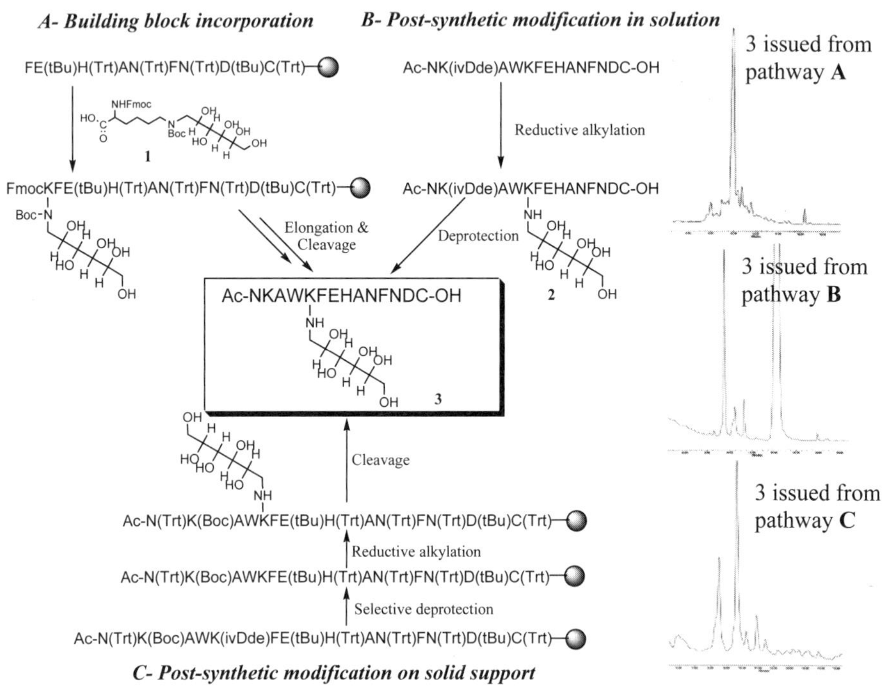

Fig. 2. Synthetic pathways for generating K^{41}-glycated CD59 antigen and the corresponding HPLC tracings of crude products (214 nm).

Conclusion

We were able to develop an orthogonally protected building block and design different synthetic approaches that led to the efficient synthesis of CD59-derived peptide antigen specifically glycated on K^{41}.

Generation of this antigen allowed us to develop rabbit polyclonal anti-glycated hCD59 antibody, which were effectively used in immunohistochemistry to demonstrate colocalization of glycated CD59 with MAC in kidneys and nerves from diabetic patients but not from non-diabetic patients.

References

1. Friedman, EA., *Diabetes Care.*, **2**, B65-71 (1999).
2. Qin, X. et al, *Diabetes*, **53**, 2653-2661 (2004).
3. Fletcher, C.M. et al, *Protein Sci.*, **2**(12), 2015-2027 (1993).

A combination of ligand SAR and receptor site-directed mutagenesis reveals requirements for antagonist activity at the melanocortin-3 receptor

Bettina Proneth, Federico P. Portillo and Carrie Haskell-Luevano

Department of Medicinal Chemistry, University of Florida, Gainesville, Fl 32610, USA

Introduction

The melanocortin system is a G-Protein coupled receptor (GPCR) pathway that contains five melanocortin receptors. The MC3R and the MC4R are involved in energy homeostasis in humans and rodents. While MC4R KO mice are extremely obese and hyperphagic, MC3R KO mice possess normal body weight with increased fat mass and decreased lean body mass (*1-3*). Extensive studies have been carried out to understand the physiological role of the MC4R but little is known about the MC3R. Identification of ligands that are selective for the MC3R can aid as *in vivo* tools to explore the complex scope of physiological functions of this receptor.

Previous structure-activity relationship (SAR) studies of the tetrapeptide Ac-His-DPhe-Arg-Trp-NH_2 identified the compound Ac-His-(pI)DPhe-Arg-Trp-NH_2 that possessed "mixed pharmacology" and resulted in a potent antagonist at the MC3R and full nM agonist at the MC4R (*4*). A rational combination of ligand SAR and *in vitro* site-directed mutagenesis studies were used in attempts to identify sub structural requirements of the ligand and the receptor that are important for selective receptor activation or inhibition of the MC3R versus the MC4R.

Results and Discussion

For the tetrapeptide ligand SAR studies the D-Phe phenyl ring of Ac-His-DPhe-Arg-Trp-NH_2 was substituted at the para position with a series of halogens (-F, -Cl, -Br and -I) to investigate this position for agonist/antagonist action at the MC3R. Peptides were synthesized using Fmoc chemistry, purified to homogeneity by RP-HPLC and analytically characterized by RP-HPLC using two solvent systems and mass spectrometry.

Functional agonist EC_{50} (nM) values and antagonist pA_2 values of the tetrapeptide ligands at the mMC3R and mMC4R were determined using a cAMP response element (CRE)/β-galactosidase reporter gene assay (Table 1). Substitution of the D-Phe phenyl ring in para position with –F and –Cl resulted into full agonist activities at both, the mMC3R and mMC4R (54 nM and 84 nM, respectively). Interestingly, upon substitution with –Br and –I, tetrapeptides were converted into potent antagonists with partial agonist activity at the mMC3R (pA_2=7.5 and 7.4, respectively) while retaining full nM agonist activity at the mMC4R (1 and 4 nM, respectively).

Homology modeling studies of the hMC4R in the active state suggested that the W258 residue (hMC4R numbering) is involved in antagonist action of the DNal(2') containing antagonist SHU9119 (Ac-Nle-c[Asp-His-DNal(2')-Arg-Trp-Lys]-NH_2) (*5*). To test if the corresponding mMC3R W255 residue is involved in differentiation of agonist versus antagonist activity of the (pI) containing tetrapeptide, Ac-His-(pI)DPhe-Arg-Trp-NH_2 was tested for functional activity at the W255A mMC3 mutant receptor. Figure 1 summarizes the pharmacology of Ac-His-(pI)DPhe-Arg-

DOI: 10.1007/978-0-387-73657-0_142, © Springer Science+Business Media, LLC 2009

Trp-NH_2 at the W255A mMC3R, resulting in full agonist activity of this tetrapeptide which is a competitive antagonist at the wild-type mMC3R.

By using a dual-pronged approach of ligand SAR studies and receptor mutagenesis this study identified new determinants for antagonist activity at the mMC3R. Findings can be used for the rational drug design of MC3R selective ligands.

Table 1. Functional Activity of tetrapeptides at the mouse MC3R and MC4R.

	Agonist EC_{50} (nM) and Antagonist pA_2 values	
	mMC3R	*mMC4R*
Ac-His-DPhe-Arg-Trp-NH_2	126	2
Ac-His-pFDPhe-Arg-Trp-NH_2	54	0.5
Ac-His-pClDPhe-Arg-Trp-NH_2	85	0.4
Ac-His-pBrDPhe-Arg-Trp-NH_2	pA_2=7.5 partial agonist	1
Ac-His-pIDPhe-Arg-Trp-NH_2	pA_2=7.4 partial agonist	4

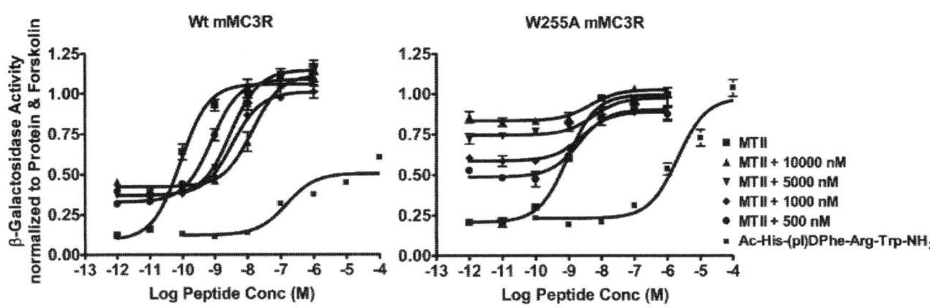

Fig. 1. Pharmacology Curves of Ac-His-(pI)DPhe-Arg-Trp-NH_2 at the Wt and W255A mouse MC3 receptors.

Acknowledgments

The work was funded by the NIH RO1DK57080 grant.

References

1. Butler, A. A., Kesterson, R. A., Khong, K., Cullen, M. J., Pelleymounter, M. A., Dekoning, J., Baetscher, M., and Cone, R. D. (2000) Endocrinology 141, 3518-21.
2. Chen, A. S., Marsh, D. J., Trumbauer, M. E., Frazier, E. G., Guan, X. M., Yu, H., Rosenblum, C. I., Vongs, A., Feng, Y., Cao, L., Metzger, J. M., Strack, A. M., Camacho, R. E., Mellin, T. N., Nunes, C. N., Min, W., Fisher, J., Gopal-Truter, S., MacIntyre, D. E., Chen, H. Y., and Van der Ploeg, L. H. (2000) Nat Genet 26, 97-102.
3. Huszar, D., Lynch, C. A., Fairchild-Huntress, V., Dunmore, J. H., Fang, Q., Berkemeier, L. R., Gu, W., Kesterson, R. A., Boston, B. A., Cone, R. D., Smith, F. J., Campfield, L. A., Burn, P., and Lee, F. (1997) Cell 88, 131-41.
4. Holder, J. R., Bauzo, R. M., Xiang, Z., and Haskell-Luevano, C. (2002) J Med Chem 45, 3073-81.
5. Pogozheva, I. D., Chai, B. X., Lomize, A. L., Fong, T. M., Weinberg, D. H., Nargund, R. P., Mulholland, M. W., Gantz, I., and Mosberg, H. I. (2005) Biochemistry 44, 11329-41.

Conformational Effects of Cyclic Hexapeptides for Ion Channel Properties

Junichi Taira[1], Satoko Imamura[1], Satoshi Osada[1], Masoud Jelokhani-Niaraki[2], Tsuguhisa Ehara[3] and Hiroaki Kodama[1]

[1]*Department of Chemistry, Faculty of Science and Engineering, Saga University, Saga, 840-8502, Japan;* [2]*Wilfrid Laurier University, Waterloo, ON N2L 3C5, Canada;* [3]*Department of Physiology, Faculty of Medicine, Saga University, Saga, 849-8501, Japan.*

Introduction

Occurrences of ion-channel functions by cyclic peptides have been reported since about two decades ago. Several ion conduct mechanisms have been suggested such as nano-tube or barrel-stave like pore structures [1, 2], although relationships between peptide structures and ion channel forming properties are unclear. We had reported ion channel activities of cyclic hexapeptides with type II' β-turn structures [3]. In this study, to investigate relationships between peptide conformation and ion channel functions, cyclic peptides possessing type I β-turn conformations were synthesized and their ion channel properties were compared to cyclic peptides having other conformations.

Results and Discussion

Structures of cyclic hexapeptides evaluated in present study were shown in Fig. 1. Previous study reported that these cyclic hexapeptides take type I β-turn conformation due to incorporated two Gly residues [4, 5]. These peptides were assembled by solution phase method with Boc strategy. Purifications of peptides were carried out by silica gel column chromatography. Peptide purities and homogeneities were confirmed by RP-HPLC and MALDI-TOF MS analysis, respectively. Peptide **3** was obtained as intermediate product of peptide **2**. Deprotection of benzyloxycarbonyl (Z) groups was carried out by catalytic reduction method by palladium black.

cyclo(Pro-Leu-Gly)$_2$	**(1)**
cyclo(Pro-Lys-Gly)$_2$	**(2)**
cyclo(Pro-Lys(Z)-Gly)$_2$	**(3)**

Fig. 1. Structures of cyclic hexapeptides. These peptides take type I β-turn conformations.

Ion channel experiments were performed by Tip-Dip technique at room temperature [6]. Electrolytes were filled by 500 mM KCl buffered with 5 mM HEPES (pH 7.4) solution and the composition was symmetrical for both sides of the diphytanoylphosphatidylcholine (DPhPC) membrane. All peptide concentrations were 1 μM, respectively. Single-channel currents were amplified using an Axpatch 1D patch-clamp amplifier. Data were filtered at 2 kHz frequencies and analyzed using Axograph 3.5. Figure 2 exhibits ion conduct patterns of cyclic hexapeptides. In our previous study, cyclo(Pro-Leu-D-Phe)$_2$ and cyclo(Pro-Leu-D-Ala)$_2$ which have type II' β-turn structures showed single level conductance patterns with clearly open-close. Peptide **1** also showed clear open-close transitions, although the patterns were composed of multi-substates. On the other hand, peptide **2** showed

S. Del Valle et al. (eds.), *Peptides for Youth: The Proceedings of the 20th American Peptide Symposium*, 321
DOI: 10.1007/978-0-387-73657-0_143, © Springer Science+Business Media, LLC 2009

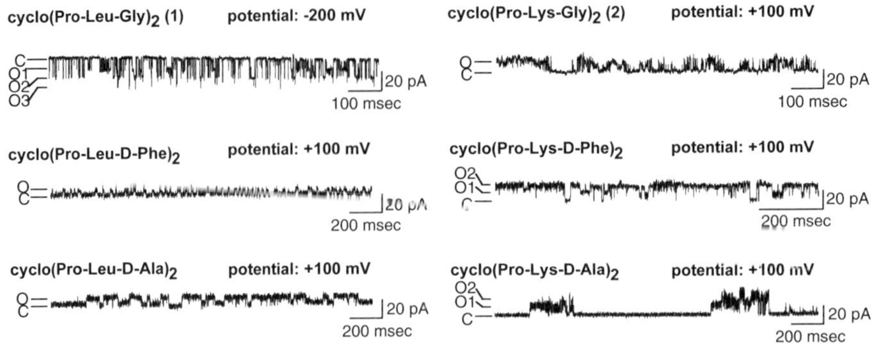

Fig. 2. Ion conduct patterns of cyclic hexapeptides. C and O represent close and open states, respectively.

single-state conductance patterns. Whereas other Lys containing cyclic hexapeptides evaluated previously exhibited multi-states ion conduct patterns. These results suggested that peptide conformation affect for ion conduct pore forming of cyclo hexapeptides. However, a further study is necessary to clarify the difference lying between Lys present and absent peptides.

We had reported that protection of ε-amino groups of Lys residues caused decrease of conductance values. Ion channel properties of peptide **3** were compared to peptide **2**. Fig. 3 illustrated ion conductance pattern of peptide **3** and histogram analysis of the protecting group present and absent peptides. Peptide **2** showed two fold excess ion conductance values than that of peptide **3**, it was suggested that amino groups contribute to occurrence of larger conductance values.

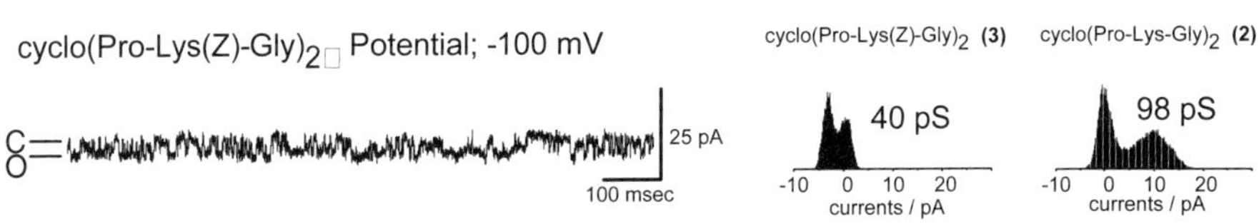

*Fig. 3. Ion conductance pattern of peptide **3** and histogram analysis of the trace. The histogram is compared to that of peptide **2**. C and O represent close and open states, respectively.*

Acknowledgments

The work was funded by JSPS Research Fellowships for Young Scientists to J. T.

References

1. Heitz, F., Jacquier, R., Kaddari, F. and Verducci, J. *Biophys. Chem.* **23**, 245-249 (1986).
2. Ghadiri, M. R., Granja, J. R. and Buehler, L. K. *Nature* **369**, 301-304 (1994).
3. Taira, J., Hayashi, R., Osada, S., Ehara, T. and Kodama, H. *Biopolomers* **80**, 581 (2005).
4. Wang, J., Osada, S., Kodama, H., Kato, T. and Kondo, M. *Bull. Chem. Soc. Jpn.* **72**, 533-540 (1999).
5. Wang, J., Osada, S., Kodama, H. and Kondo, M. *Bull. Chem. Soc. Jpn.* **73**, 1221-1226 (2000).
6. Higashimoto, Y., Kodama, H., Jelokhani-Niaraki, M., Kato, F. and Kondo, M. *J. Biochem.* **125**, 705-712 (1999).

Synthesis and Biological Activities of Cyclic Peptide, Hymenamide Analogs

Yasuhiro Shiki[1], Mika Onai[1], Daisuke Sugiyama[1], Satoshi Osada[1], Ichiro Fujita[2] and Hiroaki Kodama[1]

[1]*Department of Chemistry, Faculty of Science and Engineering, Saga University, Saga 840-8502, Japan;* [2]*Department of Pediatrics, Faculty of Medicine, Saga University, Saga 849-8501, Japan*

Introduction

A great variety of useful bioactive peptides have been isolated from various natural sources and those structures also determined. Especially, cyclic peptides have been well found as bioactive components, they were expected as seed compounds in drug design due to their metabolic stabilities and restricted conformations. Hymenamides were one class of these cyclic peptides isolated from Okinawan marine sponge *Hymeniacidon sp.* [1-4]. So far, ten peptides have been isolated. Composed of few amino acid residues, they have exhibited antifungal, cytotoxic and enzyme inhibiting activities. Napolitano *et al.* reported inhibition of human neutrophil elastase release by Hymenamide C [5]. Since excess releases of elastase cause the chronic obstructive pulmonary disease (COPD), hymenamides have been expected as the useful proto-type therapeutic tools for COPD. The development of effective syntheses of analogs and further structural-activity relationship studies are necessary for the practical usage of hymenamides in drug discovery.

In the present study to evaluate the inhibitory activities of elastase release from human neutrophils, all of the isolated analogs belonging to the hymenamide family were synthesized and their biological activities were examined. Table 1 exhibits structures reported of isolated hymenamides.

Results and Discussion

Cyclic peptides were assembled by combination of solid phase and liquid phase method. Peptide elongation was performed by Fmoc sold phase synthesis upon 2-chlorotrityl chloride resin. Fmoc groups were removed by 20% piperidine treatments. Fmoc protected amino acids were condensed by HBTU-HOBt in

Table 1. Sequences and reported activities of Hymenamides [1-4].

peptides	structures	reported activities
Hymenamide **A**	**cyclo(PPVPFWR)**	antifungal activity
Hymenamide **B**	**cyclo(PPNFVEF)**	antifungal activity and cytotoxicity
Hymenamide **C**	**cyclo(WPFGPEL)**	antifungal activity and elastase inhibitor
Hymenamide **D**	**cyclo(IPYDPLA)**	
Hymenamide **E**	**cyclo(FPTTPYF)**	antifungal activity
Hymenamide **F**	**cyclo(PPAVMLR)**	
Hymenamide **G**	**cyclo(PPYVPLIL)**	protein tyrosine kinase inhibitor
Hymenamide **H**	**cyclo(PLTPLPWV)**	cytotoxicity
Hymenamide **J**	**cyclo(PYDFWKVY)**	cytotoxicity

S. Del Valle et al. (eds.), *Peptides for Youth: The Proceedings of the 20th American Peptide Symposium*, 323
DOI: 10.1007/978-0-387-73657-0_144, © Springer Science+Business Media, LLC 2009

presence of DIEA. Linear peptides were cleaved from the resin by AcOH treatments without elimination of side chain protecting groups. The linear peptides were cyclized by HBTU and HOBt under high diluted conditions (0.4mM-0.8mM) in DMF for 2 days. Peptide purifications were achieved by preparative RP-HPLC. Homogeneities and structures of the peptides were confirmed by analytical HPLC and MALDI-TOF MS analysis, respectively.

Cytotoxicity was assessed by MTT assay using HL-60 cells. All synthetic peptides exhibited no cytotoxicity. Inhibition of elastase release by synthetic peptides was estimated using human neutrophils isolated from human peripheral blood [7, 8]. The neutrophils were stimulated by treatment of for-Met-Leu-Phe-OH (fMLP) in the presence of hymenamide peptides. After incubation, the aliquots of stimulated neutrophil supernatants were incubated with Boc-L-Ala-ONp. Absorption at 414 nm corresponded to p-nitrophenol and indicated the processing of substrates by released elastase. Figure 1 exhibits elastase inhibition activities of hymenamide peptides. Inhibition of elastase release was confirmed by treatment of hymenamide A, C, F, G, J and K. It was observed that all of the hymenamide peptides, except hymenamide B, acted as potent elastase release inhibitors. Further studies on conformation and activities of hymenamide peptides are in progress in our laboratories.

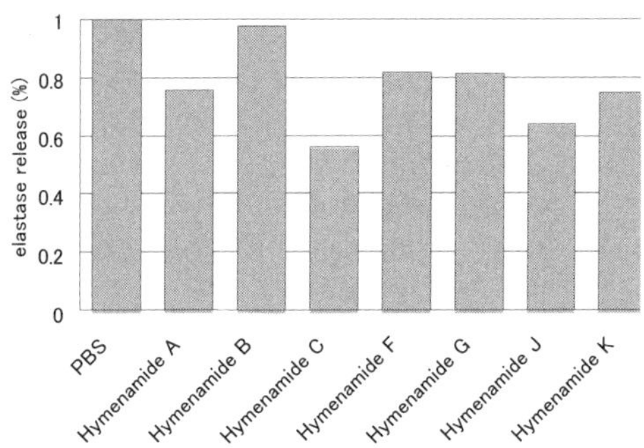

Fig. 1. Inhibition of elastase release from human neutrophils.

References

1. Kobayashi, J., Tsuda, M., Nakamura, T., Mikami, Y. and Shigemori, H. *Tetrahedron* **49**, 2391-2402 (1993).
2. Tsuda, M., Shigemori, H., Mikami Y. and Kobayashi, J. *Tetrahedron* **49**, 6785-6796 (1993).
3. Tsuda, M., Shigemori, H., Mikami, Y. and Kobayashi, J. *Tetrahedron* **50**, 4667-4680 (1994).
4. Kobayashi, J., Nakamura, T. and Tsuda, M. *Tetrahedron*. **49**, 2391-2402 (1993).
5. Napolitano, A., Bruno, I., Rovero, P., Lucas, R., Peris, M., P. Gomez-Paloma L.and Riccio, R. *Tetrahedron* **57**, 6249-6255 (2001).
6. Shiki, Y., Taira, J., Nakamura, A., Osada, S. and Kodama, H. *Biopolymers* **80**, 581-582 (2005).
7. Hayashi, R., Osada, S., Yoshiki, M., Sugiyama, D., Fujita, I., Hamasaki, Y. and Kodama, H. *J. Biochem.* **139**, 981-988 (2006).
8. Miyazaki, M. Kodama, H. Fujita, I. Hamasaki, Y. Miyazaki S. and Kondo, M. *J. Biochem.* **117**, 489-494 (1995).

Monitoring Glucagon and Glucagon Antagonist-Mediated Internalization: A Useful Approach to Study Glucagon Receptor Pharmacology

Pallavi Sachdev, Laxmi Manohar Tirunagari, Dennis Kappei and Cecilia G. Unson

The Rockefeller University, New York, NY 10021

Introduction

The impetus for studying how glucagon interacts with its receptor is to gain insight into the mechanism of glucagon action in normal physiology as well as in diabetes mellitus [1]. Continued interest in the glucagon receptor as a drug target for diabetes management is sustained by studies that show that treatment with glucagon antagonists has restored glucose homeostasis. Glucagon binding leads to internalization of glucagon receptors, a well-characterized phenomenon for GPCRs [2]. Upon activation, receptors are sequestered from the cell surface into cellular compartments where they are either recycled or targeted for degradation. Recently, studies have linked ligand-induced receptor internalization to pharmacological properties measured by conventional cell-based assays. Moreover, GPCR trafficking assays in live cells can be monitored by the use of fluorescent reagents and used to evaluate differences between agonist and antagonist effects on internalization [3, 4].

In this study we demonstrate that glucagon-mediated internalization of the glucagon receptor is a pharmacological characteristic that can be visualized, evaluated and measured. To observe internalization, we established and characterized a novel tetracycline-inducible, mammalian stable cell line for expression of human glucagon receptor fused to green fluorescent protein at the carboxy terminus (hGR-GFP). HGR-GFP expressing cells were plated on poly-L-lysine coated cover slips. Following induction with 2μg/ml tetracycline and 5 mM sodium butyrate for 24 h, the cells were stimulated with 10 nM glucagon, 10 nM glucagon antagonist desHis^{1}Nle^{9}Ala^{11}Ala^{16} glucagon amide, and 10 nM glucagon + antagonist for 1, 5, 15, and 30 min. Plasma membranes were also prepared from cells incubated with glucagon for the times indicated and analyzed on a western blot.

Results and Discussion

GFP-tagged hGR exhibited the expected pharmacological parameters in functional assays in response to glucagon and the glucagon antagonist. Membranes prepared from hGR-GFP expressing cells bound ^{125}I-glucagon with an apparent dissociation constant of 3.2 nM, while the antagonist had a competitive inhibition constant of 10 nM. The effective concentration at 50% stimulation (EC_{50}) in a typical dose-dependent assay was 6.2 nM for hGR-GFP. The antagonist had weak partial agonist activity. Glucagon stimulation of hGR-GFP triggered increased internalization of receptors over time. A fluorescence imaging system detected internalization between 5-30 min. The majority of receptors appeared as clusters inside the cells after 30 min (Fig. 1). In contrast, antagonist-bound hGR-GFP exhibited dramatically decreased kinetics of receptor internalization and most of the receptors remained on the cell membrane even after a 30 min incubation period. Western blots of plasma membranes prepared from agonist-treated cells also showed reduced receptor concentration with time, while the concentration of receptors in antagonist-treated

S. Del Valle et al. (eds.), *Peptides for Youth: The Proceedings of the 20th American Peptide Symposium*, 325
DOI: 10.1007/978-0-387-73657-0_145, © Springer Science+Business Media, LLC 2009

cell membranes appeared unchanged. The differences in the kinetics of agonist and antagonist-mediated receptor internalization corresponded well with binding affinity, adenylyl cyclase activity, and inhibition potency previously reported. With appropriate quantitative analysis, monitoring levels of receptor internalization by fluorescence microscopy can be another useful tool for the pharmacological evaluation of potential glucagon antagonists and will facilitate HT screening when automated.

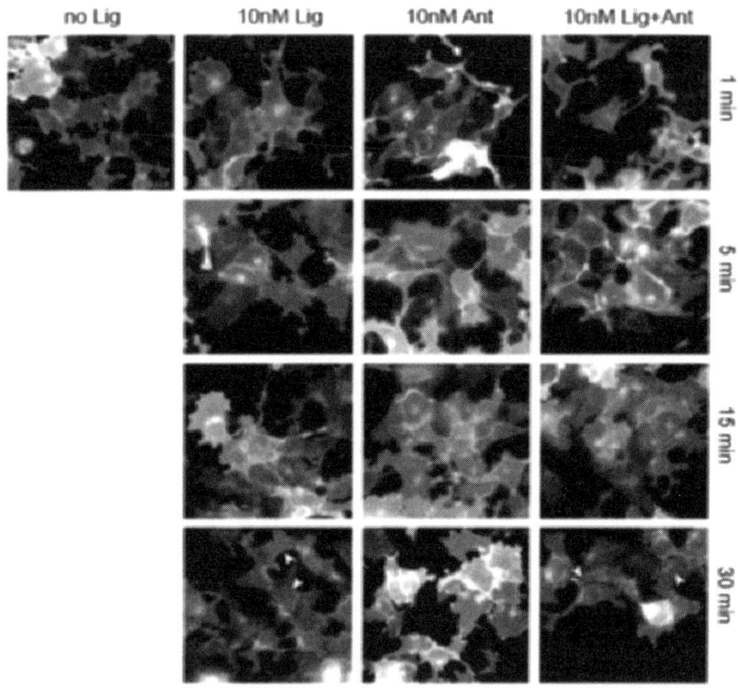

Fig. 1. Cells expressing GFP-fused human glucagon receptors were treated with either a) 10nM glucagon, b) 10nM glucagon antagonist desHis^{1}Nle^{9}Ala^{11}Ala^{16} glucagon amide, or c) 10 nM glucagon + antagonist for the indicated times. Arrowheads point to internalized hGR-GFP.

Acknowledgments

The work was funded by R01 DK024039 grant to C. Unson.

References

1. Unson, C. G. *Biopolymers* **66**, 218-235 (2002)
2. Merlen, C., Fabrega, S., Desbuquois, B., Unson, C. G., and Authier, F. *FEBS Lett.* **580**, 5697-5704 (2006).
3. Fukunaga, S., Setoguchi, S., Hirasawa, A., and Tsujimoto, G. *Life Sci.* **80**, 17-23 (2006).
4. Cai, M., Varga, E. V., Stankova, M., Mayorov, A., Perry, J. W., Yamamura, H. I., Trivedi, D., and Hruby, V. J. *Chem. Biol. Drug Des* **68,** 183-193 (2006).

Semi-Synthetic Strategies to Obtain Glucosylated MOG to Identify Antibodies as Biomarkers in Multiple Sclerosis Disease

Francesca Gori[1,2], Barbara Mulinacci[1,2], Eva Bonchi[1,2], Mario Chelli[2,3], Francesco Lolli[2,4], Anna M. Papini[2,3] and Paolo Rovero[1,2]

[1] Department of Pharmaceutical Sciences, University of Florence, Sesto Fiorentino (FI), Italy; [2] Laboratory of Peptide & Protein Chemistry & Biology, University of Florence, Sesto Fiorentino (FI), Italy; [3] Department of Organic Chemistry 'Ugo Schiff' and CNR-ICCOM, University of Florence, Sesto Fiorentino (FI), Italy; [4]Department of Neurological Sciences, University of Florence, Florence, Italy.

Introduction

Several autoimmune diseases have been associated with post-translational modifications (PTMs). In particular, glycosylation is the most important PTM of secreted proteins and plays a crucial role in several immune functions [1].

The role of autoantibodies in autoimmune diseases has been reevaluated and it is accepted that a distinct pattern of Multiple Sclerosis pathology could involve an Ab-mediated demyelination [2]. One of the proposed biomarkers, that may be objectively measured and evaluated as an indicator of normal biological processes, for MS are specific antibodies detectable by immunoassays. Previous studies showed that CSF114(Glc) [3,4], a designed glycopeptide characterized by a β-D-glucopyranosyl moiety on the tip of a I' β-turn, is able to detect and isolate specific autoAbs present in the sera of a significant number of MS patients. Since this synthetic Ag may be considered as a mimetic of aberrantly glucosylated myelin protein(s) triggering autoimmunity in MS, we focused our interest on the role of MOG (Myelin Oligodendrocyte Glycoprotein), a type I integral membrane protein specifically expressed on the outermost lamellae of myelin sheath and considered a putative autoantigen in MS [5]. In order to characterize the molecular mechanisms of the form of MS in which the demyelinization is Ab-mediated, and to design new antigenic probes to detect auto-Ab as biomarkers, the extracellular domain of MOG has been selected as putative autoantigen in these preliminary studies.

Results and Discussion

To understand the molecular mechanisms involved in Multiple Sclerosis and to get new antigenic probes to detect antibodies as biomarkers, our goal will be to obtain MOG properly glucosylated. As protein glucosilation is not under direct genetic control, the preparation of specific glycoforms may benefit from a chemical approach, such as Cysteine Ligation and Expressed Protein Ligation.

Thanks to the biorthogonality of iodoacetamide derivative with thiol free groups of Cys, we performed a selective point mutation of rMOG protein to introduce a Cys residue at the position 31, native site of glycosylation, to get a free thiol group that will be selectively modified by glucosyl iodoacetamide derivative. The mutated $rMOG_{ED}N(35)C(His)_6$ protein was obtained by over expression in *E.Coli*, purified and characterized by LC-MS.

The Expressed Protein Ligation tecnique concerns the ligation between a protein and a peptidic fragment, in which a PTM can be introduced. This strategy is an extension of the Native Chemical Ligation method, based on the reaction between a thioester

S. Del Valle et al. (eds.), *Peptides for Youth: The Proceedings of the 20th American Peptide Symposium*, 327
DOI: 10.1007/978-0-387-73657-0_146, © Springer Science+Business Media, LLC 2009

and a side chain of a Cys residue of two fully unprotected peptides to form an amide bond: so they are connected as in the native peptide backbone.

Studies performed on X-ray crystal structure of MOG showed that this protein has a β-hairpin conformation, reminiscent of the glycopeptide CSF114(Glc), in the segment 98-125 [7]. As a consequence, we decided to obtain MOG synthetically glucosylated in position 104 in order to test it in immunological assays.

This semi-synthetic bioconjugate may be obtained by ligation between $rMOG_{ED}$ (1-97) C-terminal thioester, obtained by IMPACT-TWIN system, and peptide segment 98-125 bearing a Cys residue at the N-terminus and the modifications H(103)G and S(104)N(Glc), obtained by SPPS.

The PTM may be introduced in the peptide sequence following the building-block approach: preformed glycosylated amino-acid building blocks are employed in the stepwise assembly of the peptide backbone.

The obtained homogeneously glucosylated protein carrying a glucosylated side-chain at natural glycosylation site and the semi-synthetic proteins with a well defined PTM may be tested by ELISA to study the ability to detect auto-Abs in MS patient sera, to study a cross-reactivity with CSF114(Glc) and to characterize an autoantigen in vivo.

Acknowledgments

We acknowledge Ente Cassa di Risparmio di Firenze, PRIN 2005 PROT 2005032959 and the 20th APS Travel Grant Committee for the financial support.

References

1. Daniels, M.A., *et al. Nat. Immunol.* **10**, 903-907 (2002).
2. Lucchinetti, C., *et al. Ann. Neurol.* **47**, 707-717 (2002).
3. Papini, A.M., Rovero, P., Chelli, M., & F. Lolli, PCT International Patent Application WO 03/000733.
4. Lolli, F., *et al. Proc. Natl. Acad. Sci. USA* **102**, 10273-10278 (2005).
5. Johns, T.G., Berbard, C.C.A., *J. Neurochem.* **73**, 1-9 (1999).
6. Macmillan, D., *et al. Chemistry & Biology* **8**, 133-145 (2001)
7. Breithaupt, C., *et al. Proc. Natl. Acad. Sci. USA* **100**, 9446-9451 (2003).

Photoprobe Peptides to Map the Interactions of Angiotensin II with its Receptor AT_1

D. Fillion, G. Guillemette, R. Leduc and E. Escher

Université de Sherbrooke, Faculty of Medecine & Health Sciences, Department of Pharmacology, Sherbrooke, Québec, Canada, J1H 5N4

Introduction

Seven transmembrane domains (*TMDs*) G protein-coupled receptors (GPCRs) are the largest family of cell surface receptors and are implicated in various pathologies. In fact, those proteins are privileged targets in pharmacotherapy since approximatively 50% of all current market drugs act via these receptors. The rational design of drugs targeting GPCRs requires a 3D molecular-based knowledge of those proteins that may be acquired by identification of ligand-receptor interactions through photoaffinity labeling [1]. This approach allows to directly map the ligand-receptor interface by covalent bonding of the radio-labeled photoprobe ligand within the immediate molecular surroundings of its cognate receptor. This information may then be used in conjunction with computational molecular modeling procedures, based on the X-ray crystallography of bovine rhodopsin, to build and validate homology molecular models of receptors with ligand [2]. In the present contribution, we have combined a photoaffinity labeling approach with an X→Methionine (Met) mutagenesis strategy to study the molecular structure as well as the activation mechanism of the human angiotensin II (AngII) type 1 receptor (hAT_1) following the binding of the agonist AngII hormone. AT_1 is a typical class-A, rhodopsin-like GPCR, which mediates virtually all of the known physiological action of AngII, it is responsible for cardiovascular and electrolyte homeostasis [3].

Results and Discussion

A single but systematic incorporation of both biradical-ketone-generating amino acid photoprobe *p*-benzoyl-L-phenylalanine (Bpa) or the previously synthesized [4] carbene-generating amino acid photoprobe *p*–[3–(trifluoromethyl)–3*H*–diazirin–3–yl]–L–phenylalanine (Tdf) into each of the eight positions of the $[Sar^1]$AngII sequence was carried out to produce one series of eight AngII-Bpa analogues and one series of eight AngII-Tdf analogues. These two AngII analogue series were tested for their pharmacological properties on hAT_1 as well as on its constitutively active mutant (CAM) [N111G]-hAT_1 (hAT_1-CAM). Since hAT_1-CAM had more permissive structure-activity relationship, photoaffinity labeling studies were carried out on this receptor. From all the AngII-photoprobe analogues, those substituted in position 1, 2, 3 and 5, either by Bpa or Tdf, displayed AngII-like properties on hAT_1-CAM, meaning low nM affinities with agonistic properties. Analogues substituted in position 7 and 8 were shown to be low nM affinity neutral antagonists.

In order to facilitate identification of the photoprobe contact sites in the receptor, we constructed several single X→Methionines (Met) mutants at residues 11, 21, 103, 118, 172, 195, 258, 268, 297 and 317 within the hAT_1-CAM template. All hAT_1-CAM / Met-mutants showed pharmacological properties similar to hAT_1-CAM. In photoaffinity labeling experiments, all biologically relevant AngII analogues produced an excellent photoaffinity labeling yield with the exception of ^{125}I $[Tdf^1]$AngII and ^{125}I-$[Sar^1,Bpa^7]$AngII. ^{125}I$[Bpa^1]$AngII photolabeled all Met-mutants and specific proteolytic cleavage confirmed concomitant labeling with the

S. Del Valle et al. (eds.), *Peptides for Youth: The Proceedings of the 20th American Peptide Symposium*, 329
DOI: 10.1007/978-0-387-73657-0_147, © Springer Science+Business Media, LLC 2009

extracellular loop (*ECL*) *2* (as previously described [5]), but also the *N*-terminus (*NT*) and the *ECL3* of hAT$_1$-CAM. However, ^{125}I-[Tdf^1]AngII photolabeled practically no proteinaceous material probably due to the quenching of the photogenerated carbene by extracellular water [6]. Both ^{125}I-[Sar^1,Bpa^2]AngII and ^{125}I-[Sar^1,Tdf^2]AngII presented different photolabeling profiles: ^{125}I-[Sar^1,Bpa^2] AngII labeled mostly the *ECL2* and to a lesser degree the *ECL3* of hAT$_1$-CAM, whereas ^{125}I-[Sar^1,Tdf^2]AngII labeled mostly the *NT* and to a lesser degree the *ECL3* of hAT$_1$-CAM. Both ^{125}I-[Sar^1,Bpa^3]AngII and ^{125}I-[Sar^1,Tdf^3]AngII displayed identical and exclusive labeling with *ECL2* of hAT$_1$-CAM, as previously described [7]. ^{125}I-[Sar^1,Tdf^7]AngII labeled exclusively *TMD7* of hAT$_1$-CAM, similar to the reported [Sar^1,Bpa^8]AngII [8]. Nevertheless, ^{125}I-[Sar^1,Bpa^7]AngII photolabeled no proteinaceous material for yet unknown reasons. Finally, ^{125}I-[Sar^1,Tdf^8]AngII established concomitant labeling with *TMDs3*, *5* and *6* of hAT$_1$-CAM.

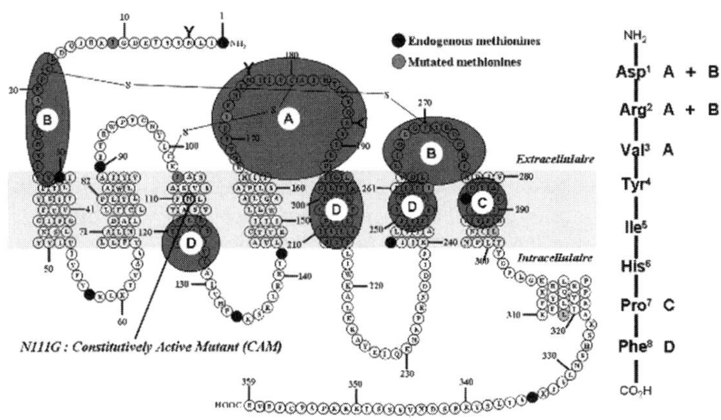

Fig. 1. Summary of the identified AngII / hAT$_1$ photolabeled regions of interactions.

Together, these results evidence the extracellular wide-ranging contact of the AngII *N*-terminal with the *NT*, the *ECL2* and the *ECL3* of AT$_1$ as well as progressively restricted contact with the *ECL2* of AT$_1$ at position 3 of AngII. The AngII *C*-terminal is in close contact with the helical bundle of the *TMDs3*, *5*, *6* and *7* of AT$_1$.

Acknowledgments

This ongoing study is funded by the Canadian Institutes of Health Research (CIHR) and the Quebec chapter of the Heart and Stroke Foundation of Canada (HSFC) to G.G., R.L. and E.E.

References

1. Vodovozova et *al.*, Biochemistry (Mosc), 2007, **72**(1): p. 1-20.
2. Patny et *al.*, Curr Med Chem, 2006, **13**(14): p. 1667-91.
3. de Gasparo et *al.*, Pharmacol Rev, 2000, **52**(3): p. 415-72.
4. Fillion et *al.*, J Med Chem, 2006, **49**(7): p. 2200-9.
5. Laporte et *al.*, Mol Endocrinol, 1999, **13**(4): p. 578-86.
6. Weber et *al.*, J Pept Res, 1997, **49**(5): p. 375-83.
7. Boucard et *al.*, Biochemistry, 2000, **39**(32): p. 9662-70.
8. Perodin et *al.*, Biochemistry, 2002, **41**(48): p. 14348-56.

Synthesis and Biological Activity of Transmembrane Peptides Derived from FPR Family Receptors

Daisuke Sugiyama[1], Ryo Hayashi[1], Satoshi Osada[1], Yuhei Hamasaki[2], Ichiro Fujita[2] and Hiroaki Kodama[1]

[1]*Department of Chemistry, Faculty of Science and Engineering, Saga University, Saga 840-8502, Japan;* [2]*Department of Pediatrics, Faculty of Medicine, Saga University, Saga 849-8501, Japan*

Introduction

Receptors, translocators and membrane enzymes exist as membrane integral proteins upon cellular surface. They are known to function as dimers or even higher oligomers. It is reported that transmembrane (TM) peptides derived from membrane integral protein modulated dimerization and consecutive function of parent proteins [1]. Neutrophils act as the first line of defense against the invasion of micro-organisms in the body. Bacterial metabolites such as N-formylmethionyl peptides are chemoattractants for neutrophils and bind to specific surface receptors, N-formyl peptide receptor (FPR) that trigger specific host defensive processes such as chemotactic migration and killing of microorganisms through superoxide and phagocytosis. Human FPR was first defined biochemically in 1976, as a high-affinity binding site on the surface of neutrophils for prototypic N-formyl peptide, formyl-Met-Leu-Phe-OH (fMLP) [2]. FPR has the structure of seven membrane-spanning α-helical domains, and three subtypes exist. However, few reports on the multimer formation of FPR were made until now. To regulate the biological activities of FPR might contribute to overcome immunologic disease. In the present study, seven transmembrane peptides (hFPRTMn, n is a number of transmembrane helix) of human formyl peptide receptor were synthesized to evaluate the interaction with receptor protein and those biological activities.

Results and Discussion

TM peptide sequences of human FPR were determined based on Miettinen's report that mapped the ligand-binding site in TM regions of human FPR [3]. High polar tag, SKSKSK sequence was attached to N- or C-terminal of TM peptides to improve solubility and to control directionality of insertion into lipid membrane. To increase the propensity for helices, TM domain peptide was acetylated and amidated at the N- and C-terminals, respectively. Cys residues were substituted by Ser residues to prevent unexpected cross-linking by disulfide formation. Trp residues were inserted in N-terminus of hFPRTM2 and C-terminus of hFPRTM3, 5 and 7 evaluate that interaction unit biomembranes by fluorescence spectroscopy (Table 1). TM peptides were synthesized through a stepwise solid phase method. Cleavage and deprotection

Table 1. Structures of synthetic peptides

Sequences	Peptides
SKSKSKIITYLVFAVTFVLGVLGNGLVI**W**VA-NH_2	hFPRTM1
Ac-**W**TISYLNLAVADFSFTSTLPFF**SKSKSK**-NH_2	hFPRTM2
SKSKSKFLFTIVDINLFGSVFLIALIALD**W**-NH_2	hFPRTM3
Ac-SLAKKVIIGP**W**VMALLLTLPVIIR**SKSKSK**-NH_2	hFPRTM4
SKSKSKRFIIGFSAPMSIVAVSYGLIATK**W**-NH_2	hFPRTM5
Ac-VLSFVAAAFFLS**W**SPYQVVALIATVR**SKSKSK**-NH_2	hFPRTM6
SKSKSKVTSALAFFNSSLNPMLYVFMGQ**W**-NH_2	hFPRTM7

S. Del Valle et al. (eds.), *Peptides for Youth: The Proceedings of the 20th American Peptide Symposium*, 331
DOI: 10.1007/978-0-387-73657-0_148, © Springer Science+Business Media, LLC 2009

were performed by TFA treatment. Purities and structures of peptides were confirmed by treatment HPLC and MALDI-TOF mass spectroscopy.

To investigate whether the hFPRTM peptides are inserted into the membrane, fluorescence measurements were carried out in phosphate buffer and SDS micelles. The sensitivity of Trp fluorescence emission to environment polarity allows us to use the Trp residues to monitor the binding of peptides to the micelles. hFPRTM peptides were resulted in increase of the fluorescence intensity and significant blue shifts of fluorescence emission maxima by ~23 nm in SDS micelles. These results suggested that hFPRTM peptides inserted into the micelles.

CD analyses were carried out to investigate the conformational properties of hFPRTM peptides in phosphate buffer and membrane-mimetic environments. To determine whether or not micelles could stabilize the folding of the peptides by providing a hydrophobic environment, CD spectra of hFPRTM4, 5 and 6 were obtained, which exhibited double minima around 205 and 220 nm and a positive band at 195 nm, indicating the α-helical property in 20 mM SDS micelles, which correlates with the micelle formation by SDS [4].

Biological activities of synthetic peptides were evaluated in human neutrophils. It was observed that pretreated human neutrophils with hFPRTM peptides were enhanced superoxide anion production by successive fMLP treatments. Neutrophils treated with hFPRTM4, 5 and 6 produced 2-4 folds superoxide anion compared with untreated cells (Fig. 1a). The effects on this superoxide production were dependent on the sequence of hFPRTM peptides. The possibility that hFPRTM peptides act as a ligand was checked using their short peptide fragments that were composed of fourth TM region of human FPR. All of short peptides did not enhance release of superoxide anion in human neutrophils (Fig. 1b). These results suggested that hFPRTM4 did not behave as a ligand. These results strongly suggested that hFPRTM4, 5 and 6 interacted with TM regions of membrane proteins via helix-helix interaction which responsible for induction of neutrophil priming.

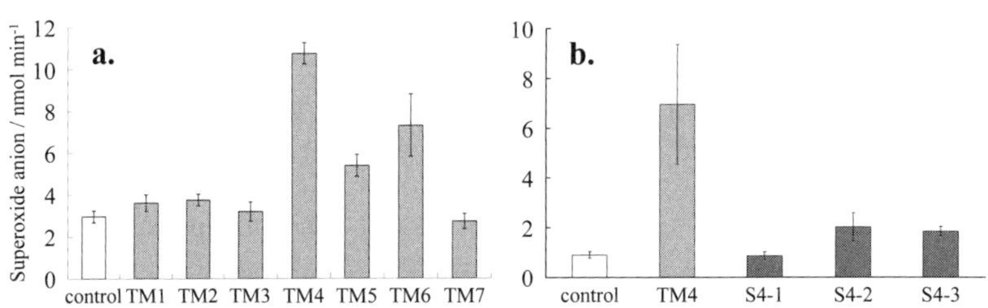

Fig. 1. Effect of hFPRTM peptides on superoxide anion production. Human neutrophils were incubated for 20 min at 37°C in the presence or absence of 10^{-6} M synthetic TM peptides and then stimulated by fMLP 10^{-7} M a) hFPRTM peptides. b) hFPRTM4 and short peptides: LAKKVIIGPW-NH_2 (S4-1), GPWVMALLLT-NH_2 (S4-2), and ALLLTLPVII-NH_2 (S4-3).

References

1. Heldin, C.H. *Cell* **80**, 213-223 (1995).
2. Showell, H. J., Freer, R. J., Zigmond, S. H., Schiffmann, E., Aswanikumar, S., Corcoran, B., Becker, E. L. *J. Exp. Med.* **143**, 1154-1169 (1976).
3. Miettinen, H.M., Mills, J.S., Gripentrog, J.M., Dratz, E.A., Granger, B.L. and Jesaitis, A.J. *J. Immunol.* **159**, 4045-4054 (1997).
4. Wang, C. and Deber, C.M. *J. Biol. Chem.* **275**, 16155-16159 (2000).
5. Hayashi, R., Osada, S., Yoshiki, M., Sugiyama, D., Fujita, I., Hamasaki, Y. and Kodama, H. *J. Biochem.* **139**, 981-988 (2006).

Incorporation of the Unnatural Amino Acid *p*-benzoyl-*L*-phenylalanine (Bpa) into a G Protein-coupled Receptor in its Native Context

George Umanah[1], Li-Yin Huang[1], Peter G. Schultz[2], Fred Naider[3] and Jeffrey M. Becker[1]

[1]Department of Microbiology, University of Tennessee, Knoxville, TN 37996; [2]Department of Chemistry. The Scripps Research Institute, La Jolla, CA 92037; [3]Department of Chemistry, College of Staten Island, CUNY, NY 10314

Introduction

Ligand binding initiates a change in the conformation of G protein-coupled receptors (GPCRs) resulting in activation of the G protein-mediated signal transduction cascade [1]. We are studying a novel approach to elucidate the dynamics of GPCR structure by the co-translational introduction of unnatural amino acids (UAAs) into the receptor.

UAAs can be synthesized to contain a variety of chemical moieties for use as photoaffinity labels, fluorescent labels, or spectroscopic probes. Orthogonal tRNA/aminoacyl-tRNA synthetase pairs evolved and expressed in the target cell have been used to incorporate UAAs into heterologously expressed protein in living cells [2]. The mutated tRNA, charged with its UAA, recognizes the amber TAG stop codon and incorporates the non-natural amino acid into the nascent polypeptide chain (See Figure below). UAAR has been widely used in the heterologous *Xenopus* oocyte expression system to insert UAAs into a variety of receptors and channel proteins. Thus far the genetic incorporation of UAAs into an integral membrane protein in its native eukaryotic host cell has not been accomplished.

Here we report the site-specific incorporation of *p*-benzoyl-*L*-phenylalanine (Bpa) into Ste2p, the prototypical yeast GPCR, using an orthologous tRNA/aminoacyl tRNA synthetase pair (See Figure Below).

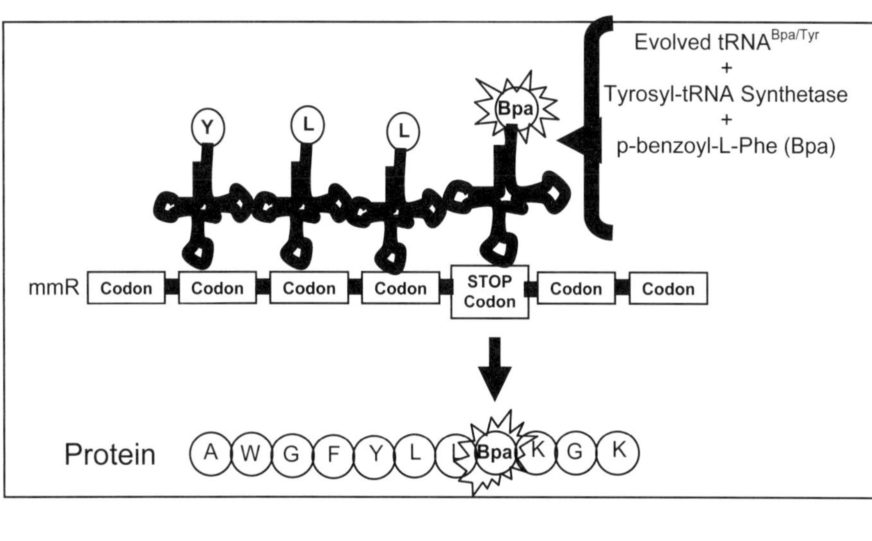

S. Del Valle et al. (eds.), *Peptides for Youth: The Proceedings of the 20th American Peptide Symposium*, 333
DOI: 10.1007/978-0-387-73657-0_149, © Springer Science+Business Media, LLC 2009

The TAG stop codon was engineered into specific sites within the *STE2* coding region. The cells accumulated Bpa from the growth medium either as the free amino acid analog or as a methionyl-Bpa peptide. Upon translation, Bpa was incorporated into the nascent Ste2p and ultimately expressed at the cell surface.

Results and Discussion

Eight different TAG stops were inserted into the *STE2* coding sequence by site-directed mutagenesis to create Ste2p-F55^{TAG}, Ste2p-S107^{TAG}, etc. (Table 1). To assay the sensitivity of the mutant receptors to α-factor, cells were grown with and without Bpa, and the amount of pheromone necessary to generate a 20 mm diameter halo was determined (Table 1).

Table 1. Pheromone required (μg) for 20 mm diameter halo for TAG mutants.

	WT	F55	S107	G115	V127	G188	Y193	F204	Y266
- Bpa	0.18	0.58	0.74	0.35	0.70	-	-	0.96	-
+ Bpa*	0.28	-	8	1.54	2.3	5.9	0.32	-	-

** Bpa concentration was 2 mm. Absence of halo is indicated by a dash (-)*

Upon binding pheromone to native Ste2p a signal transduction cascade is activated and growth arrest of the cell occurs, resulting in a clear zone or "halo"around a disk impregnated with α-factor. In the absence of Bpa, the tRNA should not be charged, resulting in a truncated Ste2p which will not respond to pheromone. In the presence of Bpa, the TAG codon would be recognized by the charged tRNA, and full-length protein should be synthesized.

The G188^{TAG} and Y193^{TAG} receptors did not respond to pheromone in the absence of Bpa, but growth arrest did occur in the presence of Bpa, although the G188 receptor did not respond as well as the Y193 receptor. (5.9 μg vs. 0.32 μg, respectively). This indicates that Bpa incorporation resulted in functional proteins Ste2p-G188Bpa and Ste2p-Y193Bpa For the S107^{TAG}, G115^{TAG} and V127^{TAG} receptors halos formed in the absence of Bpa (0.74, 0.35, and 0.70 μg per disk respectively), indicating that there was some "read-through" of the amber stop codon in the absence of Bpa. Read-through most often occurs when the context of the nucleotides flanking the stop codon is not optimal for termination [3]. Halos formed in the presence of Bpa, but the receptors were less sensitive and required more pheromone compared to the wild-type receptor (S107 ^{TAG} - 8 μg, G115^{TAG} - 1.5 μg and V127^{TAG} - 2.3 μg per disk) to generate a 20 mm halo. For the F55^{TAG} and F204^{TAG} mutants, the receptors were still active in the absence of Bpa addition, however in the presence of Bpa, no halos were observed, suggesting that under these conditions Bpa was incorporated into Ste2p, but the amino acid analog was not tolerated at those positions. The Y266^{TAG} receptor was not functional in either the presence or absence of Bpa, further supporting the essential role of the Y266 residue in signal transduction, as has been noted previously [4].

As determined by immunblot analysis wild-type Ste2p was synthesized at full length in both the presence and absence of Bpa and could be detected by antibodies directed against either the N-terminus or C-terminus of the protein. In contrast, the receptors encoded by F55^{TAG} and G188^{TAG} were detected only in the presence of Bpa. For the Y193^{TAG} mutant, a low level of expression was detected in the absence of Bpa, and was enhanced in the presence of Bpa. C-terminally truncated forms of the receptor were observed for the G188^{TAG} and Y193^{TAG} encoded receptors using antibody to the N-terminus. Truncated forms of the F55^{TAG} receptor were not detected with the N-terminal antibody. Expression of the full length F55^{TAG} Bpa-containing receptor was enhanced by increasing the Bpa concentration in the growth medium. This suggested that entry of Bpa into the cell is important in determining the efficiency of Bpa insertion into the protein. Cell surface expression of the receptors encoded by F55^{TAG}, G188^{TAG}, and Y193^{TAG} grown in the presence of Bpa was also confirmed by whole-cell saturation binding analysis using radiolabeled α-factor.

To improve delivery of Bpa into the cell we synthesized the dipeptide Met-Bpa and used it as a source of Bpa during cell growth. Small peptides enter yeast cells across the di-/tripeptide transporter Ptr2p [5, 6] and are hydrolyzed to free amino acids by intracellular peptidases. In the absence of Bpa, Y193^{TAG} mutant was minimally expressed as detemined by immunotblot, while in the presence of Met-Bpa (0.1 mM) full length protein was detected at a level exceeding the expression observed when grown on free Bpa (0.1 mM).

MALDI-TOF mass spectrum analysis of purified wild-type and receptor encoded by G188^{TAG} was performed to determine whether incorporation of Bpa was achieved in the mutant receptor. The G188^{TAG} receptor was chosen based on the CNBr cleavage profile for Ste2p which facilitated mass spectroscopy analysis. The substitution of Bpa (269.30 Da) for glycine (75.07 Da) at position 188 resulted in a mass shift of 194 Da in the 188Bpa-containing peptide. This shift confirmed incorporation of Bpa into Ste2p at position 188.

The experiments reported are the first to our knowledge to incorporate an unnatural amino acid into a GPCR in its natural environment. We believe such methodology will provide a very rich source of experimental methodologies for studying the structure and function of these important membrane proteins.

Acknowledgments

This work was supported by grants NIH GM-22086(FN) and NIH GM-22087(JMB).

References

1. Eilers, M., *et al. Biochemistry* **44**, 8959-8975 (2005).
2. Wang, L., Xie, J., and Schultz, P. G. *Annu. Rev. Biophys. Biomol. Struct.* **35**, 225-249 (2006).
3. Williams, I., *et al. Nucleic Acids Res* **32**, 6605-6616 (2004).
4. Lee, B. K., *et al. Biochemistry* **41**, 13681-13689 (2002)
5. Island, M. D., *et al. Curr. Genet.* **20**, 457-463 (1991).
6. Hauser, M., *et al. Mol. Membr. Bio*l. **22,** 215-27 (2005).

Ion Channel Formation of Dimeric Peptide Enhanced by Electrostatic Inter-Helical Interactions

Junichi Taira[1], Satoshi Osada[1], Masoud Jelokhani-Niaraki[2] Tsuguhisa Ehara[3] and Hiroaki Kodama[1]

[1]Department of Chemistry, Faculty of Science and Engineering, Saga University, Saga, 840-8502, Japan; [2]Wilfrid Laurier University, Waterloo, ON N2L 3C5, Canada; [3]Department of Physiology, Faculty of Medicine, Saga University, Saga, 849-8501, Japan.

Introduction

Helical ion channel forming peptides are simplified models of ion channel proteins. Appropriate assembly of helical peptides and stable pore structures has been pursued for artificial ion channel design. Several helix oligomerization techniques have been suggested as methods to construct stable and optimal pore structures [1, 2]. For example, peptide dimerization has been used to investigate relationships between structure and ion channel function for artificial ion channel design.

We have studied functions of 2-aminoisobutyric acid (Aib or B) containing amphiphilic helical ion channel peptides [3, 4]. In present study, heterodimer peptide composed of the Lys containing cationic helical peptide and the Glu containing anionic helical peptides were synthesized. Peptide structure, inter-, intra-molecular interactions and ion channel properties were investigated.

Ac-Cys-Gly-Gly-(Aib-Lys-Aib-Ala)$_5$-NH_2 Ac-Cys-Gly-Gly-(Aib-Glu-Aib-Ala)$_5$-NH_2	**K20E20**
Ac-(Aib-Lys-Aib-Ala)$_5$-NH_2	**BKBA20**
Ac-(Aib-Glu-Aib-Ala)$_5$-NH_2	**BEBA20**

Fig. 1. Structures of Aib containing dimeric and monomeric peptides.

Results and Discussion

Structure of the hetero dimeric model peptide K20E20, cationic peptide BKBA20 and anionic peptide BEBA20 are illustrated in Fig. 1. Electrostatic intra-molecular interactions were expected between incorporated charged residues in K20E20. Linear peptides were synthesized by solid phase method with Fmoc chemistry upon Rink amide resin. After peptide elongation, crude monomeric products were cleaved from the resins by TFA treatments. Peptide dimerization was achieved by air-oxidation at N-termini Cys side chains. Peptide purifications were carried out by RP-HPLC and confirmation of peptide homogeneities were performed by MALDI-TOF MS and amino acid analysis. To assess helix stability and peptide aggregation, CD spectra were collected in 50 mM phosphate buffer (Fig. 2). Helical contents of peptides were estimated from degrees at n-π* transitions [5]. The hetero-dimeric model peptides showed higher helical contents than those of monomeric peptides mixture. Ratios of $[\theta]_{n-\pi^*}$ / $[\theta]_{\pi-\pi^*}$ inferred helix aggregation. The ratio of the dimeric model peptide was larger than those of monomeric peptides, strong aggregation of helices in the dimeric peptide was suggested [6]. These results suggest the hetero dimerization enhance stable helix forming and helical aggregation.

S. Del Valle et al. (eds.), *Peptides for Youth: The Proceedings of the 20th American Peptide Symposium*, 337
DOI: 10.1007/978-0-387-73657-0_150, © Springer Science+Business Media, LLC 2009

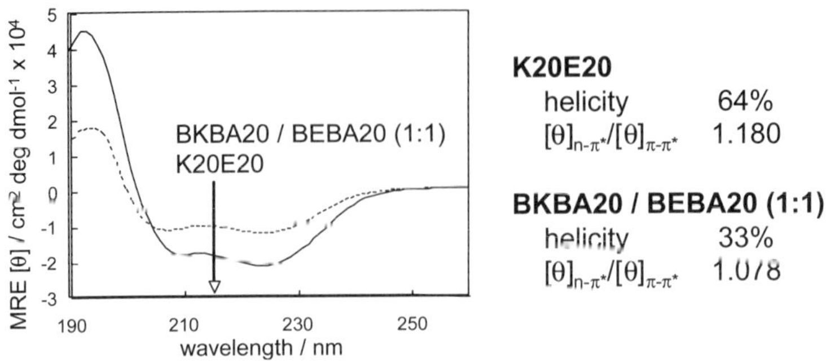

Fig. 2. CD comparison of hetero-dimeric peptides and monomeric peptides. Helicities were εστιματεδ φρομ $[\theta]_{n-\pi^*}$.

Ion channel experiments were performed by Tip-Dip method [3, 4]. Figure 3 exhibits ion conductance patterns of BKBA20 and K20E20 recorded on diphytanoyl phosphatidylcholine (DPhPC) bilayers. The model peptide showed not only large conductance values but also long channel opening durations distinguished from those of monomeric peptides. In addition, the fluctuations of conductance pattern of K20E20 were observed even in extremely low peptide concentrations, enhancement of pore formation by the hetero dimerization was suggested.

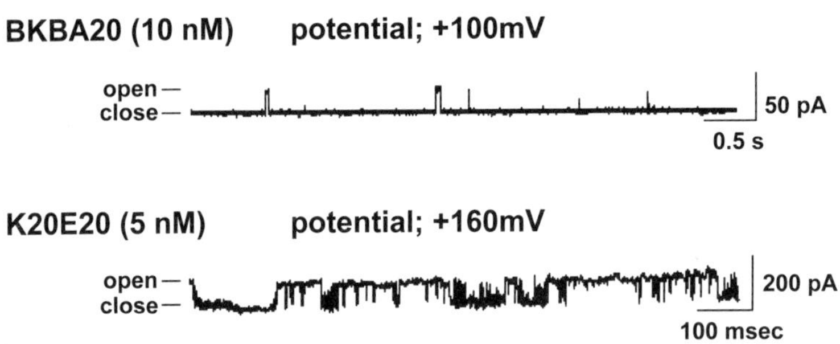

Fig. 3. Ion channel activities of BKBA20 and K20E20. The electrolyte solution was 100 mM KCl and buffered with 5 mM HEPES (pH 7.4).

Acknowledgments

The work was funded by a JSPS Research Fellowships for Young Scientists to J. T.

References

1. Regan, L. and DeGrado, W. F. *Science* **241**, 976-978 (1988).
2. Mutter, M. and Vuilleumier, S. *Angew. Chem. Int. Ed. Engl.* **28**, 535-554 (1989)
3. Higashimoto, Y., Kodama, H., Jelokhani-Niaraki, M., Kato, F. and Kondo, M. *J. Biochem.* **125**, 705-712 (1999).
4. Hara, T., Higashimoto, Y., Kodama, H., Jelokhani-Niaraki, M., Kato, F. and Kondo, M. *J. Biochem.* **130**, 749-755 (2001).
5. Chen, Y. H., Yang, J. T. and Martinez, H. M. *Biochemistry* **11**, 4120-4131 (1972).
6. Zhou, N. E., Kay, C. M. and Hodges, R. S. *J. Biol. Chem.* **267**, 2664-2670 (1992).

Temperature-induced Ligand Contact Point Variations of the hAT_1 Receptor and of the Constitutively Active Mutant N111G-hAT_1

J. Arsenault, M.P. La-Haye Renaud, M. Clément, D. Fillion, G. Guillemette, R. Leduc, P. Lavigne, and E. Escher

Département de Pharmacologie, Faculté de Médecine et des Sciences de la Santé, Université de Sherbrooke, Sherbrooke, Qc, J1H 5N4, Canada

Introduction

G-protein coupled receptors (GPCR) are crucial for intracellular signalling and primary targets for pharmacological intervention. Among typical peptidergic GPCRs, the human Angiotensin II (AngII) type 1 receptor, the hAT_1 receptor is an important pharmacological target and has been extensively scrutinized as a prototypical receptor for many years with methods such as photoaffinity labeling. In the present study, a variant of the Methionine Proximity Assay (MPA) (1) was applied to hAT_1 to investigate the effects of temperature on labeling and contact points of a neutral agonist on the receptor. Selected methionine mutants, located in transmembrane domain 6 (TMDVI); F249M, W253M and H256M of the hAT_1 and its constitutively active mutant (CAM) N111G-hAT1, were photolabeled with ^{125}I-[Sar^1, Bpa^8] AngII at specific and controlled temperatures ranging from -15°C to 37°C. These methionines were observed as contact points in previous studies (2, 4).

From a preliminary experiment, it was observed that these ligand-receptor contacts were dependent on temperature. The less accessible methionine labeled at higher temperatures implies an increased intrareceptor mobility measured at temperatures up to 37°C (3). The observed results also show such contact point changes in the CAM mutants but at higher temperatures which could indicate an altered conformational population between the WT and its CAM mutants. In addition, these results allow the assessment of the individual contributions of transmembrane residues to ligand binding and receptor activation.

Results and Discussion

This temperature-controlled version of MPA enables us to better understand the accessibility of the ligand contact points. Indeed, we show here, that there are structural differences on the WT-hAT_1 receptor compared to its CAM, the N111G-hAT_1.

The photolabeled ligand-receptor complexes were isolated and submitted to CNBr digestion. Fragments were resolved on a 16.5% tris-tricine SDS-PAGE electrophoresis. The corresponding fragment sizes were compared to previously established receptor fragments (2). Ligand contact regions were compared to its native TMDVII labeling versus the methionine specific, TMDVI, labeling. Quantified radioactive intensity was measured by densitometric analysis and γ-counting of cut gel slices. The following results are presented as the ratio of the TMDVII signal divided by the TMDVI signal.

By analyzing the photolabeling ratios on the WT (Figure 1), it appears that at -15°C the mutated residue W253M is more optimally positioned to bond covalently with the radicalized Bpa of the photolabel. F249M is in second position while H256M is the least accessible of these three mutants. Position H256 is closer to the extracellular side of the membrane than the other two residues. Interestingly the

S. Del Valle et al. (eds.), *Peptides for Youth: The Proceedings of the 20th American Peptide Symposium*,
DOI: 10.1007/978-0-387-73657-0_151, © Springer Science+Business Media, LLC 2009

order of preference for the labeling in different in the CAM: N111G/F249M is the most accessible, followed by N111G/H256M and N111G/W253M (Figure 1).

Another observation is the structural fluidity at higher temperatures, this permits the radicalized Bpa to tag the methionine. The mutated residue at H256M appears to be inaccessible at -15°C (TMDVII labeling) whereas at 37°C the majority of the photolabeling occurs at this residue (Figure 1 Right). Contrary to this, the W253M residue labeling is barely influenced by temperature, suggesting continuous accessibility within the whole temperature range. In the CAM receptor series, a different picture emerges with all three mutants having a stronger TMDVII, non-methionine labeling frequency at -15°C. TMDVI methionine-specific labeling increases proportionally with the temperature but does not reach the labeling preference as seen in the WT mutant series. This implies that in the inactive receptor conformation, the C-terminus of the ligand has better acces to TMDVI but less so in the active conformation. However, an increased conformational mobility permits a wider reach of methionine tagging to be within labeling distance of the contact residues in TMDVI. The presented method permits a wider application to the traditional photoaffinity labeling approach and further enlightens our understanding of GPCR conformations and activation.

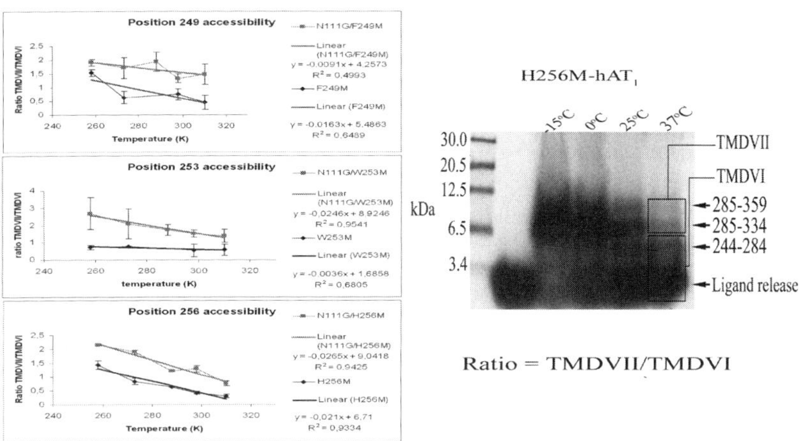

Figure 1: Temperature dependent photoaffinity labeling ratios of selected mutants and corresponding CAM double mutants of the hAT_1 receptor. On the right, SDS-PAGE gel of H256M-hAT1 to demonstrate counted receptor fragments.

Acknowledgments

We would like to thank CIHR and CHSFQ for our funding, B. Holleran, R. Hamel, E. Wieczerzak and M.-R. Lefebvre for their help and technical support.

References

1. Rihakova, L., et al., *J Recept Signal Transduct Res.*, **22**, 297-313 (2002).
2. Clément, M. et al., *J Biol Chem*, **280**, 27121-9 (2005).
3. Arsenault, J. et al., *J Peptide Science,* PAPER IN PRESS, (2007)
4. Clément, M. et al., *J Recept Signal Transduct Res.*, **26**, 417-33 (2006).

Structural changes in the binding pocket of the liganded angiotensin II receptor AT1 during activation.

Martin Clément, Gaétan Guillemette, Richard Leduc and Emanuel Escher

Department of Pharmacology, Université de Sherbrooke, Sherbrooke, Qc, J1H 5N4, Canada

Introduction

Labelling studies using benzophenone photoprobes have identified many ligand-receptor contact-points but with a surprisingly high ratio of methionine contacts [1,2]. Despite the fact that methionine represents a small proportion of the proteinogenic amino acids in most receptors, *p*-benzoyl-L-phenylalanine (Bpa) containing peptide labels have been shown to incorporate into Met residues at a disproportionate frequency [3]. Previous photochemical studies of the benzophenone radical have indicated that it exhibits strong selectivity for thioether groups due to the formation of an intermediate charge-transfer complex [4,5]. This selectivity would explain the high ratio of insertion into methionine residues of target proteins through benzophenone photoaffinity labelling. This property can be exploited to introduce Met residues into target structures as 'bait' with the goal of identifying other receptor residues that are in close proximity to the ligand label. The immediate molecular environment of this ligand residue can thus be determined.

The C-terminal residue of angiotensin II (AngII) interacts with the 7th transmembrane domain (TMD) of the AngII receptor AT_1 [6] at non-Met contacts since no endogenous Met residues is in the immediate vicinity of the ligand binding pocket. We used the methionine proximity assay (MPA) to investigate the binding environment of the C-terminal residue of AngII within this receptor with [Sar^1, Bpa^8]AngII, a well-characterized neutral antagonist [7]. We found 11 ligand contacts in TMD III, VI and VII with this approach. In the present study we applied the MPA approach also to the constitutively active hAT_1-mutant N111G-hAT_1 and compared the results with those obtained from wt-hAT_1. This would allow evidencing different contacts in the receptor and, eventually, structural changes related to receptor activation.

Results and Discussion

A further series of 54 transmembrane X→Met mutants of hAT_1 were constructed in all TMDs as double mutants on the N111G mutation (itself in TMD III and producing constitutive activity). The affinity of all double mutants was assessed, showing close to 1nM affinity for [Sar^1, Bpa^8]AngII for all, except S105M/N111G, where no binding was observed. Photolabeling of all receptor mutants was followed by isolation of the covalently labeled receptor complex, by CNBr digestion and by SDS-PAGE. Labeling of wt-hAT_1 and N111G-hAT_1 produced a 7.2 kDa fragment, corresponding to TMD VII and most of the intracellular C-terminal sequence (285-334) [6] (Fig.1). All mutant receptors in TMD I, II, IV and V produced the same 7.2kDa fragment, indicating that labeling took place on TMD VII only. Mutants L112 and Y113 (TMD III), F249, W253, H256 (TMD VI), F293, N294, N295, C296, and L297 (TMD VII) produced new fragments, indicative of ligand contacts and are identical to those of the WT-hAT_1 MPA. The only exception is the T260M/N111G mutant which was no longer MPA positive. Surprisingly, the double mutants A104M/N111G and H263M/N111G recovered ligand affinity where the single mutants were inactive.

S. Del Valle et al. (eds.), *Peptides for Youth: The Proceedings of the 20th American Peptide Symposium*, 341
DOI: 10.1007/978-0-387-73657-0_152, © Springer Science+Business Media, LLC 2009

Those three differences suggest a less hindered "entrance" of the N111G receptor close to the extracellular portion of AT_1.

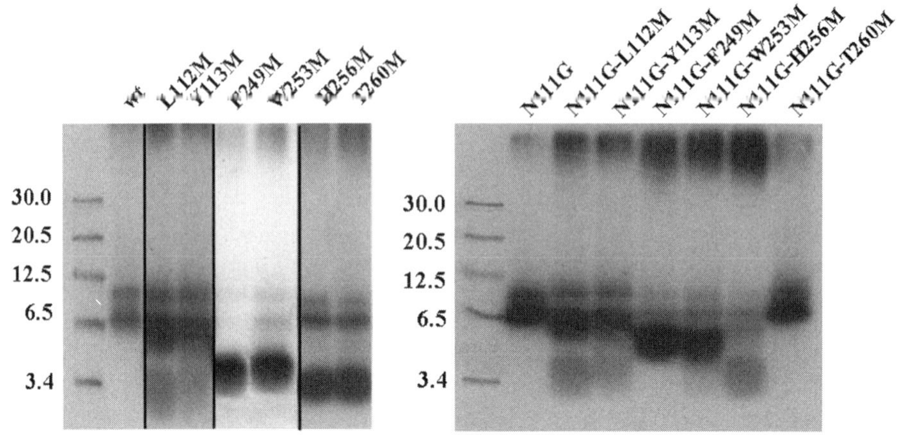

Fig. 1. CNBr cleavage of Met-mutants and Met-mutants of the N111G-hAT1 receptor.

Conclusions

These results confirm the structural model of the AngII-hAT_1 interaction where the C-terminal residue of AngII points deep into the transmembrane core towards TMD III, TMD VI and TMD VII. The small but significant differences between WT receptor and N111G-receptor further support an outward movement of TMD VII in the receptor activation mechanism without any other major structural change.

Acknowledgments

This work was supported by the Canadian Institutes of Health Research and the Quebec chapter of the Canadian Heart and Stroke Foundation.

References

1. Kage, R., et al. *J. Biol. Chem.* **271**, 25797-25800 (1996).
2. Bisello, A., et al. *J. Biol. Chem.* **273**, 22498-22505 (1998).
3. Bremer, A.A., et al. *J. Biol. Chem.* **276**, 22857-22861 (2001).
4. Bobrowski, K., et al. *J. Am. Chem. Soc.* **104**, 10279-10288 (1992).
5. Marciniak, B., et al. *J. Phys. Chem.* **97**, 11937-11943 (1993).
6. Perodin, J., et al. *Biochem.* **48**, 14348-14356 (2002).
7. Clément, M., et al. *J. Biol. Chem.* **280**, Epub (2005).

PEPTIDES AND IMMUNITY

Vaccine delivery utilizing liposaccharides

Pavla Simerska[1], Ross P. McGeary[1], Abu-Baker M. Abdel-Aal[1], Peter M. Moyle[1], Colleen Olive[2], Michael F. Good[2] and Istvan Toth[1]

[1]*The University of Queensland, St Lucia, Queensland, Australia*
[2]*Queensland Institute of Medical Research, Herston, Queensland, Australia*

Introduction

Vaccination has proven to be one of the most important and cost-effective public health interventions. Problems with conventional vaccines (such as their instability, the availability of carriers and adjuvants, and the potential danger of using live microorganisms) have led to the development of synthetic vaccines [1]. The goal of this study was to develop a synthetic peptide-based vaccine to prevent group A streptococcal (GAS) infection, a goal that is hampered by the widespread diversity of circulating GAS strains. Infection with GAS can lead to many diseases, including pharyngitis, pyoderma, invasive diseases, rheumatic fever and rheumatic heart disease.

Carbohydrates provide multiple attachment points for peptides, and the conjugation of multiple copies of a single peptide to a carrier has been demonstrated to produce higher antibody responses compared to the administration of a single peptide epitope [2]. Therefore the conjugation of a drug with lipid (adjuvant) and sugar units (carrier) represents one of the most important strategies being investigated in drug delivery.

Results and Discussion

The poly-substituted sugars, tetra-*O*-(cyanoethyl) galactopyranosyl [3] and glucopyranosyl azides, were converted into derivatives bearing linkers with terminal carboxylic acid functionalities at the anomeric position and four Boc-protected terminal amino groups. These derivatives were attached to MBHA resin previously loaded with a lipidic adjuvant (C_{12}–Gly–C_{12}–C_{12}–Gly; C_{12} = 2-amino-D,L-dodecanoic acid). After Boc-deprotection, peptide sequences (J8 or J14 antigens) were coupled onto the carbohydrate scaffolds using stepwise solid-phase peptide synthesis to form novel vaccine systems. These peptides contain a portion of the carboxy-terminal C-repeat region of the GAS cell surface M protein, a conserved sequence in 70 % of GAS strains, thereby offering the capacity to create a vaccine which could potentially prevent up to 70% of GAS infections. Finally, the products (see glucose core in Fig. 1) were cleaved from the resin using anhydrous hydrofluoric acid, then purified and characterized by mass spectrometry and HPLC. As a control, a convential lipid-core peptide system was produced which contained four copies of the J8 peptide antigen (Fig. 1).

S. Del Valle et al. (eds.), *Peptides for Youth: The Proceedings of the 20th American Peptide Symposium*, 345
DOI: 10.1007/978-0-387-73657-0_153, © Springer Science+Business Media, LLC 2009

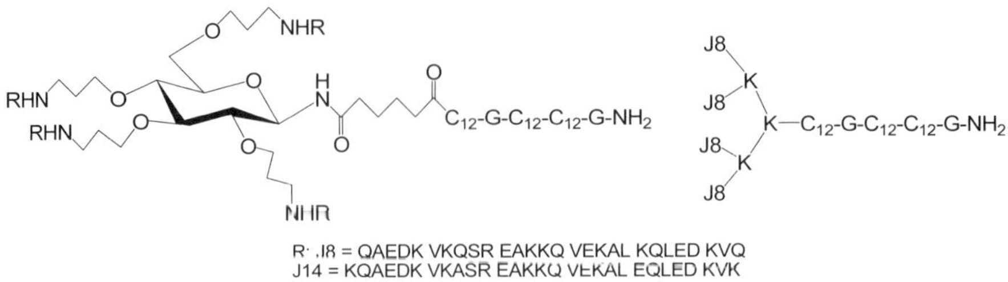

Fig. 1. Structure of the glucose-lipid core template bearing J8 or J14 antigens and poly-lysine core with J8 antigen..

In vivo experiments were performed in female B10.BR (H-2^k) mice. The mice (n=5/group) were injected subcutaneously on days 0, 21, 31 and 41 with 30 μg of these immunogens (administered with and without complete Freund's adjuvant (CFA)). Blood samples were collected one day before each injection and were analyzed by an enzyme-linked immunosorbent assay (ELISA) for total antigen-specific serum IgG antibodies elicited by immunization.

Promising preliminary results (e.g. IgG antibody responses to the galactose-core vaccine in the presence of CFA) are shown in the Table 1. The administration of these vaccines without any additional adjuvants is being tested to determine their capacity to elicit IgG antibodies against the attached peptide antigens.

Table 1. IgG antibody responses to peptide complexes (with CFA).

Tested compounds	serum antibody titre J8
J8	13 000
J8-LCP	1 100 000
J8-Gal-LCP	3 300 000

A novel drug delivery system combining lipoamino acids and carbohydrates has been developed. Immunological studies demonstrated its capacity to induce antibody responses. Moreover, these constructs are suitable for use as a template for synthesis of other vaccines.

Acknowledgments

This work was supported by the Australian National Health and Medical Research Council (NHMRC).

References

1. Olive, C., *et al. Mini Rev. Med. Chem.* I, 429 (2001).
2. Tam, J. P. *Proc. Natl. Acad. Sci. USA* **85**, 5409 (1988).
3. McGeary, R., *et al. Tetrahedron* **57**, 8733 (2001).

Vaccine Delivery: Synthesis and Investigation of a Highly Pure, Multi-Epitopic Lipopeptide Vaccine Candidate

Istvan Toth[1], Peter M. Moyle[1], Pavla Simerska[1], Yoshio Fujita[1], Colleen Olive[2] and Michael F. Good[2]

[1]*School of Molecular and Microbial Sciences, the University of Queensland, St Lucia, QLD 4072, AUSTRALIA;* [2]*Queensland Institute of Medical Research, Herston, QLD 4029, AUSTRALIA*

Introduction

Traditional vaccine development approaches (e.g. live attenuated or killed microorganisms) have produced many effective/successful vaccines. Numerous diseases exist, however, for which these methods are not suitable. Peptide vaccines, incorporating appropriate immunogenic epitopes, have advantages over traditional vaccines including a reduced risk of adverse effects, and the capacity to focus immune responses towards important antigens. In general, peptide vaccines stimulate poor or no immune responses when administered on their own, necessitating their administration with powerful adjuvants [e.g. the highly toxic Freund's complete adjuvant (CFA)]. The conjugation of bacterial or synthetic lipids to peptide antigens has been demonstrated in human clinical trials [1] to be a potentially safe means to adjuvant peptide epitopes. The lipid core peptide (LCP)-system (Figure 1) [2], incorporating a lipid-based adjuvant, a carrier, and one or more peptide epitopes, has been shown to adjuvant peptide epitopes without the need for additional adjuvants. Traditionally LCP-systems have been synthesized using solid-phase peptide synthesis (SPPS), however due to their inherent complexity, the synthesis and purification of these systems to yield a homogeneous product is difficult, if not impossible in most cases. We have therefore worked towards the development of a system [3-5], using native chemical ligation, to enable the synthesis of highly-pure, multi-epitopic LCP-systems capable of satisfying the regulatory requirements for use in human clinical trials.

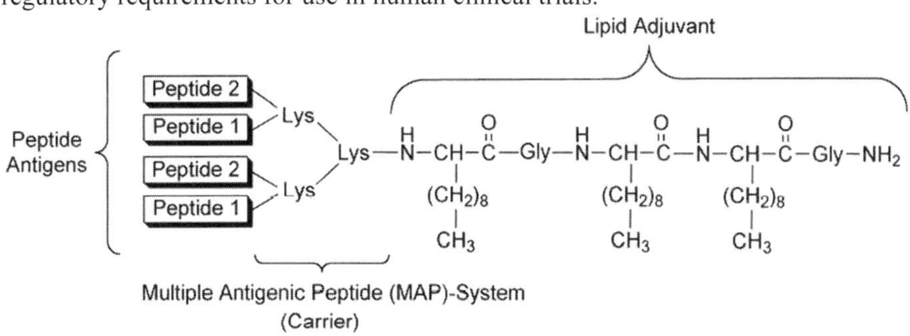

Fig. 1. Structural representation of a LCP-system.

Results and Discussion

Three peptides were synthesized using stepwise SPPS, containing peptide antigens (J8, 88/30, & PL1) from group A streptococcal strains common to Australian

S. Del Valle et al. (eds.), *Peptides for Youth: The Proceedings of the 20th American Peptide Symposium*,
DOI: 10.1007/978-0-387-73657-0_154, © Springer Science+Business Media, LLC 2009

Aboriginal populations, to enable the construction of a highly-pure, triepitopic LCP-analogue (Figure 2). Peptide **1** incorporated the lipid adjuvant portion of the LCP-system at the carboxyl (C)-terminus, followed by a glycine residue and a lysine residue at the amino (*N*)-terminus. The PL1 peptide antigen (PL1: EVLTR RQSQD PKYVT QRIS) was then conjugated to the peptides *N*-terminus, with an unprotected cysteine residue attached to the lysine side-chain in order to permit the conjugation of further peptide epitopes using native chemical ligation. Peptide **2** was designed to incorporate a thioester at the C-terminus, to enable its ligation to peptide **1**, with a glycine residue and a lysine residue attached at the *N*-terminus. Once again, a peptide antigen was attached to the lysine *N*-terminus, in this case the 88/30 peptide antigen (88/30: DNGKA IYERA RERAL QELGP), and an acetamidomethyl (Acm)-protected cysteine residue was conjugated to the lysine side-chain. Following the ligation of peptides **1** and **2**, the Acm protecting group was removed to provide a free cysteine residue onto which peptide **3** could be attached. Peptide **3** was designed to incorporate a thioester at the C-terminus, to facilitate its ligation to the di-peptide product from the first ligation reaction, followed by a glycine residue and the J8 peptide antigen (J8: QAEDK VKQSR EAKKQ VEKAL KQLED KVQ).

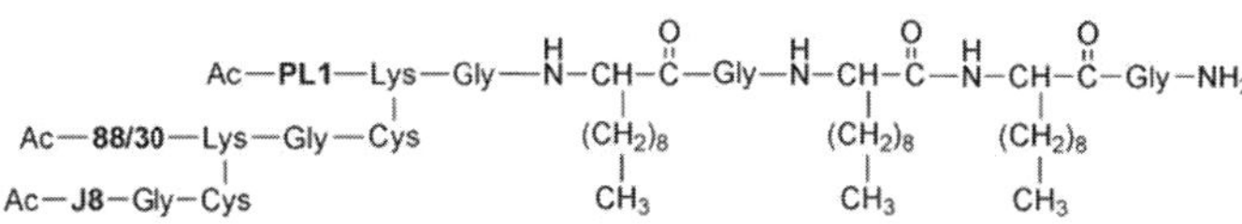

Fig. 2. Structural representation of a highly-pure, tri-epitopic LCP-analogue.

All ligation reactions were performed in 0.1M phosphate buffer pH 7.6 containing 1% (w/v) sodium dodecyl sulfate (SDS). When ligation reactions were performed without SDS, or in the presence of organic solvents, minimal to no ligation was observed. Cleavage of the cysteine Acm protecting group was performed using iodine, followed by disulfide bond reduction using tris(2-carboxyethyl)phosphine hydrochloride. The use of more traditional Acm-deprotection reagents [i.e. mercury (II) acetate, or various silver (I) salts] proved to be unsuitable in this case. Overall the product was obtained in a good yield (44.1%), and high purity (97.7%), and was readily characterized using mass-spectrometry (MS), sodium dodecyl sulfate-polyacrylamide gel electrophoresis (SDS-PAGE) and reversed-phase high-performance liquid chromatography (RP-HPLC) (see Figure 3).

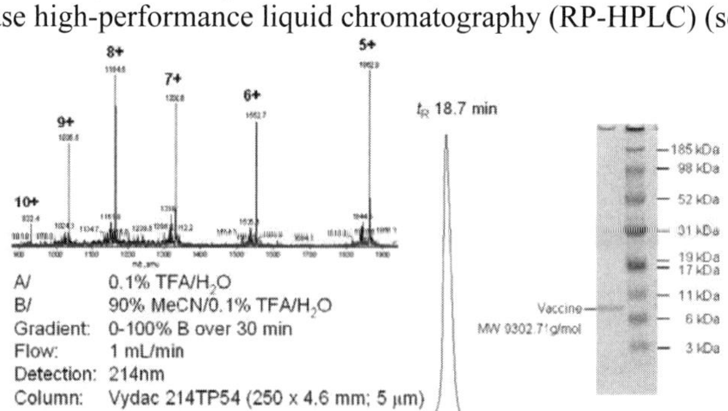

Fig. 3. Characterization data (MS, RP-HPLC, & SDS-PAGE) for the tri-epitopic LCP-analogue.

The capacity of the vaccine to elicit antigen-specific serum IgG antibodies against the PL1, 88/30 and J8 peptide antigens was assessed in B10.BR (H-2^k) mice. The mice (n = 10/group) were primed with the vaccine (30 μg) either in phosphate buffered saline (PBS, total volume 50 μL) or as a 1:1 emulsion with CFA (positive control). Three weeks later, the mice received five boosts at weekly intervals with 3 μg of the vaccine in PBS. Prior to each boost, and one week after the last boost, sera was collected to assess the levels of serum antigen-specific IgG antibodies using an enzyme-linked immunosorbent assay (ELISA). The administration of the vaccine to mice without any additional adjuvants was found to elicit serum antigen-specific IgG antibodies against each of the peptide antigens featured in the vaccine (Figure 4).

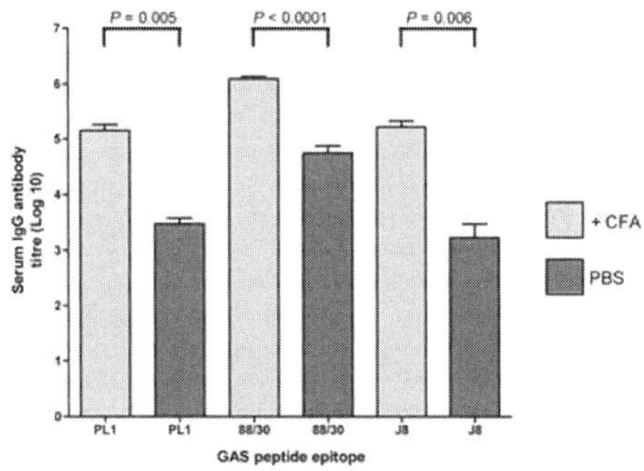

Fig. 4. Serum antigen-specific IgG antibody titers elicited following immunization of B10.BR (H-2^k) mice with the tri-epitopic LCP-analogue administered with CFA or in PBS. Antibody titers against the PL1, 88/30 and J8 antigens were measured one week after the final boost.

In conclusion, we have developed a successful technique to enable the synthesis of highly-pure, multi-epitopic LCP-system analogues, using native chemical ligation. The use of surfactants (e.g. SDS) was found to be essential for this process due to the highly lipophilic nature of the LCP-system lipid adjuvant. In this example, we were able to synthesize a tri-epitopic LCP-analogue, using this technique, in an excellent yield (44.1%) and in a highly pure state (97.7%). Immunological assessment of this analogue demonstrated its capacity to elicit high titers of serum IgG antibodies against each of the antigens included in the vaccine.

Acknowledgments

This work was funded by the National Health & Medical Research Council (Australia), the National Heart Foundation (Australia), and the Prince Charles Hospital Foundation. PMM was funded by the Queensland Government's growing the smart state PhD funding scholarship.

References

1. Durier, C., Launay, O., Meiffrédy, V., Saïdi, Y., Salmon, D., Lévy, Y., Guillet, J. G., Pialoux, G., Aboulker, J. P. *AIDS*. **20**, 1039-1049 (2006).
2. Toth, I., Danton, M., Flinn, N., Gibbons, W.A. *Tetrahedron Lett*. **34**, 3925-3928 (1993).
3. Moyle, P. M., Olive, C., Ho, M-F., Burgess, M., Karpati, L., Toth, I. *J Org Chem*. **71**, 6846-6850 (2006).
4. Moyle, P. M., Olive, C., Ho, M-F., Toth, I. *J Med Chem*. **49**, 6364-6370 (2006).
5. Moyle, P. M., Olive, C., Good, M. F., Toth, I. *J Pept Sci* **12**, 800-807 (2006).

Synthesis of Mannosylated Glycopeptides as Components for Synthetic Vaccines

Renata Kowalczyk[1], Margaret A. Brimble[1] and Rod Dunbar[2]

[1]Department of Organic and Medicinal Chemistry; [2]School of Biological Sciences, The University of Auckland, Private Bag 92019, Auckland, New Zealand

Introduction

The immune system often recognizes tumour cells and infectious agents from the unique peptides (epitopes) found on their surfaces [1] therefore, synthetic vaccines that combine many epitopes together with appropriate glycal adjuvants that stimulate Cytotoxic T Lymphocytes (CTLs) would have considerable clinical utility.

The cells responsible for initiating an immune response are APCs (antigen-presenting cells) that capture and process antigen-derived peptides for presentation to CTLs. Dendritic cells are a type of APC that express receptors capable of recognizing and internalizing foreign agents. Several of these receptors are C-type lectin receptors that bind carbohydrates [2]. The receptors we are interested in, in particular, are mannose receptors that are known in the uptake and presentation of mannosylated antigens to T cells [3].

In order to target these receptors and test their specificity for binding human skin APCs *in vitro* we have synthesized twenty mannosylated peptide derivatives, which differ in their chain length and the position of the mannosyl unit on the peptide backbone.

Results and discussion

Solid Phase Glycopeptide Synthesis (SPGS) was performed either manually or by using a CEM Microwave Peptide Synthesizer on preloaded WANG resin (0.8 mmol/g). The mannosylated building blocks are outlined in Figure 1.

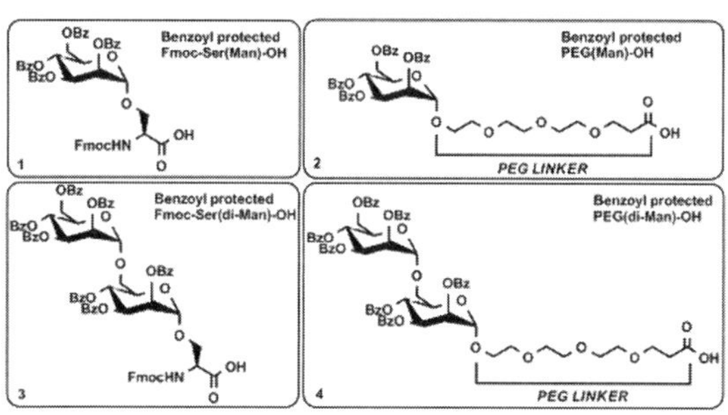

Figure 1. Mannosylated building blocks ready for incorporation into a peptide chain.

In order to prepare compounds that could be analyzed in biological screens,a fluorescent label (5,6 carboxyfluorescein) was introduced into the glycopeptides [4] via the N^{α} or the N^{ε} of the lysine residue. It was found that preparation of the glycopeptide was more facile when the peptide chain was built onto the N^{ε} of Lys (label onto N^{α}) rather than onto the N^{α} (label onto N^{ε}). In this approach the fluorescent label was coupled through the N^{α} amino group of the Lys first and then the glycopeptide chain was built up on this pre-labelled resin via the N^{ε} amino group of the Lys residue.

S. Del Valle et al. (eds.), *Peptides for Youth: The Proceedings of the 20th American Peptide Symposium*, 351
DOI: 10.1007/978-0-387-73657-0_155, © Springer Science+Business Media, LLC 2009

The SPGS was problematic, in particular when more than one glycosyl building block was involved. The low yields could be improved slightly by replacing HBTU with HATU and by using prolonged reaction times (overnight) but the results were still unsatisfactory. Based on literature reports on the use of microwave irridation for manual synthesis of glycopeptides [5] a model study of the synthesis of a glycopeptide in which mannosylated serine **1** was incorporated into an alanine scaffold peptide using a CEM microwave reactor was carried out. The use of the microwave reactor reduced the reaction time and improved the cleanliness of the reaction. A CEM Microwave Peptide Synthesizer was therefore used for all further syntheses.

The glycopeptides prepared (Figure 2) were tested for binding to monocytes using flow cytometry. Selected results are shown in Graph 1. Ratios were calculated for specific binding to monocytes (which bear mannose receptors) compared with non-specific binding to lymphocytes (no mannose receptors), both in the presence and absence of calcium. Lack of calcium abrogated specific binding, as expected for binding to C-type lectins.

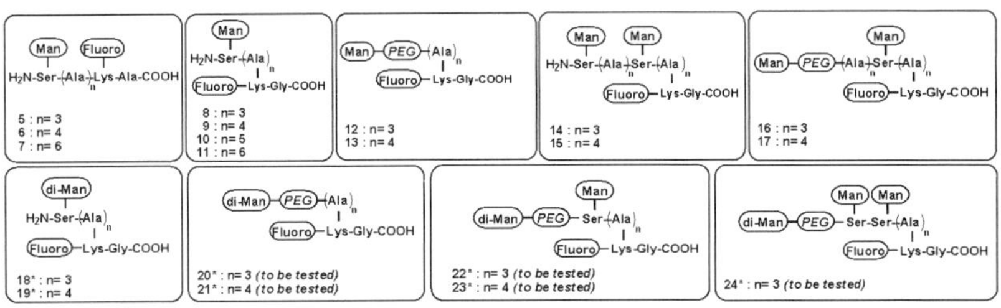

*CEM Microwave Peptide Synthesiser was used for the synthesis; **Man**= Mannose; **di-Man**= di-Mannose; **Fluoro**= 5(6) carboxyfluorescein; ***PEG***= polyethyleneglycol

Figure 2. Synthesized glycopeptides.

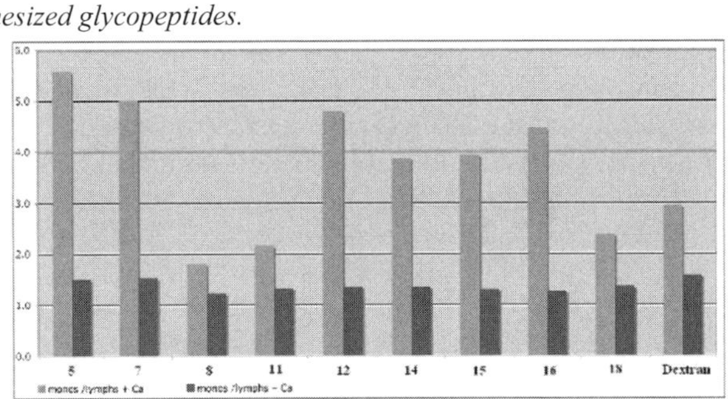

Graph 1. Glycopeptides binding activity to human peripheral blood mononuclear cells.

Acknowledgements

We thank the Maurice Wilkins Centre for Molecular Biodiscovery for financial support of this work.

References

1. Engelhard, V. H. *Sci. Amer.*, **271**, 54 (1994).
2. Figdor, C. G., van Kooyk, Y. and Adema, G. J. *Nat Rev Immunol*, **2**, 77-84 (2002).
3. Engering, A. J., Cella, M., Fluitsma, D. M., Hoefsmit, E. C. M., Lanzavecchia, A. and Pietres, J. *Adv. Exp. Med. Biol.*, **417**, 183-187 (1997).
4. Fischer, R., Mader, O., Jung, G. and Brock, R. *Bioconjugate Chem.*, **14**, 653-660 (2003).
5. Matsushita, T., Hinou, H., Kurogochi, M., Shimizu, H. and Nishimura, S. I. *Org. Lett.*, **7**, 877-880 (2005).

Bradykinin Analogues Acylated On Their N-terminus – Some Recent Development

Dariusz Sobolewski and Adam Prahl

Department of Chemistry, University of Gdańsk, Sobieskiego 18, 80-952 Gdańsk, Poland.

Introduction

Almost all cells in the majority of species express kinin receptors which mediate various physiological and pathophysiological activities of bradykinin (BK). These receptors belong to a G-protein-coupled family and their activation stimulates smooth muscle cells, sensory nerve endings, causes vasodilation and microvascular leakage and modulates the response of immunocompetent cells. There is also considerable evidence that BK contributes to the inflammatory responses. Moreover, after injection to the skin, BK produces all of the basic signs of inflammation. Due to its ability to lower blood pressure, BK has been implicated in the pathogenesis of several shock syndromes. Two receptors, B_1 and B_2, mediate biological activities of BK. While B_2 receptors require the entire BK sequence for recognition, B_1 ones recognize and bind des-Arg^9-BK only.

Our previous studies suggested that acylation of the N-terminus of several known B_2 antagonists with various kinds of bulky acyl groups consistently improved their antagonistic potency in rat blood pressure assay. On the other hand, our earlier observations also seemed to suggest that the effects of acylation on the contractility of isolated rat uterus depended substantially on the chemical character of the acyl group, as we observed that this modification might either change the range of antagonism or even transform it into agonism. Several analogues acylated on their N-terminus with various bulky groups were synthesized using SPPS method. The potencies of the analogues were assessed by their ability to inhibit vasodepressor response of exogenous BK in conscious rats and by their ability to inhibit the contractions of isolated rat uterus evoked by BK.

Results and Discussion

Summing up discovered almost twenty years ago acylation of the N-terminus of several known B_2 antagonists with various kinds of bulky acyl groups as a tool that consistently improved the antagonistic potency of analogues seems to be much more complex than was previously thought [1]. In view of reported in past years results it appears that many details including size of an acyl group attached to the N-terminus of the peptide, but also chemical features of substituent are important for B_2 antagonism [2, 3]. Different action of these acylated peptides with model blood vessels and uterus suggests a different type of interactions with B_2 receptors localized in both types of tissues. It was worth pointing out that in some cases acylation of the N-terminus of B_2 antagonists with bulky acyl groups led to the depression of analogue activity. This effect was observed for analogues substituted in the C-terminal part of molecule with conformationally restricted amino acid residues including D-Tic^7-Tic^8 sequence [4], 1-Nal or 2-Nal residues at position 8 [5], restricted dipeptide fragment D-Phe-D-Phe at positions 6-7 or 7-8 [6], 1-aminocyclopentane-1-carboxylic acid at positions 7 or 8 [7], 1-aminocyclohexane-1-carboxylic acid at positions 7 or 8 [8]. On the other hand it was reported that two of the previously used groups (acridin-9-ylacetylate and anthracen-9-ylacetylate) placed in the N-terminus of the BK molecule, transformed it activity to

S. Del Valle et al. (eds.), *Peptides for Youth: The Proceedings of the 20th American Peptide Symposium*, 353
DOI: 10.1007/978-0-387-73657-0_156, © Springer Science+Business Media, LLC 2009

antagonistic in blood pressure test and depressed agonistic activity in uterotonic test. There are the first two analogues able to antagonize the BK activity in the rat blood pressure test that does not contain any changes in the main chain (9). It is obvious that further studies are needed to confirm all these extremely interesting findings.

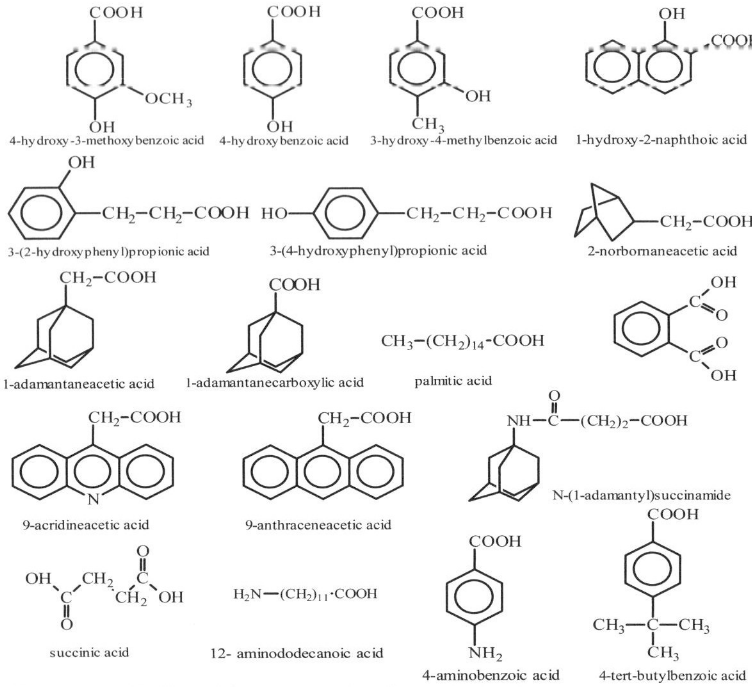

Fig. 1. Structures of bulky acid groups used as N-terminal substituents.

Acknowledgments

This work was funded by the Polish State Committee for Scientific Research, grant no. **PB 0380/H03/2006/31.**

References

1. Lammek, B. *Polish J. Chem.* **68:** 913-920 (1994).
2. Prahl, A., Wierzba, T., Wszędybył, M., Juzwa, W. and Lammek, B. *Polish J. Chem.* **71**, 915-922 (1997).
3. Trzeciak H.I., Kozik, W., Melhem, S., Kania, A., Dobrowolski, D., Prahl, A., Derdowska, I. and Lammek, B. *Peptides* **21**, 829-834 (2000).
4. Prahl, A., Derdowska, I., Winklewski, P., Wierzba, T., Juzwa, W. and Lammek, B. *Polish J. Chem.* **70**, 378-381 (1996).
5. Prahl, A., Derdowska, I., Dawidowska, O., Neubert, K., Hartrodt, B., Wierzba, T., Juzwa, W. and Lammek, B. *Polish J. Chem.* **76(10)**, 1433-1439 (2002).
6. Prahl, A., Winklewski, P., Musial, P., Juzwa, W. and Lammek, B. *Polish J. Chem.* **71**, 929-943 (1997).
7. Labudda, O., Wierzba, T., Sobolewski, D., Kowalczyk, W., Sleszynska, M., Gawinski, L., Plackova, M., Slaninova, J. and Prahl. A. *J. Pept. Sci.* **12(12)**, 775-779 (2006).
8. Labudda-Dawidowska, O., Wierzba, T., Prahl, A., Kowalczyk, W., Gawiński, Ł., Plackova, M., Slaninová, J. and Lammek, B. *J. Med. Chem.* **48(25)**, 8055-8059 (2005).
9. Prahl, A. *J. Pept. Sci.* **13(3)**, 206-210 (2006).

Rationally-designed Multivalent Architectures for Mimicking Homotrimers of CD40L, a Member of the TNF Superfamily

Gilles Guichard[1], Nathalie Trouche[1], Sébastien Wieckowski[1], Weimin Sun[1], Olivier Chaloin[1], Alberto Bianco[1], Johan Hoebeke[1], Pascal Schneider[2], and Sylvie Fournel[1]

[1]*CNRS;Institute of Molecular and Cellular Biology, Strasbourg, France;* [2]*Department of Biochemistry, University of Lausanne;Epalinges, Switzerland*

Introduction

Ligands and receptors of the TNF superfamilies play a central role in the organization and function of the immune system [1]. Ligands of the TNF-family, share a common structural motif. Monomers self-assemble around a three-fold symmetry axis to form non-covalent homotrimers that can each bind three receptor molecules. Interaction between CD40, a member of the TNF receptor superfamily, and its ligand CD40L, a 39 kDa glycoprotein is essential for the development of humoral and cellular immune responses. Selective blockade or activation of this pathway provides the ground for the development of new treatments against immunologically based-diseases and malignancies.

Synthetic multivalent ligands, owing to the presence of multiple copies of a recognition motif attached to a central scaffold, can mediate clustering of cell surface receptors and thereby function as effector molecules [2]. We have shown previously that rigid trimeric scaffolds can serve to distribute a CD40-binding motif with geometry and distances that could match that of the natural CD40L protein (Figure 1) [3,4]. A short sequence from the surface of CD40L, encompassing three "hot-spot" residues (Lys143, Tyr145 and Tyr146) critical for binding to CD40, was chosen as a CD40 binding motif.

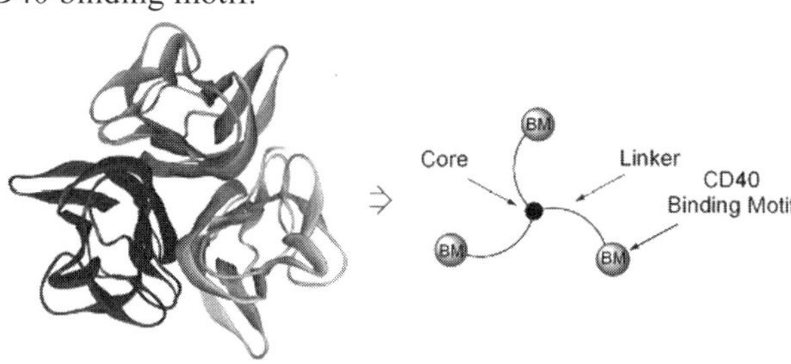

Fig. 1. X-ray structure of CD40L homotrimer (PDB 1ALY) viewed down the C_3 axis and schematic representation of the strategy used to design CD40L mimetics.

Two trimeric molecules **1** and **2** built on the rigid cyclo(Lys-D-Ala)$_3$ and cyclo(β^3-HLys) peptide templates, respectively, with amino hexanoic acid (Ahx) residues as additional spacer arms were found to bind in a specific manner to CD40 and to compete with the binding of CD40L homotrimers. In various cell-based assays, **1** and **2** displayed some effector functions of the much larger recombinant CD40L homotrimers, i.e. induction of apoptosis of Burkitt B-lymphomas and maturation of dendritic cells (DCs). Both **1** and **2** were shown recently to be effective *in vivo* and to promote the control of *T. cruzi* (i.e. the etiological agent of the Chagas'disease) infection in an experimental mouse model [5].

S. Del Valle et al. (eds.), *Peptides for Youth: The Proceedings of the 20th American Peptide Symposium*, 355
DOI: 10.1007/978-0-387-73657-0_157, © Springer Science+Business Media, LLC 2009

wever, the molecular determinants of the design of CD40L mimetics with optimal CD40 binding properties and effector functions are not yet fully delineated. To assess the contribution of each individual component of the trivalent architecture, we have undertaken a detailed study by systematically varying the shape and the valency of the core structure as well as the length of the spacer arm.

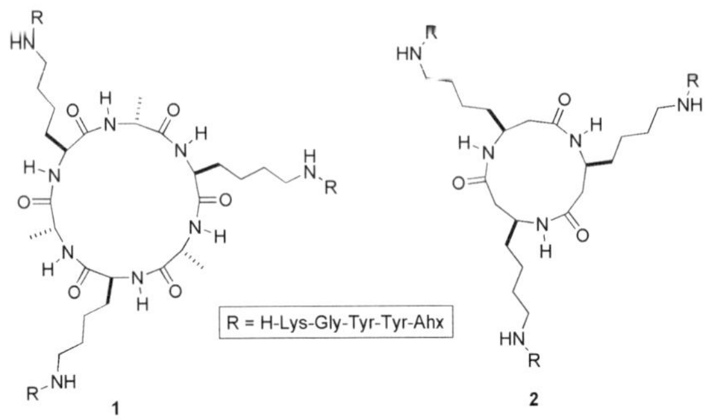

Results and Discussion

All CD40L mimetics described in this study were synthesized according to Scheme 1 by fragment coupling in solution of a fully protected peptide fragment encompassing the receptor binding motif and a spacer arm, to the corresponding amino-functionalized core structure.

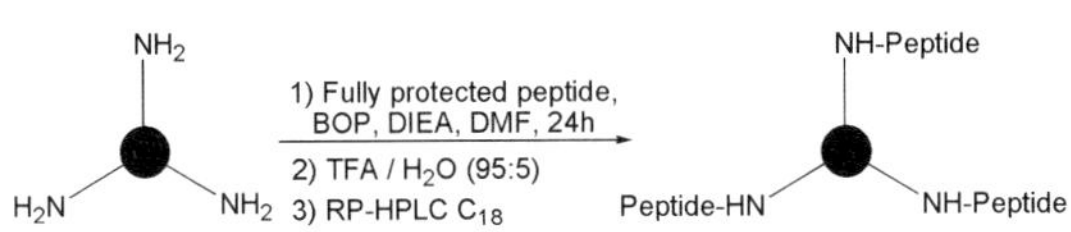

Scheme 1. General synthetic approach to trimeric CD40L ligands.

The protected CD40-binding peptides with their linker were prepared on solid support starting from a 2-chlorotrityl chloride resin. Peptide assembly on core structures was generally performed with BOP as the coupling agent in DMF for 24 h, after which the crude fully protected multimeric construct recovered by filtration after precipitation with a saturated $NaHCO_3$ solution was washed extensively with ethyl acetate and dried under high vacuum. All protecting groups were removed by treatment with TFA to afford crude ligands. Finally, ligands were purified by C_{18} RP-HPLC and recovered in overall yields ranging from 6-37%. It is noteworthy that the synthesis of **1** was routinely achieved on > 100 mg scale with an overall yield exceeding 25%.

The importance of a C_3-symmetric trivalent display was assessed by systematically varying the number of CD40 recognition elements appended to the cyclic D,L-peptide core. We found that removing one CD40-binding motif while maintaining a radial distribution resulted in a strong decrease of the CD40 binding capacity and biological effects, i.e. apoptosis of human B-lymphoma cells and maturation of mouse dendritic cells D1. Not surprisingly, an analogue of **1** with only one CD40-binding motif appended to a cyclic D,L-hexapeptide core was inactive.

To investigate how the length of the linker can modulate the activity of CD40L mimetics, we have prepared a series of variants of **2** with oligomethylene [–$(CH_2)n$–] tethers of length varying from n = 3 to n = 7. Compound **3**, in which the oligomethylene chain is reduced by two carbons (n = 3), is no longer able to in apoptosis of human B-lymphoma BL41 cells and exhibits drastically reduced binding to CD40. Significant binding to CD40 is retained when the length of the spacer is decreased by only one carbon (**4**), but the capacity to induce apoptosis is not restored. In contrast, analogues of **2** with one and two additional methylene groups (**5**, n = 6 and **6**, n = 7), respectively induced significantly higher levels of apoptosis compared to **2**. Further increments in chain length from the 7-carbon chain to the 10-carbon chain led to a compound (**7**) with lower solubility that could not be evaluated in the two assays.

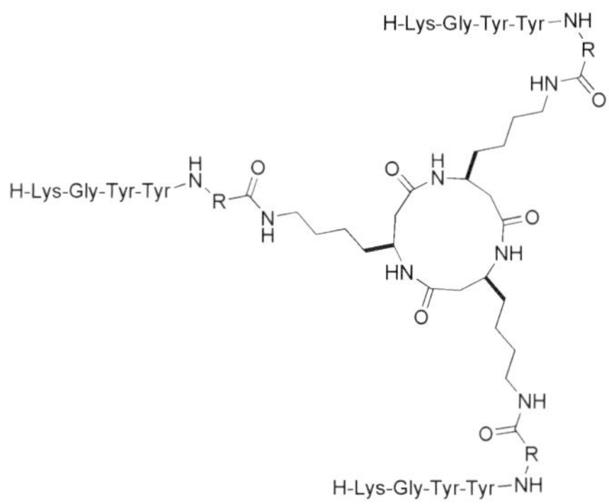

3 R = $(CH_2)_3$ (spacer Abu)
4 R = $(CH_2)_4$ (spacer Apa)
2 R = $(CH_2)_5$ (spacer Ahx)
5 R = $(CH_2)_6$ (spacer Ahp)
6 R = $(CH_2)_7$ (spacer Aoc)
7 R = $(CH_2)_{10}$ (spacer Aud)

The data reported here i) suggest that radial distribution of CD40-binding units and C_3-symmetry are preferred for optimal binding to CD40 and signaling; ii) underscores the importance of choosing an appropriate linker to connect the receptor binding motif to the central scaffold; and iii) show the versatility of planar cyclic α- and β-peptides as templates for the design of CD40L mimetics. The use of *miniCD40Ls* such as those reported here is complementary to other approaches (recombinant ligands, agonistic anti-receptor antibodies) and may find interesting therapeutic applications. Furthermore, the results disclosed in this paper provide the basis for future design of other TNF family member mimetics.

Acknowledgments

This work was supported by the CNRS, the Ministère de la Recherche (ACI « Jeunes Chercheurs »), « La Ligue contre le Cancer » and INCA. WS and SW were supported by grants from the Ministère de la Recherche and from the Fondation pour la Recherche Médicale (FRM). NT was supported by « La Ligue contre le Cancer »

References

1. Bodmer, J.L., Schneider, P., Tschopp, J. *Trends Biochem. Sci.* ***27***, 19-26 (2002).
2. Kiessling, L. L., Gestwicki, J. E., Strong, L. E., *Angew. Chem. Int. Ed. Engl.* ***45***, 2348-2368 (2006).
3. Fournel, S., Wieckowski, S., Sun, W., Trouche, N., Dumortier, H., Bianco, A., Chaloin, O., Habib, M., Peter, J.-C., Schneider, P., Vray, B., Toes, R. E., Offringa, R., Melief, C. J. M., Hoebeke, J., Guichard, G. *Nature Chem. Biol.* ***1***, 377-382 (2005).
4. Wieckowski, S., Trouche, N., Chaloin, O., Guichard, G., Fournel, S., Hoebeke, J. *Biochemistry* ***46***, 3482-3493 (2007).
5. Habib, M., Chamekh , M., Noval Rivas, M., Wieckowski, S, Sun, W., Bianco, A., Trouche, N., Chaloin, O., Dumortier, H., Goldman, M., Guichard, G., Fournel, S., Vray, B. *J. Immunol.* ***178***, 6700-6704 (2007).

A Strategy for Selectively Shielding Portions of a Peptide/Protein from Immune Response while Maintaining Immunogenicity of Contiguous Epitopes

Elisabetta Bianchi[1], Marco Finotto[1], Paolo Ingallinella[1], Mike Citron[2], Renee Hrin[2], Meiqing Lu[2], Romas Geleziunas[2], Michael D. Miller[2], Xiaoping Liang[2] and Antonello Pessi[1]

[1]*IRBM-MRL Peptide Centre of Excellence, Via Pontina Km 30.600, 00040 Pomezia (Rome) Italy;* [2]*Merck Research Laboratories, Sumneytown Pike P.O. Box 4, West Point, PA, USA*

Introduction

Considerable effort is being dedicated to devise vaccination strategies to elicit a humoral response against human immunodeficiency virus (HIV). Recently we validated the fusogenic gp41 subunit of the envelope glycoprotein as a vaccine target by discovering a neutralizing antibody, D5 that inhibits HIV entry by preventing formation of the 6-helix bundle [1]. IgG D5 binds to a highly conserved hydrophobic pocket in the groove formed by the trimeric coiled coil of the gp41 N-helices.

We designed peptides that are mimetics of the gp41 trimeric N-helix coiled coil. These chimeric peptides, such as $(CCIZN17)_3$, comprise a designed coiled coil scaffold with a covalent stabilizing moiety and a region of the HIV N-helix sequence [2]. Our aim is to use peptide vaccines of this class to raise D5-like antibodies that target the HIV N portion. However, when using our chimeric peptides, a substantial amount of antibodies are directed against the scaffold IZ. To circumvent this, we have developed a strategy to selectively mask, or "shield", an antigenic epitope within a peptide or protein, without affecting the antigenicity and immunogenicity of contiguous epitopes. Here we show that immunofocusing towards the HIV sequence portion is achieved by attachment of small molecular weight polyethylene glycol (SMWPEG) chains at suitable location(s) within the scaffold peptide sequence.

Results and Discussion

Starting from $(CCIZN17)_3$ [2], we designed a new chimeric peptide $(CCIPN17)_3$ (shown in Fig. 1 a), where we introduced polyethyleneglycol chains of the nominal size of 15 **Å** at two exposed positions of each peptide chain of the trimeric coiled coil scaffold IZ of the $(CCIZN17)_3$ immunogen. The SMWPEGs were positioned in such a way, not to perturb accessibility to the HIV sequence, that contains 17 N-helix residues encompassing the D5 neutralizing epitope.

The binding to the IgG D5 was used as a guiding criterion for the integrity of the conformational epitope of the designed immunogens. $(CCIPN17)_3$ binds D5 with a 5-fold higher efficiency with respect to the non-PEGylated $(CCIZN17)_3$ suggesting that:

a) the introduction of the SMWPEG chains at suitable locations on the IZ scaffold does not preclude accessibility to the HIV sequence portion and it is not detrimental for the structural integrity of the D5 conformational epitope.
b) Because of the higher affinity for D5, the insertion of the hydrophilic PEG chains at suitable positions of the scaffold, also has a beneficial effect on the stabilization of the coiled coil structure, through a classical

S. Del Valle et al. (eds.), *Peptides for Youth: The Proceedings of the 20th American Peptide Symposium*, 359
DOI: 10.1007/978-0-387-73657-0_158, © Springer Science+Business Media, LLC 2009

"hydrophobic effect" that translates in enhanced presentation of the conformational epitope.

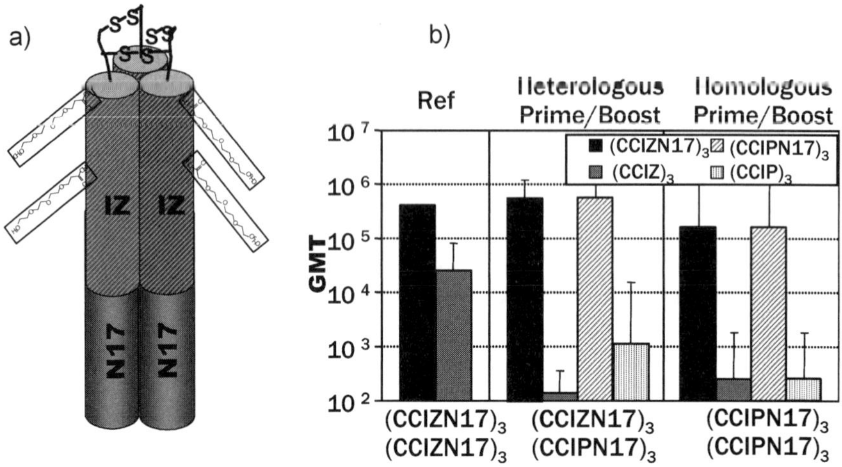

Fig. 1. a) Schematic model of the trimeric coiled coil (CCIPN17)$_3$ with insertions of small MW PEGs on the IZ scaffold. b) Antibody titers against the intact antigens, (CCIZN17)$_3$ and (CCIPN17)$_3$ and against the scaffolds (CCIZ)$_3$ and (CCIPN17)$_3$ with two different modalities, homologous and heterologous injections.

(CCIPN17)$_3$ was used as an immunogen in guinea pigs using two different modalities: an homologous vaccination strategy, (prime and boost injections with the same antigen) and an heterologous strategy (prime injection with the non-PEGylated antigen (CCIZN17)$_3$, followed by two boost injections with (CCIPN17)$_3$. Antisera were assayed by ELISA for antibody titers against the intact antigens (CCIZN17)$_3$ and (CCIPN17)$_3$ and against the scaffolds, (CCIZ)$_3$ and (CCIP)$_3$ as shown in Fig. 1 panel B. For reference, in the left panel are also reported the ELISA data obtained for the (CCIZN17)$_3$ immunogen for which much of the total antibody response was against the scaffold IZ. The immunizations with (CCIPN17)$_3$ showed that while the antibody titers for the intact antigens are quite high in both modalities, we observe very low responses against both the two scaffolds, (CCIZ)$_3$ and (CCIP)$_3$. The data demonstrate that by shielding the scaffold IZ by means of small MW PEGs it was possible to selectively suppress the antibody immune recognition of that region of the immunogen while retaining the immunogenicity of the epitope of choice, the conformational HIV epitope.

We conclude that it is possible to selectively mask, or "shield" a portion of a peptide or a protein. By selective masking we were able to eliminate an unwanted immune reactivity to the peptide antigen, without affecting the antigenicity and immunogenicity of contiguous epitopes.

References

1. Miller, M.D. *et al. Proc. Natl. Acad. Sci. USA 102*, 14759 (2005).
2. Bianchi, E. *et al. Proc. Natl. Acad. Sci. USA 102*, 1290 (2005).

Synthesis of Peptide-Based Vaccines

Zhanna Potetinova and Gordon E. Willick

Institute for Biological Sciences, National Research Council, Ottawa, ON K1A 0R6, Canada

Introduction

Lipopeptides are potent and non-toxic adjuvants. Peptides by themselves are poor immunogens, even when incorporated in liposomes. For effective interaction with both the innate and adaptive immune responses, peptides must be conjugated with one of several known lipid moieties.

Tri- or dipalmitoyl-S-glyceryl-Cys-Ser-Ser, based on a bacterial lipoprotein, was found to be an effective adjuvant [1] that was later replaced by the more readily synthesized dipalmitoyl-Lys-Ser-Ser [2]. The purification of lipopeptides remains a problem, however, due to low aqueous solubility and the loss of resolution using traditional HPLC technology. Moreover, difficulties in their analysis hinder the use of such peptide-based vaccines in human clinical experiments. Purification of the unprotected peptides prior to conjugation of whatever lipid modification being used is highly desirable. Methods now available rely on exploitation of Cys residues for conjugation. Maleimide-based conjugation is simple, but is not applicable if there is a Cys within the epitope sequence, unless it can be left blocked until the final purification. Thioester-based coupling, commonly used in native chemical ligation of peptides is very attractive because it is specific for an N-terminal Cys and results in a normal peptide bond. A protocol of choice would provide specific lipid conjugation to any peptide.

We synthesized lipopeptides containing epitopes from ovalbumin ($OVA_{257-264}$, SIINFEKL), tyrosine-related protein ($TRP2_{180-188}$, SVYDFFVWL), and a melanoma antigen ($MART_{26-35}$, EAAGIGILTV) by Fmoc-protocol. After purification, these were lipidated by chemoselective native ligation using either maleimide [3] or thioester [4] conjugation.

Results and Discussion

Three different strategies for lipopeptide synthesis were used and compared in this work: direct SPPS, thiol-active maleimide, and thioester. In direct SPPS of Pam_2KSS-epitopes, the peptides were analyzed before lipidation by RP-HPLC and MALDI methods to evaluate purity. Lipidation was performed with HATU/DIEA/DMF at 40° C and monitored by Kaiser test. However, this approach yields lipopeptides only as pure as the crude synthesis product. For example, HPLC purity of KSS-MART crude reached 53%. The Cys-epitopes used in chemoselective ligation were synthesized on an Applied Biosystems continuous-flow peptide synthesizer with FmocLeu or FmocVal NovaSyn TGT resin (0.22 mM/g, NovaBiochem) as support by standard Fmoc SPPS protocols (TBTU/HOBt/DIEA/DMF; 20% piperidine/1% triton X100/DMF). After treatment with reagent K, unprotected peptides with N-terminal Cys were purified by HPLC (Vydac C_{18}, 218 x 22 mm, 0.1% TFA/MeCN gradient in aqueous 0.1% TFA). For comparison, C-MART was 95% purity that improved quality of the final lipopeptide. The lipopeptide adjuvant sequence was activated with N-(2-aminoethyl) maleimide by two different pathways, either before or after deprotection. Better solubility in organic solvents of the protected Pam_2KS(tBu)S(tBu), than the deprotected Pam_2KSS, resulted in a quantitative yield of the activation stage and

S. Del Valle et al. (eds.), *Peptides for Youth: The Proceedings of the 20th American Peptide Symposium*, 361
DOI: 10.1007/978-0-387-73657-0_159, © Springer Science+Business Media, LLC 2009

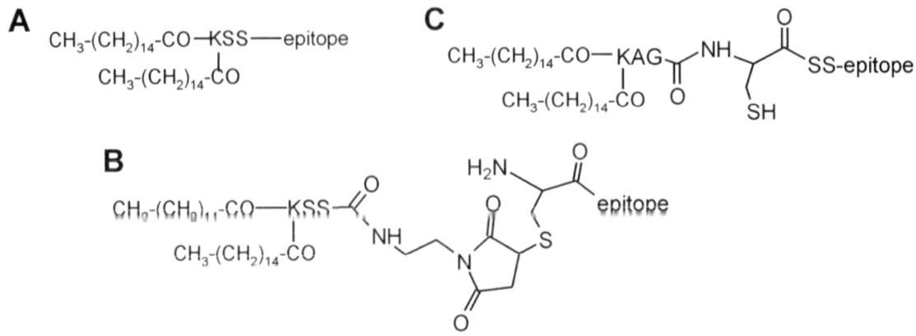

Fig. 1. Structures of lipidated epitopes: (A) direct SPPS, (B) maleimide, (C) thioester methods.

Simplified purification of the resulting maleimide derivatives by extraction. For the thioester approach, the thiophenol derivative, Pam_2KAG-SPh, was obtained using HOBt/EDC/DIEA in DCM/DMF. Native ligation of maleimide or thioester lipopeptides to the HPLC purified Cys-epitope was carried out in 30% MeCN/ 0.1 M phosphate buffer, pH 7.2. Excess Cys-epitopes were washed off with aqueous-organic solutions (OVA, MART - 0.1% TFA/50% MeCN/H_2O; TRP2 – 0.1% TFA/16% MeOH/60% iPro/H_2O). The unreacted maleimide derivative can be removed with thiomethylpolymer. Lipopeptides were characterized by MS, 2D-NMR, and amino acid analysis. To confirm structure and evaluate purity, unlipidated and lipidated peptides were analyzed by HPLC, MS, and α-chymotrypsin (data not shown).

Table 1. Molecular mass (Da) of synthesized lipopeptides.

Pathway*	Lipopeptide	MALDI-TOF	Calculated
A	Pam_2KSS-$OVA_{257\text{-}264}$	1741.2	1742.3
A	Pam_2KSS-$TRP2_{180\text{-}188}$	1952.7	1954.5
A	Pam_2KSS-$MART_{26\text{-}35}$	1721.1	1722.3
B	Pam_2KSSMal-C-$OVA_{257\text{-}264}$	1986.5	1985.6
B	Pam_2KSSMal-C-$TRP2_{180\text{-}188}$	2196.4	2197.8
B	Pam_2KSSMal-C-$MART_{26\text{-}35}$	1961.3	1965.6
C	Pam_2KAG-CSS-$OVA_{257\text{-}264}$	2019.4	2017.6 (+2Na)

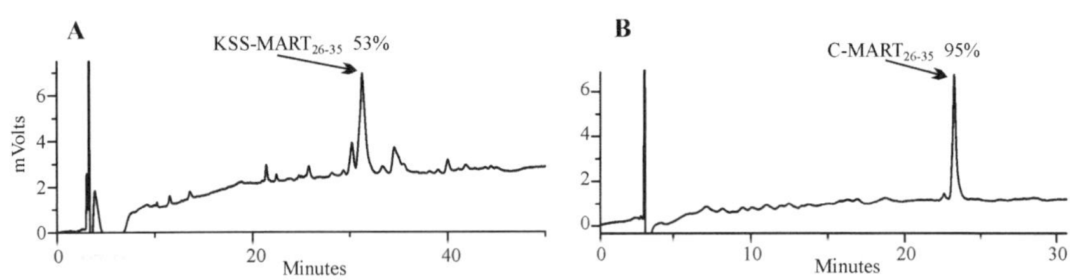

Fig. 2. HPLC of $MART_{26\text{-}35}$ before lipidation by: (A) direct SPPS, (B) maleimide method.

References

1. Hoffmann, P., *et al. Immunobiology* **177**, 158 (1988).
2. Huang, W., *et al. Mol. Immunol.* **31**, 1191 (1994).
3. Roth, A., *et al. Bioconjugate Chem.* **15**, 541 (2004).
4. Tam, J.P., *et al. Proc. Natl. Acad. Sci. USA* **92**, 12485 (1995).

A Peptide Vaccine Based on Retro-Inverso β-Amyloid Sequences Fails to Elicit a Cross-reactive Immune response

Elisabetta Bianchi[1], Paolo Ingallinella[1], Marco Finotto[1], Xiaoping Liang[2], Gene G. Kinney[3] and Antonello Pessi[4]

[1]*IRBM-MRL Peptide Centre of Excellence, Via Pontina Km 30.600, 00040 Pomezia (Rome) Italy;* [2]*Department of Vaccine and Biologics Research;* [3]*Department of Alzheimer's Reserach; Merck Research Laboratories, Sumneytown Pike P.O. Box 4, West Point, PA, USA*

Introduction

Retro-inverso peptides as vaccines. In a retro-inverso (RI) peptide, the direction of the amino acid sequence is reversed, and the chirality of each amino acid residue is inverted, resulting in inversion of each peptide bond within the peptide sequence [1].

Several studies have documented that the RI analogs of linear peptide epitopes can be useful as vaccines, the most convincing example being the retro-inverso analog of the immunodominant epitope of Foot-and-Mouth disease virus [2]. In general, the likelihood of a RI peptide successfully mimicking a given epitope depends on how much the epitope is defined by the overall topology of the side-chains, rather than by interactions involving the peptide backbone, and successful mimicry is more often observed for epitopes in random coil, loop, or cyclic conformations, rather than in specific secondary structures. Conversely, the available studies indicate a low likelihood of a RI peptide eliciting a T-cell response cross-reactive with the native T-cell epitope [3]. This is because maintenance of the side chains topology, which is required for binding the T-cell receptor, requires orienting the peptide backbone in the opposite direction with respect to the L-peptide, thus effectively inverting the position of all the hydrogen bond donors/acceptors to the MHC molecule. Absence of cross-reactivity is especially true for class II restricted T-epitopes, whose binding is largely influenced by the peptide backbone, rather than the side chains.

Retro-inverso peptides for vaccination against Alzheimer's Disease. For Alzheimer's Disease (AD), the possibility of eliciting a cross-reactive antibody response to amyloid beta (Aβ) peptides in the absence of a cross-reactive T-cell response, is particularly attractive. A human vaccination trial with full-length Aβ(1-42) peptide showed some indication of efficacy in terms of plaque clearance and cognitive benefit, but was halted when some patients showed symptoms consistent with meningoencephalitis: although several factors may have contributed, it is widely assumed that Aβ mediated T-cell activation was central to this adverse event. A RI Aβ vaccine could thus be safer than any vaccine based on natural Aβ fragments.

Results and Discussion

We systematically explored the entire Aβ(1-40) peptide. Within its sequence, the critical target region for a therapeutically useful antibody response is the N-terminus [4], which is exposed and devoid of secondary structure [5,6], thus representing an ideal target for the RI modification.

The size of our RI peptide scan window was kept to 8 amino acids, i.e. smaller than the size of a typical T-cell epitope, to further increase the safety margin. The immunodominant Aβ B-epitope, ^{4}FRHDSGY^{10}, is just seven amino acids. In our

S. Del Valle et al. (eds.), *Peptides for Youth: The Proceedings of the 20th American Peptide Symposium*, 363
DOI: 10.1007/978-0-387-73657-0_160, © Springer Science+Business Media, LLC 2009

RI vaccine, the T-cell help for elicitation of an antibody response to the RI peptide is provided by the carrier, as a peptide-carrier conjugate.

Table 1. Immunological data on RI peptide conjugate vaccines. The antibody response against each immunizing peptide is shown as Geometric Mean Titer (GMT) in ELISA against a biotinylated form of the RI peptide, while the antibody response cross-reactive with the native Aβ sequence is shown as GMT to biotinylated Aβ(1-40).

RI Peptide conjugate	*GMT[1] (RI peptide)*	*GMT (Aβ1-40)*	RI Peptide conjugate	*GMT[1] (RI peptide)*	*GMT (Aβ1-40)*
1-8	409,600[1]	<100	9-16	102,400[1]	<100
2-9	102,400[1]	<100	10-17	409,600[1]	<100
3-10	40,637[2]	<100	11-18	204,800[1]	<100
4-11	102,400[1]	<100	17-24	102,400[1]	<100
5-12	25,600[1]	<100	21-28	204,800[1]	1600
6-13	51,200[1]	<100	1-18	102,400[2]	<100
7-14	40,637[2]	<100	1-40	102,400[1]	<100
8-15	51,200[1]	<100			

[1]Guinea Pig antiserum; [2]Rhesus macaque antiserum

The conjugates used in this study are shown in Table 1. They all had high peptide/carrier molar ratio, a feature that we found to be important for eliciting high-titer immune response. Accordingly, the antibody response against all vaccines was very high. However, when we measured the level of cross-reactivity of the RI antiserum against the native Aβ(1-40) sequence, only one antiserum [RI Aβ(21-28)] was found to be weakly cross-reactive with Aβ(1-40). Moreover, when the cross-reactive antiserum was tested on AD-positive tissue sections, it failed to react.

We conclude that despite the promising rationale, the RI approach is not a viable option for the rational design of a safe and efficacious anti-Aβ vaccine for the treatment of Alzheimer's disease.

Acknowledgments

The authors thankfully acknowledge the intellectual contributions to this work of G. Ciliberto, R. Cortese, G. Seabrook, and J. Shiver.

References

1. Chorev, M., and Goodman, M., *Acc. Chem. Res.* 26:266-273 (1993).
2. Briand, J. P., Benkirane, N., Guichard, G., Newman, J. F., Van Regenmortel, M. H., Brown, F., and Muller, S., *Proc. Natl. Acad. Sci. U.S.A.* 94:12545-50 (1997).
3. Apostolopoulos, V., and Lazoura., E., Expert Rev. Vaccines 3:151-162 (2004).
4. Bard, F., Barbour, R., Cannon, C., Carretto, R., Fox, M., Games, D., *et al., Proc. Natl. Acad. Sci. U.S.A.* 100:2023-8 (2003).
5. Morimoto, A., Irie, K., Murakami, K., Masuda, Y., Ohigashi, H., Nagao, M., *et al., J. Biol. Chem.* 279:52781-8 (2004).
6. Urbanc, B., Cruz, L., Yun, S., Buldyrev, S.V., Bitan, G., Teplow, D.B., and Stanley, H.E., *Proc. Natl. Acad. Sci. U.S.A.* 101:17345-50 (2004).

Synthesis and Application of (*Z*)-Alkene- and (*E*)-Fluoroalkene-Dipeptide Isosteres as *cis*-Amide Equivalents

Kenji Tomita, Tetsuo Narumi, Ayumu Niida, Shinya Oishi, Hiroaki Ohno and Nobutaka Fujii

Graduate School of Pharmaceutical Sciences, Kyoto University, Sakyo-ku, Kyoto 606-8501, Japan

Introduction

Alkene-type dipeptide isosteres have been developed as nonhydrolyzable peptide-bond bio-isosteres that mimic the planar structure of peptide bonds. Restricted peptide bond (ω-angle) rotations in alkene-type isosteres enable one to investigate the *cis* or *trans* conformation of the peptide bonds. The precedented synthetic methods reported for alkene-type dipeptide isosteres preferentially provide *trans*-amide bond mimics, such as (*E*)-alkene dipeptide isosteres (ADIs) and (*Z*)-fluoroalkene dipeptide isosteres (FADIs). On the other hand, there have only been a few reports on the synthesis of *cis*-amide bond mimics possessing (*Z*)-alkene or (*E*)-fluoroalkene, which limits their application.

Recently, we established a widely applicable synthetic method for the synthesis of (*Z*)-ADI and (*E*)-FADI using organocopper-mediated reduction of certain α,β-unsaturated-δ-lactams possessing a leaving group(s) at the γ-position, followed by ring-opening of the lactams [1].

Herein, we report the use of Phe-Gly-type (*Z*)-ADI and (*E*)-FADI in the Fmoc-based solid-phase peptide synthesis (SPPS) of peptide derivatives of a GPR54 agonist **1** [H-Amb-Phe-Gly-Leu-Arg-Trp-NH_2; Amb = 4-(aminomethyl)benzoic acid], to probe the conformational requirement of its Phe-Gly peptide bond.

Results and Discussion

The synthetic route for the preparation of Fmoc-protected dipeptide isosteres is shown in Scheme 1. Key substrates **4** and **5** were prepared from allylic alcohol **2** and β-amino ester **3**, respectively. Lactams **4** and **5** were treated with organocopper reagents, affording β,γ-unsaturated-δ-lactams **6** and **7**. Sequential hydrolysis and N^{α}-Fmoc-protection yielded the desired dipeptide isosteres **8** and **9** containing *cis*-amide equivalences.

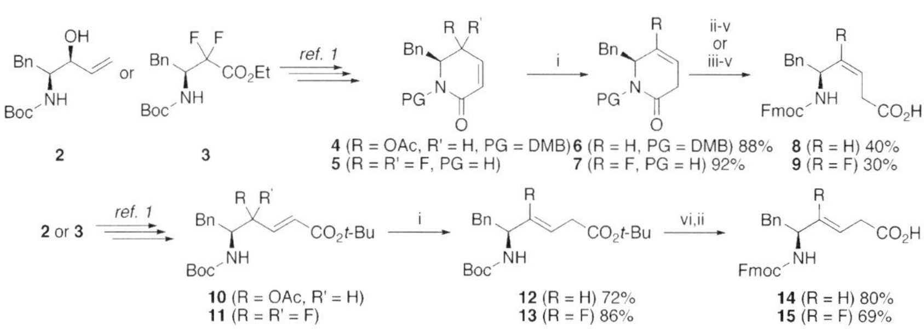

Scheme 1. Synthesis of Fmoc-protected Phe-Gly-type (E/Z)-ADIs and (E/Z)-FADIs. Reagents and conditions: (i) $Me_3CuLi_2{\cdot}LiI{\cdot}_3LiBr$; (ii) TFA; (iii) $Me_3O{\cdot}BF_4$; (iv) 0.5 N HCl aq.-THF (4:1); (v) Fmoc-OSu, Et_3N; (vi) 4 N HCl-dioxane.

S. Del Valle et al. (eds.), *Peptides for Youth: The Proceedings of the 20th American Peptide Symposium*, 365
DOI: 10.1007/978-0-387-73657-0_161, © Springer Science+Business Media, LLC 2009

*Table 1. GPR54 agonistic activities of pentapeptide analogues **1** and **16-19**.*

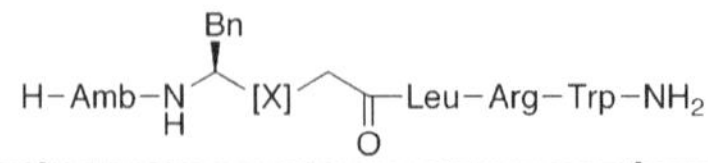

Peptide	X	EC_{50} (nM)[a]
1	CO NH	2.9
16	–[(*Z*)–CH=CH]–	> 100
17	–[(*E*)–CF=CH]–	> 100
18	–[(*E*)–CH=CH]–	7.8
19	–[(*Z*)–CF=CH]–	> 100

[a]*EC_{50} values were evaluated using the FLIPR assay.*

For the comparative structure-activity relationship study, the *trans*-amide dipeptide isosteres were constructed from acyclic substrates **10** and **11** using an organocopper reagent in essentially the same manner as previously reported [1]. N^{α}-Boc-deprotection of **12** and **13** and subsequent Fmoc-protection provided Fmoc-protected (*E*)-ADI **14** and (*Z*)-FADI **15**, respectively.

The resulting four Fmoc-protected dipeptide isosteres **8**, **9**, **14** and **15** were applied to Fmoc-based SPPS, affording dipeptide isostere-containing pentapeptide analogues **16-19** derived from a reported GPR54 agonist **1** [2].

Among four of the isostere-containing peptides, the (*E*)-ADI-containing peptide **18** exerted potent GPR54 agonistic activity (EC_{50} = 7.8 nM) in the FLIPR assay, comparable to the parent peptide **1** (EC_{50} = 2.9 nM). In contrast, the *cis*-amide mimetic-containing peptides **16** and **17** induced extremely low receptor activation (EC_{50} = > 100 nM), indicating that the *trans*-amide conformation of the Phe-Gly peptide bond is favorable for the agonistic activity [3]. In contrast to our expectation, peptide **19**, possessing (*Z*)-FADI as a *trans*-amide bond mimetic, was not active, although the loss of its bioactivity has yet to be clarified.

Acknowledgments

We thank Takeda Pharmaceutical Company Limited for the biological activity evaluation of a series of compounds. This work was supported by a Grant-in-Aid for Scientific Research from the Ministry of Education, Culture, Sports, Science, and Technology of Japan, and Health and Labour Sciences Research Grants (Research on HIV/AIDS). K.T., T.N. and A.N. are grateful for Research Fellowships from the JSPS for Young Scientists.

References

1. Niida, A., Tomita, K., Mizumoto, M., Tanigaki, H., Terada, T., Oishi, S., Otaka, A., Inui, K. and Fujii, N. *Org. Lett.* **8**, 613-616 (2006).
2. Tomita, K., Niida, A., Oishi, S., Ohno, H., Cluzeau, J., Navenot, J.-M., Wang, Z., Peiper, S. C. and Fujii, N. *Bioorg. Med. Chem.* **14**, 7595-7603 (2006).
3. Tomita, K., Narumi, T., Niida, A., Oishi, S., Ohno, H. and Fujii, N. *Biopolymers* **88**, 272-278 (2007).

Gadd45β dimerization does not affect MKK7 binding

Laura Tornatore[1,2], Simona M. Montil[1], Daniela Marasco[1], Nina Dathan[1], Rosa M. Vitale[1], Ettore Benedetti[1,2], Carlo Pedone[1,2], Salvatore Papa[3], Guido Franzoso[3] and Menotti Ruvo[1]

[1]Istituto di Biostrutture e Bioimmagini - CNR, via Mezzocannone, 16, Napoli, Italy; [2]Università Federico II di Napoli, via Mezzocannone, 16, 80134, Napoli, Italy; [3]The Ben May Institute for Cancer Research and the University of Chicago, 24 East 57th Street, Chicago, IL60637.

Introduction

The Gadd45 family comprises three small acidic proteins, α, β, and γ, largely involved in cellular response to genotoxic stress and DNA damage. They exert and mediate a large number of functions, including regulation of cell cycle, indeed, it has been demonstrated that, depending on cell type or tissues, Gadd45α and Gadd45γ have strong pro-apoptotic properties, whereas, Gadd45β is a potent mediator of the protective effects of NF-kB against the TNFα induced apoptosis [1,2]. Gadd45β fulfils this function by strongly interacting with MKK7, an essential JNK activator, and blocking its catalytic activity [1,2] and this mechanism has been largely investigated elucidating key residues on both the kinase [1] and the Gadd45 protein [2]. The regions 132-156 of MKK7, containing the ATP-binding K149, and 60-86 of Gadd45β, containing a large acidic stretch, participate in this interaction and the corresponding synthetic peptides strongly block the Gadd45β/MKK interaction and promote apoptosis when administered to *RelA*$^{-/-}$ cells [1,2]. Based on large biochemical and site-directed mutagenesis data, a 3D model of the MKK7/Gadd45β has been proposed, corroborating the view that the acidic loop D61-I69 of Gadd45β enters the kinase catalytic pocket and prevents access to the ATP, thereby inactivating the enzyme. However, Gadd45 proteins interact with several other partners and in addition, it has been reported that they all can homo- and/or hetero-dimerize or oligomerize [3]. With the aim of studying the influence that Gadd45β oligomerization can have on MKK7 binding and, therefore, on the mechanism by which NF-kB counteracts the TNFα-mediated programmed cell death, we have here investigated the Gadd45β regions involved in self-association.

Results and Discussion

Human Gadd45β was obtained as a recombinant protein and purified to homogeneity. Analyses carried out by gel filtration showed that, in vitro, Gadd45β exists predominantly as a non covalent dimer, similarly to the highly homologous Gadd45α. Furthermore CD studies in the far-UV region suggested the presence of a folded structure with a predominantly high α-helical content. The self-association affinity constant, as estimated by an ELISA-like assay, was around 100 nM. In this assay, soluble biotinylated Gadd45β was bound in a dose-dependent and saturable way to the unlabelled protein coated on a microtiter plate. To identify regions involved in homo-dimerization-oligomerization, we next carried out competition assay, whereby trypsin-derived Gadd45β fragments were used as competitors. By these experiments, it emerged that fragments 16-32 and 132-146 blocked Gadd45β self-association, while other peptides did not. Notably, by looking at the protein predicted structure (Fig. 1) [2], we observed that these peptides covered most part of

S. Del Valle et al. (eds.), *Peptides for Youth: The Proceedings of the 20th American Peptide Symposium*,
DOI: 10.1007/978-0-387-73657-0_162, © Springer Science+Business Media, LLC 2009

the Gadd45β 1st (Helix 1) and 5th (Helix 5) putative helices. The corresponding synthetic peptides, including N- and C-terminus amino acids necessary to complete the helices (Gadd45β 12-35 and 129-148), were thereby prepared by SPPS and tested in the same assay, using the peptide 91-104, corresponding to the putative Helix 4 as negative control. As expected, Helix 1 and Helix 5 blocked the protein homo-dimerization in a dose-dependent way with estimated IC_{50} of 76 and 178 nM, respectively, whereas Helix 4, MKK7(132-156) and Gadd45β (60-86) were ineffective. Importantly, the capacity of the single peptides to abolish protein self-association, suggests that recognition occurs in a cooperative way and the block of one site (Helix 1 or Helix 5) prevents complex formation. To investigate whether this regions could be somehow involved in MKK7 inactivation, we set up a binding assay with the kinase domain of this protein. To this aim, the region (120-380) of MKK7 was produced as recombinant protein and purified. Remarkably, Gadd45β bound in a dose-dependent and saturable way to this domain with an estimated K_D of about 90 nM. Confirming previous observations [1,2], this interaction was completely abolished by peptides MKK7(132-156) and Gadd45β(60-86), whereas Helices 1 and 5 of Gadd45β were totally ineffective.

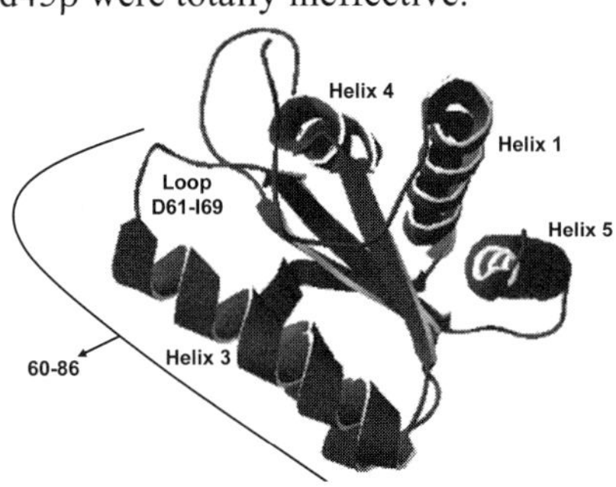

Fig. 1. Predicted structure of Gadd45β [2]: evidenced are the Helix 1 and Helix 5 involved in protein self-association and the region 60-86 responsible of MKK7 binding and inactivation.

Data confirm the notion that the 60-86 region of Gadd45β is involved in MKK7 binding and that the protein self-association, mediated by regions corresponding to the predicted Helix 1 and Helix 5, does not interfere with this interaction. We therefore propose that the Gadd45β-MKK7 interaction occurs within the contest of a complex that at least contains the MKK7/Gadd45β:Gadd45β/MKK7 oligomer. Indeed, as also the kinase reportedly is able to dimerize, the formation of higher order oligomers cannot be ruled out.

Acknowledgments

Financial support from project FIRB RBNE03PX83_005 is acknowledged.

References

1 Papa, S., F. Zazzeroni, et al. *Nat. Cell. Biol.* **6**, 146-153 (2004).
2 Papa, S., S. M. Monti, et al. *J. Biol. Chem.* **282**, 19029-19041 (2007).
3 Kovalsky, O., F. D. Lung, et al. *J. Biol. Chem.* **276**, 39330-39339 (2001).

A Glycopeptide-based Technique for Selective Antibody Purification

Elisa Peroni[1,2], Marta Tontini[1,3], Barbara Mulinacci[1,3], Mario Chelli[1,2], Francesco Lolli[1,4], Paolo Rovero[1,3] and Anna M. Papini[1,2]

[1]*Laboratory of Peptide & Protein Chemistry & Biology;* [2]*Dept. of Organic Chemistry "Ugo Schiff", CNR-ICCOM, University of Florence;* [3]*Dept. of Pharmaceutical Sciences, University of Florence;*[4]*Dept. of Neurological Sciences, University of Florence.*

Introduction

We demonstrated, for the first time, that an aberrant post-translational modification (PTM, *N*-glucosylation) is possibly triggering autoantibody response in Multiple Sclerosis. This was possible because of an innovative "chemical reverse approach" we followed to develop the best synthetic glycopeptide antigen for fishing out circulating autoantibodies from sera of Multiple Sclerosis patients. In fact CSF114(Glc), a structure-based designed glycopeptide, is the first Multiple Sclerosis Antigenic Probe accurately measuring high affinity autoantibodies as biomarkers of disease activity in sera of a statistically significant patients' population, compared to other autoimmune diseases and normal blood donors (NBD) [1].

In order to further investigate the molecular mechanism triggering an autoimmune response in MS, antibodies specifically recognized by CSF114(Glc) were produced in rabbit and purified directly from rabbit serum.

With this aim we investigated the immunogenic activity of the glycopeptide immunizing rabbits with the glycopeptide CSF114(Glc) without any protein carrier. We further studied and developed a glycopeptide-based immunoaffinity technique for selective antibodies purification.

Results and Discussion

For immunizations, HICR rabbits were used. On day 0 they were injected with CSF114(Glc), methylated BSA and Freund's Complete Adjuvant in H_2O. Subsequent booster immunizations were repeated after 4 weeks, 8 weeks, and 12 weeks with Freund's Incomplete Adjuvant. Immunization of rabbit with CSF114(Glc) produced a polyclonal sample containing antibodies able to recognize both the glycopeptide CSF114(Glc) and the unglycosylated peptide CSF114.

Immunoaffinity chromatography based on CSF114(Glc)-Sepharose and CSF114-Sepharose was used to isolate and purify antibodies from rabbit sera. This purification technique allows to isolate an antibody family recognizing specifically CSF114(Glc).

By ELISA and BIAcore analysis we tested the specificity of purified IgG for CSF114(Glc) and CSF114. The peptides were immobilized on the wells of the ELISA plates, or linked by the biotin/streptavidin interaction on BIAcore chip.

The antibodies purified on CSF114-Sepharose do not recognize in ELISA the glycosylated peptide antigen CSF114(Glc), moreover by inhibition ELISA we demonstrated that both isolated antibodies and immune serum are specifically inhibited by CSF114(Glc).

S. Del Valle et al. (eds.), *Peptides for Youth: The Proceedings of the 20th American Peptide Symposium*, 369
DOI: 10.1007/978-0-387-73657-0_163, © Springer Science+Business Media, LLC 2009

BIAcore analysis confirm that the isolated antibodies are specifically recognized by CSF114(Glc) and not by CSF114 (Table 1).

All these data demonstrated for the first time that the glycopetide CSF114(Glc) is able to produce antibodies in rabbit without any carrier protein.

Table 1. BIAcore data.

Purified IgG	***CHANEL 1***	***CHANEL 2***	***relative signal***
	IgG versus CSF114	***IgG versus CSF114(Glc)***	
Sample 1	96.6	401.4	304.8
Sample 2	38.2	147.3	109.1
Sample 3	24.2	116.9	92.7

Moreover thanks to the CSF114(Glc)-based chromatography we were able to isolate and purify antibodies specifically recognizing CSF114(Glc) directly from rabbit serum.

These properties suggested that the use of our glycopeptide as a synthetic antigen (Ag) might represent a simple experimental system, not only to investigate auto-reactive Ab response to follow-up MS disease activity, but also to explore new immuno-therapies for this autoimmune disease. Affinity separation by Antibody-Antigen interaction, where either the Ag or the Ab are immobilized to various matrices, can be used for the purification or removal of the corresponding Ab or Ag, respectively from MS patients sera. Synthetic peptides to specifically remove Abs, have not been yet so much used as affinity ligands. Moreover, no selective apheresis technique was performed up to now, particularly for MS, because no selective Ag was identified for this disease. On the base of the results obtained in this study, our challenge is to investigate the possibly development of a selective apheresis based on synthetic antigen

Acknowledgments

We thank Dr J. Hoebeke from the CNRS of Strasbourg for the fruitful discussion. The work was funded by MIUR FIRB 2005 RBIN04TWKN

References

[1] (a) Lolli F, Mulinacci B, Carotenuto A, Bonetti B, Sabatino G, Mazzanti B, D'Ursi AM, Novellino E, Pazzagli M, Lovato L, Alcaro MC, Peroni E, Pozo-Carrero MC, Nuti F, Battistini L, Borsellino G, Chelli M, Rovero P, Papini AM. *Proc Natl Acad Sci U S A.*, **102**(29) 10273-8, 2005.

(b) Lolli F, Mazzanti B, Pazzagli M, Peroni E, Alcaro MC, Sabatino G, Lanzillo R, Brescia Morra V, Santoro L, Gasperini C, Galgani S, D'Elios MM, Zipoli V, Sotgiu S, Pugliatti M, Rovero P, Chelli M, Papini AM. *J Neuroimmunol.*, **167**(1-2), 131-7, 2005.

(c) Papini, A.M. *Nat. Med.*, **11**, 13, 2005.

(d) Carotenuto A., D'Ursi A. M., Mulinacci B., Paolini I., Lolli F., Papini, A. M., Novellino E., Rovero P., *A. J. Med. Chem.*, **49**, 5072-5079, 2006.

(e) Papini, A.M., Rovero P. Chelli M.; Lolli F., Granted U.S.A. Patent & PCT Application WO 03/000733 A2.

C-Terminal Amidation On Aryl Hydrazine Resin

Witold A. Neugebauer, Amélie Parent, Xue Wen Yuan and Robert Day

Department of Pharmacology, Faculty of Medicine, University of Sherbrooke, Sherbrooke, Québec, J1H 5N4, Canada

Introduction

Carboxy-terminal 7-amino 4-methyl coumarin (AMC) peptide amides are useful chromogenic compounds for studying proteolytic enzymes. Solid phase synthesis of these in a classical form could not be applied. However, a new strategy to build a peptide on special resin used by Kwon et al [1] and couple C-terminally to peptide with resin cleavage (see Figure 1) seems to be an ideal method to make Pyr-Arg-Thr-Lys-Arg-AMC [2]. Fmoc-aryl hydrazine resin (Novabiochem) is used for peptide synthesis in a continuous flow system peptide synthesizer.

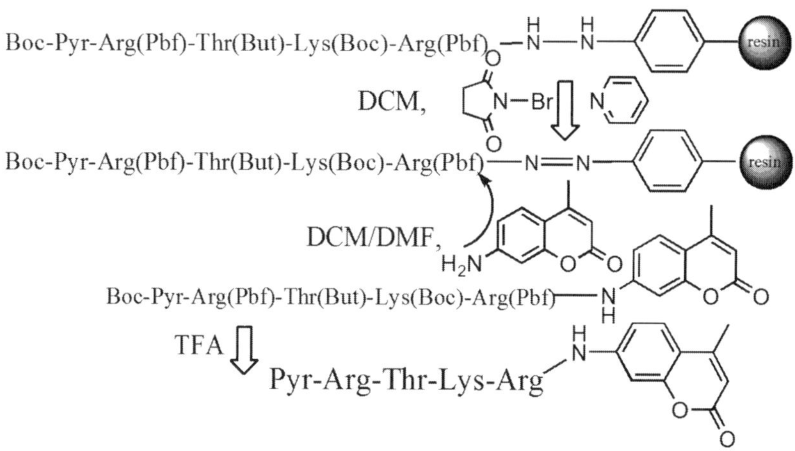

Fig. 1. Scheme of peptide-4-methylcoumarin-7-amide synthesis.

Then aryl hydrazine resin is oxidized by NBS for 10 min. to diazene, which is displaced by coumarin overnight. The peptide amide is then deprotected with TFA:TIPS:water/95:2.5:2.5 and purified on silica and C18 column chromatography. The final product was identified by mass spectrometry (MALDI) and its purity tested on HPLC and TLC.

Results and Discussion

Pyr-Arg-Thr-Lys-Arg peptide was synthesized using Fmoc chemistry on aryl-hydrazine resin in a continuous flow system peptide synthesizer. The last Pyr in sequence was coupled as Boc-Pyr-OH. Peptide-resin was suspended in dry DCM in amber glass and was oxidized in 10 min with N-bromosuccinimide (NBS) while formed HBr was trapped by dry pyridine. The N-terminal Boc-group was left on during the amidation removal of synthesized peptide in DCM/DMF with coumarin overnight in amber glass. Then resin was filtered off and solvent evaporated in vacuum. The protecting groups were then cleaved by TFA /triisopropylsilane (TIPS) / water and precipitated in ethyl ether. The crude peptide was freeze-dried and purified on silica gel followed

S. Del Valle et al. (eds.), *Peptides for Youth: The Proceedings of the 20th American Peptide Symposium*, 371
DOI: 10.1007/978-0-387-73657-0_164, © Springer Science+Business Media, LLC 2009

by reversed phase column chromatography. Yield of crude peptide in this amidation/ cleavage was ~50% and purified product ~16%. Peptide identity was confirmed with MALDI mass spectrometry. Peptide purity was verified by HPLC and biochemical test.

The peptide activity was tested in 96 well plate assay, and fluorescence was measured using a Molecular Devices spectrofluorometer, as previously described [3]. Each AMC-peptide was tested with and without furin, a proprotein convertases that requires the sequence RXKR for cleavage activity [4]. As shown in Figure 2, the newly synthesized AMC peptide Pyr-Arg-Thr-Lys-Arg-MCA was compared to a control peptide, of the same structure but obtained commercially (Bachem Inc.). Both peptides were incubated under identical conditions, with and without furin. Lack of furin in the reaction reveals the background levels of the peptide. The data shows that the newly synthesized peptide has an identical behavior to the commercially available peptide, synthesized by standard means.

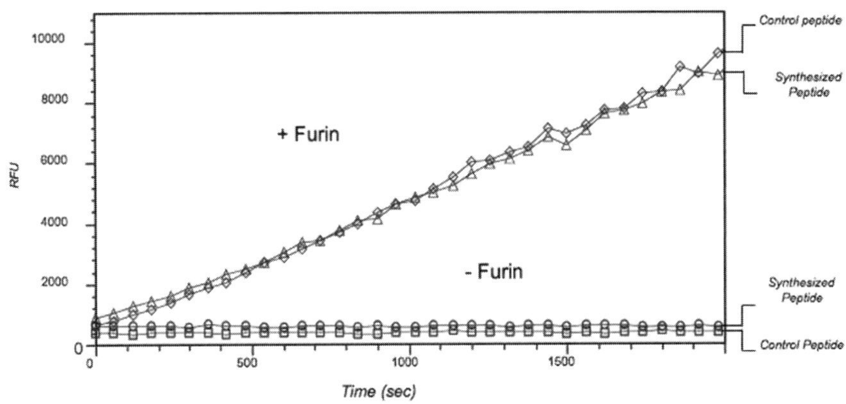

Fig 2. Fluorescence activity of newly synthesized peptide in comparison to control peptide, in the presence and absence of furin. In the absence of furin the background levels are very low, while identical progress curves are obtained in the presence of furin.

Conclusions

This new procedure greatly simplifies the synthesis of C-terminal AMC peptides, at a reduced cost. The described method could be used as general way to synthesize peptide amino-coumarin amides.

Acknowledgments

The work was funded by a research grant from the Canadian Instituts of Health Research (CIHR) to RD and WAN.

References

1. Kwon, Y., *et al.* J.A. *Organic Letters* **6(21)**, 3801-3804 (2004).
2. Hatsuzawa, K., *et al. J. Biol. Chem.* **267**, 16094-16099 (1992).
3. Fugère, M., *et al. Molecular Pharmacology* **71**, 323-332 (2007).
4. Fugère, M. and Day, R. *Trends in Pharmacological Sciences* **26**, 294-301 (2005).

PEPTIDES AND INFECTIOUS DISEASE

Synergy Between a Lead Proline-rich Antibacterial Peptide Derivative and Small Molecule Antibiotics

Laszlo Otvos, Jr[1], Marco Cassone[1], Vanessa de Olivier Inacio[2], Paul Noto[1], John J. Rux[3], John D. Wade[4] and Predrag Cudic[2]

[1]Sbarro Institute for Cancer Research and Molecular Medicine, Temple University, Philadelphia, PA 19122, USA; [2]Department of Chemistry and Biochemistry, Florida Atlantic University, Boca Raton, FL 33431, USA; [3]The Wistar Institute, Philadelphia, PA 19104; [4]Howard Florey Institute, University of Melbourne, Victoria 3010, Australia

Introduction

Proline-rich antimicrobial peptides kill bacteria by binding to the C-terminal D-E helix of DnaK, the 70 kDa bacterial heat shock protein and inhibiting chaperone-assisted protein folding [1]. Indeed, the strain selectivity of two family members, drosocin and pyrrhocoricin, and the D-E helix homology of Enterobactericeae can be fully correlated [2,3]. However, mechanistically it is still unclear how binding to the multihelical lid region would allosterically prevent substrate-DnaK interactions taking place 200 residues upstream. If a proper theoretical explanation can be found, the process can be utilized for the development of standalone peptide antibiotics, or for reversing resistance mechanisms that are developed against small molecule antibiotics where failure to kill microorganisms involves the activation of properly folded bacterial proteins, including enzymes. Newly introduced functional enzymes are responsible for the emerging resistance against once powerful antibiotics such as amoxicillin (inactivated by β-lactamase), trimethoprim (inactivated by altered dihydrofolate reductase), chloramphenicol (inactivated by chloramphenicol acetyltransferase) and sulfonamide (inactivated by altered tetrahydropteroic synthetase). Earlier we documented that a lead dimeric proline-rich antibacterial peptide analog, A3-APO given in sublethal doses, is able to reverse β-lactamase induced resistance against amoxicillin in clinical *Escherichia coli* isolates [4]. In general, peptide A3-APO has a series of attractive pharmacological parameters that warrant further preclinical development [5]. In the current study, we investigated the in vitro scope and limitations of peptide A3-APO therapy for the design of appropriate in vivo mouse models of infection control.

Results and Discussion

A frequently overseen physical property of biologically active Hsp70 molecules is their dimeric form where oligomerization is related to the intrinsic ATPase activity of the enzymes [6]. Dimerization of Hsp70 chaperones occurs at their C-terminus [7]. Thorough examination of the dimer structure reveals that one of the interacting surfaces is exactly the D-E helix, where the proline-rich antimicrobial peptides bind (Figure 1a). Thus, the most probable mode of action of pyrrhocoricin, drosocin and analogs is the competitive inhibition of the dimerization process of DnaK, leading to diminished ATPase capacity, a secondary function that provides energy for the primary role of protein refolding activity. We used a combination of liquid broth microdilution, fluorescence microscopy and dual fluorochrome healthy/damaged membrane assays to show that the isolated D-E helix fragment is inefficient in killing bacteria because this neutral peptide is unable to interact with the negatively

S. Del Valle et al. (eds.), *Peptides for Youth: The Proceedings of the 20th American Peptide Symposium*,
DOI: 10.1007/978-0-387-73657-0_165, © Springer Science+Business Media, LLC 2009

charged bacterial membrane and penetrate into bacterial cells. Nevertheless, truly active proline-rich antibacterial peptides exhibit both DnaK antagonistic and membrane-destroying properties [8].

In this regard, the sequence of peptide A3-APO is a fortunate combination of the two requirements without interacting with mammalian Hsp70 or influencing the growth and homeostasis of eukaryotic cells. A3-APO kills multidrug resistant bacteria that are insensitive to most currently used small molecule antibiotics including the classical β-lactams or fluoroquinolones, the preferred first-line defense compounds in hospitals (Figure 1b). In addition, this peptide shows remarkable stability to proteolytic enzymes present in pooled mouse serum in vitro or whole mouse blood ex-vivo. In 25% diluted serum as much as 78%, 66%, 47% and 26% peptide remained intact after 45 min, 2h, 4h and 8h incubation periods, respectively. The peptide was injected into mice either intravenously or subcutaneously, blood and urine were taken in a 5 min – 2h period, and processed similar to the in vitro serum stability assay protocol [9]. MALDI-MS failed to identify the intact molecule or early degradation products in either biological fluid. This finding supports our earlier observation that amphipathic antimicrobial peptides frequently stick to biological and physical surfaces, and thus HPLC/MS-based pharmacokinetics studies may not be applicable to these molecules [5]. True absorption and stability data can only be obtained from in vivo efficacy studies. The activity measurements of A3-APO in comparison with ciprofloxacin in mouse models of bacteremia induced by fluoroquinolone-resistant bacteria are in progress.

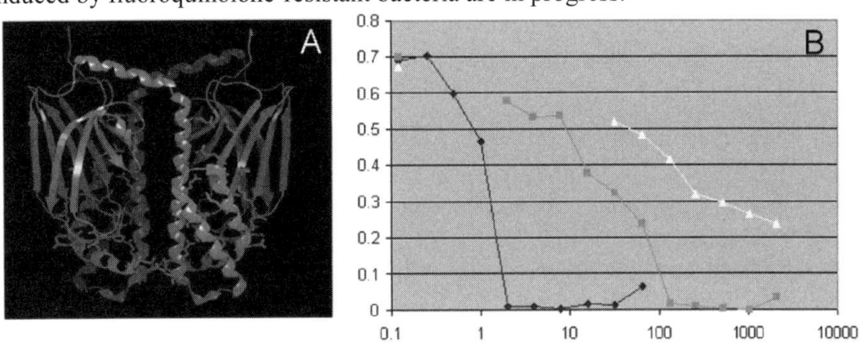

Fig.1. Panel A: Dimeric structure of the C-terminal third of E. coli *DnaK. The figure also shows interacting residues of the D-E helix and an upstream discontinuous region. Panel B: Antibacterial activity of peptide A3-APO (black diamonds), ciprofloxacin (grey squares) and amoxicillin (white triangles) against the multidrug-resistant clinical isolate* E. coli *HK179. The concentration of the antibiotics (X axis) is given in μg/mL. The Y axis shows bacterial growth as absorbance at 600 nm.*

Synergy between A3-APO and amoxicillin, trimethoprim, chloramphenicol or sulfonamide was tested with two different methods. Either the concentration of both types of antibiotics varied (classical checkerboard assay) indicating full synergy, or with fixed, sublethal doses of A3-APO and varying small molecule antibiotics concentration indicating a one-dimensional (peptide to small molecule) interaction. In both cases peptide A3-APO was added to growing bacterial colonies 1h prior the addition of conventional antibiotics to cease protein folding in bacteria. The two techniques yielded identical results (Table 1). A3-APO successfully reversed resistance developed against all four antibiotics and in all Enterobactericeae species tested: *E. coli*, *Klebsiella pneumoniae*, *Salmonella typhimurium* and *Haemophilus influenzae*. Thus, A3-APO may give renewed life to conventional antibiotics for which large stockpiles are currently available. However, the activity was strain- and small molecule-dependent: not all strains of the same species could be killed by the antibiotics after prior A3-APO treatment suggesting that additional resistance factors, or unrelated cell compositional differences maintain the resistant phenotype.

Table 1. Synergy between peptide A3-APO and small molecule antibiotics against strains that are resistant to the conventional antimicrobial compounds when these are added alone. The two types on antibacterial molecules were either applied in a checkerboard fashion at 0.25-512 mg/l for both compounds (CB), or A3-APO was used in a fixed, sublethal dose and the concentration of the small molecule antibiotics varied between 0.1-128 mg/l (SD).

Bacterial Strains	Small Molecule Antibiotics			
	Amoxicillin	Trimethoprim	Chloramphenicol	Sulfonamide
E. coli SEQ102	CB:synergy SD:synergy	CB:synergy	CB:synergy	
E. coli BF1023	CB:synergy SD:synergy			
E. coli S4362		CB:no synergy		
E. coli S5081				SD:no synergy
K. pneumoniae K6	SD:no synergy	CB:synergy	CB:synergy SD:synergy	
K. pneumoniae RP1	CB:no synergy SD:no synergy		CB:synergy	
K. pneumoniae MH4	CB:synergy		CB:synergy	
S. typhimurium G10215	CB:no synergy SD:no synergy		CB:synergy	CB:synergy
H. influenzae R387	SD:synergy		SD:synergy	

Acknowledgment

This work was supported by the Sbarro Health Research Organization.

References

1. Kragol, G., *et al. Biochemistry* **40**, 3016-3026 (2001).
2. Bikker, F.J., *et al. Chem. Biol. Drug Des.* **68**, 148-153 (2006).
3. Kragol, G., *et al. Eur. J. Biochem.* **269**, 4226-4237 (2002).
4. Otvos, L., Jr., *et al. Antimicrob. Agents Chemother.* **50**, 3146-3149 (2006).
5. Otvos, L. Jr., *et al. J. Med. Chem.* **48**, 5349-5359 (2005).
6. Richarme, G. and Kohiyama, M. *FEBS Lett.* **322**, 277-279 (1993).
7. Chou, C.-C., *et al. J. Biol. Chem.* **278**, 30311-30316 (2003).
8. Otvos, L., Jr., *et al. Int. J. Pept. Res. Ther.* **11**, 29-42 (2005).
9. Powell, M.F., *et al. Pharm. Res.* **10**, 1268-1273 (1993).

Natural mutants of HIV CTL epitopes with enhanced immunogenicity as potential vaccine candidates

Sylvie E. Blondelle[1], Rosa Moya[1], Kim Schroder[1], Keiko Osawa[1] and Darcy B. Wilson[1,2]

[1]*Mixture Sciences, Inc., San Diego, CA 92121, USA.* [2]*Torrey Pines Institute for Molecular Studies, San Diego, CA 92121, USA*

Introduction

HIV/AIDS remains a serious, ultimately lethal pandemic disease in much of the developing world. Significant treatment advances have been made with anti-retroviral agents, but major problems remain. These treatments are expensive and difficult to apply especially for developing world nations that lack adequate finances and infrastructure, do not eliminate all virus and therefore require life long treatment regimens, carry significant toxic side effects and in some quarters becoming less effective due to viral mutation. There remains a major need to identify effective measures to prevent new HIV infections, especially in areas of the world where it is a pandemic. Based on the history of vaccine development, it remains a strong consensus in the biomedical research community that immunization will provide the most effective and affordable long term-approach to controlling the spread of HIV/AIDS.

Strong cytotoxic T-lymphocyte (CTL) responses are thought to be important for the development of an effective human immunodeficiency virus (HIV) vaccine; their benefit has been established in simian immunodeficiency virus (SIV) vaccination studies and in natural infections [1,2]. One of our major goals is to identify optimized, highly immunogenic peptides for the design of prophylactic or therapeutic vaccines for HIV infection. Our strategy derives from the hypothesis that the rarer mutant CTL epitopes of HIV may be more immunogenic than the common ones and that these rarer mutant sequences of HIV epitopes might be more effective in generating cross reactive anti-HIV CTL responses against both the rare and common mutant sequences.

Results and Discussion

Epitope based vaccines

In comparing various vaccine approaches designed to induce $CD8^+$ CTL responses, epitope-based vaccines offer advantages over vaccines encoding whole protein antigens. Thus, epitope-based vaccines are capable of inducing more potent responses than whole protein vaccines [3]. By design, it is possible to control qualitative aspects of the immune response by simultaneously targeting multiple dominant and subdominant epitopes [4,5]. Inclusion of both may be important for prophylactic and therapeutic HIV vaccines, to provide protection even after emergence of escape variants. The use of epitopes can overcome potential safety concerns associated with some vaccinating proteins or viral vectors. For example, the multifunctional Vpr protein induces apoptosis in neuronal cells [6]. Finally, it is substantially easier to produce altered peptide sequences than proteins having enhanced immunogenicity.

S. Del Valle et al. (eds.), *Peptides for Youth: The Proceedings of the 20th American Peptide Symposium*,
DOI: 10.1007/978-0-387-73657-0_166, © Springer Science+Business Media, LLC 2009

Table 1. Frequency of conserved and non-conserved amino acid changes in common and rare CTL epitope mutants.

CTL epitope	Among common mutants No. of entries with:		Among rare mutants No. of entries with:	
	conservative mutation	**non-conservative mutation**	**conservative mutation**	**non-conservative mutation**
p17 77-85	119 (70%)	0 (0%)	17 (23%)	10 (14%)
p24 19-27	161 (96%)	0 (0%)	4 (26%)	1 (6%)
p24p2p7 230-7	17 (9%)	0 (0%)	12 (31%)	4 (10%)
RT 309-317	54 (57%)	0 (0%)	13 (30%)	10 (23%)
gp41 591-599	181 (83%)	6 (3%)	16 (21%)	11 (14%)
Nef 92-100	279 (96%)	5 (2%)	19 (25%)	10 (13%)
Nef 190-198	100 (22%)	58 (12%)	29 (10%)	109 (36%)

Number (percent) calculated based on the number of entries of a given mutant in the Los Alamos HIV Molecular Immunology Database [http://www.hiv-lanl.gov – 7].

CTL escape

Results of numerous recent studies have shown that evasion of the immune response via CTL escape is an important factor in HIV and SIV pathogenesis [8]. CTL escape has been well documented in both acute and chronic HIV infection. An average of 60% of HIV isolates has mutated CTL epitope sequences. Of note, CTL epitope sequence mutations are represented by a few common mutations (less than 10 separate sequences) found in 75% of mutated viral isolates and by numerous different rare mutations (more than 50 separate sequences) found only in single or very few individuals. We hypothesized that the common mutants represent sequence variations that have most successfully evaded the selective pressure of CTL recognition, and that the rare mutants have not.

The pattern of CTL escape mutations depends on a balance of functional constraints on viral proteins and immune pressure [9,10]. A significant proportion of non-conservative mutants were found among the rare mutants, while most of the common mutants have none or few non-conservative substitutions (Table 1). Mutations with conservative amino acid changes would have a lesser effect on viral fitness and therefore are more common than mutations with non-conservative residues; on the other hand, those mutants with non-conservative substitutions compatible with viral integrity and survival occur in rare case of extreme immune pressure.

Immunogenicity of various mutants

The immunogenicity of mutants from three separate CTL epitopes (Gag $p24_{19\text{-}27}$ TV9, Gag $p17_{77\text{-}85}$ SL9, and $RT_{309\text{-}317}$ IV9) was assessed in the HLA-A2 transgenic mouse Cy HLA-A2/Kb model system. CD8 responses against the immunizing peptides were determined in murine IFN-γ ELISPOT assays 8 days post-immunization. As anticipated, the common mutants of the three CTL epitopes studied, all having conservative substitutions, were less or similarly immunogenic than the native sequences (illustrated in Table 2). Similar results were observed for rare mutants

Table 2. Immunogenicity of CTL mutants determined by ex vivo ELISPOT assay.

	Immunizing peptide									ELISPOT responses	
										% positive response	**Mean SFU/500th LN cells**
TV9	T	L	N	A	W	V	K	V	V	89	65
TV9 C1	–	–	–	–	–	–	–	–	I	44	35
TV9 R1	–	–	–	–	–	–	–	***A***	–	100	185
TV9 R2	–	–	–	–	–	–	–	–	L	17	15
SL9	S	L	Y	N	T	V	A	T	L	22	30
SL9 C1	–	–	F	–	–	–	–	–	–	31	25
SL9 R1	–	–	F	–	–	–	–	***V***	–	0	0
SL9 R2	–	–	–	–	**L**	–	–	–	–	100	105
SL9 R3	–	–	F	–	**L**	I	–	–	–	100	100
SL9 R4	–	I	–	–	**L**	–	–	–	–	100	65
IV9	I	L	K	E	P	V	H	G	V	80	70
IV9 C1	–	–	–	D	–	–	–	–	–	42	40
IV9 R1	–	I	–	–	–	–	–	–	–	0	0
IV9 R2	–	–	–	–	–	–	–	**R**	–	100	80

Non-conservative mutations are in bold and boxed. Semi-conservative mutations are in italics and underlined. Common mutants are referred as "C" and rare mutants are referred as "R". The results are expressed as expressed as percent positive from at least ten responses. and as the mean number of spot forming units (SFU) per 0.5 million input cells for all positive responses after subtracting the number of spots obtained with no peptide.

containing only conservative substitutions. In contrast, rare mutants containing non-conservative substitution were in most cases among the most immunogenic peptides. Mixed results were observed for those rare mutants containing semi-conservative substitutions.

Cross-recognition to the native sequence and other mutants was then investigated by performing similar IFN-γ ELISPOT assays following immunization. Cross recognition was observed for mutants TV9 R1, SL9 R4, and IV9 R2 (100%, 72%, and 73% positive responses to each respective native peptide). Of note substitutions resulting in cross-recognition are located in all three cases at a potential TCR contact position (position 5 or 8). These observations suggest that TCR/peptide interactions may play an important role towards high cross-reactive immunogenicity.

Table 3. Individual peptide ELISPOT responses following multi-peptide immunization.

	Immunizing peptide cocktail									ELISPOT responses	
										% positive response	**Mean SFU/500th LN cells**
TV9	T	L	N	A	W	V	K	V	V	100	95
TV9 R1	–	–	–	–	–	–	–	***A***	–	100	105
TV9 R2	–	–	–	–	–	–	–	–	L	33	35
SL9	S	L	Y	N	T	V	A	T	L	83	20
SL9 R1	–	–	F	–	–	–	–	***V***	–	33	25
SL9 R4	–	I	–	–	**L**					100	45

Multi-epitope approach

While a rapid emergence of CTL escape occurs during acute infection with HIV, i.e., when few epitopes are targeted, a broader T cell response, associated with better control of viral replication, is observed after acute infection. This can be explained by a T cell response directed at multiple epitopes. An immune response targeted at multiple epitopes therefore makes the probability of simultaneous escape at all epitopes in a single virus very unlikely. Furthermore, it is probable that the virus would rapidly mutate away from an epitope used in a vaccine to elicit immunity. A "multi-mutant" vaccine that elicits multiple T cell clonotypes against several HIV epitopes may therefore provide immunity that can best tolerate amino acid mutations.

To investigate the multi-epitope immunization strategy, mice were immunized with a cocktail of three TV9 mutants and three SL9 mutants, and CD8 response to each individual peptides was determined by IFN-γ ELISPOT assay. The cocktail included the native sequences, a highly cross-reactive mutant and a week immunogen for each CTL epitope. As shown in Table 3, weak immunogens do not interfere with the response to strong immunogens, strong immunogens do not enhance the response to weaker immunogens, and multiple different strong immunogen epitopes do not compete with one another.

In summary, the results presented here confirm our premise that naturally occurring sequences among the reported rare mutants represent potential candidate vaccines since they can generate better immune responses than the consensus native epitopes against the consensus sequence itself, as well as against common mutants of that sequence and against multiple rare mutant sequences. Furthermore, we have demonstrated that peptide pools can be used as the basis of an effective multi-epitope vaccine.

Acknowledgments

The work was funded by the National Institutes of Health grants R43 DA021175 to SEB and R44 DA015212 to DBW.

References

1. Sewell, A. K., Price, D. A., Oxenius, A., Kelleher, A. D. and Phillips, R. E. *Stem Cells* **18**, 230-244 (2000).
2. Johnston, M. I. and Fauci, A. S. *N. Engl. J. Med.* **356**, 2073-2081 (2007).
3. Ishioka, G. Y., *et al. J. Immunol.* **162**, 3915-3925 (1999).
4. Oukka, M., *et al. J. Immunol.* **157**, 3039-3045 (1996).
5. Tourdot, S., *et al. J. Immunol.* **159**, 2391-2398 (1997).
6. Patel, C. A., Mukhtar, M. and Pomerantz, R. J. *J. Virol.* **74**, 9717-9726 (2000).
7. Korber, B., Walker, B., Brander, C., Koup, R., Moore, J., Haynes, B., Meyers, G. HIV Molecular Immunology Database. Los Alamos National Laboratory: Theoretical Biology and Biophysics, Los Alamos, NM (1996).
8. Goulder, P. J. and Watkins, D. I. *Nat. Rev. Immunol.* **4**, 630-640 (2004).
9. Wagner, R., *et al. J. Immunol.* **162**, 3727-3734 (1999).
10. Walker, B. D. and Korber, B. T. *Nat. Immunol.* **2**, 473-475 (2001).

Lipodepsipeptide antibiotic fusaricidin and its analogues. Total solid-phase and biological activity.

Maciej Stawikowski and Pedrag Cudic

Department of Chemistry and Biochemistry, Florida Atlantic University, 777 Glades Rd, 33431 Boca Raton, FL, USA

Introduction

Infectious diseases due to antibiotic-resistant bacteria pose a serious threat to public health [1]. The Centers for Disease Control and Prevention has estimated that nearly two million people a year in the U.S. develop nosocomial infections, and nearly 90,000 of those people die. For the past three decades the antibiotic vancomycin has been the last line of defense against bacterial strains that are resistant to most of antibiotics. However, the utility of this drug has been limited because of the emergence of vancomycin-resistant bacterial strains, most notably in *enterococcal* and *staphylococcal* nosocomial pathogens. Therefore, developing new antibiotics capable of treating infections from drug-resistant bacteria is of vital importance.

Promising candidate for reverting multi-drug resistant bacteria is naturally occurring cyclic lipodepsipeptide antibiotic fusaricidin A. This natural product was isolated from *Bacillus polymyxa* KT-8 strain, and exhibits strong activity against various Gram-positive bacteria including MRSA [2]. Here, we report the total solid-phase synthesis of fusaricidin A, solid-phase synthesis of its analogs, their structural characterization and biological activities.

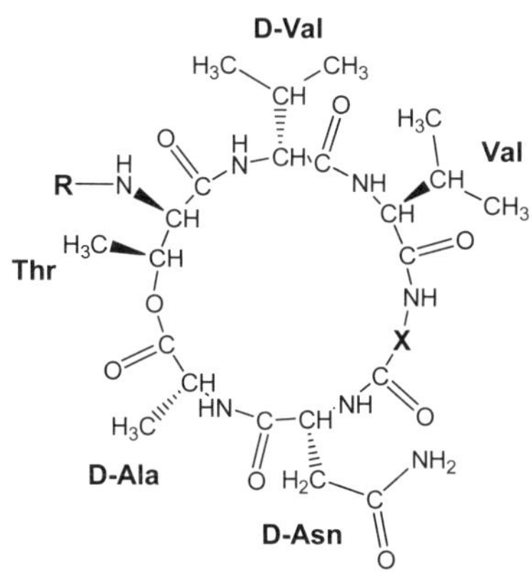

Compound	X	R
Fusaricidin A	D-*allo*-Thr	15-guanidino-3-(*R*)-hydroxypentadecanoic acid
1	D-Thr	12-guanidinododecanoic acid
2	D-Thr	12-aminododecanoic acid
3	D-*allo*-Thr	acetyl
4	D-Thr	acetyl

Fig. 1. Fusaricidin A and its analogs.

S. Del Valle et al. (eds.), *Peptides for Youth: The Proceedings of the 20th American Peptide Symposium*, 383
DOI: 10.1007/978-0-387-73657-0_167, © Springer Science+Business Media, LLC 2009

Results and Discussion

Fusaricidin's exploitation as a lead compound for the development of new antibiotics strongly depends on gaining unlimited synthetic access to this natural product. Fusaricidin A and its analogs were prepared using standard solid-phase Fmoc chemistry according to the strategy previously described by our group. A key building block in fusaricidin A's synthesis, Fmoc protected 15-amino-3-(*R*)-hydroxypentadecanoic acid, was prepared over nine steps starting from 11-bromoundecanol. Fmoc protected 12-aminododecanoic acid, acetic acid, and D-Thr, were used to prepare analogues **1-4**. In all cases the fatty acid was incorporated into linear lipidic peptide precursor prior the depsipeptide ester bond formation and on-resin cyclization reaction. In the very last step fatty acid's amino group was converted into desired guanidino group using Mukaiyama's guanylating reagent [4]. The structural studies of analogs **1-4** in water using CD spectroscopy showed that all depsipeptides possess random-coil structure, and that the secondary structure of the cyclic depsipeptide fragment is not influenced by the presence and character of the lipid tail nor by the D-Thr/D-*allo*-Thr residue. Fusaricidin A analogs **1-4** were screened for antibacterial activity against various multidrug resistant Gram-positive and Gram-negative bacteria. As indicated by determined MIC, analogue **1** showed most promising activity against Gram-positive bacteria including methicillin-resistant *S. aureus* (16 μg/mL), methicillin-resistant *S. epidermidis* (16 μg/mL) and vancomycin-resistant *E. faecium* (64 μg/mL). Analogs **2-4** did not show activity against these strains even at high concentration (128 μg/mL). None of analogs **1-4** were active against Gram-negative bacteria. The antimicrobial assays revealed that depsipeptide's guanidino lipid tail is crucial for peptide antibacterial activity. To test whether synthesized peptides were toxic, the hemolytic activity of each analogue was determined using human RBCs (0.5%). PBS and 1% Triton controls were used to determine 0% and 100% hemolysis. At a final concentration of 16 μg/mL the highest hemolytic activity showed analogue **2** (50%), followed by analogue **1** (27%), analogue **3** (8.5%), and analogue **4** (14%), indicating that lipid tail contributes significantly to the toxicity of the lipodepsipeptide antibiotics. All peptides displayed remarkable stability in 25% human serum. After depsipeptide incubation for 6 h at 37 °C, concentration of **1-4** in human serum remained unchanged. However, 50% change in concentration was observed after 24 h of incubation. Since the same concentration change was observed after incubation of peptides only in PBS buffer (pH 7.6), we assumed that rather water than enzymatic hydrolysis of depsipeptide bond occurred. This assumption was confirmed by MALDI-TOF MS analysis of crude reaction mixture after incubation of **1** in human serum and PBS buffer ($[M+H]^+$, **1**, m/z = 825.4956, hydrolysis product, m/z = 845.1799). The investigation of the mode of action of the fusaricidin antibiotics family is currently underway and will be reported elsewhere.

Acknowledgments

This work was supported by NIH (1S06-GM073621-01) and AHA (0630175N) grants to P. Cudic.

References

1. Lipsitch, M. *Trends in Microbiology*, **9**, 438-444 (2001)
2. Kajimura, Y., Kenda, M. *J. Antibiot.*. **49,** 129-135 (1996)
3. Stawikowski, M., Cudic. P. *Tetrahedron Lett.* **47**, 8587-8590 (2006)

Progress toward Total Solid-Phase Synthesis of Depsipeptide Antibiotic Katanosin B

Roma Stawikowska, Maciej Stawikowski and Predrag Cudic

Department of Chemistry and Biochemistry, Florida Atlantic University, 777 Glades Road, Boca Raton, FL 33431, USA

Introduction

The peptidoglycan synthesis pathway is an attractive target for antibacterial and antifungal agents in terms of specificity for microorganisms. Drugs with different mode of action from vancomycin should be promising candidates against vancomycin resistant strains. Katanosin B is naturally occurring cyclic depsipeptide antibiotic isolated from strain related to the genus *Cytophaga*.[1] According to recent literature data it inhibits transglycosylation and transpeptidation steps of bacterial cell wall biosynthesis by an unidentified mechanism that differs from those of vancomycin. Katanosin B is composed of eleven amino acids from which nine form cyclic structure joined with one ester linkage. In addition, this antibiotic contains three unusual L-β-hydroxy amino acids: phenylserine (PhSer), hydroxyleucine (HyLeu) and hydroxyasparagine (HyAsn). Lack of commercial sources of katanosin B producing organism PBJ-5356 and, particularly, unlimited access to its synthetic analogs hampered katanosin B utilization as a lead compound for development of new antibiotics. In addition to this, isolation of the katanosin B from *Cytophaga sp.* is laborious process which results in low quantities of the final product. In order to obtain milligram quantities of katanosin B it is required to use large volumes (~10L) of the culture broth, organic solvents such as butanol, methanol and petroleum ether (necessary for the extraction of katanosin B from the culture broth) to perform centrifugation, decolorization with large amounts of active charcoal, and in final step preparative HPLC. As an alternative, total solid–phase peptide synthesis (SPPS) of this important natural product and combinatorial chemistry approach can be employed to elucidate structure-activity relationship and to find new potent compounds of this class. Since our goal is to develop a general Fmoc solid-phase methodology for the synthesis of katanosin B and its analogs, preparation of the Fmoc-protected L-β-hydroxy amino acids represent a challenging synthetic task. Chemo-enzymatic approach was used to prepare these unusual amino acids allowing therefore katanosin B synthesis on a solid-support according to the strategy already developed in our group.[2] Advantages of this synthetic approach include facile assembly, high purity, and potential for combinatorial diversity afforded by SPPS of protected amino acid modules.

Results and Discussion

Our strategy for katanosin B total SPPS include amide resin attachment of L-HyAsp *via* side chain, use of combination of four quasi-orthogonal removable protecting groups, stepwise Fmoc solid-phase synthesis of a linear precursor peptide, followed by the last amino acid coupling *via* ester bond and on-resin head-to-tail macrolactamization. As we demonstrated previously, this strategy allows complete suppression of undesired $O \rightarrow N$ acyl shift, and efficient automated solid-phase synthesis of cyclic depsipeptides.[2] Therefore, preparation of suitably protected

S. Del Valle et al. (eds.), *Peptides for Youth: The Proceedings of the 20th American Peptide Symposium*, 385
DOI: 10.1007/978-0-387-73657-0_168, © Springer Science+Business Media, LLC 2009

L-*β*-hydroxy amino acids represents the first step toward total Fmoc SPPS of katanosin B. Although several enantioselective procedures for the L-HyAsn, L-PheSer and L-HyLeu synthesis have been reported,[3] the Sharpless asymmetric aminohydroxylation (AA) was most frequently used as the most practical and shortest route to these unusual amino acids.[4] Since D,L-PhSer (MP Biomedicals), and D,L-HyAsp (TCI America) are commercially available at an affordable price, we chose to prepare orthogonally protected L-*β*-hydroxy amino acids suitable for Fmoc SPPS from these commercially available racemic mixtures. Fmoc-L-PhSer was obtained in pure form by enantiomeric resolution of D,L-PhSer with *Aspergillus melleus* acylase,[5] followed by Fmoc protection of α-NH_2 group of desired L-enantiomer. An enzymatic resolution of D,L-HyAsp could not be performed because HyAsp is not a substrate for acylase.[5] In this case an alternative approach has been developed and suitably protected Fmoc-D,L-HyAsp(*O*-THP)-OAllyl ready for coupling to the resin *via* side-chain was prepared. Fischer-type esterification of β-CO_2H by benzyl alcohol, followed by Fmoc protection of α-NH_2 group, allyl protection of α-CO_2H, and β-OH protection with THP using standard protocols resulted in synthesis of fully protected D,L-HyAsp (yield 75%). In the final step Bzl protecting group was removed with 20% K_2CO_3 in $EtOH/H_2O$ (1:1 v/v, yield 35%). Optimization of pseudoephedrine method for enentiomeric resolution is currently underway.

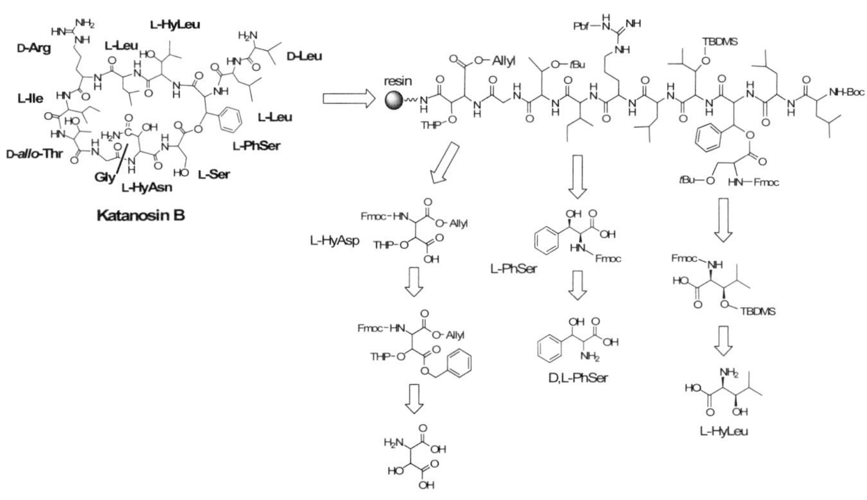

Fig. 1. Retrosynthetic analysis of katanosin B.

Acknowledgment

The work was funded by an AHA-SDG grant (0630175N) to P. Cudic.

References

1. Maki, H., Miura, K., Yamano, Y., *Antimicrob. Agents Chemother.* **45**, 1823-1827 (2001)
2. Stawikowski, M., Cudic, P. *Tetrahedron Lett.* **47**, 8587-8590 (2006)
3. Jung, M. E., Jung, Y., H. *Tetrahedron Lett.*, **30**, 6637-6640 (1989)
4. Boger, D. L., Lee, R. J., Bounaud, P. Y., Meier, P. *J. Org. Chem.* **65**, 6770-6772 (2000)
5. Chenault, H. K., Dahmer, J., Whitesides, G. M. *J. Am. Chem. Soc.* **111**, 6354-6364 (1989)

Determination of the minimal fusion peptide of HIV, SIV and BLV fusion glycoproteins

Aurélien Lorin[1], Benoît Charloteaux[1], Laurence Lins[1], Vincent Stroobant[2] and Robert Brasseur[1]

[1] *Gembloux Agricultural University, Centre de Biophysique Moléculaire Numérique, 2 Passage des Déportés, B-5030 Gembloux;* [2]*Ludwig Institute, 74 av. Hippocrate, B-1200 Brussels, Belgium*

Keywords : membrane fusion, fusion peptide, tilted peptide, HIV, SIV, virus, lipid-mixing, leakage, liposome, membrane insertion, molecular modeling

Introduction

The entry of enveloped viruses into target cells requires the fusion between the viral envelope and the target cell membrane. In the case of many viruses like HIV, SIV and BLV, the fusion is mediated by class 1 fusion glycoproteins located on the viral envelope. These fusion glycoproteins contain a region at their N-terminal extremity called the "fusion peptide", which interact with the target membrane. Many mutagenesis studies showed that this region is required for mediating membrane fusion [1]. Moreover, synthetic peptides corresponding to the fusion peptide of many glycoproteins induce membrane fusion *in vitro*. Despite the large number of studies on synthetic fusion peptides, the region necessary and sufficient to induce optimal membrane fusion is not known. To determine this minimal fusion peptide, we used the "tilted peptide" theory. According to this theory, a helical peptide inserting obliquely into membranes induces fusion [2]. Moreover, the more tilted the peptide is, the more important the fusion is. Then, we postulate that the minimal fusion peptide corresponds to the shortest helical fragment able to insert into the membrane with an angle close to 45°. This peptide was predicted using the IMPALA algorithm, which allow to predict peptide-membrane interactions [3]. Fusogenicity of this peptide was then assessed in liposome lipid-mixing and leakage assays and compared to the fusogenicity of smaller and longer peptides to check the validity of the prediction. This methodology was used to determine successfully the minimal fusion peptide of three viruses, HIV, SIV and BLV.

Results and Discussion

For each virus, fusion peptides of different lengths were constructed as α-helices using Hyperchem 6.0 (Hypercube Inc.) and analyzed with IMPALA. For HIV, models derived from the Protein DataBank structure (PDB code: 1ERF) proposed by Gordon et al. [4] were also used [5]. The *in silico* analysis indicated that the shortest peptide that inserts into the membrane with an optimal tilted orientation corresponds to the 12, 11 and 15 first residues of the fusion glycoprotein of HIV, SIV and BLV, respectively (Figure 1). These peptides are thus the predicted minimal fusion peptides. These predictions were challenged by analyzing the fusogenicity of these peptides with liposomes. Lipid-mixing assays confirmed that these peptides induce liposome fusion to the same extent as longer peptides but more than shorter peptides (Figure 1). The predicted peptides are thus the minimal fusion peptides [5,6]. The "tilted peptide" theory, in association with IMPALA, allows thus to predict successfully the minimal fusion peptides and can be used to find the minimal fusion peptide of other viruses.

S. Del Valle et al. (eds.), *Peptides for Youth: The Proceedings of the 20th American Peptide Symposium*, 387
DOI: 10.1007/978-0-387-73657-0_169, © Springer Science+Business Media, LLC 2009

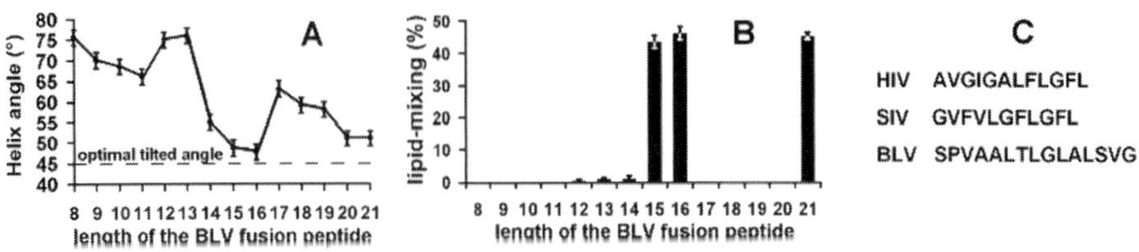

Figure 1. A. Angle with respect to the membrane plane predicted by IMPALA for BLV fusion peptides of different length. The 15-residue fusion peptide is the shortest helical peptide with an optimal tilted angle. B. Percentage of lipid-mixing of liposomes induced by BLV fusion peptides of different length. The 15-residue fusion peptide induce lipid-mixing to the same extent as longer peptides but more than shorter peptides. It is thus the minimal fusion peptide. C. Sequences of the minimal fusion peptide of HIV, SIV and BLV.

IMPALA was also used to study the impact of mutations known to have an effect on the fusogenicity of the whole glycoproteins. Results show that the loss of the fusogenicity is correlated with the loss of the obliquity of the corresponding fusion peptide (Table 1). For BLV, *in vitro* analyses with synthetic peptides confirmed the relationship between the angle of FPs in the membrane and membrane fusion : the more tilted the peptide is, the more destabilized the membrane is [6]. These results indicate a correlation between the ability of fusion peptides to adopt a tilted orientation, their ability to induce *in vitro* membrane fusion and the ability of the whole glycoproteins to induce membrane fusion. This is of particular interest to help in the development of new antiviral drugs or vaccines.

Table 1. Effect of mutations on the fusogenicity of whole glycoproteins (gp) and on the insertion angle of the fusion peptide (FP).

mutation	gp fusogenicity *	FP angle
HIV		
Wild-type	100%	56°
A1E	8%	61°
SIV		
Wild-type	100%	55°
F10SL11R	6%	83°
BLV		
Wild-type	100%	48°
S1LA4L	15%	70°

* *characterized by the % syncitia (see 5,6,7)*

Acknowledgments

The work was funded by a Grant FNRS-Televie No. 7.4.527.05.F to A. Lorin.

References

1. Martin, I., Ruysschaert, J. M. and Epand, R. M. *Adv. Drug. Deliv. Rev.* **38**, 233-255 (1999).
2. Brasseur, R. *Mol. Memb. Biol.* **17**, 31-40 (2000).
3. Ducarme, P., Rahman, M., Brasseur, R. *Proteins* **30**, 357-371 (1998).
4. Gordon, L.M., *et al. Biochim. Biophys. Acta.* **1559**, 96-120 (2002).
5. Lorin, A., *et al. J. Mol. Biol.* **359**, 597-609 (2006).
6. Lorin, A., *et al. Biochem. Biophys. Res. Commun.* **355**, 649-653 (2007).
7. Horth, M., *et al. EMBO J.* **10**, 2747-2755 (1991).

Development of a Novel Fusion Inhibitor against T-20-resistant HIV-1

Shinya Oishi[1], Saori Ito[1], Hiroki Nishikawa[1], Michinori Tanaka[1], Hiroaki Ohno[1], Akira Otaka[1], Kazuki Izumi[2], Eiichi Kodama[2], Masao Matsuoka[2] and Nobutaka Fujii[1]

[1]Graduate School of Pharmaceutical Sciences, [2]Institute for Virus Research, Kyoto University, Kyoto 606-8501, Japan.

Introduction

T-20 (also known as Fuzeon and enfuvirtide) is a potent human immunodeficiency virus-1 (HIV-1) fusion inhibitory peptide derived from a C-terminal heptad repeat (C-HR) of envelope glycoprotein gp41 [1]. Although even multi-drug-resistant HIV variants are effectively suppressed by T-20-containing regimens in the treatment of HIV/AIDS, it has been reported that T-20-resistant HIV strains can emerge after prolonged therapy [2].

Another C-HR-derived peptide C34 also exhibits potent anti-HIV activity by inhibiting the fusogenic six-helical bundle formation. It has been revealed that binding of C34 with the coiled-coil structure of an N-terminal heptad repeat (N-HR) occurs in α helix conformations, while the binding mode of T-20 remains ambiguous [3]. Previously, we have demonstrated that replacement of the i, $i + 4$ residues at solvent-accessible sites of C34 with Glu or Lys leads to stabilization of the six-helical bundle by potential salt-bridge formation of Glu–Lys pairs [4]. The designed peptides, SC34, SC34EK and SC35EK, demonstrated potent anti-HIV activity against T-20- or C34-resistant, as well as wild-type viruses. SC35EK exerted most potent inhibitory activity, in which α-helix-inducible X-EE-XX-KK (X, original amino acid residues; E, Glu; K, Lys) motifs were applied.

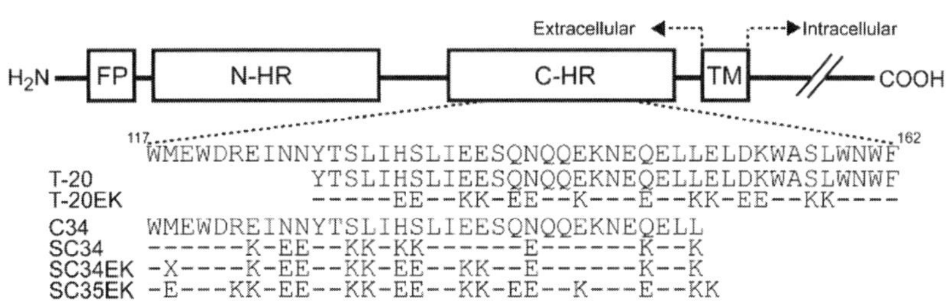

Fig. 1. Schematic view of gp41 and the sequences of C-HR peptide derivatives. The locations of the fusion peptide (FP), N-HR, C-HR, transmembrane domain (TM) and amino acid sequence of C34, T-20 and the derivatives are shown. X indicates norleucine. The residue numbers correspond to their positions in gp41 of NL4-3 strain.

In order to enhance the anti-HIV activity of T-20 by stabilization of the α helix structure, and to identify the interactive surface with N-HR, we designed a novel T-20 derivative T-20EK, which has four X-EE-XX-KK repeats. Structure–activity relationship (SAR) studies of these T-20EK derivatives and anti-HIV activity against T-20-resistant HIV-1 strains are reported.

S. Del Valle et al. (eds.), *Peptides for Youth: The Proceedings of the 20th American Peptide Symposium*, 389
DOI: 10.1007/978-0-387-73657-0_170, © Springer Science+Business Media, LLC 2009

Table 1. Anti-HIV activity of peptides.

Peptide	EC_{50} (nM) [a]
T-20	15 ± 3.9
T-20EK	1.8 ± 0.4
T-20EK W155A	5.8 ± 1.0
T-20EK W159A	49 ± 8.6
T-20EK W161A	24 ± 3.8
T-20EK F162A	27 ± 6.8

[a] Anti-HIV activity was determined with MAGI assay using HIV-1 NL4-3 strain. All data represent means±standard deviation.

Table 2. Anti-HIV activity of T-20 and T-20EK against wild-type and T-20-resistant strains.

Peptide	EC_{50} (nM) [a]	
	T-20	T-20EK
NL4-3 (wild-type)	15 ± 3.9	1.8 ± 0.4
D36G	2.3 ± 0.5	0.9 ± 0.2
D36G V38A	22 ± 7.6	3.3 ± 1.0
D36G N43D	46 ±9.6	1.7 ± 0.3

[a] Anti-HIV activity was determined with the MAGI assay. All data represent means±standard deviation.

Results and Discussion

All T-20 derivatives were manually synthesized using standard Fmoc chemistry on Rink-amide resins. Final deprotection and cleavage using TFA/thioanisole/*m*-cresol/1,2-ethandithiol/water, followed by HPLC purification (Cosmosil 5C18 AR-II column) with a linear gradient of CH_3CN in aqueous 0.1% TFA, yielded the expected peptides. Anti-HIV activity was evaluated by MAGI assay using HIV-1 NL4-3 strain as a wild-type virus. Two HIV-1 strains containing a V38A or N43D mutation at N-HR, which is relevant to resistance against T-20, were also employed for bioassays. Since the replication of these variants is significantly reduced, an additional mutation D36G was experimentally added to these variants as well as the wild-type virus.

T-20EK showed more potent bioactivity against wild-type strain (EC_{50}, 1.8 nM, Table 1) compared with T-20 (EC_{50}, 15 nM). The increased bioactivity of T-20EK could be rationalized by pre-ordering of the α helix structure, which was demonstrated by the CD spectrum of T-20EK (Fig. 2a). In addition, it seems that the residues replaced with Glu or Lys were not relevant to the direct interaction with N-HR of viral gp41. Alanine scanning of hydrophobic residues in the C-terminal Trp-rich region of T-20EK was performed. Anti-HIV activity of W155A-substituted peptide was retained, while the other substitutions led to a significant decrease in activity, even with the stabilized α helix structure (Fig. 2b), indicating that hydrophobic residues downstream of the helix region are critical for bioactivity. Interestingly, T-20EK retained its anti-HIV activity against T-20-resistant variants V38A (EC_{50}, 3.3 nM) and N43D (EC_{50}, 1.7 nM) (Table 2), although T-20 was significantly less potent against these strains [EC_{50}, 22 nM (V38A); EC_{50}, 46 nM (N43D)]. The impaired interaction with viral mutated N-HR could be overcome by the stabilized α helix structure.

As such, a novel fusion inhibitor T-20EK was identified, which was potent against T-20-resistant as well as wild-type HIV-1 strains. The interactive surface of T-20 with the viral component was also characterized by SAR studies, with inner surface residues of a six-helical bundle and three hydrophobic residues at the C terminus. T-20EK could be a lead compound for highly potent fusion inhibitors and a chemical tool to understand membrane fusion of HIV-1.

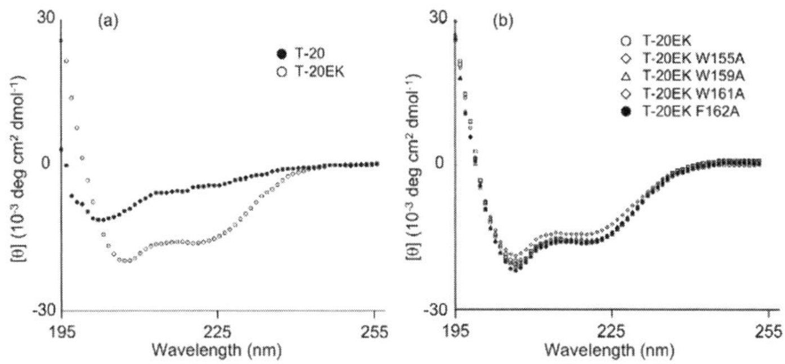

Fig. 2. CD spectra of (a) T-20, T-20EK and (b) Ala-substituted analogues of T-20EK. CD spectra were acquired at 25°C after incubation at 37°C for 30 min (the final concentrations of peptides were 10 mM in phosphate buffered saline (PBS), pH 7.4, containing 0.4% TFE).

Acknowledgments

This work was supported by a Grant-in-Aid for Scientific Research from the Ministry of Education, Culture, Sports, Science, and Technology of Japan, and Health and Labour Sciences Research Grants (Research on HIV/AIDS). H.N. is grateful for Research Fellowships from the JSPS for Young Scientists.

References

1. Matthews, T., Salgo, M., Greenberg, M., Chung, J., DeMasi, R., Bolognesi, D. *Nat. Rev. Drug. Discov.* **3**, 215-225 (2004).
2. Greenberg, M. L., Cammack, N. *J. Antimicrob. Chemother.* **54**, 333-340 (2004).
3. Liu, S., Jing, W., Cheung, B., Lu, H., Sun, J., Yan, X., Niu, J., Farmar, J., Wu, S., Jiang, S. *J. Biol. Chem.* **282**, 9612-9620 (2007), and references therein.
4. Otaka, A., Nakamura, M., Nameki, D., Kodama, E., Uchiyama, S., Nakamura, S., Nakano, H., Tamamura, H., Kobayashi, Y., Matsuoka, M., Fujii, N. *Angew. Chem. Int. Ed. Engl.* **41**, 2937-2940 (2002).

ε-Peptide Chimeras as Novel Antimicrobials

Yi-An Lu[1], Jin-Long Yang[2] and James P. Tam[1]

[1]*Department of Biochemistry, The Scripps Research Institute, Jupiter, FL 33458, USA;*
[2]*Vanderbilt Eye Institute, Vanderbilt University, Nashville TN 37232, USA*

Introduction

The prevalence of antibiotic resistance of bacterial pathogens requires development of new antibacterials to overcome this limitation. Our previous work has demonstrated that cationic antimicrobial peptides with unusual architecture are often resistant to proteolytic degradation [1]. An example is a cascade-type of dendrimeric peptide $(RLYR)_4K_2K$ that shows high *in vitro* antibacterial activity. It contains a trilysine core (K_2K) that radiates four tetrapeptide RLYR arms with its sequence derived from the repeating topological motif of a known and potent antimicrobial peptide, protegrin-1 [2].

Here, we report our continuing effort in developing new antimicrobial peptides with novel branched architecture, using the ε-amine of lysine for the ε-peptide backbone and functionalized α-amines as pendant side chains to give a branched α,ε-peptide containing two RLYR motifs. Since ε-peptides are found naturally as poly-ε-lysines which are generally stable to proteolytic enzymes, but exhibit extended conformation [3], we have designed an α,ε-peptide chimera ε(RLY)-R-R-ε(YLR) with a Arg-Arg dipeptide to facilitate a reverse turn that could mimic the two-β-stranded protegrin-1 (PG-1). Our results show that the new α,ε-peptide displays broad-spectrum antimicrobial activity against Gram-positive bacteria, Gram-negative bacteria and fungi.

Results and Discussion

Our prototypic design of the α,ε-peptide chimera ε(RLY)-R-R-ε(YLR) was an open-chain mimetic of protegrin (Figure 1) by rearranging the architecture of the cascade-type $(RLYR)_4K_2K$ to a branch-type peptide dendrimer. The chimera is symmetrical in design with two parallel extended strands of ε-peptides ε(RLY) and an α-dipeptide Arg-Arg as the reverse turn. The sequence of RLYR (basic-hydrophobic- hydrophobic-basic amino acid) [1] was derived from PG-1 (RGGRLCYCRRRFCV- CVGR) containing 18 α-amino acids that give its α-peptide backbone 54 atoms in length [2]. In contrast, the prototypic α, ε chimera, ε(RLY)-R-R-ε(YLR) with ε-lysyl peptide as the main backbone has a length of 48 atoms contributed by only 8 amino acids. The side-chain residues R, L, and Y were attached onto the α-amino ends of the ε-peptide chain to mimetic the side chain of PG-1.

The α,ε-peptide was assembled by the stepwise solid-phase method using both Boc- and Fmoc-chemistry. Boc-Bzl chemistry was used to elongate ε-peptide chain to afford an α,ε-octapeptide chimera and Fmoc-Bzl chemistry used for attaching N^{α}-amino acid (R, L, Y) on N^{α}-lysine. The synthesis of ε(RLY)-R-R-ε(YLR) was initiated on a Fmoc-Lys(Boc)-MBHA-resin, a TFA-stable benzhydrylamine resin. Functionalization of the α-amine of the Fmoc-Lys(Boc)-MBHA-resin was mediated by coupling with Ac-Arg(Tos) after removing its N^{α}-Fmoc protecting group by 20% piperidine/DMF. Continuing elongation of the ε-peptide chain was performed by removing Boc-group on ε-Lys, followed by coupling with Fmoc-Lys(Boc), and then a repeating deprotection/coupling cycle on the N^{α}-amino terminal to append the

S. Del Valle et al. (eds.), *Peptides for Youth: The Proceedings of the 20th American Peptide Symposium*, 393
DOI: 10.1007/978-0-387-73657-0_171,

© Springer Science+Business Media, LLC 2009

commercially available Ac-Leu and Ac-Tyr(Br-Bzl) as side chains to complete the first half of the ε-peptide chimera. In the putative turn sequence, Boc-chemistry to couple two Arg in a straight forward manner, followed by strategy described above to complete the second half of the ε-peptide chimera to afford final peptide chain ε[Ac-Arg(Tos)-Ac-Leu-Ac-Tyr(Br-Bzl)]-Arg(Tos)-Arg(Tos)-ε[Ac-Tyr(Br-Bzl)-Ac-Leu-Ac-Arg(Tos)]-MBHA-resin which was cleaved by HF to give 40% yield of expected peptides after purification.

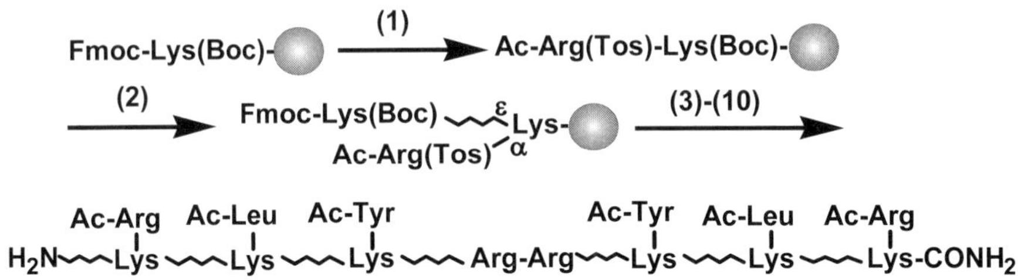

(1) 20% piperidine/DMF, coupling Ac-Arg(Tos)-OH
(2) 50%TFA/DCM, coupling Fmoc-Lys(Boc)-OH
(3) Repeat (1) coupling Ac-Leu-OH, then repeat (2)
(4) Repeat (1) to assemble Ac-Tyr(Br-Bzl)-OH
(5) Boc-chemistry coupling consecutive two Boc-Arg(Tos)-OH
(7) Repeat (2), then repeat (1) to assemble Ac-Tyr(Br-Bzl)-OH
(8) Repeat (1) to assemble Ac-Leu-OH, then repeat (2)
(9) Repeat (1) to assemble Ac-Arg(Tos)-OH
(10) HF

Fig. 1. Synthesis of ε(RLY)-RR-ε(YLR) and its structure.

A two-stage radial diffusion assay [4] was employed to test the antimicrobial activity of ε(RLY)-R-R-ε(YLR) which shows a broad-spectrum retained activity and potency comparable to protegrin-1 under low-salt conditions.

In conclusions, this study provides insights to simplify peptidyl architectures and to learn rules for designing α,ε-peptide as β-strand mimetics found in antimicrobial peptide such as protegrins. Our solid-phase syntheses of α,ε-peptide also represent the first step toward an exploitation the potentials of α,ε-peptide as novel antibacterials.

Acknowledgments

The work was in part supported by US public Health Service NIH Grant EB001986.

References

1. Tam, J. P., Lu, Y-A. and Yang J-L. *Eur. J. Biochem.* **269**, 923-932 (2002).
2. Fahrner, R. L., *et al. Chem. Biol.* **3**, 543-550 (1996).
3. Shima, S., *et al. J. Antibiotics* **37**, 1449-1455 (1984).
4. Lehrer, R. L., *et al. J. Immunol. Med.* **137**, 167-173 (2000).

Chemical modifications of short antimicrobial peptides from insects and vertebrates to fight multi-drug resistant bacteria

Daniel Knappe[1], Christin Stegemann[1], Ariane Nimptsch[1], Alexander Kolobov, Jr[2,3], Ekaterina Korableva[2,3], Olga Shamova[2], Vladimir N. Kokryakov[2,3] and Ralf Hoffmann[1]

[1] *Institute of Bioanalytical Chemistry, Center for Biotechnology and Biomedicine, Faculty of Chemistry and Mineralogy, Leipzig University, 04103 Leipzig, Germany;* [2] *Institute of Experimental Medicine, Russian Academy of Medical Sciences, 197376 Saint-Petersburg;* [3] *Saint-Petersburg State University, 199034 Saint-Petersburg, Russia*

Introduction

The presence of antimicrobial substances in secretions, blood and leukocytes has been known since the end of the 19th century. Later different classes of antimicrobial peptides (AMPs) with specific activities against bacteria, fungi and viruses were isolated from vertebrates as well as from invertebrates. These gene-encoded peptides differ in length from about 10 to more than 100 amino acid residues showing a great variety in structure and antimicrobial modes of action, targeting either the cell membrane or intracellular targets.

The emergence of bacterial and fungal pathogens resistant to small molecule antimicrobial drugs demands the development of new antibiotics with novel modes of action. These new antibacterial molecules are urgently needed to treat patients in hospitals and nursing homes infected by resistant or multidrug resistant pathogens that become increasingly live threatening for patients with an already weakened immune system. Many native antimicrobial peptides kill bacteria by mechanisms different than those employed by current antibiotics such as quinolones and tetracyclines.

Results and Discussion

Here we studied the antibacterial activities of drosocin (*Drosophila melanogaster*) [1], formaecin (*Myrmercia gulosa*) [2], and pyrrhocoricin (*Pyrrhocoris apterus*) [3], which are members of the short proline-rich AMPs, as well as histone and haemoglobine sequences newly identified cationic AMPs in dog (*Canis familiaris*) and European pond turtle (*Emys orbicularis*).

Oncopeltus 4	H-VDKPPYLPRP(X^{11})PPRRIYN(NR)-OH
Drosocin	H-GKPRPYSPRPTSHPRPIRV-OH
Haemoglobine	H-VYPKTKTYFPHFDLHHGSAQVRTHGKKVLNALG-OH
Histone H4	H-RDNIQGITKPAIRRLARRGGVKRISGLIY-OH

Eight peptides were synthesized at a 30 µmol scale on solid phase using Fmoc/^{t}Bu-strategy and DIC/HOBT activation and a MultiSynTech Syro 2000 synthesizer with a 48 well reaction block. All peptides were obtained in high yields and purities using single coupling and a 8 fold reagent excess, except the haemoglobine sequence. Due to its high aggregation potential Fmoc-Lys(Boc)-Thr($\Psi^{Me,Me}$pro)-OH and Fmoc-Gly-Ser($\Psi^{Me,Me}$pro)-OH pseudoproline dipeptides were incorporated yielding the peptide at 30 % in the crude product. All peptides were purified to homogeneity by reversed

S. Del Valle et al. (eds.), *Peptides for Youth: The Proceedings of the 20th American Peptide Symposium*, 395
DOI: 10.1007/978-0-387-73657-0_172, © Springer Science+Business Media, LLC 2009

phase HPLC and their masses were confirmed by MALDI-MS. The minimal inhibitory concentrations (MIC) for *Bacillus subtilis*, *Escherichia coli,* and *Micrococcus luteus* were determined in broth microdilution efficacy assays (Tab. 1).

Table 1. In vitro antimicrobial activity of peptides.

	Minimum Inhibitory Concentration (μg/mL)[a]			
Peptide	*E. coli* BL 21AI	*E. coli* DSM 10233	*M. luteus* DSM 1790	*B. subtilis* DSM 347
Oncopeltus 4 Pro[11]	125	n.d.	62.5	n.d.
Oncopeltus 4 Thr[11]	125	n.d.	62.5	n.d.
Oncopeltus 4 His[11]	62.5	n.d.	31.3	n.d.
Oncopeltus 4 Lys[11]	15.6	n.d.	15.6	n.d.
Oncopeltus 4 Arg[11]	7.8	>125	15.6	>125
Drosocin	3.9	3.9	0.5	>125
Haemoglobine α chain fragment	>125	>125	15.6	125
Histone H4 fragment	31.3	>125	7.8	31.3

[a]MIC is the lowest concentration that inhibit bacterial growth in 1% tryptic soy broth after 24h incubation at 37°C

As the sequence of the antibacterial peptide oncopeltus 4 isolated from the Milkweed bug (*Oncopeltus fasciatus*) was only partially sequenced by Edman degradation [4], i.e., position 11 is unknown, five different amino acids were tested (Table 1) at this position. Among those Arg and Lys showed the best activities. Arg completes the PRPRP motif, which is characteristic for proline-rich AMPs and is essential for its antibacterial activity.

The haemoglobine and histone fragments were active against *B. subtilis* in an agar diffusion assay. This bacterial strain was not susceptible to a majority of proline-rich AMPs including drosocin. In our experiments the histone H4 fragment exhibited an higher activity than the haemoglobine fragment against all tested strains and showed MIC values comparable to those of Oncopeltus 4. In comparison to the difficult synthesis of haemoglobine sequences histone H4 fragment and oncopeltus 4 are relatively easy to synthesize and may therefore be interesting targets for further investigations.

Acknowledgements
Financial support by the European Fond for Regional Structure Development (EFRE, European Union and the Free State Saxonia) is gratefully acknowledged. We thank Laszlo Otvos Jr. for helpful discussions.

References
1. P. Bulet, J. L. Dimarcq, C. Hetru et al., *J. Biol. Chem.* **268**, 14893-7 (1993).
2. J. A. Mackintosh, D. A. Veal, A. J. Beattie, et al., *J. Biol. Chem.* **273**, 6139-43 (1998).
3. S. Cociancich, A. Dupont, G. Hegy, et al., *Biochem. J.* **300**, 567-75 (1994).
4. M. Schneider, A. Dorn, *J. Invert. Path.* **78**, 135-140 (2001).

PROTEOMICS AND PEPTIDOMICS

Isolation of Peptides From Rat Tissues: Peptidomics vs. Degradomics

Oleg N. Yatskin[1], Olga V. Sazonova[1], Dmitry P. Khachin[1], Elena Y. Blishchenko[1], Andrei A. Karelin[1] and Vadim T. Ivanov[2]

[1]Group of Regulatory Peptides; [2]Laboratory of Peptide Chemistry, Shemyakin-Ovchinnikov Institute of Bioorganic Chemistry RAS, Miklukho-Maklaya 16/10, 117997, Moscow, Russia

Introduction

Earlier we have isolated about 200 functional protein fragments from acidic rat tissue extracts [1,2]. The conditions of their isolation (treatment time, t°C, absence of protease inhibitors) could indicate that at least part of those peptides resulted from post-mortem protein degradation. Such sets of protein fragments, or peptide pools, were shown to be tissue-specific and highly reproducible in individual animals, their composition changing upon metabolic stress [3]. Here we report isolation of peptides from rat brain, heart, lung and spleen extracts prepared in the presence of protease inhibitors and compare the resultant peptide sets with the earlier described sets of peptides.

Results and Discussion

The isolation of rat organs was performed within 1 min with immediate liquid nitrogen freezing. Tissues were homogenized in 10% acetic acid, in the presence or in absence of protease inhibitors (10^{-6}M of pepstatin A, 10^{-4}M of phenylmethylsulfonylfluoride (PMSF), 2 мM EDTA). The extracts were separated by size-exclusion chromatography (4 fractions) and then by RP-HPLC in a linear water-acetonitrile gradient. Fractions corresponding to major chromatographic peaks were analyzed by MS, peptides corresponding to major MS signals were identified on the basis of secondary ion spectra (MALDI TOF/TOF) using Mascot search. For some peptides, the structure was confirmed by Edman sequencing.

In inhibitor-treated extracts, 111 peaks common for 4 tissues were identified by comparison of retention times, MWs and structures. MW was determined for 102 compounds; 32 peptides were sequenced. They corresponded to the N- and C-terminal fragments of abundant functional proteins and had average length 21 a.a.r. 27 of them were fragments of hemoglobin, other precursors were: serum albumin precursor, ferritin, fibrinogen precursor. Those peptides are with high probability endogenous, i.e., can be regarded as tissue peptidome components. The content of hemoglobin fragments per g of tissue (hundreds pmol/g) was comparable with the content of hemoglobin fragments in acidic extract of erythrocytes obtained with protease inhibitors (pmol/mL of packed erythrocytes) [5]. As erythrocytes constitute less than 10% of tissues, the hemoglobin fragments found in tissue extracts cannot be of intra-erythrocyte origin.

In non-treated extracts, 36 common peaks were identified, for 43 compounds MW was measured, 10 sequences were determined, including 5 hemoglobin fragments. Thus, the majority of peptidome components (isolated with the inhibitors) were common in 4 tissues. In contrast, large share of peptides isolated in the absence of protease inhibitors, i.e., degradome products, were tissue-specific.

Earlier we have reported on the analysis of peptides from rat brain acidic extract [2]. The conditions of peptide extraction (absence of protease inhibitors, duration, temperature) favored more pronounced protein degradation than the conditions

S. Del Valle et al. (eds.), *Peptides for Youth: The Proceedings of the 20th American Peptide Symposium*, 399
DOI: 10.1007/978-0-387-73657-0_173, © Springer Science+Business Media, LLC 2009

employed in the absence of inhibitors in the present study. One of size exclusion fractions of this extract was separated in the standard RP-HPLC system. Fig. 1 compares the RP-HPLC elution profiles obtained for the extract treated with protease inhibitors, and 2 profiles for the non-treated extracts obtained in the present study (early stage of post-mortem protein degradation) and in the previously applied conditions (later stage of post-mortem degradation). In the treated extract, the number of peaks is comparatively small. Upon protein degradation, new peaks appear in addition to peaks seen with suppressed proteolysis, the total number of peaks growing by 40-50%. The composition of components strongly differs between stages 1 and 2 of degradome products formation. However, 2 hemorphins were identified at both stages, significant amounts of those peptides being produced at the later stage of protein degradation.

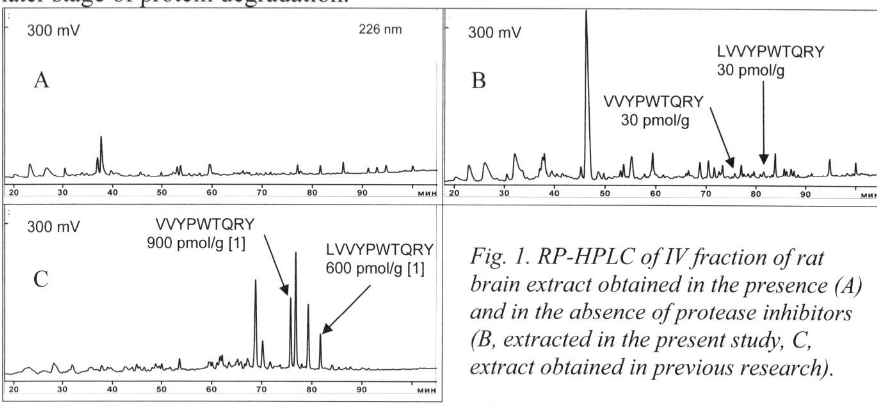

Fig. 1. RP-HPLC of IV fraction of rat brain extract obtained in the presence (A) and in the absence of protease inhibitors (B, extracted in the present study, C, extract obtained in previous research).

To summarize, the main characteristics of sets of peptidome components and degradome products are given in Table 1. Analysis of degradome products seems to be useful in diagnostics and in the search for novel active sequences. Study of endogenous peptidome components is of major importance in the framework of basic research of regulatory networks.

Table 1. Comparison of tissue peptide sets obtained with and without protease inhibitors.

Parameters	Peptidome components	Late degradome products
Reproducibility between individuals	Yes	Yes
Tissue specificity	Low	High
Correlation of composition with tissue state	Not studied	Yes
Presence of bioactive peptides	Yes	Yes
Number of major peaks, % of peptidome	100	150-200
Average length of peptides, a.a.r.	21±8	10±7

Acknowledgments

The work was funded by RAS Presidium grant “Molecular and Cellular Biology”.

References

1. Yatskin, O.N., *et al. FEBS Lett.* **428**, 286-290 (1998).
2. Karelin, A.A., *et al. J. Pept. Sci.* 2000, **6**, 345-354 (2000).
3. Karelin, A.A., *et al. J. Pept. Sci.* 2000, **6**, 168-174 (2000).
4. Ivanov, V.T., *et al. Biopolymers* **80**, 332-346 (2005).

PEPTIDES AS DIAGNOSTICS, PROBES, AND BIOMARKERS

Multivalent peptidomimetics for tumor targeting

Nadine Pannier[1], Valerie Humblet[2], Preeti Misra[2], John V. Frangioni[2] and Wolfgang Maison[1]

[1]*Institute of Organic Chemistry, University Giessen, Heinrich-Buff-Ring 58, 35392 Giessen, Germany;* [2]*Beth Israel Deaconess Medical Center, 330 Brookline Avenue, Boston, MA 02215*

Introduction

Multivalency is a common phenomenon in nature to increase affinity and specifity of receptor-ligand interactions, especially on the cell surface. Chemists have tried to make use of multivalent interactions in different context and have synthesized a lot of scaffolds for the assembly of multivalent receptor ligands [1].

We designed suitable scaffolds for the assembly of multiple targeting ligands and contrast agents on the basis of adamantane in the hope that multimerization would improve the performance of our cancer specific ligands. The adamantane scaffold provides a unique tripodal recognition motif for cell surface receptors. We suggest that our scaffold might be generally useful for affinity maturation of small molecules targeting cell surface epitopes.

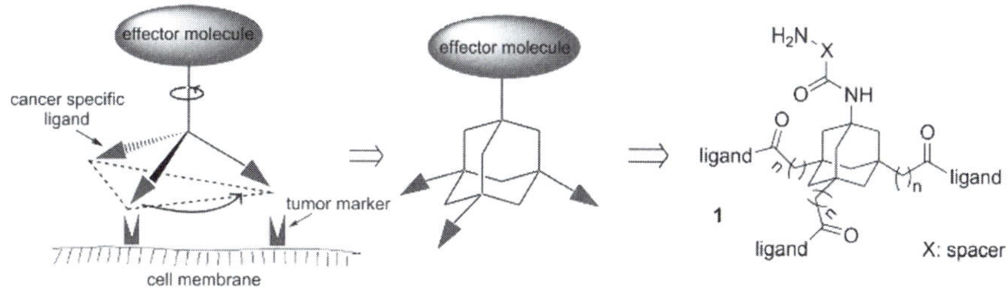

Fig. 1. Tripodal recognition motif for cell surface receptors.

Results and Discussion

One of the major challenges in current cancer therapy is the sensitive detection of small tumors. None of the known methods is sufficiently precise for high resolution cancer imaging. In this context, cancer specific receptors, which are over expressed ideally on the surface of cancer cells, have attracted considerable attention as tumor markers [2]. If conjugated to contrast reagents or dyes, ligands for cancer specific receptors can guide marker moieties to a tumor site and allow the sensitive detection of cancer [3]. A suitable marker for malignant prostate tumor cells is the prostate specific membrane antigen (PSMA) [4].

In this context, we have previously described the development of a high affinity (10 nM), small molecule ligand for PSMA [5]. Conjugates of GPI and contrast agents like **3** were therefore thought to be valuable tools for prostate tumor imaging. However, it turned out that GPI can not effectively compete with high concentrations of endogenous phosphate, limiting its use *in vivo*.

To improve binding affinity and off-rates of our ligand in order to improve its performance in biological media we used the multivalency approach. The tetravalent scaffold **1** can be used for tumor imaging, because it can be attached to tumor

S. Del Valle et al. (eds.), *Peptides for Youth: The Proceedings of the 20th American Peptide Symposium*, 403
DOI: 10.1007/978-0-387-73657-0_174, © Springer Science+Business Media, LLC 2009

specific ligands and a contrast agent to give conjugate **5-7**. The synthesis of the tetravalent adamantane derivatives **5-7** was realized in 12 steps with an overall yield of 10% [6,7].

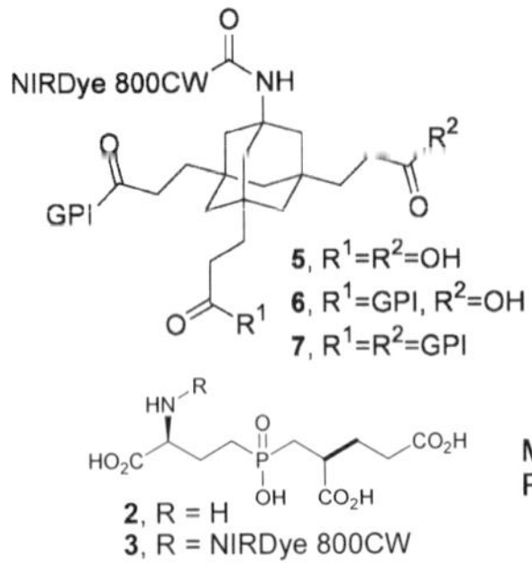

Molecule	K_d [nM]/TBS	K_d [nM]/PBS	K_d [nM]/Serum
GPI (2)	10.5 ± 1.4	N.D.	N.D.
[800CW]GPI (3)	0.0 ± 0.0	N.D.	N.D.
[800CW]Adam GPI monomer (5)	5.4 ± 1.1	N.D.	N.D.
[800 CW]Adam GPI dimer (6)	1.3 ± 0.2	1.9 ± 0.5	1.7 ± 0.2
[800 CW]Adam GPI trimer (7)	0.5 ± 0.1	0.5 ± 0.1	0.4 ± 0.1

Mean ± S.D. / TBS: Tris-(hydroxymethyl)-aminomethane buffered saline / PBS: Phosphate buffered saline / N.D.: None detected / serum: 100% calf serum

Fig. 2. Live cell binding affinity of PSMA ligands in different media.

In vitro assays were run with LNCaP cells (express PSMA in high surface density) and for comparison PC3 cells (PSMA deficient). In contrast to the monomeric GPI **2**, dimeric GPI-conjugates **6** and their trimeric homologues **7** show almost the same binding affinities to LNCAP cells in two different buffer systems and even in calf serum. The affinities of PSMA ligands are increased by an order of magnitude upon trimerization leading to picomolar cell surface binders in serum.

We also studied the binding of NIR-labelled compounds by *in vitro* NIR fluorescence imaging and the binding of ^{99m}Tc radiolabelled adamantane derivatives by SPECT (data not shown) [7,8]. We are currently transferring the system into animal models and trying to understand the improved binding characteristics of PSMA-ligands by multimerization.

Acknowledgments

This work was funded by NIH grant R01-CA-115296, grants of the Lewis Family Fund, Ellison Foundation and DFG MA 2519/3.

References

1. M. Mammen, S. K. Chio, G. M. Whitesides, *Angew. Chem., Int. Ed.* **1998**, *37*, 2755.
2. W. Maison, J. V. Frangioni, *Angew. Chem. Int. Ed.* **2003**, *42*, 4726.
3. M. Harada, M. Noguchi, K. Itoh, *Int. J. Clin. Oncol.*, **2003**, 8, 193.
4. J. Konvalinka, R. Hilgenfeld *et al.*, *EMBO* **2006**, *25*, 1375.
5. V. Humblet, R. Lapidus, L. R. Williams, T. Tsukamoto, C. Rojas, P. Majer, B. Hin, S. Ohnishi, A. M. De Grand, A. Zaheer, J. T. Renze, A. Nakayama, B. S. Slusher, J. V. Frangioni, *Mol. Imaging* **2005**, *4*, 448.
6. W. Maison, J. V. Frangioni, N. Pannier, *Org. Lett.* **2004**, *6*, 4567.
7. P. Misra, V. Humblet, N. Pannier, W. Maison, J. V. Frangioni, *J. Nucl. Med.* **2007**, *48(8)*, 1379-1389.
8. V. Humblet, P. Misra, Y. Ko, N. Pannier, W. Maison, J. V. Frangioni, *Angw. Chem.* **2007**, *submitted for publication.*

Comparison of [^{18}F]FBA and [^{18}F]FPyMe as peptide radio-labeling agents of PEPHC1 for PET imaging of EGFRvIII

Charlotte L. Hansen[1], Bertrand Kuhnast[3], Franciose Hinnen[3], Frédéric Dollé[3], Nic Gillings[1], Paul R. Hansen[2], Andreas Kjær[1]

[1] PET & Cyclotron Unit, Copenhagen University Hospital, Denmark, [2] Department of Natural Science, University of Copenhagen, Denmark, [3] Départment de Recherche Médicale, Service Hospitalier Frédéric Joliot, France

Introduction

Receptor binding peptides labeled with positron-emitting nuclides for Positron Emission Tomography (PET) are useful targeting agents for diagnostic imaging of various types of cancer. Among the positron emitting nuclides, fluorine-18 is often the radionuclide of choice for labeling of peptides, due to the physical and nuclear characteristics of the isotope [1]. A two-step ^{18}F-labeling of resin-bound linear peptides can be achieved by acylation of the N-terminus with 4-[^{18}F]fluorobenzoic acid ([^{18}F]FBA) [2], Figure 1a. Alternatively, ^{18}F-labeling may be performed via the thiol group of cysteine using 1-[3-(2-[^{18}F]fluoropyridin-3-yloxy)propyl]pyrrole-2,5-dione ([^{18}F]FpyMe) [3], Figure 1b. In this study, we compare [^{18}F]FBA and [^{18}F]FPyMe for the ^{18}F-labeling of PEPHC1 (HFLIIGFMRRALCGA), a peptide which is selective towards the cancer specific epidermal growth factor tyrosine kinase receptor mutation (EGFRvIII) [4].

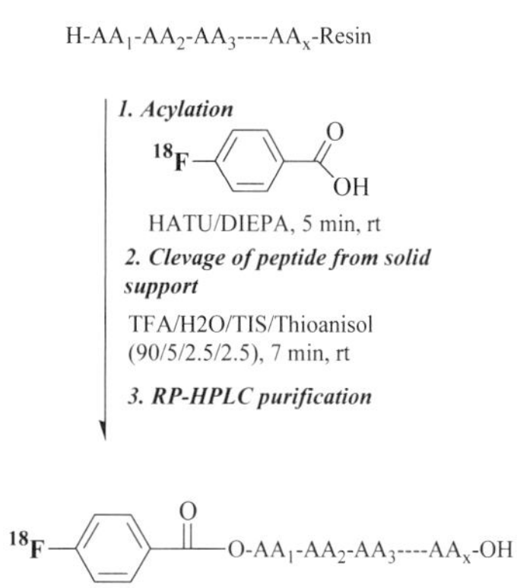

Figure 1a. ^{18}F labeling of peptides using [^{18}F]FBA

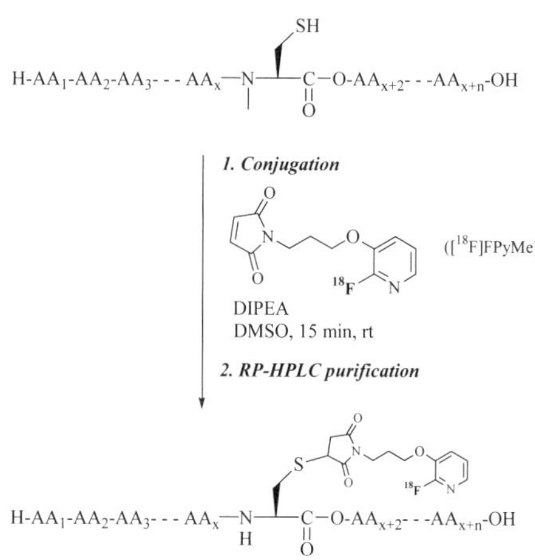

Figure 1b. ^{18}F labeling of peptides using [^{18}F]FPyMe

Results and Discussion

The ^{18}F-labeling of PEPHC1 with [^{18}F]FPyMe was obtained in 15 min with a decay-corrected (d.c.) radiochemical yield of 40-50% and a radiochemical purity of 96-98% prior to HPLC.

DOI: 10.1007/978-0-387-73657-0_175, © Springer Science+Business Media, LLC 2009

The ^{18}F-labeling of PEPHC1 using $[^{18}F]FPyMe$ was obtained in 15 min with a decay-corrected (d.c.) radiochemical yield of 40-50% and with a radiochemical purity of 96-98% prior to HPLC.

Radio-labeling of resin-bound PEPHC1 with $[^{18}F]FBA$, followed by cleavage was achieved in 120 min with <5% radiochemical yield. The product appeared to contain a mixture of many different labeled compounds. To improve the radiochemical yield different coupling reagents and various resins were tested. HATU/DIEPA as coupling reagents gave the highest yield. Moreover, the different types of resin had no influence on the yield.

The identity of the radio-labeled peptides was confirmed using RP-HPLC, by co-injection with authentic reference samples.

The most efficient and rapid radio-labeling of PEPHC1 was achieved by using $[^{18}F]FPyMe$. The drawback of this method is that the peptide must contain a cysteine residue. Using $[^{18}F]FBA$ as a radio-labeling agent is not limited to peptides with a certain amino acid sequence. However, the method is time consuming and in our experience results in a low radiochemical yield. Alternative labeling strategies may also be more appropriate.

Acknowledgment

The Danish Cancer Society for funding.

References

1. Kilbourn, M.R., Dence, C.S., Welch, M.J., Mathias, C.J. The Journal of Nuclear Medicine, 1987, **28**: 462-470
2. Sutcliffe-Goulden, J.L., O'Doherty, M.J., Marsden, P.K., Hart, I.R., Marshall, J.F., Bansal, S.S.,European Journal of Nuclear Medicine and Molecular Imaging, 2002, **29**(6): 754-759
3. de Bruin, B., Kuhnast, B., Hinnen, F., Yaouancq, L., Amessou, M., Johannes, L., Samson, A., Boisgard, R., Tavitian, B., Dollé, F., Bioconjugate Chemistry, 2005, **16**: 406-420
4. Campa, M.J., Serlin, S.B. and E.F. Patz, Academic Radiology, 2002, **9**(8): 927-932

Now back to the sample manuscript… *Vibrio anguillarum* infections result in huge economic losses in the aquaculture industry annually. Pleurocidin, an antimicrobial peptide (AMP) isolated in 1997 from winter flounder, exhibited MICs in the micromolar range against a wide variety of fish and human pathogens [1]. Studies have shown a correlation between antimicrobial activity and the charges, hydrophobicities, and hydrophobic moments of amphipathic peptides that are capable of adopting an α-helical structure [2]. Although the C-terminal acid is the naturally-occurring form, synthetic pleurocidin amide is more potent and has been shown to protect coho salmon from vibriosis *in vivo* [3]. No additional increase in potency was obtained by adding a lysine at the N-terminus of pleurocidin amide to increase the positive charge on the peptide [3]. To characterize the structural determinants required for antimicrobial activity and to discover minimal sequences for genetic engineering, we synthesized pleurocidin amide and two new pleurocidin analogs LM4-1 and LM4-2 (see Figure 1). The analogs were responsible for helix formation in cecropin A [4]. AMPs were screened against *E.coli, V. anguillarum*, and *V.carchariae*.

Cecropin A	H-WKLF**K**KIEKVGQNIRDGIIKAGPAVAVVGQATQIAK-NH_2
LM4-1	H-GWGSFKKAAHVGKHVGK-NH_2
Pleurocidin	H-GWGSF**F**KKAAHVGKHVGK**AALTHYL**-OH
LM4-2	H-GWGSFFKKAAHVGKHVGKAAL-NH_2

Fig. 1. Sequences of cecropin A, native pleurocidin and synthetic pleurocidin analogs

Results and Discussion

Synthetic peptides pleurocidin amide (25 residues), LM4-1 ([des-F^5]Pleurocidin 1-18 amide) and LM4-2 (pleurocidin 1-21 amide) were synthesized using Boc chemistry on MBHA (1.1 mmol/g, 1g) resins in a CS-Bio peptide synthesizer. The N-terminal Boc-group was left on during removal of Dnp from His residues using thiophenol in DMF. Following low-high HF deprotection and cleavage with thiocresol and cresol scavengers, the crude peptides were extracted from the resin with 10% aqueous HOAc; material excluded from a Sephadex® G-25 column was pooled and lyophilized, then dissolved in water, and analyzed by C_{18} RP-HPLC (Vydac 218TP54) with a gradient of 5-85% CH_3CN in aq. 0.1% TFA.

Table 1. Antibacterial activity of peptides.

Peptide	Minimum Bactericidal Concentration (μM)[a]			
	CSMHB		CSMHB + 1.5% NaCl	
	V. anguillarum	*E.coli* (ZK4)	*V. anguillarum*	*V. carchariae*
Pleurocidin amide	25	100	100	100
LM4-1	>100	>200	>200	>100
LM4-2	100	200	>200	>100
Tachyplesin (control)	25	50	50	25

[a]MBC is the minimum peptide conc. where bacterial growth was not observed at 21h

Antibacterial assays were performed by evaluating the ability of synthetic peptides to inhibit bacterial growth in cell cultures (Table 1). Just to show how to insert another type of Figure, stereo picture is shown in Figure 2. Synthetic pleurocidin amide exhibited similar antibiotic activity to tachyplesin, with the desired activity against marine pathogens *V. anguillarum* and *V. carchariae*. Pleurocidin was not as potent against *E. coli* as it was against *V. anguillarum*. The pleurocidin analogs were an order of magnitude less active than the parent peptide, and this result indicated that the C-terminus of pleurocidin might play a role in stabilizing the helix. LM4-2 was more active than LM4-1, possibly indicating that Phe^6 is required for optimal helix formation and activity.

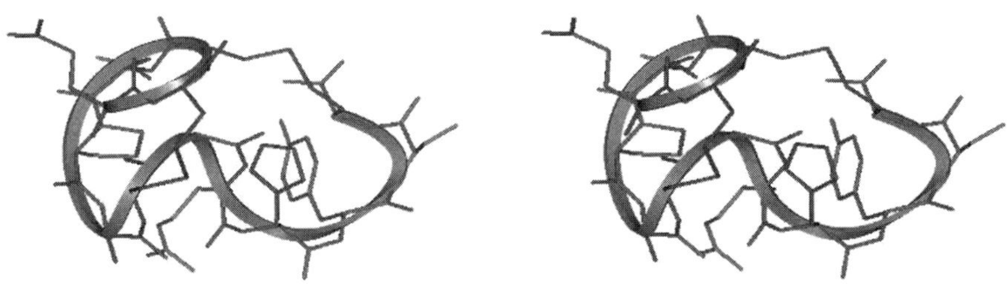

Fig. 1. A representative structure of GIDY. The ribbon shows the peptide backbone.

Acknowledgments

The work was funded by an USDA-NRICGP grant to L. Martin.

References

1. Cole, A. M., Weis, P. and Diamond, G. *J. Biol. Chem.* **272**, 12008-12013 (1997).
2. Tossi, A., Sandri, L. and Giangaspero, A. *Biopolymers* **55**, 4 30 (2000).
3. Jia, X., *et al. Appl. Environ. Microbiol.* **66**, 1928-1932 (2000).
4. Merrifield, R. B., Vizioli, L. D. and Boman, H. G. *Biochem.* **21**, 5020-5031 (1982).

Carbohydrated $[^{99m}Tc(CO)_3](N^{\alpha}His)Ac$-Bombesin(7-14) analogs

Veronique Maes[1], Luc Brans[1], Christian Schweinsberg[2], Elisa García-Garayoa[2], Peter Bläuenstein[2], Peter A. Schubiger[2] and Dirk Tourwé[1]

[1]*Vrije Universiteit Brussel, Department of Organic Chemistry, Pleinlaan 2, 1050 Brussels, Belgium;* [2]*Center for Radiopharmaceutical Science, Paul Scherrer Institute, CH-5232 Villigen, Switzerland*

Introduction

Bombesin (BN) is a tetradecapeptide with high affinity for the Gastrin Releasing Peptide (GRP) receptor. Overexpression of this receptor subtype on a variety of human tumors, e.g. breast and prostate, makes radiolabeled bombesin analogs attractive targets for in vivo imaging and therapy of these carcinoma types. The retro-N-α-carboxymethyl-histidine moiety, short $(N^{\alpha}His)Ac$, functions as an efficient chelator for the $^{99m}Tc(CO)_3$ core and allows labeling of the peptides with very high specific activity[1]. The biological active BN(7-14) sequence (Gln-Trp-Ala-Val-Gly-His-Leu-Met-NH_2) could be stabilized against enzymatic cleavage in plasma by replacing Leu^{13} by the constrained Cha (cyclohexylalanine), and Met^{14} was substituted by the Nle (norleucine) residue which is less prone to oxidation ($t_{1/2} > 16$ h). The introduction of spacers (βAla-βAla) between the chelator and the BN(7-14) sequence improved the accumulation of the radiolabeled analogs in the BN-receptor expressing tissues and resulted in the first bombesin analogs with target-to-non target ratios that increase with time [2]. In order to further increase the tumor uptake, and to enhance the excretion kinetics and blood clearance, the ^{99m}Tc-labeled bombesin analogs were made more hydrophilic by conjugation of a carbohydrate.

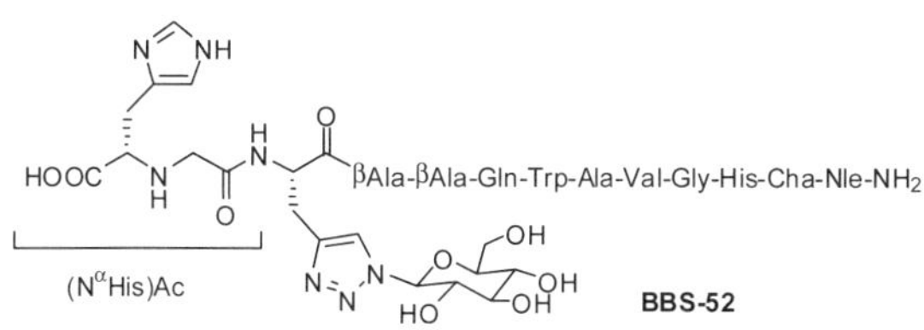

Fig. 1. Carbohydrated $(N^{\alpha}His)Ac$-Pra(Glu)-Bombesin(7-14) analog.

Results and Discussion

A drawback of the $(N^{\alpha}His)Ac$ chelator is its unusual reactivity, preventing straightforward carbohydration of $(N^{\alpha}His)Ac$ containing peptides by the Maillard reaction[3]. The conjugation of a sugar derivative through a linker to the bombesin sequence by means of the Tornøe-Sharpless 1,3-dipolar cycloaddition [4,5] is a very selective glycation method for peptides containing the $(N^{\alpha}His)Ac$ chelator. The $(N^{\alpha}His)Ac$-containing bombesin analog BBS-52 (Figure 1) was synthesized on a Rink Amide resin using the Fmoc-strategy. Protected azidoglucose was conjugated to the bombesin sequence containing a propargylglycine (Pra) residue using the Cu(I) catalyzed 1,3-dipolar cycloaddition. The reaction was performed on the resin-bound peptide before the construction of the chelator and resulted in the peptide analog BBS-52 (Figure 2).

S. Del Valle et al. (eds.), *Peptides for Youth: The Proceedings of the 20th American Peptide Symposium*, 409
DOI: 10.1007/978-0-387-73657-0_176, © Springer Science+Business Media, LLC 2009

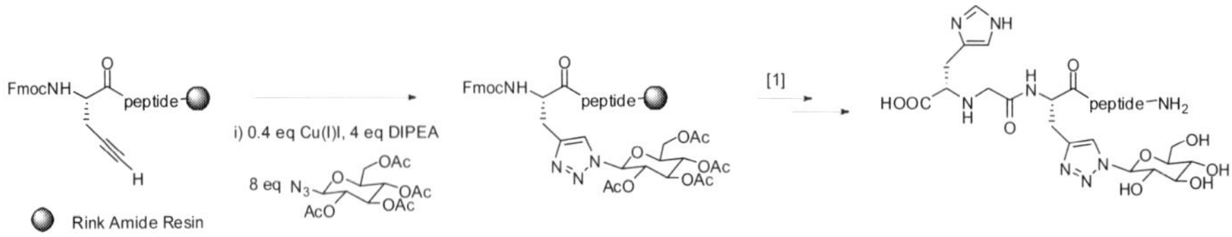

Fig. 2. Synthesis of the carbohydrated (N^{α}His)Ac-Pra(Glu)-Bombesin(7-14) analog (BBS-52).

Biological studies

After labeling of the BN-analog with [^{99m}Tc(CO)$_3$], the affinity of BBS-52 for the GRP receptor on PC-3 cells was determined (Table 1) and biodistribution studies were performed in nude mice bearing PC-3 tumors. (Figure 3) The introduction of the carbohydrate moiety Pra(Glu) did not significantly change the binding affinity for the GRP receptor on the human PC-3 tumor cells compared to the previous investigated BBS-38 ((N^{α}His)Ac-βAla-βAla-Gln-Trp-Ala-Val-Gly-His-Cha-Nle-NH_2). [2] However, an increased tumor uptake was observed with BBS-52, and also the uptake in the other GRP receptor positive tissues (pancreas and colon) was higher than with BBS-38. The activity wash-out from the tumor is much slower than from healthy organs, resulting in better tumor-to-background ratios. Due to the presence of the glucose moiety, the radiolabeled peptide is mainly excreted via the kidneys. The excretion kinetics from the kidneys is very fast and at 5 h p.i. the accumulation in the kidneys is about 1% ID/g.

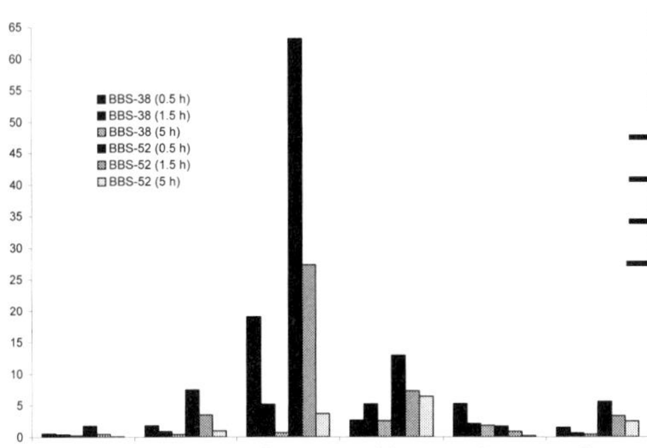

Fig. 3. Biodistribution of ^{99m}Tc-labeled BBS-38 and BBS-52 in nude mice bearing PC-3 tumor cells.

Table 1. Binding affinity of unlabeled (IC_{50}) and labeled (K_D) analogs for GRP receptor on PC-3 cells.

	IC_{50} (nM)	K_D (nM)
BBS-52	4.2 ± 0.1	0.3 ± 0.1
BBS-38	3.9 ± 1.1	0.2 ± 0.1

Acknowledgments

V. Maes is a Postdoctoral Researcher of the Fund for Scientific Research-Flanders (Belgium), which we also acknowledge for financial support.

References

1. Maes, V., *et al. J. Med. Chem.* **49**, 1833-1836 (2006).
2. Garcia-Garayoa, E., *et al. Nucl. Med. Biol.* **34**, 17-28 (2007).
3. Maes, V. and Tourwé, D. *Int. J. Pep. Res. Therapeutics* **12**, 197-202 (2006).
4. Tornøe, C. W. and Meldal, M. *Proc. of the 17th APS, San Diego* (2001).
5. Kolb, H. C., Finn, M. G. and Sharpless, K. B. *Angew. Chem. Int. Ed.* **40**, 2004-2021 (2001).

A Synthetic Peptide Approach for Elucidating the Points of Natural Auto-Antibody Reactivity to Proteolytic Fragments of Human IgG

A. Schmidt, D. DeRiggi, R.E. Jordan, B. Whitaker, G.A. Heavner and M. Kruszynski

Protein Engineering, Centocor R&D, Inc., Radnor, PA 19087 U.S.A.

Introduction

There are naturally occurring human antibodies to the F(ab')$_2$ portion of IgG [1]. Extracellular proteases and bacterial enzymes can cleave the IgG hinge region and release the Fc domain (with its effector functions. The human immune repertoire contains auto-antibodies that are directed against endogenous Fab and F(ab')2 fragments [2]. The anti-fragment reactivity is at newly exposed epitopes in the cleaved IgG hinge region.

By systematically constructing a bi-directional approach of 14-mer biotinylated peptides through the hinge region with either a free C-termini or free N-termini for each position of the hinge, one can study other single heavy chain terminal positions (eg. Fab, F(ab')$_2$ and Fc) at which no proteases that cleave IgG have been identified. By exposing streptavidin-captured peptides to human serum, immune reactivity could be identified at various free carboxy termini sites. This mapping approach will aid in identifying pre-existing auto-antibodies that recognize the newly-generated epitopes which correspond to protease cleavage sites in IgG and possibly to other non-IgG proteins.

Results and Discussion

The peptides were synthesized on a Rainin Symphony Multiple Peptide Synthesizer SMPS-110 using Fmoc chemistry. The α-amino groups of N-peptides and α-carboxylic groups of C-peptides were biotinylated either with Ac-Lys(Biotin) or with Glu(Biotin-PEG$_3$)-amide The peptides were deprotected and cleaved from the resin then purified by preparative reversed-phase HPLC Peptide identification and purity was determined by analytical HPLC, capillary electrophoresis, SELDI MS and/or MALDI MS. Each analyzed cleavage site generated the free α-carboxylic group of N-peptide and free α-amino group of C-peptide that were left unmodified. I.e. the cleavage site K^{223} - T^{224} generated the Ac-Lys(Biotin)-hinge(210-223)-acid and N-α-amino- hinge(224-237)-Glu(Biotin-PEG$_3$)-amide.

We analyzed the possible enzymatic cleavage sites of IgG starting at Lys^{223} – Thr^{224} and ending at Pro^{239}–Ser^{240}. The 17 overlapping 14-mer pairs of peptides, each shifted by one amino acid along the sequence, covering the 210 - 253 hinge area of human IgG1 were synthesized.

When tested in streptavidin-based ELISA format with serum from healthy human volunteers, antibodies were detected that bound to varying extents to different peptides with free carboxy termini, but minimally to peptides with free amino termini. Antigenic peptides correspond with a number of previously identified sites of protease cleavage that yield Fab and F(ab')$_2$ fragments. The most antigenic of these fragments (and peptide analogs) are those that are generated by human extracellular proteases such as MMP-3, MMP-12, cathepsin G, neutrophil elastase, plasmin and Cruzipain and the bacterial protease, IdeS The peptide map of

S. Del Valle et al. (eds.), *Peptides for Youth: The Proceedings of the 20th American Peptide Symposium*, 411
DOI: 10.1007/978-0-387-73657-0_177, © Springer Science+Business Media, LLC 2009

autoimmune reactivity predicted sites in the hinge for which no known enzyme cleaved. In another case however, peptide antigenicity predicted a $F(ab')_2$ antigen terminating at Glu^{233} – an enzyme that generates such a fragment was later identified as cathepsin G. Thus, this peptide mapping approach yielded results that paralleled the immune reactivity of cleaved antibody fragments and allows for a systematic analysis of IgG1 hinge antigenicity at each potential position. The relationship of the presence of auto-antibodies to protease cleavage sites in IgG with autoimmune diseases is under investigation.

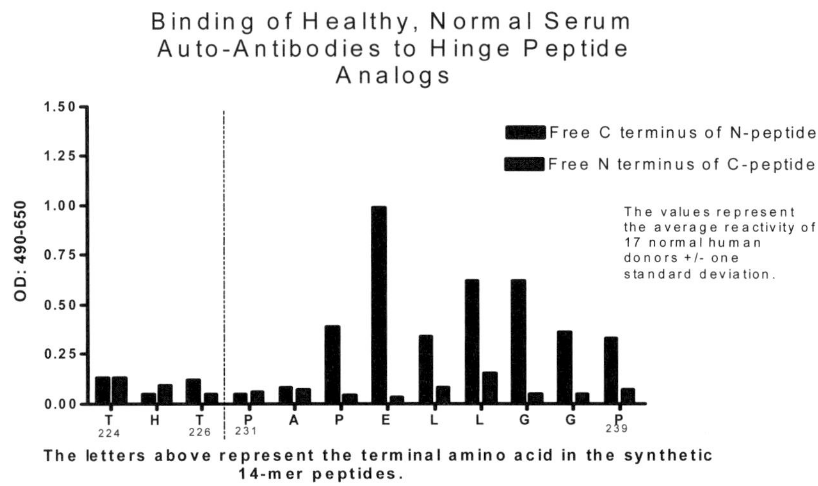

Fig. 1. "Map" of an overlapping series of 14-mer peptides which demonstrates the reactivity to the hIgG1 hinge domain.

References

1. Nasu, H., Chia, D. S., Knutson, D. W. and Barnett, E. V. *Clin. Exp. Immunol.* **42**, 378-386 (1980).
2. Terness, P., Kohl, I., Hübener, Gerd., Battistutta, R., Moroder, L., Welschof, M., Dufter, C., Finger, M., Gain, C., Jung, M., and Opelz, G. *J. Immun.* **154**, 6446-6452 (1995).

Heterobivalent Ligands Crosslink Multiple Cell-Surface Receptors – A Step Towards Personal Medicine

Jatinder S. Josan[1], Rajesh Sankaranarayanan[1], Heather L. Handl[3], Steve Fernandes[1], Liping Xu[4], Josef Vagner[1,4], Robert J. Gillies[2,3,4] and Victor J. Hruby[1,3,4]

[1]Departments of Chemistry, [2]Radiology, [3]Biochemistry & Molecular Biophysics, and [4]BIO5 Institute, University of Arizona, Tucson, AZ 85721, USA

Introduction

Effective treatment of tumor malignancies depends upon identifying targets – molecular markers that differentiate cancer cells from healthy cells. Current cancer therapies involve targeting either differential metabolism, especially nucleic acid biosynthesis pathways, or overexpressed specific gene products. We propose an alternative approach – to specifically target combinations of cell-surface receptors using heteromultivalent ligands (htMVLs). We envision that a three cell-surface protein combination that is expressed on a cancer cell but not on a normal cell could be targeted with heteromultivalent ligands displaying cognate binding motifs of weak affinities. These constructs should bind with high avidity and specificity to cancer population *in vivo* [1, 2]. As a proof-of-concept, we have synthesized series of heterobivalent constructs of MSH, CCK and δ-opioid receptors (Fig. 1). The constituent binding motifs in these constructs were moderately potent (see Fig. 1 legend) and were connected via a semi-rigid poly(Pro-Gly) linker flanked by flexible polyethylene glycol based PEGO chains.

Results and Discussion

Heterobivalent ligands were synthesized using Fmoc/*t*Bu chemistry on Tentagel Rink amide resin. A synthetic scheme for MSH-CCK series is given in Fig. 2. Using a parallel solid-phase synthesis approach, the first ligand (either CCK-6 or MSH-7) was assembled on the resin followed by a PEGO linker attachment [3]. Subsequently, proline and glycine residues were added to build the poly(Pro-Gly) linker followed by a second PEGO addition. Finally, the sequence of the second ligand (either MSH-7 or Deltorphin-II) was assembled. Compounds were cleaved

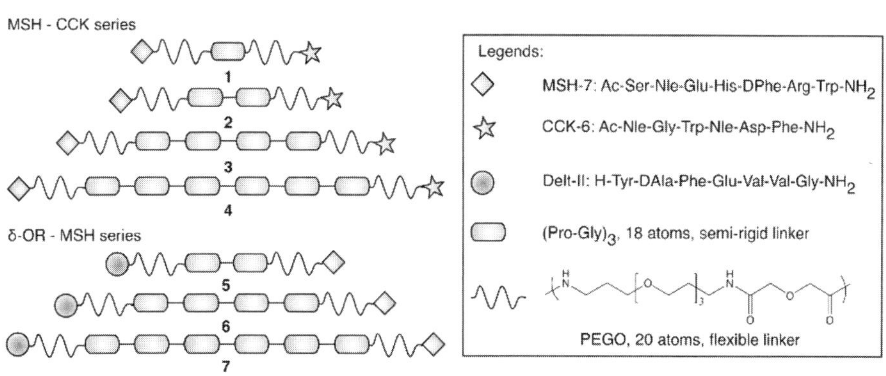

Fig. 1. Heterobivalent ligands of melanocortin receptor (hMC4R), cholecystokinin receptor (CCK-2R) and delta-opioid receptor (δOR).

S. Del Valle et al. (eds.), *Peptides for Youth: The Proceedings of the 20th American Peptide Symposium*, 413
DOI: 10.1007/978-0-387-73657-0_178, © Springer Science+Business Media, LLC 2009

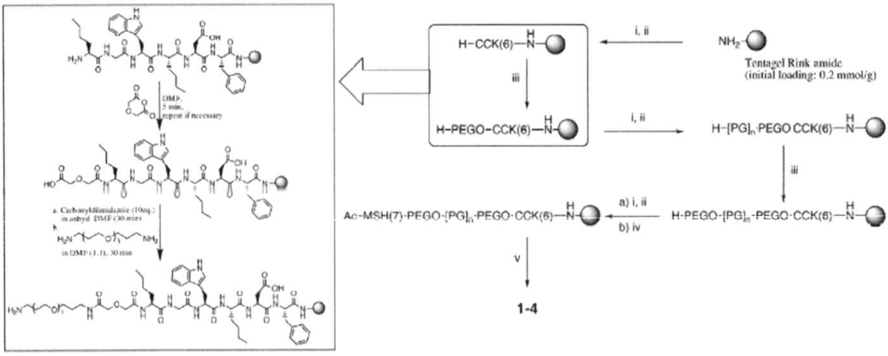

Fig. 2. Synthesis of MSH-CCK heterobivalent ligands. Reagents: i. 3eq. each of Fmoc-amino acid-OH, Cl-HOBt, DIC; ii. 20% Piperidine/DMF; iii. The PEGO linker can be introduced by reacting the free amino groups with diglycolic anhydride (10eq) in DMF followed by activation with carbonyldiimidazole (CDI, 20eq) in DMF for 30 min and substitution with diamine 4,7,10-trioxa-1,13-tridecaneamine (20eq) in DMF; iv. 50% Ac_2O in Pyridine; v. 91% TFA, 3% water, 3% thioanisole, 3% triisopropylsilane, 4 h. n is 3, 6, 12 or 18.

with TFA cocktail, purified by C_{18} RP-HPLC, and characterized by ESI-MS and/or MALDI-TOF and/or FT-ICR.

The heterobivalent constructs were evaluated for bivalent binding in Hek293 cells lines that co-expressed either hMC4R and CCK-2R or δOR and hMC4R. Monovalent binding was tested on cells lines that expressed only one of these receptors. In a preliminary study, these compounds were shown to have enhanced bivalent binding. For example, compound **1** exhibited a monovalent IC_{50} of 150 nM and a bivalent IC_{50} of 6.3 nM for CCK-6 binding. Thus, the construct exhibited a roughly 25-fold enhancement in binding affinity when targeted to a two-receptor combination on the cell-surface. Similarly, up to 50-fold decrease in IC_{50} for bivalent binding of MSH-7 ligand in δOR-hMC4R series was noticed for compound **6**. Clearly, this level of enhancement results from crosslinking of heterologous receptors. It is our contention that a high degree of specificity can be achieved in the context of heterovalent affinity enhancement. We envision that in the near future when a patient tissue expression profile could be easily identified, it will become possible to tailor heteromultivalent ligands to a combination of receptors using these expressed patterns. Thus, a high degree of specificity in multivalent targeting coupled with the ability to identify unique receptor combinations will provide a novel and revolutionary platform technology with which to direct therapeutics to defined cell populations. The work presented here is a step towards this goal of personal medicine in cancer.

Acknowledgments

The work was supported by NIH R33 CA95944, R01 CA97360 and the Arizona Biomedical Research Commission.

References

1. Caplan M. R., Rosca E. V. *Ann. Biomed. Eng.* **33**, 1113-1124 (2005).
2. Gillies R. J., Hruby V. J. *Expert Opin. Ther. Targets*. **7**, 137-139 (2003).
3. Vagner J., Handl H. L., Gillies R. J., Hruby V. J. *Bioorg. Med. Chem. Lett.* **2004**, 211-215 (2004).

Development of Polyvalent Peptide-Conjugated Magnetic Nanoparticles for Targeted In-Vivo Imaging of Micro-Thrombi

Rana Filfil[1], Albert Jin[2], Dmitri Tolkatchev[1], Kenji Tonan[1], Richard Gingras[1], Philip A. Barber[2] and Feng Ni[1]

[1]*Bio-NMR and Protein Research Laboratory, Biotechnology Research Institute, National Research Council of Canada, 6100 Royalmount Avenue, Montreal, Quebec, Canada H4P 2R2,* [2]*Clinical Neurosciences, University of Calgary, Calgary, Alberta, Canada*

Introduction

The need for potent and more selective diagnostic and therapeutic agents directed significant attention to the design of targeted nanoparticle-based pharmaceuticals [1]. Nanoparticles confer two specific advantages: 1) unique physicochemical properties due to the arraying of thousands of atoms (gold and iron) in the core and targeting peptides or antibodies on the surface, and 2) surface-mediated multivalency effects due to simultaneous interactions between peptide- or antibody-conjugated nanoparticles and the cell surface [2]. Furthermore, conjugation of bivalent polypeptides can create nanoparticles as efficient cross linkers and aggregators to mimic and induce the formation of biological assemblies [3].

To this end, we array polypeptides to magnetic nanoparticles (MNPs) to target a number of validated biomarkers and molecular mediators of human cancers. MNPs have been coupled to monovalent and bivalent polypeptide systems with specific binding to thrombin and P-selectin immobilized on the surface of fibrin clots or enriched on the surface of platelets; as both platelets and fibrin aggregates are two integral components of micro-thrombi [4]. Functional binding of the peptide-MNP conjugates was characterized by use of well-defined ex-vivo fibrin clots

As targeting efficacy is an overriding concern for peptide-based agents, an integral component of this work is to determine the in-vivo bioavailability of MNP-assembled peptides using animal models in conjunction with MR imaging studies tracking the accumulation of target-specific MNPs.

Results and Discussion

Coupling between the coagulation system and angiogenesis has promoted significant recent interests in the use of anticoagulant agents as part of new treatment strategies for devastating human cancers [5]. We use the peptide fragment Bbs-Arg-dPip-G-C, which mimics the anticoagulant drug, Argatroban, which was tested recently for the treatment of gliomas [5]. Multiple copies of this peptide were covalently linked to MNPs (50 nm, Mw 53000 kDa, with an NH_2-terminated dextran coating, Corpuscular Inc, New York) to create a multivalent molecular assembly with sufficient surface density (~62%) of conjugated peptides, corresponding to ~1000 peptide per particle, to target clot-bound thrombin.

Ex-vivo fibrin clots were used to characterize the targeting capacity of peptide-MNP conjugates. Polymerized fibrin matrices were loaded with three doses of thrombin ranging from 0.1 to 5 μM. The nanoparticles functionalized by Bbs-Arg-dPip-G-C binds specifically to the active site of thrombin, with an inhibition potency of IC_{50}= 2.78 + 0.19 μM. The release of the conjugated MNPs from the fibrin polymer was induced by an irreversible thrombin inhibitor D-phenylalanyl-L-prolyl-L-arginine

S. Del Valle et al. (eds.), *Peptides for Youth: The Proceedings of the 20th American Peptide Symposium,* 415
DOI: 10.1007/978-0-387-73657-0_179, © Springer Science+Business Media, LLC 2009

chloromethyl ketone (PPACK). UV absorbance spectra (Fig. 1) reveal an increase in absorption with increases in thrombin-loading levels, suggesting that conjugating particles are specific for clot-bound thrombin and that the release of these conjugated MNPs is dose-dependent on the density of surface-bound thrombin.

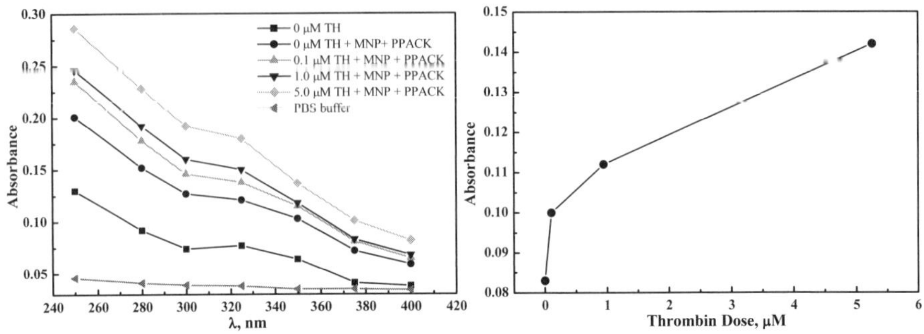

Fig. 1. Release of conjugated-MNPs from polymerized fibrin matrices loaded with doses of thrombin (TH) ranging from 0.1-5 μM.

To test the efficacy of peptide conjugated MNPs, we coupled a bivalent polypeptide (GSIQPRPQIHNDGDFEEIPEEYLQ-GGSS-LVSVLDLEPLDAAWL) targeting both thrombin (GSIQPRPQIHNDGDFEEIPEEYLQ, IC_{50} = 10 nM) [6] and P-selectin (LVSVLDLEPLDAAWL , IC_{50}=150μM) [7], linked by a -GGSS- linker. The P-selectin-specific contrast agent MNP-Psel appears to accumulate in the stroke region to a greater extent than is seen with the non-specific agent Feridex (Fig. 2).

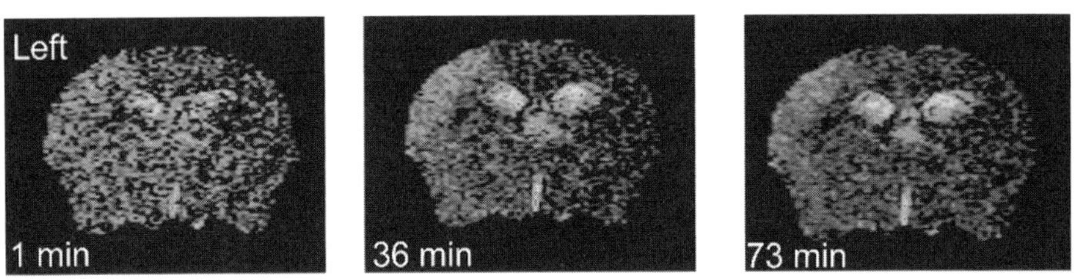

Fig. 2. Accumulation of MNP-Psel in the infarct and pre-infarct regions in the mouse model of stroke via pre-post injection T2 map difference MR images.

Acknowledgments

This work was supported by the Genomics and Health Initiative (GHI) of the National Research Council of Canada (NRCC Publication No. 49518), sponsored by the Government of Canada.

References

1. Ferrari, M. *Nat. Rev. Cancer* **5**, 161-171 (2005).
2. Mammen, M., *et al. Angew Chem., Int. Ed.* **37**, 2754-2794 (1998).
3. Mulder, A., Huskens, J. and Reinhoudt, D. N. *Org. Biomol. Chem.* **2**, 3409-3424 (2004).
4. Im, J.H., *et al. Cancer Res.* **64**, 8613-8619 (2004).
5. Hua, Y., *et al. J. Thromb. Haemost.* **9**, 1917-1923 (2005).
6. Su, Z., *et al. Protein Eng. Des. Sel.* **17**, 647-657 (2004).
7. Tom, J.M., *et al. Biessen. Biochem. Pharmacol.* **66**, 859-866 (2003).

Design, Selection and Binding Mechanism of Bivalent Miniproteins Targeting Human Thrombin

Andy Ng[1], Jamshid Tanha[2], Saran Narang[2] and Feng Ni[1]

[1]Biomolecular NMR and Protein Research, Biotechnology Research Institute, National Research Council, Montreal, Quebec, Canada H4P 2R2; [2]Institute for Biological Sciences, National Research Council, Ottawa, Ontario, Canada K1A 0R6

Introduction

Multilvalent molecular designs have emerged as an important tool for protein engineering and drug discovery complementary to the traditional methods of medicinal chemistry and recombinant antibody technology. Like antibodies, this new generation of molecules is particularly suited for the inhibition of difficult pharmaceutical targets such as protein-protein interactions. Unlike antibody engineering, the design of effective multivalent (or minimally bivalent) molecules is still wrought with trial and error, especially with regard to the selection of individual binding moieties and the choice of linking strategies. Using the thrombin system as an example, we aim to evolve a general approach for the design of protein-targeting bivalent polypeptides. Our design scheme involved the selection of binding moieties through screenings of relatively unsophisticated libraries, which often yield low-affinity, fast-dissociating hits. Through bivalency, however, we showed that these weak but specific hits can be transformed into higher affinity protein-targeting molecules. We found that bivalent constructs consisting of these fast-dissociating binding moieties can discourage the inevitable and often undesirable intermolecular cross-linking interactions, an important property not commonly present in other poly- or multivalent systems.

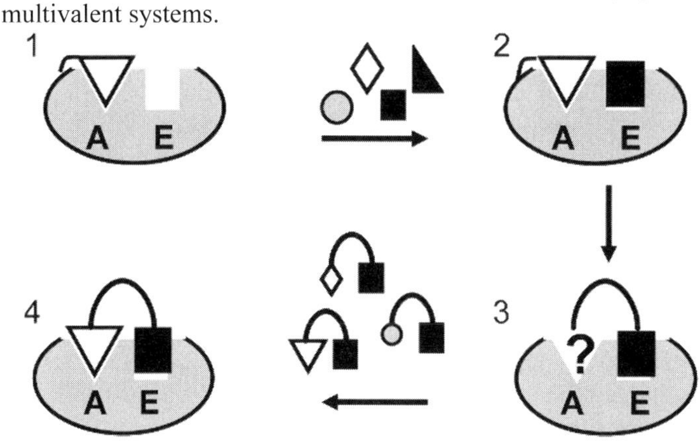

Fig. 1. Design of bivalent miniproteins targeting the active site (A) and exosite I (E) of human thrombin.

Results and Discussion

Novel thrombin exosite I (a binding surface on thrombin away from the catalytic active site) –binding miniproteins were selected from a small, naïve VH fragment phage-display library [1]. To mimic bivalent binding, thrombin that is blocked by an

S. Del Valle et al. (eds.), *Peptides for Youth: The Proceedings of the 20th American Peptide Symposium*, 417
DOI: 10.1007/978-0-387-73657-0_180, © Springer Science+Business Media, LLC 2009

irreversible active site inhibitor, PPACK, was used as the immobilized target (Fig. 1, steps 1 and 2). Screening of this relatively unsophisticated library led to the identification of weak, but specific-binding ligands targeting exosite I, evidenced by inhibition assays, fluorescence and NMR binding experiments. In addition, this 'monovalent' ligand was found to exhibit 10-fold higher affinity to active-site-blocked thrombin, apparently taking advantage of the positive allostery between exosite I and the active site (Table 1).

To mimic the effect of the inhibitor PPACK and to introduce bivalent avidity, the selected miniprotein was extended with a polypeptide carrying a randomized sequence reaching to the active site of thrombin (Fig. 1, steps 3 and 4). Phage-display selection was employed to characterize the enhancement of inhibitory activity as compared to the monovalent miniprotein alone (Table 1).

Table 1. Thrombin binding actions of the selected binding moieties and the designed bivalent miniprotein.

	IC_{50} (μM)	K_i (μM)	K_D (μM)
Exosite-I-binding miniprotein	80 ±5	N/A	85 ±5 [a] 6.5 ±0.7 [b]
Active-site-binding sequence	~2000	~2000	N/A
Bivalent miniprotein	0.70 ±0.1	0.80 ± 0.1	N/A

[a] Native thrombin; [b] PPACK-thrombin

Dissection of the novel bivalent miniprotein revealed that the activity enhancement was conferred by an approximately equal contribution of positive cooperativity and the avidity effect of bivalency up to a factor of 10 for both effects. It is intereseting to note that, linkage of an exosite I ligand and an active-site-targeting sequence with a proper linker often result in a 10-fold affinity enhancement conferred by linkage alone [2]. We found that similar lengths and compositions of the linker sequences represent the underlining determinants for this common degree (a factor of ten) of affinity enhancement in different bivalent constructs.

It is also interesting to note that, the weak affinities (Table 1) and the fast-dissociation rates of the individual binding moieties promote 1:1 binding geometry with thrombin, despite the rather long (8 residues) and flexible linker moiety. Bivalent molecules that can discourage the intermolecular cross-linking are indispensable in systems where cross-linking can lead to undesirable protein aggregations, for example, in the design of retractable enzyme inhibitors [3].

This work (NRC publication No. 49519) was supported by a "Health and Genomics Initiative of the National Research Council of Canada, sponsored by the Government of Canada."

References

1. Tanha, J., Dubuc, G., Hirama, T., Narang, S.A., and MacKenzie, C.R. *J. Immunol. Methods* **263**, 97-109 (2002).
2. Su, Z., Vinogradova, A., Koutychenko, A., Tolkatchev, D. and Ni, F. *Protein Eng. Des. Sel.* **17**, 647-657 (2004).
3. Tolkatchev, D., Vinogradova, A. and Ni, F. *Bioorg. Med. Chem. Lett.* **15**, 5120-5123 (2005).

Quenched Fluorescent Peptide Substrates as Tools for the Discovery of Novel Cardiovascular Disease Biomarkers

A. Ian Smith[1], Fiona J. Warner[1], Rebecca A. Lew[1], Mike Yarski[2], Barry McGrath[3] and Louise M. Burrell[4]

[1]*Dept. Biochemistry and Molecular Biology, Monash University, Clayton, Vic 3800;* [2]*Baker Institute, Commercial Road, Prahran, Vic 3181;* [3]*Monash Medical Centre, Monash University, Clayton, Vic 3800;* [4]*Dept Medicine, Austin Hospital, Heidelberg, Australia*

Introduction

The generation of angiotensin II (Ang II) within the renin-angiotensin system (RAS) is recognised as a critical point in the regulation of cardiovascular function The final step in the production of Ang II is catalysed by the membrane-bound ectoenzyme angiotensin converting enzyme (ACE), and inhibitors of ACE are the most widely prescribed therapy in the treatment of hypertension, heart failure, and myocardial infarction Six years ago, the homologous enzyme ACE2 was discovered (1, 2) Evidence thus far suggests that ACE2 may work in a counter-regulatory role to ACE, via the inactivation of Ang II, and the formation of the putative vasodilator, Ang 1-7 (see 3 for review) This hypothesis is supported by an increasing number of published reports, including the development of an ACE2 knock-out mouse displaying severe defects in cardiac contractility which were prevented by concomitant ablation of ACE (4) Within the cardiovascular system, ACE2 expression is normally relatively low and restricted to endothelial cells of the coronary and renal circulation, and epithelial cells of the renal distal tubules (2). However, we and others have shown that ACE2 levels in these and other tissues increase markedly in a number of pathologies, including myocardial infarction (5) suggesting a role for the enzyme in limiting the damage associated with RAS activation in these conditions.

It has long been recognized that in addition to its localization on endothelial cell membranes, ACE is also present in biological fluids, including plasma. This soluble ACE arises from proteolytic cleavage of the membrane-bound enzyme, a process referred to as "shedding". Although ACE2 has been shown to be proteolytically shed from cells in culture (6), there have been no previous reports that ACE2 circulates in human plasma. Indeed, our initial attempts to measure ACE2 activity directly in plasma were unsuccessful; however, during the course of these studies, we made the observation that plasma itself potently inhibits the activity of purified recombinant ACE2, suggesting the presence of an endogenous inhibitor of ACE2. In this report, we describe the quantitation of plasma ACE2 activity and present preliminary evidence that plasma ACE2 is elevated in heart disease.

Results and Discussion

Preparation of plasma samples for ACE2 activity assay and Western blot analysis:

Given that plasma contains an endogenous ACE2 activity inhibitor, the inhibitor and other contaminating components have to be removed from the plasma by anion

S. Del Valle et al. (eds.), *Peptides for Youth: The Proceedings of the 20th American Peptide Symposium*, 419
DOI: 10.1007/978-0-387-73657-0_181, © Springer Science+Business Media, LLC 2009

exchange chromatography. Briefly, plasma (0.25 mL) was diluted into low ionic strength buffer (20 mM Tris-HCl, pH 6.5), and added to 200 μl ANX Sepharose 4 Fast-Flow resin (Amersham Biosciences, GE Healthcare, Uppsala, Sweden). Following binding and washing, proteins were eluted with high salt buffer (20 mM Tris-HCl, 1 M NaCl, pH 6.5) and the resulting eluate was assayed for ACE2 catalytic activity and the ACE2 characterised by Western blot analysis.

Western immunoblot analysis
Following anion exchange extraction, plasma proteins were concentrated using a 30 kDa molecular weight cut-off Nanosep centrifugal concentrator (Pall Sciences) prior to separation by SDS-PAGE. A portion of the sample was deglycosylated using peptide:N-glycosidase F (PNGase F; New England Biolabs) for 4 hr, according to the manufacturer's instructions. Western immunoblot analysis was performed using a polyclonal anti-ACE2 antiserum (R&D Systems, catalog # AF933, 1:100).

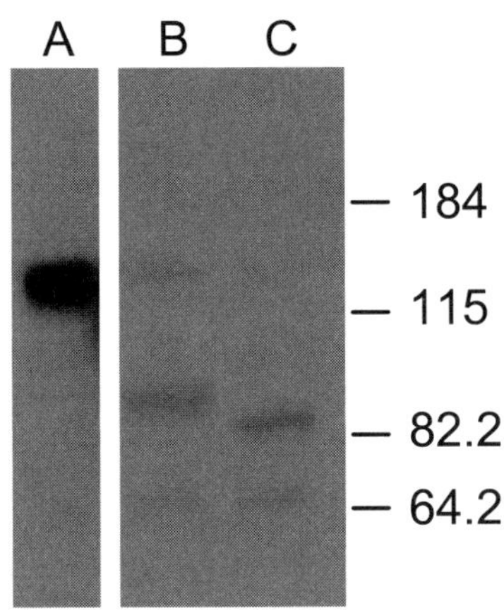

Figure 1 shows a Western blot analysis of immunoreactive ACE2 in human plasma Lane A is full-length (membrane-bound) human ACE2 purified from transiently transfected HEK 293T cells (300 μg total protein); Lane B is derived from human plasma following anion exchange chromatography and concentration (300 μg protein); and Lane C is the same as Lane B following treatment with PNGase F (4 hr).

Using Western immunoblot analysis, ACE2 was undetectable in straight plasma samples, even after concentration (data not shown); however, following both anion exchange and concentration, a faint band was visible at ~90 kDa which was reduced to 83 kDa following PNGase F treatment (Figure 1). An additional faint band was seen at ~130 kDa, the same size as full-length membrane-bound recombinant protein; this band was not present in the PNGase-treated sample (Figure 1).

ACE2 activity assay

ACE2 catalytic activity was measured as described previously (5), using an ACE2 quenched fluorescent substrate (QFS) = (7-methoxycoumarin-4-yl)acetyl-Ala-Pro-Lys(2, 4-dinitrophenyl). The rate of substrate cleavage was expressed as pmoles substrate cleaved/min//ml plasma.

Plasma ACE2 activity was measured in 17 samples from apparently healthy post-menopausal women (50-70 years of age); of these, 16 samples ranged from 0.633 – 1.67 pmoles substrate cleaved/ml/min (mean ± s.e.m. = 1.07 ± 0.07. A single sample measured 5.61 pmol/ml/ml (approximately 12 ng ACE2/ml), more than 5-fold greater than the mean of the other 16 samples.

ACE2 activity levels in plasma from patients admitted to the coronary care unit were also determined, and ranged from 0.53 – 7.57 pmol/min/ml plasma (Figure 2; mean ± sem = 1.99 ± 0.34 pmol/min/ml, n=25), corresponding to approximately 1 – 18 ng ACE2/ ml plasma.

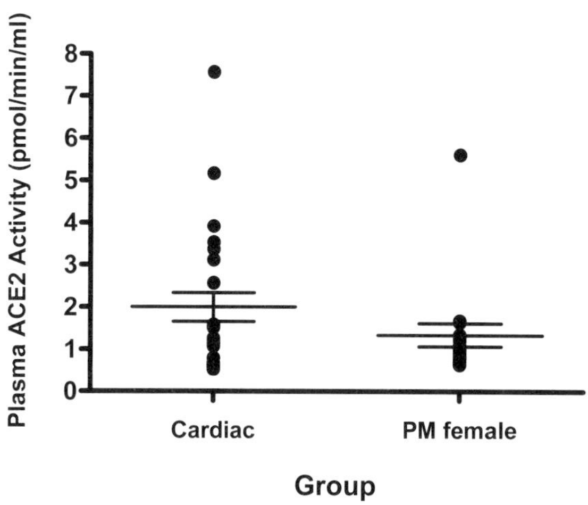

Figure 2 – ACE2 activity in plasma samples from coronary care patients ("Cardiac", n=25) and healthy post-menopausal women ("PM female", n=17); each value is represented by a circle, with the mean (thick line) and s.e.m. (thin lines) shown.

These data were not normally distributed, as determined by the Kolmogorov-Smirnov test (GraphPad Prism software), with an apparent gap between low and high ACE2 "clusters," which may reflect differences in the clinical state of these patients. The relatively small numbers included in this initial study do not allow us to determine if ACE2 levels are associated with particular coronary/cardiac syndromes; however, inspection of the data reveals two "clusters" of plasma ACE2 values, the lower of which overlaps substantially with those in the post-menopausal cohort. Given our previous finding that ACE2 expression in the heart is markedly up-regulated following myocardial infarction (5), it is possible that those patients with a high plasma ACE2 have some level of ischemic disease. Thus, the appearance of ACE2 in plasma may serve as a biomarker of ischemic heart disease. Further studies with increased numbers of coronary care patients may reveal a correlation between plasma ACE2 levels and the extent of myocardial injury, the time since injury, and prognosis.

Acknowledgments

The work was funded by an NH&MRC project grant awarded to A. I. Smith.

References

1. Tipnis SR, *et al. J. Biol. Chem.* 2000;275:33238-33243.
2. Donoghueet *et al. Circ. Res.* 2000;87:E1-E9.
3. Ferrario CM, Trask AJ, Jessup JA., *Am J Physiol Heart Circ Physiol.* 2005;289: H2281-2290.
4. Crackower MA, *et al. Nature* 2002;417:822-828.
5. Burrell, *et al. Eur Heart J.* 2005;26:369-375.
6. Lambert DW, *et al. J. Biol. Chem.* 2005;280:30113-3011.

Solid Phase Synthesis and Analysis of Amadori Peptides

Andrej Frolov, David Singer, Thomas Zauner and Ralf Hoffmann

Institute of Bioanalytical Chemistry, Center for Biotechnology and Biomedicine, Faculty of Chemistry and Mineralogy, Leipzig University, Deutscher Platz 5, Leipzig, 04103, Germany

Introduction

Glycation (or non-enzymatic glycosylation) is a common modification in living organisms formed by the reaction of carbohydrates or reactive carbonyl-compounds with amino groups in peptides and proteins. Increased glucose levels in organs and tissues, for example, yield fructosamines, also called Amadori products. These are early markers of several diseases including diabetes mellitus, atherosclerosis, Parkinson disease, and Alzheimer's disease. In organisms these Amadori products can undergo further irreversible reactions, such as oxidations, yielding various advanced glycation end products (AGEs), which might be important for the pathogenesis of several diseases [1]. To further study the molecular processes well defined glycated peptide and protein models are essential tools. However, the typical approach to modify peptides in aquatic buffers at elevated temperatures yields the Amadori products only at low yields often contaminated with many by-products difficult to separate on RP-HPLC. Thus we compare two recently developed synthetic strategies to obtain site-specifically modified peptides on solid-phase by Fmoc/^{t}Bu-chemistry using either direct glycation with D-glucose or reductive amination with 2,3:4,5-di-O-isopropylidene-aldehydo-ß-D-arabino-hexos-2-ulo-2, 6-pyranose [2].

Results and Discussion

Lysine residues to be modified were protected with the very acid labile methyltrityl (Mtt) protection group. After completion of the peptide synthesis the Mtt group was cleaved with 1% TFA in DCM and this position glycated either with 0.25 M D-glucose (110°C, 40 min, DMF, nitrogen atmosphere; method 1) or by reductive amination with 2,3:4,5-di-O-isopropylidene-aldehydo-ß-D-arabino-hexos-2-ulo-2, 6-pyranose in the presence of $NaCNBH_3$ in a mixture of methanol, isopropanol and

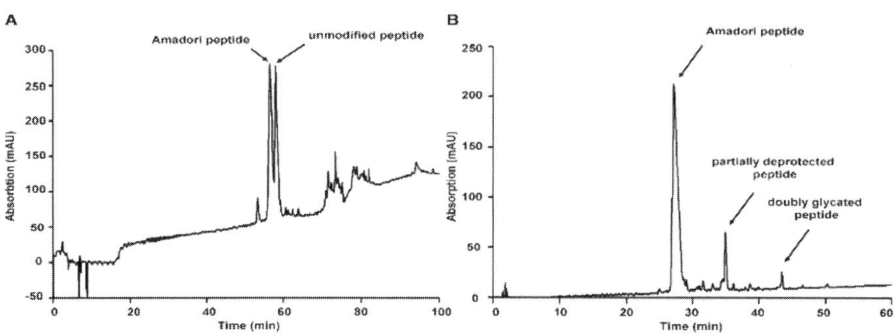

Fig. 1. Reversed-phase chromatograms of the crude H-ASK$_{Amadori}$ASKFL-NH$_2$ peptide after glycation with D-glucose (A) and via reductive amination (B).

S. Del Valle et al. (eds.), *Peptides for Youth: The Proceedings of the 20th American Peptide Symposium*, 423
DOI: 10.1007/978-0-387-73657-0_182, © Springer Science+Business Media, LLC 2009

water (2:2:1 by vol.) at 70°C for 18 h (method 2). The peptides were cleaved with 5% H_2O in TFA for two hours (method 1) or in two steps for 1 h first and then again for 2 h in fresh cleavage solution (method 2) to deprotect the sugar moiety effectively.

The crude peptides contained the targeted Amadori products at a content of 30% (method 1) and 80% (method 2) according to the peak areas on RP-HPLC using a linear aqueous acetonitrile or methanol gradient in the presence of 0.1% trifluoroacetic acid (TFA) as ion pair reagent (Figure 1). As for some sequences the Amadori product coeluted with the unmodified peptide, it was necessary to replace TFA by heptafluorobutyric acid (HFBA) to achieve base line separation. The main by-products of the direct glycation approach (method 1) were the unmodified peptide and several oxidation products of minor intensities. In contrast, the reductive amination (method 2) yielded the desired Amadori peptide as main compound contaminated with only two minor by-products (Figure 1B). The higher hydrophobicity of these partially protected by-products allowed an efficient purification by RP-HPLC using a fast acetonitrile gradient (0.1% TFA). Thus this approach appears superior to the direct glycation despite the longer reaction times.

The correct structure of the Amadori-peptides was confirmed by both matrix-assisted laser desorption/ionization (MALDI-) and electrospray ionization (ESI-) tandem mass spectrometry (MS/MS) using collision induced dissociation. The sequences were retrieved from the pyrilium- (-54 u) and furylium- (-84 u) ion series dominating the spectra (Figure 2).

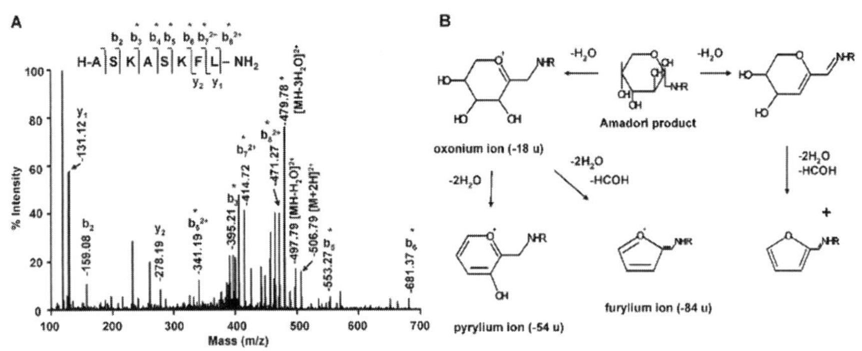

Fig. 2. ESI-MS/MS of peptide H-ASK$_{Amadori}$ASKFL-NH$_2$. Fragmentation pattern of the glucose-derived Amadori peptide. Furylium ions are marked with an asterisk.

Acknowledgments

Financial support by the Deutsche Forschungsgemeinschaft (DFG, Graduiertenkolleg 378) and the Free State Saxony is gratefully acknowledged.

References

1. Baynes, J.W. and Thorpe, S.R. *Diabetes.* **48**, 1-9 (1999).
2. Ledesma, M.D., Bonay, P. and Avila, J. *J. Neurochem.* **65**, 1658-1664 (1995).
3. Jakas, A. and Horvat, Š. *J. Chem. Soc., Perkin Trans.* **255**, 789-794 (1996).

Design of Hetero-Bivalent Polypeptides as Kinetics-Based Sensors for High-Sensitivity Detection of Weak Molecular Interactions

Kenji Tonan, Dmitri Tolkatchev, Ping Xu, Ping Wang, Josee Plamondon, Anna Vinogradova, Richard Gingras and Feng Ni

Biomolecular NMR and Protein Research Group, Biotechnology Research Institute, National Research Council of Canada, 6100 Royalmount Avenue, Montreal, H4P2R2, Canada

Introduction

Multivalent molecular interactions found throughout biological systems provide stable high-affinity complexes by increasing the statistical probability of re-binding of constituent binding moieties to the receptor sites at supramolecular surfaces [1] and in solution [2]. Not only its thermodynamic stability, kinetics behavior of the multivalent system is also important. Transient exposure of unoccupied binding sites, which ocuurs in the course of stepwise dissociation of the multivalent complex, allows for fast displacement of high-affinity ligands with competing molecules [3]. To understand thermodynamic and kinetic aspects of dynamic control of naturally-occurring or artificially-designed multivalent protein inhibitors, a detailed analysis of molecular events accompanying the process of stepwise dissociation is required.

We used one of the best-characterized multivalent (bivalent) molecular systems, i.e. the complexes formed between human thrombin and polypeptides derived from a natural protein inhibitor hirudin. We constructed heter-bivalent thrombin inhibitors by tethering two peptide fragments that target two distinct binding sites on thrombin, namely, the active site (A-site) and the fibrinogen-recognition exosite (E-site). Through competitive displacement of bivalent peptide-protein complexes by monovalent competitor, our NMR study aim to evaluate quantitatively the dissociation pathways engaged by bivalenet molecular complexes.

TWE1	Bbs-R-dPip-GC-*SS*-CGSGSGSGSGSGSGSGSGDFEEIPEEYLQ
TWE2	Bbs-R-dPip-GC-*SS*-CPHYEKVSGDFEEIPEEYLQ
MH2	IRFTDGEGTPNPESHNNGDFEEIPEEYLQ

Fig. 1. Sequences of hetero-bivalent peptide inhibitors for human thrombin. The SS stands for a disulfide bond. Bbs is ter-butylbenzenesulfonyl, and dPip is (D)pipecolyl.

Results and Discussion

The inhibitory constants (K_I) determined were 2.2+/-0.4 nM (TWE1) and 4.8+/-0.8 nM (TWE2), 10+/-2 nM (MH2), 12+/-2 nM (MH2GS) were much lower than those previously determined for isolated segments that target the active site of thrombin, Bbs-R-dPpi-NH_2 (2.0 μM) and IRFTDG (107 μM) [3] and that for isolated fragment from the C-terminus of hirudin binds to the exosite (E-site) of thrombin, GDFEEIPEEYLQ (~1.5 μM) [4-5]. Such enhanced binding of to thrombin was accomplished through synergestic interactions of the two binding moieties of the bivalent peptides with the A- and E-sites of thrombin and form thermodynamically stable complex, which was also confirmed by NMR spectra characteristics supporting a very slow exchange between the free and bound states of the peptides (Fig. 2).

S. Del Valle et al. (eds.), *Peptides for Youth: The Proceedings of the 20th American Peptide Symposium*,
DOI: 10.1007/978-0-387-73657-0_183, © Springer Science+Business Media, LLC 2009

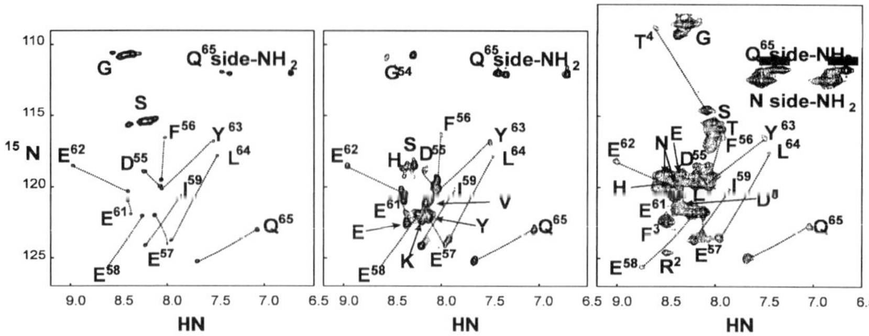

Fig. 2. 800 MHz [^{15}N,^{1}H]-HSQC spectra of TWE1 (left), TWE2 (middle), and MH2 (right) in mixtures at concentration ratios of peptide to thrombin are 211/155, 123/68, 150.90 (μM/μM), respectively, measured at 25°C and pH 5.5. The assigned free and thrombin-bound signals are connected by dashed line for each residue.

Addition of much weaker monovalent competitors (K_I~1 μM) to the mixture of TWE1 and thrombin promotes NMR-line broadening only for residues of the free peptide. Such line-broadening behaviour can be interpreted by switching of exchanging kinetics to the path on which free molecular species can exchange with a generating competitor-displaced monovalent complex of the bivalent peptide with thrombin at a rate intrinsically fast in the NMR time scale. A reduced 3-site exchange system has been formulated to quantify the observed line-broadening and the stepwise dissociation mechanisms whereby the two binding moieties of the bivalent peptides dissociate from thrombin independently. Such quantitative analysis also showed it will be possible to modify the two binding moieties and the linker segments of bivalent sensor peptides to enable NMR (line-broadening)-based screening of even weaker (K_I ~mM) protein-ligand interactions.

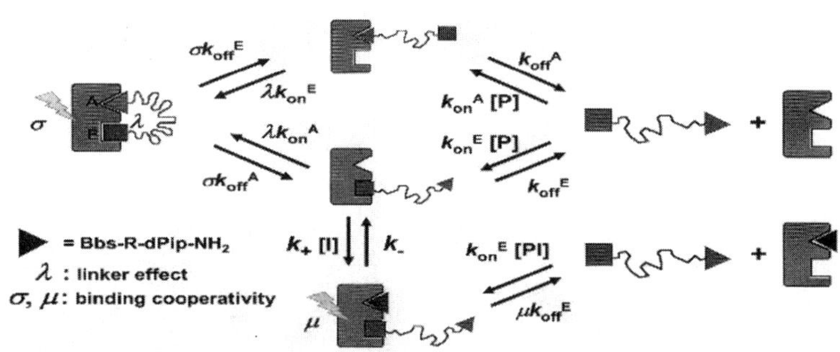

Fig. 3. Possible chemical exchange pathway for a system with bivalent peptide and thrombin and monovalent A-site competitor.

References

1. Huskens, J., *et al. J. Am. Chem. Soc.* **126**, 6784-6787 (2004).
2. Zhou. H. X. *J. Mol. Biol.* **329**, 1-8 (2003).
3. Filfil, R., *et al. in preparation.*
4. DiMaio, J., *et al. J. Biol. Chem..* **265**, 21698-21703 (1990).
5. Anderson, P. J., *et al. J. Biol.Chem.* **275**, 116428-16434 (2000).

Monitoring of native chemical ligation by surface plasmon resonance

Ewa Wieczerzak[1], Raymond Jr Hamel[1,2], Vincent Chabot[2], Vincent Aimez[2], Paul G. Charette[2], Michel Grandbois[1] and Emanuel Escher[1]

[1]Department of Pharmacology; [2]Department of Electrical and Computer Engineering, Université de Sherbrooke, Sherbrooke, QC, Canada

Introduction

Native chemical ligation (NCL) [1] is being widely used for the chemical synthesis of large peptides and entire proteins. This reaction involves a chemoselective reaction in aqueous solution of an unprotected peptide α-thioester with a second unprotected peptide containing a free N-terminal cysteine and yields a product with a native peptide bond at the ligation site.

In the present contribution we show the feasibility and monitoring of peptide synthesis and NCL by surface plasmon resonance (SPR). The application of SPR as a detection method is based on the utilization of a noble metal layer to generate a surface plasmon electromagnetic field, which can be used to probe the changes in refractive index associated with the binding or adsorption reactions occurring in the vicinity of the surface. In our work we used an SPR device that was developed with chemical compatibility for the synthetic procedures.

Results and Discussion

In our studies we used gold surfaces with 5 nm layer of silicon oxide deposited by means of plasma-enhanced chemical vapour deposition which allowed chemical functionalization with 3-aminopropyltrimethoxysilane (APS). After attachment of Boc-Cys(Boc)-OH and subsequent removal of Boc-protecting groups, native chemical ligation of a thioester of a model peptide – bradykinin (synthesized by solid phase method [2]), was carried out (Figure 1).

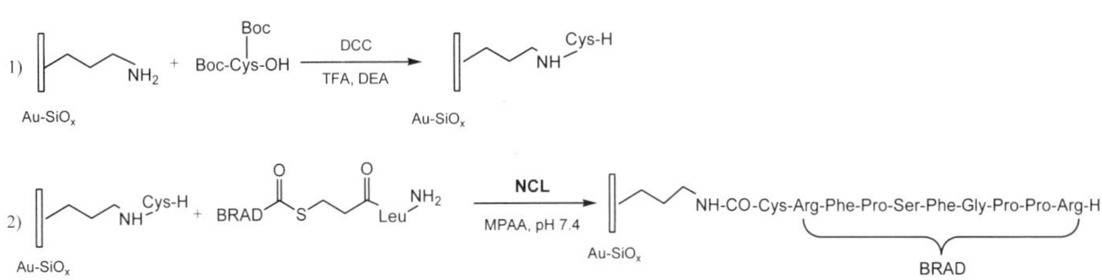

Fig. 1. Reaction scheme showing the NCL immobilization of peptides on the modified SPR surface.

Since the synthesized alkyl thioester of bradykinin is relatively unreactive under native chemical ligation conditions, we conducted our experiments in the presence of a thiol catalyst – 4-mercaptophenylacetic acid (MPAA) that proved to be particularly useful for the facile conversion of a peptide alkyl thioester to the corresponding highly reactive aryl thioester [3].

S. Del Valle et al. (eds.), *Peptides for Youth: The Proceedings of the 20th American Peptide Symposium*, 427
DOI: 10.1007/978-0-387-73657-0_184, © Springer Science+Business Media, LLC 2009

All chemical reaction steps were monitored by SPR and were accompanied by the expected resonance angle shifts. To independently prove NCL taking place, we carried out this reaction with iminobiotin-labeled bradykinin thioester. The iminobiotin group was used to anchor avidin which can be detected by a reversible shift in the reflectance curve. After attachment, specific avidin-iminobiotin interaction was reversed by acetic acid solution (pH3) and/or 2 mM solution of biotin in pH3 buffer, as shown on Figure 2, left. A part of the resonance angle shift caused by avidin attachment was stable in the conditions mentioned above; this part corresponds to non-specific avidin binding. This was confirmed by NCL of bradykinin thioester that did not contain iminobiotin and subsequent avidin addition, as presented on Figure 2, right.

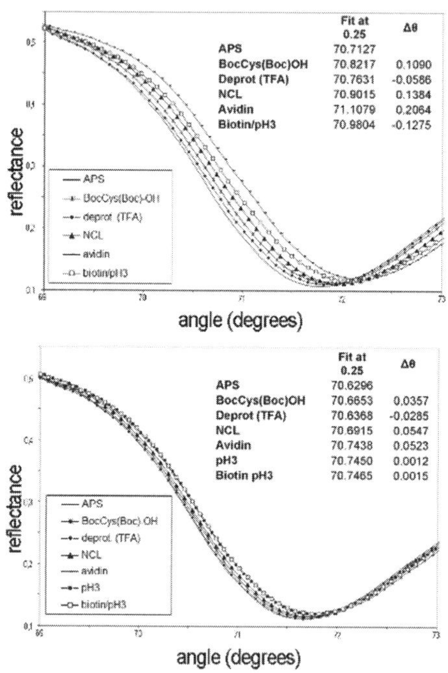

Fig. 2. SPR curves corresponding to NCL of iminobiotinylated bradykinin thioester and bradykinin thioester, respectively.

The time-dependent evaluation of the NCL-induced SPR signal shift allows a direct determination of NCL reaction kinetics and the reaction speed in function of thioester nature and thiol catalyst.

In this contribution we presented real-time evaluation of peptide synthesis and native chemical ligation by means of surface plasmon resonance. This SPR monitoring gives synthetic access to a new range of SPR bioanalytical procedures.

Acknowledgments

The work was supported by funds from Canadian Institutes of Health Research and Fonds de la Recherche en Santé du Québec.

References

1. Dawson, P.E., Muir, T.W., Clark-Lewis, I. and Kent, S. B. H. *Science* **266**, 776-779 (1994).
2. Hackeng, T. M., Griffin J. H. and Dawson, P. E. *Proc. Natl. Acad. Sci. USA* **96**, 10068-10073 (1999).
3. Johnson, E. C. B. and Kent, S. B. H. *J. Am. Chem. Soc.* **128**, 6640-6646 (2006).

Studies for Identification of the Minimal Epitope(s) mimicked by the Synthetic Glucopeptide CSF114(Glc)

Francesca Nuti[1,2], Barbara Mulinacci[1,3], Elisa Peroni[1,2], Maria C. Alcaro[4], Ilaria Paolini[4], Francesco Benedetti[1,3], Alfonso Carotenuto[5], Francis Ciolli[1,2], Francesco Lolli[1,6], Mario Chelli[1,2], Paolo Rovero[1,3] and Anna M. Papini[1,2]

[1]*Laboratory of Peptide & Protein Chemistry & Biology;* [2]*Dept. of Organic Chemistry "Ugo Schiff" & CNR-ICCOM, University of Florence;* [3]*Dept. of Pharmaceutical Sciences, University of Florence;* [4]*Toscana Biomarkers S.r.l., Siena, Italy;* [5]*Dept. of Pharmaceutical and Toxicological Chemistry, University of Naples "Federico II", Napoli, Italy;* [6]*Dept. of Neurological Sciences, University of Florence.*

Introduction

Molecules present in biological fluids, such as antibodies, can be identified as biomarkers and used to set up diagnostic/prognostic tools and for monitoring the effects of a therapeutic treatment. In the case of autoimmune disorders, such as Multiple Sclerosis (MS), antibodies have shown to reflect the presence, nature and intensity of the immune response. In particular, an antibody-mediated disease pattern seems to correlate with the relapsing-remitting form of Multiple Sclerosis. Antibodies may therefore be potentially useful as biomarker for the diagnosis, prognosis and classification of disease activity in MS and other immune-related diseases. Antibodies should be specific and they have to be detected by simple immunoassays, but unfortunately up to now there is no biomarker available fulfilling the criteria of a surrogate endpoint for Multiple Sclerosis. In order to discover more efficient tools for Multiple Sclerosis diagnosis, recent studies have focused on the use of synthetic peptides. In our laboratory, we developed CSF114(Glc), a structure-based designed glucosylated peptide, characterized by a β-D-glucopyranosyl moiety on the tip of a I' β-turn, as the first Multiple Sclerosis Antigenic Probe (MSAP) accurately measuring autoantibodies, as specific biomarkers of disease activity, by ELISA experiments on sera of a statistically significant number of patients [1]. Moreover, we demonstrated that Asn(Glc) moiety is the minimal epitope able to recognize MS autoantibodies in a solution-phase competitive ELISA experiment on patients' sera. Aim of this work is to investigate the role of the amino acid residues surrounding the minimal epitope in antibody recognition and therefore to study the size of the antibody binding groove.

Results and Discussion

In previous results, we synthesized glycopeptides derived from the original sequence of CSF114(Glc), shortened at both termini and containing an Asn(Glc) residue to identify the minimal glycopeptide sequence to be used in SP-ELISA (Solid Phase ELISA). We demonstrated that the longest is the synthetic antigen, the highest is the efficiency of its coating to the polystyrene of the plate [2]. In this study, we synthesized the following CSF114-glycopeptide fragments focusing on the amino acids surrounding the minimal epitope: [Asn^7(Glc)]CSF114(6-9) (**I**), [Asn^7(Glc)]CSF114(5-10) (**II**), [Asn^7(Glc)]CSF114(4-11) (**III**) to identify the

S. Del Valle et al. (eds.), *Peptides for Youth: The Proceedings of the 20th American Peptide Symposium*, 431
DOI: 10.1007/978-0-387-73657-0_185, © Springer Science+Business Media, LLC 2009

minimal glycopeptide sequence able to recognize MS autoantibodies in a solution-phase competitive ELISA experiment on MS patients' sera (Table 1).

Moreover, in previous studies the CSF114(Glc)-sequence was modified introducing different amino acids to facilitate the formation of several β-turn types to study any possible correlation between the antibody affinity and the β-turn types. [Pro^7,Asn^8(Glc),Thr^{10}]CSF114 [3] was obtained introducing Pro in position 7 instead of a Gly present in the original CSF114(Glc), Thr in position 10 to generate an *N*-glycosylation consensus sequence, not present in CSF114(Glc). [Pro^7,Asn^8(Glc),Thr^{10}]CSF114 was demonstrated to be a type II β-turn and to be more efficient and selective than CSF114(Glc) [*Carotenuto A. et al.* unpublished results].

Starting from these promising results, we synthesized and tested fragments of the new glycopeptide, [Pro^7,Asn^8(Glc),Thr^{10}]CSF114: [Pro^7,Asn^8(Glc),Thr^{10}]CSF114 (6-10) (**IV**), [Pro^7,Asn^8(Glc),Thr^{10}]CSF114(5-11) (**V**), [Pro^7,Asn^8(Glc),Ser^{10}] CSF114(5-11) (**VI**), [Pro^7,Asn^8(Glc),Gly^{10}]CSF114(5-11) (**VII**), [Pro^7,Asn^8(Glc),Thr^{10}]CSF114(4-11) (**VIII**).

The glycopeptide **VII** contains Gly at position 10 to favor β-turn structure.

Table 1. Shortened peptides sequences.

CSF114	[Pro^7,Asn^8(Glc),Thr^{10}]CSF114
[Asn^7(Glc)]CSF114(6-9) (**I**)	[Pro^7,Asn^8(Glc),Thr^{10}]CSF114(6-10) (**IV**)
[Asn^7(Glc)]CSF114(5-10) (**II**)	[Pro^7,Asn^8(Glc),Thr^{10}]CSF114(5-11) (**V**)
[Asn^7(Glc)]CSF114(4-11) (**III**)	[Pro^7,Asn^8(Glc),Ser^{10}]CSF114(5-11) (**VI**)
	[Pro^7,Asn^8(Glc),Gly^{10}]CSF114(5-11) (**VII**)
	[Pro^7,Asn^8(Glc),Thr^{10}]CSF114(4-11) (**VIII**)

The autoantibody recognition by glycopeptides **I-VIII** was evaluated by competitive ELISA on Multiple Sclerosis patients' sera. The glycopeptide **V** displayed an antibody affinity higher than CSF114(Glc).

Biological results showed that the modifications introduced in the original sequence of CSF114(Glc) is also useful in the case of shortened peptides sequences. Therefore, it is possible that the seven amino acids sequence with Asn(Glc) at the centre is possibly the minimal length important for an optimal interaction with the antibodies specific for MS.

Acknowledgments

We thank FIRB 2005 RBIN04TWKN, PRIN 2005 PROT 2005032959 and the University of Florence for "*Assegno Premiante*" for Francesca Nuti.

References

1. (a) Lolli, F. *et al. P.N.A.S.* U.S.A., **102**(29), 10273-10278, (2005); (b) Lolli, F. *et al. J. Neuroimmunology*, **167**, 131-137, (2005); (c) Papini, A.M. *Nat. Med.* **11**, 13, (2005); (d) Papini, A.M. *et al.* Granted U.S.A. Patent & PCT Application WO 03/000733 A2.
2. Mulinacci B., *et al.* In: Peptides 2004. Bridges Between Disciplines. M. Flegel, M. Fridkin, C. Gilon, J. Slaninova (Eds), Kenes International, Geneva, Switzerland, 593-594, (2005).
3. Alcaro M.C., *et al.* In: Peptides 2006. K. Rolka, P. Rekowski, J. Silberring (Eds.), Kenes International, Geneva, Switzerland (**2007**), pp 684-685.

Synthesis of Bivalent MSH Ligands and Evaluation of their Binding to hMC4R using MSH Lanthaligand

Steve M. Fernandes[1], Rajesh Sankaranarayanan[2], Liping Xu[1], Jatinder Josan[2], Josef Vagner[2], Robert Gillies[1] and Victor J. Hruby[1,2]

[1]Department of Biochemistry and Molecular Biophysics; [2]Department of Chemistry, University of Arizona, Tucson, AZ 85721, USA

Introduction

Membrane proteins especially GPCRs are interesting and important therapeutic targets since many of them serve in intracellular signaling critical for all aspects of health and disease. Here, the potential utility of designed bivalent ligands as targeting agents for cancer diagnosis and/or therapy is evaluated by determining their binding to the cell surface protein, the hMC4R receptor. Based on modeling data, the optimal linker lengths of 35±10Å is required to link the two binding pharmacophores within a bivalent construct. The heptapeptide MSH(7), agonist homo-bivalent ligands (EC_{50} 15nM) were observed to bind with 15-fold higher affinity than the corresponding monovalent ligand (1). Accordingly, we have evaluated the influence of the non-selective antagonist, SHU9119 (EC_{50} 80pM) within bivalent constructs on receptor binding using a lanthanide-based [DTPA(Eu^{3+})-NDP-α-MSH] (2) competition binding assay on HEK293/hMC4R.

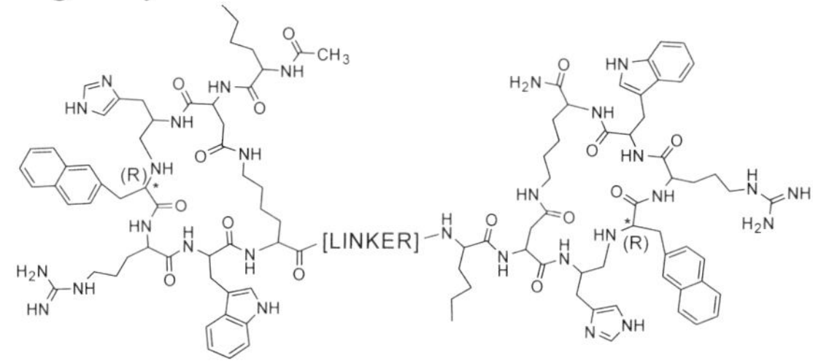

Fig. 1. Structure of a non-selective Melanocortin class4 receptor bivalent antagonist.

Results and Discussion

The MSH bivalent analogues I-VI syntheses were carried out through stepwise solid-phase peptide synthesis utilizing Fmoc-chemistry on Rink-amide Tentagel resin. Individual amino acids were activated with 3 eq. HBTU-HOBt in the presence of 6 eq. DIEA. The coupling reaction was carried out using 3 eq. of the amino acid and allowed to react for 2 hr. The Fmoc group was removed by 20% piperidine in DMF. Aloc and OAllyl side chain protecting groups for Lysine and Aspartic acid respectively, were deprotected using 3 eq. $Pd(PPh_3)_4$/ AcOH/ NMM in $CHCl_3$ (3). Within analogues I, II, III; lactam cyclization within the C-terminal SHU9119 was carried out prior to peptide chain elongation and subsequent second lactam cyclization at the N-terminal SHU9119. The linear (t_R 15 min) and cyclized product (t_R 19 min) were monitored by analytical RP-HPLC (gradient 10-90% CH_3CN over 30 min). The linker PEGO was synthesized on the solid support using 20 eq. diglycolic anhydride, 20 eq. carbonyl diimidazole and 40 eq. of tridecanediamine. After chain elongation, the N-terminal was finally acetylated using acetic anhydride/ pyridine (1:1) for 10 mins and the resin treated with TFA. The peptide was precipitated

S. Del Valle et al. (eds.), *Peptides for Youth: The Proceedings of the 20th American Peptide Symposium*, 433
DOI: 10.1007/978-0-387-73657-0_186, © Springer Science+Business Media, LLC 2009

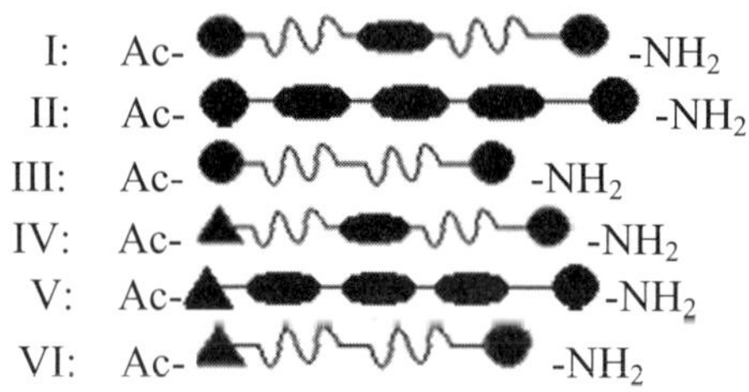

Fig. 2. Schematic bivalent sequence analogs with varied linker lengths I-VI;
SHU9119 - Nle-c[Asp-His-D-Nal(2')-Arg-Trp-Lys]; - $(Pro\text{-}Gly)_3$
MSH(7) - Ser-Nle-Glu-His-D-Phe-Arg-Trp; - *PEGO moiety*

Table 1. Physiochemical and competition binding data for the bivalent constructs.

Analogs	Molecular Weight	[a]Retention time (t_R) (min)	# atoms present in the linker	(Å) Estimated Linker Length	[c]EC_{50} (nM)
I	3187	13.6[b]	20	13-46	2.2
II	3477	13.2[b]	54	15-30	1.6
III	2726	13.9[b]	58	8-36	1.9
IV	3129	12.2[b]	20	13-46	1.3
V	3417	12.0[b]	54	15-30	2.0
VI	2665	12.5[b]	58	8-36	1.7

[a]HPLC eluents: A- 0.1% TFA/H_2O, B- 100% CH_3CN, [b]gradient of 10-40%B over 30min; [c]Determined from Dissociation Enhanced Lanthanide FluoroImmuno Assay (DELFIA) using optimized 10nM standard agonist, DTPA(Eu^{3+})-NDP-α-MSH chelate.

precipitated by cold ether and was purified by semi-preparative HPLC. Homogeneity of the bivalent ligands was confirmed by analytical RP-HPLC and MALDI-MS.

The competition binding assay using the lanthaligand was performed to evaluate the binding affinity of the bivalent analogues 1-VI to the HEK293 cells expressing hMC4R. The EC_{50} values are summarized in Table 1. The incorporation of the potent non-selective antagonist SHU9119 within the bivalent construct showed upto 200-fold increase in the binding affinity to whole cells compared to the monovalent analogue. As proof of concept, these potent agonist/ antagonist bivalent ligands can be used in targeting cell surface receptors to study internalization and the increased specificity will help achieve our goal of establishing diagnostic agents for targeted therapy.

Acknowledgments

The work was funded by grants from USPHS, NCI and the AZ Biological Research Commission.

References

1. Sankaranarayanan, R., Hruby, V.J., *et al. Bioconjugate Chemistry*, **18(4)**, 1101-1119 (2007).
2. Handl, H.L., Hruby, V.J., Gillies, R.J., *et al. Analytical Biochemistry* **343**, 299-307 (2005).
3. Albericio, F., Thieriet, N., *et al. Tetrahedron Letters* **38(41)**, 7275-7278 (1997).

Synthesis Of Organometallic Glycopeptides And Electrochemical Studies To Detect Autoantibodies In Multiple Sclerosis Patients' Sera.

Feliciana Real-Fernández[1], Amélie Chamois-Colson[2], Jérome Bayardon[2], Francesca Nuti[1], Elisa Peroni[1], Maria R. Moncelli, Rita Meunier-Prest[2], Sylvain Jugè[2] and Anna Maria Papini[1]

[1]*Laboratory of Peptide & Protein Chemistry & Biology, Polo Scientifico, University of Florence, I-50019 Sesto Fiorentino (FI), Italy;* [2]*Laboratoire de Synthese et d'Electrosynthese Organometalliques (LSEO), Université de Bourgogne, 21068, Dijon, France*

Introduction

Measurement of disease specific biomarkers, such as antibodies, is an important tool for diagnosis, follow-up, and prevention of autoimmune diseases. Identification of autoantibodies is achieved up to now using native protein antigens in immunoenzymatic assays. Unfortunately, in autoimmune diseases only very low specific antibodies are detected in serum, possibly because protein antigens used in the assays contain more than one epitope, of which only few are involved in the disease. Moreover, the use of recombinant protein antigens doesn't allow reproduction of aberrant post-translational modifications (PTMs) involved in triggering autoantibodies. These observations prompted us to set up an innovative "chemical reverse approach" to select specific synthetic peptide antigens containing the minimal epitope with the correct PTM to fishing out specific biomarkers in biological fluids [1].

As a proof-of-concept, we have recently developed CSF114(Glc), a structure-based designed glucosylated peptide as the first Multiple Sclerosis Antigenic Probe (MSAP) [2], accurately measuring high affinity autoantibodies in sera of a statistically significant patients' population [3]. The ELISA diagnostic/prognostic test MSPepKit [4], based on CSF114(Glc), has been developed to recognize specific autoantibodies in MS patients' sera and follow up disease activity [5].

The aim of our study is to evaluate the specificity and sensitivity of Cyclic Voltammetry (CV) measurements in the analysis of antibody profiles in MS patients' sera, possibly improving detection of a panel of antibodies as specific biomarkers for different forms of the disease. Development of sensitive, quantitative, and specific electrochemical devices, based on a simple, rapid, and reproducible protocol, can dramatically improve the sensitivity of the traditional immunoassays (e.g. ELISA).

Results and Discussion

In this context, our MSAP CSF114(Glc) has been properly modified in N-terminus with new organometallic amino acids as "Electrochemical Probes" to perform CV measurements in solution and/or grafting peptides on a gold electrode. A small library of organometallic glycopeptides was synthesized modifying CSF114(Glc) and the corresponding unglucosylated sequence with a series of specifically designed organometallic amino acids (Fig. 1).

We tested MS patients' sera and Normal Blood Donors' sera (NBDs) in ELISA and competitive ELISA versus the modified peptides. CSF114(Glc) modified with organometallic amino acid **1**, **2**, or **3** is always able to detect and inhibit

S. Del Valle et al. (eds.), *Peptides for Youth: The Proceedings of the 20th American Peptide Symposium*, 435
DOI: 10.1007/978-0-387-73657-0_187, © Springer Science+Business Media, LLC 2009

autoantibodies in MS patients' sera. Therefore, the autoantibody recognition is not affected by these modifications. On the other hand, all the peptides are not able to detect any antibody titre in NBDs' sera.

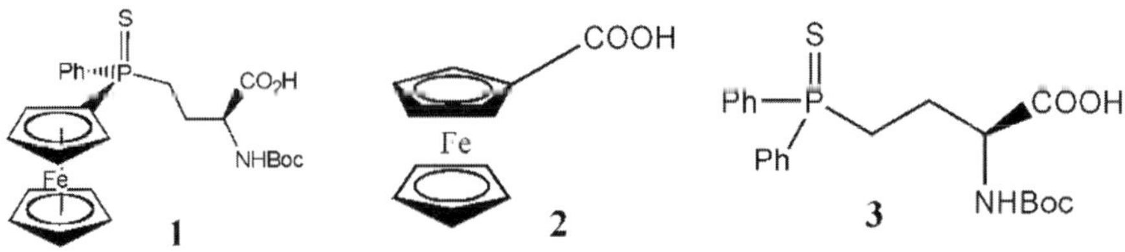

Fig. 1. Organometallic amino acids.

The glycopeptides were used on gold electrodes covered with a proper polymer monolayer of mercaptoundecanoic acid, to perform CV measurements in the presence of antibodies. An increase in intensity of the current (faradic and capacitive) and a shift to more positive potentials have been observed after addition of purified anti-CSF114(Glc) antibodies in a saline solution of the glycopeptide modified with the amino acid **2**.

The same behaviour was observed in the case of the glycopeptide modified with the amino acid **1** (including characteristics of both amino acids **2** and **3**). Moreover, the presence of thiophosphine function in the N-terminus allows glycopeptide immobilization on the gold electrode. CV measurements grafting the glycopeptide antigen directly on the gold electrode showed similar electrochemical response both in solution and on the electrode surface.

The new peptides containing unnatural amino acid **1** or **2** were able to detect specific anti-CSF114(Glc) antibodies in CV. Moreover, thanks to the presence of thiophosphine, CSF114(Glc) modified with **1** was able to build simple monolayers on gold surfaces. The possibility of grafting synthetic probes directly on the electrode surface will enable to obtain innovative strategies to develop new and more reliable techniques for antibody detection and quantitative determination.

Acknowledgments

PAI Galilée N°11686QA (2006), FIRB Internalization 2005-2008 RBIN04TWKN, MIUR (Italy), Ente Cassa Risparmio di Firenze and 20th APS travel grant committee are gratefully acknowledged.

References

1. Lolli F, Rovero P, Chelli M, Papini AM, *Expert Rev. Neurotherapeutics,* **6** (5), 781-794, 2006
2. Lolli F, Mulinacci B, Carotenuto A, Bonetti B, Sabatino G, Mazzanti B, D'Ursi AM, Novellino E, Pazzagli M, Lovato L, Alcaro MC, Peroni E, Pozo-Carrero MC, Nuti F, Battistini L, Borsellino G, Chelli M, Rovero P, Papini AM. *Proc Natl Acad Sci U S A.*, **102** (29), 10273-8, 2005.
3. Carotenuto A., D'Ursi A. M., Mulinacci B., Paolini I., Lolli F., Papini, A. M., Novellino E., Rovero P., *A. J. Med. Chem.*, **49**, 5072-5079, 2006.
4. Papini, A.M. et al. *Nat. Med.*, **11**, 13, 2005.
5. Papini, A.M., Rovero P., Chelli M., Lolli F., Granted U.S.A. Patent & PCT Application WO 03/000733 A2.

Tumor Imaging With Tetrabranched Neurotensin

Monica Fabbrini, Chiara Falciani, Barbara Lelli, Alessandra Cappelli, Alessandro Pini, Luisa Lozzi and Luisa Bracci

Department of Molecular Biology, University of Siena, ITALY

Introduction

The finding that neurotensin (NT) receptors are expressed on the plasma membrane of a number of human tumors opened the way for peptide receptor targeting using NT sequence. The main drawback in the use of NT, or any other endogenous peptide, as a drug, is its extremely short half-life due to very rapid cleavage by different peptidases.

We previously demonstrated that NT, when synthesized as a dendrimer, retains biological activity and becomes resistant to proteases [1-2].

Here we describe the use of the neurotensin functional fragment NT(8-13) in a tetra-branched form linked to different units for tumor imaging.

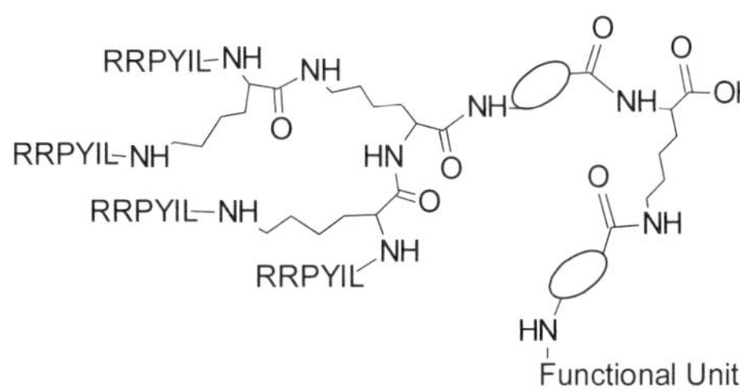

Fig. 1. Structure of tetrabranched functionalized neurotensin derivatives.

Tumor specific therapy by peptide-receptor targeting depends on the presence and concentration of target receptors on neoplastic tissues. Specific diagnostic tools to be used on biopsies, would be extremely useful to predict the efficacy of the treatment.

Branched peptides conjugated with different chromophore can be tested on human specimens from tumors known to express the target receptors. Our goal is to produce peptide molecules that can be used for two purposes: for a specific diagnostic test (i.e, for testing the presence of the specific target in primary tumor or methastasis), and for specific therapy or in vivo imaging by delivering radio or chemotherapeutic moieties to the same tumor cells. Branched peptide molecules can be used for both purposes with no modification of the tumor targeting moiety, but only by addition of different effector units to the conserved branched core.

Results and Discussion

Tetrabranched NT4-Biotin (NT4-Bio) was synthesized and tested for receptor binding on the human colon adenocarcinoma cell line HT29. Once verified that the conjugation of the branched peptide did not reduce its ability to bind the receptor, we tested our peptide on samples from surgical resection of human colon adenocarcinoma, in comparison with healthy tissues from the same patients.

Samples from 15 individuals were submitted to histological investigation and the tumors were classified and staged. Sections (15μm) from tumor and healthy tissues were incubated with NT4-Bio (1 μg/ ml) for 1h at 37°C and after washing incubated with avidine-FITC (0.5 μg/ml). Images were obtained with a confocal microscope.

S. Del Valle et al. (eds.), *Peptides for Youth: The Proceedings of the 20th American Peptide Symposium*, 437
DOI: 10.1007/978-0-387-73657-0_188, © Springer Science+Business Media, LLC 2009

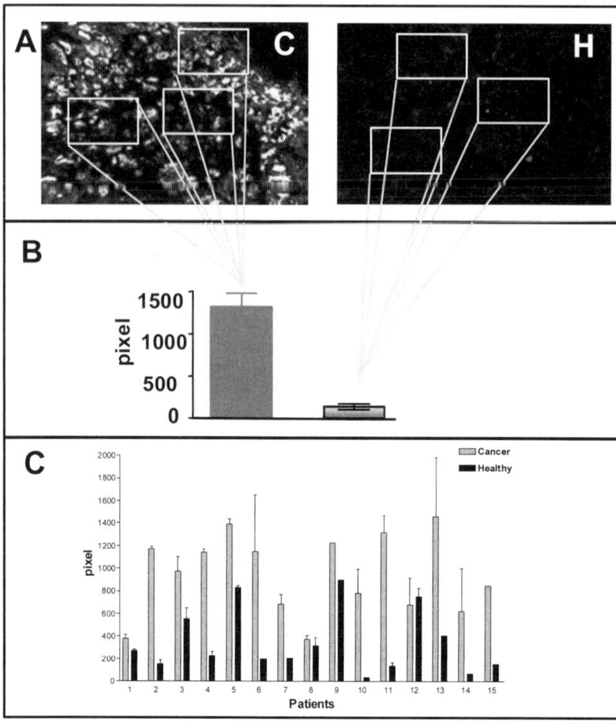

Fig. 2. NT4-biotin binding to human colon biopsies. ***A)*** *Confocal microscopy images of colon carcinoma (C) and healthy colon (H) sections. Frozen sections of human biopsies were fixed with 4% formaldehyde, saturated with FCS and incubated with NT4-biotin (1 µg/ml in TBS-3% BSA) for 30' at 37°C. After washing the sections were incubated with avidin-FITC (0.5 µg/ml in TBS-BSA) for 30' at 37°C, washed and mounted on a glass slide.* ***B)*** *Quantification of the fluorescence intensity: green pixels are counted over three different significant areas of tumor (C) or normal mucosa (H).* ***C)*** *Chart of the quantified fluorescence signals of cancer and healthy sections coming from 15 different patients that underwent tumor resection.*

Confocal images show a general good staining of tumor cells by the tetrabranched NT peptide in most tumor samples and a much lower staining in healthy tissue from the same patient.

Acknowledgments

This work was supported by grants from AIRC.

References

1. Bracci, L., Falciani, C., Lelli, B., Lozzi, L., Runci, Y., Pini, A., De Montis, M.G., Tagliamonte, A., Neri, P. *J Biol Chem* **278**, 46590-46595 (2003).
2. Falciani C, Lozzi L, Pini A, Corti F, Fabbrini M, Bernini A, Lelli B, Niccolai N, Bracci L. *Chem Biol Drug Des.* **69**, 216-221 (2007).

A new and selective radiolabeled $\alpha_V\beta_3$ peptide antagonist as tracer in tumor diagnosis

Stefania De Luca[1], Annarita Del Gatto[1], Menotti Ruvo[1], Maria Rosaria Panico[2], Antonella Zannetti[2], Silvana Del Vecchio[2], Marco Salvatore[2], Carlo Pedone[1], Michele Saviano[1] and Laura Zaccaro[1]

[1] *Institute of Biostructures and Bioimaging, CNR, CIRPEB, Via Mezzocannone 16, 80134 Naples, Italy;*
[2] *Institute of Biostructures and Bioimaging, CNR, via S. Pansini 5, 80121 Naples, Italy.*

Introduction

Angiogenesis, the process whereby new capillaries are formed by outgrowth from existing microvessels, is required for tumor growth and metastasis [1,2].

The transmembrane cell-surface receptor $\alpha_v\beta_3$ has recently received increasingly attention, because of the critical role in tumor associated angiogenesis and metastasis formation. The restricted expression of integrin $\alpha_V\beta_3$ during tumor growth, invasion, and metastasis presents an interesting target for both detection and treatment of solid tumors [3]. Targeting $\alpha_V\beta_3$ with radiolabelled ligands may provide information about the receptor status and enable the planning and the monitoring of therapeutic approaches.

Recently we developed a novel $\alpha_V\beta_3$ antagonist that showed a high selectivity for the receptor [4]. Adhesion assays, competitive binding assays and cross-linking experiments performed in human erythroleukemia K562 cells, stably cotransfected with cDNA of α_v or α_{IIb} and β_3, demonstrated the high selectivity for $\alpha_v\beta_3$ integrin [4].

Starting from these evidences RGDechi was covalently bound to the chelating agent DTPA (Diethylene Triamine Pentaacetic Acid) able to give stable complexes of radionuclides such as ^{111}In and ^{90}Y. The final goal has been to obtain a radiolabelled compound to be used in nuclear medicine as diagnostic and therapeutic agent. In particular DTPA-RGDechi has been labelled with ^{111}In and used in SPECT (Single Photon Emission Computed Tomography) for diagnostic purpose.

Moreover, the ligand RGDechi has been labelled with ^{18}F for micro-PET in order to evaluate the use of this peptide for *in vivo* imaging.

Results and Discussion

The synthesis of RGDechi was performed by solid phase method using Fmoc chemistry and standard protocol. The conjugation of RGDechi with DTPA for SPECT imaging was carried out on solid phase using orthogonal aminoacidic side chain protecting group.

RGDechi with ^{18}F for micro-PET imaging was obtained by one-step procedure in solution using ^{18}F-N-succinimidyl 4-fluorobenzoate to label the peptide. Both RGDechi analogs were purified by RP-HPLC and identified by LC/MS.

To develop xenografts in nude mice, U87MG human glioblastoma cells and A431 human epidermoid cells endogenously expressing high levels $\alpha_v\beta_3$ and $\alpha_v\beta_5$ respectively were used. U87MG tumor xenografts showed a high tumor uptake of ^{111}In-labeled and ^{18}F-labeled RGDechi assessed by SPECT and microPET, respectively. No tumor uptake of radiolabeled RGDechi could be observed A431 tumor xenografts overexpressing $\alpha_v\beta_5$.

S. Del Valle et al. (eds.), *Peptides for Youth: The Proceedings of the 20th American Peptide Symposium*, 439
DOI: 10.1007/978-0-387-73657-0_189, © Springer Science+Business Media, LLC 2009

In summary, our findings indicate that RGDechi represents a new tracer in tumor imaging that may offer insight into molecular processes during tumor development and metastasis and be a helpful tool in the planning and the monitoring of therapeutic approaches.

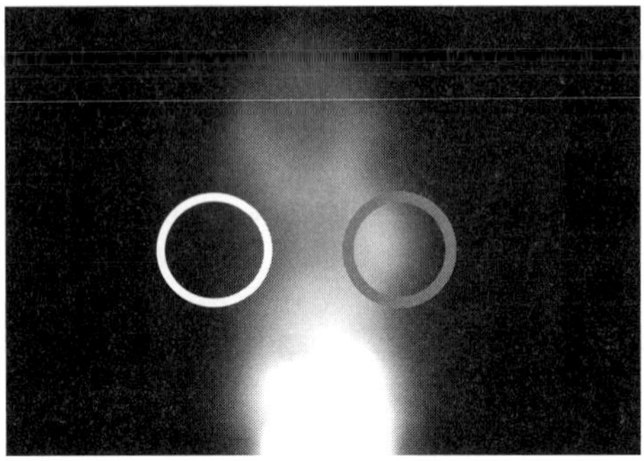

Fig. 1. Representative microPET images of nude mice bearing U87MG tumor on right shoulder and A431 tumor on left shoulder with ^{18}F-RGDechi.

Acknowledgments

This study was supported by the "Centro Regionale di Competenza in Diagnostica e Farmaceutica Molecolari" of the Regione Campania, by Italian C.N.R. Dr Giuseppe Perretta is also acknowledged for technical assistance.

References

1. Hood J.D. and Cheresh D.A. *Nat. Rev. Cancer* **2**, 91-100 (2002).
2. Ruoslathi E. *Nat. Rev. Cancer* **2**, 83-90 (2002).
3. Walton H.L., Corjay M.H., Mohamed S.N., Mousa S.A., Santomenna L.D. and Reilly T.M. *J Cell Biochem.* **78**, 674-680 (2000).
4. Del Gatto A., Zaccaro L., Grieco P., Novellino E., Zannetti A., Del Vecchio S., Iommelli, F., Salvatore M. and Saviano, M. *J. Med. Chem.* **49**, 3416-3420 (2006).

Ribose Building Block For The Synthesis Of Glycopeptides For Fishing Out Antibodies In Autoimmune Diseases

Mª Angeles Bonache[1,2], Alexandra Le Chevalier[1,2], Francesca Nuti[1,2], Elisa Peroni[1,2], Mario Chelli[1,4,5], Paolo Rovero[1,3,4] and Anna M. Papini[1,2,4]

[1]*Laboratory of Peptide & Protein Chemistry & Biology;* [2]*Dept. of Organic Chemistry "Ugo Schiff", University of Florence;* [3]*Dept. of Pharmaceutical Sciences, University of Florence;* [4]*Toscana Biomarkers S.r.l., Siena, Italy;* [5]*CNR-ICCOM, University of Florence.*

Introduction

It is well known that ribose (Rib) is produced *in vivo* from glucose and it plays a fundamental role in cellular energy metabolism and cellular signaling. Mono (ADP-ribosyl)transferases transfer single ADP-ribose moieties to acceptor proteins. Following our hypothesis that an aberrant glucosylation triggers autoimmunity in Multiple Sclerosis (MS) (because of a bacterial or viral infection?), we can assume that a ribose moiety may be also transferred to protein components of organs or tissue target of autoimmune diseases.

For the first time, by a structure-based design, we selected CSF114(Glc) as a synthetic glycopeptide, containing a β-glucosyl moiety linked to an Asparagine residue at position 7. CSF114(Glc) can be used as a specific Multiple Sclerosis Antigenic Probe (MSAP). We demonstrated that the glucosylasparagine Asn(Glc) is fundamental for autoantibody recognition in a form of MS [1].

In previous studies, we synthesized and tested in Enzyme-Linked Inmuno Sorbent Assay (ELISA) a large CSF114-type glycopeptide library based on N-glycosyl amino acid diversity (Glc, Man, GlcβGlc, Gal, GlcNAc, etc) and up to now, we have never investigated the role of ribose [2]. Considering that different families of autoantibodies to proteins aberrantly glycosylated with different sugar moieties could correlate with disease activity, we synthesized Fmoc-Asn(Rib)-OH. This building block will be introduced in the CSF114 β-turn scaffold to test this new ribosylated peptide as synthetic antigen contributing to the library of glycopeptides for fishing out families of autoantibodies specific for different forms of Multiple Sclerosis or other autoimmune diseases.

Results and Discussion

We synthesized Fmoc-Asn[Rib(OAc3)]-OH (**5**) from β-D-ribofuranose **1**.

The amine **3** was obtained by azidation of β-ribose **1** using Me_3SiN_3, catalysed by $SnCl_4$, followed by reduction of β-azide **2** with $Pd(OH)_2$.

Coupling of the amine **3** with Fmoc-Asp-O*t*Bu, using TBCR/BF_4 [4-(4,6-Dimethoxy-[1,3,5]triazin-2-yl)-4-methylmorpholinium tetrafluoroborate] [3] allows to obtain the protected building block **4**. The deprotection of the carboxyl function by acidic treatment afforded the new deprotected building block **5** (Scheme 1).

Unfortunately, the yield of the reduction is very low (6%), probably due to the ring opening of ribose.

S. Del Valle et al. (eds.), *Peptides for Youth: The Proceedings of the 20th American Peptide Symposium*, 441
DOI: 10.1007/978-0-387-73657-0_190, © Springer Science+Business Media, LLC 2009

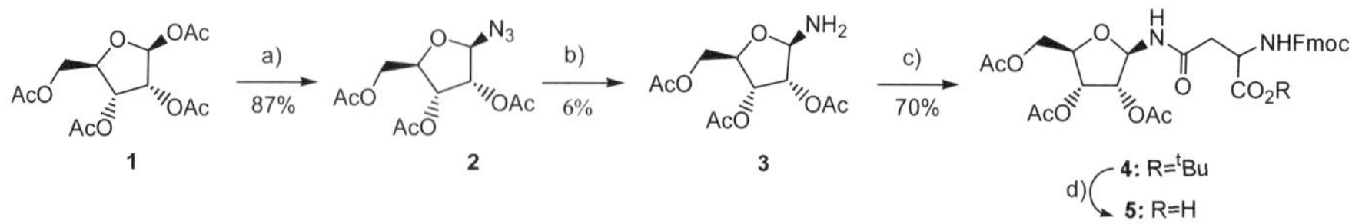

a) Me_3SiN_3, $SnCl_4$, CH_2Cl_2. b) H_2, $Pd(OH)_2$, MeOH. c) Fmoc-Asp-O*t*Bu, TBCR/BF_4, CH_3CN. d) TFA, CH_2Cl_2.

Scheme 1. Synthetic pathway to Fmoc-Asn[Rib(OAc3)]-OH (5).

In order to improve yield of the reduction reaction, we have investigated an alternative synthetic strategy. The new building block was synthetized from ribose derivative **7** obtained by condensation of D-ribose with acetone, followed by acetylation (Scheme 2) [4].

The azide reaction was catalysed by $SnCl_4$ in presence of Me_3SiN_3 to afford a mixture of anomers **8α/8β** (1:1.6), which were easily separated by FCC (Scheme 2).

The reduction of the azide **8β** in presence of Pd black or triphenylphosphine as reagents afforded the amine **9**. Using Pd black as catalyst, we obtained the amine **9** with better yield (89%) (Scheme 2).

Fmoc-Asn[Rib(*O*-2,3-Isopropiliden,5-Ac)]-OH **11** was obtained directly by coupling the amine **9** to Fmoc-Asp-OAll using TBCR/BF4 as coupling reagent, followed by the removal of the OAll group with $Pd(PPh_3)_4$ and triphenylsilane with moderate yield.

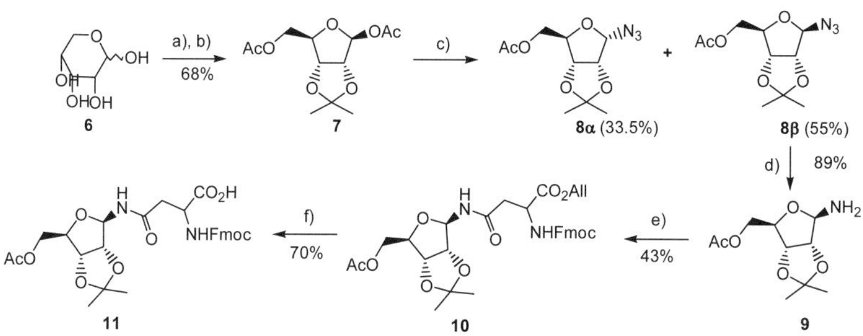

a)Acetone, $CuSO_4$, H_2SO_4. b) Ac_2O, Pyridina. c) Me_3SiN_3, $SnCl_4$, CH_2Cl_2. d) H_2, Pd black, EtOH. e) Fmoc-Asp-OAll, TBCR/BF_4, CH_3CN. f) $Pd(PPh_3)_4$, Ph_3SiH, CH_2Cl_2.

Scheme 2. Synthetic pathway to Fmoc-Asn[Rib(*O*-2,3-Isopropiliden,5-Ac)]-OH (11).

Acknowledgments

This work was supported by Ente Cassa Risparmio di Firenze and FIRB 2005 RBINOYTWKN,MIUR (Italy). M.A.B. thanks Alfonso Martín Escudero Foundation (Spain) and the 20th APS travel grant committee for the financial support.

References

1. (a) Lolli, F. *et al. P.N.A.S. U.S.A.* **102** (29), 10273-10278 (2005), (b) F. Lolli *et al. J. Neuroimmunology* **167**, 131-137 (2005), (c) A.M. Papini, P. Rovero, M. Chelli, F. Lolli. Applicant: University of Florence, Italy. PCT International application (2003) WO 03000733 A2. Brevetto italiano n. 0001327122 (27/04/2005). Granted USA Patent (02/08/2005).
2. Nuti, F. *et al. Bioorg. Med. Chem.* **15**, 3965-3973 (2007).
3. Kaminski, Z.J. *et al. J. Am. Chem. Soc.,* **127**, 16912-16920 (2005). Z. Kaminski, A.M. Papini, B. Kolesinska, J. Kolesinska, K. Jastrzabek, G. Sabatino, R. Bianchini. PCT/EP2005/055793 (2005).
4. Cruickshank, *et al. Synthesis* **3**, 199-200(1983).

Design, Synthesis and Pharmacological Characterization of Fluorescein-Derived Endothelin Antagonists Bq-123 And Bq-788

Jean-Philippe Brosseau[1], **Witold A. Neugebauer**[2], **Fernand Gobeil Jr**[2] **and Pedro D'Orléans-Juste**[2]

[1]*Laboratoire de Génomique Fonctionnelle de l'Université de Sherbrooke, Centre de développement des biotechnologies (CDB) de Sherbrooke, Sherbrooke, J1E 4K8, Canada;* [2]*Département de Pharmacologie, Université de Sherbrooke, Sherbrooke, J1H 5N4, Canada*

Introduction

Endothelin-1 (ET-1) is a 21 amino acid peptide involved in the cardiovascular system. ET-1 binds to two G-protein coupled receptors namely, endothelin receptor A (ET_A) and endothelin receptor B (ET_B). Our goal is to develop fluorescent ET_A and ET_B analogues towards the study of the ET system in situ (*in vivo* imaging). To date, there is no published data on fluorescent and selective probe for those receptors. We design a strategy to generate FITC-derivative (fluorescein isothiocyanate-derivative) of the known BQ-123 *cyclo*(Pro-D-Val-Leu-D-Trp-D-Asp) and BQ-788 (N-(cis-2,6-dimethylpiperidinocarbonyl-γ-methylleucyl-D-1-(methoxycarbonyl)-tryptophanyl-D-norleucine), two potent and selective antagonists for ET_A and ET_B, respectively. We also provide an *in vitro* pharmacological profile for each new compound.

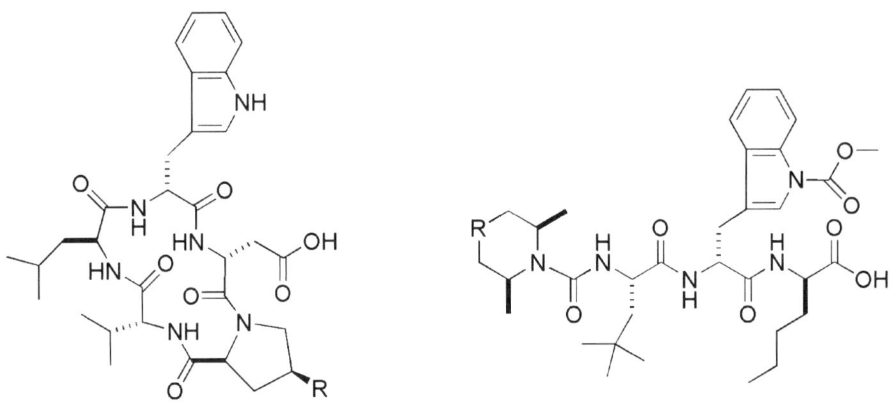

BQ-123; R = H | BQ-788; R = CH_2

BDN-123N; R = NH_2 | BDN-788N; R = NH

BDN-123NF; R = fluoresceinthiourea | BDN-788NF; R = fluoresceinthiourea

Fig. 1. Structures of BQ-123, BQ-788 and their novel analogs.

S. Del Valle et al. (eds.), *Peptides for Youth: The Proceedings of the 20th American Peptide Symposium*, 443
DOI: 10.1007/978-0-387-73657-0_191, © Springer Science+Business Media, LLC 2009

Table 1. Apparent affinity of agonist (pD_2) and antagonist (pA_2) in rabbit carotid arteries (rca) and rabbit pulmonary arteries (rpa)

	rca		rpa		
Agonist	pD_2	*Max Response (g)*	pD_2	*Max Response (g)*	
ET-1	8.98	2.62	-	-	
IRL 1620	-	-	8.65	1.04	
Antagonist	pA_2		pA_2		(r^a)
BQ-123	6.77		*n.d.*		0.99
BDN-123N	inactive		n.d.		-
BDN-123NF	6.42		inactive		0.99
BQ-788 [b]	inactive		8.24		0.97
BDN-788N	inactive		7.07		0.99
BDN-788NF	inactive		6.52		0.98

[a]Correlation coefficient; [b]Values from [2]

Results and Discussion

Since, there is no obvious anchoring site for labeling in neither BQ-123 nor BQ-788 (except carboxylic acid which is critical for activity), we designed new molecules (see figure 1) with suitable amino groups and synthesized them according to previously described strategies [1,2]. Subsequently, those molecules were individually labeled with FITC (3eq) in presence of pyridine (1eq) and DIPEA (cat) in MeCN/H_2O (50/50) and were purified by preparative TLC with DCM/MeOH/ AcOH (16/2/1) to yield BDN-123NF or BDN-788NF.

New derivatives were tested on the rabbit carotid (ET_A-rich preparation) and pulmonary arteries (ET_B-rich preparation) [3]. Results from Schild plot regression analysis show that the BDN-123NF fully retains potency and selectivity of the parent compound (BQ-123). In contrast, the BDN-788NF fully retains selectivity albeit it is about 50 fold less potent when compared to the parent compound (BQ-788) (see table 1). All novel compounds are specific to ET receptors as they were found inactive against the vasoconstrictive responses to Angiotensin II. Our results suggest that our proposed strategy has allowed the development of ET receptor-specific probes towards a better identification of endothelin receptors in various tissues.

Acknowledgments

We acknowledge Guillaume Desmarais for providing specificity characterization and the Canadian Institutes for Health Research for financial support.

References

1. Neugebauer, W.A., Gratton, J.P., Ihara, M., Bkaily, G. and D'Orléans-Juste, P. *Peptides 1996, Proceedings of the 24th European Peptide Symposium,* 677-678 (1996).
2. Brosseau, J.P., D'Orléans-Juste, P. and Neugebauer, W.A. *Peptides* **26**, 1441-1453 (2005).
3. Calo, G., Gratton, J.P., Telemaque, S., D'Orléans-Juste, P. and Regoli, D. *Mol. Cell. Biochem.* **154**, 31-37 (1996).

Peptide Antagonists of the PED-hPLD1 Binding

F. Viparelli[1], N. Doti[1,3], S.M. Monti[1], D. Marasco[1], N. Dathan[1], C. Pedone[1], C. Miele[2], P. Formisano[2], F. Beguinot[2], M. Ruvo[1]

[1]*Istituto di Biostrutture e Bioimmagini - CNR, via Mezzocannone 16, Napoli, Italy;* [2]*DBPCM, Università degli Studi di Napoli Federico II, Via S. Pansini 5, 80131, Napoli, Italy;* [3]*Dipartimento di Biochimica e Biofisica, Seconda Università di Napoli, via SM Costantinopoli, 16, 80138, Napoli, Italy*

Introduction

PED (Phosphoprotein Enriched in Diabetes) is an ubiquitously expressed 15 kDa cytosolic protein with recognized multiple functions. It has been demonstrated that PED has broad anti-apoptotic properties[1,2] and, by altering insulin secretion, has a proven implication in diabetes[3]. In cultured cells and in transgenic mice, PED overexpression induces insulin-resistance and impairs glucose tolerance. Recent evidence indicates that increased interaction of PED with Phospholipase D1 (hPLD1) is a key event leading to these abnormalities *in vivo* and that this interaction is mediated by a C-terminal domain of PLD1 comprising residues 712-1070, D4[4]. To further study this interaction, we prepared the recombinant proteins/sub-domains and after purification and preliminary characterization by CD, we measured the K_D by SPR. Furthermore, following an approach of protein fragmentation and peptide fractionation we have been able to identify regions on PED that are involved in the interaction with the phospholipase sub-domain.

Results and Discussion

Recombinant expression of D4 yielded low amounts of the protein fused to Trx-His6 (hereafter, D4-Trx-His6), with a strong tendency to degradation; however, the amount was enough for our purposes. Two different isoforms of D4-Trx-His6 were purified: a monomeric (D4M) and a dimeric one (D4D). Notably, dimerization was observed also for purified PED. Interaction between PED and D4-Trx-His6 was assessed by SPR using a BIACORE 3000 system. Recombinant PED was efficiently immobilized on the sensor surface, then increasing concentration of D4-Trx-His6 were injected on it. Assays using solutions of Trx-His6 showed no binding to immobilized PED. Fractions containing the dimeric protein did not exhibit any binding to PED. On the other hand, fractions containing the D4M exhibited dose-dependent binding curves with a K_D of $2\pm1*10^{-7}$ M. To delineate PED sub-domains involved in D4 binding, we extensively hydrolyzed PED with trypsin; protein fragments were then fractionated by RP-HPLC in 6 separate fractions and subsequently lyophilized and dissolved in H_2O. Peptides within the fractions were identified by LC-MS/MS. Then, in a first approach (INCUBATION), peptide competitors were pre-incubated with soluble D4M and injected on the chip; results were evaluated by observing the RUmax for a given competitor compared to RUmax determined without competitors. In a second approach (COINJECTION), peptides were instead injected at the end of the association phase, evaluating the capability of increasing the K_{off} by competing with the soluble ligand. In both cases, PED(1-24) was the strongest competitor for PED-D4 interaction. We then prepared a panel of synthetic peptides corresponding to the most active tryptic fragments and to several PED secondary structure elements[5] (Table I). Synthetic peptides were then used in the SPR-based competition assay: PED(1-24) reduced the signal up to 50%, but also

S. Del Valle et al. (eds.), *Peptides for Youth: The Proceedings of the 20th American Peptide Symposium*, 445
DOI: 10.1007/978-0-387-73657-0_192, © Springer Science+Business Media, LLC 2009

other peptides, mostly PED(1-15), did. The capability of all synthetic peptides to associate to immobilized PED was also evaluated by SPR. Remarkably, peptide 71-92 was able to bind PED, suggesting that this region could be involved in PED dimerization. To further clarify these data, we set up a new competition experiment based on an ELISA-like assay. Again, the synthetic PED(1-24) consistently reduced the binding, while neither the shorter PED(1-15), previously identified as a strong competitor, or other peptides were as effective as PED(1-24), except for peptide PED(71-92), that surprisingly blocked PED-D4 association at an extent comparable to PED(1-24). Peptides PED(1-24) and PED(71-92) were finally tested in a dose-dependent competition assay, determining an IC_{50} of about 3 μM for both molecules.

Table 1: Bolded peptides are those derived from tryptic digestion.

Peptide	Secondary structure
^{1}MAEYGTLLQDLTNNITLEDLEQLK^{24}	α1+part of α2
^{1}MAEYGTLLQDLTNNI^{15}	α1
^{16}TLEDLEQLKSACKED^{30}	α2
^{72}RPDLLTMVVDYR^{83}	part of α6
^{71}RRPDLLTMVVDYRTRVLKISEE^{92}	α6
^{36}SEEITTGSAWFSFLESHNKLDK^{54}	part of α3+α4
^{33}SEKSEEITTGSAWFSFLESHNKLD^{56}	α3+α4
^{40}TTGSAWFSFLESHNKLD^{56}	α4
^{53}NKLDKDNLSYIEHIFEISRRPDLLT^{77}	α5+part of α6
^{79}VVDYRTRVLKISEEDELDTKLTRIPSAKKYKDII^{112}	part of α6
^{114}QPSEEEIIK^{122}	unstructured

In conclusion, we identified the segment 1-24 of PED (containing the first helix and few residues of the second helix) as a region strongly involved in hPLD1 recognition. A second region encompassing residues 71-92 (corresponding to PED sixth helix) also appears involved in the interaction with the phospholipase, but it also has a role in PED dimerization. Remarkably, α1 and α6 are exposed on the same side of PED and form a large flat surface. Regulation of glucose transport by PED-hPLD1 system, in the end, might be due to several equilibria involving PED dimerization and PED-hPLD1 interaction. In fact, preliminary experiments on PED overexpressing cells treated with the synthetic peptide PED(1-24), show that the peptide blocks PED binding to hPLD1 and that insulin stimulated glucose transport can be reverted to basal level.

Acknowledgments

The work was funded by the italian Ministry of University and Research (MUR), legge 297. Financial support also from project FIRB RBNE03PX83_005 is acknowledged.

References

1. Trencia A., Perfetti A., Cassese A., Vigliotta G., Miele C., Oriente F., Santopietro S., Giacco F., Condorelli G., Formisano P., Beguinot F. *Mol. Cell. Biol.* **45**, 11-21 (2003).
2. Arajou E., Danziger N., Cordier J., Glowinski J., Chneiweiss H. *J. Biol. Chem.* **268**, 5911-5920 (1993).
3. Vigliotta G., Miele C., Santopietro S., Portella G., Perfetti A., Maitan M.A., Cassese A., Oriente F., Trencia A., Fiory F., Romano C., Tiveron C., Tatangelo L., Troncone G., Formisano P., Beguinot F. *Mol Cell Biol*, **24**, 5005-15 (2004).
4. Zhang Y., Redina O., Altshuller Y.M., Yamazaki M., Ramos J., Chneiweiss H., Kanaho Y., Frohman M.A. *J. Biol. Chem.* **275**, 35224-32 (2000).
5. Hill J.M., Vaidyanathan H., Ramos J.W., Ginsberg M.H., Wemer M.H. *EMBO J.* **21**, 6494-504 (2002).

In vivo Imaging Using Tissue Specific Near Infrared fluorescent Peptide Conjugate, c[RGDyK(HiLyte Fluor™ 750)]

Jose I. Rey[2], Haihong Huang[1], Ling Sheng[1], Xiao Fen Zhong[1], Rich Meyer[1], Xiaohe Tong[1], Francis Moussy[2] and Anita Hong[1]

[1]AnaSpec, Inc. 2149 O'Toole Ave. San Jose, CA 95131 and [2]Univ. of South Florida, Tampa, FL 33606, USA

Introduction

Extracellular matrix proteins that contain the Arg-Gly-Asp (RGD) sequence, and integrin receptors which bind this sequence, constitute a major recognition system for cell migration and adhesion processes. In fibronectins and other proteins, the RGD binding sequence is found at the apex of a loop; such conformation has been found to allow for high affinity selectivity to integrin receptors. Cyclic peptides have been shown to be more stable than linear peptides; in the case of RGD cyclic peptide c(RGDyK), its structure also confers increased affinity and selectivity for integrin $\alpha v\beta_3$ both in cell culture and in living subjects. We report here an *in vivo* testing of this peptide labeled with a proprietary near infrared fluorescent dye, HiLyte Fluor™ 750-labeled RGD peptide, c[RGDyK(HiLyte Fluor™ 750)], with excitation and emission wavelengths at 750 and 780 nm. We found that for our animal model, this conjugate binds specifically to some tissues in organs that are known to be rich in integrin $\alpha v\beta_3$.

Results and Discussion

Preparation of c[RGDyK(HiLyte Fluor™ 750)]: Cyclic RGD peptide c(RGDyK) (cat# 61183, AnaSpec Inc., San Jose, CA) dissolved in $NaHCO_3$ buffer (pH = 8.5) was mixed with HiLyte Fluor™ 750 OSu (cat# 81266, AnaSpec Inc., San Jose, CA) in DMF. The solution was stirred in the dark at r.t. for 1 hour. The conjugate mixture was analyzed and purified by RP-HPLC using 0.1% TFA in water (solvent A) and 0.1%. TFA in Acetonitrile (solvent B). Pure conjugate was confirmed by MS.

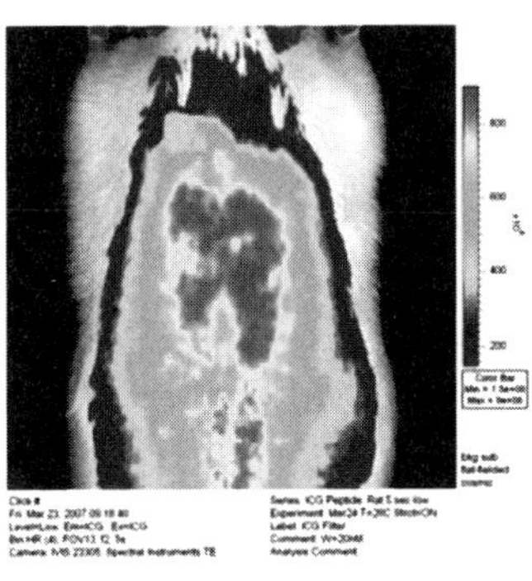

Fig. 1. In vivo fluorescent imaging of rat injected with 20 nmol c[R-GDyK(HiLyte FluorTM 750)] conjugate 3 h post injection.

In vivo Imaging: 20 nmol of the c[RGDyK(HiLyte Fluor™ 750)] was diluted in 200 μL saline solution. The saline solution was injected intravenously (IV) into a Sprague-Dawley (Harlan, Indianapolis, Indiana) inbred rat in two separate instances (experiments). A control solution of 20 nmol of HiLyte Fluor™ 750 Acid (cat# 81265, AnaSpec Inc., San Jose, CA) was injected in the Sprague-Dawley rat in two separate instances (controls). *In vivo* imaging was made using a Xenogen IVIS® Imaging System 200 (Figures 1 and 2); the animal was imaged at 0, 3, 6, and 24 hours post injection using a Indocyanine Green (ICG) Filter set (excitation 710-760nm, emission 810-875 nm). Organs were dissected and imaged 3 hours after injection of conjugate, and imaged (Figure 3).

S. Del Valle et al. (eds.), *Peptides for Youth: The Proceedings of the 20th American Peptide Symposium*, 447
DOI: 10.1007/978-0-387-73657-0_193, © Springer Science+Business Media, LLC 2009

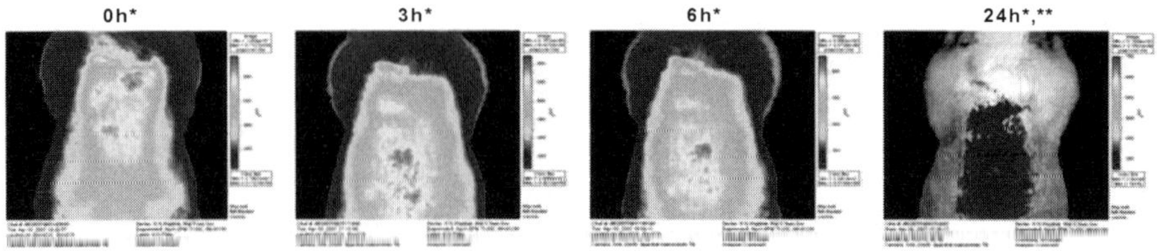

*Fig. 2. Progression of conjugate clearance over a 24 hour period. *Different scales for Photon Flux (p/s). **48h image not included here, very similar to 24h image and to background.*

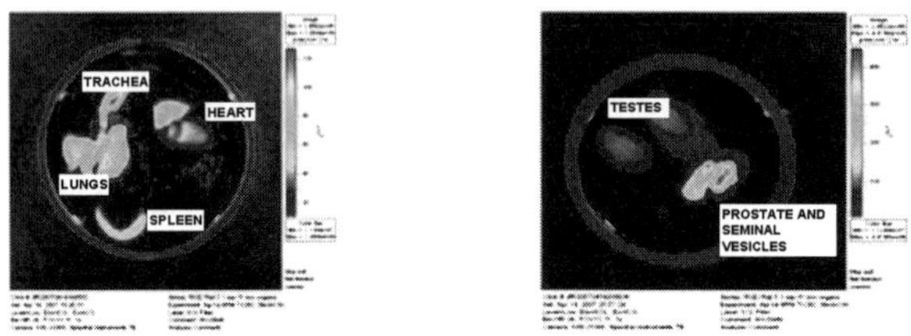

Fig. 3. Dissected organs imaged 3 hours post conjugate injection.

Conjugate Compared to Dye Only: The RGD-HiLyte750 conjugate shows increased maximum fluorescence at 0, 3, and 6 hours when compared to the controls (HiLyte Fluor™ 750 dye only) confirming preferential accumulation of the conjugate in certain organs (Fig. 4). The RGD-HiLyte Fluor™ 750 conjugate shows increased fluorescence in the lung, and prostate and seminal vesicles tissues. High fluorescence in the liver and kidney is expected as part of the conjugate clearance process in the animal (Fig. 5).

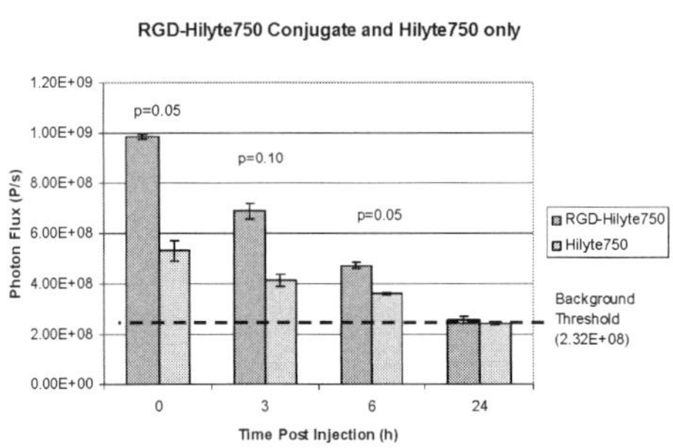

Fig. 4. Comparison of fluorescence in the RGD conjugate injections versus controls

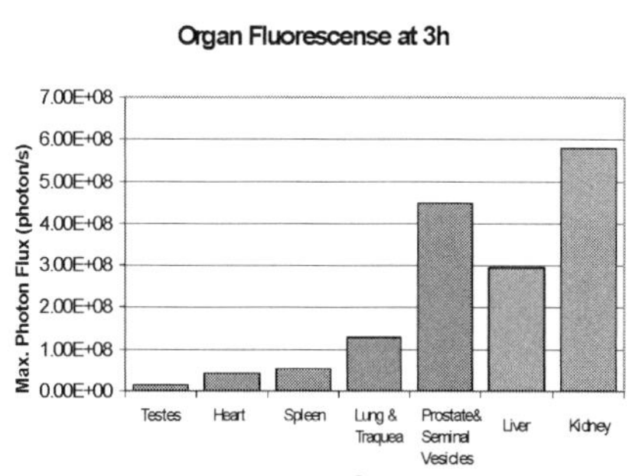

Fig. 5. Organ maximum fluorescence.

Acknowledgments

Financial Support provided by the NIH-NBIB grant #1R01 EB001640.

References

1. Chen, X., P. Conti, and R. Moats, *In vivo Near-Infrared Fluorescence Imaging of Integrin $\alpha v\beta_3$ in Brain Tumor Xenografts.* Cancer Res, 2004. **64**, 8009-8014.

Photoinduced Intramolecular Covalent Bond Formation in Structurally Rigid -Bpa-(spacer)-Met Hexapeptides

Alessandro Moretto[1], Simona Oancea[1], Fernando Formaggio[1], Claudio Toniolo[1], Laurence A. Huck[2], Dino Mangion[2] and William J. Leigh[2]

[1]*ICB, Padova Unit, CNR, Department of Chemistry, University of Padova, 35131 Padova, Italy;* [2]*Department of Chemistry, McMaster University, Hamilton, ON, Canada L89 4M1*

Introduction

The *para*-benzoylphenylalanyl (Bpa) residue [1] is extensively used as a photoaffinity label, particularly for studies of *intermolecular* (peptide)ligand-receptor(protein) interactions, where it is believed to most frequently remove a hydrogen atom from the side chain of a Met residue followed by covalent C-C bond formation of the resulting radical pair [2].

We are currently carrying out a detailed investigation of this photochemical reaction in a series of five, backbone rigidified, hexapeptides of general sequences Boc-$A_xBA_yMA_z$-OMe and Boc-$A_xMA_yBA_z$-OMe, where B=(S)-Bpa, A= Aib, M=(S)-Met, and A_x+A_y+A_z=4. We aim at determining the effects induced by the spacer length (A_y=1-3) on the rate of the *intramolecular* excited state reaction (Yang photocyclization), and the chemical and 3D-structures of the resulting products.

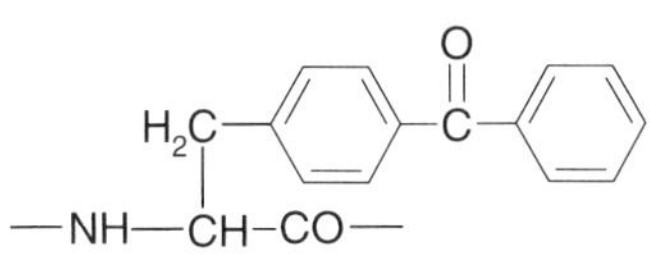

Bpa

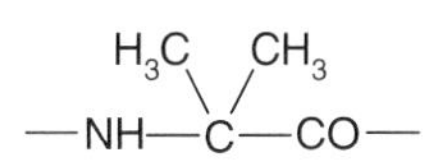

Aib

Results and Discussion

The rank order of the triplet state lifetimes of the Bpa residues in the five terminally-protected hexapeptides, determined in a deoxygenate, dilute, acetonitrile solution by laser flash photolysis, is as follows: ABA_2MA, τ=60 ns; AMA_2BA, τ=190 ns; $ABAMA_2$, τ=350 ns; $ABMA_3$, τ=430 ns; and ABA_3M, τ=920 ns. In addition to the information these data provide on the structural requirements for intramolecular excited state quenching in the molecules, they also serve to define the conditions necessary for optimal intramolecular reaction in preparative photolysis experiments.

S. Del Valle et al. (eds.), *Peptides for Youth: The Proceedings of the 20th American Peptide Symposium*, 449
DOI: 10.1007/978-0-387-73657-0_194, © Springer Science+Business Media, LLC 2009

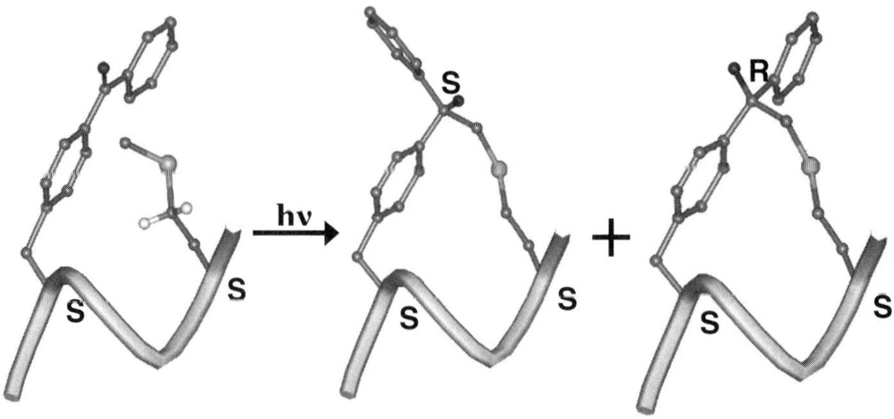

Fig. 1. Cartoon showing the two diastereomeric products afforded by the Yang photocyclization reaction on the terminally-protected hexapeptide ABA_2MA. Hydrogen abstraction took place only from the Met ε-CH_3 group.

So far, we have isolated (by HPLC or flash chromatography) and analyzed (by mass spectrometry and NMR) the only two products resulting from the reaction involving each of the ABA_2MA and $ABAMA_2$ hexapeptide substrates. In both cases an investigation of the chemical structures has clearly indicated that they are couples of diastereomers arising exclusively from the Bpa diradical attack on the Met ε-methyl function (Fig. 1). Configurational assignments (by X-ray diffraction) of all of the four products are in progress by taking advantage of a single crystal already obtained from one of the two diastereomers in each reaction.

References

1. Kauer, J.C., Erickson-Viitanen, S., Wolfe, Jr., H.R. and DeGrado, W.F. *J. Biol. Chem.* **261**, 10695-10700 (1986).
2. Pérodin, J., Deraët, M., Auger-Messier, M., Boucard, A.A., Rihakova, L., Beaulieu, M.-E., Lavigne, P., Parent, J.-L., Guillemette, G., Leduc, R. and Escher, E. *Biochemistry* **41**, 14348-14356 (2002).

Evaluation of chemically and enzymatically prepared cross-linked BNP fragments

Ganga Beligere* Huaiqin Wu, Jim Fino and Reika Campbell

Core R and D, Abbott Diagnostic Division, 100 Abbott Park Rd, Abbott Park, IL 60064-6016 USA

Introduction

B-type natriuretic peptide (BNP) is a member of the family of hormones known as natriuretic peptides. BNP is initially synthesized as a 134-amino acid pre-prohormone that is processed to a 108-amino acid polypeptide, proBNP. proBNP is subsequently cleaved to yield a 76-amino acid N-terminal peptide (NT-proBNP) and the biologically active 32-amino acid peptide BNP [1]. The measurement of BNP levels in patient plasma is known to be an effective diagnostic biomarker for the assessment of heart failure. However, active BNP is susceptible to proteolytic cleavage at the two termini and/or within the ring structure of the cyclic peptide [2]. The products of these cleavage reactions can be debated but play a particularly important role during antibody selection for immunoassay development. One proposed cleavage product includes a ring cleaved BNP product containing the two linear peptide fragments held together by a single disulfide linkage. We describe here preparation and characterization of this proposed BNP ring cleavage product by synthetic (CNBr) and enzymatic (endopeptidase Arg-C) methods. The cross-linked fragments were isolated and were compared to full length cyclic BNP in a prototype assay using selected anti-BNP mAbs.

Results and Discussion

BNP peptide 1-32 is known to undergo fast proteolytic degradation in some patient samples. The arginine residues 30 and 29 are extremely susceptible to proteolytic cleavage. Arg-C digestion of the BNP peptide yielded not only ring-cleaved product. The arginine and lysine on the ring were cleaved due to proteolysis and this lead to mixture of two BNP peptides attached through disulphide bonds as shown in Fig 1. To selectively cleave the ring residue on the peptide, the methionine group at position 16 was selected. It was cleaved in acetic acid solution using CNBr.

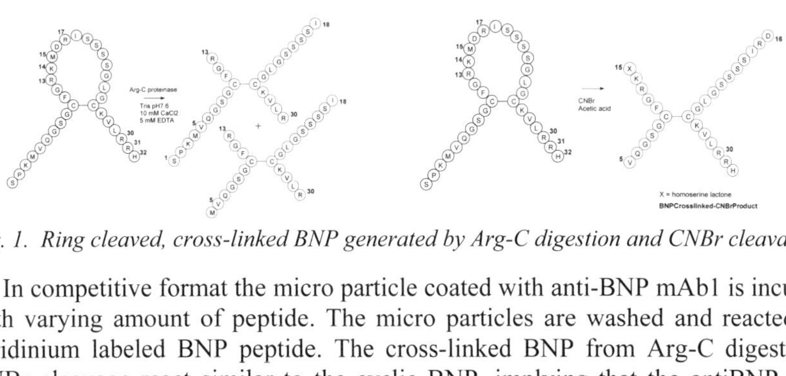

Fig. 1. Ring cleaved, cross-linked BNP generated by Arg-C digestion and CNBr cleavage.

In competitive format the micro particle coated with anti-BNP mAb1 is incubated with varying amount of peptide. The micro particles are washed and reacted with acridinium labeled BNP peptide. The cross-linked BNP from Arg-C digestion or CNBr cleavage react similar to the cyclic BNP, implying that the antiBNP mAb1 does not differentiate the ring cleaved, cross-linked peptide and the cyclic BNP.

S. Del Valle et al. (eds.), *Peptides for Youth: The Proceedings of the 20th American Peptide Symposium*, 451
DOI: 10.1007/978-0-387-73657-0_195, © Springer Science+Business Media, LLC 2009

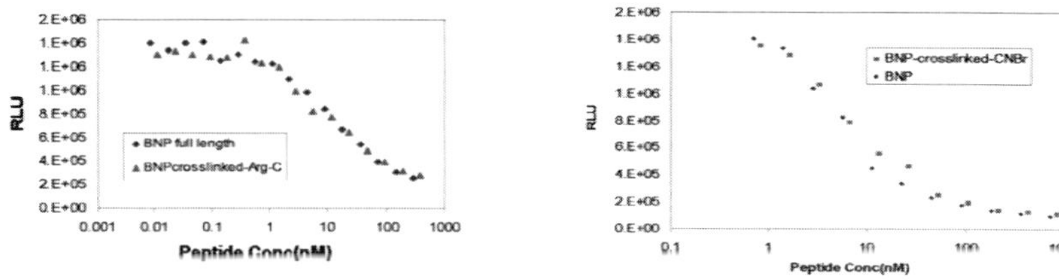

Fig. 2. Evaluation of BNP-cross-linked peptide from ArgC digestion and CNBr binding to anti-BNP mAb1, using competitive format.

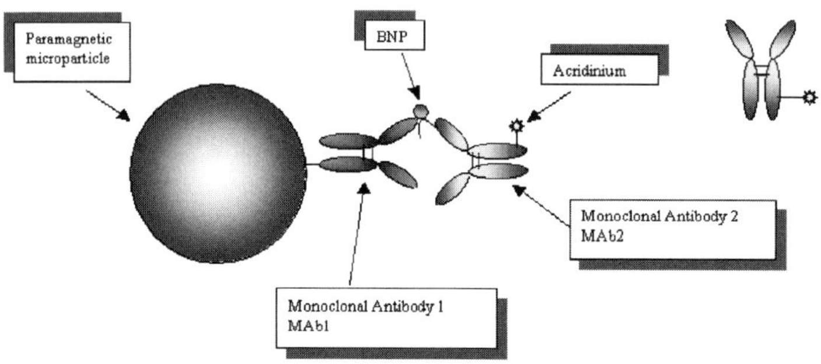

In a sandwich assay, one of the antibodies is coated on the solid phase (paramagnetic microparticle) and the other mAb recognizing the same antigen, carries acridinium labeled antibody. The BNP peptide and ring cleaved peptides are separately incubated with the microparticles and the amount of BNP or cross-linked BNP captured by the micro particle is directly read by the second antibody carrying the acridinium label. The CNBr cleaved product does not show any differentiation in its binding to the two antibodies. It is observed that the Arg-C cleaved cross-linked peptide losses binding to the second antibody due to loss of epitope

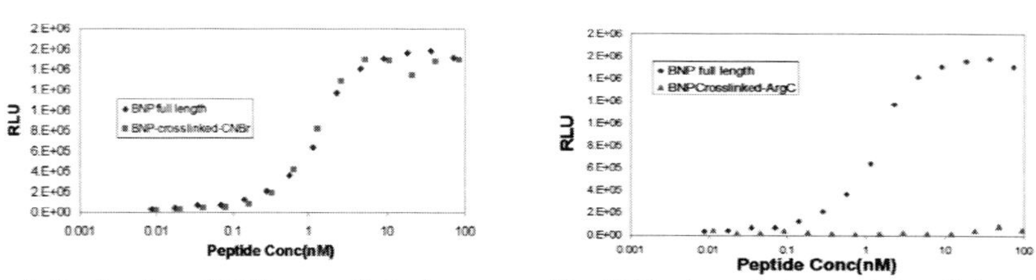

Fig. 3. Evaluation of BNP cross-linked generated by CNBr degradation and Arg-C digestion binding to antiBNP mAb2 using antiBNP mAb1 coated micro particle in sandwich format.

The proteolytic cleavage of BNP to form ring cleaved fragment was demonstrated and the cleavage products were isolated. These ring cleaved cross-linked BNP fragments were recognized by antibodies implying that the structural feature or conformation restriction due to disulphide ring formation does not affect the epitope.

References

1. Valli et al., *J. Lab. Clin. Med.*, Vol 134(5) **1999**: 437-444.
2. Shimizu et al., *Clinica Chimica Acta*, Vol 316 **2002**: 129-135.

Monitoring of MMPs activity *in vivo*, non-invasively, using solubility switchable MRI contrast agent

Beata Jastrzebska[1,2], Rejean Lebel[1,2], Oliver J. McIntyre[3], Benoit Paquette[2], Witold Neugebauer[4], Emanuel Escher[4] and Martin Lepage[1,2]

[1]Sherbrooke Molecular Imaging Centre; [2]Department of Nuclear Medicine and Radiobiology, Université de Sherbrooke, Sherbrooke, QC, Canada; [3]Department of Cancer Biology and Vanderbilt-Ingram Cancer Center, Vanderbilt University, Nashville, TN 37232, USA; [4]Department of Pharmacology, Université de Sherbrooke, Sherbrooke, QC, Canada

Introduction

The matrix metalloproteinases (MMPs) belong to a family of extracellular zinc-dependent proteinases with important roles in normal biological processes and pathological states. MMPs are known to be overexpressed in tumor cells and secreted into the microenvironment of a variety of tumours. Their activity is associated with the tumour metastatic process, notably through breaching of the basement membrane [1].

Non-invasive imaging techniques provide an extraordinary opportunity for "molecular imaging" of tumours to increase the sensitivity of detecting early-stage tumours, to identify tumours that require particularly aggressive therapy, to identify tumours appropriate for specific anti-protease therapeutics, and to address the issue of target modulation and dose selection.

We have developed a novel peptide-based contrast agent (CA) to image the activity of matrix metalloproteinases (MMPs) *in vivo,* non-invasively, using magnetic resonance imaging (MRI). The agent is composed of a paramagnetic gadolinium chelate (Gd-DOTA) attached to a peptide sequence made hydrophobic by an 12-aminododecanoic acid. The peptide is cleavable by a specific MMP (MMP-2). The aqueous solubility of the compound is restored by linking a PEG chain on the C-terminal of the peptide together with an Arg residue. Upon cleavage of the MMP substrate sequence, the Gd-DOTA moiety becomes sufficiently hydrophobic to precipitate or to associate to cell membranes, modifying its pharmacokinetic properties detected by MRI.

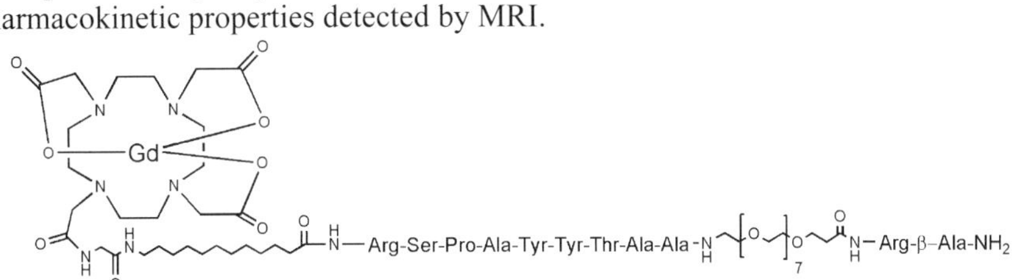

Fig. 1. Sequence of MRI contrast agent (JB-P2).

Results and Discussion

Synthesis. The peptide JB-P2 (Fig. 1) was prepared by solid phase methodology on Rink amide resin using Fmoc strategy. Final cleavage and deprotection of the peptide was carried out with TFA and H_2O (90:5) in the presence of thioanisole and 1,2-ethanedithiol as scavengers at room temperature for 3.5 h. The Gd(III) complex was obtained by adding 1.5-fold excess of $GdCl_3$ to the aqueous solution of the

S. Del Valle et al. (eds.), *Peptides for Youth: The Proceedings of the 20th American Peptide Symposium*, 453
DOI: 10.1007/978-0-387-73657-0_196, © Springer Science+Business Media, LLC 2009

ligand at pH = 6.5-7.0. The crude product was purified by preparative C_{18} glass column medium pressure chromatography. The pure contrast agent was characterized by analytical C_{18} RP-HPLC and MALDI-TOF mass spectrometry.

Mouse model. The murine MC7-L1 breast cancer cell line was stably knocked-down for MMP-2 (MC7-L1*). MC7-L1 and MC7-L1* cells were injected subcutaneously respectively on the right and left hind limb of Balb/c mice. After 5 weeks, the animals were imaged using MRI.

Magnetic resonance imaging. Dynamic T_1-weighted images were continually recorded before and after injection of the agents. JB-P2 (2.05 μmoles) or a control compound Magnevist (Gd-DTPA, 5.8 μmoles) was injected as a bolus over one minute at time 8.5 min via a tail vein catether. The change in signal intensity was calibrated in terms of a change in relaxation rate (ΔR_1), which is related to the CA concentration.

Figure 2 shows that ΔR_1 in the tumours is increased by both Magnevist and JB-P2. The control Magnevist experiment confirms that both tumours (MC7-L1 and MC7-L1*) have similar vascular properties, as the intensity reaches similar levels with comparable time curves. Both tumours initially incorporate similar amounts of JB-P2 agent, but there is an increased uptake in the MMP-2 expressing tumours after 30 minutes, while the amount of JB-P2 is stable in the MMP-2 knockdown tumour. This suggests a preferential accumulation of the cleaved compound inside the tumour expressing MMP-2. The signal intensity decreases towards the background at longer times. Our data show a very marked signal increase in the bladder, suggesting JB-P2 is eliminated via the kidneys.

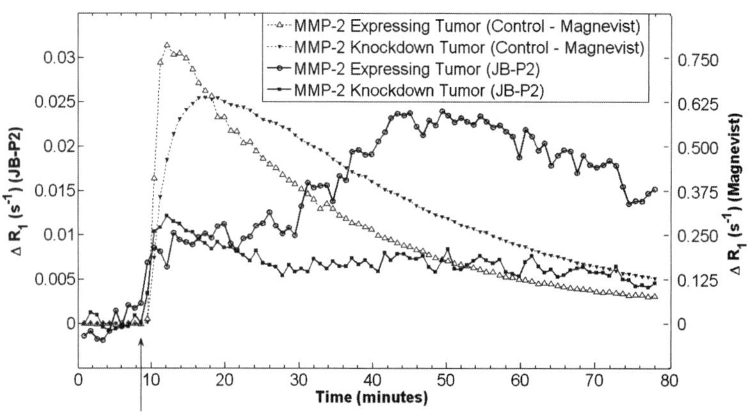

Fig. 2. Variation in relaxation rate (ΔR_1) in two different tumors as a function of time after the injection of JB-P2 or Magnevist. The arrow indicates the time of injection.

Acknowledgments

This work was supported by the National Cancer Institute of Canada. We acknowledge the help of H. Therriault, M. Archambault, C. Dubuc and V. Dumulon. ML is the Canada Research Chair in magnetic resonance imaging.

References

1. Liotta, L. A., Tryggvason, K., Garbisa, S., Hart, I., Foltz C. M., and Shafie S. *Nature* **284**, 67-68 (1980).

^{99m}Tc-labeled Ac-DEVD Peptides as a Substrate for Measuring Caspase Activity

Chiyi Xiong, Zhi Yang, Rui Zhang, William Tong, Juri Gelovani and Chun Li

Department of Experimental Diagnostic Imaging, The University of Texas M. D. Anderson Cancer Center, 1515 Holcombe Blvd., Texas 77030

Introduction

Caspases are a structurally related group of cysteine proteases that cleave peptide bonds following specific recognition sequences. The activation of caspases is a hallmark of apoptosis. Objective and quantitative noninvasive imaging of apoptosis would be a significant advance for rapid and dynamic screening as well as validation of experimental therapeutic and diagnostic agents.

Historically, the detection of the caspases activity was accomplished by utilizing a synthetic tetrapeptide, Asp-Glu-Val-Asp (DEVD), labeled with either a fluorescent molecule, or a colorimetric molecule as substrates. DEVD-dependent protease activity is assessed by detection of the free fluorescent or colorimetric molecule cleaved from the substrates. However, the current state of the art does not allow detection of specific caspases in vivo.

Nuclear imaging techniques are especially suited to track radio-labeled therapeutic or diagnostic agents and to investigate specific molecular interaction with their target. Here, we report the synthesis and radio-labeling efficiency of single amino acid chelate and its DEVD derivatives, $^{99m}Tc(CO)_3$-DEVD-SAAC-AAn, and, $^{99m}Tc(CO)_3$-SAAC-DEVD-AAn, (Fig 1.) that can be cleaved by caspase 3 to generate $^{99m}Tc(CO)_3$-SAAC fragment. Peptides were screened by evaluating differential cellular retention between the substrates and the corresponding fragment in both vial tumor cells and apoptotic cells.

DEVD-SAAC-*d*-Tyr-NH_2
DEVD-SAAC-*d*-Lys-NH_2
DEVD-SAAC-*d*-Leu-NH_2
DEVD-SAAC-*d*-Phe-NH_2
DEVD-SAAC-*d*-Cys-NH_2
Ac-DEVD-SAAC-Leu-OMe
Ac-DEVD-SAAC-Phe-OMe

SAAC-DEVDK
SAAC-DEVDF
SAAC-DEVDC
Y-SAAC-DEVDG
SAAC-DEVDG

SAAC =

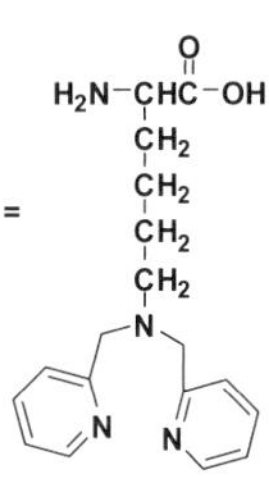

Fig. 1. Single amino acid chelates (SAAC) and its incorporated DEVD peptides

Results and Discussion

The single amino acid chelate (SAAC) as a building block for technetium and rhenium-based radio-labeled peptides was conveniently synthesized in high yields. The reaction was through direct reductive N-alkylations of N_α-Fmoc-Lys-OH and pyridine-2-carboxyaldehyde using $NaBH(OAc)_3$ as an efficient reducing reagent. Dialkylated N_α-Fmoc-Lys-OH was prepared in one-pot synthesis.

Synthetic DEVD peptide amides or methyl esters were synthesized using Fmoc/tBu chemistry on rink amide or oxime resins in a *Prelude* peptide synthesizer

S. Del Valle et al. (eds.), *Peptides for Youth: The Proceedings of the 20th American Peptide Symposium*, 455
DOI: 10.1007/978-0-387-73657-0_197, © Springer Science+Business Media, LLC 2009

(Protein Technologies, Inc). Removal from the resin was accomplished using standard TFA cleavage cocktails in the presence of scavengers. The crude peptides were purified by reverse-phase HPLC. All structures were then characterized and confirmed by electrospray ionization (ESI) mass spectrometry.

Preparation of $[^{99m}Tc(CO)_3(OH_2)_3]^+$ was accomplished by boiling 1 mL of sodium pertechnetate ($Na^{99m}TcO_4$) with Isolink carbony reaction kit for 20 min. After cooling on an icebath, the alkaline solution was neutralized by the addition of 120 µL I N HCl.

An aliquot (100 µL) of peptides (1 mg/ml) in saline were added to the reaction vial containing 1.0 mCi $[^{99m}Tc(CO)_3(OH_2)_3]^+$ by syringe and the mixture heated to 70° C for 45 min. The reaction mixture was cooled in an ice bath and the product isolated by filtration with high radiochemical yield and radiochemical purity. (Fig 2.). The radio-labeled peptides were efficiently cleaved by human recombinant caspase-3 in enzyme assays.

Further work is in progress to focus on differentiating the peptides cell permeability and the corresponding fragment retention in both tumor cells and apoptotic cells.

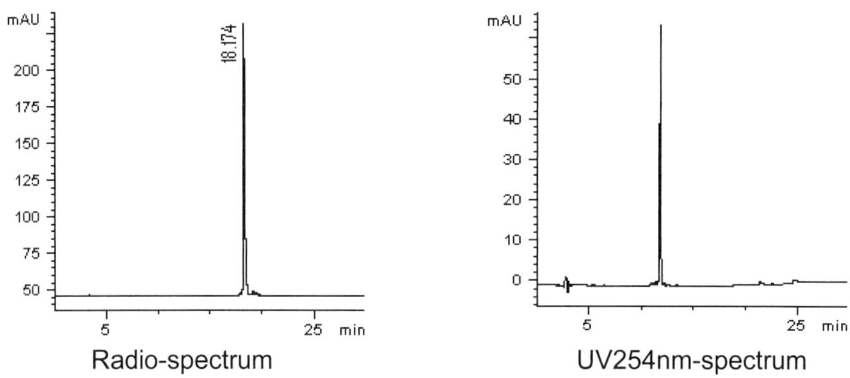

Fig. 2. γ-Radiochromatograms and HPLC chromatogram (UV-detection, λ = 254 nm) of peptide $^{99m}Tc(CO_3)$*-DEVD-SAAC-d-Leu-*NH_2

Acknowledgments

The work was funded by John S. Dunn Foundation.

References

1. Stephenson, K.A, Banerjee, S. R, Sogbein, O.O, Levadala, M. K, McFarlane, N, Boreham, D. R, Maresca, K. P, Babich, J. W, Zubieta, J, Valliant, J. F. *Bioconjugate Chem.* **16**, 1189-1195 (2005).
2. Wang, Z-Q, Liao, J and Diwu, Z. *Bioorg. Med. Chem. Chem. Lett.* **15**, 2335-2338 (2005).
3. Zhang, H-Z, Kasibhatla, S, Guastella, J, Tseng, B, Drewe, J, Cai, S. X. *Bioconjugate Chem.* **14**, 458-463 (2005).

PEPTIDES LEADS TO THE DRUGS OF TOMORROW

PepLook: An innovative in silico tool for determination of structure, polymorphism and stability of peptides

Annick Thomas [1], Sébastien Deshayes[1], Marc Decaffmeyer[1], Marie-Hélène Van Eyck[2], Benoit Charloteaux B[1] and Robert Brasseur[1]

[1]Centre de Biophysique Moléculaire Numérique (CBMN) – Faculté des Sciences Agronomique de Gembloux – Belgium; [2]Biosiris SA – Gembloux – Belgium

Introduction

Peptides are selective biological tools with an increasingly important place in therapeutic and diagnostic fields. Despite a promising future, characterization of their structure and stability can be very difficult. Structures can be approached by spectroscopy and NMR techniques but data from these approaches frequently diverge. Structures can also be calculated in silico from primary sequences using different algorithms. We recently developed PepLook **[1]**. All in silico algorithms provide 3D model structures from sequences. PepLook also gives information on the polymorphism, stability and binding ability of peptides.

Method

PepLook is based on a Boltzman-stochastic algorithm **[2]** and explores by an iterative calculation the conformational space for a peptide sequence using couples of phi/psi angles derived from the work of Etchebest et al. **[3] (Fig. 1)**. PepLook provides the most stable 3D structure, namely the Prime and 98 other low energy 3D structures. A stability score of residues of the Prime structure, based on MFP calculations, points out which residues are candidates as binding sites. A polymorphism score is also provided, based on the dispersion of the 98 lowest energy structures mapped on the sequence on which residues should be involved in the binding site. This method gives a fairly good approach of experimental structures and predicts which peptide will lead to controversial experimental data obtained because of polymorphism.

Results and Discussion

The method has been firstly used for the study of several peptides such as magainin 2, a helical CPP (Cell Penetrating Peptide), transportan, a mixed helical/ramdom coil CPP, HA2, the fusion peptide of influenza hemaglutinin which is a random coil peptide, and hCT (9-22) (segment 9-32 of the human calcitonin) which is polymorphic. We have shown that the structures calculated using PepLook fit nicely with the NMR structures for magainin, transportan and HA2. For hCT(9-32), experimental data are more controversial, suggesting structural polymorphism, which is also suggested by the PepLook analysis. The polymorphism scores highlight that magainin2 is monomorphic, transportan is N-disordered, HA2 is C-disordered and hCT9-32 is fully disordered. The stability scores indicate that the MFP values of NMR models and of PepLook models are in the same range, excepted for HA2 where NMR structures are more stable for this peptide.

We also used PepLook to propose, in addition of the length, the positive net charge and the amphipaticity, that the structural variability of cell penetrating peptides (CPPs) is related to their cellular internalization properties. Indeed calculations carried out on transportan and penetratin and on their permeant and non-permeant mutants highlighted that both wild type CPPs are polymorphic

S. Del Valle et al. (eds.), *Peptides for Youth: The Proceedings of the 20th American Peptide Symposium,* 459
DOI: 10.1007/978-0-387-73657-0_198, © Springer Science+Business Media, LLC 2009

allowing a modification of their apparent hydrophobicity when their non-permeant mutants are monomorphic **[4]**.

In another field we showed, using PepLook, that the hydrophobic segment (18-42) of the epsilon isoform of diacyl glycerol kinase (DGKε) can adopt two different conformations (straight helical and U bent) in membranes and that these confirmations are related to the binding of cholesterol and to dimerization **[5]**.

Because we know the need of pharmaceutical companies for more and more of biological stability for peptides, PepLook can also be dedicated to the prediction of structures for peptides including non natural or D amino acids.

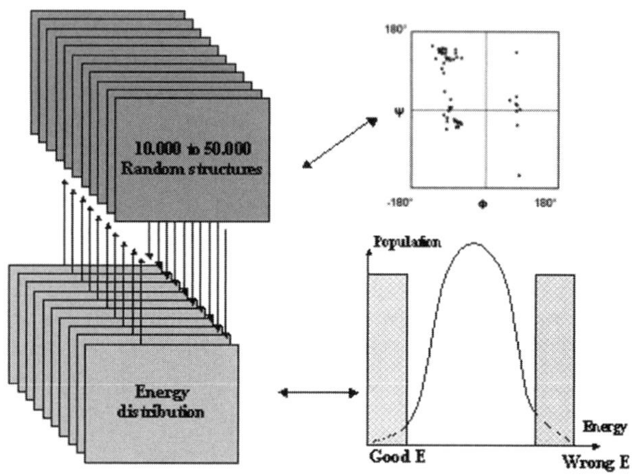

Fig. 1. Schematic view of the procedure used by PepLook.

Acknowledgments

R.B. is Director of Research at the Belgian Fund for Scientific Research (FNRS). A.T. is Research Director at the French Institute for Medical Research (INSERM). M.D. is supported by Walloon Region (DGTRE,PepSein grant).

References

1. Thomas A., Deshayes S., Decaffmeyer M., Van Eyck MH., Charloteaux B., Brasseur R. *Proteins* **65(4)**, 889-97 (2006).
2. Glick M, Rayan A, Goldblum A, *Proc. Natl. Acad. Sci. U.S.A* **99**, 703-708 (2002).
3. Etchebest C, Benros C, Hazout S, de Brevern AG. *Proteins* **59(4)**, 810-27 (2005).
4. Deshayes S, Decaffmeyer M, Brasseur R and Thomas A. (in correction).
5. Decaffmeyer M, Dicu A, Thomas A, Topham MK, Brasseur R and Epand RM. (in correction).

The Effects of the EW - Peptide Optical and Chemical Isomers on the Hemopoietic Stem Cells in Intact and Irradiated Mice

Vladislav I. Deigin[1,#], Tamara N. Semenets[2], Irina A. Zamulaeva[2], Yana V. Maliutina[2], Elena I. Selivanova[2], Alexander S. Saenko[2] and Olga V. Semina[2]

[1]*Department of peptide chemistry, Shemyakin and Ovchinnikov Institute of Bioorganic Chemistry RAS, (Moscow, Russia);* [2]*Medical Radiological Research Center, RAMS (Obninsk, Russia);* [#]*Immunotech Developments Inc. (Mississauga, Ontario, Canada L5K 1B3)*

Introduction

The influence of Glu-Trp (EW) synthetic dipeptide isomers on hemopoietic stem cells and basic immune response reactions is determined by their optical and chemical properties. L-amino acid containing dipeptides lE-lW (l-l) and l-γiE-lW (γ-l-l) have no effect on proliferation of intact bone marrow cells. The optical isomerism of the Glu residue is the essential determinant of the EW dipeptides biological activity. The inversion of the amino acid optical form imparts suppressor properties: dE-dW (d-d), d-γE-dW (γ-dd), dE-lW (d-l) and d-γE-lW (γ-d-l) inhibit proliferation of hemopoietic stem cells in intact bone marrow. The type of the peptide bond between l-Glu and Trp is another important factor for the biological activity of the l-Glu-containing peptides. Unlike lE-dW (l-d) with α-peptide bond, the dipeptide l-γE-dW (γ-l-d) with γ-peptide bond stimulates CFU-S-8 proliferation in intact bone marrow. In animals, ***pre-irradiation*** injection of dE-dW (d-d), d-γE-dW (γ-d-d), dE-lW (d-l), d-γE-lW (γ-d-l), or ***post-irradiation*** injection of lE-lW (l-l), l-γiE-lW (γ-l-l) promoted regeneration of the hemopoietic stem cells population.

Results and Discussion

There have been relatively few studies on the effects of synthetic peptides consisting of d- or l-amino acid residues on the development of immune response. Practically unexplored are the effects of d- or mixed l- and d-peptides at the initial stages of hemopoiesis, i.e. on stem cells that are the progenitors of all morphologically identifiable blood and immune cells.

Using the Glu-Trp dipeptides as an example, we have demonstrated the dependence of peptide biological activity on the optic form of each constituent amino acid residue. In our earlier publications, we reported that l-Glu-l-Trp (EW) and its structural and optical d-isomer d-γGlu-d-Trp (γEW) have different effects on hemopoietic stem cells *in vivo*: the l-l-peptides have hemostimulating properties, while the d-d-peptides — hemoinhibiting ones [1]. This brief publication is devoted to demonstrate the effects of synthetic d-d, l-l, d-l and l-d isomers of the dipeptides EW (α-peptide bond) or γEW (γ-peptide bond) on the number, proliferating capacity and other properties of hemopoietic stem cells in normal animals, as well as upon exposure to ionizing radiation. We have studied the activity of dE-dW (d-d); dE-lW (d-l); lE-dW (l-d); lE-lW (l-l); d-γE-dW (γ-d-d); d-γE-lW (γ-d-l); l-γE-dW (γ-l-d) and l-γE-lW (γ-l-l). The relative number (per 10^5 injected bone marrow cells) of CFU-S in the bone marrow of donor mice, injected with all tested dipeptides in a dose of 1, 10, 100, 1000 or 10 000 μg/kg 18 hours before the organ extraction has

S. Del Valle et al. (eds.), *Peptides for Youth: The Proceedings of the 20th American Peptide Symposium*, 461
DOI: 10.1007/978-0-387-73657-0_199, © Springer Science+Business Media, LLC 2009

been determined. The effects of the EW and γEW dipeptide isomers on the colonies forming cells (CFU-S) proliferation in normal and irradiated donor mice were studied using the "thymidine suicide" method in two test systems, with intact and irradiated donors (Figures 1A and 1B). In the first case, intact donor mice received peptide injections 48 hrs ***prior*** to bone marrow extraction. As evident from the results (Fig. 1A), only l-γE-dW stimulated proliferation of CFU-S in intact donors. The other peptides had no effect on the number of CFU-S in the S-phase of the cycle in this test system. The specific effect of the peptides on proliferation in ***intact*** system visible only for 3 to 10% CFU-S cells in the S-phase, the remaining cells being at "rest". For capturing more pronounced effects, the cells in the S-phase must account for at least 40%. This is the case with the mice bone marrow on day 7 ***after irradiation*** with the dose of 4 Gy. In the alternative test system donor mice were irradiated with 4 Gy; on day 5 post irradiation the animals received an injection of one of the peptides under study, and 2 days later (on day 7 post irradiation) their bone marrow was analyzed for CFU-S in the S-phase using the method of "thymidine suicide". The results (Fig. 1B) show that the dGlu-containing peptides (dE-dW, dE-lW and d-γE-dW, d-γE-lW) had similar effects on the actively proliferating bone marrow: the peptides reduced the percent of CFU-S in the S-phase of the cell cycle to the control level. The effects of l-Glu-containing peptides (lE-dW, lE-lW and l-γE-dW, l-γE-lW) were also similar: they neither reduced, nor increased statistically significant the number of CFU-S in the cycle.

In vivo studies of the biological effects of chemical and optical isomers of the natural EW dipeptide (d-d, l-l, d-l and l-d isomers) showed that a change of the optical orientation of even one of the constituent amino acids causing new biological properties. l-dipeptides (lE-dW, l-γE-lW) are inert to committed and pluripotent CFU-S of intact bone marrow, but restore the cells population affected by ionizing irradiation. d-Glu-containing dipeptides (dE-dW, dE-lW d-γE-dW, d-γE-lW) have an inhibiting effect on the pool of partially committed CFU-S-8, and only dE-dW or d-γE-dW can increase the population of pluripotent hemopoietic progenitors (CFU-S-12) increasing the number of the hemopoietic stem cells. The effects of the mixed dipeptides lE-dW and l-γE-dW on the CFU-S population differ: lE-dW restores the number of CFU-S collected from irradiated bone marrow and at the same time, like lE-lW, has no effect on *in vivo* colony formation by intact bone marrow cells. The proliferation of CFU-S-8 stimulates l-γE-dW, increasing the number of colonies by over 40% as compared to control. Thus, the change of the α-peptide bond for the non-natural γ-peptide bond in the EW dipeptide imparts a new property to the l-γE-dW peptide — the capacity to increase the number of CFU-S *in intact* bones marrow.

Flow cytometry was used to evaluate the effects of EW isomers on the relative number of CD34+ cells (hemopoietic progenitors) and the surface marker density in the bone marrow of mice. The peptides were injected to donor mice. On the second day following the injection, the relative number of CD34+ bone marrow cells has been evaluated. Each of the tested peptides statistically significantly decreased the parameter as compared to control. The only dE-dW or d-γE-dW increased the marker density ($p<0.05$) and the percent of pluripotent progenitors in the bone marrow.

Based on the effects of the EW dipeptides on the initial stages of hemopoiesis, it was evaluated their potential use as agents reducing the detrimental effect of ionizing radiation on hemopoiesis. According to our data, the d-Glu-containing peptides have a radioprotective activity, while the l-Glu- ones — radiotherapeutical properties. The injection of dE-dW, d-γE-dW or dE-lW and d-γE-lW ***prior to***, and lE-lW, lE-dW, l-γE-dW or l-γE-lW ***after*** irradiation intensified the regeneration of CFU-S.

The d-Glu- containing dipeptides are capable to inhibit the proliferation of CFU-S and have a radioprotective effect on the cell population. The peptides reduce the number of progenitors in the cycle, and hence the radiation acts on "resting" cells that more resistant to damage. The l-Glu-containing dipeptides l-γE-dW and l-γE-lW are only effective when used after irradiation: they intensify the regeneration of the population of hemopoietic progenitors as compared to irradiated control.

A change in the spatial orientation of a diastereomer can alter not only the magnitude, but also the direction of its biological effects. Our findings of such changes in the fine structure of the EW dipeptide isomers suggests that at least the EW dipeptides could be considered as an integral organic molecule, whose general chemical structure, optical and spatial orientation determines the magnitude and the nature of its biological effects. Based on the obtained data, the dipeptide molecules can be considered as a sort of unique "bridge" between polypeptides, on the one hand, and small organic molecules, that are the active substances of most modern synthetic drugs, on the other hand. We believe that the established relationships between the optical and chemical structures of the EW dipeptides and their biological properties [2] will help in the search for new peptide drug candidates.

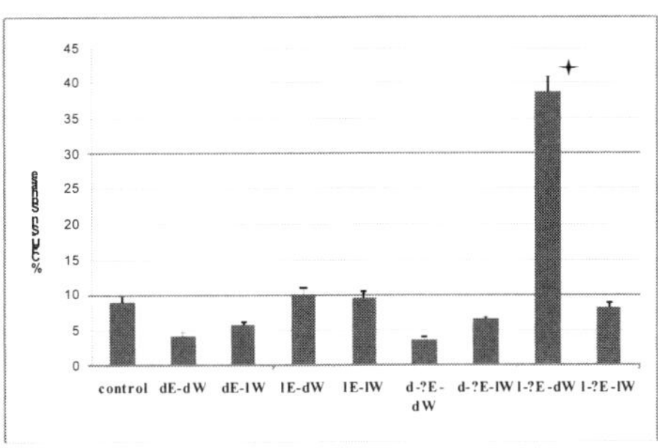

Fig.1 A

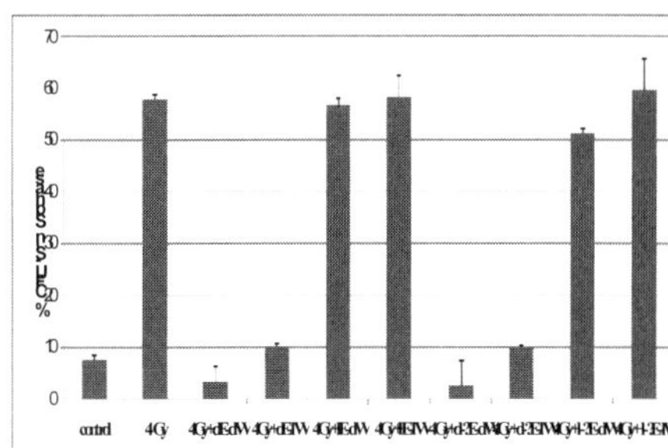

Fig.1 B

Fig. 1A *The effects of the EW dipeptide isomers on the percentage of CFU-S-8 in the S-phase in intact donor bone marrow (10μg/kg). *Significantly different (p<0.001) from control.*

Fig. 1B *The effects of the EW dipeptide isomers on the percentage of CFU-S-8 in the S-phase in irradiated donor (4 Gy) bone marrow (10μg/kg). *Significantly different (p<0.001) from 4 Gy.*

References

1. Deigin V.I., Poverenny A.M., Semina O.V. and Semenets T.N. Immunol. Lett **67**, 1: 41-46 (1999)
2. Deigin V.I., Semenets T.N., Zamulaeva I.A, Maliutina Ya.V., Selivanova E.I, Saenko A.S, Semina O.V. Int. Immunopharmacol., **7**, 375-382 (2007)

N-(Fluorenyl-9-methoxycarbonyl)amino Acid Amide Derivatives as a New Class of Anti-cancer Agents

Lajos Gera[1,3], Daniel C. Chan[2,3], Vitalija Simkeviciene[1], Paul A. Bunn, Jr[2,3] and John M. Stewart[1,3]

[1]*Department of Biochemistry and Molecular Genetics, University of Colorado Health Sciences Center, Denver, CO 80262;* [2]*Cancer Center, University of Colorado Health Sciences Center;* [3]*ApopLogic Pharmaceuticals LLC, Aurora, CO 80010, USA*

Introduction

We have developed highly potent, orally active peptide bradykinin B2 (B10238), B1 (B10324) and peptide dimer B1/B2 (B9870) receptor antagonists with excellent anti-cancer activity [1, 2]. Based on these studies a small molecule mimetic, BKM-570 (F5c-OC2Y-Atmp, F5c: 2,3,4,5,6-pentafluorocinnamoyl; OC2Y: (O-2,6-diclorobenzyl)-tyrosine, Atmp: 4-amino-2,2,6,6-tetramethylpiperidine) was developed by Gera at the Stewart laboratory [3]. This acyl-tyrosine amide derivative inhibited growth of small cell lung cancer (SCLC) *in vitro* and *in vivo* (68-91%). In a recent study we incorporated a variety of nonsteroidal anti-inflammatory drugs (NSAIDs) as the acylating group of BKM-570. BKM-1394 (Ktlc-OC2Y-Atmp, Ktlc: ketorolac) showed high growth inhibition (70%) *in vivo* against non-SCLC (NSCLC) line A-549 and inhibited growth (44%) of PC3 prostate cancer xenografts [2]. The N-(fluorenyl-9-methoxycarbonyl) amino acids (Fmoc-AAs) also are a class of anti-inflammatory agents [4], therefore we designed and synthesized new Fmoc-analogs of our potent anti-cancer compounds, BKM-570 and BKM-1394, and tested them against lung and prostate cancer.

Results and Discussion

The Fmoc-amino acid amides and the Fmoc-peptides were synthesized in DMF solution using BOP for coupling protected amino acids or for acylations of Atmp. The compounds were purified by preparative HPLC and characterized by TLC, analytical HPLC and laser desorption mass spectroscopy (LDMS).

After the discovery of the anti-cancer action of the N-terminal dimer and the N-terminal acylated bradykinin monomers (B10238, B10324), our goal was to develop nonpeptide small molecule anti-cancer compounds. Our lead compounds, BKM-570 and BKM-1394 have the same tyrosine amide template (OC2Y-Atmp) and their acylating groups are bulky and hydrophobic F5c and Ktlc. Therefore, we gave preference to the incorporation of a variety of bulky and hydrophobic Fmoc-amino acids which had anti-inflammatory activity [4]. BKM-1794 (Fmoc-Leu-OC2Y-Atmp) showed the same high growth inhibition *in vitro* of SCLC as BKM-570 (Table 1). Based on this discovery we designed two shorter structural analogs of Fmoc-Leu-OC2Y-Atmp, BKM-1798 (Fmoc-Leu-Atmp) and BKM-1800 (Fmoc-OC2Y-Atmp). Interestingly, BKM-1800 with bulky and significantly hydrophobic O-(2,6-dichloro-benzyl)-tyrosine (Fig. 1) showed a remarkable high growth inhibition both *in vitro* and *in vivo* (Table 1). This simple Fmoc-tyrosine amide derivative inhibited growth of PC3 prostate cancer both i.p. (78 %) and orally (55 %). However, this newly developed, inexpensive and non-toxic BKM-1800 showed higher growth inhibition (55-78 %) than the widely used but highly toxic chemotherapeutic drug cisplatin (39 %) in our PC assays.

S. Del Valle et al. (eds.), *Peptides for Youth: The Proceedings of the 20th American Peptide Symposium*, 465
DOI: 10.1007/978-0-387-73657-0_200, © Springer Science+Business Media, LLC 2009

Table 1. Structures and activities of selected compounds.

Number	Structure	SHP-77[a]	SHP-77[b]	PC3[b,d]	PC3[b,e]
		in vitro	*in vivo %*	*in vivo %*	orally %
BKM-570	F5c-OC2Y-Atmp	1.8	91[c]	65	21
BKM-1794	Fmoc-Leu-OC2Y-Atmp	1.8	--	2	--
BKM-1798	Fmoc-Leu-Atmp	7.6	--	-54	--
BKM-1800	Fmoc-OC2Y-Atmp	1.7	--	78	55
BKM-1824	Fmoc-Leu-Bip-Atmp	3.4	--	36	--
BKM-1852	Fmoc-Bip-Atmp	7.7	--	46	--
Cisplatin		--	60	39[f]	--

[a]IC_{50} (μM)for cytotoxicity by MTT test. [b]Inhibition obtained in nude mice. [c]Compounds were injected i.p. at 5 mg/kg daily for 28 days or [d]every 4th day at 10 mg/kg. [e]Compounds were given 50 mg/kg every 2nd day. [f]Cisplatin was given as 4 weekly injections of 10 mg/kg. Abbreviations: Bip, β-(4-biphenylyl) alanine

These studies suggest that Fmoc-amino acid amide derivatives may be valuable therapeutic agents for cancer treatment.

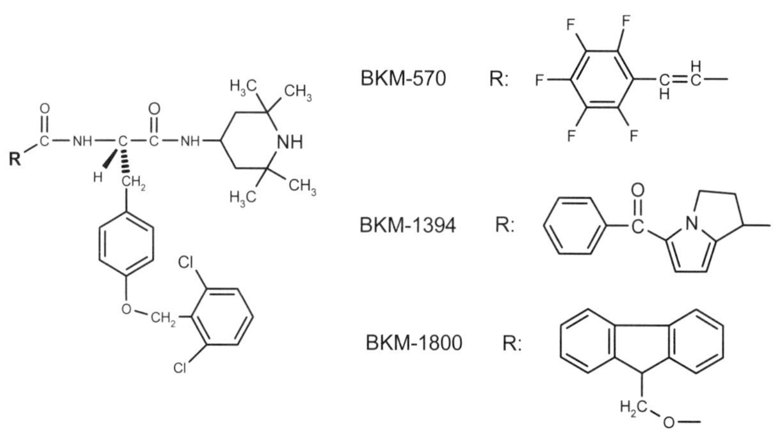

Fig. 1. Structures of BKM-570, BKM-1394 and BKM-1800.

References

1. Gera, L., Stewart, J.M., Whalley, E., Burkard, M. and Zuzack, J. Immunopharmacol. **33,** 178-182 (1996).
2. Gera, L., *et al.* In Flegel, M., Fridkin, M., Gilon, C. and Slaninova, J. (Eds.) *Peptides 2004, Proceedings of the 3rd International and 28th Eur. Peptide Symp.* International, Geneva, 2005, pp. 846-847.
3. Gera, L., *et al.* In Martinez J. and. Fehrentz, J.-A. (Eds.) *Peptides 2000, Proceedings of the 26th Eur. Peptide Symp.* Edk, Paris, 2001, pp. 637-638.
4. Birch, R.M., *et al. Proc. Nat. Acad. Sci. USA.* **88**, 355-359 (1991).

Opportunities and Challenges of Developing Peptide Drugs in the Pharmaceutical Industry

Waleed Danho[1], Joseph Swistok[1], Wajiha Khan[1], Xin-Jie Chu[1], Adrian Cheung[1], David Fry[1], Hongmao Sun[1], Grazyna Kurylko[1], Leonid Rumennik[1], Joseph Cefalu[2], Gretchen Cefalu[2] and Philip Nunn[2]

[1]*Roche Research Center; Hoffmann-La Roche Inc. Nutley, New Jersey 07110, USA;*
[2] *Roche Bioscience, Palo Alto, California 94304, USA*

Introduction

The peptide market is growing nearly ***twice*** as fast as the overall pharmaceutical market due to an increased number of therapeutic targets and improved delivery methodologies. There are 67 therapeutic peptides on the market, 150 in clinical phases, 400 in advance pre-clinical stages. Over 100 pharmaceutical and biotech companies are active in the peptide field. The market size for peptide drugs in 2007 is estimated to be 3.01 billion US$. [1]

Important factors and questions need to be considered before starting peptide based projects. They include: is there an unmet medical need, and an opportunity to define clinical differentiation for the indication of interest? What is the market size? What about the competitive landscape in the clinic and in the market? Is the target amenable to alternate approaches, such as small molecules or monoclonal antibodies? What are the de-risking approaches vs. small molecules and monoclonal antibodies? What are the possible delivery methods? Are the starting materials available in the amount needed? What will the cost of goods be? To illustrate these points, Male erectile dysfunction (MED) and obesity indications will be used.

Results and Discussion

Medical need for MED: There are approximately 70 million men in seven major pharmaceutical markets diagnosed with MED, due to aging population, lifestyle and increased prevalence of chronic diseases that predispose men to ED drive increase prevalence. Current oral agents (PDE-5 inhibitors) target peripheral erectogenic mechanisms have 20-30% non responders, and face issues related to headaches, dyspepsia and flushing. A melanocortin (MC-4) receptor agonist may restore the natural function of neuro-hormonal pathways in mediating sexual response and offer a non-PDE-5 choice to effectively and safely restore erectile and sexual function. [2]

Medical need for treatment of obesity: in the US there are 42.5 million people estimated to be obese (BMI 30+), and 105.5 million are estimated to be overweight and obese (BMI 25+) which represent 38.7% of the US population. Current approaches for treatment of obesity include dietary, surgical intervention and pharmacotherapy. Pharmacological and genetic evidence suggests that MC-4R is a key regulator of energy homeostasis and feeding behavior regulation, and therefore an MC-4 agonist should be effective as a pharmaceutical anti-obesity agent. [3]

Small molecule vs. peptides: Are these targets amenable for small molecules approaches? We developed a strategy for de-risking such projects by using a designed series of peptides, for example: N- and C-terminal truncation, systematic

S. Del Valle et al. (eds.), *Peptides for Youth: The Proceedings of the 20th American Peptide Symposium*, 467
DOI: 10.1007/978-0-387-73657-0_201, © Springer Science+Business Media, LLC 2009

substitutions with Ala, (D) and N-methyl amino acids; helix stabilization, cyclo scan using side chain to side chain, and end to end cyclization, ring size variation, all of which allow definition of the size and nature of the active site pharmacophore, and a judgment as to whether it can be replicated by a small molecule.

Delivery strategies: Advances in nasal, inhalable, bucal, sublingual and oral formulations and delivery systems have been made recently which enable non intravenous administration of peptides. In addition sustained release formulation using Poly(lactide-co-glycolide). PLGA, made it possible to have once weekly administration of a peptide drug by SC route of administration.

Identification of highly selective linear and cyclic peptides for human MC-4 Receptor: Extensive structure activity studies suggest that [His-(D)Phe-Arg-Trp] is regarded as a "pharmacophore core" essential to the interaction of the ligand to its receptor, and *His* is the most critical position responsible for the selectivity toward of MC-4R *versus* the other MC- receptors. Replacement of His with conformationally constrained amino-acids resulted in peptides highly selective for MC-4R [4]. Details of the synthesis and agonist activities of these linear and cyclic peptides are reported elsewhere [5].

In vivo efficacy of a Roche MC-4R specific agonist in the ex copula rat model: MC-4R activation stimulates erectile activity in a variety of species, including man [6]. We tested a Roche MC-4R specific agonist in the ex copula rat model previously described [7], and compared the effects to the reference compound Melanotan-II. Intranasal (IN) administration of the Roche compound produced a significant number of observed erections compared to vehicle-treated animals, and was equipotent to Melanotan-II (Figure 1). The Roche compound also exhibited a rapid onset of action: 37 % of the total erections occurred within 15 minutes, and an additional 43 % within 30 minutes.

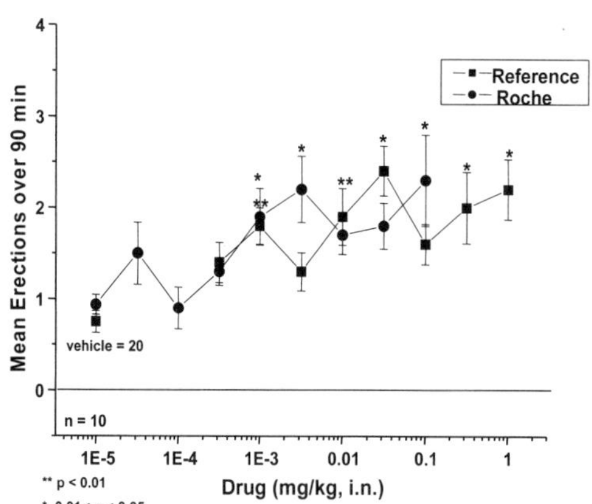

Fig. 1. Erectile events in the ex copula rat model. Comparison between Melanotan-II, and the Roche MC-4R specific agonist.

In vivo efficacy of Roche MC4-R specific agonist on food intake and body weight by IN route of administration in DIO mice model: The role of the MC-4R agonist in controlling feeding behavior and body weight is well supported by several studies. We tested our MC-4 agonist in DIO mouse model. By IN route of administration, Roche compound acutely reduced food intake 66% @ 5 mg/Kg. as shown in (Figure 2a). Moreover, the compound caused a significant reduction of food intake and 19 % reduction of body weight by sub-chronic (21 days), IN route of administration @ 5 mg/Kg, as shown in (Figure 2b). In addition, the compound produced a significant dose-dependent (19%) reduction in plasma glucose and (18%) reduction in plasma cholesterol levels after sub-chronic treatment due to reduction of body weight.

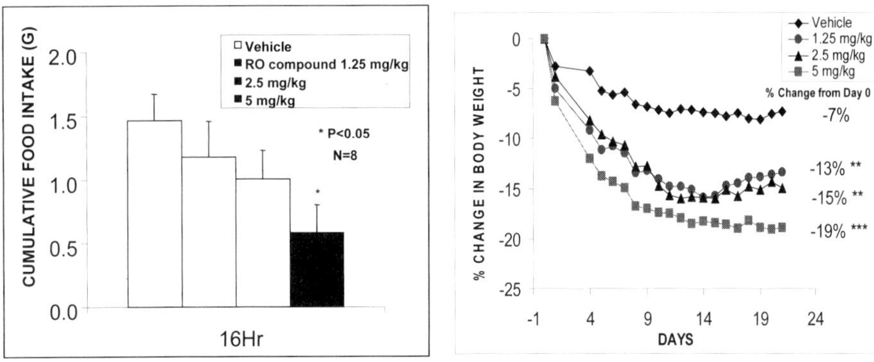

Fig. 2. (a) Acute reduction of food intake in DIO mice following (IN) administration of Roche compound .(b) Sub-chronic treatment with Roche compound (21 days) produced reduction of body weight in DIO mice.

Acknowledgments

The authors wish to thank Drs. Lee Babiss, Nader Fotouhi, Jefferson Tilley, Rebecca Taub, Louis Renzetti, Michael Brunda, Waseem Malick, and Lisa Chao from Roche (Nutley) and Drs. Anthony Ford and Sally Meadows from Roche (Palo Alto) for stimulating discussions, their support and encouragement during the course of this research.

References

1. D&MD publications. Peptides 2006. New Application in Discovery, Manufacturing, and Therapeutics. Sehgal, A., Author. One Research Drive, suite 400A. Wesborough, MA 01581.
2. Giuliano, F., *J. Androl* **25**, 683-691 (2004).
3. MacNeil D. J. et al., *Eur. J. Pharmaco*logy, **450**, 93-101 (2002).
4. Danho, W. et al., in *Peptides*: The Wave of the Future. M. Lebel and R. Houghten (Editors) American Peptide Society, 701-703 (2001).
5. US Patens 6,600.015 and 7,045,591.
6. Martin, W.J. et al., *Eur. J. Pharmacology*, **454**, 71-79 (2002).
7. Cefalu, G. et al., Presented at Western Pharmacology Socirty Conference (2005).

Inhibition of pathogenically-related morphologic transition in *Candida albicans* by disrupting Cdc42 binding to its effectors

Hongjian Li[1], Yang Li[2], Feng Ni[2,*] and Zhengding Su[2,3,*]

[1]*Department of Biotechnology, Jinan University, Guangzhou, Guangdong 510632 China,* [2]*Health Sector, Biotechnology Research Institute, National Research Council of Canada, Montreal, Quebec H4P 2R2 and* [3]*Department of Chemistry, University of Waterloo, Waterloo, Ontario, N2L 3G1, Canada (*z2su@uwaterloo.ca)*

Introduction

Signal transduction pathways are typically regulated by protein–protein interaction networks. As many human diseases exhibit dysfunctional aspects in these networks, there has been a great deal of enthusiasm for the prospect of identifying novel drug candidates targeting key signal transduction components in human diseases.

Candida albicans is an opportunistic fungal pathogen that is the cause of most "yeast infections" or candidiasis in humans. The pathogenicity of C. albicans is contingent upon its ability to switch from yeast to hyphal growth in response to environmental signals. Invasive hyphal growth requires signal transduction through the interaction of *Candida* Cdc42 (CaCdc42), a small GTPase of the Rho-family, with the CRIB (Cdc42/Rac Interactive Binding) domains of its two downstream effectors, Cst20 and CaCla4, under all conditions examined [1]. Thus, these protein-protein interactions are potential targets for curbing candidiasis and that the CRIB peptides may function as potent anti-fungal agents against the CaCdc42. In this work, we have investigated the intracellular activities of the CRIB peptides transduced by a peptide transduction domain (i.e. NPF: VLTNENPFSDP [2]), in inhibiting the morphologic transition of *C. albicans*.

Results and Discussion

In order to achieve intracellular inhibition of CaCdc42, the NPF peptide was arbitrarily fused to the N-terminus of all the examined fusion proteins and two NPF conjugated CRIB peptides were thereby defined as NPF-eCaCla4 and NPF-eCst20, respectively. These peptides were readily over-produced as fusion proteins in *Escherichia coli* cells through a recombinant peptide expression system [3]. Importantly, two NPF-CRIB peptides retained their high-affinity binding capacity for CaCdc42 as the CRIBs peptides alone (*Data not shown*).

To examine whether the NPF-CRIB peptides could penetrate *C. albicans* cells, a GFP protein was fused to the C-terminus of the NPF-CRIB peptides (**Figure 1A**). The internalization of these GFP-fusion proteins into mammalian cells and *C. albicans* cells were monitored by quantitative GFP fluorescence. Apparently, these GFP fusion proteins were not internalized into mammalian cells, such as *HeLa* cells shown in **Figures 1B** & **1C**. When the GFP fusion proteins were incubated with *C. albicans* cells composed of both yeast and hyphal form, the fusion proteins exhibited specific targeting to the hyphal cells (**Figures 1D** & **1F**). The GFP fusion proteins not only penetrated into the hyphal cells, but also accumulated on the surface of hyphal cells (**Figure 1F**). Interestingly, both fusion proteins had not been observed to penetrate into *C. albicans* cells in the yeast state under the current experimental conditions (**Figure 2C**). In contrast, the GFP protein alone was unable to penetrate into *C. albicans* cells in either the yeast or the hyphal state under similar conditions.

S. Del Valle et al. (eds.), *Peptides for Youth: The Proceedings of the 20th American Peptide Symposium*, 471
DOI: 10.1007/978-0-387-73657-0_202, © Springer Science+Business Media, LLC 2009

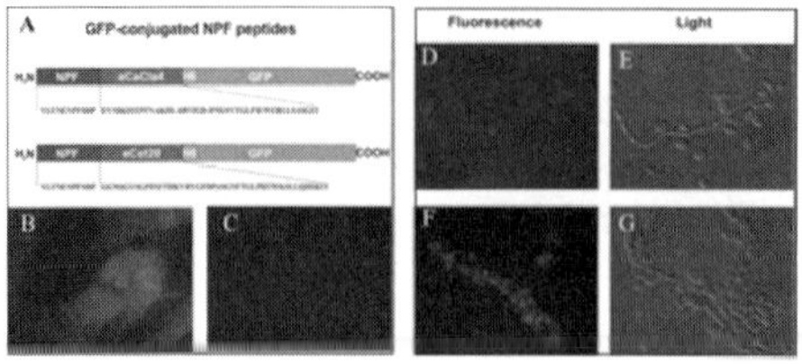

Fig. 1. NPF-conjugated CRIB peptides preferentially bind to Candida cells in the hyphal state. (A) Schematic representations of two NPF-CRIB-GFP fusion proteins. Phase-contrast micrographs were taken for HeLa cells after being incubated with NPF-eCla4-GFP protein (B & C); for C. albicans cells after incubated with NPF-eCst20-GFP protein (D & E) or NPF-eCaCla4-GFP protein (F & G).

C. albicans cells in the yeast state were incubated with different concentrations of peptides (0 ~ 100 μM) at room temperature before the temperature was switched into 37°C. After 5 hrs, no apparent inhibition was observed for those with lower concentrations of peptides (< 25 μM). When peptide concentration increased to 25 μM, both the peptides exhibited significant inhibition on the yeast-to-hyphal transition. When the peptide concentration further increased (> 25 μM), two peptides exhibited different effects on cell growth. The NPF-eCaCla4 peptide was prone to aggregation in the culture medium and adhered to cell surfaces whereas the NFP-eCst20 peptide showed no aggregation. Nevertheless, the population of yeast cells increased with increasing concentration of the peptide. In control experiments, neither CRIB peptide alone exhibits inhibition at a concentration of 100 μM.

After 16 hrs, the inhibitory effects of peptides on hyphal growth became more significant with low concentrations of peptides. In cell cultures where the initial peptide concentration was 12 μM, the majority of *C. albicans* cells became the yeast form (Figures 2C & 2G). This postponed inhibition observed at the low peptide concentration could be a result of low efficiency of NPF-mediated membrane crossing transduction into *C. albicans*. Nevertheless, it is consistent with a previous observation that the inhibition by NPF-mediated peptides exhibits long-lasting effects in cultures [2]. Given the complexity of the *C. albicans* cell wall, the significant inhibition with low micromolar concentration of peptides (e.g. 12 μM) indicates that such peptides are promising candidates of potent antifungal agents. Similar inhibitions were also observed in other culture media. Thus, the inhibition of *C. albicans* hyphal growth by both NPF-CRIB peptides is a time- and dose-dependent process and appears to be mediated by intracellular transduction.

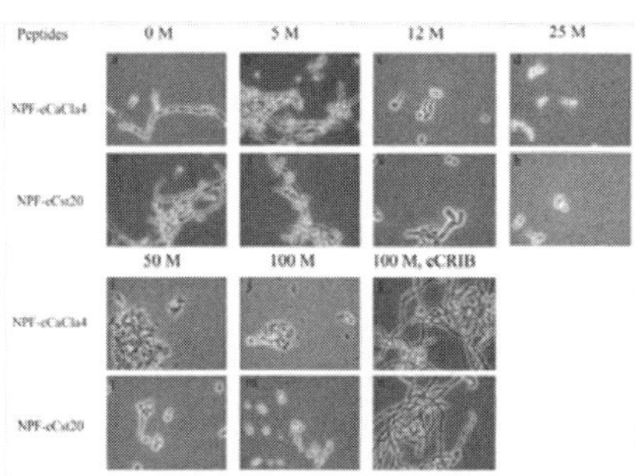

Fig. 2. Concentration dependence of peptide inhibition on Candida yeast to hyphal transition.

Acknowledgment: This work was supported in part by a Genomics and Health Initiative of the National Research Council of Canada (NRCC Pulbication No. 49514), sponsored by the Government of Canada.

References

1. Su, Z., Osborne, M. J., Xu, P., Xu, X., Li, Y. & Ni, F. (2005) *Biochem.* **44**, 16461-16474.
2. Nekhotiaeva, N., Elmquist, A., Rajarao, G. K., Hallbrink, M., Langel, U. & Good, L. (2004). *FASEB J.* **18**, 394-396.
3. Su, Z., Vinogradov A., Koutychenko, A., Tolkatchev, D. & Ni, F. (2004). *Protein Eng. Des. Sel.* **17**, 647-657.

Synthesis and Pharmacological Evaluation of Highly Potent [Dmt[1]]DALDA Analogs

P. Anantha Reddy[1], Anita H. Lewin[1] and Peter W. Schiller[2]

[1]*Center for Organic and Medicinal Chemistry, Research Triangle Institute, P. O. Box 12194, Research Triangle Park, NC 27709-2194, USA;* [2]*Laboratory of Chemical Biology and Peptide Research, Clinical Research Institute of Montreal, 110 Pine Avenue West, Montreal, Quebec, Canada H2W 1R7*

Introduction

Structure variation of the highly potent and μ-selective, peripheral analgesic [D-Arg[2],Lys[4]]dermorphin-(1-4)-amide (DALDA) [1] led to [Dmt[1]]DALDA, with 12-fold higher affinity than DALDA at μ-receptors and potency180-fold higher in the guinea pig ileum (GPI) assay. Receptor selectivity was maintained as well (K_i^{δ}/ K_i^{μ} = 14700 relative to 11400 for DALDA) [2]. [Dmt[1]]DALDA is a potent, systemically active peripheral opioid analgesic [3]. We had access to protected analogs of [Dmt[1]]DALDA which were utilized in a fragment condensation approach to the synthesis of gram-amounts of three antioxidants (including [Dmt[1]]DALDA) [4]. Encouraged by the recent report on the use of fluorescent analogs of [Dmt[1]]DALDA for confocal laser scanning microscopy (CLSM) and flow cytometry studies [5], these synthetic intermediates which may be blood-brain barrier(BBB)-permeant were examined for their bioactivity. Surprisingly, the protected analogs of [Dmt[1]]DALDA exhibited higher receptor selectivity compared to the parent peptide. The details of the synthetic approach and biological evaluation of these intermediates, having the masked side chain charges, is described.

Results and Discussion

Synthesis of the three antioxidant peptides Dmt-D-Arg-Phe-Lys-NH_2 (SS-02) (**1**), D-Arg-Dmt-Lys-Phe-NH_2 (SS-31) (**5**), and Phe-D-Arg-Phe-Lys-NH_2(SS-20) (**6**) involved routinely used side chain protecting groups for amino acid building blocks. Thus, the guanidino group was protected with NO_2 and the ε-NH_2 of Lys was protected by Cbz or 2-Cl-Cbz. This afforded the protected analogs Dmt-D-Arg(NO_2)-Phe-Lys-NH_2 (**2**), Dmt-D-Arg(NO_2)-Phe-Lys(Cbz)-NH_2 (**3**) and Dmt-D-Arg(NO_2)-Phe-Lys(2-Cl-Cbz)-NH_2 (**4**), and Boc-D-Arg-Dmt-Lys-Phe-NH_2 (**7**). These peptides were synthesized using Boc/Cbz chemistry and BOP reagent coupling as exemplified by the synthesis of **4**. Starting with the C-terminal Lys residue protected as H-Lys(2-Cl-Cbz)-NH_2, (prepared from the commercially available Boc-Lys(2-Cl-Cbz)-OH in two steps by amidation with NH_4HCO_3 in the presence of DCC/HOBt following a literature procedure [6], followed by exposure to TFA). Selective removal of the 2-Cl-Cbz in the presence of the NO_2 group was accomplished using catalytic transfer hydrogenolysis (CTH) [7]. Opioid activities were determined *in vitro* using the GPI and mouse vas deferens (MVD) assays as previously described [2]. The results are summarized in Table I.

In agreement with the literature, [Dmt[1]]DALDA (**1**) has nanomolar potency at μ receptors with 5 fold selectivity over δ receptors. Masking the guanidine side chain of D-Arg with NO_2 (**2**) led to a 2-fold potency enhancement at μ receptors and a similar potency increase at δ receptors. Masking of **2** by protection of the ε-NH_2 of Lys with Cbz (**3**) had no effect on potency at the μ receptor and produced only slightly enhanced potency at the δ receptor. Addition of a 2-Cl substituent to the Cbz

S. Del Valle et al. (eds.), *Peptides for Youth: The Proceedings of the 20th American Peptide Symposium*,
DOI: 10.1007/978-0-387-73657-0_203, © Springer Science+Business Media, LLC 2009

protecting group on ε-NH_2 of Lys in **3** to give **4** further increased μ agonist potency with δ agonist potency similar to that of **1**. Thus, **4** has significantly higher μ *vs.* δ selectivity and is 3 fold more potent than **1**. As expected, the other related peptides (**5**, **6** and **7**) were found to be inactive.

Table I. Guinea pig ileum (GPI) and mouse vas deferens (MVD) assay of [Dmt[1]]DALDA and related compounds.

		GPI	MVD	MVD/GPI
No.	***Compound***	IC_{50}[nM][a]	IC_{50}[nM][a]	IC_{50} ratio
1	Dmt-D-Arg-Phe-Lys-NH_2 (SS-02)	1.03 ± 0.04	5.38 ± 0.91	5.22
2	Dmt-D-Arg(**NO_2**)-Phe-Lys-NH_2	0.534 ± 0.028	2.92 ± 0.49	5.46
3	Dmt-D-Arg(**NO_2**)-Phe-Lys(**Cbz**)-NH_2	0.513 ± 0.066	2.19 ± 0.10	2.19
4	Dmt-D-Arg(**NO_2**)-Phe-Lys(**2-Cl-Cbz**)-NH_2	0.340 ± 0.053	5.79 ± 0.77	17.0
5	D-Arg-Dmt-Lys-Phe-NH_2 (SS-31)	4900 ± 210	inactive	-
6	Phe-D-Arg-Phe-Lys-NH_2 (SS-20)	inactive	inactive	-
7	Boc-D-Arg-Dmt-Lys-Phe-NH_2	inactive	inactive	-

[a] *Mean of 3-6 determinations ± SEM.*

Acknowledgments

The work was funded by the National Institute on Drug Abuse (NIDA), Contract No. NO1DA-3-7736 and DA-004443, and the Canadian Institutes of Health Research (FRN 5655).

References

1. Schiller, P. W., Nuguyen, T. M. D., Chung, N. N. and Lemieux, C. *J. Med. Chem.* **32**, 698-703 (1989).
2. Schiller, P. W., Nuguyen, T. M. D., Berezowska, I. and Dupuis, S. *J. Med. Chem.* **35**, 895-901 (2000).
3. Zhao, G.-M., Wu, D., Soong, Y., Shimoyama, M., Berezowska, I., Schiller, P.W. and Szeto, H.H. *J. Pharmacol. Exp. Ther.*, **302**, 188-196 (2002).
4. Zhao, K., Zhao, G.-M., Wu, D., Soong, Y., Birk, Schiller, P. W. and Szeto, H. H. *J. Biol Chem.*, **279** (33), 34682-34690 (2004).
5. Zhao, K., Zhao, G.-M., Wu, D., Soong, Y., Birk, Szeto, H. H. and Schiller, P. W. *Peptides,* **24**, 1195-1200 (2003).
6. Ueyama, N, ueno, S., Nakamura, A., Wade, K., Matsubara, H., Kumagai, S.-I., Sakakibara, S. and Tsukihara, T. *Biopolymers*, **32**, 1535-1544 (1992).
7. Gowda, D. C. and Abiraj, K. *Lett. Pept. Sci.* **9**, 153-165 (2002).

Protease-resistant glucagon like peptide-1 analogs with long-term anti-diabetes type 2 activity

Hongjian Li[1] Cindy X Zhou[2] and Zhengding Su[2,3,*]

*[1]Depart. of Biotechnology, Jinan University, Guangzhou, 510632, China, [2]Amersino Biodevelop, Kitchener ON N2M 2K7 and [3]Department of Chemistry, University of Waterloo, waterloo, ON N2L 3G1, Canada (*z2su@uwaterloo.ca)*

Introduction

Type 2 diabetes (also known as non-insulin-dependent diabetes, NIDDM) is increasing in prevalence worldwide. It is well accepted that NIDDM therapeutic approaches should rely on peptides because of the large interaction interface between glucagon like peptide-1 (GLP-1) and its cognate receptor. The size of the interface limits the efficacy of small molecule antagonists [1]. Human GLP-1 is an endogenous 30 or 31 amino acid residue peptide that occurs in two forms; GLP-1(7-36) amide and GLP-1(7-37). Both peptides have limited therapeutic utility owing to rapid degradation in the circulation by DPP IV peptidase and susceptibility to degradation by neutral endopeptidases (e.g. NEP24.11) [2,3]. As such, the drug discovery challenge is to make a stable GLP-1-like compound with a long half-life. In last decades, GLP-1 analogs have been extensively designed using mutagenesis-based side-chain replacement, but significant progress has not occurred until Extindin-4 was developed [4]. However, these GLP-1 analogues are still rapidly cleared by kidneys and thus *in situ* longevity must be ameliorated.

In the current study, we aim to modify the GLP-1 scissile peptide bonds susceptible to increase the bioavailability of therapeutic peptides by conferring protease resistance. The approach described in this current work is substitution protease-sensitive peptide bonds with pseudopeptide bonds (i.e. ethyl bond or $-CH_2-NH-$) (**Table 1**) so that the side-chains of amino acid residues will not be affected.

Results and Discussion

Peptide bonds susceptible to DPP IV and NEP24.11 peptidases are located at Ala_8-Glu_9, Asp_{15}-Val_{16}, Ser_{18}-Tyr_{19}, Tyr_{19}-Leu_{20}, Glu_{27}-Phe_{28}, Phe_{28}-Ile_{29} and Trp_{31}-Leu_{32}, (see **Table 1**). In this study, these peptide bonds are replaced with the ethyl bonds during peptide synthesis using protected *di-* or *tri-*peptide analogs as precursors. These *di-* or *tri-*peptide precursors are Fmoc-amino acid analogs with protected

Table 1. GLP-1 peptide and its pseudopeptides.

Peptide	Sequence	τ*
wtGLP-1	$H_7A\ EG_{10}TFTSD_{15}\ VSS\ \ Y\ L_{20}EGQAA_{25}KE\ F\ \ IA^{30}W\ \ LVKG_{35}R\text{-}NH_2$	<3.2 min
ABI-001	$H^7A_\Psi EG^{10}TFTSD_{15}\ VSS\ \ Y\ L_{20}EGQAA_{25}KE\ F\ \ IA_{30}W\ \ LVKG_{35}R\text{-}NH_2$	> 5 days
ABI-043	$H_7A_\Psi EG_{10}TFTSD_{15\Psi}VSS_\Psi Y_\Psi L_{20}EGQAA_{25}KE_\Psi F_\Psi IA_{30}W_\Psi LVKG_{35}R\text{-}NH_2$	> 5 days

*Ψ represents pseudopeptidyl bond with structure of $-CH_2-NH-$, replacing peptide bond at corresponding position in wtGLP-1. * τ refers to stability against proteolysis in vitro.*

S. Del Valle et al. (eds.), *Peptides for Youth: The Proceedings of the 20th American Peptide Symposium*, 475
DOI: 10.1007/978-0-387-73657-0_204, © Springer Science+Business Media, LLC 2009

ups and are prepared by a step-wise protocol using Fmoc-amino acid aldehyde as substrate. The resultant GLP-1 peptide analog with pseudopeptide bonds between Ala_8 and Glu_9 was named ABI-001 and the analog with all sensitive peptide bonds replaced was designated ABI-043 (**Table 1**). As expected, both peptidomimetics of GLP-1 (i.e. ABI-001 and ABI-043) significantly resist DPP IV- and NEP 24.11-mediated proteolysis (**Table 1**).

Importantly, both analogs (i.e. ABI-001 and ABI-043) exhibit binding to human GLP-1 receptor (hGLP-1R) expressed in CHO cells. The analog binding affinities are similar to that of native GLP-1 (i.e. IC_{50} = 0.34 nM, 0.53 nM for ABI-001 and ABI-043, respectively versus IC_{50} = 0.30 nM for wtGLP-1). Both peptide analogs also increased the accumulation of cAMP in the CHO cells with EC_{50} values similar to that of the wtGLP-1 (3.2 nM). Both the GLP-1 analogs (i.e. ABI-001 and ABI-043) are under consideration as therapeutic agents for the treatment of NIDDM. Therefore, additional studies were conducted to establish their effectiveness in animal models. Basal glycemia was markedly elevated in these mice (4.8 ± 0.3 mM in CD1 mice), as was the glycemic response to an oral glucose challenge (270 ± 18 mM·120 min in the CD1 mice). Administration of ABI-001 or ABI-043 to these mice resulted in reduction of glycemia for the duration of the 2 hr study period (**Figure 1**). The response to these analogues was markedly greater than observed for the wtGLP-1 peptide ($P < 0.05$-0.001 for all doses tested).

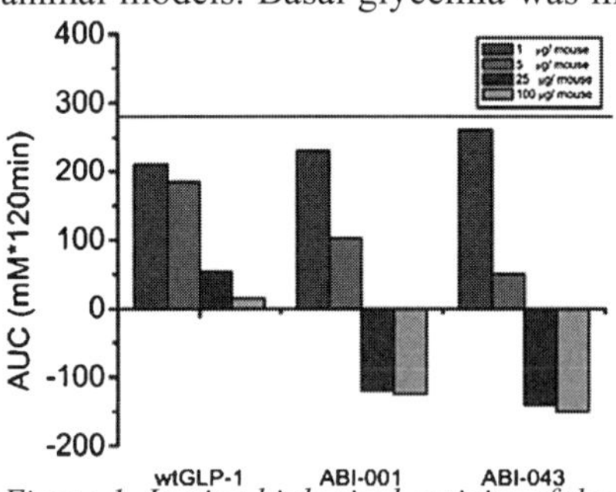

Figure 1. In vivo biological activity of the wtGLP-1 and its two pseudopeptide analogs in CD1 mice.

In summary, GLP-1 analogs with substituted pseudopeptide bonds were tested for both their binding affinities to GLP-1 receptors *in vitro* and their physiological activity *in vivo*. Results indicated that these analogs have longer lifetime in blood than wild-type GLP-1 due to protease resistance and are superior to previous GLP-1 analogs with substituted amino acid residues or the modified amino acid side-chains. Therefore, the novel GLP-1 analogs developed in the present work could be used to enhance the bioavailability of endogenous GLP-1. The result would be a GLP-1-based NIDDM therapy possible through the delivery of a potent biologically active GLP-1 analog such that therapeutic serum levels are achieved.

Acknowledgment

This work was partially supported by Guangdong Key Laboratory of Biopharmacy (2006B60119) and Guangzhou Scientific and Technological Fund (2006Z3-E4152) to H. J. L and Z.D.S.

References

1. Gallwitz, B. (2006) *Minerva Endocrinol.* **31**, 133-147.
2. Arulmozhi, D. K. & Portha, B. (2006) *Eur J Pharm Sci.* **28**, 96-108.
3. Trebbien, R.; Klarskov, L.; Olesen, M.; Holst, J. J.; Carr, R. D. & Deacon, C. F. (2004) *Am. J. Physiol. Endocrinol. Metab.*, **287**, 431-438.
4. Eng, J.; Kleinman, W. A.; Singh, L.; Singh, G. & Raufman, J. P. (1992) *J. Biol. Chem.* **267**, 7402-7408.

The discovery and development of a natural combinatorial peptide template: the cyclotides

David C. Ireland[1,2], Michelle L. Colgrave[1], Norelle L. Daly[1] and David J. Craik[1]

[1]*Institute for Molecular Bioscience;* [2]*UQ Business School, University of Queensland, Brisbane, 4053, Australia*

Introduction

The cyclotides are a group of macrocyclic peptides from the Violaceae, Rubiaceae and Curcurbitaceae plant families. They are typically comprised of 27-38 amino acids and are defined by their N-C cyclized peptide backbone and a knotted arrangement of their six conserved cysteine residues. This complex topology is termed the cyclic cystine knot (CCK) motif (1). Although their natural role in plants is believed to be one of defense, cyclotides display a broad spectrum of biological activities, including insecticidal, anti-HIV and anti-bacterial activity. Cyclotides are also completely resistant to thermal, chemical, and enzymatic degradation (2). To date, over 100 cyclotides have been isolated and characterized, making them a large family of cyclic peptides. Their interesting protein topology, proposed native function as a combinatorial template for targeting pests and pathogens, as well as their remarkable chemical properties, all suggest application of cyclotides in drug design programs.

Results and Discussion

To further our understanding of the natural diversity of cyclotides and the implications of sequence variation on structure, function and stability, we set out to characterize cyclotides expressed in five different plants from the Rubiaceae and Violaceae families. 17 novel sequences were found from two of the species. These are presented in Table 1.

Table 1. Sequences of novel cyclotides discovered in this study.

Peptide	Amino acid sequence (Cys I, II, III, IV, V, VI boxed; loops 1–6)
hyfl A	S I S . . . . C G E S C V Y I P C T V T A L V G . C T C K D . . . K V C Y . L N
hyfl B	G . S . P I Q C A E T C F I G K C Y . T E E L . G . C T C T A . . . F L C M K . N
hyfl C	G . S . P R Q C A E T C F I G K C Y . T E E L . G . C T C T A . . . F L C M K . N
cycloviolacin O13	G I . . P . . C G E S C V W I P C I . S . A A I G . C S C K S . . . K V C Y R . N
cycloviolacin O14	G . S I P A . C G E S C F K G K C Y . T . P . . G . C S C S K . . Y P L C A K . N
cycloviolacin O15	G L . V P . . C G E T C F T G K C Y . T . P . . G . C S C S . . . Y P I C K K . N
cycloviolacin O16	G L . . P . . C G E T C F T G K C Y . T . P . . G . C S C S . . . Y P I C K K I N
cycloviolacin O17	G I . . P . . C G E S C V W I P C I . S . A A I G . C S C K N . . . K V C Y R . N
cycloviolacin O18	G I . . P . . C G E S C V Y I P C T V T . A L A G . C K C K S . . . K V C Y . . N
cycloviolacin O19	G . T L P . . C G E S C V W I P C I . S S . V V G . C S C K S . . . K V C Y K . D
cycloviolacin O20	G I . . P . . C G E S C V W I P C L . T S A I . G . C S C K S . . . K V C Y R . D
cycloviolacin O21	G L . . P V . C G E T C V T G S C Y . T . P . . G . C T C S . . . W P V C T R . N
cycloviolacin O22	G L . . P I . C G E T C V G G T C N . T . P . . G . C T C S . . . W P V C T R . N
cycloviolacin O23	G L . . P T . C G E T C F G G T C N . T . P . . G . C T C D S S . W P I C T H . N
cycloviolacin O24	G L . . P T . C G E T C F G G T C N . T . P . . G . C T C D . . P W P V C T H . N
cycloviolacin O25	D I . F . . . C G E T C A F I P C I . T H V P . G T C S C K S . . . K V C Y . F N
violacin A	S A . I . S . C G E T C F K F K C Y . T . P . . R . C S C S . . . Y P V C K . . .

DOI: 10.1007/978-0-387-73657-0_205, © Springer Science+Business Media, LLC 2009

Figure 1 presents the hemolytic activities of a selection of the new cyclotides. It is clear that variations in the sequences cause significant changes in the hemolytic activity. We used hemolytic activity as a simple marker for bioactivity given that the insecticidal activity of cyclotides has been proposed to result from damage to membranes within the insect gut (3). All cyclotides were significantly less active than melittin. Since all cyclotides contain the CCK motif, the variations in activity likely reflects variations in the surface-exposed residues in the various backbone loops. It is thought that the hydrophobic patch formed by the residues of loops 2 and 5 plays an important role in producing the hemolytic activity.

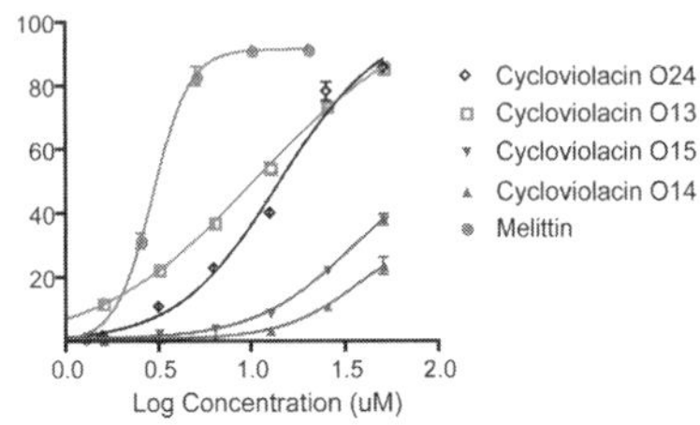

Figure 1. Hemolytic activity of a selection of cyclotides.

To test their stability, we subjected a selection of the above cyclotides to various endo- and exoproteases. The naturally occurring linear derivative, violacin A, was used as a natural negative control since it has a cystine knot, but not a cyclic backbone (4). All cyclic cyclotides were resistant to degradation by endo- and exoproteases. Although violacin A was also resistant to degradation by endoproteases, it was susceptible to degradation by exoproteases, but only residues between the termini and the nearest Cys of the cystine knot were able to be cleaved. This suggests that an intact cystine knot provides the structural stability of the cyclotides.

Using the enhanced sequence diversity and structure activity relationship knowledge of cyclotides gained from this study, we sought to exploit the stability, plasticity, and natural bioactivities of cyclotides for drug design and development applications. Two broad strategies for this were chosen. The first involved grafting therapeutically relevant peptide epitopes onto the cyclotide, which acts as a scaffold. Seven grafted cyclotide analogs were synthesized and one was amenable to folding into the native disulfide connectivity. This analog is currently being tested for activity and stability. The second strategy involves modifying the amino acid sequence of a cyclotide to enhance or suppress a bioactivity. This strategy will be used to improve the HIV inhibitory activity of cyclotides and to reduce their cytotoxicity to uninfected cells.

Acknowledgments

The work was funded by an NHMRC and ARC grant to DJC.

References

1. Craik, D. J. *et al. J. Mol. Biol.* **294**, 1327-1336 (1999).
2. Colgrave, M. L. and Craik, D. J. *Biochemistry* **43**, 4965-4975 (2004).
3. Jennings, C. *et al. Proc. Natl. Acd. Sci. USA.* **98**, 10641-10619 (2001).
4. Ireland, D. C. *et al. J. Mol. Biol.* **357**, 1522-1535 (2006).

Cyclic peptides as VEGF receptor antagonist

Victor Goncalves[1,2], Benoit Gautier[1,2], Christiane Garbay[1,2], Michel Vidal[1,2] and Nicolas Inguimbert[1,2]

[1]*Université Paris Descartes, Laboratoire de Pharmacochimie Moléculaire et Cellulaire, 45 rue des Saints Pères, Paris, F-75006 France;* [2]*INSERM, U648, Paris, 75006 France*

Introduction

Click chemistry finds more and more application as attested by the growing number of publications [1]. It covers a broad field from materials science to medicinal chemistry. This chemistry exploited mainly the copper I catalyzed Huisgen cycloaddition, which allows the formation of an anti-1,2,3-triazole by reaction of an alkyne with an azide [2]. Consequently, we sought to extend Huisgen cycloaddition chemistry to the synthesis of cyclic peptides antagonist of the VEGFR receptor. Indeed, on the basis of the structural data and mutagenesis studies [3], we designed a cyclic peptide **1** mimicking the interaction zones of the VEGF with its receptor (Figure 1). The peptide **1** behaves as an antagonist and it's able on cells to disrupt the signal transduction induced by VEGF. Such binding site antagonists could find an application in the inhibition of the tumoral growth by blocking the development of the blood vessels ensuring the nutrition of the tumor [4].

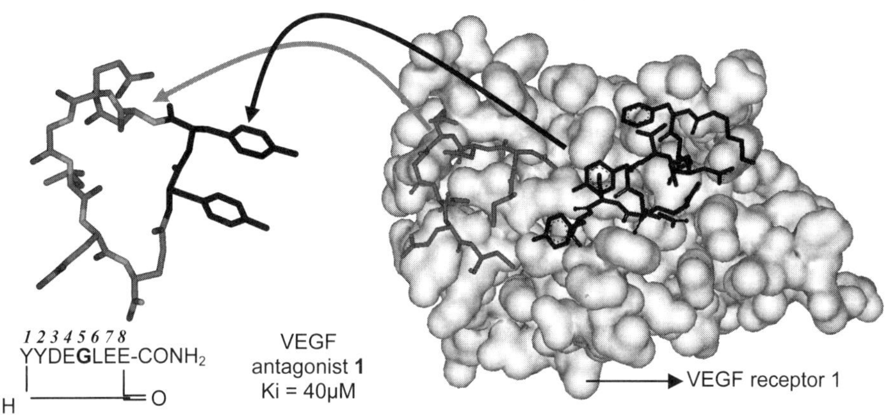

Fig. 1. VEGFR-1 in interaction with VEGF parts, synthetic VEGF antagonist.

In order to test the replacement in peptide **1** sequence of the amide bond by the isosteric 1,2,3 triazole moiety as well as the tyrosine *2* by different amino acids and so on to refine our comprehension of the structure-activity relationships of the antagonist, we synthesized peptides **3** by click chemistry on solid support (Figure 2).

Results and Discussion

Up to now, cyclization of peptides by Huisgen cycloaddition, are performed either in solution or on-resin [5]. Unexpectedly, on resin cyclization led to the formation of cyclodimeric products despite the pseudo-dilution effect of the resin [6]. In our case, the peptides **3** were synthesized using Fmoc chemistry on Rink amide resin (Figure 2). The N-terminal azide was introduced on the resin by coupling an azido acid or by

S. Del Valle et al. (eds.), *Peptides for Youth: The Proceedings of the 20th American Peptide Symposium*, 479
DOI: 10.1007/978-0-387-73657-0_206, © Springer Science+Business Media, LLC 2009

diazo transfert on the N-terminal amino acid [7]. The alkyne moiety was introduced as a L-propargylglycine (Pra) introduced in the C-terminal position. The cyclization was performed on resin and its progress was monitored by IR. After deprotection and cleavage with TFA, TIPS and water as scavengers, the crude peptides were purified using a Vydac® C-18 column and identified by means of ESI mass spectroscopy. Surprisingly, we did not observe any dimeric compounds. Replacement of the glycine in peptide **1** by a proline residue was shown to be necessary in order to introduce a bend which brings in close proximity the alkyne and azide parts thus allowing an easier cyclization. Furthermore, this cyclization is relatively general as attested by the variety of the R substituents.

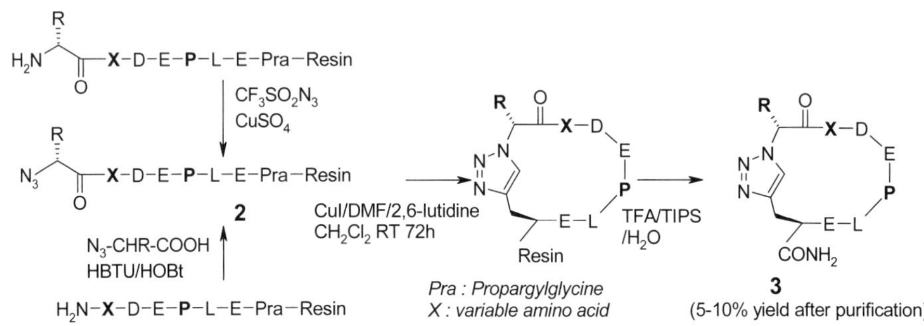

Fig. 2. Monomeric cyclic peptide containing the 1,2,3 triazole ring as amide bound isostere.

Antagonist assays were performed by evaluating the ability of synthetic peptides to inhibit binding of biotinylated VEGF to its receptor (Table 1) [8]. None of the synthesized peptides was found more active than peptide **1.** This is certainly due to the structure rigidification induced by the triazole ring. Nevertheless, these results highlight the usefulness of Huisgen cycloaddition to specifically produce monomeric cyclic peptide on resin.

Table 1. Antagonist activity of peptides.

Peptide	**1**	**2a**	**3a**	**2b**	**3b**	**2c**	**3c**	**2d**	**3d**	**2e**	**3e**	**2f**	**3f**
R	-	H		4-OH-Bz		Bz		$(CH_2)_2CONH_2$		$(CH_2)_2CO_2H$		$(CH_2)_4NH_2$	
X	-	hF[a]		Y		Y		Y		Y		Y	
Activity[b] at 100 µM (%)	99	0	29	0	0	55	61	0	0	0	6	27	28
IC50 (µM)	40	ND	ND	ND	ND	96	93	ND	ND	ND	ND	151	162

[a]hF: homophenylalanine; [b]The inhibitory activity corresponds to the percentage of biotinylated $VEGF_{165}$ displaced by 100 µM of peptide on VEGFR1; N.D.: not determined; In this set of experiments, recombinant human $VEGF_{165}$ displayed an IC_{50} of 387 ± 60 pM.

References

1. Wu, P.; Fokin, V. V. *Aldrichimica Acta*, **40**, 7-17, (2007).
2. Kolb, H.C.; Finn M. G.; Sharpless, K.B. *Angew. Chem. Int. Ed.* **40**, 2004-2021, (2001).
3. Wiesmann, *et al Cell*, **91**, 695-704, (1997).
4. Risau, W. *Nature* **386**, 671-674, (1997).
5. Tornøe, C.W., Christensen, C., Meldal, M. *J. Org. Chem*, **67**, 3057-3064, (2002).
6. Punna, S.; *et al. Angew. Chem. Int. Ed.* **44**, 2215-2220, (2005).
7. Lundquist, J. T.; Pelletier, J. C. *Org. Lett.* **3**, 781-783, (2001).
8. Goncalves, V.; *et al. Anal. Biochem.* **366**, 108-109, (2007).

Facile Synthesis of Hydrocarbon-Stapled Peptides

Introduction

The intracellular protein-protein interactions that govern many biological pathways are frequently mediated by α-helix structure of protein. Theoretically, helical peptides also can interfere with or stabilize protein-protein interactions, but native helical peptides have major shortcomings as experimental or therapeutic agents because of low potency, instability, and inefficient delivery to cells. Verdine's group [1-2] has shown that these problems could be overcome by a chemical modification of α-helical peptides they termed hydrocarbon stapling. They used (S)-α-(2'-pentenyl)alanine containing olefin-bearing tethers to generate an all-hydrocarbon "staple" by ruthenium-catalyzed olefin metathesis. The (S)-α-(2'-pentenyl)alanine peptides were made to flank three (substitution positions l and $l + 4$) or six (l and $l + 7$) amino acids within the peptide, so that reactive olefinic residues would reside on the same face of α-helix. The modified hydrocarbon-stapled peptides are helical, relatively protease-resistant, and cell-permeable peptides that bind with increased affinity for its target, and may provide a useful strategy for experimental and therapeutic modulation of protein-protein interactions in many signaling pathways.

Here we report a versatile synthetic method for hydrocarbon-stapled peptides. Asymmetric synthesis of (S)-Fmoc-α-(2'-pentenyl)alanine was successfully accomplished via an Ala-Ni (II)-BPB-complex [3] in three steps with 40% total yield. The 12-mer peptide containing two α-pentenyl-alanines on positions 4 and 8 was synthesized by the Fmoc solid phase synthesis method. After olefin metathesis and cleavage, the peptide was purified by HPLC to obtain the hydrocarbon-stapled peptide.

Results and Discussion

In contrast with Verdine's method [2] for (S)-Fmoc-α-(2'-pentenyl)-alanine, we chose Ala-Ni (II)-BPB-complex method [3] for asymmetric synthesis. The Ala-Ni (II)-BPB-complex [4] was reacted with 5-bromo-1-pentene in acetone under basic conditions to give a mixture of a Ni(II) complex of Schiff base of (S)-BPB-(S)-trans-α-(2'-pentenyl)alanine [α-(S)-**2**] and Ni(II) complex of Schiff base of (S)-BPB-(R)-trans-α-(2'-pentenyl)-alanine [α-(R)-**2**] with ratio 6:1. After separation with silica gel column, diastereo-pure α-(S)-2 complexes were obtained in 44% yield.

The α-(S)-**2** complexes were decomposed with 3N HCl/MeOH to afford (S)-α-(2'-pentenyl)alanine (**3**) as well as chiral ligand which was extracted out with DCM. After work up, (S)-α-(2'-pentenyl)alanine (**3**) was protected with Fmoc-OSu to give the (S)-Fmoc-α-(2'-pentenyl)alanine (**4,** see Figure 1) with yield 93% (two steps). The peptide **1** was synthesized manually by Fmoc solid phase synthesis methods using Rink amide MBHA resin. For the normal amino acids, the couplings were performed with fourfold excess of amino acids. Fmoc-amino acids were activated using the ratio of Fmoc-amino acid:HBTU:HOBt:DIEA, 1:1:1:2. For (S)-Fmoc-α-(2'-pentenyl)alanine , the coupling was performed with twofold excess of amino acid which was activated with DIC:HOAt (1:1). For peptide olefin metathesis, the peptide resin with N-terminal protected by Fmoc group was treated with degassed 1, 2-dichloroethane containing Bis(tricyclohexyl-phosphine)-benzylidine ruthenium

S. Del Valle et al. (eds.), *Peptides for Youth: The Proceedings of the 20th American Peptide Symposium*,
DOI: 10.1007/978-0-387-73657-0_207, © Springer Science+Business Media, LLC 2009

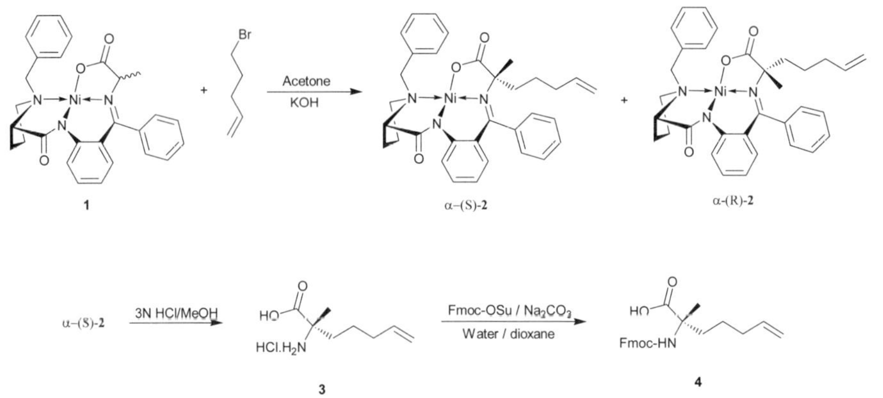

Fig. 1. Synthesis of (S)-Fmoc-α-(2'-pentenyl)alanine.

(IV) dichloride at room temperature for two hours and the reaction was repeated once for completion (see Figure 2). After Fmoc removal, resin bound peptide was cleaved using standard protocols (95% TFA, 2.5% water, 2.5% TIS). The cleaved peptide was purified by RP-HPLC using 0.1% (v/v) TFA/water and 0.1% (v/v) TFA/acetonitrile. Chemical composition of the pure product was confirmed using MS.

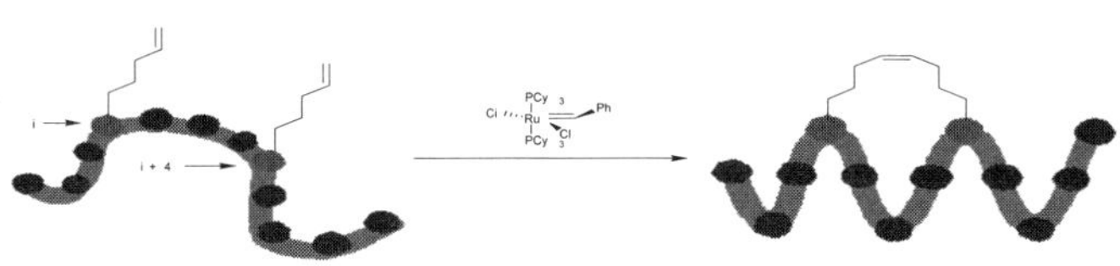

*The Sequence of Peptide **1**.*	*XXFZDLLZYYGX*
*The Sequence of Peptide **2**.*	*FITC-(βA)XXFZDLLZYYGX*
Z = (S)-α-(2'-pentenyl)alanine	

Fig. 2. Strategy for hydrocarbon-stapled peptide with enhanced α helix structure.

For fluorescently labeled peptide **2,** the N-terminal group of peptide **1** was further derivatized with β-Ala followed by FITC (DMF/DIEA) on the resin before the cleavage. The other cleavage, purification and confirmation steps were same as above. The peptide **1** not only showed enhanced α-helicity and resistance to proteolysis, but also had antiviral activity (Manuscript in preparation).

References

1. Walensky, L.D., et al., *Science* **305**, 1466-1470, (2004).
2. Schafmeister, C.E., et al., *J. Am. Chem. Soc.* **122**, 5891-5892, (2002).
3. Qiu, W., et al., *Tetrahedron* **56**, 2577-2582, (2000).
4. Belokon, Y.N., et al., *Tetrahedron: Asymmetry*, **9**, 4249-4252, (1998).

Isolation of angiotensin I converting enzyme inhibitory peptides from a fish by-products hydrolysate Tensideal®

Véronique Lahogue[1,2], Karine Vallée-Rehel[1], Mylène Cuart[1], Dominique Haras[1] and Patrick Allaume[2]

[1]Laboratoire de Biotechnologie et Chimie Marines – Université de Bretagne Sud – BP 92116 – 56321 Lorient - France; [2]IDMER – 2 rue Batelière – 56100 Lorient - France

Introduction

Many ACE inhibitory peptides have been isolated from various protein hydrolysates such as casein, soybean or fish protein. The purpose of this work is to isolate ACE inhibitory peptides derived from an enzymatic hydrolysate of fish by-products named Tensideal®, which is actually produced industrially by IDMER.

The measurement of ACE inhibitory activity was performed using a HPLC method consisting on measuring the levels of the substrate (FAPGG (N-(3-[2-furylacryloyl)-Phe-Gly-Gly)) or the product formed (FAP) with and without inhibitor. By this method, the hydrolysate Tensideal® showed an IC50 of 65µg of peptides while the reference Captopril® showed an IC50 of 0.3ng.

To isolate ACE inhibitory peptides from the hydrolysate, different chromatographic methods were successively applied to select step by step the most active fraction. Gel filtration has demonstrated that major peptides have molecular weight less than 1000 Da and some tests realised by C18 HPLC have showed that active peptides were very polar molecules. After gel filtration, SPE and LC-MS, we succeeded in identifying an active fraction with an interesting mass profile. The interpretation of the mass spectrum is actually in progress.

Measurement of ACE inhibitory activity

The compound FAP resulting of the FAPGG hydrolysis by ACE is quantified with and without inhibitor and results are compared to determine the ACE inhibition degree.

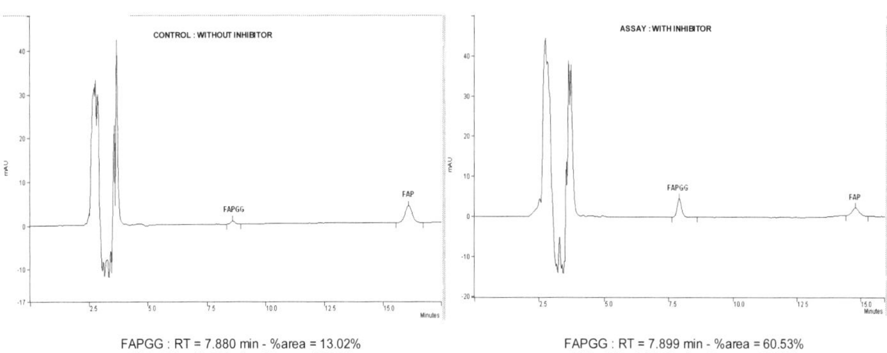

Fig. 1. HPLC Measurement of ACE inhibitory activity : % ACE inhibition = 54.6%.

S. Del Valle et al. (eds.), *Peptides for Youth: The Proceedings of the 20th American Peptide Symposium*, 483
DOI: 10.1007/978-0-387-73657-0_208,

© Springer Science+Business Media, LLC 2009

Hydrolysate Gel Filtration and ACE inhibition

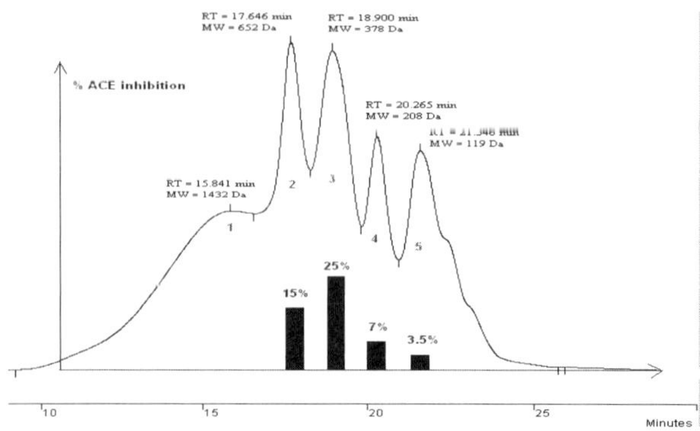

Fig. 2. Hydrolysate Gel filtration chromatogram and ACE inhibition degree of the different fractions showing the 3rd one is the most active.

Solid Phase Extraction

The 3rd peak which showed the best ACE inhibitory activity was loaded onto C18 and NH2 SPE cartridges to go on separation of active peptides. Extractions were performed with different solvents (water, acetonitrile and methyl alcohol) and each fraction (non retained and extracts) was tested in terms of ACE inhibitory activity. In fact, the non retained fractions, especially the one obtained with the C18 cartridge, were the most active confirming that active peptides are very polar molecules. The acetonitrile and methyl alcohol fractions obtained with the C18 cartridge showed also a low activity. So, it appeared to be interesting to analyse the different active fractions by LC-MS to compare their mass spectra and to identify the active peptide.

LC-MS analysis

The active fractions obtained after SPE were analysed by LC-MS with a Pursuit PentaFluoroPhenyl column fitted to very polar molecules.

All the active fraction showed the same peak at the beginning of the chromatogram corresponding to a similar mass profile (major peak at m/z 258 uma). The interpretation of this mass spectrum is actually in progress.

Conclusion

At this time, we have demonstrated that the hydrolysate Tensideal® have an ACE inhibitory activity due to small and very polar peptides. A peptidic fraction have been isolated after several chromatographic separations. The corresponding mass spectrum is composed of intense peaks and will probably allow a structural characterization of the active peptide.

Novel Melanocortin-4 Receptor Agonists That Decrease Food Intake and Body Weight

Jesse Z. Dong[1], Daniel B. DeOliveira[1], Heather A. Halem[1], John E. Taylor[1], Pierre Roubert[2], Pascale Plas[2], Michael D. Culler[1] and Jundong Zhang[1]

IPSEN Group, [1]Milford, Massachusetts 01757, U.S.A. and [2]5 avenue du Canada, 91966 Les Ulis, France

Introduction

Melanocortin-4 (MC4) receptor is considered as a viable target for developing anti-obesity drugs. Alpha-melanocyte stimulating hormone (α-MSH), a natural ligand for the melanocortin-4 (MC4) receptor, has been demonstrated to inhibit food intake and reduce body weight. In an effort to search for therapeutic agents for the treatment of obesity, we have developed a novel series of MC4 receptor agonists. One of the unique structural features of this new class of MC4 agonists is the presence of a D-amino acid residue in position 5. Such an unnatural amino acid substitution contributes to not only the selectivity but also the binding affinity at the target receptor.

Results and Discussion

Ac-Nle-c(Cys-DAla-His-DPhe-Arg-Trp-Cys)-NH_2 is a representative peptide of this new series of MC4 ligands that contain D-amino acids in position 5. Its binding affinity (Ki) for the human MC4 (hMC4) receptor is 7.6 nM versus 41.4 nM for α-MSH. As demonstrated by measurement of dose-related cyclic AMP (cAMP) production in hMC4-transfected CHO-K1 cells, Ac-Nle-c(Cys-DAla-His-DPhe-Arg-Trp-Cys)-NH_2 is a full agonist at the hMC-4 receptor with an EC_{50} of 78.9pM (peptide 1, Table 1). Its EC_{50}s on hMC1, hMC3, and hMC5 receptors, however, are approximately 31, 4, and 5385-fold higher, demonstrating that the peptide is a selective agonist at the hMC4 receptor subtype. The structural modification of Ac-Nle-c(Cys-DAla-His-DPhe-Arg-Trp-Cys)-NH_2 by replacing the Cys residue in position 10 with a penicillamine (Pen) amino acid further improves the selectivity at hMC4 receptor (peptide 2, Table 1). Peptide 1 is resistant to plasma enzymatic degradation with a half-life of 5.3 hours in rat plasma, in part due to the D-Ala amino acid residue in position 5.

Table 1. Stimulation of cAMP production in hMC4-transfected CHO-K1 cells.

Peptide	Sequence	hMC1R EC_{50}(nM)	hMC3R EC_{50}(nM)	hMC4R EC_{50}(nM)	hMC5R EC_{50}(nM)
1	Ac-Nle-c(Cys-DAla-His-DPhe-Arg-Trp-Cys)-NH_2	2.44	0.331	0.0789	417
2	Ac-Nle-c(Cys-DAla-His-DPhe-Arg-Trp-Pen)-NH_2	14.3	2.03	0.183	2240

S. Del Valle et al. (eds.), *Peptides for Youth: The Proceedings of the 20th American Peptide Symposium,* 485
DOI: 10.1007/978-0-387-73657-0_209, © Springer Science+Business Media, LLC 2009

To determine the ability of peptide 1 to reduce food intake *in vivo*, fasted male Sprague-Dawley rats were injected subcutaneously (s.c.) with peptide 1 (0.8 or 8 μmole/kg) or with vehicle. Rats treated with peptide 1 consumed ~11% less with the 0.8 μmole/kg dose, and ~63% less food with the 8 μmole/kg dose as compared with vehicle-treated control animals during the 6 hour period (Fig. 1).

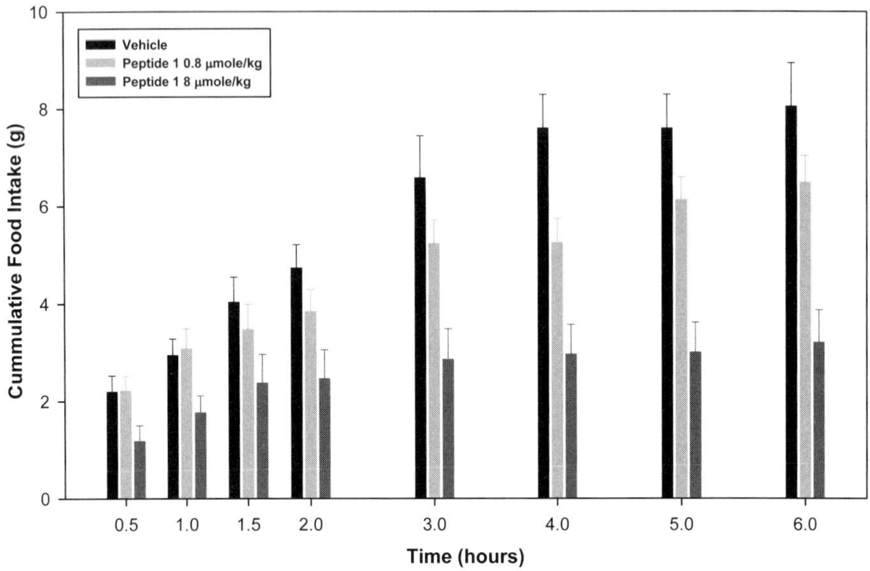

Fig. 1. Peptide 1 suppresses acute food intake in male, Sprague-Dawley rats.

The effect of peptide 1 on body weight was studied in male, Sprague-Dawley rats. At doses from 1 to 16 μmole/kg (3x/day, s.c), peptide 1 induced a progressive, dose-related decrease in body weight that reached 1-10% less than that of vehicle-treated controls by day 7.

These data demonstrate that peptide 1 is potent, selective agonist of the hMC4 receptor and is highly effective in inhibiting food intake and body weight gain with peripheral administration to normal rats.

GHS-1a Agonists That Effectively Stimulate Food Intake and Body Weight Gain

Jesse Z. Dong, Jundong Zhang, John E. Taylor, Heather A. Halem, Rakesh Datta, Michael D. Culler and John Eynon

IPSEN Group, 27 Maple Street, Milford, MA 01757, USA

Introduction

Ghrelin, a 28 amino acid octanoylated peptide, is an endogenous ligand for the growth hormone (GH) secretagogue (GHS-1a) receptor. In addition to its GH secretion effect, ghrelin enhances food intake and induces body weight gain. Due to its unique biological activities on appetite and positive energy balance, ghrelin is considered as a viable target for developing therapeutic agents for the treatment of cachexic and anorexic disorders. Previously, we reported the discovery of a novel pentapeptide that has higher GHS-1a receptor binding affinity and longer plasma half-life than human ghrelin (h-ghrelin) [1]. Based on the discovery, recently we have synthesized new peptide agonists of the GHS-1 receptor.

Results and Discussion

Screening of a series of peptides against the GHS-1a receptor revealed that the pentapeptide Inp-D-Bip-D-Trp-Phe-Lys-NH_2 is a ligand for the receptor with moderate binding affinity (peptide 1, K_i=46.8nM, Table 1). A significant improvement in the receptor binding was achieved when the first residue isonipecotic acid (Inp), the second residue D-diphenylalanine (D-Dip) and the fourth residue Phe were replaced with 4-amino-4-carboxypiperidine (Apc), D-1Nal and 4-thiazolylalanine (Taz), respectively. The resulting pentapeptide Apc-D1Nal-DTrp-Taz-Lys-NH_2 has sub-nanomolar binding affinity to the GHS-1a receptor (peptide 2, K_i=0.45nM), which is ~2-fold more potent than h-ghrelin (K_i=0.45nM vs. 0.89nM, respectively).

Table 1. GHS-1a receptor binding affinity.

Peptide	Sequence	GHS-1a Receptor Binding Affinity, K_i (nM)[a]
1	Inp-D-Dip-D-Trp-Phe-Lys-NH_2	46.8
2	Apc-D1Nal-DTrp-Taz-Lys-NH_2	0.45
3	Apc-DBal-DTrp-Taz-Lys-NH_2	0.50
4	Apc-D1Nal-DTrp-Taz-NH_2	1.41
human ghrelin		0.89

[a] *Membranes prepared from GHS-1a/CHO-K1 cells were incubated with [^{125}I]ghrelin, with or without unlabeled competing test peptides. The membrane-bound radioactivity was counted by gamma spectrometry, which was used in the K_i calculation.*

Further replacement of D-1Nal residue in peptide 2 with its derivative D-3-benzothienyl-alanine (D-Bal) did not significantly change the receptor binding affinity (peptide 3, K_i=0.50nM). The deletion of the C-terminal Lys amino acid of peptide 2 decreased the receptor binding affinity by approximately 3-fold (peptide 4,

S. Del Valle et al. (eds.), *Peptides for Youth: The Proceedings of the 20th American Peptide Symposium*, 487
DOI: 10.1007/978-0-387-73657-0_210, © Springer Science+Business Media, LLC 2009

K_i=1.41nM), indicating that the Lys residue is important for the ligand-receptor interaction.

The plasma stability of the peptides was studied in comparison with that of h-ghrelin. The *in vitro* half-lives of peptides 2 and 3 in rat plasma were ~ 8.3 and 6.0 hours respectively, compared to ~1.9 hours of h-ghrelin. The longer plasma half-lives of the pentapeptides are likely due to the presence of D-amino acids and Apc residue that are resistant to enzymatic cleavage.

To examine the ability of peptide 2 in stimulating body weight gain, male, Sprague- Dawley rats were injected 3x/day, ip, with the peptide at the dose of 40 nmole/kg or vehicle for 7 days. Treatment with the peptide induced a progressive, dose-related increase in body weight gain that reached ~6g over that of vehicle-treated controls by day 7 (Fig. 1).

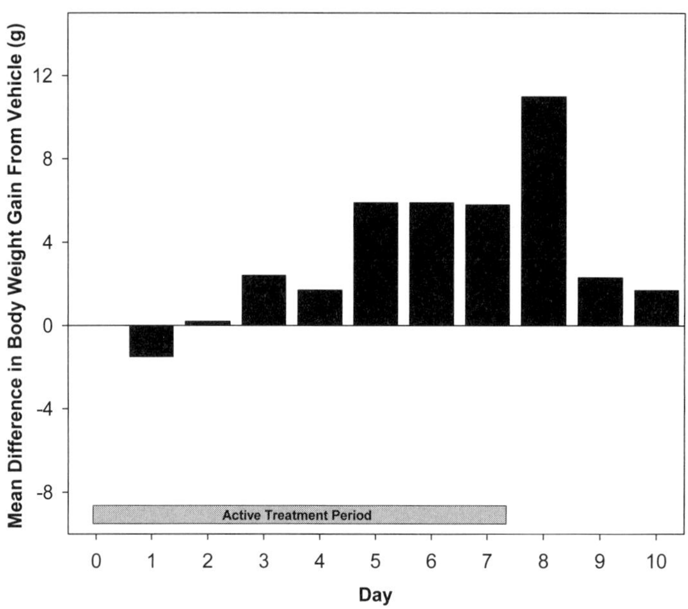

Fig. 1. Body weight gain induced by peptide 2 in male, Sprague-Dawley rats.

Taken together, due to its high GHS-1a receptor binding affinity and long plasma half-life, the pentapeptide Apc-D1Nal-DTrp-Taz-Lys-NH_2 is highly efficacious in stimulating body weight gain in rats.

Reference

1. Dong, J.Z., Eynon, E., Zhang, J., Taylor, J.E., Halem, H.A., Datta, R. and Culler, M.D. *Understanding Biology Using Peptides, Proceedings of the 19th American Peptide Symposium* (Blondelle S.E. editor) San Diego, 587-588 (2005).

Conformation-activity relationships of *cyclo*-constrained μ/δ opioid agonists derived from the N-terminal tetrapeptide segment of dermorphin/deltorphin

Jerzy Ciarkowski[1], Cezary Czaplewski[1], Sylwia Łuczak[1], Sylwia Rodziewicz-Motowidło[1], Irena Berezowska[2] and Peter W. Schiller[2]

[1]*Faculty of Chemistry, University of Gdańsk, 80-952 Gdańsk, Poland;* [2]*Clinical Research Institute of Montreal, Montreal, Quebec, Canada H2W 1R7*

Introduction

The N-terminal tetrapeptide segments of dermorphin (Tyr-D-Ala-Phe-Gly-Tyr-Pro-Ser-NH_2) and deltorphin (Tyr-D-Ala-Phe-Asp/Glu-Val-Val-Gly-NH_2) are agonists at the opioid receptors μ and δ, respectively. (D-Arg^2,Lys^4)-dermorphin(1-4) amide (Tyr-D-Arg-Phe-Lys-NH_2, DALDA) and [Dmt^1]DALDA (Dmt is 2′,6′-dimethyl-tyrosine) are among the most potent and selective μ-agonists reported to date, both *in vitro* (the latter one having picomolar μ receptor affinity) and *in vivo*. In this communication, conformation-activity studies of cyclic tetrapeptide analogs of dermorphin/deltorphin are presented and discussed. They include the peptide Tyr-c[D-Cys-Phe-Cys]NH_2, constrained via an $S^{\gamma}S'^{\gamma}$ disulfide between Cys^2 and Cys^4, and its dicarba analogs, the $C^{\gamma}C'^{\gamma}$-saturated and -olefinic ones. They are potent non-selective or moderately μ-selective opioid agonists in vitro [1] (Table 1). With a major structural constraint imposed by the 11-membered ring spanning residues 2-4, they are expected to manifest well-defined conformations of the backbone. Given the small size and confirmed bioactivity of the peptides [1], there is a good chance that their conformations determined in solution would correspond to the receptor-bound ones. We used 2D-NMR in H_2O/D_2O supported with molecular dynamics (MD) in this study.

Table 1. The analogs studied and their biological ativities [a].

Tyr-D-Bbn[b]-Phe-Bbn[b]-NH_2 (CH$_2$-X-X-CH$_2$)	No	**GPI** (μ-selective) **EC_{50} [nM]**	**MVD** (δ-selective) **EC_{50} [nM]**	**MVD/GPI** **EC_{50} ratio**
– S – S –	**1**	64.7 ± 11.9	740 ± 187	11.4
-HC=CH-(*cis*)	**2**	162 ± 17	444 ± 36	2.74
-HC=CH-(*trans*)	**3**	436 ± 97	460 ± 44	1.05
-CH_2-CH_2-	**4**	76.5 ± 5.5	655 ± 68	8.56

[a] *Reference [1];* [b] *Bbn – backbone.*

Results and Discussion

NMR spectra were recorded in H_2O/D_2O solution at 305 K with peptide concentrations about 3 mM, on a Varian Unity 500 Plus spectrometer with sodium 3-trimethylsilyltetradeuteriumpropionate, TSP, as internal standard. The assignment of proton chemical shifts was accomplished using 2D proton spectra TOCSY (80 ms), ROESY (150 ms), and DQF-COSY. Due to problems with solubility, productive sets of 2D spectra with usable cross-peaks could be collected only for

S. Del Valle et al. (eds.), *Peptides for Youth: The Proceedings of the 20th American Peptide Symposium,* 489
DOI: 10.1007/978-0-387-73657-0_211, © Springer Science+Business Media, LLC 2009

analogs **1** and **3**. Data were processed using ACDLabs [2] and XEASY [3] software. The $^3JH_{N\alpha H}$ coupling constants were obtained from the DQF-COSY spectra. The structures of **1** and **3** were refined using 2.9 ns productive MD runs, utilizing the Time-Averaged Constraints procedure (TAV-MD) dedicated to structure determinations from NMR of small flexible peptides [4]. Prior to the productive MD, routine operations, including parameterization of new moieties [5], TAV-constrained model-building, energy minimization, thermalization, etc., were performed. MD simulations were carried out using the AMBER, ver. 8.0 software [5]. Each set of conformations from the TAV-MD trajectories was clustered into 5 to 6 families of conformations. Analogs **2** and **4** were submitted to a similar 2.9 ns MD, non-restrained by NMR data, and with starting structures taken at random. Remarkably, the structures of these two essentially diverse MDs converged into two clearly different types of structures: analogs **1** and **4** merged into one of these and analogs **2** and **3** into the other, as shown in Figure 1.

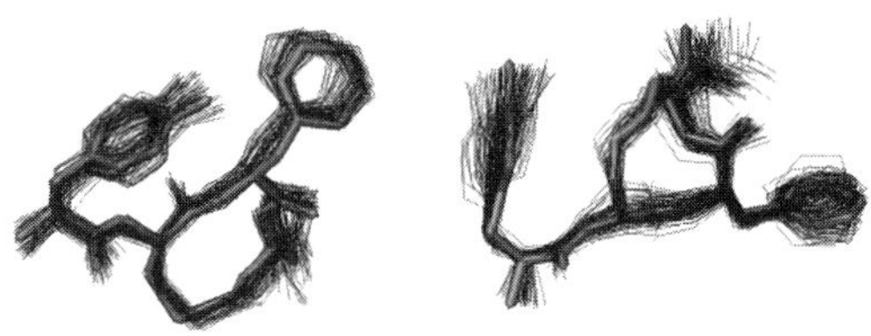

*Fig. 1. Left: 103 refined structures of **1**; Right: 109 structures of **3**.*

The aromatic rings of Tyr[1] and Phe[3] in **1** and **4** are on the same side of the heterodetic ring at an interacting distance, while in **2** and **3** the two rings are located far apart from one another. The remarkable merge of analogs **1** and **4** into one and analogs **2** and **3** into the other structural type correlates with the higher μ agonist activities and μ vs. δ selectivities of the former as compared to the latter. Thus, it is tempting to hypothesize that the enhanced μ agonist potency and μ receptor selectivity of analogs **1** and **4** may be attributed to the location of the two aromatic rings in close proximity and engaged in a stacking interaction. Of course, the exocyclic Tyr[1] residue and the Phe[3] side chain still enjoy considerable orientational freedom and it is possible that they may change their orientation upon binding to the receptor.

Acknowledgments

This work was funded by Polish Ministry of Science and Higher Education (grants BW/8000-5-0376-7, BW/8000-5-0058-7) and by grants from the CIHR and NIH.

References

1. Berezowska, I., Chung, N.N., Lemieux, C., Wilkes, B.C. and Schiller, P.W. *Acta Biochim. Polon.* **53**, 73-76 (2006).
2. ACD/Labs, Toronto, Ontario, Canada M5C 1T4, Copyright © 1994-2004.
3. Bartles, C., Xia, T., Billeter, M., Günter, P. and Wütrich, K. *J. Biomol NMR* **6**, 1-10 (1995).
4. Pearlman, D.A. *J. Biomol. NMR* **4**, 1-16 (1994).
5. Case, D.A., Darden, T.A., Cheatham, III, T.E., Simmerling, J.C.L., *et al.*.. and Kollman, P.A., AMBER 8, University of California, San Francisco (2004).

Deltorphin analogs restricted via a urea bridge: structure and opioid activity

J. Ciarkowski[1], A. Zieleniak[2], S. Rodziewicz-Motowidło[1], Ł. Rusak[1], N.N. Chung[3], C. Czaplewski[1], E. Witkowska[2], P.W. Schiller[3] and J. Izdebski[2]

[1]Faculty of Chemistry, University of Gdańsk, 80-952 Gdańsk, Poland; [2]Peptide Laboratory, Faculty of Chemistry, Warsaw University, 02-093 Warsaw, Poland; [3]Clinical Research Institute of Montreal, Montreal, Quebec, Canada H2W 1R7

Introduction

The deltorphins (Tyr-D-Ala-Phe-(Asp/Glu)-Val-Val-Gly-NH_2) are the most selective natural opioid agonists for δ receptors. To understand the molecular basis of opioid bioactivity and selectivity, we synthesized eight deltorphin analogs with major conformational restraints imposed by cyclization. Linear peptides containing dibasic amino acid residues (Lys, Orn, Dab, Dap) in positions 2 and 4 were prepared and cyclization was achieved by a urea closure through reaction of bis(4-nitrophenyl)carbonate with the side chain amino groups [1]. Molecular weights were confirmed using LSIMS mass spectrometry. Cyclic deltorphin analogs having 14-18-membered depsipetide ring structures were obtained. The peptides were tested in the guinea-pig ileum (GPI) and mouse vas deferens (MVD) assays [2] (Table 1). DEL1, DEL3 and DEL8 showed significantly enhanced potency relative to [Leu]Enk at the δ receptor and, in the case of DEL1 and DEL8, also high selectivity towards δ receptors, as determined in the MVD and GPI assays.

Table 1. The analogs studied and their biological activities.

Analog	$(CH_2)_nNHCONH(CH_2)_m$ / Tyr-*D*-Bbn[a]-Phe-Bbn[a]-Val-Val-Gly-NH_2		GPI μ-selective IC_{50}[nM]	MVD δ-selective IC_{50}[nM]
	n	m		
DEL1	4	2	65.4± 9.6	0.640 ± 0.043
DEL2	3	2	460± 9	106 ± 7
DEL3	4	1	25.4±2.0	0.483±0.065
DEL4	3	1	>10,000	27.1±3.1
DEL5	4	4	888±99	11.8±1.6
DEL6	3	4	1020±250	2.73±0.028
DEL7	4	3	>10,000	67.0±6.9
DEL8	3	3	159±23	0.814±0.054
[Leu[5]]enk	H-Tyr-Gly-Gly-Phe-Leu-OH		246 ± 39	11.4 ± 1.1

[a] *Bbn – backbone*

Results and Discussion

NMR spectra were recorded in H_2O/D_2O (9:1), with addition of CD_3COOD (DEL5-DEL8) at 305 K with peptide concentrations 3-9 mM, on a Varian Unity 500 Plus spectrometer with sodium 3-trimethylsilyltetradeuteriumpropionate, TSP, as internal standard. The assignment of proton chemical shifts was accomplished based on 2D

S. Del Valle et al. (eds.), *Peptides for Youth: The Proceedings of the 20th American Peptide Symposium*,
DOI: 10.1007/978-0-387-73657-0_212, © Springer Science+Business Media, LLC 2009

proton spectra TOCSY (80 ms), NOESY (100 ms), ROESY (200 ms), and DQF-COSY. Due to poor solubility, spectra of DEL3 and DEL4 could not be collected. NMR data were processed using the XEASY [3] software. The $^{3}JH_{N\alpha H}$ coupling constants were obtained from the DQF-COSY spectra. The structures of all but DEL3 and DEL4 were refined using 4.0 ns productive MD runs, utilizing the Time-Averaged Constraints procedure (TAV-MD) dedicated to structure determinations from NMR of small flexible peptides [4]. Prior to the productive MD, routine operations, including parameterization of new moieties [5], TAV-constrained model-building, energy minimization, thermalization, etc., were done. MD simulations were carried out using the AMBER, ver. 8.0 software [5]. Each set of conformations from the TAV-MD trajectories was clustered into 5 to 6 families of conformations. Analogs DEL3 and DEL4 were submitted to a similar 4.0 ns MD, not restrained by NMR data and with starting structures taken at random.

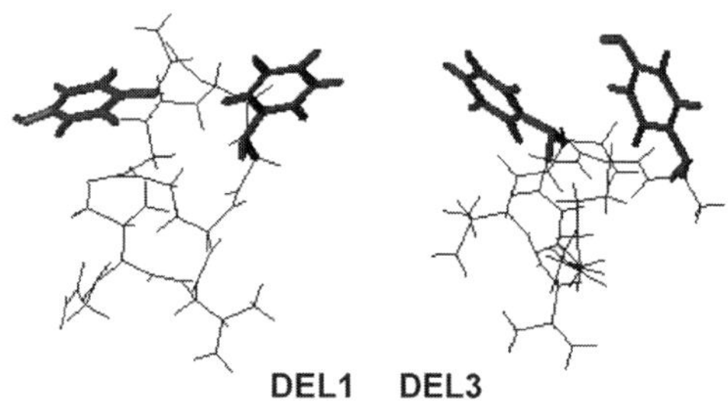

Fig. 1. Refined conformations of DEL1 (NMR data-supported) and DEL3 (NMR not supported; see above) with Tyr[1]-Phe[3] interacting aromatic rings exposed. The ring-ring centroid distances are equal to 5.8 and 6.5 Å, respectively.

Both NMR-constrained and non-constrained MD results indicate that all studied analogs tend to take up well defined structures in solution. The highly potent (DEL1, DEL3 and DEL8) and δ-selective (DEL1 and DEL3) compounds share the following common features. 1) They have a moderate size 15-16-membered ring spanning residues 2 and 4. 2) The Tyr[1] and Phe[3] aromatic rings are preferably close to each other to enable a ring stacking interaction, e.g. they are on the average 5.8, 6.5 and 9.8 Å apart in DEL1, DEL3 and DEL8, respectively. 3) The structures of DEL1, DEL3 and DEL8 are compact, showing mean distances between the N- and the C-termini of 9.4, 6.7 and 9.9 Å, respectively.

Acknowledgments

Supported by Polish Ministry of Science grant 3T09A02328 (JI, AZ, EW), BW8000-5-0058-7 (JC, SR-M) and by grants from the CIHR and NIH (NNC and PWS).

References

1. Filip, K., Oleszczuk, M., Pawlak, D., Wójcik, J., Chung, N.N., Schiller, P.W. and Izdebski, J. *J. Peptide Sci,* **9**, 649-657 (2003).
2. DiMaio, J. and Schiller, P.W. *Proc. Natl Acad. Sci* **77**, 7162-7166 (1980).
3. Bartles, C., Xia, T., Billeter, M., Günter, P. and Wütrich, K. *J. Biomol NMR* **6**, 1-10 (1995).
4. Pearlman, D.A. *J. Biomol. NMR* **4**, 1-16 (1994).
5. Case, D.A., *et al.* and Kollman, P.A., AMBER 8, Univ of California, San Francisco (2004).

Synthesis and Characterization of Dimeric Venom Peptides

J. Vizzavona, A. Zufferey and K. Rose

[1] *Department of Structural Biology and Bioinformatics. University of Geneva, 1 rue Michel Servet, 1211 Geneva 4 (Switzerland*

Introduction

Venom peptides represent a major source of bioactive compounds [1] and have led to the development of several new drugs, including Prialt (FDA approved), a peptide originating from the venom of a cone snail. A large project involving about 20 European and US groups, aims to identify a wide variety of peptide toxins of potential therapeutic interest. Each identified peptide will be synthesized, its structure determined and will be biologically tested in a variety of assays. Attempts will be made to improve on their activity.

One of the ways investigated to improve activity will be dimerization. We have investigated the dimerization of alpha-conotoxin MII [2], a 16 residue peptide containing 2 disulfide bridges and isolated from the marine snails (Conus Margus). This toxin has been found to be a highly selective inhibitor of neuronal acetylcholine receptors (nAChRs) of the subtype α3/β2. It is also able to bind to its receptor at several sites, suggesting the possibility to increase the avidity of the toxin-receptor interaction by a dimerization approach.

We have made alpha-conotoxin dimers directly using oxime chemistry and we have also incorporated the biocompatible polymers (Peg-Suc units) [3] with 2 defined lengths. The synthesis and characterization of these dimeric compounds will be described.

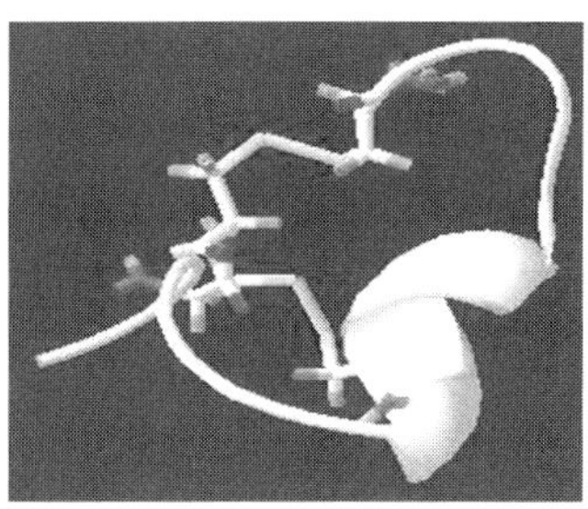

Fig. 1. 3D representative structure of an α-conotoxin.

Results and Discussion

The synthetic peptide α-conotoxin (MII) was synthesized using Boc chemistry on MBHA resin (0,9 mmol/g, 0,2 mmol scale), using the in situ neutralization approach [4], HBTU/DIEA coupling reagent using a modified ABI 433A synthesizer. Following HF deprotection with p-cresol scavenger, the crude peptide was isolated in a standard manner to give a clean HPLC profile (not shown). Two different linkers, derived from the biocompatible polymers PegSuc units, were synthesized according to the procedure described in the literature [3]. All the compounds were analyzed and purified by C_8 RP-HPLC, and characterized by MALDI-TOF mass spectrometry. The reduced form of α-conotoxin was simply oxidized at 0,2 mg/ml in

S. Del Valle et al. (eds.), *Peptides for Youth: The Proceedings of the 20th American Peptide Symposium*, 493
DOI: 10.1007/978-0-387-73657-0_213, © Springer Science+Business Media, LLC 2009

Table 1. List of compounds.

Peptides	Length of linker(A)	Expected mass	Experimental mass
α-conotoxin (MII)	na	1710,63	1711,13
MII-PegSucEDASucPeg-MII[a]	50	4308,70	4309,67
MII-(PegSucPeg)$_2$Peg(SucPeg)$_2$-MII	108	5074,18	5074,83
MII-PegSucEDASucPeg-CH_3 (control)	108	2571,82	2572,84

[a]EDA: ethylene diamine. Peg: NH-$(CH_2)_3$-$(OCH_2)_3$-CH_2NH. Suc: succinic acid.

presence of NH_4CO_3 100 mM, 15h, pH 7,8 at room temperature. The table 1 summurizes the list of molecules made in this project.

The chemistry employed for dimerization was the oxime bond that is specifically formed between an aldehyde group (introduced on the two extremities of the linkers with a periodate-mediated-oxidation of a Ser residue) and the aminooxyacetyl function [5]. The general synthetic route of MII dimer is shown in the figure 2.

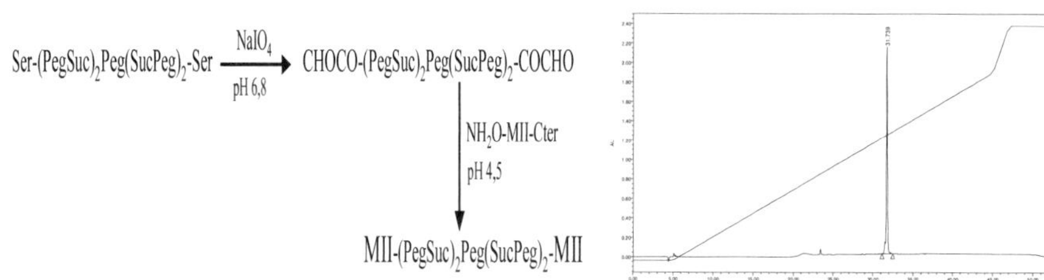

Fig. 2. example of synthetic route of MII (PegSucPeg)$_2$ Peg (SucPeg)$_2$ -MII. Right: HPLC of purified material.

All the compounds are being tested on the neuronal acetylcholine receptors of the subtype α3/β2.

Acknowledgments

The work was supported by the Swiss National Science Foundation.

References

1. Lewis Richard J. and Garcia Maria L., G. Nature, 2, 790-802 (2003).
2. Blanchfiled Joanne T. et al. J. Med. Chem. 46, 1266-1272 (2003).
3. Rose K. and Vizzavona J. J. Am. Chem. Soc. 121, 7034-7038 (1999).
4. Schnolzer M. *et al.* Int. J. Pept. Protein Res. 40, 180-193 (1992).
5. Rose K. J. Am.Chem. Soc. 116, 30-33 (1994).

Glycosyl-Enkephalins: Synthesis and Binding at the Mu, Delta & Kappa Opioid Receptors. Antinociception in Mice

Charles M. Keyari[1], Brian I. Knapp[2], Jean M. Bidlack[2], John Lowey[3], Edward J. Bilsky[3], and Robin Polt[1]

[1]*Department of Chemistry, The University of Arizona, Tucson AZ 85721;* [2]*University of New England;* [3]*University of Rochester*

Introduction

Opioid peptides are unstable toward enzymatic biodegradation, resulting in poor bioavailability in tissues and organs, and hence are not effective analgesic agents. Glycosylation of small peptides increases the hydrophilicity, stability, and bioavailability of peptides and hence transport across the blood-brain barrier (BBB) [1-3]. BBB penetration studies of glycopeptides have indicated up to a 3-fold increase in the rate of brain delivery of these compounds compared with the unglycosylated parent peptides [4,5]. Also glycosylation of the naturally occurring μ-selective dermophin and δ-selective deltorphin opioid agonists resulted in analogues with higher brain permeability and greater analgesic actions [6,7]. Therefore, glycosylation seems to be an ideal strategy for the development of drug candidates for analgesia.

The carbohydrate moiety plays important roles in both the mechanism of transport of glycosylated peptides across the BBB, as well as stability towards proteolysis. Since the glycopeptides are amphipathic in nature the concept of *"membrane hopping"* was proposed [8,9]. By NMR, CD, spin labels and molecular modeling studies it was determined that the 6-residue glycosylated enkephalin analogues adopt turn conformations at the C-terminus in the presence of micelles.

Therefore, our design strategy involves introduction of a carbohydrate moiety in the address segment of an enkephalin analog by glycosylation through the side chain of serine. This increases hydrophilicity of the peptide sequence and also increases stability towards enzymatic degradation. Second, the second position of the enkephalin sequence in the message segment will be substituted with different D-amino acids having lipophilic and hydrophilic side chains. Studies have shown that the 2nd and 5th positions can be altered to yield peptides with potent activity [10-12].

Results and Discussion

The glycopeptides were synthesized according to established solid phase procedures using Rink amide MBHA resin (substitution: 0.65 mmole/g, 1% DVB) on an automated solid phase peptide synthesizer (PS3, Protein Technologies, Inc.). The acetate protecting groups were removed under 80% H_2NNH_2/MeOH and final cleavage of the product from the resin effected by a cocktail of 90% TFA, anisole, TES-H and water. The crude peptide was precipitated from $(C_2H_5)_2O$ and purified by RP-HPLC using CH_3CN- 0.1% TFA/H_2O gradient and lyophilized to obtain white powders. The glycopeptides were characterized and analyzed by ^{1}H NMR, DQF-COSY, TOCSY, ROESY and NOESY in 90% H_2O/D_2O at pH 4.5. For structural analysis, the NMR experiment was repeated by adding 100 mg d_{25} SDS micelles to 90%H_2O/D_2O, pH 4.5.

S. Del Valle et al. (eds.), *Peptides for Youth: The Proceedings of the 20th American Peptide Symposium,*
DOI: 10.1007/978-0-387-73657-0_214, © Springer Science+Business Media, LLC 2009

Table 1. Binding affinity and antinociception of glycopeptides.

Opioid	μ	δ	κ	A_{50} *i.c.v.*	A_{50} *i.v.*
	*K_i (nM)	*K_i (nM)	*K_i (nM)	(nmol)	μmol/Kg
DAMGO	0.53	990	270	----	----
Morphine	0.79	290	12	2.38	7.84
Y-t-G-F-L-[β-Xyl]S-CONH$_2$	1.6	7.6	23	0.092	9.15
Y-t-G-F-L-[β-Fuc]S-CONH$_2$	2.3	8.4	16	0.125	4.54
Y-t-G-F-L-[α-Fuc]S-CONH$_2$	1.4	5.3	33	----	----
Y-t-G-F-L-[β-Lac]S-CONH$_2$	5.0	9.2	42	0.018	4.07
Y-t-G-F-L-S-CONH$_2$	1.4	4.1	34	0.068	46.4
Y-a-G-F-L-[β-Glc]S-CONH$_2$	2.1	6.2	340	0.034	----
Y-a-G-F-L-[β-Lac]S-CONH$_2$	4.0	15	470	0.021	5.00
Y-s-G-F-L-[β-Glc]S-CONH$_2$	2.4	6.4	270	----	----
Y-s-G-F-L-[β-Lac]S-CONH$_2$	6.0	23	480	0.223	----
Y-v-G-F-L-[β-Glc]S-CONH$_2$	9.7	180	†58%	0.064	----
Y-v-G-F-L-[β-Lac]S-CONH$_2$	19	430	†43%	0.206	----

* *Versus* μ = [^{3}H]DAMGO; δ = [^{3}H]Naltrindole; κ = [^{3}H]U69,593
† %-Inhibition at 10 μM concentration of drug
i.c.v. = Intracerebroventricular; *i.v.* = intravenous

The peptides were tested using human receptors expressed in CHO cells (Table 1). The *i.c.v.* and *i.v.* A_{50} values were calculated from dose-response curves in the 55°C tail-flick assay. The *i.c.v.* and *i.v.* administration tests show a dose-dependent antinociception. The glycopeptides with DThr^2 shows little selectivity for μ, δ or κ-opioid receptors whereas replacement with DSer^2, DAla^2 or DVal^2 resulted in the loss of κ-binding probably due to the peptide adopting a conformation that is less favorable for κ-binding. The κ-binding receptor affinity for the sequences tested follow the trend D-Thr^2 > D-Ser^2 > D-Ala^2 > D-Val^2.

Acknowledgments

We thank the Office of Naval Research (N00014-05-1-0807 & N00014-02-1-0471) and the National Science Foundation (CHE-607917) for support of these studies.

References

1. Polt, R., *et al. Pro. Natl. Acad. Sci. USA.* **91**, 7114-7118 (1994).
2. Bilsky, E. J., *et al. J. Med. Chem.* **43**, 2586-2590 (2000).
3. Albert, R., *et al. Life Sci.* **53**, 517-525 (1993).
4. Egleton, R. D., *et al. J. Pharmacol. Exp. Ther.* **299**, 967-972 (2001).
5. Egleton, R. D., *et al. Brain Res.* **881**, 37-46 (2000).
6. Negri, L., *et al. J. Med. Chem.* **42**, 400-404 (1999).
7. Negri, L., *et al. Br. J. Pharmacol.* **124**, 1516-1522 (1998).
8. Dhanasekaran, M., *et al. J. Am. Chem. Soc.* **127**(15), 5435-5448 (2005).
9. Egleton, R. D., *et al. Tetrahedron: Asymmetry* **16**, 65-75 (2005).
10. Gorin, F. A., *et al. J. Med. Chem.* **23**(10), 1113-1122 (1980).
11. Roemer, D. and Pless, J. *Life Sci.* **24**(7), 621-624 (1979).
12. Ling, N., *et al. Pept., Proc. Am. Pept. Symp.* 96-99 (1977).

Agonists and Partial Antagonists Acting on the Leptin – Leptin Receptor Interface

Laszlo Otvos, Jr[1], Marco Cassone[1], Marianna Terrasi[1], Sandra Cascio[1], George D. Mateo[1], Daniel Knappe[2], Ralf Hoffmann[2], Predrag Cudic[3], John D. Wade[4] and Eva Surmacz[1]

[1]Sbarro Institute for Cancer Research and Molecular Medicine, Temple University, Philadelphia, PA 19122, USA; [2]Center for Biotechnology and Biomedicine, Leipzig University, Leipzig D-04103 Germany; [3]Department of Chemistry and Biochemistry, Florida Atlantic University, Boca Raton, FL 33431, USA; [4]Howard Florey Institute, Melbourne, Victoria 3010, Australia

Introduction

Leptin is a neurohormone regulating energy balance and food intake acting in the hypothalamus [1]. Leptin can also modulate immune response, fertility, and hematopoiesis, acting as a mitogen, metabolic regulator, or pro-angiogenic factor. Both leptin and the leptin receptor (ObR) are overexpressed in breast and colorectal cancer [2]. An ever increasing wealth of evidence suggests that leptin is a novel pharmaceutical target for diseases related to leptin overabundance (cancer) as well as deficit (lipodystrophy or anorexia-related infertility). The full form of ObR is 1,165 amino acids long and contains the extracellular, transmembrane, and intracellular domains. The extracellular domain binds ligand, whereas the intracellular tail recruits and activates signaling substrates. Leptin appears to interact with its receptor at three discontinuous surfaces, at each with two arms [3]. Interfering with these surfaces may decrease or increase the efficiency of downstream ObR signaling in target tissues. Recombinant leptin mutants (all alanine) were suggested to act as true antagonists on STAT3 activation [4,5]. The mutated residues were as follows: in Site I Leu39-Asp40, in Site III Tyr119, and in Site III Ser120-Thr121.

The goals of our study were: i) Verify the activity of leptin fragments on cells expressing ObR and the lack of toxicity of cells lacking ObR;
ii) Analyze the antagonist activities of the alanine mutants at the peptide level;
iii) Attempt to develop novel peptide antagonists and agonists;
iv) Characterize the activity of analogs designed to improve *in vivo* stability and blood-brain-penetration (for agonists).

Results and Discussion

Our *in vitro* models were MCF-7 breast and DU-45 prostate cancer cells that express ObR, with the negative control normal mammary epithelial cell line MCF-10 that lacks the leptin receptor. *In vivo* stability was modeled by stability studies in mouse serum. The following leptin fragments and analogs were synthesized and tested:

Site I: 36-49; Site I Ala39, Ala40; Site II fragment a: 3-18; Site II fragment b: 70-89; Combined Site II construct: 3-18-Gly-Gly-Gly-70-89; Site III: 117-132; Site III Ala120, Ala121; F2: 118-129 119Tyr(I_2), 122D-Gla, 125D-Ala; D12: 119-129 119Tyr(Me), 122D-Gla, 125D-Ala; A4: 118-129 119Tyr(I_2), 125βAla, A11: 119-129 119Tyr(Me), 125βAla; E1: 119-129 119Tyr(I_2), 120Serβ($GlcAc_4$), 130Dap(Ac); D6: 118-129 119Tyr(I_2), 120Serβ($GlcAc_4$), 130Dap(Ac)

S. Del Valle et al. (eds.), *Peptides for Youth: The Proceedings of the 20th American Peptide Symposium*, 497
DOI: 10.1007/978-0-387-73657-0_215, © Springer Science+Business Media, LLC 2009

Table 1. Activity of leptin-derived peptides on MCF-7 and DU-45 (ObR+) as well as MCF-10 (ObR-) cell growth.

Peptide (10 nM – 1 μM)	Stimulation of Cell Growth After 5-day Incubation			
	MCF-7			
	No Leptin	10 nM Leptin Added	DU 45	MCF 10
Site I Ala39, Ala40	No Activity	No Activity	No Activity	Weakly Toxic
Combined Site II construct	Strong Agonist	Antagonist	Agonist	No Activity
Site III Ala120, Ala121	Agonist	Antagonist	Agonist	No Activity
E1	Strong Agonist	Weak Agonist	Not Tested	No Activity

We found that the N-terminal arm of Site I did not bind ObR and had no activity on ObR expressing cells. Alanine mutations in of Site I failed to improve ObR binding. The other two sites measurably bound to the leptin receptor. At the peptide level, all natural fragments that bound ObR and their alanine mutants appeared to be partial agonists rather than true antagonists (i.e inhibited cell growth in the presence of exogenous full-sized leptin but stimulated cell growth when full-sized leptin was unavailable). A chimera of the two independent arms of Site II was more powerful agonist than any of the constituents alone, yet it also exhibited partial antagonist activities. From the three sites, Site III seemed most suitable for medicinal chemistry manipulations. Indeed, the binding domain of Site III could be reduced to the undecapeptide fragment 119-129. Incorporation of iodinated tyrosine to the N-terminal position, a glycosylated serine to position 120 and an acetylated diaminopropionic acid moiety into the C-terminus in derivative E1 resulted in a true agonist, the only one-dimensionally active derivative we found so far. The unnatural amino acid replacements at the exopeptidase cleavage sites (N- and C-termini) improved the stability in serum. The D6 analog remained without any amino acid loss after 90-minute incubation in 25% mouse serum. The only degradation products that could be detected corresponded to analogs without the glucose protecting acetyl groups. Our data on the Site III designer glycopeptide warrant further *in vitro* studies for measuring the ability to cross the blood-brain-barrier *via* carbohydrate-mediated transport and *in vivo* studies evaluating activity in mouse models of lipodystrophy.

Acknowledgments

This work was supported by the Formula Fund of the PA Department of Health and by the Sbarro Health Research Organization.

References

1. Wauters, M., Considine, R.V. and Van Gaal, L.F. *Eur. J. Endocrinol.* **143**, 293-311 (2000).
2. Garofalo, C. and Surmacz, E. *J. Cell. Physiol.* **207**, 12-22 (2006).
3. Isarentant, H., *et al. J. Cell. Sci.* **118**, 2519-2527 (2005).
4. Niv-Spector, L., *et al. Biochem. J.* **391**, 221-230 (2005).
5. Peelman, F., *et al. J. Biol. Chem.* **279**, 41038-41046 (2004).

Effect of 4-Fluoro-L-proline on the SH3 Binding Affinity

Paolo Ruzza[1], Chiara Rubini[1], Giuliano Siligardi[2], Rohanah Hussain[2], Andrea Calderan[1], Andrea Guiotto[1], Luca Cesaro[3], Anna M. Brunati[3] and Arianna Donella-Deana[3]

[1]CNR Institute of Biomolecular Chemistry, Padua Unit, 35131 Padua, Italy; [2]Diamond Light Source, Diamond House, Chilton, Didcot, Oxfordshire, OX11 0DE, UK; [3]Department of Biological Chemistry, University of Padua, 35121 Padua, Italy.

Introduction

Short proline-rich peptides corresponding to well-known SH3 ligands exhibit little or no secondary structure before their binding to the cognate protein-targets. Under these conditions the association of a proline-rich peptide with the SH3 domain indicates unfavorable binding entropy, likely resulting from a loss of rotational freedom on the formation of the PPII helix.

With the aim of stabilizing the PPII helix conformation in SH3 binding motifs we replaced the proline residues of the HPK1 proline-rich decapeptide, PPPLPPKPKF (**P2**), either with the 4-R-fluoro-L-proline (**FPro**) or with the 4-S-fluoro-L-proline (**fPro**) [1] at different *i, i+3* positions (Table 1). The interactions of the fluoro-proline peptides with the SH3 domain of the protein cortactin were analyzed quantitatively by non-immobilized ligand interactions assay by circular dichroism (NILIA-CD). The conformation of each peptide in both aqueous and organic solvents was investigated as a function of temperature using CD spectroscopy.

Results and Discussion

The peptides were synthesized by manual solid-phase method using Fmoc chemistry. Coupling to the secondary amino group of fluoro-proline was performed in DMF using TFFH (5 eq) as coupling reagent in presence of the carboxyl component (5 eq) and DIEA (10 eq).

The conformational properties of fluoro-proline containing peptides were studied by CD spectroscopy in 25 mM Tris-HCl, pH 7.0 buffer. The CD spectra of **P2**, FPro-containing peptides and **S2** (Table 1) at 5°C showed a positive band at about 225 nm and a negative band at 200 nm, which are hallmarks of the PPII conformation. The more intense CD maximum at 225 nm for **R3** indicates that this peptide has higher PPII content. The temperature dependence showed that in all peptides a decreased PPII content was observed on increasing the temperature. However, this trend was smaller for **R3** and **R5** as their positive CD band, though smaller was still present at 45°C. In addition, the presence of an isosbetic point at about 210 nm is also indicative of an equilibrium between two conformational states: PPII and the irregular structure.

CD spectra of all peptides were also acquired in n-propanol (95% v/v), a solvent that tend to promote PPI helices [2]. As the PPII→PPI conversion is a rather slow process, the peptides were incubated in n-propanol for 6 days prior measurements. In 95% n-propanol, the shape of the CD spectra of 4(R)-analogues is characterized by the absence of the positive band at 220-230 nm range, and the reduction of the negative band at about 200 nm, suggesting the presence of an irregular structure. The CD spectra of 4(S)-analogues is characterized by the presence of a negative band at 205 nm and a shoulder (**S2** and **S3**) or a negative band (**S5**) at about 225 nm.

S. Del Valle et al. (eds.), *Peptides for Youth: The Proceedings of the 20th American Peptide Symposium*, 499
DOI: 10.1007/978-0-387-73657-0_216, © Springer Science+Business Media, LLC 2009

The binding of peptides with the SH3 domain of cortactin was monitored by the CD changes of the Trp side-chain of the recombinant fusion protein GST-SH3 upon peptide addition. The dissociation constants K_d were determined analyzing the CD data at 294 nm using a non-linear regression method [3].

Table 1. Effect of 4-fluoro-proline on the K_d values of binding to the GST-SH3$_{cort}$ protein.

Peptide	*Sequence*	K_d *(μM)*
P2	H-Pro-Pro-Pro-Leu-Pro-Pro-Lys-Pro-Lys-Phe-OH	0.8
R2	H-Pro-Pro-**FPro**-Leu-Pro-**FPro**-Lys-Pro-Lys-Phe-OH	0.9
R3	H-Pro-**FPro**-Pro-Leu-**FPro**-Pro-Lys-**FPro**-Lys-Phe-OH	5.4
R5	H-Pro-**FPro-FPro**-Leu-**FPro-FPro**-Lys-**FPro**-Lys-Phe-OH	3.9
S2	H-Pro-Pro-**fPro**-Leu-Pro-**fPro**-Lys-Pro-Lys-Phe-OH	4.7
S3	H-Pro-**fPro**-Pro-Leu-**fPro**-Pro-Lys-**fPro**-Lys-Phe-OH	18.0
S5	H-Pro-**fPro-fPro**-Leu-**fPro-fPro**-Lys-**fPro**-Lys-Phe-OH	60.0

FPro: 4-R-fluoro-L-proline; fPro: 4-S-fluoro-L-proline.

The introduction of fPro residues into the **P2** peptide chain decreases the affinity for the SH3 domain, as expected from the conformational properties of the synthetic peptides (contents of PPII helix) previously described. The fPro peptides are less structured than **P2** peptide and consequently these analogues undergo a change from almost non-structured to partially structured in a PPII helix. This process is likely to produce a significant entropic penalty and undoubtedly accounts for some of the unfavorable contributions to the observed K_d values. Indeed, the K_d value determined for the peptide **S2** (4.7 μM) is 6-fold higher than that of the parent peptide **P2**, while peptides **S3** and **S5**, characterized by the absence of PPII helix also at low temperature (5°C), display high K_d values (18.0 and 60.0 μM, respectively).

Unexpectedly, the peptides containing the FPro residues and adopting a PPII helix conformation are also characterized by K_d values that are higher than that the displayed by the parent peptide **P2** (see Table 1), with the only exception of **R2** peptide that shows K_d values comparable with that of **P2** peptide.

These results provide two different conclusions: i) the replacement of Pro residues by fluoro-proline confirms that its ability to stabilize the PPII-like conformation is strictly related to the residue position as well as to the configuration of the γ-substitution; ii) the stabilization of the peptide conformation is not sufficient to increase the binding affinity, suggesting that the interaction of the Pro-rich peptides to the cortactin-SH3 domain is the result of a delicate balance of mutually compensating contributions and that different factors play a role in the binding energetics.

References

1. Horng, J. C. and Raines, R. T. *Protein Sci.* **15**, 74-83 (2006).
2. Mutter, M., *et al. Biopolymers* **51**, 121-128 (1999).
3. Siligardi, G., *et al. J. Biol. Chem.* **277**, 20151-20159 (2002).

Analogues of arginine vasopressin modified in the N-terminal part of the molecule with pipecolic acid isomers

Dariusz Sobolewski[1], Adam Prahl[1], Jiřina Slaninová[2] and Bernard Lammek[1]

[1]*Faculty of Chemistry, University of Gdańsk, Sobieskiego 18, 80-952 Gdańsk, Poland;* [2]*Institute of Organic Chemistry and Biochemistry, Academy of Sciences of the Czech Republic, Flemingovo sq. 2, 16610 Prague, Czech Republic*

Introduction

One of the most straightforward approaches for peptide modification is to introduce changes into the side chains of single amino acids. At this level, a multitude of possibilities for the synthesis of non-proteinogenic amino acids already exists, and useful preparative routes for the asymmetric synthesis of many derivatives have been developed. This strategy enabled incorporation of amino acids with side chains that do not or uncommonly occur naturally in peptides or proteins, with the aim to introduce special functional groups, either to restrict conformational flexibility of a peptide or to enhance its metabolic stability. Furthermore, D-configured amino acids, N^{α}-alkylated amino acids, or C^{α}-dialkylated amino acids can be employed.

We have designed ten new analogues of arginine vasopressin (AVP) modified in the N-terminal part of the molecule with pipecolic acid (Pip) isomers, a non-proteinogenic α-imino acid, also know as homoproline. Eight of the peptides were designed by substitution of Tyr^{2} or Phe^{3} of the AVP and [3-mercaptopropionic acid (Mpa)^{1}]AVP (dAVP) with L- or D-pipecolic acid. In the case of a potent V_2 agonist, [Mpa^{1},Val^{4},D-Arg^{8}]VP (dVDAVP), two similar analogues have been prepared by substituting position 3 with L- or D-Pip. This modification, apart from reducing the flexibility, also changed the character of a fragment of the molecule from aromatic to aliphatic. We designed the following analogues: [L-Pip^{2}]AVP (**I**), [Mpa^{1},L-Pip^{2}]AVP (**II**), [D-Pip^{2}]AVP (**III**), [Mpa^{1},D-Pip^{2}]AVP (**IV**), [L-Pip^{3}]AVP (**V**), [Mpa^{1},L-Pip^{3}]AVP (**VI**), [D-Pip^{3}]AVP (**VII**), [Mpa^{1},D-Pip^{3}]AVP (**VIII**), [Mpa^{1},L-Pip^{3},Val^{4},D-Arg^{8}]VP (**IX**), [Mpa^{1},D-Pip^{3},Val^{4},D-Arg^{8}]VP (**X**).

Results and Discussion

Ten new analogues of AVP (**I-X**) were synthesized manually, using Fmoc chemistry. Surprisingly enough, peptides **I-IV** demonstrated a high sensitivity to trifluoroacetic acid (TFA) used for deprotection and cleavage of the synthesized peptides from the resin. The products of fragmentation are presented in Table 1.

Table 1. TFA-catalyzed hydrolysis of peptides I-IV during TFA cleavage from the resin.

Peptide		Molecular ion[a] *calc.*	Molecular ion[a] *found*	Products of cleavage
[L-Pip^{2}]AVP	**I**	1032.2	931.2	H-Cys + L-Pip-Phe-Gln-Asn-Cys-Pro-Arg-Gly-NH_2
[Mpa^{1},L-Pip^{2}]AVP	**II**	1017.2	820.2	Mpa-L-Pip + Phe-Gln-Asn-Cys-Pro-Arg-Gly-NH_2
[D-Pip^{2}]AVP	**III**	1032.2	931.2	H-Cys + D-Pip-Phe-Gln-Asn-Cys-Pro-Arg-Gly-NH_2
[Mpa^{1},D-Pip^{2}]AVP	**IV**	1017.2	820.3	Mpa-D-Pip + Phe-Gln-Asn-Cys-Pro-Arg-Gly-NH_2

[a] *Mass spectra of the peptides were recorded on a MALDI TOF mass spectrometer*

S. Del Valle et al. (eds.), *Peptides for Youth: The Proceedings of the 20th American Peptide Symposium*, 501
DOI: 10.1007/978-0-387-73657-0_217, © Springer Science+Business Media, LLC 2009

The purity of compounds **V**-**X** was determined by HPLC and the molecular ions values were as expected. The activities of these analogues were determined in the *in vitro* rat uterotonic test in the absence or in the presence of 1mM magnesium ions, in the rat pressor test, and in the antidiuretic assay on conscious rats (for details see ref. [1]). None of the six analogues showed any activity in these assays.

For analogues **II** and **IV**, we propose a mechanism of cleavage *via* formation of an oxazolone-like intermediate (Fig. 1) from a species protonated at the Pip residue by nucleophillic attack of an adjacent carbonyl group [2]. Apparently, the important factor here is the charge density on the carbonyl oxygen. The inductive (I) effect of the alkyl side chain of the residue adjacent to the carbonyl group enhances the carbonyl nucleophillicity.

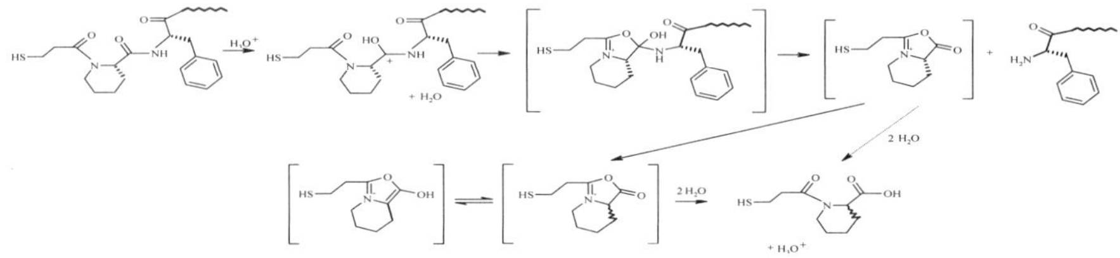

Fig. 1. Proposed mechanism of acidic hydrolysis of analogue II (similar to analogue IV).

A comparison of the hydrolysis products of peptides with Mpa[1] (**II** and **IV**) and Cys[1] (**I** and **III**) residues reveals importance of the N-terminal amino group on the mechanisms of hydrolysis. Protonation of the free N-terminal amino group of peptides **I** and **III** results in both decreased electron density (-I effect) on the carbonyl oxygen of the Cys[1] residue and in formation of a hydrogen bond between the N-terminus and that oxygen. Consequently, the acid-promoted hydrolysis of the amides can serve as a model for cleavage of the peptide bond in analogues **I** and **III** (Fig. 2) [3].

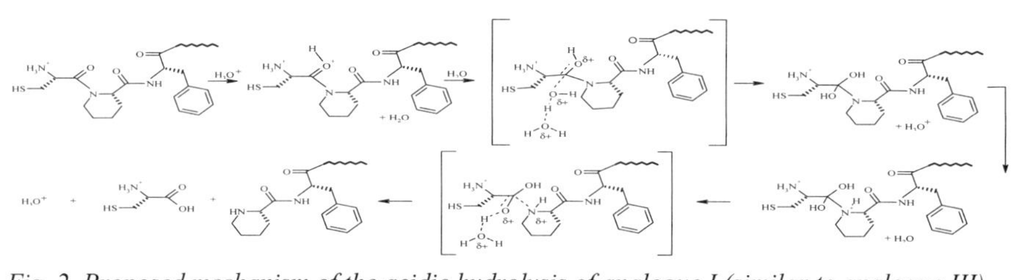

Fig. 2. Proposed mechanism of the acidic hydrolysis of analogue I (similar to analogue III).

Our results, while not impressive in terms of biological activities of the reported analogues, offer important information on structure – activity relationships and response of the L- or D-Pip-containing peptides to standard TFA cleavage conditions.

Acknowledgments

Supported by grants 0066/H03/2006/30 and 1312/T09/2005/29 and by research project Z4055905.

References

1. Kowalczyk, W., *et al. J. Med. Chem.*, **47**, 6020-6024 (2004).
2. Urban, J., Vaisar, T., Shen, R. and Lee, M. S. *Int. J. Pept. Protein Res.* **47**, 182-189 (1996).
3. Brown, R.., Bennet, A. and Ślebocka-Tilk, H. *Acc. Chem. Res.* **25**, 482-488 (1992).

Analogues of arginine vasopressin modified at position 2 with proline derivatives: selective antagonists of oxytocin *in vitro*

Dariusz Sobolewski[1], Adam Prahl[1], Jiřina Slaninová[2] and Bernard Lammek[1]

[1]*Faculty of Chemistry, University of Gdańsk, Sobieskiego 18, 80-952 Gdańsk, Poland;* [2]*Institute of Organic Chemistry and Biochemistry, Academy of Sciences of the Czech Republic, Flemingovo sq. 2, 16610 Prague, Czech Republic*

Introduction

In mammals, arginine vasopressin (AVP) is mainly synthesized and released into the circulation by the magnocellular neurons of the supraoptic and paraventricular hypothalamic nuclei, with axons projecting to the pituitary [1]. Physiological effects of AVP are mediated by at least three distinct vasopressin receptor subtypes: V_{1a}, known to mediate the contractile action of AVP on vascular smooth muscles and stimulate glycogenolysis in the liver; V_{1b} (V_3), involved in the release of ACTH from pituitary; and V_2, receptors mediating antidiuretic action in the kidney [2]. AVP also interacts with oxytocin (OT) receptors which are responsible for the galactobolic and uterotonic effects [3].

Many of the arginine vasopressin (AVP) agonists and antagonists have been designed and synthesized in the course of extensive investigations of the structure – activity relationships [4,5]. All these efforts resulted in probably the best understanding of the relationships among peptide hormones. However, the design of analogues that are very active and truly selective for AVP receptors, still remains an area of great interest.

We have designed, synthesized and determined some pharmacological properties of eight new analogues of AVP and some of its agonists modified at position 2 with conformationally constrained amino acids, (2S,4S)-4-cyclohexyl-pyrrolidine-2-carboxylic acid (Cpc) and (2S,4S)-4-phenyl-pyrrolidine-2-carboxylic acid (Ppc). We assumed that these modifications reduce flexibility of the peptides and force their backbones and side chains to adopt specific orientations. We synthesized the following analogues: [Cpc^2]AVP (**I**), [Ppc^2]AVP (**II**), [Mpa^1,Cpc^2]AVP (**III**), [Mpa^1,Ppc^2]AVP (**IV**), [Cpc^2,D-Arg^8]VP (**V**), [Ppc^2,D-Arg^8]VP (**VI**), [Mpa^1,Cpc^2,D-Arg^8]VP (**VII**), and [Mpa^1,Ppc^2,D-Arg^8]VP (**VIII**).

Results and Discussion

The eight new analogues of AVP (**I-VIII**) were synthesized by Fmoc strategy, purified and characterized. Their molecular ion values were as expected and their purity was better than 97%. The activities of the new analogues were determined in the *in vitro* rat uterotonic test in the absence of magnesium ions, the rat pressor test, and in the antidiuretic assay the conscious rats were used (for details see ref. [6]). The results of pharmacological evaluation of peptides **I-VIII**, together with relevant data for AVP and some related peptides, are presented in Table 1. Compounds modified at position 2 with Cpc and Ppc exhibited either weak (**I**, **II**, **V** and **VI**) or moderate (**III**, **IV**, **VII** and **VIII**) antiuterotonic activities. They display no pressor activity. They showed either low (**III**, **IV**, **VII** and **VIII**) or negligible (**I**, **II**, **V** and **VI**) potency in the antidiuretic test; the dose-response curve of peptides containing Mpa^1 residue (**III**, **IV**, **VII** and **VIII**) was steeper than that of AVP.

S. Del Valle et al. (eds.), *Peptides for Youth: The Proceedings of the 20th American Peptide Symposium*, 503
DOI: 10.1007/978-0-387-73657-0_218, © Springer Science+Business Media, LLC 2009

Table 1. Pharmacological properties of the new AVP analogues (IU/mg or pA_2).

Analogue		Activity		
		Oxytocic Uterus *in vitro* test no Mg^{2+}	Pressor IU/mg	Antidiuretic[b] IU/mg $t_{1/2}$ 60 ($t_{1/2}$ 200)
AVP[a]		17	412	465
[Mpa[1]]AVP[a]		27-63	346-370	1300-1745
[D-Arg[8]]VP[a]		0.4	4.1	114-257
[Mpa[1],D-Arg[8]]VP[a]		1.5-5.1	~0.39	800-50000
[Cpc[2]]AVP	**I**	pA_2=6.87±0.14	0	<0.45 (<0.45)
[Ppc[2]]AVP	**II**	pA_2=6.73±0.17	0	<0.45 (<0.45)
[Mpa[1],Cpc[2]]AVP	**III**	pA_2=7.67±0.11	0	2.5 (40)
[Mpa[1],Ppc[2]]AVP	**IV**	pA_2=7.36±0.15	0	4.5 (20)
[Cpc[2],D-Arg[8]]VP	**V**	pA_2=6.42±0.18	0	<0.45 (<0.45)
[Ppc[2],D-Arg[8]]VP	**VI**	pA_2=6.18±0.18	0	<0.45 (<0.45)
[Mpa[1],Cpc[2],D-Arg[8]]VP	**VII**	pA_2=7.17±0.30	0	4.5 (20)
[Mpa[1],Ppc[2],D-Arg[8]]VP	**VIII**	pA_2=7.28±0.10	0	4.5 (20)

[a] values taken from [5];
[b]the activities obtained by comparing doses of the analogues and AVP resulting in an antidiuresis time of $t_{1/2}$=60 min in arbitrary units; in parentheses, the activities obtained by comparing doses of the analogues and AVP resulting in an antidiuresis time of $t_{1/2}$=200 min.

In general, the results presented here confirm previous knowledge about the significant role of the Tyr residue at position 2 in initiating the pressor response [7]. The results imply that our modifications are incompatible with the binding sites of the V_{1a} and V_2 receptors and also disrupt the interaction with OT receptors. All the analogues are selective oxytocin antagonists. Summing up, our studies provide new, useful information about structure-activity relationships and open up new possibilities for designing potent OT antagonists.

Acknowledgments

Partial funding for this work was provided by Polish Ministry of Education and Science under Grant No. 0066/H03/2006/30 and 1312/T09/2005/29 and by research project No. Z4055905 of the Academy of Sciences of the Czech Republic.

References

1. Barberis, C. and Tribollet, E. *Crit. Rev. Neurobiol.* **10**, 119-154 (1996).
2. Jard, S. *Adv. Exp. Med. Biol.* **449**, 1-13 (1998).
3. Mouillac, B., *et al. J. Biol. Chem.* **270**, 25771-25777 (1995).
4. Manning, M. and Sawyer, W. H. *J. Receptor Res.* **13**, 195-214 (1993).
5. Lebl, M., Jošt, K. and Brtník, F., In Jošt, K., Lebl, M. and Brtník, F. (Eds.) Handbook of Neurohypophyseal Hormone Analogs, Vol II, Part 2: CRC Press Inc., Boca Raton, Florida, 127-267 (1987).
6. Kowalczyk, W., *et al. J. Med. Chem.*, **47**, 6020-6024 (2004).
7. Hlavacek, J., In Jošt, K., Lebl, M. and Brtník, F. (Eds.) Handbook of Neurohypophyseal Hormone Analogs, Vol I, Part 2: CRC Press Inc., Boca Raton, Florida, 109-129 (1987).

Multimeric peptides as agonists of the erythropoietin receptor

Oscar Vadas and Keith Rose

Department of Structural Biology and Bioinformatics, University Medical Center (CMU), CH-1211 Geneva 4, Switzerland.

Introduction

Erythropoietin (EPO) is a hormone that controls the proliferation and differentiation of red blood cells in bone marrow. The recombinant protein (rEPO) and other erythropoiesis stimulating agents are administered for the treatment of anaemia resulting from renal failure. EPO specifically interacts with EPO Receptor (EPOR) by inducing the reorientation of two receptor chains on the membrane, initiating signal transduction.

Small Erythropoietin Mimetic Peptides (EMP) that share no sequence similarities with the endogenous hormone were previously reported to activate EPOR [1]. Crystal structure of EMP1 bound to EPOR demonstrated that two peptides were necessary to activate the dimeric receptor. Synthesis of EMP dimers to increase there biological activity by multivalency were synthesized, either with a commercial polyethylene glycol (PEG) as linker or using a C-terminal lysine residue as branchpoint [2]. In both cases, the avidity for EPOR and the biological activity tested on cell proliferation assays was increased for the dimer compared to the monomeric peptide. To test the influence of linker length and of peptide attachment side on receptor activation, we synthesized a panel of EMP dimers with various linker lengths. Monomers were linked by a dialdehyde PEG-based polyamide linker of defined length, allowing homogeneous molecules of various lengths to be synthesized [3]. Using the chemoselective oxime chemistry, peptides could be attached through their N- or C-terminal side, via the coupling of an aminooxy moiety. Biological activity of peptides, measured by cell proliferation assays, showed that dimers had higher potency than monomers, whereas linker length as well as site of peptide attachment did not have much influence on biological activity.

N-EMP	AoA-GGLYACHMGPMTWVCQPLRG-NH_2
C-EMP	GGLYACHMGPMTWVCQPLRGK(AoA)-NH_2

Fig. 1. Sequences of EMP monomers with the aminooxyacetyl (AoA) functionality on the N-terminus (N-EMP) or on the C-terminus (C-EMP).

Results and Discussion

Peptide dimers were synthesized using Boc/Bzl chemistry on MBHA resin (see figure 1 for sequences). For N-terminal attachment of linker (N-EMP) to the peptide, Boc-aminooxyacetyl (Boc-AoA) functionality was manually coupled as the N-hydroxysuccinimide ester (Boc-AoA-OSu) after chain elongation. For the C-terminal attachment of linker (C-EMP) to the peptide, Boc-Lys(Fmoc)-OH was coupled as the first residue. After completion of the synthesis, Fmoc protecting group was removed by treatment with 20% piperidine and Boc-AoA-OSu was coupled on the ε-amine of the lysine side chain. Certain protecting groups were removed (Dnp with 20% 2-mercaptoethanol and 10% DIEA; formyl with 20% piperidine; Boc with neat TFA) prior to acid cleavage with HF containing 5%

S. Del Valle et al. (eds.), *Peptides for Youth: The Proceedings of the 20th American Peptide Symposium*, 505
DOI: 10.1007/978-0-387-73657-0_219, © Springer Science+Business Media, LLC 2009

p-cresol for 60 minutes at 0°C. After refolding step, peptides were purified by reverse-phase HPLC to be ready for oximation reaction with a dialdehyde linker.

Dialdehyde linkers were synthesized by successive addition of diacid (succinic anhydride) and diamine (4,7,10-trioxa-1,16-tridecanediamine) as described in [3]. Three sizes of linkers were synthesized; short (dialdehydes are separated by 170 bonds), medium (246 bonds) and long (322 bonds). Purification was done by reverse phase HPLC. Reaction of pure aminooxyacetyl peptide with dialdeyde linker gave the expected dimeric molecule through the formation of an oxime bond, which was characterized by analytical HPLC and MALDI-TOF MS.

Cell proliferation assays were performed with an EPO-responsive cell line: UT-7/EPO [4]. Cells were grown in 96-well plates in presence of increasing concentrations of agonist for 4 days. Total number of living cells was measured by a colorimetric assay with tetrazolium salt [5]. All experiments were done in triplicate, and the same experiment was repeated at least 3 times for each compound. Determination of the EC_{50} by measuring the absorbance at 570 nm allowed to compare the potency of the molecules, using rEPO as standard (table 1).

Table 1. EC_{50} of molecules measured by proliferation assay on UT-7/EPO cells (pM).

Peptide	Monomer	Short dimer	Medium dimer	Long dimer
N-EMP	45062	136	108	98
C-EMP	20658	152	164	284
EPO	23			

N-EMP: N-terminally linked peptide; C-EMP: C-terminally linked peptide

Activity of dimers, given by EC_{50} values, was much greater than activity of monomers. Regarding the side of attachment of the linkers, we saw that N-terminal linked dimers and C-terminal linked dimers had similar potencies, although monomeric C-EMP was two times more potent than N-EMP. Minor differences in potency between short, medium and long dimers were observed, showing that the linker length does not affect the activation of the receptor. Finally, acting on the way of attachment of two EMPs resulted in molecules with comparable potencies, with N-EMP dimers showing EC_{50} values only four times higher than that of rEPO.

Acknowledgments

This work was funded by the Swiss National Research Foundation (FNRS), the University of Geneva and the Schmidheiny Fondation.

References

1. Wrighton, N.C., *et al. Science* **273**, 458-463 (1996).
2. Johnson, D.L., *et al. Chem Biol.* **4**, 939-950 (1997).
3. Rose, K., *et al. J. Am. Chem. Soc.* **121**, 7034-7038 (1999).
4. Komatsu, N., *et al. Blood* **82**, 456-464 (1993).
5. Mosmann, T. *J. Immunol. Methods.* **65**, 55-63 (1983).

Synthesis and *in vitro* Pharmacological Profile of Potent and Selective Peptidic V1a Receptor Agonists.

Kazimierz Wisniewski, Robert Galyean, Hiroe Taki, Sudar Alagarsamy, Glenn Croston, Regent Laporte, Claudio D. Schteingart and Pierre J-M. Rivière

Ferring Research Institute Inc., 3550 General Atomics Court, San Diego, CA 92121, USA

Introduction

The neurohypophyseal hormone, 8-arginine-vasopressin (AVP, **1**), produces arterial vasoconstriction by activation of the V1a vasopressin receptor (V1aR) located in vascular smooth muscle. This vasopressor effect can be used to treat vasodilatory hypotension in critical care and other conditions [1]. AVP is the endogenous ligand for the V1aR and the two additional vasopressin receptors (V1bR and V2R), and also binds to the oxytocin receptor (OTR). When used as a vasopressor, the lack of selectivity of AVP could possibly generate unwanted side effects such as V2 receptor mediated antidiuresis and release of coagulation factors. In order to find new, more V1a selective agonists, analogs of **1** of the general structure [X^2,Ile^3,Y^4,Z^8]VP have been designed, synthesized and tested *in vitro*.

Results and Discussion

Enhanced selectivity towards V1aR in a rat *in vivo* model for [Phe^2,Ile^3,Orn^8]VP, **2**, was first reported by Huguenin [2]. The peptide was tested in our *in vitro* RGA assays and showed improved selectivity versus related receptors as compared to **1** (see table 1). It was also demonstrated that the Ferring compound, [Hmp^1,Phe^2,Ile^3, Hgn^4,$Orn(Dab)^8$]VP (F-180), is a potent and selective V1a receptor agonist [3]. Therefore, all new analogs were designed with the Ile^3 residue and positions 2, 4 and 8 were selected for modifications.

The analogs with an N-ω-alkylated diamino acid residue in position 8 were prepared by Fmoc SPPS. The Mtt group was employed for the side chain amino group protection. After peptide assembly the Mtt group was removed with 1% TFA/DCM. Reductive alkylation with acetone/$NaBH(OAc)_3$ provided the N-isopropyl peptides. To obtain N-alkyl analogs with non-branched alkyl groups, the o-nitrobenzenesulfonyl group (o-NBS) [4] was transiently introduced and the intermediate sulfonamide was alkylated with an appropriate alcohol under Mitsunobu reaction conditions. The o-NBS group was removed with 5% potassium thiophenolate in DMF, and the peptides were cleaved from the resin with TFA.

The analogs containing ω-amides of Asp or Glu in position 4 were synthesized by Boc strategy. The dicarboxylic amino acid residue was introduced in the sequence as Boc-Aaa(OFm)-OH. After peptide assembly the side chain protection was removed with 30% piperidine in DMF. The resulting free carboxylic group was converted to the desired amide by coupling with an appropriate amine mediated by PyBOP/DIEA. The N-terminal Boc group was then removed and the peptides were cleaved with HF. All linear analogs were cyclized with iodine and purified by HPLC.

Analogs, where X is a residue containing a small unsubstituted heteroaromatic ring (e.g. 3-Thi, **6**), do not show an improved pharmacological profile over the Phe^2 peptides, and the 2-furylalanine^2 (Ala(2-Fur)) compounds (i.e. **3**) are less potent than

DOI: 10.1007/978-0-387-73657-0_220, © Springer Science+Business Media, LLC 2009

Table 1. Structures and in vitro profile of selected peptides.

Peptide[a]				EC50 [nM]	Selectivity[b]		
#	X^2	Y^4	Z^8	hV1aR	hV2R	hOTR	hV1b
1 (AVP)	Tyr	Gln	Arg	0.24	0.2	92	18
2	Phe	Gln	Orn	0.27	840	30	46
3	Ala(2-Fu)	Gln	Orn	1.40	>71	>71	160
4	Phe	Gln	Dab	0.50	130	11	13
5	3-Pal	Gln	Dab	29.9	>3	>3	7
6	3-Thi	Gln	Dab	0.61	>160	44	82
7	Phe	Hgn	Orn	0.90	1800	>11000	51
8	Phe	Asn(Me_2)	Orn	1.65	>6000	>6000	38
9	Phe	Asn(Et)	Dab	2.70	600	>3700	8
10	Phe	Gln	Orn(Me)	0.26	360	140	280
11	Phe	Hgn	Orn(iPr)	2.40	1100	440	140

*[a]AVP, **1**, is the Phe[3] peptide and all other analogs are the Ile[3] peptides; [b]EC50 (receptor)/EC50 (hV1aR); [c]Highest concentration tested – 100 nM; [d]Highest concentration tested – 10000 nM*

their Phe[2] counterparts. When the heteroaromatic ring in position 2 contains nitrogen (e.g. X = 3-Pal, compound **5**), there is a significant loss of potency at the hV1aR as compared to **1** or **2**. Position 4 seems to be the key one to attain high selectivities versus both hV2R and hOTR. The much desired V2 selectivity can be tuned by changing the position 4 side chain lengths as exemplified by analogs **7** and **11** or by alkylating the primary amide functionality (compounds **8** (Y = Asn(Me_2) and **9** (Y = Asn(Et)). Replacing the Orn[8] residue with diaminobutyric acid (Z = Dab) appears to have a deteriorating effect on the overall selectivity as exemplified by analogs **4** and **6**. The N-alkylation of the Orn side chain (Z = Orn(Me) or Orn(iPr), compounds **10** and **11**) provides improved selectivity towards hV1bR that couldn't be achieved by any other modifications described in this paper. Combining selected modifications in positions 2, 4 and 8 resulted in potent and selective peptidic V1a agonists exemplified by compounds **7** through **11**.

In conclusion, new V1a selective AVP analogs with *in vitro* pharmacological profile superior to AVP are described.

Acknowledgments

The authors wish to thank Marlene Brown and Monica Wolfe for their excellent technical assistance.

References

1. Landry, D. W., Levin, H. R., Gallant, E. M., Ashton, R. C. Jr., Seo, S., D'Alessandro, D., Oz, M. C. and Oliver, J. A., *Circulation*, **95**, 1122-1125 (1997).
2. Huguenin, R. L., *Helv. Chim. Acta*, **47**, 1934-1941 (1964).
3. Andres, M., Trueba, M. and Guillon, G., *Br. J. Pharm.* **135**, 1828-1836 (2002).
4. Fukuyama, T., Jow, C.-K. and Cheung, M., *Tetrahedron Lett.* **36**, 6373-6374 (1995).

Discovery of a Ligand that Compensates for Decreased Endogenous Agonist Potency of Melanocortin-4 Receptor Polymorphisms Identified in Obese Humans

Carrie Haskell-Luevano[1], Zhimin Xiang[1], Andrzej M. Wilczynski[1], Kimberly R. Haskell[1], Amy M. Andreasen[1], Sally A. Litherland[2], William J. Millard[3], Ira D. Pogozheva[4], Henry I. Mosberg[4] and Nicholas B. Sorenson[1]

Departments of [1]Medicinal Chemistry, [2]Pahtology, Immunology and Laboratory Medicine, and [3]Pharmacodynamics, University of Florida, Gainesville, FL 32610 USA; [4]Department of Medicinal Chemistry, University of Michigan, Ann Arbor, MI 48109 USA

Introduction

Genetic studies of morbidly obese human patients and normal weight control patients have resulted in the discovery of over 70 human melanocortin-4 receptor (MC4R) polymorphisms observed as both heterozygous and homozygous forms. A number of laboratories have been studying these hMC4R polymorphisms in attempts to understand the molecular mechanism(s) that might explain the obese human phenotype. Herein, we have studied polymorphic hMC4Rs that have been identified to possess statistically significant decreased endogenous agonist potency [1] in attempts to identify ligands that can pharmacologically "rescue" the polymorphic hMC4R agonist response. The hMC4R polymorphisms included in this study are the S58C, N97D, I102S, L106P, S127L, T150I, R165Q, R165W, G252S, C271Y, and I301T. We have discovered that the peptide template (AMW3-130) possessing nM to sub nM agonist potency, can rescue the functional endogenous agonist defect at these human MC4R polymorphisms.

AMW3-130	Tyr-c[**Cys**-His-DPhe-Arg-Trp-Asn-Ala-Phe-**Cys**]-Tyr-NH_2
AMW632	Tyr-c[**Asp**-His-DPhe-Arg-Trp-Asn-Ala-Phe-**Dpr**]-Tyr-NH_2

Fig. 1. Sequences of AMW3-130 and AMW632.

Results and Discussion

Our laboratory [2,3], and another [4], have reported on the concept of agonist ligands that possess amino acid residues from the endogenous melanocortin agonists incorporated into various hAGRP antagonist based templates that result in the conversion of the endogenous AGRP antagonist ligands into agonists. These chimeric ligands appear to possess unique melanocortin receptor pharmacology distinct from either the synthetic melanocortin agonists or antagonists and AGRP based antagonists [3,4]. The AMW3-130 ligand is comprised of the AGRP(108-119) antagonist template incorporating the agonist His-DPhe-Arg-Trp residues instead of the antagonist Arg-Phe-Phe(111-113) amino acids. These studies resulted in the AMW3-130 ligand possessing nM to sub nM agonist potency at all the hMC4R polymorphisms examined (Table 1). Thus, this ligand could generically rescue the potency and stimulatory response of the abnormally functioning hMC4Rs studied. Attempts to dissolve AMW3-130 peptide (containing a disulfide bridge) in saline for use in *icv* food intake studies resulted in a flocculent at the desired nmol concentrations. The lactam bridge containing derivative AMW632 (Figure 1) .

S. Del Valle et al. (eds.), *Peptides for Youth: The Proceedings of the 20th American Peptide Symposium*, 509
DOI: 10.1007/978-0-387-73657-0_221, © Springer Science+Business Media, LLC 2009

Table 1. Peptide agonist potency EC_{50}(nM) at the hMC4R Polymorphisms.

hMC4R Mutation	α-MSH	β-MSH	ACTH(1-24)	AMW3-130
WT hMC4R	0.65	0.42	0.65	0.21
S58C	18	15	24	0.53
N97D	>10	660	>10	4
I102S	260	150	480	0.44
L106P	50% @ 1μM	356	40% @ 1μM	0.62
S127L	5	2	11	0.40
T150I	8	6	14	0.90
R165Q	18	6	39	0.094
R165W	21	8	58	0.25
G252S	6	0.53	10	0.13
C271Y	48% @ 1μM	154	42% @ 1μM	0.80
I301T	8	2	16	0.13

The values indicated represent the mean±SEM (at least three independent experiments). >10 indicates that an EC_{50} value was not reportable at up to 10 μM concentrations. A percentage value indicates that some stimulatory agonist pharmacology resulted at up to 10 μM.

however (that possessed the same receptor pharmacology potency profile at the mouse melanocortin receptors [2] as AMW3-130), was completely soluble in the saline vehicle. Central administration of AMW632 into *ad libitum* cannulated mice resulted in a significant reduction in cumulative food intake (Figure 2). These studies demonstrate that chimeric melanocortin-AGRP ligands can functionally rescue endogenous agonist potency at dysfunctional hMC4R polymorphisms identified in obese individuals and that this template can affect food intake in mice.

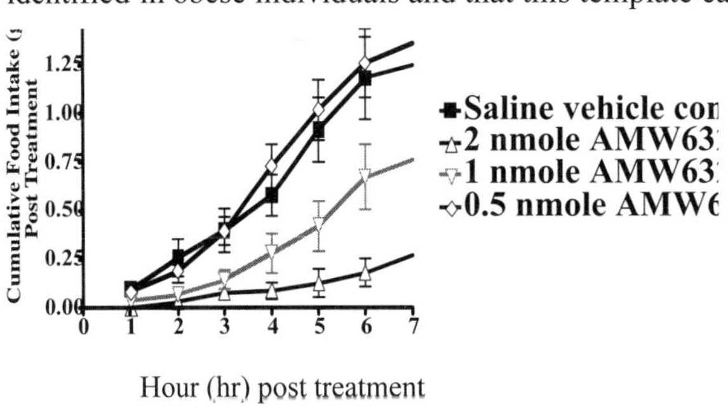

Fig. 2. Cumulative food intake of AMW632 in mice over time (hr) post treatment.

Acknowledgments

This work has been supported by NIH Grants to CHL(DK063974, DK57080, DK64250) and to HIM (DA003910) and an ADA Research Award (CHL).

References

1. Xiang *et al. Biochemistry.* **45**, 7277-7288 (2006).
2. Wilcxynaki *et al. J.Med.Chem.* **47**, 2194-2207 (2004).
3. Wilczynski *et al. J.Med.Chem.* **48**, 3060-3075 (2005).
4. Jackson *et al. Peptides.* **26**, 1978-1987 (2005).

High-affinity FGFR2 Binding Peptides Derived from the Native Epitope of the KGF Ligand

Yijun Pan, Yvonne M. Angell, Jennifer M. Green, Kathy L. Wright, Joy Miguel, Peter J. Schatz and Christopher P. Holmes

Affymax, Inc., 4001 Miranda Avenue, Palo Alto, CA 94304, USA

Introduction

The fibroblast growth factor receptors (FGFRs) are cell surface tyrosine kinase receptors that are implicated in numerous processes during cell growth and development. KGF (keratinocyte growth factor, FGF-7) is a member of the FGF family that binds the FGF receptor splice variant: FGF-R2(IIIb), i.e., KGF-R. KGF mediates proliferation of epithelial cells and has potential for treatment of oral mucositis, venous ulcers and ulcerative colitis. A series of high affinity peptides based on the native sequence of the RTQ loop (residues 65-80) of KGF were synthesized and tested for their ability to inhibit the binding of KGF to its receptor, KGFR (FGFR2IIIb). The structures of those synthetic peptides were optimized primarily by varying their length at the *N*- and *C*-termini, mutating some of the key residues at the *N*- and *C*-termini, and introducing an intramolecular disulfide constraint with various loop sizes. In addition, the position of the native RTQ epitope was varied within the peptide sequence. The peptide monomers were dimerized via a lysine at their *C*-terminus using a small molecular bi-functional linker to give the corresponding peptide dimers. Both the monomers and dimers were tested for their ability to inhibit the KGF/KGFR interaction. The most potent peptide dimers inhibited KGF binding with an IC50 of 4~8 nM, while the most potent peptide monomers inhibited KGF binding with an IC50 of 15~97 nM. These results demonstrate that these native epitope derived synthetic peptides bind directly to the KGFR with high affinity and can effectively compete with KGF for the KGFR binding site.

Results and Discussion

Synthetic peptides (Table 1) were synthesized using Fmoc chemistry on TentaGel R RAM (0.18 mmol/g, 400 mg) resins in a PTI Symphony peptide synthesizer. The *N*-terminal amine groups were capped with Ac_2O capping agent. Following deprotection and cleavage with 85% TFA, 10% TIPS, 2.5% H_2O and 2.5% thioanisole, the crude peptides were precipitated using ether, purified by preparative RP-HPLC using linear gradients of acetonitrile (containing 0.1%TFA) in H_2O (containing 0.15%TFA) on Waters RCM Delta-Pak (C18, 200 Å, 10 mm, 40 x 200 mm) columns and lyophilized.

An initial peptide monomer was synthesized and tested in a competition binding assay, and was found to inhibit KGF binding with an IC50 of 629 nM. The peptide sequences were optimized by varying their length at the *N*- and *C*-termini (Table 1, entries 1-15). The peptide monomers were dimerized via a lysine at their *C*-terminus via a bi-functional linker (Scheme 1). The lead peptide dimers inhibited KGF binding with an IC50 of 4~8 nM, while the most potent peptide monomers inhibited binding with IC50 of 15~97 nM. The peptide architectures were also modified by introducing a disulfide constraint with various loop sizes in their sequences (Table 1,

S. Del Valle et al. (eds.), *Peptides for Youth: The Proceedings of the 20th American Peptide Symposium*, 511
DOI: 10.1007/978-0-387-73657-0_222, © Springer Science+Business Media, LLC 2009

entries 16-27). The most potent peptide with the disulfide constraint inhibited KGF binding with an IC50 of 8 nM.

Table 1. Sequence Optimization of the Peptide Monomers.

Entry	Sequences
1	Ac I R V R R L F S R T Q W Y L R I D R R K NH_2
2	Ac R V R R L F S R T Q W Y L R I D R R K NH_2
3	Ac V R R L F S R T Q W Y L R I D R R K NH_2
4	Ac R R L F S R T Q W Y L R I D R R K NH_2
5	Ac R L F S R T Q W Y L R I D R R K NH_2
6	Ac I R V R R L F S R T Q W Y L R I D R K NH_2
7	Ac R V R R L F S R T Q W Y L R I D R K NH_2
8	Ac V R R L F S R T Q W Y L R I D R K NH_2
9	Ac R R L F S R T Q W Y L R I D R K NH_2
10	Ac R L F S R T Q W Y L R I D R K NH_2
11	Ac I R V R R L F S R T Q W Y L R I D K NH_2
12	Ac R V R R L F S R T Q W Y L R I D K NH_2
13	Ac V R R L F S R T Q W Y L R I D K NH_2
14	Ac R R L F S R T Q W Y L R I D K NH_2
15	Ac R L F S R T Q W Y L R I D K NH_2
16	Ac I R V R R C F S R T Q W Y C R I D R K NH_2
17	Ac R V R R C F S R T Q W Y C R I D R K NH_2
18	Ac V R R C F S R T Q W Y C R I D R K NH_2
19	Ac I R V R C L F S R T Q W Y C R I D R K NH_2
20	Ac R V R C L F S R T Q W Y C R I D R K NH_2
21	Ac R V R C L F S R T Q W Y C R I D R K NH_2
21	Ac R V R C L F S R T Q W Y C R I D Q K NH_2
22	Ac I R V R R C F S R T Q W Y L C I D R K NH_2
23	Ac R V R R C F S R T Q W Y L C I D R K NH_2
24	Ac V R R C F S R T Q W Y L C I D R K NH_2
25	Ac I R V R C L F S R T Q W Y L C I D R K NH_2
26	Ac R V R C L F S R T Q W Y L C I D R K NH_2
27	Ac V R C L F S R T Q W Y L C I D R K NH_2

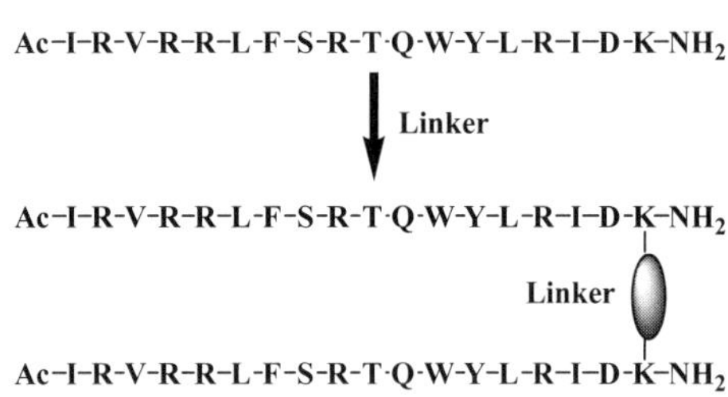

Scheme 1. Dimerization of the Peptide Monomers.

References

1. William, J. LaRochelle, *et al. Biochemistry,* **38**, 1765 (1999).
2. Timothy, D. Osslund, *et al. Protein Science,* **7**, 1681-1690 (1998).
3. Sheng, Ye., *et al. Biochemistry,* **40**, 14429-14439 (2001).
4. Donald, P. Bottaro, *et al. J. Biol. chem.,* **268**, 9180-9183 (1993).
5. Luca, Pellegrini, *et al. Nature*, **407**, 102(2000).

Application of Intramolecular Migration Reaction in Peptide Chemistry to Chemical Biology, Chemical Pharmaceutics and Medicinal Chemistry

Yoshio Hayashi,[1,2] Yuri Yamazaki,[1,2] Mariusz Skwarczynski,[2] Youhei Sohma,[2] Atsuhiko Taniguchi,[2] Mayo Noguchi,[2] Tooru Kimura,[2] Saskia Neuteboom,[3] Barbara Potts,[3] G. Kenneth Lloyd[3] and Yoshiaki Kiso[2]

[1]School of Pharmacy, Tokyo University of Pharmacy and Life Sciences, Hachioji, Tokyo 192-0392, Japan; [2]Department of Medicinal Chemistry, Kyoto Pharmaceutical University, Kyoto 607-8412, Japan; [3]Nereus Pharmaceuticals, San Diego, CA 92121, USA
yhayashi@mb.kyoto-phu.ac.jp

Introduction

Nucleophilic reactions such as N–O/O–N intramolecular acyl migration in Ser- or Thr-containing peptides and succinimide formation in Asp derivatives by cleavage of ester or amide bonds are commonly used in Peptide Chemistry. These reactions are of great interest in Peptide Chemistry, Chemical Pharmaceutics and Medicinal Chemistry. In regard to O–N intramolecular acyl migration, new synthetic methods, designated as "O-acyl isopeptide methods" have been developed for making difficult sequence containing peptides, as well as for application as water-soluble prodrugs of HIV-1 protease inhibitors and anti-cancer drugs such as paclitaxel and other taxoid analogs possessing α-hydroxy-β-amino acid moieties. !Moreover, new peptidic self-cleavable spacers that release a parent drug via succinimide formation in both water-soluble prodrugs and the oligoarginine-based cargo-transporter (OACT) system were developed as means to potentially enhance intestinal absorption of the parent drug.

Results and Discussion

In O–N intramolecular acyl migration, a new peptide synthetic method designated "isopeptide method" was developed for the synthesis of difficult sequence containing peptides [1]. Successful synthesis of an isopeptide of amyloid β peptide 1-42 (Aβ1-42) at Gly25-Ser26 by this method brought about a quick and pH dependent one-way migration to Aβ1-42 under physiological conditions [1c-f]. This controllable conversion can produce intact monomeric Aβ1-42 formation for biological evaluation systems and contribute to the investigation of actual Aβ1-42 oligomerization-related pathogenic factors of Alzheimer disease. By introducing a photo-cleavable group on the alpha-amine of the O-acyl isopeptide, we have recently developed a photo-triggered O-acyl isopeptide, Aβ1-42 Isopeptide [1g]. The photo-triggered cleavage of a protective group at the N-terminus of the isopeptide part of the non-aggregative Aβ1-42 isopeptide resulted in the formation of intact Aβ1-42 in a quick and one-way manner.

Much like conventional pharmaceutics, Chemical Pharmaceutics deals with drug delivery. However, contrary to conventional Pharmaceutics that focused on formulation techniques, Chemical Pharmaceutics delivers drugs using techniques based on Organic Chemistry. We applied the aforementioned nucleophilic reactions to prodrugs to enhance water-solubility of injectable drugs, or intestinal absorption of drugs with low intestinal permeability. These studies included water-soluble

S. Del Valle et al. (eds.), *Peptides for Youth: The Proceedings of the 20th American Peptide Symposium*, 513
DOI: 10.1007/978-0-387-73657-0_223, © Springer Science+Business Media, LLC 2009

prodrugs of taxoids [2] and HIV-1 protease inhibitors [3] as well as a novel self-cleavable spacer strategy [4]. As examples, we developed a novel water-soluble taxoid prodrug isotaxel. This prodrug, a 2'-O-isoform of paclitaxel, showed promising results with enhanced water solubility (0.8-1.1 mg/mL) and proper kinetics for parent drug release via a simple pH-dependent O-N intramolecular acyl migration mechanism. No additional functional auxiliaries were released during the conversion to the parent drug [2]. We recently developed a new peptidic self-cleavable spacer that released a parent drug via succinimide formation. The conversion time can be controlled by an amino acid side-chain structure next to the succinyl moiety on the spacer. This self-cleavable spacer is a promising development for the OACT system as a mean to enhance intestinal absorption of the parent drug [4].

As a target of our water-soluble prodrug strategy, we developed a new cyclic dipeptidic microtubule targeting agent, NPI-2358 [5], which is in Phase I clinical trial in the USA as a new anticancer drug. This agent acts as a "vascular disrupting agent (VDA)" to induce tumor-selective vascular collapse. In the field of Chemical Biology, the use of photoaffinity probes derived from this-type of VDA is also being intensively investigated to understand its precise tubulin binding and microtubule depolymerization mechanisms.

References

1. (a) Sohma, Y., Sasaki, M., Ziora, Z., Takahashi, N., Kimura, T., Hayashi, Y. and Kiso, Y. *Peptides, Peptide Revolution: Genomics, Proteomics & Therapeutics* Kluwer Academic: Netherlands, pp 67-68 (2003); (b) Sohma, Y., Sasaki, M., Hayashi, Y., Kimura T. and Kiso, Y. *Chem. Commun.* 124-125 (2004); (c) Sohma, Y., Sasaki, M., Hayashi, Y., Kimura, T. and Kiso, Y. *Tetrahedron Lett.* **45**, 5965-5968 (2004); (d) Sohma, Y., Hayashi, Y., Skwarczynski, M., Hamada, Y., Sasaki, M., Kimura and T., Kiso, Y. *Biopolymers* **76**, 344-356 (2004); (e) Sohma, Y., Hayashi, Y., Kimura, M., Chiyomori, Y., Taniguchi, A., Sasaki, M., Kimura, T. and Kiso, Y. *J. Peptide Sci.*, **11**, 441-451 (2005); (f) Sohma, Y., Chiyomori, Y., Kimura, M., Fukao, F., Taniguchi, A., Hayashi, Y., Kimura, T. and Kiso, Y. *Bioorg. Med. Chem.* **13**, 6167-6174 (2005); (g) Taniguchi, A., Sohma, Y., Kimura, M., Okada, T., Ikeda, K., Hayashi, Y., Kimura, T., Hirota, S., Matsuzaki, K. and Kiso, Y. *J. Am. Chem. Soc.* **128**, 696-697 (2006). (h) Sohma, Y., Taniguchi, A., Skwarczynski, M., Yoshiya, T., Fukao, F., Kimura, T. and Hayashi, Y., Kiso, Y. *Tetrahedron Lett.* **47**, 3013-3017 (2006).
2. (a) Hayashi, Y., Skwarczynski, M., Hamada, Y., Sohma, Y., Kimura, T. and Kiso, Y. *J. Med. Chem.* **46**, 3782-3784 (2003); (b) Skwarczynski, M., Sohma, Y., Kimura, M., Hayashi, Y., Kimura, T. and Kiso, Y. *Bioorg. Med. Chem. Lett.*, **13**, 4441-4444 (2003); (c) Skwarczynski, M., Sohma, Y., Noguchi, M., Kimura, M., Hayashi, Y., Hamada, Y., Kimura, T. and Kiso, Y. *J. Med. Chem.* **48**, 2655-2666 (2005); (d) Skwarczynski, M., Noguchi, M., Hirota. S., Sohma, Y., Hayashi, Y., Kimura, T. and Kiso, Y. *Bioorg. Med. Chem. Lett.* **16**, 4491-4495 (2006); (e) Skwarczynski, M., Sohma, Y., Noguchi, M., Kimura, T., Hayashi, Y. and Kiso, Y. *J. Org. Chem.* **71**, 2542-2545 (2006).
3. Sohma, Y., Hayashi, Y., Ito, T., Matsumoto, H., Kimura, T. and Kiso, Y. *J. Med. Chem.* **46**, 4124-4135 (2003).
4. Hayashi, Y., Takayama, K., Suehisa, Y., Fujita, T., Nguyen, J.-T., Futaki, S., Yamamoto, A. and Kiso, Y. *Bioorg. Med. Chem. Lett.* **17**, 5129-5132 (2007).
5. Nicholson, B., Lloyd, G. K., Miller, B. R., Palladino, M. A., Kiso, Y., Hayashi, Y. and Neuteboom, S. T. C. *Anti-Cancer Drugs* **17**, 25-31 (2006).

Synthesis & Biological Evaluation of PYY(3-36) Analogs Substituted with Alanine

John S. Ahn, Ramina Nazarbaghi, Lawrence J. D'Souza, Soumitra Ghosh, Carolyn M. Jodka, Aung N. Lwin, Odile E. Levy

Amylin Pharmaceuticals, Inc., 9360 Towne Centre Drive, San Diego CA 92121, USA

Introduction

PYY is a 36-residue peptide first isolated from porcine intestine (Figure 1) [1,2]. Two endogenous forms of PYY, PYY(1-36) and the post-DPPIV [3,4] activated PYY(3-36), arc released into the circulation following a meal [1,5]. PYY(3-36) appears to be the predominant secreted form. PYY has been known to inhibit gastric, [6] pancreatic and intestinal secretions [7]. PYY binds and activates at least four receptor subtypes (Y_1, Y_2, Y_4 and Y_5) [8-10] in rats and humans. These Y receptor subtypes display different patterns of affinity and activation for PYY, PYY(3-36) and synthetically modified PYY analogs. PYY(3-36) is a selective ligand for Y_2 and Y_5 receptors, implicated in food intake and feeding behavior, respectively [11]. In this study, PYY(3-36) analogs, where each residue of the natural sequence is replaced by L-alanine, and analogs with multiple alanine substitutions were synthesized. The three alanines at positions 7, 12 and 22 were replaced by a D-alanine. The affinity of each analog to the Y family of receptors and the ability of the analogs to decrease acute food intake in mouse are presented. Additionally, the efficacy of a single equimolar dose of PYY(3-36) and a subset of the analogs to reduce body weight in the chronic weight loss assay in the mouse is summarized.

PYY(1-36) H-YPIKPEAPGEAASPEELNRYYASLRHYLNLVTRQRY-NH_2
PYY(3-36) H-IKPEAPGEAASPEELNRYYASLRHYLNLVTRQRY-NH_2

Fig. 1. Sequence of PYY(1-36) and PYY(3-36).

Results and Discussion

At the N-terminus, disruption of Pro residues at position 5 and 8 resulted in loss of efficacy in the mouse food intake assay when dosed at 10 nmol/kg (Table 1). With the exception of Tyr^{20}, Ala substitutions were well tolerated in the middle region of PYY (residues 9-22). However, with the exception of Tyr^{21}, Ala^{22} and His^{26}, Ala substitutions at the C-terminus (residues 24-36) were not tolerated and led to a great loss of efficacy in the acute food intake assay. Although the Y receptor binding profiles of most peptides were similar to those of PYY (3-36), Ala substitutions at the C-terminus were detrimental to the Y_2 receptor binding affinity and led to loss of activity in the food intake assay, thus confirming the importance of C-terminus residues for biological activity. Beck-Sickinger et al., have also shown an analogous alanine scan effect in the Y_2 receptor binding assay in their study on NPY(1-36) [12]. In general, both studies demonstrated extreme sensitivity to alanine substitution at the C-terminus, and identified a peptide region where alanine or glycine substitutions are allowed from positions 6 to 18. The Pro^{14} to Ala^{14}, Asn^{18} to Ala^{18} and His^{26} to Ala^{26} substitutions, and combinations thereof, were used in the design of the next set of molecules. These novel substitutions led to no loss of

S. Del Valle et al. (eds.), *Peptides for Youth: The Proceedings of the 20th American Peptide Symposium*, 515
DOI: 10.1007/978-0-387-73657-0_224, © Springer Science+Business Media, LLC 2009

efficacy in the primary and secondary *in vivo* assays when compared to PYY(3-36) and have been used extensively in the SAR program that followed this study.

The scope of the alanine scan study determined that the C-terminal end of PYY(3-36) was the most sensitive to modifications in both the Y_2 receptor binding affinity and the food intake assay. This investigation was the genesis to help the progression of new molecular entities to further our SAR studies in PYY and potential anti-obesity therapeutics.

Table 1. Mouse Food Intake (10 nmol/kg) [a].

	30 min	60 min	120 min		30 min	60 min	120 min
PYY(3-36)	-33	-40	-36	PYY(1-36)	---	-40	---
[Ala^3]PYY(3-36)	-16	-25	-21	[Ala^{20}]PYY(3-36)	5	-3	-6
[Ala^4]PYY(3-36)	-45	-60	-62	[Ala^{21}]PYY(3-36)	-13	-19	-11
[Ala^5]PYY(3-36)	-15	-20	-15	[Ala^{22}]PYY(3-36)	-18	-21	-17
[Ala^6]PYY(3-36)	-28	-39	-33	[Ala^{23}]PYY(3-36)	-19	-28	-20
[Ala^7]PYY(3-36)	-5	-7	4	[Ala^{24}]PYY(3-36)	-15	-12	-14
[Ala^8]PYY(3-36)	18	4	-5	[Ala^{25}]PYY(3-36)	-6	0	5
[Ala^9]PYY(3-36)	-14	-33	-34	[Ala^{26}]PYY(3-36)	-29	-47	-49
[Ala^{10}]PYY(3-36)	-20	-30	-25	[Ala^{27}]PYY(3-36)	-6	-6	2
[Ala^{11}]PYY(3-36)	-32	-41	-38	[Ala^{28}]PYY(3-36)	-13	-3	5
[Ala^{12}]PYY(3-36)	-16	-27	-24	[Ala^{29}]PYY(3-36)	-5	2	4
[Ala^{13}]PYY(3-36)	-15	-31	-31	[Ala^{30}]PYY(3-36)	-25	-30	-18
[Ala^{14}]PYY(3-36)	-28	-34	-32	[Ala^{31}]PYY(3-36)	-15	-16	-3
[Ala^{15}]PYY(3-36)	-27	-31	-16	[Ala^{32}]PYY(3-36)	-2	-6	1
[Ala^{16}]PYY(3-36)	-17	-28	-17	[Ala^{33}]PYY(3-36)	-8	-4	-2
[Ala^{17}]PYY(3-36)	-14	-22	-10	[Ala^{34}]PYY(3-36)	-6	-7	-5
[Ala^{18}]PYY(3-36)	-29	-41	-46	[Ala^{35}]PYY(3-36)	0	-4	2
[Ala^{19}]PYY(3-36)	-18	-29	-22	[Ala^{36}]PYY(3-36)	8	5	-2

[a] *Values represent amount (%) consumed relative to vehicle treated mice.*

References

1. Balasubramaniam A., *et al.*, *J. Med. Chem.* **43,** 3420-3427 (2000).
2. Tatemoto K., *Proc. Natl. Acad. Sci. USA* **79,** 2514-2518 (1982).
3. Keire D.A., Bowers C.W., Solomon T.E., Reeve Jr. J.R., *Peptides* **23,** 305-321 (2002).
4. Mentlein R., Dahms P., Grandt D., Kruger R., *Regul. Pept.* **49 (2),** 133-144 (1993).
5. Adrian T.E., Ferri G.L., Bacarese-Hamilton A.J., Fuessl H.S., Polak J.M., Bloom SR., *Gastroenterology* **89,** 1070-1077 (1985).
6. Savage A.P., Adrian T.E., Carolan G., Chatterjee V.K., Bloom S.R., *Gut* **28,** 166-170 (1987).
7. Lundberg J.M., *et al.*, *Proc. Natl. Acad. Sci. USA* **79,** 4471-4475 (1982).
8. Keire D.A., Mannon P., Kobayashi M., Walsh J.H., Solomon T.E., Reeve Jr. J.R., *Am J Physiol. Gastrointest. Liver Physiol.* **279,** G126-G131 (2000).
9. Beck-Sickinger A.G., Jung G., *Biopolymers* **37,** 123-142 (1995).
10. Bard J.A., Walker M.W., Branchek T.A., Weinshank R.L., *J. Biol. Chem.* **270,** 26762-27765 (1995).
11. Batterham R.L., *et al.*, *Nature* **418,** 650-654 (2002).
12. Beck-Sickinger A.G., Wieland H.A., Wittneben H., Willim K-D., Rudolf K., Jung G., *Eur. J. Biochem.* **225,** 947-958 (1994).

Development of μ/δ Opioid Ligands: Enkephalin Analogues Containing 4-Anilidopiperidine Moiety

YeonSun Lee[1], Ravil Petrov[1],Vinod Kulkarni[1], Byoung J. Min[1], Shou-wu Ma[2], Peg Davis[2], Janice Oyarzo[2], Todd Vanderah[2], Josephine Lai[2], Frank Porreca[2], Ruben Vardanyan[1] and Victor J. Hruby[1]

[1]*Department of Chemistry;* [2]*Department of Pharmacology, University of Arizona, Tucson, AZ 85721, USA*

Introduction

It has been known that modulation of both μ and δ opioid receptors can give therapeutic benefit for the treatment of pain without the serious side effects caused by morphine [1-2]. Thus, our group developed a series of enkephalin analogues for the μ and δ receptors in which enkephalin-related tetrapeptide structures were linked to three different 4-anilidopiperidine moieties [3]. Among them, one analogue with a 2,6-dimethyltyrosine-containing tetrapeptide structure and *N*-phenyl-*N*-piperidin-4-yl propionamide moiety, **LYS540**, showed highly potent biological activities at both receptors (Table 1) (Figure 1) [3]. Further modifications of this ligand, mainly at the 2 and 4 positions of the tetrapeptide, were performed and found to increase both opioid activities in the binding and functional assays. We report here structure-activity relationships in detail, along with in-vivo assay results.

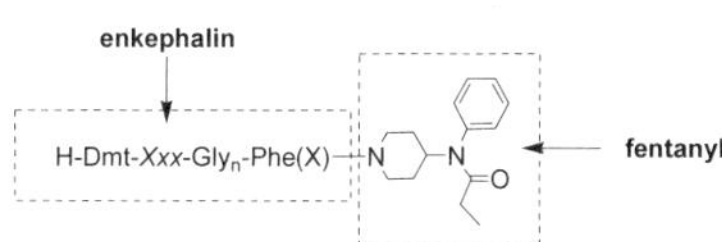

(Xxx = *D*Ala, *D*Nle, *D*Tic; n = 0, 1; X = H, Cl, F)

Fig. 1. Design of enkephalin analogues.

Results and Discussion

Synthetic enkephalin analogues were prepared by stepwise solution phase peptide synthesis using N^{α}-Boc chemistry starting from *N*-phenyl-*N*-piperidin-4-yl propionamide in 40-50% overall yields [3]. After chain elongation, crude peptides were isolated by preparative C_{18} RP-HPLC (Vydac 218TP1010) with a gradient of 10-50% CH_3CN in aq. 0.1% TFA.

Opioid binding affinities were determined by radioligand competition analyses against [^{3}H]DPDPE (δ) and [^{3}H]DAMGO (μ) using membrane preparations from transfected HN9.10 cells. Agonist activities for opioid receptors were evaluated by classical assays in stimulated mouse vas deferens (MVD, δ) and guinea pig isolated ileum (GPI, μ) [4].

The ligands were designed to have good agonist activities for both μ and δ opioid receptors with increased potential cell permeability due to the liphophilic character of the C-terminal 4-anilidopiperidine moiety. Ligating the moiety to the enkephalin-related tetrapeptide structure increased their hydrophobicities from aLogP = -0.54 to 2.8 [5]. **LYS644** in which *D*-Ala^2 was replaced by *D*-Nle^2 increased opioid agonist activities in the MVD (IC_{50} = 0.70, 3 fold) and GPI (IC_{50} = 1.6, 5 fold) assays (Table 1). In our study, we observed that all truncation of Gly^3, for example **LYS711**,

S. Del Valle et al. (eds.), *Peptides for Youth: The Proceedings of the 20th American Peptide Symposium*, 517
DOI: 10.1007/978-0-387-73657-0_225, © Springer Science+Business Media, LLC 2009

Table 1. Bioactivities of the opioid ligands.

ligand	*K*i (nM)[a]		IC$_{50}$ (nM)[b]	
	hDOR [^{3}H]DPDPE	rMOR [^{3}H]DAMGO	MVD(δ)	GPI(μ)
LYS540 (Xxx = *D*Ala; n = 1; X = H)	0.38	0.30	1.8	8.5
LYS644 (Xxx = *D*Nle; n = 1; X = H)	0.18	0.39	0.70	1.6
LYS711 (Xxx = *D*Ala; n = 0; X = H)	1.7	0.15	16	3.6
LYS707 (Xxx = *D*Ala; n = 1; X = *p*-Cl)	0.14	0.14	0.70	2.6
H-Tyr-*D*Ala-Gly-Phe-NH$_2$	300	2.8	120	47

[a]*Competition analyses against radiolabeled ligand.* [b]*Concentration at 50% inhibition of muscle contraction at electrically stimulated isolated tissues.*

decreased δ opioid activities in the binding and functional assays, but nonetheless retained μ opioid activities. Shortening the backbone length might change the topographical structure of the essential two aromatic groups for the opioid activities and result in decreased δ opioid activity. Substitution on the phenyl ring with chlorine (**LYS707**) increased both opioid activities 2-3 fold in the binding and functional assays. The ligand showed very good antihyperalgesic and antiallodynic effects in in-vivo tests using radiant heat and von frey filaments, respectively, in L5/L6 SNL SD rats (Figure 2).

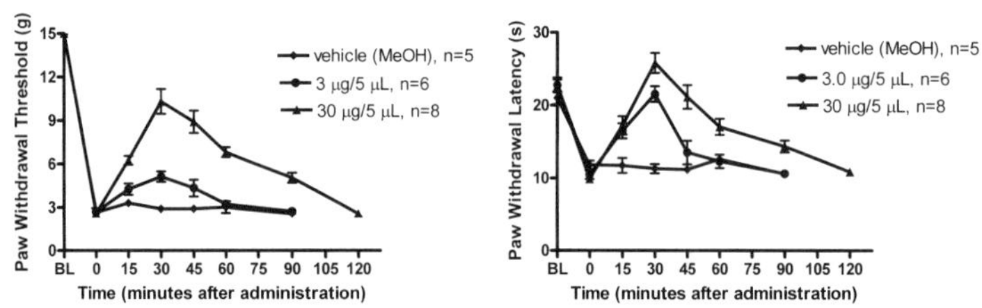

*Fig. 2. Antiallodynic (left) and antihyperalgesic (right) effects of **LYS707** (i.th) in L5/L6 SNL SD rats.*

Acknowledgments

This work was supported by grants from the USDHS, National Institute on Health (DA12394 and DA06284).

References

1. Zhao, G-M, *et al. J. Pharmacol. Exp. Ther.* **302**, 188-196 (2002).
2. Horan, P. J., *et al. J. Pharmacol. Exp. Ther.* **265**, 896-902 (1993).
3. Lee, Y.S., *et al. Bioorg. Med. Chem. Lett.* **17**, 2161-2165 (2007).
5. Lee, Y.S., *et al. J. Med. Chem.* **49**, 1773-1780 (2006).
6. Tetko, I. V., *et al. J. Comput. Aid. Mol. Des.* **19**, 453-63 (2005).

Factors that Affect Dissociation Degree of Strong Aggregated Peptide Chains when Bound to a Polymer or Free in Solution

Luciana Malavolta and Clóvis R. Nakaie

Department of Biophysics, Universidade Federal de São Paulo, Sao Paulo, Brazil.
e-mail: luciana@biofis.epm.br

Introduction

In a conceptual departure from the innumerable approaches that have been applied for studying solute-solvent interaction, we have focused on examining the solvation effect of polymeric material attaching or not peptide chains selected as solute models in their interactions with different solvent systems [1]. As one of the consequences of these studies, the sum (1:1) of the Gutmann´s electron acceptor (AN) and electron donor (DN) values [2] was previously proposed as a novel polarity parameter [3]. Following with these investigations, the search for rules which might govern solubilization processes of peptide sequences in solution has already been initiated [4-5]. The present report has aimed to further evaluate the dissociation phenomenon of peptide sequences (attached or not to a solid support) having different levels of propensity to aggregation, some of them of physiological relevance.

Results and Discussion

i) *Dissocitation of peptide chains attached to resins*: the solvation degree of (1-42) β-amyloid (DAEFRHDSGYDVHHQKLVFFAEDVGSQKGAIIGLMVGGVVIA) peptide involved in the Alzheimer disease [6] and the fragments containing 7, 14, 21, 28, 35 residues bound to resin were studied in different solvents. The results were interpreted in the light of the (AN) and (DN) values of the solvent. The strong nucleophilic DMSO depicted greater solvation of peptide-resins in comparison with the strong electrophilic TFE whereas MeCN, likely by having similar AN and DN numbers (intermolecular neutralization effect) was unable to disrupt peptide chains association within polymer matrix. In addition, the peptide chain mobility in these solvents was also estimated through electron paramagnetic resonance (EPR) method by labeling these compounds with the amino acid-type spin label Fmoc-TOAC (2,2,6,6-tetramethylpiperidine-1-oxyl-4-amino-4-carboxylic acid) [7]. Faster chain motion was verified in the order DMSO > TFE > MeCN as the smallest EPR central field peak W_0 value was verified in DMSO and the largest in MeCN. These findings are thus in close accordance with the solvation results when the swelling degree of the peptide-resins beads were measured with a microscope in these solvents.

ii) *Solubilization of insoluble peptides*: the relevance of applying the concept of solvent AN and DN values was also verified in the explanation of the dissociation process of the following aggregated peptide segments in solution: the (1-42) β-amyloid peptide and, their (12-24) sequence including three analogues with variations of residues at positions 15 and 22, the short aggregating sequence LPFFD [8] and lastly, the fragment (66-97) transmembrane

S. Del Valle et al. (eds.), *Peptides for Youth: The Proceedings of the 20th American Peptide Symposium*, 519
DOI: 10.1007/978-0-387-73657-0_226, © Springer Science+Business Media, LLC 2009

VAEIYLGNLAGAKLILASGLPFWAITIANNFD sequence of the bradykinin B2 receptor [9]. The solubilization experiments of peptides were designed to encompass single and mixed solvents, including the strong electrophilic water (AN= 54.8) in this process. In agreement with previous report [5], improved solubilization for all peptides was observed in solvents having the highest possible (AN-DN) values (in positive or negative mode). Amongst the single solvents, the highly electrophilic HFIP (hexafluorisopropanol) or the highly nucleophilic DMSO induced greater solubilization of peptides. The solubility effect induced by the addition of water to other organic solvents is highly dependent upon the type and amount of organic solvent to be cosolvated. Care should be taken when the electrophilic water is mixed with strong nucleophilic solvent such as DMSO or DMF as an internal self-neutralizing effect can occur in the mixture. In this case, each component tends to attract to the other of the mixture and fails to disrupt solute-solute aggregation. The same precaution is valid when the chaotropic urea is added in the medium. An inverse relationship between solubility percentage and urea concentration (0.5 to 6.0 M) was observed for the β-amyloid and its segments. Otherwise, the influence of the pH of the medium was also evaluated. With exception for the β-amyloid peptide and its (12-24) segment, the LPFFD and the B2 receptor transmembrane sequences displayed much greater solubilization in basic pH. The CD spectroscopy was also applied in an attempt to correlate secondary structure characteristics of peptides with their solubility degree in selected solvent systems. No clear relationship was detected in this case as improved solubilizations were observed in α-helix or disordered structures, depending on the peptide sequence. Therefore, in light of the Lewis acidity and basicity properties of solvent systems, some relevant rules could be established for the complex processes of peptide dissolution and peptide-resin solvation. However, the exact effect of other parameters upon the peptide solubility degree such as the medium pH, presence of chaotropic agent or the type of peptide conformation is still far from a more clear and consistent explanation.

Acknowledgments

Grants from the Brazilian FAPESP and CNPq agencies are gratefully acknowledged.

References

1. Cilli, E.M., Oliveira, E., Marchetto, R. and Nakaie, C.R. *J. Org. Chem.* **61,** 8992-9000 (1996).
2. Gutmann, V. The Donor-Acceptor Approach to Molecular Interactions; Plenum: New York, 1978.
3. Malavolta, L., Oliveira, E., Cilli, E.M.and Nakaie, C.R. *Tetrahedron* **58**, 4383-4394 (2002).
4. Malavolta, L. and Nakaie, C.R. *Tetrahedron* **60**, 9417-9424 (2004).
5. Malavolta, L., Pinto, M.R.S., Cuvero, J.H. and Nakaie, C.R. *Prot. Sci.* **15**, 1476-1488 (2006).
6. Gordon, D.J. and Meredith, S.C. *Biochemistry* **42**, 475-485 (2003).
7. Marchetto, R., Schreier, S. and Nakaie, C.R. *J. Am. Chem. Soc.* **115,** 11042-11043 (1993).
8. Adessi, C., Frossard, M.J., Boissard, C., Fraga, S., Bieler, S., Ruckle, T., Vilbois, F., Robinson, S.M., Mutter, M., Banks, W.A. and Soto, C. *J. Biol. Chem.* **278,** 13905-13911 (2003).
9. McEarchen, A.E., Shelton, E.R., Obernolte, R., Bach, C., Zuppan, P., Fujisaki, J., Aldrich, R.W. and Jarnagin, K. *Proc.Natl. Acad Sci. USA.* **88,** 7724-7728 (1991).

Development of AEZS-115 (ZEN-019) by Optimization of Structurally Unique, Orally Active, Peptidomimetic GnRH Antagonists

K.W. Paulini*[$], S. Baasner, M. Gerlach, E.E. Polymeropoulos, P. Schmidt, T. Schuster, M. Teifel and E.G. Guenther

Æterna Zentaris GmbH, Discovery and Preclinical Development, Weismuellerstr. 50, 60314 Frankfurt am Main, Germany

Introduction

While the LHRH (GnRH) receptor is an established target for the treatment of various sex hormone dependent benign and malignant conditions such as breast and prostate cancer, uterus myoma, endometriosis, benign prostate hyperplasia (BPH), precocious puberty or in vitro fertilization, drug discovery has recently been focused on the development of both new peptidic as well as orally bioavailable, antagonistic compounds [1-6].

Major advantages for LHRH receptor antagonists as compared to corresponding (super)agonists are:

- No flare-up effect
- Fast onset of action/shorter treatment period
- Dose-dependent reduction of sexual hormones
- No castration necessary
- Intermittent/chronic therapy possible
- Reduced side-effects

Medicinal Chemistry and Pharmacology

Screening and docking experiments led to the discovery of a non-natural tetra-peptide as a suitable starting point for optimization based on privileged structures. Variation of sequence, C- and N-termini, substitution patterns and side chains, aided by molecular modelling and a proprietary receptor model [7], led to AEZS-115 (ZEN-019), a promising, peptidomimetic candidate for clinical development. Based on a reliable structure-activity-relationship (SAR), successful optimization for improved receptor binding affinity, oral efficacy and higher metabolic stability was achieved by medicinal chemistry. A binding mode deep down in highly conserved regions of the LHRH receptor‘s active site, leads to low species selectivity as seen for the decapeptide anatgonists.

The competitive antagonist AEZS-115 has high affinity towards the LHRH receptors of various species (^{125}I-triptorelin displacement assay):

human LHRH-R (cells)	IC_{50}	2.4 nM
human LHRH-R (membranes)	IC_{50}	1.7 nM
rat LHRH-R (cells)	IC_{50}	0.7 nM
rat LHRH-R (membranes)	IC_{50}	0.3 nM
mouse LHRH-R (cells)	IC_{50}	27.5 nM

A single oral administration of AEZS-115 (20mg/kg) in female or in male rats leads to efficient and reversible suppression of either plasma luteinizing hormone (LH) levels or plasma testosterone levels for 12 hours.

S. Del Valle et al. (eds.), *Peptides for Youth: The Proceedings of the 20th American Peptide Symposium*, 521
DOI: 10.1007/978-0-387-73657-0_227, © Springer Science+Business Media, LLC 2009

ADME, Safety, Tox

Evaluation of ADME parameters revealed:

- Satisfactory liver microsomes and hepatocyte stability across species (human, monkey, dog and rat)
- Very good stability in human plasma

Early, orienting safety pharmacology and toxicology assessment revealed:

- Very good selectivity of AEZS-115 against a variety of receptors and enzymes (low side effect potential)
- No concerns regarding hERG inhibition in vitro by AEZS-115
- No influence on QTc intervals in guinea pigs
- No adverse effects on cardiovascular and respiratory parameters like blood pressure and heart rate in rats
- No behavioural changes observed in Irwin test with rats
- No mutagenic potential observed in AMES tests
- No toxicity in *in vitro* cell proliferation assays in various cancer cell lines
- No relevant histamine release in vitro up to 400µM concentration
- No relevant findings in a dose range finding (DRF) study up to 2000mg/kg single doses in rats
- No relevant findings in repeated dosing study (8 x 1000mg/kg) in rats

Conclusions

- Clear structure activity relationships (SAR) established
- Low nM binding affinity against LHRH receptor of various species
- Excellent selectivity (receptors, enzymes and transporters)
- No issues in early toxicity assessment
- Very good safety profile *in vivo* and *in vitro*
- Sufficient metabolic stability for p.o. administration
- >15% oral bioavailability in rats and dogs depending on formulation
- Efficient kg-scale synthesis developed and optimized
- Three patent applications filed, one already granted

AEZS-115 is a promising candidate for clinical development!

Acknowledgements

The authors would like to thank A. Bork, R. Eckhard, B. Heil, A. Hofmann, E. Jaschob, B. Krönert, T. Kynast, C. Lang, A. Linsner, G. Müller, J. Schumann and K. Zimmermann for excellent technical assistance.

References

1. Zhu, Y.-F., et al., C., Exp. Opinion on Therapeutic Patents **14**(2), 187-199 (2004).
2. Armer, R.E., et al., Curr. Med. Chem. **11**(22), 3017-3028 (2004).
3. Betz, St.F., et al., J. Med. Chem. **49**(2), 637-647 (2006).
4. Sasaki, S., et al., J. Med. Chem. **46**, 113-124 (2003).
5. Cui, J., et al., Molecular Endocrinology **14**, 671-681 (2000).
6. Anderes, K.L., et al., J. Pharm. Exp. Therapeutics **305**, 688-695 (2003).
7. Soederhaell, J.A., et al., Biochem. and Biophys. Res. Commun. **333**(2), 568-582 (2005).

Synthesis and solid-phase application of suitably protected γ-hydroxyvaline building blocks

Mare Cudic, Frank Marí and Gregg B. Fields

Department of Chemistry and Biochemistry, Florida Atlantic University, Boca Raton, FL 33431, USA

Introduction

Hydroxylated amino acids constitute an important modification in proteins. The best characterized of the post-translationally hydroxylated amino acids are γ-hydroxy-Pro (Hyp) and δ-hydroxy-Lys (Hyl), which are commonly found in collagen. Recently, an unexpected modified residue, γ-hydroxy-D-valine (D-Hyv), was identified within ribosomally expressed polypeptide chains of four conopeptides from the venoms of *Conus gladiator* and *Conus mus*.[1] These conopeptides were the first known examples of a naturally occurring polypeptide chain containing Hyv. In general, γ-hydroxyamino acids are not that common in nature (except γ-hydroxy-Pro) since a hydroxyl group in the γ-position can undergo intramolecular cyclization to form a lactone, cleaving the peptide bond. The stability of Hyv within conopeptides has been explained by the D-configuration at the α-carbon in conjunction with specific interactions with the surrounding L-amino acids. Given the wide range of neurophysiological responses these conopeptides elicit, there is much interest in developing efficient methods for their synthesis.

Results and Discussion

The goals of this work were to develop a convenient method for the synthesis of Hyv derivative suitably protected for Fmoc solid-phase chemistry and its subsequent incorporation into a conopeptide, gld-V* (Ala-Hyp-Ala-Asn-Ser-D-Hyv-Trp-Ser-NH_2), isolated from *Conus gladiator*.

The synthesis of Hyv derivative is complicated by the fact that Hyv contains two asymmetric centers and therefore exists as four stereoisomers: L-iso [(*S*,*S*)], D-iso [(*R*,*R*)], L-allo [(*S*,*R*)], and D-allo [(*R*,*S*)]. Since prior analysis indicated that gld-V* corresponds to the Hyv (3*S*,4*S*) configuration[1] we utilized a variation of the previously published $K_2PtCl_4/CuCl_2$ oxidative method for the efficient regio- and stereoselective production of the (3*S*,4*S*) isomer of D-Hyv starting from D-Val.[2] One troublesome step, use of gaseous H_2S for removal of excess of $CuCl_2$, has been optimized and potassium ferrocyanide successfully applied for copper precipitation. The challenging step in the synthesis of appropriately functionalized D-Hyv building blocks was suppression of intramolecular lactone formation. We successfully resolved this potential problem previously during the preparation of the suitably protected Hyl derivative. An α-amino group protection scheme was developed that included prior hydroxyl protection of copper-complexed D-Hyl with the *tert*-butyldimethylsilyl (TBDMS) group. Fmoc-Hyl(ε-Boc,*O*-TBDMS) was obtained in 67% overall yield starting from Hyl in 5 distinct steps.[3]

A similar approach was applied in the synthesis of an Fmoc-protected D-Hyv derivative. Copper-complexed Hyv **5a** and **5b** was prepared as previously described.[3] The hydroxyl group of Hyv was protected by treatment with TBDMSOTf in the presence of DMAP in pyridine. The desired copper complexed D-Hyv(*O*-TBDMS) **7a**

S. Del Valle et al. (eds.), *Peptides for Youth: The Proceedings of the 20th American Peptide Symposium*, 523
DOI: 10.1007/978-0-387-73657-0_228, © Springer Science+Business Media, LLC 2009

and **7b** was separated from copper complexed Val by extraction with EtOAc-H_2O. Disruption of the copper complex was achieved using Na^+ Chelex 100 resin in a pyridine-H_2O mixture. The product **8a** and **8b** was reacted with Fmoc-OSu and the resulting residue purified by flash chromatography to provide Fmoc-D-Hyv (*O*-TBDMS) **9a** and **9b** in 56% yield starting from **8a** and **8b**. NMR and mass spectra confirmed the purity and composition of Fmoc-D-Hyv (*O*-TBDMS). The overall yield of **9a** and **9b** was 26% starting from D-Val. RP-HPLC analysis of Fmoc-D-Hyv(*O*-TBDMS) revealed the presence of 2 peaks (at 17.7 and 17.9 mins) whose areas corresponded well to the observed diastereoselective ratio of 3:1 (3*S*,4*S*) to (3*S*,4*R*) for the D-Hyv mixture of **1a** and **1b**.

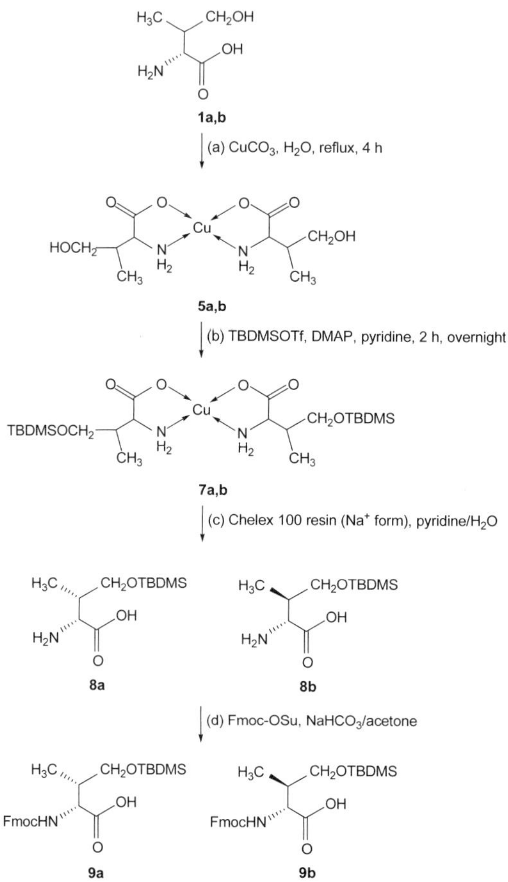

The mixture of the major and minor isomers of Fmoc-D-Hyv(*O*-TBDMS) **9a** and **9b** was used for the solid-phase synthesis of the D-Hyv containing conopeptide gld-V̲* (Ala-Hyp-Ala-Asn-Ser-D-Hyv-Trp-Ser-NH_2).

NMR spectroscopic analysis found that the most prominent isolated product corresponded to naturally occurring gld-V̲*. Analysis of the Fmoc-Hyv (*O*-TBDMS) derivative used for peptide synthesis indicated 75% of the (3*S*,4*S*) configuration present, and thus it is logical that the most abundant synthetic peptide diastereomer is the one containing the Hyv (3*S*,4*S*) configuration (gld-V̲*).

Acknowledgments

This work was supported by the Florida Sea Grant College Program (R/LR-MB-18), the National Institutes of Health (GM 066004 to F.M., CA 77402 to G.B.F.), and the FAU Center of Excellence in Biomedical and Marine Biotechnology.

References

1. Pisarewicz, K.; Mora, D.; Pflueger, F. C.; Fields, G. B.; Marí, F. *J. Am. Chem. Soc.* **2005**, *127*, 6207-6215.
2. Dangel, B. D.; Johnson, J. A.; Sames, D. *J. Am. Chem. Soc.* **2001**, *123*, 8149-8150.
3. Cudic, M.; Lauer-Fields, J. L.; Fields, G. B. *J. Peptide Res.* **2005**, *65*, 272-283.

Synthesis and Structure-based Dissection of Cyclic Peptide Chitinase Inhibitors: New Leads for Antifungal and Anti-Inflammatory Drugs

Mark J. Dixon[1], Amit Nathubhai[1], Ole A. Andersen[2], Daan M. F. van Aalten[2] and Ian M. Eggleston[1*]

[1]*Department of Pharmacy and Pharmacology, University of Bath, Bath BA2 7AY, UK;* [2]*Division of Biological Chemistry and Drug Discovery, College of Life Sciences, University of Dundee, Dundee DD1 5EH, UK;*

Introduction

Family 18 chitinases play key roles in the life cycles of many organisms that are pathogenic to humans, such as fungi, bacteria, and parasitic nematodes. The onset or transmission of several major human diseases (e.g. malaria, filariasis) has been linked with the activity of chitinases from such organisms, while the overexpression of endogeneous mammalian chitinases, notwithstanding the absence of chitin from mammalian physiology, is a significant feature in asthma and lipid storage disorders such as Gaucher's disease [1]. Selective inhibitors of these enzymes are of great interest as new drug leads or biochemical probes, and in this context, we recently reported the first syntheses of two cyclic peptide natural products, argifin [2] and argadin [3], that show nanomolar inhibition of a range of family 18 chitinases.

The structure of argifin in complex with various family 18 chitinases shows that the carbamoylated Arg moiety (Arg(MC)) of the cyclic peptide interacts with conserved catalytic residues in the active site of these enzymes, and also occupies the i+1 position of a β-turn. Critically, the peptide bond to the adjacent MePhe residue has a *cis*-configuration in the enzyme-bound structure, as is also observed by NMR in solution. We speculated that linear peptides based around the Arg(MC)-MePhe motif might possess sufficient conformational rigidity to adopt this same bioactive conformation, and with this in mind, we synthesised three argifin-derived fragments and evaluated them as chitinase inhibitors.

Tetrapeptide 1: Ac-D-Ala-Arg(MC)-MePhe-βAsp-OH
Tripeptide 2: Ac-Arg(MC)-MePhe-βAsp-OH
Dipeptide 3: Ac-Arg(MC)-MePhe-OH
Argifin: cyclo[βAsp-D-Ala-Arg(MC)-MePhe-βAsp]

Fig. 1. Argifin fragment peptides investigated.

Results and Discussion

1-3 were assembled by manual Fmoc SPPS on 2-chlorotrityl chloride resin, using Pmc and ^{t}Bu ester protection for Arg and Asp respectively. Following cleavage from the resin with TFA/thioanisole/DCM/H_2O(16:2:1:1) [2], the three Arg-containing precursors were treated with MeNHCO$_2$Su/DBU/DMF at 40°C, in order to introduce the required Arg(MC) unit. Following purification by HPLC, argifin fragments **1-3** were evaluated against the fungal chitinase *Af*ChiB1 from *Aspergillus fumigatus*, human chitotriosidase (HCHT) and human acidic mammalian chitinase (*h*AMCase). The latter two enzymes have been implicated in the onset of Gaucher's disease and asthma respectively, and are therefore potential targets for the development of novel

S. Del Valle et al. (eds.), *Peptides for Youth: The Proceedings of the 20th American Peptide Symposium*, 525
DOI: 10.1007/978-0-387-73657-0_229, © Springer Science+Business Media, LLC 2009

peptide-derived therapeutics. Inhibition was determined using a standard fluorescence-based assay [1] with 4-methylumbelliferyl-β-*N',N''*-diacetylchitobiose as substrate (Table 1).

*Table 1. IC_{50} values for inhibition of family 18 chitinases by **1-3**.*

Peptide	*h*AMCase	HCHT	*Af*ChiB1
1	0.94 μM	28 μM	4.3 μM
2	1.5 μM	68 μM	5.1 μM
3	12 μM	190 μM	12 μM
Argifin	33 nM	1.1 μM	29 nM

1-3 were all found to retain substantial activity against fungal and mammalian chitinases. X-ray structures obtained for **1-3** in complex with *h*AMCase (and *Af*ChiB1) reveal an identical binding mode to that seen for the natural product, with a *cis* Arg(MC)-MePhe peptide bond. The Arg(MC) moiety again reproduces key hydrogen bonding interactions with key conserved catalytic residues, and for **1** and **2**, the turn conformation is reinforced by an intramolecular hydrogen bond, as observed in the natural product. These results confirm that dissecting the argifin scaffold using X-ray structural data, is a useful strategy for identifying simplified inhibitor leads, and opens the way towards the design of effective peptidomimetic chitinase inhibitors that incorporate the essential features of the Arg(MC)-MePhe dipeptide.

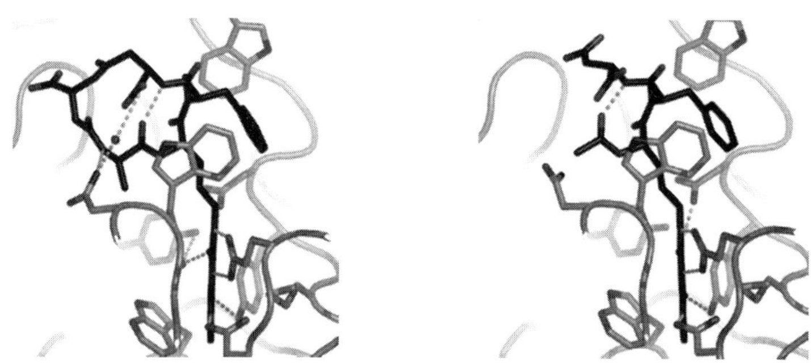

*Fig. 2. X-ray structures of argifin (left) and tripeptide **2** (right) in complex with hAMCase.*

Acknowledgments

We thank the Wellcome Trust and the Biotechnology and Biological Sciences Research Council for financial support.

References

1. Andersen, O. A., Dixon, M. J., Eggleston, I. M., van Aalten, D. M. F. *Nat. Prod. Rep.* **22**, 563-569 (2005).
2. Dixon, M. J., Andersen, O. A., van Aalten, D. M. F., Eggleston, I. M. *Bioorg. Med. Chem. Lett.* **15**, 4717-4721 (2005).
3. Dixon, M. J., Andersen, O. A., van Aalten, D. M. F., Eggleston, I. M. *Eur. J. Org. Chem.* **24**, 5002-5006 (2006).

Synthesis of Biotin-Tagged Diketopiperazine-Based Anti-Microtubule Agents and Tubulin Photoaffinity Labeling

Yuri Yamazaki[1,4], Kyoko Kohno[2], Hiroyuki Yasui[3], Yoshiaki Kiso[1], Saskia Neuteboom[5], Ana M. Barral[5], Barbara Potts[5], G. Kenneth Lloyd[5] and Yoshio Hayashi[1,4]

[1]Department of Medicinal Chemistry; [2]Department of Molecular Genetics; [3]Department of Analytical and Bioinorganic Chemistry, Kyoto Pharmaceutical University, Yamashina-ku, Kyoto 607-8412, Japan; [4]Department of Medicinal Chemistry, School of Pharmacy, Tokyo University of Pharmacy and Life Sciences, Hachioji, Tokyo 192-0392, Japan; [5]Nereus Pharmaceuticals, San Diego, CA92121, USA

Introduction

The introduction of anti-microtubule agents such as taxanes and *vinca* alkaloids has revolutionized cancer treatment and improved patient survival time. However, tumors become resistant to these drugs after long-term clinical treatment. Hence, there is a significant need to develop novel anti-microtubule agents. As one of such candidates, we have focused on a natural diketopiperazine (DKP), phenylahistin (PLH, halimide) that exhibits colchicine-like anti-microtubule activity [1, 2]. From structure-activity relationship (SAR) study [3], a highly potent cytotoxic derivative NPI-2358 (**1**, KPU-2, IC_{50} 15 nM against HT-29 cells) was developed. Additionally, it was recently shown that NPI-2358 functions as a strong "vascular disrupting agent (VDA)" to induce tumor-selective vascular collapse [4]. Therefore, NPI-2358 is now in Phase I clinical trial as a promising anticancer drug in the US.

Although it is known that PLH recognizes the colchicine binding site on tubulin, precise binding mode of NPI-2358 and its microtubule depolymerization mechanism have not well been investigated and three-dimensional structure of NPI-2358 failed to favorably superimpose with that of colchicine in molecular modeling study. In the SAR studies based on compound **1**, a more potent benzophenone derivative KPU-244 (**2**, IC_{50} 3.9 nM against HT-29 cells) was discovered. Since the benzophenone structure is often used for protein photoaffinity labeling, in the present study, based on the structure of compound **2**, we designed and synthesized biotin-tagged derivatives, KPU-244-B1 (**3**), which can be detected at the labeled position by an avidin-peroxidase system. Then, tubulin photoaffinity labeling was performed.

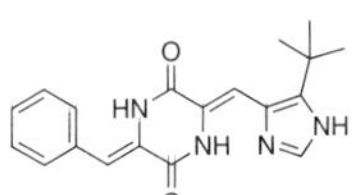

NPI-2358 (**1**)

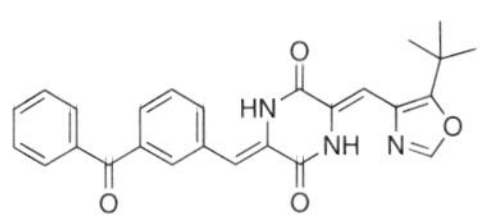

KPU-244 (**2**)

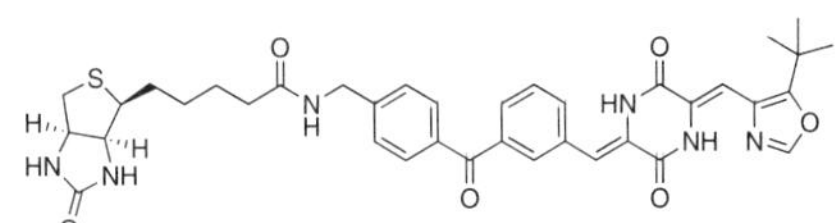

KPU-244-B1 (**3**)

Fig. 1. Structure of NPI-2358, KPU-244 and biotin-tagged photoaffinity probe KPU-244-B1.

Results and Discussion

In the synthesis of compound **3**, a biotin-tag was successfully connected at the 4' position of the benzophenone moiety of KPU-244 via an additional aminomethyl group. To evaluate whether compound **3** can recognize tubulin, we performed the binding assay to tubulin based on fluorescence quenching and the dissolution

S. Del Valle et al. (eds.), *Peptides for Youth: The Proceedings of the 20th American Peptide Symposium*, 527
DOI: 10.1007/978-0-387-73657-0_230, © Springer Science+Business Media, LLC 2009

constant (K_d) of compound 3 to tubulin was calculated to be 7.95 μM. Additionally, compound 3 (5 μM) exhibited a 30 % inhibition of the tubulin polymerization and cytotoxic activity against HT-29 human colon cancer cell lines with a IC_{50} value of 0.91 μM. These results suggest that compound 3 is able to function as a photoaffinity probe.

Since compound **3** has an anti-microtubule activity, tubulin was photo-irradiated at 365 nm in the absence or presence of compound **3** under the physiological condition. An aliquot of sample solution was applied to SDS-PAGE and electrically blotted onto nitro-cellulose membrane, followed by detection using the avidin-biotin system. A specific and irradiation-time-dependent labeling was observed (Fig.2A). This labeling was also dose-dependently inhibited by colchicine (Fig.2B). Moreover, the labeling was relatively selective to beta-tubulin over alpha-tubulin. These results suggest that compound **3** specifically and covalently binds to near the colchicine binding site on beta-tubulin. Therefore NPI-2358 and KPU-244 as well as compound **3** would also recognize the similar site, resulting in the potent cytotoxic effect which can be exploited as an anti-cancer effect.

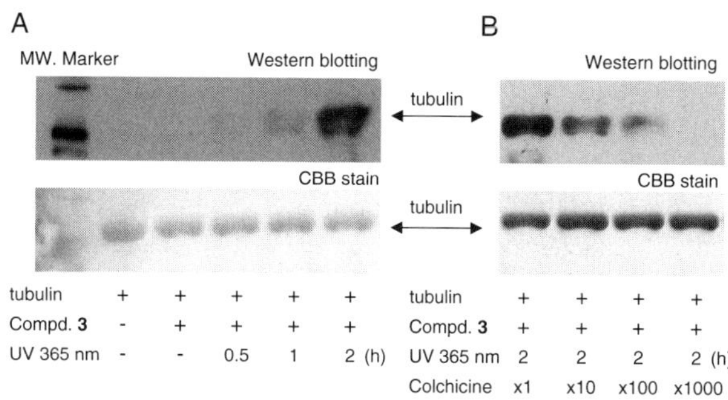

Fig. 2. Photoaffinity labeling of KPU-244-B1 to tubulin.

Acknowledgments

This research was supported by various grants from MEXT (Ministry of Education, Culture, Sports, Science and Technology), Japan, including the 21st Century COE Program.

References

1. Kanoh, K., Kohno, S., Katada, J., Hayashi, Y., Muramatsu M., and Uno, I. *Biosci. Biotechnol. Biochem.*, **63**, 1130-1133 (1999).
2. Kanoh, K., Kohno, S., Katada, J., Takahashi, J., Uno, I., and Hayashi, Y. *Bioorg. Med. Chem.*, **7**, 1451-1457 (1999).
3. Hayashi, Y., Orikasa, S., Tanaka, K., Kanoh, K., and Kiso, Y. *J. Org. Chem.*, **65**, 8402-8405 (2000).
4. Nicholson, B., Lloyd, G. K., Miller, B. R., Palladino, M. A., Kiso, Y., Hayashi, Y., and Neuteboom, S. T. C. *Anti-Cancer Drugs*, **17**, 25-31 (2006).

Study on Interactions of Oligo(tyrosine sulfate)s with Synthetic Heparin-binding Peptides by Affinity Chromatography and MALDI-TOF-MS

Miyuki Yamaguchi and Masaaki Ueki

Department of Applied Chemistry, Tokyo University of Science, Tokyo, 162-8601, Japan

Introduction

The heparin/heparan sulfate-like glycosaminoglycans (HLGAGs) play important roles in the regulation of many biological processes. However, a controlled application has been limited because of their heterogeneity and difficulty in synthesis. To solve the problems, many synthetic mimics including sulfated peptides have been prepared. Since the discovery of anti-HIV activity in oligo(tyrosine sulfate)s in our laboratory [1, 2], we have been interested in their potential as HLGAG mimics. In this study, we investigated their interactions with synthetic heparin-binding peptides using affinity chromatography and matrix-assisted laser desorption/ionization time-of-flight mass spectrometry (MALDI-TOF-MS).

Results and Discussion

Binding affinities of the heparin-binding peptides were preliminarily compared by retention on a NaO_3S-[Tyr(SO_3Na)]$_9$-immobilized affinity column and a H-[Tyr(SO_3Na)]$_9$-immobilized affinity column in the presence of increasing NaCl concentrations. Four heparin-binding peptides were used in this study. Three were derived from heparin-binding domains of human antithrombin III (hAT III) [3] and a HIP peptide is derived from heparin-interacting protein (HIP) [4]. While hAT III (123-139) and the HIP peptide bound to both columns, hAT III (39-54) and hAT III (286-301) were not retained at all.

Based on these results, relative binding affinities of oligo(tyrosine sulfate)s to hAT III (123-139) and the HIP peptide were investigated using surface non-covalent affinity mass spectrometry (SNA-MS), which enables affinity isolation of specific binders on MALDI surface [5]. The sample preparation procedure consists of immobilization of biotinylated heparin-binding peptides on MALDI surface, addition of oligo(tyrosine sulfate)s, low salt wash (0.1 M Na_2SO_4) to remove non-specific binders, and addition of 2,4,6-trihydroxyacetophenone (THAP) as matrix and tetrabutylphosphonium bromide (TBP·Br) as co-matrix [6]. In this study, *N*-sulfated oligomers NaO_3S-[Tyr(SO_3Na)]$_n$-ONa (n = 5-12) and *N*-nonsulfated oligomers H-[Tyr(SO_3Na)]$_n$-ONa (n = 4-12) were used as samples.

We compared the relative binding affinity between oligo(tyrosine sulfate)s and other sulfated heparin mimics to hAT III (123-139). In the analysis of mixtures of each oligo(tyrosine sulfate) and a heparin-derived sulfated hexasaccharide ($\Delta U_{2S}H_{NS,6S}I_{2S}H_{NS,6S}I_{2S}H_{NS,6S}$), oligo(tyrosine sulfate)s from pentamer to dodecamer were detected selectively. This result suggests that oligo(tyrosine sulfate)s bind to hAT III (123-139) more strongly than the hexasaccharide.

Next, we compared the relative binding affinity between oligo(tyrosine sulfate)s and a pentapeptide, SY(SO_3)DY(SO_3)G [Y(SO_3) denotes tyrosine *O*-sulfate]. This disulfated short peptide was identified as the top binder to $VEGF_{165}$ among the library of tetrapeptides using a fluorescence assay, and shown to bind to $VEGF_{165}$ 100-fold more strongly than suramin by surface plasmon resonance spectroscopy [7].

S. Del Valle et al. (eds.), *Peptides for Youth: The Proceedings of the 20th American Peptide Symposium*, 529
DOI: 10.1007/978-0-387-73657-0_231, © Springer Science+Business Media, LLC 2009

In the analysis of mixtures of each *N*-nonsulfated oligomer shorter than the hexamer ($n = 6$) and the pentapeptide, none of them was retained after a low salt wash. When these mixtures were analyzed without a low salt wash, the oligo(tyrosine sulfate) was detected selectively. In the cases of *N*-sulfated oligomers and the *N*-nonsulfated hexamer and the longer oligomers, only the oligo(tyrosine sulfate)s were detected. These results suggest that oligo(tyrosine sulfate)s have a much higher binding affinity to hAT III (123-139) than the pentapeptide.

Then, dependence of binding affinity on chain-length was definitively determined. When an equimolar mixture of two oligo(tyrosine sulfate)s was analyzed, the oligo(tyrosine sulfate) with higher affinity could be detected selectively. For example, when a mixture of two *N*-nonsulfated oligomers with different chain length was analyzed, the longer oligomer was bound and detected selectively. The same chain length dependence was observed in the analysis of mixtures of two *N*-sulfated oligomers, but in some cases both oligomers were detected. From these results, it seems that the difference in binding affinity between the two *N*-sulfated oligomers is relatively small. Thus, we succeeded in observing chain-length dependence in non-covalent complex formation of oligo(tyrosine sulfate)s with hAT III (123-139).

To evaluate the effects of *N*-sulfation on the binding affinity to hAT III (123-139), *N*-sulfated oligo(tyrosine sulfate)s were compared with *N*-nonsulfated oligomers containing the same or one more sulfate groups. In the analysis of mixtures of *N*-sulfated oligomer with tyrosine residue number n, NaO_3S-[Tyr(SO_3Na)]$_n$-ONa, and *N*-nonsulfated oligomer with residue number n+1 or n+2, H-[Tyr(SO_3Na)]$_{n+1}$-ONa or H-[Tyr(SO_3Na)]$_{n+2}$-ONa, signals of both oligomers were observed. These results suggest that the effect of *N*-sulfation is more important than chain-length elongation in enhancing the binding affinity of oligo(tyrosine sulfate)s to hAT III (123-139).

In the case of the HIP peptide, binding affinity of oligotyrosine sulfate)s was almost the same as that to hAT III (123-139).

In conclusion, we succeeded in observing specific binders in non-covalent complex formation of oligo(tyrosine sulfate)s with heparin-binding peptides. Further investigation is now in progress in our laboratory.

Acknowledgments

One of us (MY) is indebted to support from the Hayashi Memorial Foundation for Female Natural Scientists.

References

1. Ueki, M., *et al. Bioorg. Med. Chem.*, **9**, 477-486 (2001).
2. Ueki, M., *et al. Bioorg. Med. Chem.*, **9**, 487-492 (2001).
3. Onoue, S., *et al. Peptides*, **24**, 821-826 (2003).
4. Liu, S., *et al. Proc. Natl. Acad. Sci. USA*, **94**, 1739-1744 (1997).
5. Kaiser, N., *et al. Nature Med.*, 7, 123-128 (2001).
6. Ueki, M. and Yamaguchi, M. *Rapid Commun. Mass Spectrom.*, **20**, 1615-1620 (2006).
7. Maynard, H. D. and Hubbell, J. A. *Acta Biomaterialia*, **1**, 451-459 (2005).

Fallaxin analogues with improved antibacterial activity

S.L. Nielsen[1], N. Frimodt-Møller[2] and P.R. Hansen[1]

[1]*Department of Natural Sciences, Faculty of Life Sciences, University of Copenhagen, 1871 FC, Denmark.* [2]*National Center for Antimicrobials and Infection Control, Statens Serum Institut, 2300 S, Denmark.*

Introduction

The antibacterial peptide "fallaxin" H-GVVDILKGAAKDIAGHLASKVMNKL-NH_2, recently isolated from the West-Indian mountain chicken frog "*leptodactylus fallax*", have been shown to inhibit the growth of a number of Gram-negative bacteria including *E.coli, P.aeruginosa and K.pneumoniae* [1]. Antibacterial activity and selectivity has in studies shown to depend on various parameters i.e. the net charge, mean hydrophobicity $\langle H \rangle$, mean hydrophobic moment $\langle \mu_H \rangle$ and the ability to fold into amphipatic structures upon interaction with membrane models [2]. To gain further insight into the structure-activity relationship of fallaxin, we performed a complete alanine-scan of fallaxin. The peptides were tested for antibacterial activity against methicilin-resistant *S.aureus* (MRSA), vancomycin-intermediate resistant *S.aureus,* (VISA), *E.coli* and *K.pneumoniae*. Finally, the cytotoxicity of the fallaxin analogues against human erythrocytes was assessed in a hemolytic activity assay

Results and Discussion

Peptide synthesis was performed manually using standard Fmoc-chemistry on a TentaGel S RAM resin. Protected amino acids were coupled in three-fold excess employing a protocol using HATU/DIEA (1:1.5) activation, and NMP as solvent. Final cleavage and deprotection were carried out using reagent L: TFA, TIS, DTT and water (88:2:5:5) for 2h. Following synthesis, the peptides were purified by RP-HPLC and masses verified by MALDI-TOF MS. The peptides were tested for antibacterial activity against *S.aureus* ATCC 33591 (MRSA), 700699MU50 (VISA), *E.coli* ATCC 25922 and *K.pneumoniae* ATCC 700603. The hemolytic activities were determined against human erythrocytes (ORh+).

The antibacterial activity of fallaxin was shown to depend on overall charge. Sub- stitution of Asp^{12} with the electrically neutral alanine (FA12) resulted in the best activity towards *E. coli* with an observed MIC value of 3.12 μM compared to that of fallaxin of 100 μM. However, this analogue showed a 32 % increase in hemolytic activity. Replacing the positively charged lysine with alanine in FA11 and FA20 resulted in a decreased activity profile [3]. The alanine-scan revealed some analogues with increased antibacterial activity and retained hemolytic activity. The analogues FA4 and FA19 showed an activity towards *E.coli* of 50 and 25 μM respectively and very low hemolytic activity. The analogues FA9, 10, 12, 14 and 18 showed an increased activity towards MRSA and VISA, compared to that of the parent peptide [3]. We observed that the majority of the active fallaxin analogues had a mean hydrophobicity ranging from -0,0256 to -0,0088 and a hydrophobic moment ranging from 0,00766 to 0,1164 showed MIC values of 25 μM or better against at least one bacterial strain. This is true for FA8, FL9, FL10, FA12, FL14

S. Del Valle et al. (eds.), *Peptides for Youth: The Proceedings of the 20th American Peptide Symposium*, 531
DOI: 10.1007/978-0-387-73657-0_232, © Springer Science+Business Media, LLC 2009

and FA15. The exceptions being FA1 and FA12. The less active analogues had values outside this range [3].

Characteristics of fallaxin and analogues

Peptide	Minimum Inhibitory Concentration (MIC)				Hem. Activity	<H>	$<\mu_H>$
	K.pneumoniae	*E.coli*	*VISA*	*MRSA*			
Fallaxin	100	100	>100	>100	0	-0.0292	0.1055
FA25	>100	>100	>100	>100	0	-0.0400	0.1030
FA23	100	50	>100	>100	0	0.0064	0.0900
FA20	>100	100	>100	>100	2	0.0248	0.1368
FA19	100	25	>100	>100	2	-0.0088	0.0858
FL18	50	12,5	100	100	38	-0.0180	0.1053
FA15	50	12,5	>100	>100	14	-0.0256	0.1023
FL14	25	6,25	50	100	55	-0.0180	0.1124
FA12	25	3,12	50	100	32	0.0096	0.0766
FA11	>100	100	>100	>100	5	0.0248	0.0968
FL10	50	12,5	50	100	43	-0.0180	0.1164
FL9	50	12,5	50	50	27	-0.0180	0.1069
FA8	50	25	>100	>100	11	-0.0256	0.1019
FA4	50	50	100	>100	1	0.0096	0.0832
FA3	100	>100	>100	>100	0	-0.0408	0.0961
FA1	>100	>100	>100	>100	0	-0.0256	0.1020

FAX or FLX corresponds to a fallaxin analogue with an alanine or a leucine in position X with respect to the primary structure of fallaxin. The hemolytic activity is in procentage.

In conclusion, we have identified several fallaxin analogues which show an increased activity profile towards MRSA, VISA, *E.coli* and *K.pneumoniae.* As expected the antibacterial activity was shown to be dependent upon peptide charge. We found a correlation between antibacterial activity and mean hydrophobicity <H> and mean hydrophobic moment $<\mu_H>$ and we were able to group the active and inactive analogues according to mean hydrophobicity and mean hydrophobic moment [3]. This study indicates that fallaxin analogues are promising lead-structures for the development of new antibacterial agents.

Acknowledgments

SLN gratefully acknowledges a scholarship from Novozymes/Novo-Nordisk.

References

1. Rollins-Smith, L.A., et al. (2005) An Antimicrobial peptide from the skin secretions of the mountain chicken frog *Leptodactulys fallax* (*Anura: Leptodactylidae*) *Regulatory Peptides* 124: 173-179
2. Hancock, R.E.W., and Sahl, H.G., (2006). Antimicrobial and host-defense peptides as new anti-infective therapeutic strategies. *Nature Biotechnology* 24: 1551-1557.
3. Nielsen, S.L., et al. (2007) Structure-activity study of the antibacterial peptide fallaxin. *Protein Science* 16: 1969-1976.

Alkene/Alkane-Bridged Mimics of the Lantibiotic Nisin: Toward Novel Peptide-Based Antibiotics

Rianne A. G. Harmsen[1], Nourdin Ghalit[1], Johan Kemmink[1], Hans W. Hilbers[1], Eefjan Breukink[2], Dirk T. S. Rijkers[1] and Rob M. J. Liskamp[1]

[1]*Medicinal Chemistry & Chemical Biology, Utrecht Institute for Pharmaceutical Sciences, Department of Pharmaceutical Sciences;* [2]*Chemical Biology & Organic Chemistry, Department of Chemistry, Faculty of Sciences, Utrecht University, Utrecht, The Netherlands*

Introduction

Nisin belongs to a class of natural antimicrobial peptides, the lantibiotics, which is produced by a broad range of bacteria e.g. *Bacillus*, *Lactococcus*, *Streptomyces* and *Staphylococcus* species [1]. A general feature of these peptides is the presence of a lanthionine moiety (thioether- or sulfide bridge) which gives these peptides their specific bioactive conformation.

Nisin binds with its N-terminus to lipid II thereby enabling the C-terminus to form pores in the phospholipid membrane which ultimately leads to cell leakage and causes a collapse of the vital ion gradients across the cell membrane. Lipid II is an important target for many antimicrobial peptides such as vancomycin, ramoplanin and nisin and serves as a general target for the development of new antibiotics [2].

Recently, the molecular mechanism of the nisin – lipid II interaction has been unraveled by NMR studies [3]. It has been found that nisin binds the pyrophosphate moiety in the pyrophosphate cage which is formed by the peptide backbone of the N-terminal AB-ring system (see Figure 1).

Binding of the AB-fragment to lipid II also displays antimicrobial activity, although to a lesser extent than full length nisin [4]. It is thought that binding of the AB-fragment to lipid II interferes with the cross-linking of the peptidoglycan network in the bacterial cell wall and the formation of this complex implies that any molecule with binding affinity toward lipid II can act as a potential antibiotic.

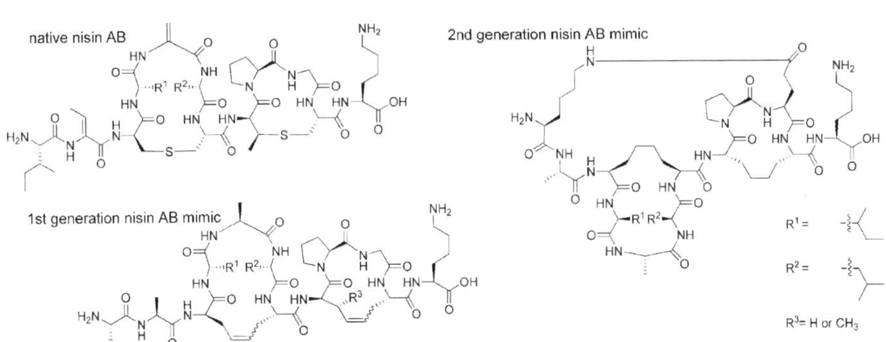

Fig. 1. Chemical structures of the native nisin AB fragment and its corresponding 1st and 2nd generation alkene/alkane-bridged mimics.

The lanthionine moiety as a natural constraint in bioactive peptides can be replaced by an alkene/alkane bridge in order to increase the metabolic stability of the newly designed peptide derived antibiotics [5]. In this contribution we describe the

S. Del Valle et al. (eds.), *Peptides for Youth: The Proceedings of the 20th American Peptide Symposium,* 533
DOI: 10.1007/978-0-387-73657-0_233, © Springer Science+Business Media, LLC 2009

design, synthesis and biochemical evaluation of the first and second generation nisin AB mimics.

Results and Discussion

A potential isostere of the thioether constraint is an alkene bridge which can be introduced by ring-closing metathesis (RCM) using amino acid residues with a terminal alkene moiety in the side chain. Since the (3-methyl)lanthionine is formed by a combination of a D- and L-amino acid, the alkene/alkane bridge mimic is synthesized using two L-allylglycine residues, a combination of a D- and L-allylglycine and a β-methyl substituted D-allylglycine together with an L-allylglycine residue, respectively. The synthesis of the nisin AB mimics consisted of the linear synthesis of the individual ring fragments, either in solution or on the solid phase, followed by RCM in solution using the 2nd generation Grubbs' catalyst. Then, the AB mimics were assembled by coupling of the cyclized fragments in solution.

The tricyclic nisin AB mimic has been designed based on the solution structure of the nisin – lipid II complex as determined by NMR [3]. In this structure, the δ-CH_3 moiety of the L-isoleucine residue at position 1 points toward the pro-*S* hydrogen atom of the glycine residue at position 10. Thus, a third covalent constraint, connecting both side chains, might result in an increased lipid II binding affinity. Therefore, a lactam bridge was introduced using an L-lysine – L-glutamic acid combination at position 1 and 10, respectively.

Both nisin AB mimics were tested on their lipid II binding affinity by evaluating their potency to inhibit nisin-induced carboxyfluorescein (CF) release from large unilamellar vesicles loaded with lipid II. In a typical experiment, preincubation of CF-loaded vesicles containing lipid II with the nisin AB mimics was performed to test their ability to occupy the binding site of nisin on lipid II. Subsequent addition of full length nisin will result in a reduced amount of pore formation depending on the affinity of the AB fragments toward lipid II. Native nisin AB as well as both AB mimics were not able to form pores in the phospholipid bilayer. However, in the presence of these AB fragments, nisin-induced CF-release was reduced, indicating that these compounds were capable of competing with nisin for the pyrophosphate binding site. At 100 μM, a 14% inhibition of nisin-induced CF-release was observed with the first generation AB mimic (all L-residues). A significant increase in affinity was found with the second generation AB mimic (all L-residues) since an inhibition of 52% was observed at the same concentration. Native nisin AB was still found to be most active compound, showing an inhibition of 37% at 20 μM.

These data imply that an alkene/alkane moiety is a suitable thioether mimic. It is expected that fine-tuning of the backbone stereochemistry in combination with a defined configuration of the double bond will further increase the affinity of the nisin AB mimics toward lipid II.

References

1. Chatterjee, C., Paul, M., Xie, L. and van der Donk, W. A. *Chem. Rev.* **105**, 633-683 (2005).
2. Breukink, E. and de Kruijff, B. *Nat. Rev. Drug Disc.* **5**, 321-332 (2006).
3. Hsu, S.-T. D., Breukink, E., Tischenko, E., Lutters, M. A. G., de Kruijff, B., Kaptein, R., Bovin, A. M. J. J. and van Nuland, N. A. J. *Nat. Struct. Mol. Biol.* **11**, 963-967 (2004).
4. Hasper, H. E., Kramer, N. E., Smith, J. L., Hillman, J. D., Zachariah, C., Kuipers, O. P., de Kruijff, B. and Breukink, E. *Science* **313**, 1636-1637 (2006).
5. Ghalit, N., Reichwein, J. F., Hilbers, H. W., Breukink, E., Rijkers, D. T. S. and Liskamp, R. M. J. *ChemBioChem* **8**, 1540-1554 (2007).

Peptide-Lisinopril Conjugates: Design, Synthesis and Biological Activities

Josue Alfaro-Lopez, Chris Soares, Soumitra Ghosh, Alexander Kravchuk, Eugene Coates, Lynn Jodka, Amy Carroll and Que Liu

Research Chemistry, Amylin Pharmaceuticals Inc. 9360 Towne Centre Drive, San Diego, CA 92121, USA

Introduction

The proteolytic cleavage of peptide bonds by regulatory and non-specific peptidases is the principal mechanism of inactivation and clearance of bioactive peptide drugs *in vivo*. When the peptidase itself is a therapeutic target in the disease of interest, focusing on the peptidase and a peptide receptor simultaneously could provide dual action to improve the overall therapeutic outcome. We have explored this new approach by covalently attaching lisinopril, a potent angiotensin converting enzyme (ACE) inhibitor, to the carboxy terminus of a series of bioactive peptides. ACE is a major factor determining systemic blood pressure through its action to release the potent vasoconstrictor angiotensin II. Lisinopril is now widely used to treat systemic hypertension, heart failure and the nephrotoxic effects of diabetes (see Figure 1).

The mapping of bound inhibitors at the ACE active site has been validated for known experimental compounds [1]. The bound conformations of the chemically relevant atoms were accurately deduced from the geometry of the ligand and it was evident, in the case of lisinopril, that the Lys side chain was not important in the ACE active site model. Based on these and other results [2] we decided to use this group as the linkage point in our conjugates.

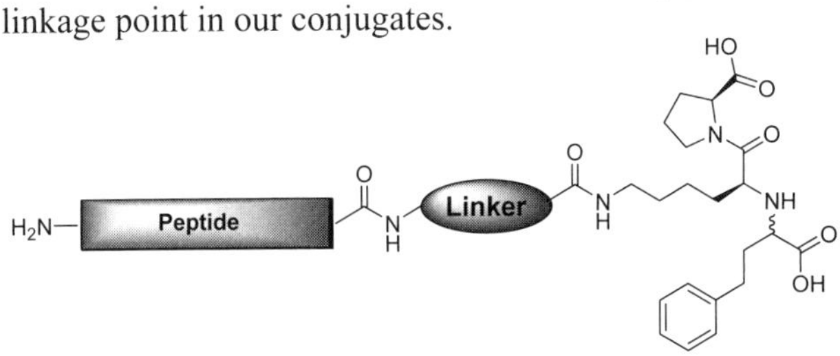

Fig. 1. Structure of Peptide-Linker-Lisinopril Conjugates.

Results and Discussion

We have developed an efficient solid phase synthesis of lisinopril (see Figure 2). Starting with commercially available H-Pro-Trt resin (0.68 mmol/g), coupling of Fmoc-Lys(alloc)-OH using HOBt and HBTU yielded 100% loading of resin 1. Fmoc deprotection of 1 was followed by reductive amination using 20 equivalents of the α-oxoester, which was then treated under neutral conditions of alloc deprotection to give resin 2. Seven different linkers, ranging from 5 to 21 atoms in size, were then coupled and the resin was prepared for peptide synthesis in a Symphony synthesizer (100 μmol scale). The crude peptides were purified by LCMS-PrepExpress (mass triggered collection) (C_{18}, 25-60% CH_3CN in 0.1% TFA/H_2O over 30 min gradient) to yield the titled compounds as white powders. Peptide-ACE peptidase inhibitor conjugate compounds shown in Table 1 were assessed in an ACE inhibition assay using purified ACE enzyme and chromogenic substrates

S. Del Valle et al. (eds.), *Peptides for Youth: The Proceedings of the 20th American Peptide Symposium*, 535
DOI: 10.1007/978-0-387-73657-0_234, © Springer Science+Business Media, LLC 2009

The conjugates were also tested for GLP-1 cyclase activity using a whole cell assay which measures increases in cAMP in a cell line via the peptide-induced activation of the endogenously expressed GLP-1 receptor. After SAR analysis, a selected compound (AC164786) was tested in a congestive heart failure (CHF) *in vivo* assay and the results are also shown in Table 1.

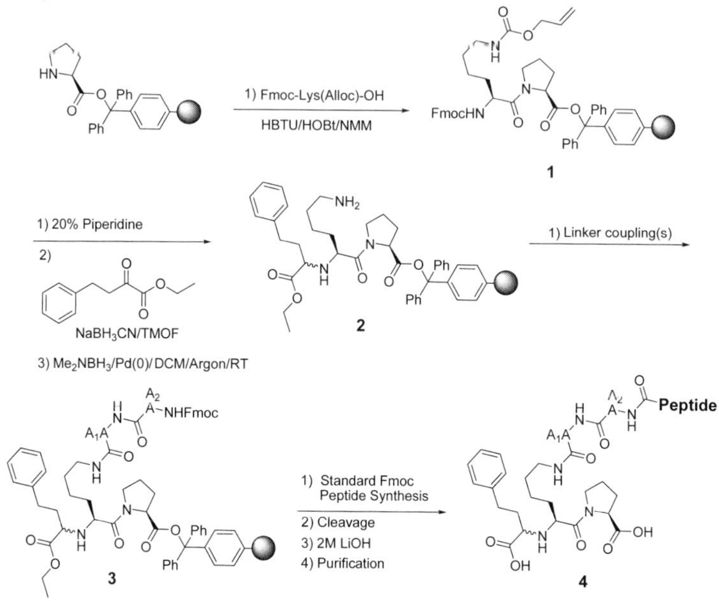

Fig. 2. Solid phase synthesis of peptide-linker-lisinopril conjugates.

Table 1. Biological activities of selected conjugates.

Compound	Linker size	GLP-1 cyclase EC_{50}	ACE IC_{50}	Glucose Lowering	In vivo CHF
AC164678	21	0.55 nM	100 nM	GLP-1-like[a]	ND
AC164681	9	2.0 nM	50 nM	Exenatide-like[a]	ND
AC164787	9	0.14 nM	140 nM	Not active	ND
AC164786	21	0.06 nM	260 nM	Exenatide-like[a]	Active

[a]*Similar in-vivo activity at 2 nmol/kg dose, ND= Not determine.*

The *in vitro* and *in vivo* results shown in Table 1 provide evidence that these conjugates present a dual mechanism of action. Further investigation of the potential of these types of compound to form a new therapeutic class is warranted.

References

1. Kuster, D.J., Marshall, G.R. *J. Comp. Aid. Mol. Des.* **19**, 609-615 (2005).
2. Bernstein, K.E., Inman, J.K. *Biochem. Biophys. Res. Commun.* **167**, 310-316 (1990).

Novel Bifunctional Peptides as Opioid Agonists and NK-1 Antagonists

Padma Nair[1], Takashi Yamamoto[1], Vinod Kulkarni[1], Sharif Moye[2], Edita Navratilova[2], Peg Davis[2], Tally Largent[2], Shou-wu Ma[2], Henry I.Yamamura[2], Todd Vanderah[2], Josephine Lai[2], Frank Porreca[2] and Victor. J. Hruby[1]

[1]*Departments of Chemistry and* [2]*Pharmacology, University of Arizona, Tucson, AZ 85721, U.S.A.*

Introduction

Inadequate treatment of pain is an important and urgent problem, which needs to be addressed. Opiates are still the drugs of choice for the treatment of moderate to severe acute pain; however their use are generally accompanied by undesired side effects like analgesic tolerance. Several pharmacological studies have suggested the role of substance P, in pain transmission [1]. It has been observed that co-administration of neurokinin-1 receptor antagonists and opioid agonists augmented the acute effects of opioids and also prevented opioid induced tolerance [2]. In view of these and other observations we have designed novel bifunctional peptides with mixed opioid μ/δ agonist activity and neurokinin-1 antagonist activity [3]. Earlier reports of molecules with opioid μ/δ agonist activity and neurokinin-1 antagonist activity did not have balanced activity at both the receptor's systems [4]. In our studies the opioid pharmacophore chosen has the sequence H-Tyr-DAla-Gly-Phe which is a substructure of biphalin, a highly potent agonist at both δ and μ opioid receptors [5]. A 3,5-(bistrifluoromethyl) benzyl ester of *N*-acylated tryptophans was chosen as the NK1 pharmacophore [6]. These two pharmacophores were combined with amino acids acting as possible address moieties they could be part of the two pharmacophores.

Sequences of some of the bifunctional peptides synthesized are represented below (Fig. 1)

NP30 H-Tyr-DAla-Gly-Phe-**Gly**-Trp-O-3,5-Bzl$(CF_3)_2$
NP32 H-Tyr-DAla-Gly-Phe-**βAla**-Trp-O-3,5-Bzl$(CF_3)_2$
NP34 H-Tyr-DAla-Gly-Phe-**γAbu**-Trp-O-3,5-Bzl$(CF_3)_2$
NP40 H-Tyr-DAla-Gly-Phe-***D*Ala**-Trp-O-3,5-Bzl$(CF_3)_2$

Fig. 1. Sequences of the peptides with opioid agonist and NK-1 antagonist activity.

Results and Discussion

The bifunctional peptide ligands were synthesized using solution phase chemistry. The synthetic strategy was started with coupling of tryptophan 3,5-(bistrifluoromethyl)benzyl ester with the respective Boc protected amino acid followed by subsequent chain elongation using BOP/HOBT/NMM method. The N^{α}-Boc groups were deprotected by 100% TFA. The final crude peptides were purified by C_{18} RP-HPLC (Vydac 10 mm x 250 mm, 10 μM) with a gradient of 30-70% CH_3CN in aq. 0.1% TFA.

NP30, with Gly as the linker showed excellent binding affinity at both δ and μ opioid receptors and rat NK1 receptors (K_i = 4.7 nM, δ opioid; 0.29 nM, μ opioid;

DOI: 10.1007/978-0-387-73657-0_235, © Springer Science+Business Media, LLC 2009

Table 1. Binding affinities.

Peptide	Binding affinities			Functional assay		
	K_i(nM)			IC_{50} (nM) (agonist)		K_e (nM)
	hDOR	rMOR	rNK1	MVD	GPI	SP(GPI) (antagonist)
NP30	4.7	0.29	4.2	21	26	59
NP32	58	11	1.6	13	430	250
NP34	25	2.2	7.4	68	200	0.96
NP40	3.9	2.8	2.4	58	530	610

4.2 nM, rNK1). It also showed potent δ and μ opioid agonist efficacies in the MVD and GPI assays binding assays with the IC_{50} values of 21 and 26 nM, respectively.

GPI assay clearly showed that NP30 was an antagonist against substance P stimulation with K_e = 59 nM. NP34 with γAbu (Aminobutyric) at position 5 turned out to be the most potent at the GPI assay (K_e = 0.96nM) among all the compounds synthesized. On the other hand substitution with DAla (NP40) at the same position resulted in drastic lowering of antagonistic activity in the GPI assays (K_e = 610nM). However both NP34 and NP40 were highly potent in the binding affinity studies at the opioid and rat NK1 receptors.

Conclusion

Several bifunctional peptides with δ/μ opioid agonist pharmacophore and NK1 antagonist pharmacophore were designed and synthesized. Binding assays performed showed potent activities as an opioid agonist and as NK1 antagonists thus exhibiting the potential of these compounds to act as bifunctional ligands (Table 1).

Acknowledgments

The work was funded by U. S Public Health Service, National Institute of Drug Abuse.

References

1. Khasabov, S.G., Rogers, S.D., Ghilardi, J.R., Peters, C.M., Mantyh, P.W., Simone, D.A., *J. Neurosci.* **22**, 9086–98 (2002).
2. Powell, K.J., Quirion, R., Jhamandas, K. *Neuroscience*, **18**, 1572-83 (2003).
3. Yamamoto, T., Nair, P., Davis, P., Ma, S.W., Moye, M., Largent-Milnes, T., Vanderah, T.W., Lai, J., Porreca, F.; Yamamura, H.I., Hruby, V.J. *J. Med Chem.*, **50**, 2779-86 (2007).
4. Bonney, I.M., Foran, S.E., Marchand, J.E., Lipkowski, A.W., Carr, D.B. *Eur. J. Pharmacol.*, **488**, 91-99 (2004).
5. Horan, P. J., Mattia, A., Bilsky, E.J., Weber, S., Davis, T.P., Yamamura, H.I., Malatynska, E., Appleyard, S.M., Slaninova, J., Misicka, A., Lipowski, A.W., Hruby, V.J., Porreca, F. *J. Pharmacol Exp. Ther.* **265**, 1446–54 (1993).
6. Macleod, A.M., Merchant, K.J., Cascieri, M.A., Sadowski, S., Ber, E., Swain, C.J., Baker, R. *J. Med Chem.*, **14**, 2044-5 (1993).

.

Anti-HIV dendrimeric peptides

Qitao Yu, Ling Li and James P. Tam

Department of Biochemistry, The Scripps Research Institute in Florida, Jupiter, FL 33458, USA

Introduction

Novel dendrimeric peptides were developed in our laboratory as new architectural peptides to facilitate the peptide-based drug design [1]. Previously, we reported a tetrapeptide RLYR to be a motif that contributes to the antimicrobial activity of a cascade-type peptide dendrimer. Our studies have demonstrated that dendrimeric peptides containing several copies of this motif on their surface were found to possess broad antimicrobial activity [1]. Here we report the development of a branch-type of RLYR-containing dendrimeric peptide that displays potent HIV-1 inhibition. Comparing with a control α-peptide consisting of tandemly repeating RLYR, this dendrimer is 50-fold more resistant to proteinase K than the corresponding linear α-peptide. Our results show that the unusual architectural design of dendrimeric peptide allows us to integrate multiple copies of functional motifs in parallel while improving stability and simplifying the synthetic process.

Results and Discussion

Our prototypic design of a branched dendrimeric peptide contains an α-peptide backbone α-$(Lys)_4$ and four pendant branches of RLYR tethered to the ε-amines of α-$(Lys)_4$. This is based on our original design of the cascade-type $(RLYR)_4K_2K$ which we now rearranged the backbone to an α-peptide and the architecture to a branch-type peptide dendrimer. For synthetic expedience, the N-terminal α-amine is capped with a Fmoc protecting group. For comparison, we also prepared a 16-residue of linear peptide containing four tandemly repeating RLYR (Figure 1).

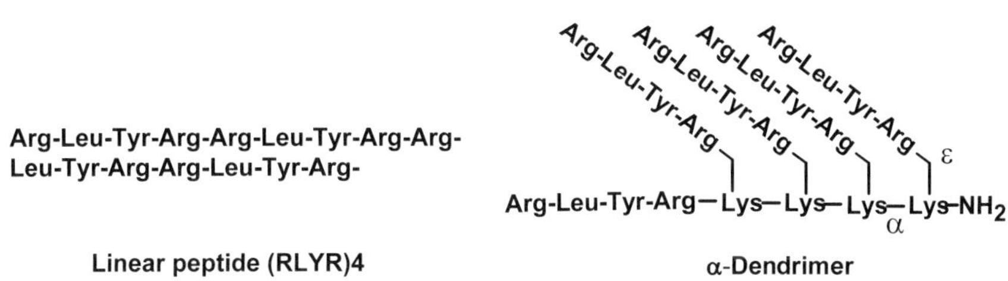

Fig. 1. Schematic representations of linear and α-dendrimeric peptides.

The α-dendrimer was prepared by a combination of Boc- and Fmoc-chemistries, with the α-peptide backbone being assembled by the Fmoc chemistry and the side-chain RLYR peptides by the Boc chemistry (Figure 2).

To determine effect of peptide architecture on proteolytic stability, we compared the rates of proteinase K digestion on both linear and branched peptide. Using 0.05mg/ml proteinase K, the half life for linear peptide $(RLYR)_4$ was <2 min whereas half life for α-dendrimer was 150 min, a 75 fold increases in resistance to proteinase K digestion comparing the corresponding linear peptide.

S. Del Valle et al. (eds.), *Peptides for Youth: The Proceedings of the 20th American Peptide Symposium*, 539
DOI: 10.1007/978-0-387-73657-0_236, © Springer Science+Business Media, LLC 2009

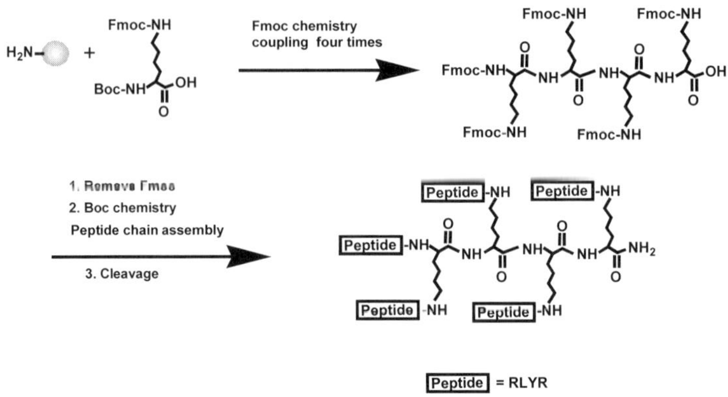

Fig. 2.Synthetic scheme of α-dendrimer.

The effect of our prototype dendrimer on HIV -1 was examined by Magi assay. Briefly, HIV-1 NL4-3 viral supernatant was harvested from transient transfection of 293T. The reporter cell line, P4-R5 magi cells were seeded $5x10^3$ cells per well in 96 well plates. 24 hours later, cell media were replaced with viral supernatant containing serial diluted dendrimers. 48 hours later, the β-galactosidase activity was assayed using β-glo kit (Promega, Madison, WI).

Our result shown that dendrimer inhibits the infection of HIV NL4-3 with IC_{50} at ~0.15 μM level. However, α-dendrimer failed to inhibit VSV-G pseudo-typed NL4-3 luciferase in the same assay system, indicating that α-dendrimer inhibited NL4-3 replication at the entry level.

We found the prototype dendrimer has no harmful effect of on HIV-1 susceptible cells. Cell toxicity study was performed using three cell lines, P4R5 Magi cell, MT4 and SupT 1 cells, which are often used for the study of HIV-1 replication. Under our experimental conditions, no toxicity was observed with all three cell lines using concentration up to 1μM by MTT assay (Invitrogen, Carlsbad, CA) and Cell glo kit (Promega, Modison, WI).

In conclusions, our prototypic dendrimeric peptide displays potent anti-HIV activity, low cell toxicity, and enhanced stability. While the mechanism of their anti-HIV activity is still being investigated, the novel architectural design and easy preparation (requiring fewer chemical steps for synthesis) of dendrimers offer strong potential for developing "druggable" peptides.

Acknowledgments
The work was in part supported by US public Health Service NIH Grant EB001986 . P4R5 cells and plasmids, NL4-3 Luc were obtained from Dr. Nathaniel Landau through the AIDS Research and Reference Reagent Program, Division of AIDS, NIAID, NIH.

References

1. Tam, J. P., Lu, Y-A. and Yang J-L. *Eur. J. Biochem.* **269**, 923-932 (2002).

Ago-Antagonists for G Protein-Coupled Peptide Hormone Receptor by Modifying the Agonist's Signalling Domain

Michael Beyermann, Nadja Heinrich, Klaus Fechner, Jens Furkert, Wei Zhang, Oliver Kraetke, Michael Bienert and Hartmut Berger

Department of Peptidchemistry/Biochemistry, Leibniz-Institute of Molecular Pharmacology, Berlin, D-13125, BRD

Introduction

Most of the pharmaceuticals today target at G protein-coupled receptors (GPCRs) [1] that transmit extracellular signals into cells. GPCRs are promiscuous, that means a single receptor can activate different signalling events, presumably via different G protein subtypes. Functional selectivity of GPCR ligands has been observed and points at a new orientation in pharmaceutical research. However, the structural characteristics of ligands that produce this selectivity are far from known [2].

The polypeptide hormone urocortin I (Ucn) activates both Gs and Gi proteins via the corticotropin-releasing factor receptor type 1 (CRF_1). We have recently developed an easy method for the separate measurement of Gs and Gi activation at HEK 293 cells stably transfected with cDNA coding for CRF_1 [3,4]. The aim of this study was to search for structural determinants of the peptide agonist Ucn (DDPPLSIDLT FHLLRTLLEL ARTQSQRERA EQNRIIFDSV-NH_2) that direct the signalling to Gs and Gi, respectively. For this purpose, the effect of single replacements by bulky amino acids, benzoyl-phenylalanine (Bpa) and naphthyl alanine (Nal), on Gs and Gi signalling pathways was measured in HEK-CRF_1 cells, using receptor binding, [^{35}S]-GTPγS binding stimulation, and cAMP accumulation assays.

Results and Discussion

From former structure-activity relationship studies the two-domain binding model was derived for CRF receptor activation by peptide agonists, whereupon the receptor N-terminus captures the ligand C-terminus, representing the high affinity binding site, and the ligand N-terminus then activates the receptor, presumably interacting with the juxta-membrane domain of receptor. The activating function of the ligand N-terminus (the first 9-11 residues) was mainly deduced from the fact that corresponding N-terminal truncations of the natural agonists convert them into potent antagonists. For G protein coupling of CRF_1 our screening results for Bpa and Nal analogues of Ucn do show a particular role of residues (6-15).

For the natural peptide ligands, such as Ucn and sauvagine, we have recently shown that Gs activation at HEK-CRF_1 cells correlates with a high affinity receptor state (for 3-I-Tyr(0)-Svg; $EC_{50(1)}$: 3.24E-11 M/ K_{d1}: 3.85E-11 M) and Gi activation with a low affinity state ($EC_{50(2)}$: 3.23E-09 M/ K_{d2}: 1.47E-08 M) [3]. In principle, we observed the same for Ucn analogues substituted outside of the domain (6-15); a decrease in potency comes along with a loss of affinity for both receptor states. For analogues substituted at positions 6-15, with the exception of position 12, the results were completely different, in that we have seen a much more reduced or even no Gi activity, whereas the Gs coupling was much less affected as exemplarily demonstrated for Bpa(7)-Ucn (Fig. 1). The domain 6-15 of Ucn represents the signalling domain, modifications of which yielded analogues that fully activate Gs

S. Del Valle et al. (eds.), *Peptides for Youth: The Proceedings of the 20th American Peptide Symposium*, 541
DOI: 10.1007/978-0-387-73657-0_237, © Springer Science+Business Media, LLC 2009

and thus stimulate cAMP production (Fig. 1), but completely failed in Gs binding stimulation (Fig. 1) and, moreover, antagonize sauvagine stimulated Gi activation (Fig. 2). Thus, we could show that ligands actually can be at the same time both agonists and antagonists of different G proteins at a single receptor, what we name in this case Gs-ago-Gi-antagonism.

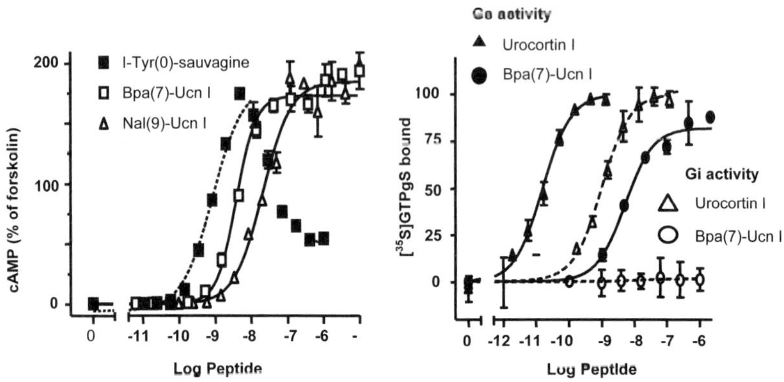

Fig. 1. Dose-response curves for general agonists (I-Tyr(0)-sauvagine, urocortin) and Gs-ago-Gi-antagonists (Bpa(7)Ucn, Nal(9)-Ucn): cAMP production (left) and GTPγS binding stimulation in HEK-CRF$_1$ cells [5].

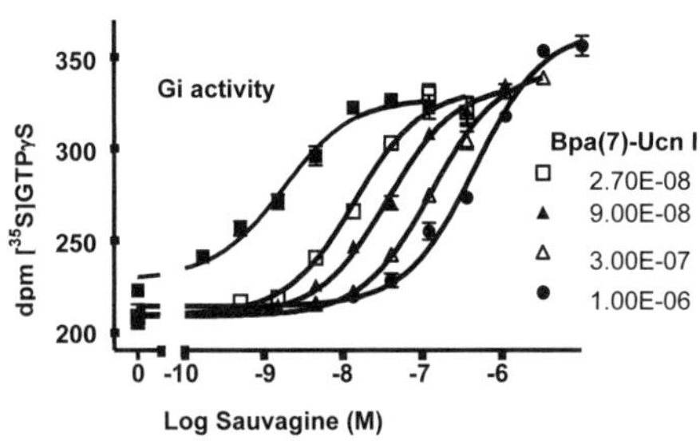

Fig. 2. Right-shift of dose-response curve for Gi binding stimulation of the general agonist I-Tyr(0)-sauvagine by increasing concentration of the Gs-ago-Gi-antagonist Bpa(7)Ucn at CRF$_1$: GTPγS binding stimulation in HEK-CRF$_1$ cells [5].

The ago-antagonism cannot be explained by a single active receptor state, but presupposes significant differences between active conformations of a single receptor when coupled to different G proteins. Assuming a similar structural encoding of signalling information in other polypeptide hormones, a new concept for the development of signalling-selective drug candidates on the basis of polypeptides would turn out.

References

1. Davey, J. *Expert Opinion on Therapeutic Targets* **8**, 165-170 (2004).
2. Urban, J. D. et al. *J. Pharmacol. Exp. Ther.* **320**, 1-13 (2007).
3. Wietfeld, D. et al. *J. Biol. Chem.* **279**, 38386-38394 (2004).
4. Berger, H. et al. *British J. Pharmacol.* **149**, 942-47 (2006).
5. Beyermann, M. et al. *British J Pharmacol.*, doi 10.1038 (2007).

Model of Intermolecular Interactions between High Affinity Phosphopeptides and Stat3

John S. McMurray[1], Zhiyong Ren[2], Pijus K. Mandal[1] and Xiaomin Chen[2]

[1]Department of Experimental Therapeutics; [2]Department of Biochemistry and Molecular Biology, The University of Texas M.D. Anderson Cancer Center, Houston, TX 77030, USA

Introduction

Signal transducer and activator of transcription (Stat3) is constitutively active in several cancer types and it participates in the increased transcription of cell cycling, survival, and angiogenesis genes. Therefore Stat3 is a target for anti-cancer drug design [1]. Compounds targeted to the SH2 domain of Stat3 would uncouple this protein from its aberrant activity by preventing recruitment to receptors, blocking reciprocal pTyr-SH2 domain dimerization, translocation to the nucleus and DNA binding, thus preventing transcription of the cell-cycling, survival, and angiogenesis genes. To date there are no crystal or NMR structures of high affinity phosphopeptides complexed with the SH2 domain of Stat3 to aid in inhibitor design.

Stat3 binds phosphopepties in the context of pYXXQ. Structure activity studies suggest that Gln at pY+3 is optimal [2]. A model of our lead peptide, pYLPQTV, would be helpful in revealing the interactions between the specificity determinant Gln and the SH2 domain of Stat3.

Examination of the crystal structures of Stat1 dimer-DNA [3], Stat1 bound to pYDKPH [4], Stat3 dimer-DNA [5], and Stat5 [6] revealed remarkable overlap in the main chain conformation of the SH2 domain backbone. Furthermore, there is a new helix, named αB', that immediately follows strand βD. This is unique to the Stat proteins. A survey of crystal or NMR structure of 20 non-Stat SH2 domains revealed that βE, the EF loop, and βF follow strand βD. The crystal structure of the Stat1-pYDKPH complex revealed a heretofore unseen mode of phosphopeptide binding to SH2 domains. The specificity determinant, His, at position pY+4 resides in a unique pocket at the juncture of strand βD and the new helix αB'. This pocket exists in the mammalian Stats. Given the unique structure of the Stat SH2 domains, the differences with non-Stat, proteins, and the novel phosphopeptide binding mode in the Stat1 structure, we hypothesize that, in addition to the usual pTyr interactions shown in the crystal structure, the side chain of the Stat3 specificity determinant, Gln and pY+3, binds in the unique pocket at the juncture of strand βD and helix αB'.

Results and Discussion

To examine this hypothesis, a model of Ac-pTyr-Leu-Pro-Gln-NHBn [2] docked to the SH2 domain of Stat3 was constructed using molecular modeling. The coordinates of residues 585-688 from the crystal structure of Stat3β (PDB code 1BG1) [5] were used for the SH2 domain. The coordinates of pTyr705 and Leu706 were used as the starting point for the phosphopeptide. Proline and glutamine were appended to the leucine and their ϕ and ψ angles were rotated to maximize interaction energy, as measured using the Docking module of InsightII (Accelrys, Inc.). The χ angles of Gln were altered to place the side chain of Gln into the pocket bounded by residues of βD and αB'. The bonds of the C-terminal benzylamide were rotated to place the benzene ring into the same pocket occupied by Phe611 in the

S. Del Valle et al. (eds.), *Peptides for Youth: The Proceedings of the 20th American Peptide Symposium*, 543
DOI: 10.1007/978-0-387-73657-0_238, © Springer Science+Business Media, LLC 2009

Stat3β dimer structure [5]. The complex was subjected to extensive energy minimization, was soaked with water, and re-minimized. The resulting structure is depicted in Figure 1.

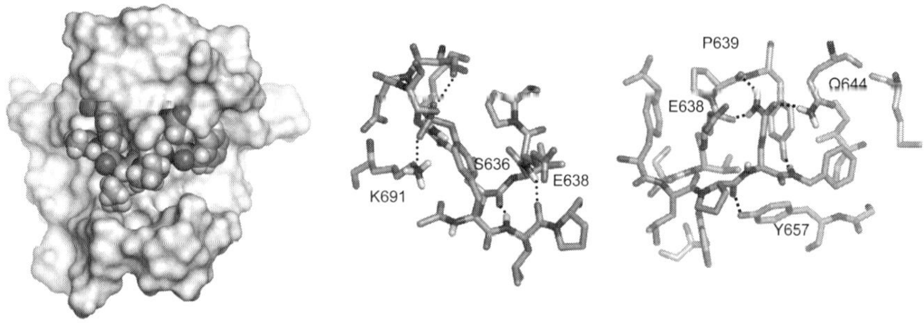

Fig. 1. Model of Ac-pYLPQ-NHBn dock to the SH2 domain of Stat3. Left: Space filling model of phosphopeptide and surface rendering of protein. Center: Interactions of pTyr and Leu. Right, interactions of Pro, Gln, and NHBn.

In addition to the "standard" phosphotyrosine and pY+1 interactions, the model reveals unique peptide backbone-protein hydrogen bonds between Leu C=O and Glu638 NH, Pro C=O and Tyr657 OH, and the benzylamine NH and Tyr640 OH. The $CONH_2$ of Gln resides in the pocket between β strand D and helix αB' and forms hydrogen bonds with main chain carbonyl groups of Glu638 and Pro639 and the side chain of Gln644. The phenyl group of NHBn fits into the hydrophobic pocket occupied by Phe611 in the Stat3 dimer stabilizing the helix. Substitution of Glu for Gln was shown to reduce affinity 40-fold [7], suggesting coulombic repulsion of the side chain carboxyl with that of Glu638, this supporting the pocket at the βD-αB' juncture. Gln H-bonding is supported by the observation that mono- and di- *N*-methylation of the side chain severely impairs affinity, as does substitution of Met(O) at this position [2]. The tight fit of glutamine is supported by the decrease in affinity of Thr($CONH_2$) in relation to Ser($CONH_2$) indicating that β-substitution is not tolerated well [8].

This model is being used to optimize affinity of our peptidomimetics.

Acknowledgments

This work was supported by the National Cancer Institute (CA096652) and the M. D. Anderson Cancer Center University Cancer Fund. The NCI Cancer Center Support Grant CA016672 is gratefully acknowledged for the support of our molecular modeling software and hardware.

References

1. Yu, H. and Jove, R. *Nature Rev. Cancer.* **4**, 97-105 (2004).
2. Coleman, IV, D.R. et al. *J. Med. Chem.* **48**, 6661-6670 (2005).
3. Chen, X.; et al. *Cell*, **93**, 827-839 (1998).
4. Mao, X.; et al. *Mol. Cell* **17**, 761-771 (2005).
5. Becker, S. et al. *Nature* **394**, 145-151 (1998)
6. Neculai, D. et al. *J. Biol. Chem.* **280**, 40782-40787 (2005).
7. Ren et al. *Bioorg. Med. Chem. Lett.* **13**, 633-636 (2003).
8. Mandal, P.K. et al. *Bioorg. Med. Chem. Lett.* **17**, 654-656 (2007).

Inhibition of Stat3 by Cell-Permeable Peptidomimetic Prodrugs Targeted to its SH2 Domain

John S. McMurray[1], Pijus K. Mandal[1], Warren S. Liao[2], Zhiyong Ren[2], and Xiaomin Chen[2]

[1]*Department of Experimental Therapeutics;* [2]*Department of Biochemistry and Molecular Biology, The University of Texas M.D. Anderson Cancer Center, Houston, TX 77030, USA*

Introduction

Signal transducer and activator of transcription 3 (Stat3) is a cytosolic transcription factor that transmits signals directly from the cytokine or growth factor receptors to the nucleus [1]. Stat3 is constitutively activated in several cancers and participates in the expression of cell cycling, survival, and angiogenesis genes. It is recruited to receptors via it SH2 domain, becomes phosphorylated on Tyr705, and dimerizes via reciprocal pTyr-SH2 domain interactions. The dimer translocates to the nucleus where it initiates gene transcription. We are targeting the Stat3 SH2 domain to create new anti-cancer chemotherapeutic agents.

We found a high affinity lead peptide, Ac-Y(p)LPQTV-NH_2, **1**, [2], and have converted it into two phosphatase-stable, cell-permeable peptidomimetics that inhibit Stat3 activity and inhibit tumor cell growth.

Results and Discussion

We performed structure-affinity relationship studies on truncated **2** to determine the pharmacophores and to gain information on the conformation in the bound state [3,4]. Peptide **7** was culmination of that effort (mPro = cis-3,4-methanoProline) (Table 1).

Table 1. Affinity of phosphopeptides for Stat3.

Peptide	Sequence	IC_{50} (nM)
1	Ac-Y(p)LPQTV-NH_2	290 ± 63[a]
2	Ac-Y(p)LPQT-NH_2	739 ± 31
3	pCinn-LPQT-NH_2	136 ± 23
4	pCinn-LPQ-NHBn	138 ± 8[b]
5	Ac-Y(p)-Haic-QT-NH2	231 ± 22
PM6	pCinn-Haic-Q-NHBn	162 ± 19
7	$PhCH_2CH_2CO$-Y(p)L-mPro-Q-NHBn	125 ± 19[c]
PM7	pCinn-Y(p)-Nle-mPro-Q-NHBn	50 ± 8

[a] *Affinity measured by fluorescence polarizarion [3]*

[b] *From reference [4]*

[c] *From reference [3].*

Replacement of pTyr with 4-phsophoryloxycinnamate increased activity of **2** 5-fold , (Peptide **3**). The C terminal Thr was replaced with benzyl amide [3] and **4**

S. Del Valle et al. (eds.), *Peptides for Youth: The Proceedings of the 20th American Peptide Symposium*, 545
DOI: 10.1007/978-0-387-73657-0_239, © Springer Science+Business Media, LLC 2009

was equally active. Replacement of the central Leu-Pro unit with Haic (**5**) increased the activity of **2** to 231 nM but decreased peptide character. Combining pCinn, Haic, and benzylamide gave **PM6**, a peptidomimetic. The same strategy with the mPro sequence gave **PM7**. Both have IC_{50} values in the low nM range.

To impart phosphatase stability to **PM6** and **PM7**, the phosphate was converted to a difluoromethyl phosphonate [5]. The negatively charged phosphonate oxygens were blocked with the esterase-labile pivaloyloxymethyl (POM) groups [6] to impart cell penetratability (**BP-PM6** and **BP-PM7**, Figure 1).

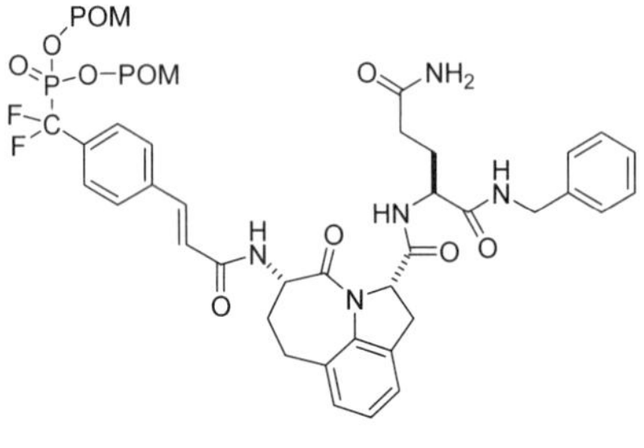

BP-PM6

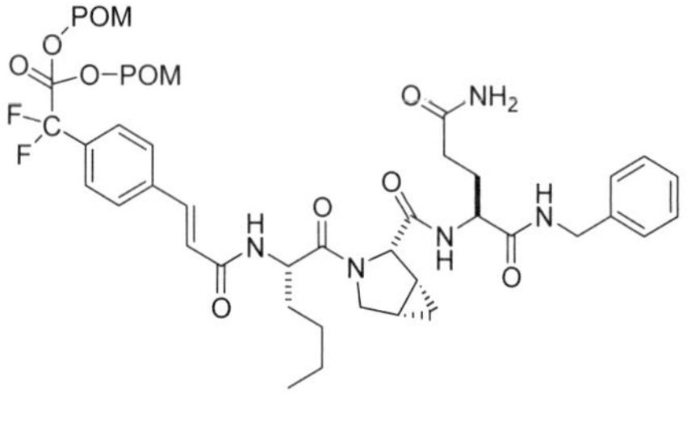

BP-PM7

Fig. 1. Prodrug inhibitors of Stat3.

BP-PM6 and **BP-PM7** inhibited constitutively phosphorylated Stat3 in BT-20, MDA MB 231, and MDA-MB 468 breast tumor cell lines with IC_{50}'s of 1-5 μM. These compounds also inhibited Stat3 phosphorylation in ovarian and head and neck cell lines. IC_{50}'s for inhibition of a luciferase reporter gene were 5 and 1 μM, respectively. **BP-PM6** was tested more extensively and was found to be selective for Stat3 over Stat1 and Stat5, and did not inhibit an NFκB reporter gene, suggesting that it does not bind to the SH2 domains of p85 PI3K. IC_{50}'s for growth inhibition of BT20 and multiple myeloma 1 cells were 5 and 25 μM, respectively. **BP-PM7** inhibited BT20 growth with an IC_{50} of 25 μM.

In conclusion, we have converted a phosphohexapeptide into cell-permeable, phosphatase stable peptidomimetics that inhibit their target in the low μM range. Further studies are in progress to optimize delivery and potency of these compounds.

Acknowledgments

This work was supported by the National Cancer Institute (CA096652) and the M. D. Anderson Cancer Center University Cancer Fund. The NCI Cancer Center Support Grant CA016672 is acknowledged for the support of our NMR facility.

References

1. Yu, H. and Jove, R. *Nature Rev. Cancer.* **4**, 97-105 (2004)
2. Ren, Z. et al. *Bioorg. Med. Chem. Lett.* **13**, 633-636 (2003).
3. Coleman, IV, D.R. et al. *J. Med. Chem.* **48**, 6661-6670 (2005).
4. Mandal, P.K. et al. *Bioorg. Med. Chem. Lett.* **17**, 654-656 (2007).
5. Burke, T.R., Jr et al *Biochemistry*, **33**, 6490-6494 (1994).
6. Farquhar, D. et al. *J. Med Chem.* **37**, 3902-3909 (1994).

Synthesis and Biological Activity of Conformational Restricted Galanin Fragments

Marcus Marra, Marcos A. Fázio and Antonio Miranda

Department of Biophysics, Federal University of São Paulo, São Paulo, SP, 04044-020, Brazil

Introduction

Since the discovery of galanin (Gal) by Tatemoto et al [1], this neuropeptide has obtained increasing interest, mainly motivated by its widespread distribution and involvement in a multitude of physiological actions, such as central cardiovascular control, control of appetite and seizures, effects on pain and depression. Pharmacological tools to fully examine the actions of galanin and its receptors are still lacking and need to be better explored [2]. Galanin consists of 29 amino acid (30 in humans) residues and the N-terminal part (1-16) of the peptide has proven crucial for Gal's bioactivity [3]. A number of attempts have been made to design chimeric peptides as ligands for the receptors. One of the limitations of the pharmacological use of peptide ligands is due to their short half-lives and poor bioavailability in the CNS. Inclusion of a lactam bridge in the peptides sequence is one strategy employed in an attempt to promote peptide degradation resistance and also structure stabilization [2].

Our goal is to investigate the SAR of a series of Gal-related fragments of restricted conformation by the insertion of different sizes, chirality and positions of lactam bridges (see Fig. 1). Peptides were synthesized by SPPS, purified by HPLC, and characterized by LC/MS and AAA [4]. Biological activities were determined in colonic smooth muscle strips of Wistar rats of both sexes. Conformational studies were performed by CD spectroscopy in different environments.

Nº	Peptide	Sequence[1]
1	Gal	GWTLNSAGYLLGPHAIDNHRSFHDKYGLA-NH_2
2	Gal_{1-15}	GWTLNSAGYLLGPHA-NH_2
3	Gal_{1-16}	GWTLNSAGYLLGPHAI-NH_2
4	$hGal_{1-16}$	GWTLNSAGYLLGPHAV-NH_2
5	$c[E^4, K^8]Gal_{1-16}$	GWT[ENSAK]YLLGPHAI-NH_2
6	$c[D^4, K^8]Gal_{1-16}$	GWT[DNSAK]YLLGPHAI-NH_2
7	$c[D^4, Orn^8]Gal_{1-16}$	GWT[DNSAB]YLLGPHAI-NH_2
8	$c[D^4, Dap^8]Gal_{1-16}$	GWT[DNSAX]YLLGPHAI-NH_2
9	$c[d^4, K^8]Gal_{1-16}$	GWT[dNSAK]YLLGPHAI-NH_2
10	$c[D^4, k^8]Gal_{1-16}$	GWT[DNSAk]YLLGPHAI-NH_2
11	$c[d^4, k^8]Gal_{1-16}$	GWT[dNSAk]YLLGPHAI-NH_2
12	$c[d^1, K^5]Gal_{1-16}$	[dWTLK]SAGYLLGPHAI-NH_2
13	$c[d^3, K^7]Gal_{1-16}$	GW[dLNSK]GYLLGPHAI-NH_2
14	$c[d^5, K^9]Gal_{1-16}$	GWTL[dSAGK]LLGPHAI-NH_2
15	$c[d^7, K^{11}]Gal_{1-16}$	GWTLNS[dGYLK]GPHAI-NH_2
16	$c[d^9, K^{13}]Gal_{1-16}$	GWTLNSAG[dLLGK]HAI-NH_2
17	$c[d^{11}, K^{15}]Gal_{1-16}$	GWTLNSAGYL[dGPHK]I-NH_2

[1]Lowercase letters correspond to D-amino acid.

Fig. 1. Sequences of Gal and its fragments.

S. Del Valle et al. (eds.), *Peptides for Youth: The Proceedings of the 20th American Peptide Symposium*, 547
DOI: 10.1007/978-0-387-73657-0_240, © Springer Science+Business Media, LLC 2009

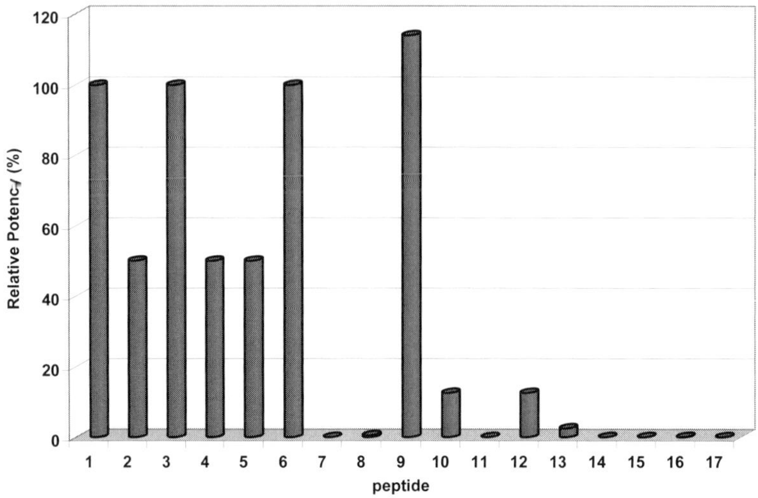

Fig. 2. Relative potency of Gal and its fragments on rat colonic smooth muscle strips.

Results and Discussion

As expected, **1** and its short fragment **3** were equipotent (Fig. 2). **2** and **4** (human sequence) showed relative potency of 50% in comparison to Gal. Concerning the bridge-head amino acids pair (E/K, D/K, D/Orn and D/Dap) the best activity was observed with **6**, in fact fragments, **5** was just 50% active and **7** and **8** were completely inactive. Systematic chiral optimization (**6, 9-11**) of the bridge heads at positions 4 and 8 resulted in the most potent galanin fragment **9**. We also have scanned the whole gal sequence with an i-(i + 4) bridge consisting of the DAsp-X-Y-Z-Lys scaffold. Most analogues (**13-17**) were inactive, with exception of fragment **12** that presents 12.5% of relative potency. No clear correlation was found between CD and the biological activity results. From our results we concluded that the conformational restriction was important to the maintenance of the biological activity and that the best size and chirality for the cycle was obtained by the incorporation of the D-Asp and Lys residues as lactam bridge-heads components in the sequence region from 4 to 8 in agreement with Carpenter et al. [2].

Acknowledgments

The work was supported by FAPESP, CNPq and UNIFESP/FADA.

References

1. Tatemoto, K., Rokaeus, A., Jornvall, H., Mcdonald, T.J. and Mutt, V. *FEBS Letters* **164**, 124-128 (1983).
2. Sollenberg, U. E., Lundstrom, L., Bartfai, T. and Langell, U. *Int. J. Pep. Research Therapeutics* **12**, 115-119 (2006).
3. Carpenter, K. A., Schmidt, R., Yue, S. Y., Hodzic, L., Pou, C., Payza, K., Godbout, C., Brown, W. and Roberts, E. *Biochemistry* **38**, 15295-15304 (1999).
4. Fázio, M. A., Oliveira, V. X., Bulet, P., Miranda, M. T. M., Daffre, S., Miranda, A. *Biopolymers* **84**, 205-218 (2006).

Interaction of membrane mimetic vesicles with the antimicrobial peptide gomesin and its analogues

Marcela S. Carmona[1], Luis G. M. Moraes[1], Marcos A. Fázio[1], Rosemaire A. S. Lucca[2], Maria L. V. Oliva[3], Sirlei Daffre[4] and Antonio Miranda[5]

[1]*Department of Biophysics, UNIFESP, São Paulo, SP;* [2]*Center of Engineering and Accurate Sciences, UNIOESTE, Toledo, PR;* [3]*Department of Biochemistry, UNIFESP, São Paulo, SP and* [4]*Department of Parasitology, ICB-USP, São Paulo, SP, Brazil*

Introduction

Gomesin (*Gm*) is an antimicrobial peptide very effective against several pathogenic microorganisms. The peptide contains four cysteines performing two disulfide bridges in positions 2/15 and 6/11 (pGlu-CRRLCYKQRCVTYCRGR-NH_2)[1]. NMR data showed that *Gm* consists in a well-resolved two-stranded antiparalell β-sheet connected by a noncanonical β-turn [2].

To better understand the mechanisms of action, we performed studies on the interactions of the peptides with large unilamellar vesicles composed of L-α-phosphatidic acid (PA) and L-α-phosphatidylethanolamine (PE)(1:1 mol ratio), by evaluating the release of the fluorescence probe calcein, trapped inside of the vesicles. Conformational behavior of the peptides was also evaluated by CD and fluorescence spectroscopies in different environments. Studies were done with *Gm*, [Trp^9]-*Gm*; [$Ser^{2,6,11,15}$, Trp^9]-*Gm*; [$Thr^{2,6,11,15}$, D-Pro^9]-*Gm*; and [Trp^1, $Thr^{2,6,11,15}$, D-Pro^9]-*Gm*. Their antimicrobial activities were evaluated against *M. luteus*, *E. coli*, and *C. albicans*. Peptides hemolytic activities were determined against human erythrocytes.

Results and Discussion

Peptides were synthesized by SPPS on a MBHA resin. After HPLC purification, they were characterized by AAA, CE and LC/ESI-MS. Antimicrobial activities were measured using liquid growth inhibition assays. MICs were expressed as the lowest peptide concentration that causes total microbial growth inhibition (Table 1). As planed, no significant differences between the biological activities of the *Gm* and its

Table 1. Antimicrobial activities of the gomesin analogues studied.

Peptide	MIC (μM)		
	M. luteus[1]	*E. coli*[1]	*C. albicans*[2]
Gm	0.64	1.28	0.64
[Trp^9]-*Gm*	1.28 (2)	1.28 (1)	0.64 (1)
[$Ser^{2,6,11,15}$, Trp^9]-*Gm*	10.24 (16)	10.24 (8)	10.24 (16)
[$Thr^{2,6,11,15}$, D-Pro^9]-*Gm*	2.56 (4)	5.12 (4)	1.28 (2)
[Trp^1, $Thr^{2,6,11,15}$, D-Pro^9]-*Gm*	0.64 (1)	2.56 (2)	1.28 (2)

[1]*PB (217 mOsM; 1.0 g Peptone + 86 mM NaCl in 100 mL of H_2O).* [2]*PDB (79 mOsM; 1.2 g potato dextrose in 100 mL of H_2O). Numbers in parenthesis mean: analogue MIC/Gm MIC.*

S. Del Valle et al. (eds.), *Peptides for Youth: The Proceedings of the 20th American Peptide Symposium*, 549
DOI: 10.1007/978-0-387-73657-0_241, © Springer Science+Business Media, LLC 2009

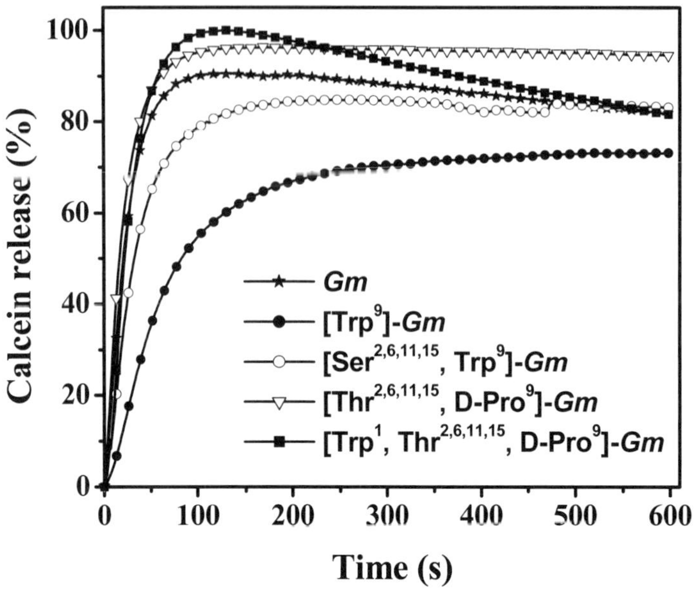

Fig. 1 - Releasing of calcein trapped inside of large unilamellar vesicles of PA:PE.

analogues were observed with exception of the linear analogue [Ser^{2,6,11,15},Trp^{9}]-*Gm* that was 8-16-fold less active than gomesin against all the strains tested and was used as a negative control. These findings indicated that the incorporation of Trp, in positions 1 and 9, caused no reduction on the antimicrobial activity of the cyclic and linear analogues. Almost all analogues presented a beta-hairpin conformation identical to that of the native gomesin in all environments tested, while the linear analogue [Ser^{2,6,11,15}, Trp^{9}]-*Gm* presented an unordered conformation. All analogues strongly interact with the phospholipids vesicles and induce leakage of their content in a pH-dependent manner (Fig. 1). Fluorescence spectra of all peptides showed a blue-shift of emission from around 350 nm to as low as 325 nm in the presence of SDS. No similar changes were detected in the presence of LPC. These results corroborate with previous results and indicate that the first step of the gomesin killing mechanism on bacteria is an electrostatic interaction with the lipid bilayer causing the disruption of the internal membrane.

Acknowledgments

The work was supported by FAPESP, CAPES, CNPq and UNIFESP/FADA.

References

1. Silva Jr., P. I., Daffre, S., Bulet, P. *J. Biol. Chem.* **275**, 33464-33470 (2000).
2. Mandard, N., Bulet, P., Caille, A., Daffre, S., Vovelle, F., *Eur. J. Biochem.* **269**, 1190-1198 (2002).
3. Moraes, L. G., Fázio, M. A., Vieira, R. F. F., Nakaie, C. R., Miranda, M. T. M., Schreier, S., Daffre, S. and Miranda, A. *BBA Biomembranes* **1768**, 52-58 (2006).
4. Fázio, M. A., Jouvensal, L., Vovelle, F., Bulet, P., Miranda, M. T. M., Daffre, S. and Miranda, A. *Biopolymers*, **88**, 386-400 (2007).

Limited Structure-Activity Relationships for Carboxy and AminoTerminal Substitutions of Peptide YY 25-36

David H. Singleton, James G. Boyd, Nancy A. Nardone, Brett S. Robinson, Lydia C. Pan, John R. Hadcock and A. Zane Wenzel

Pfizer Global Research and Development, Eastern Point Rd., Groton, CT 06340 USA

Introduction

PYY is a naturally occurring peptide hormone of 36 amino acids in length with high homology to NPY. It is secreted from the gut (released by endocrine cells) in direct response to feeding and is active in the central nervous system. PYY binds to the Y family of GPCR receptors of which there are five sub-types (Y1-Y5). PYY participates in regulation of numerous processes including blood pressure, circadian rhythm, sexual behavior, and appetite suppression. N-terminal processing of PYY 1-36 by dipeptidyl peptidase (DPP) produces PYY 3-36, which is selective for and activates the Y2 receptor. Because PYY 3-36 is efficacious at reducing food intake in humans[1] and rodents[2] the appetite suppression indication is believed to be due to Y2 agonism.

Previous literature reports demonstrate that truncation of the 3-36 amide to the 13-mer acetyl PYY 24-36 amide provides a Y2R full agonist with 3-5 fold reduction in potency[3]. Using this as our starting point, series of discreet analogs exploring the C-terminal portion of PYY were synthesized, screened and assessed for their binding and functional activities. Sets of amino terminal truncated, amino terminal modified, and carboxamide terminal modified analogs have afforded SAR in binding and functional *in vitro* assays. Modified truncation mutants incorporating the C-terminal 12 amino acids exhibit nanomolar binding and agonist activity providing promise as potential therapeutic agents in obesity intervention therapy.

Results and Discussion

Initial syntheses focused on downsizing PYY 3-36 while maintaining activity. A set of N-terminal acetylated peptide amides, progressively N-terminal truncated by 4 residues was synthesized. Consistent with literature results an acute drop off in Y2R binding and functional activity was observed for NT peptides shorter than 13 residues (NT 24-36). 16-mer peptide **4** (Ac PYY 21-36 amide) retained almost all of the binding and functional potency of NT. Peptide **3** (Ac PYY 25-36 amide) was a full agonist but at reduced potency to 3-36 or 24-36 PYY. This compound provided an activity suitable for an SAR starting point.

Additional SAR seems to define an activity drop off point for 12-mer peptides. Removal of the N-terminal acetyl group from peptide **3** (**13**) significantly reduces potency in both binding and functional assays. We sought to recoup some or all of the activity observed for acetyl PYY 24-36 by examining simple aliphatic replacements for the acetyl in (**3**). N-butyryl capping (**11**) provides the largest increase in binding affinity.

Finally, carboxy terminal modifications were explored. Regarding PYY 3-36, literature suggests that C-terminal amidation is required for activity[4], however, a modified amide would benefit from metabolic stability countering de-amidation to the acid and thereby prolonging t1/2 *in vivo*. Analogs were prepared via two synthetic routes either from the acid by HATU activation of the peptide and

S. Del Valle et al. (eds.), *Peptides for Youth: The Proceedings of the 20th American Peptide Symposium*, 551
DOI: 10.1007/978-0-387-73657-0_242, © Springer Science+Business Media, LLC 2009

Table 1. Analogs of PYY 25-36.

Sequence	X-Sequence-Z *Cpd. Number*	 *Binding IC50 (nM)*	X-Sequence-Z *Functional EC50(nM)*
Ac-RQRY	1	>5uM	ND
Ac-NLVTRQRY	2	>5uM	ND
Ac-RHYLNLVTRQRY	3	56	11
Ac-YASLRHYLNLVTRQRY	4	1.8	7.4
Ac-LNRYYASLRHYLNLVTRQRY	5	6.2	1.6
X=4-Methylvaleramido	6[a]	64	97
X=Phenylacetamido	7[a]	51	74
X=Hydrocinnamido	8[a]	46	ND
X=Cyclohexane propionamido	9[a]	118	137
X=Propionamido	10[a]	44	8
X=Butyramido	11[a]	24	2.5
X=Valerimido	12[a]	77	81
X=H	13[a]	622	48
Z=OH	14[a]	>5uM	>10uM
Z=N-CH3	15[a]	>5uM	>10uM
Z=N-CH2CH3	16[a]	>5uM	>10uM
Z=Morpholine	17[a]	149	5000
Z=Benzylamine	18[a]	631	10000

[a] *Peptide Sequence is that of compound 3, X and Z N and C-terminal modifiers respectively*

amination by either morpholine or benzyl amine or by reductive amination of BAL-aldehyde resins[5] with N-methyl or N-ethyl amine followed by peptide synthesis displaying modest, but measurable μM activities. Peptides switched from agonist to antagonist activities and true estimation of EC50's were difficult to determine as full potency requires concentrations higher than the assay permits.

References

1. Bloom, S.R. *et al.*, Nature Clinical Practice Endocrinology & Metabolism, (2006), 2(11), 612-620.
2. Stanley, B. G., *et al.*, PNAS, (1985), 3940-43.
3. Potter, E.K. *et. al.,* European J. Pharmacology, (1994), 253-262.
4. Small, C.J. *et al.* Proc. Nat. Acad. Sciences, (1997), 94, 11686-11691.
5. Jensen, K.A. *et al*, JACS, (1998), 120, 5441-5452.

Cysteine Based PNA (CPNA): Design and Synthesis of Novel CPNA Monomers

Sung Wook Yi,[1,2] Priyesh Jain,[1,2] Mehul Ajmera,[1,2] Sridhar Reddi Kaulagari,[1,2] Melissa Topper,[1,2] Laura Anderson,[1,2] and Mark L. McLaughlin*[1,2]

[1]*Drug Discovery Program, H. Lee Moffitt Cancer Center & Research Institute, 12902 Magnolia Drive, Tampa, FL 33612-9416, USA* [2]*Department of Chemistry, University of South Florida, 4202 E. Fowler Avenue, Tampa, FL 33620. USA*

Introduction

Peptide nucleic acids (PNAs), a class of pseudo-peptide DNA mimics, were first introduced by Nielsen and his coworkers in 1991 [1]. PNA oligomers form very stable sequence-specific duplexes with complementary DNA and RNA strands through Watson-Crick base paring, and can bind to duplex DNA by helix invasion [2]. These intriguing properties of PNA implicated great potential for medical and biotechnical applications. Not surprisingly, tremendous efforts on the synthesis of PNAs and their applications have been reported from various scientific groups since their discovery. However, owing presumably to the nature of the simple neutral backbones, most of PNA oligomers have shown poor water-solubility and cell permeability, and these problems have been the main stumbling blocks for PNA application in biomedical fields [3]. To improve these properties, scientists have been attaching cell-membrane-permeabilizing peptide sequences and trying to modify the backbone of PNA, and only a few examples have shown some positive improvement to date [4]. In this report, we introduce our synthesis of novel cysteine based PNA monomers with various alkyl chains attached as shown in **Figure 1**. Unlike other amino acids, the cysteine possesses the thiol group which complicates the PNA synthesis and that is presumably why cysteine based PNAs have never been reported. We have envisioned that the thiol group of cysteine could take the role as an attachment site for groups designed to improve the physical properties of the PNA oligomers.

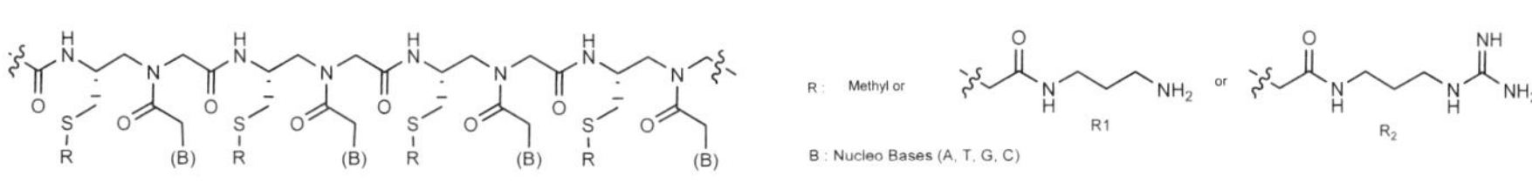

Figure 1. Structure of the Cysteine Based PNA.

Results and Discussion

While the conventional protocols for the PNAs utilize the reductive amination to prepare the monomer unit, we developed a S_N2 route to prepare the monomer unit (**Scheme 1**). This procedure eliminates the possibility of epimerization, which could occur during the conventional reductive amination using chiral aldehyde intermediates. Synthesis of the Fmoc- or Boc-protected monomers started with commercially available **1**, as shown in **Table 1**, which was smoothly converted to the primary amide **2** mediated by DCC and HOBt followed by addition of ammonium hydroxide [5], and borane reduction of the amide **2** gave free amine **3** in

S. Del Valle et al. (eds.), *Peptides for Youth: The Proceedings of the 20th American Peptide Symposium*, 553
DOI: 10.1007/978-0-387-73657-0_243, © Springer Science+Business Media, LLC 2009

moderate yield. Nucleophilic substitution reaction of **3** with alkyl bromoacetate, as shown in **Table 1**, afforded the protected sub-monomer **4**. Selective trityl group cleavage of **4** was smoothly carried in 10 % TFA in DCM in the presence of triethylsilane and the thiol was coupled with alkyl halides in situ as shown in **Table 1**. HATU activation of Cbz-protected nucleic acids [6] in the presence of TEA provided our monomers with excellent yields (**Scheme 1**). In summary, novel cysteine based PNA monomers are prepared for both Fmoc and Boc-solid phase synthesis. This synthetic protocol provides PNAs with a very broad range of possible side chains expected to enhance the cell membrane permeability of the PNAs. We are enthusiastically optimizing the solid phase synthesis of CPNA oligomers.

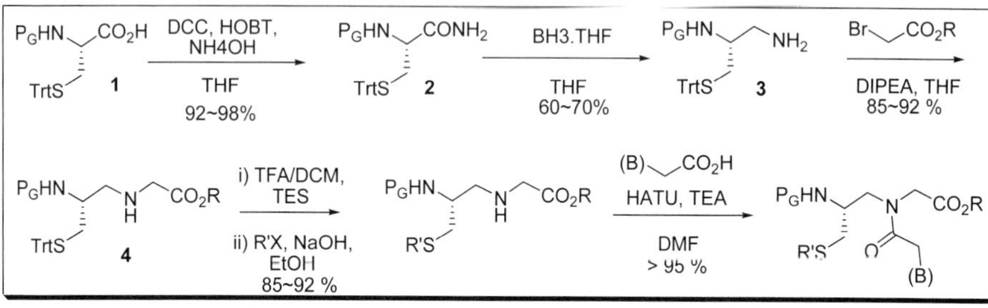

Scheme 1. General Procedures for Synthesis of CPNA Monomers.

Table 1. Protecting groups and Nucleic Based coupled to the CPNA backbones.

P_G: Protecting group	R	R'X	Nucleic Bases on the Monomers
Fmoc	tBu	MeI	T, G^{Cbz}, C^{Cbz}, A^{Cbz}
Fmoc	tBu	R_1Br (Fig 1)	T, G^{Cbz}, C^{Cbz}, A^{Cbz}
Fmoc	tBu	R_2Br (Fig 1)	T, G^{Cbz}, C^{Cbz}, A^{Cbz}
Boc	Me	MeI	T, G^{Cbz}, C^{Cbz}, A^{Cbz}

Acknowledgments

We thank the Cancer Research and Prevention Foundation, and the NIH (R21GM072772) and for financial support through grants.

References

1. Nielsen, P. E., Egholm, M., Berg, R. H. and Buchardt, O. *Science,* **254**, 1497-1500 (1991).
2. Nielsen, P. E. *Peptide Nucleic Acids: Protocols and Applications* (**2nd Edition**). Wymondham, Norfolk: Horizon Bioscience (2004). Egholm, M., et al. Nature, 365, 566-568 (1993).
3. Nielsen, P. E. Quarterly Reviews of Biophysics, 38, 4, 345-350 (2005).
4. Ganesh, K. N., Nielsen, P. E. Current Organic Chemisty, 4, 931-943 (2000). Zhou, P., Wang, M., Du, L., Fisher, G. W., Waggoner, A., and Ly, D. H. J. Am. Chem. Soc. 125, 6878-6879 (2003).
5. Nakamura, Y., Okumura, K., Kojima, M. and Takeuchi, S. Tetrahedron Letters, 47, 239-243 (2006).
6. Kofoed, T., Hansen, H. F., Ørum, H., and Koch, T. J. Peptide Sci. 7, 402–412 (2001). Thomson, S. A., et al. Tet. Lett. 51, 6179-6194 (1995).

Synthesis and Biological Activity of Nicotianamine and Analogues

M'Barek Bouazaoui[1], Stéphane Mari[2], Pierre Czernic[2], Catherine Curie[2], Jean Martinez[1] and Florine Cavelier[1]

[1]*Institut des Biomolécules Max Mousseron, UMR 5247 CNRS-UM1-UM2, Place E. Bataillon, CC19, 34095 Montpellier, France*
[2]*INRA, UMR 5004, Laboratoire de Biochimie et Physiologie Moléculaire des Plantes, 2 pl. Viala 34060 Montpellier, France*

Introduction

Nicotianamine (NA), which was first isolated from the leaves of *Nicotiana Tabacum L.*[1], is a key biosynthetic precursor of phytosiderophores. It is known as the normalizing factor for the tomato mutant *Chloronerva L.* NA has been shown to act as an iron chelating agent and to promote the transport of iron [2] required for chlorophyll biosynthesis. To obtain synthetic NA, two main strategies are described in the literature: reductive amination of intermediate protected aldehydes [3] and more recently reduction of amide bonds via thioamide [4]. However, in practice, these two strategies failed to furnish sufficient amount of material for further biological investigations on the role of NA in metal circulation.

Therefore, we have developed a new strategy based on the nucleophilic substitution of iodine by an amine as the key reaction.

Results and Discussion

Our synthetic strategy is illustrated in Figure 1. Benzyl 2-(S)-(tert-butoxycarbonyl amino)-4-iodobutanoate **6**, the key precursor of this new method, was prepared in three steps using L-tert-butyl aspartate as starting materiel. Treatment of **4** with benzyl chloroformate under usual conditions ($NaHCO_3$, H_2O/Dioxane) yielded the Z derivative, which was readily reduced using BOP/$NaBH_4$ to afford the desired alcohol **5** in 80 % overall yield. The use of tert-butyl as protecting group significantly increased the yield of the desired alcohol, by preventing lactonisation. Then, alcohol **5** was converted into iodide **6** using iodine, triphenylphosphine and imidazole [5]. The halide synthon was N-alkylated with L-azetidine tert-butyl carboxylate. Alkylation conditions were optimized: DIEA (2 eq) in acetonitrile at 60°C provided **8** in good yield. Hydrogenolysis of **8** using 10% Pd/C and H_2 occurred quantitatively. The (2-trimethylsilyl)ethanesulfonyl group (SES) was then introduced to activate [6] the intermediate primary amine to give sulfonamide **9** easily purified by column chromatography. Several alkylating conditions were then tested. Optimal yield was obtained by using cesium carbonate (2 eq) in acetonitrile at 55° C. Higher temperature and large excess of cesium carbonate led to decomposition of the iodide. The targeted compound **10** was obtained in 55 % yield. Removal of SES, Z and tert-butyl protecting groups proceeded without difficulties in a HF single step to afford NA after lyophilisation as a white powder, in high purity.

We then extended this strategy to produce two synthetic analogues, by replacing the carboxy-azetidine ring with silaproline or proline to yield the new NA analogues **2** and **3**.

S. Del Valle et al. (eds.), *Peptides for Youth: The Proceedings of the 20th American Peptide Symposium*, 555
DOI: 10.1007/978-0-387-73657-0_244, © Springer Science+Business Media, LLC 2009

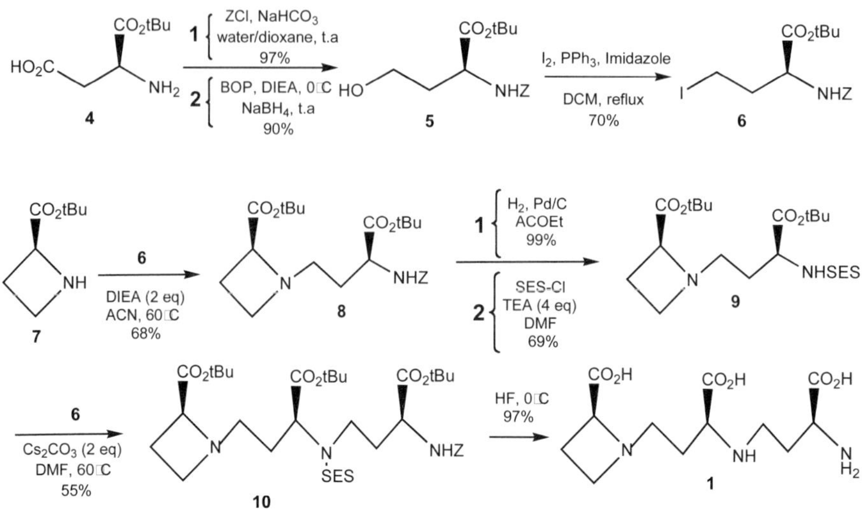

Fig. 1. Synthesis of nicotianamine.

Biological activity:

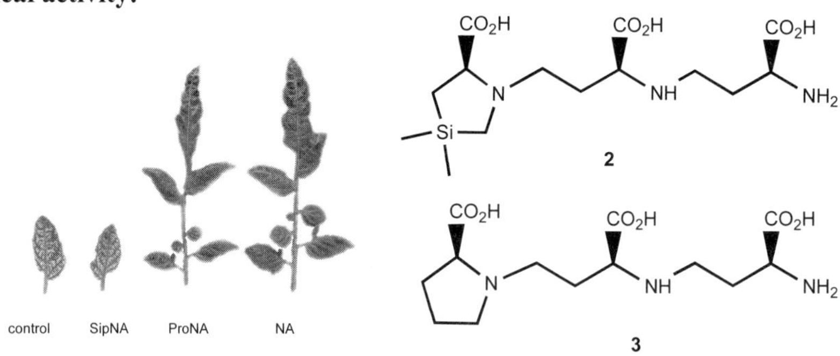

*Fig. 2. Biological activity of NA **1** and analogues **2** and **3**.*

The biological test of the different compounds was realized on the NA-free tomato mutant *chloronerva*. Leaves of *chloronerva* plants were infiltrated with 30 μM solutions of each molecule on a daily basis, for seven days. A biologically active molecule induces the re-greening and restores the growth of the leaves. The biological test has revealed clearly that both the synthetic NA and ProNA could revert the chlorotic phenotype and restore normal growth whereas the SipNA was totally inactive, suggesting that hydrophilicity of the molecule may be crucial for its activity.

In conclusion, this new synthetic pathway of NA and NA analogues is practical, straightforward and occurred with no racemisation, good overall yield and reproducibility.

Acknowledgments

The financial support of CEA, INRA and CNRS in the frame of the ToxNuc programme is gratefully acknowledged.

References

1. *Noma M. et al. / Tetrahedron Lett,* 22 (1971), 2017.
2. *Stephan, U.W. et al. / Physiol. Plant.* 88 (1993), 522.
3. *Shioiri T. et al. / Heterocycles,* 44 (1997), 519.
4. *Miyakoshi K. et al. / Tetrahedron,* 57 (2001), 3355.
5. *Richard F.W. et al. / J. Org. Chem.,* 63 (1998), 7875.
6. *André N. et al. / Letters in Peptides Sciences,* 6 (1999), 239.

The active site specificity of angiotensin II converting enzyme 2 investigated through single and multiple residue changes and β-amino acid substrate analogs.

Daniel Clayton[1], Iresha Hanchapola[1], Patrick Perlmutter[2], A. Ian Smith[1] and Marie-Isabel Aguilar[1]

[1]*Department of Biochemistry & Molecular Biology, Monash University, Clayton Vic 3800, Australia,* [2]*School of Chemistry, Monash University, Clayton Vic 3800, Australia,*

Introduction

Angiotensin converting enzyme 2 (ACE2) is a recently discovered zinc carboxypeptidase involved in the renin angiotensin system (RAS). This enzyme inactivates the potent vasopressive peptide angiotensin II (AngII) by removing the C-terminal phenylalanine residue. ACE2 is also found to be upregulated in the heart in human and animal models of cardiovascular disease and specifically in the human fibrotic liver. Given the role of ACE2 in the RAS and its suspected involvement in other organ systems, a stable and specific inhibitor of low toxicity would be useful in further elucidating the precise role that ACE2 plays in the pathology of heart and liver disease and as a possible therapeutic for the hypotension associated with chronic liver failure.

Approach & Methods

Several approaches were taken to identify leads for a high affinity ACE2 inhibitor: 1) single and multiple residue substitutions were made to the peptide encompassing the four C-terminal residues (IHPF) of AngII (DRVYIHPF), yielding a library of tetrapeptides, 2) a set of full-length AngII β-amino acid analogs were made, 3) optimal C-terminal sequences were incorporated into full-length AngII 'chimeras' having the native N-terminal sequence, 4) β-amino acid (C3) substitution was used to stabilise the scissile bond in order to generate a high affinity inhibitor resistant to proteolytic cleavage by ACE2. Manual and microwave-assisted Fmoc solid-phase peptide synthesis (SPPS) were used to make the AngII peptides. Peptides were assayed for ACE2 inhibition by Quenched Fluorescence Substrate (QFS) assays, while ACE2 cleavage of the peptides was assessed by RP-HPLC mass spectrometry (MS) analysis.

Results

Substitution of isoleucine for proline or valine at the first position in the C-terminus of AngII (IHPF), tyrosine for histidine at the second position, or valine for proline at the third position result in significant increases in the ability of the C-terminus to inhibit ACE2. Combinations of these favourable single residue substitutions generally resulted in further increases to inhibition. N-terminal acetylation of the most potent tetrapeptide inhibitors resulted in small increases to inhibition for many peptides while C-terminal amidation greatly diminished inhibition of all but one of these tetrapeptides, VHVF. The best tetrapeptide inhibitors had IC_{50} values in the

DOI: 10.1007/978-0-387-73657-0_245, © Springer Science+Business Media, LLC 2009

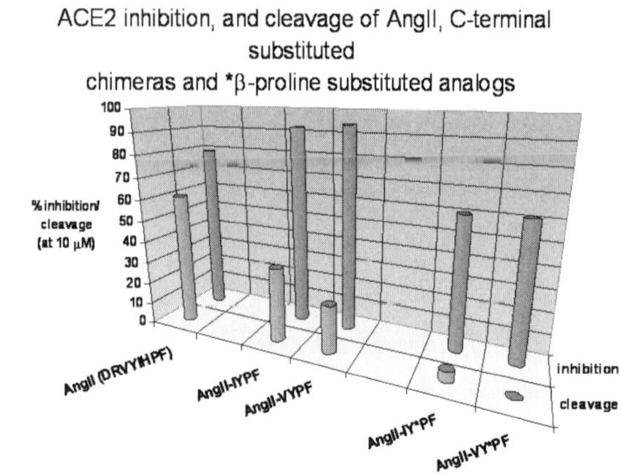

Fig. 1.

5-20 μM range (AngII IC_{50}, 5-10 μM). β-amino acid (Fig. 1) substitutions to AngII had variable effects on ACE2 inhibition; β-D and β-V substitution had little effect while all other substitutions resulted in decreases to ACE2 inhibition. Interestingly, many of the substitutions that resulted in decreases to inhibition resulted in increases in the extent of ACE2 cleavage. In contrast, β-substitution to residues at the scissile bond (β-P & β-F) showed little or no cleavage by ACE2 yet retained approximately 40% inhibition when assayed at 10 μM. Combining the optimal C-terminal tetrapeptides with the native N-terminal AngII amino acid sequence resulted in chimeras that exhibited increased ACE2 inhibition (92% vs. 78% AngII, 10 μM) and decreased ACE2 cleavage as compared to native AngII (Fig. 2). β-substitution to the proline residue at the scissile bond of these chimeras resulted in decreases to ACE2 inhibition, however these analogs showed minimal ACE2 cleavage (5 and 0%) as compared to AngII (60%) (Fig. 2).

Conclusions

Screening of a library of AngII C-terminal analogs identified a number of tetrapeptides with increased ACE2 inhibition, and identified residues critical to the binding of AngII to the active site of ACE2. Screening a β-amino acid library of AngII analogs identified substitutions (β-P & β-F) that stabilise the scissile bond against ACE2 cleavage, albeit with reduced inhibition. Combining successful features from the different libraries into chimeric AngII analogs yielded high affinity inhibitors having reduced ACE2 cleavage. Subsequent studies will determine the enzyme specificity and AngII receptor binding of these analogs.

Acknowledgments

The support of the Australian Research Council is gratefully acknowledged, and many thanks to Dr. Rebecca Lew for discussions on ACE2 QFS assay analysis.

Effects of net charge and the number of positively charged residues on the biological activity of amphipathic α-helical cationic antimicrobial peptides

Ziqing Jiang[1], Adriana I. Vasil[2], John Hale[3], Robert E. W. Hancock[3], Michael L. Vasil[2] and Robert S. Hodges[1]

[1]Department of Biochemistry & Molecular Genetics and [2]Department of Microbiology, University of Colorado at Denver and Health Sciences Center, Aurora, CO, 80045, USA. [3]Department of Microbiology and Immunology, University of British Columbia, Vancouver, British Columbia V6T 1Z3, Canada

Introduction

The widespread use of classical antibiotics has resulted in the emergence of many antibiotic-resistant strains. Cationic antimicrobial peptides (AMPs) have become important candidates as potential therapeutic agents. Cationic AMPs of the α-helical class have two unique features: a net positive charge of at least +2 and an amphipathic character, with a non-polar face and a polar/charged face.

We previously designed a peptide V13K from our original α-helical antimicrobial peptide V681, which had excellent antimicrobial activity but also exhibited high toxicity [1]. D-V13K showed greater antimicrobial activity, no toxicity and excellent stability to proteolytic digestion compared to D-V681. The valine to lysine substitution in the center of the non-polar face is the first report of a specificity determinant in α-helical antimicrobial peptides between eukaryotic and prokaryotic cells. The hydrophobic residues on the non-polar face of the helix played a more important role in stabilizing peptide secondary structure than residues on the polar face and are an essential requirement for antimicrobial activity.

L-V13K was used as the framework to study the effects of net charge and the number of positively charged residues on the hydrophilic/polar face to determine the optimum charge for the highest therapeutic index.

Table 1. Peptides used in this study.

Peptide Name	Sequence[a] (1 … 13 … 26)	Substitution	Number of positively charged residues	Number of negatively charged residues	Net charge
-5	Ac-E-W-E-S-F-L-E-T-F-E-S-A-K-E-T-V-L-H-T-A-L-E-A-I-S-S-amide	V13K, K1E, K3E, K7E, K10E, K14E, K22E	1	6	**-5**
+1	Ac-K-W-K-E-F-L-K-E-F-K-E-A-K-K-E-V-L-H-E-A-L-K-A-I-S-E-amide	V13K, S4E, T8E, S11E, T15E, T19E, S26E	7	6	**+1**
+4S	Ac-S-W-K-S-F-L-K-T-F-S-S-A-K-S-T V L-H-T-A-L-K-A-I-S-S-amide	V13K, K1S, K10S, K14S	4	0	**+4**
+4E	Ac-K-W-K-E-F-L-K-T-F-K-E-A-K-K-E-V-L-H-T-A-L-K-A-I-S-S-amide	V13K, S4E, S11E, T15E	7	3	**+4**
V13K	Ac-K-W-K-S-F-L-K-T-F-K-S-A-K-K-T-V-L-H-T-A-L-K-A-I-S-S-amide	V13K **(Lead Compound)**	7	0	**+7**
+8	Ac-K-W-K-S-F-L-K-T-F-K-S-A-K-K-K-V-L-H-T-A-L-K-A-I-S-S-amide	V13K, T15K	8	0	**+8**
+9	Ac-K-W-K-S-F-L-K-T-F-K-S-A-K-K-K-V-L-H-K-A-L-K-A-I-S-S-amide	V13K, T15K, T19K	9	0	**+9**
+10	Ac-K-W-K-S-F-L-K-T-F-K-S-A-K-K-K-V-L-H-K-A-L-K-A-I-S-K-amide	V13K, T15K, T19K, S26K	10	0	**+10**

Peptide sequences are shown using the one-letter code for amino acid residues; Ac- denotes N^{α}-acetyl and -amide denotes C-terminal amide. Histidine is considered neutral at pH7.

Results and Discussion

The sequences of peptide analogs are shown in Table 1. We kept the hydrophobic/ non-polar face unchanged and decreased net charge by decreasing the number of positively charged residues or increasing the number of negatively charged residues (-5, +1, +4S, +4E), or increased net charge by increasing the number of positively charged residues (+8, +9, +10). The net charge varied between -5 and +10 and the number of positively charged residues varied from 1 to 10.

S. Del Valle et al. (eds.), *Peptides for Youth: The Proceedings of the 20th American Peptide Symposium*, 561
DOI: 10.1007/978-0-387-73657-0_246, © Springer Science+Business Media, LLC 2009

Table 2. Biological activity of V13K analogs against Pseudomonas aeruginosa strains.

Peptide Name	Number of positively charged residues	Number of negatively charged residues	Net Charge	Hemolytic activity		Antimicrobial activity								Therapeutic index[f]	
				MHC(μg/ml)[a]	Fold[b]	MIC(μg/ml)[c]							Fold[e]	MHC/MIC	Fold[g]
						PAO 1	PAK	CP 204	PA 14	WR 5	M 2	GM[d]			
-5	1	6	-5	>1000.0	8.0	>500.0	>500.0	>500.0	>500.0	>500.0	>500.0	1000.0	0.05	2.0	0.4
+1	7	6	+1	>1000.0	8.0	>500.0	>500.0	>500.0	>500.0	>500.0	>500.0	1000.0	0.05	2.0	0.4
+4S	4	0	+4	125.0	0.5	>500.0	>500.0	>500.0	>500.0	>500.0	>500.0	1000.0	0.05	0.1	0.02
+4E	7	3	+4	250.0	1.0	500.0	250.0	250.0	125.0	250.0	250.0	250.0	0.2	1.0	0.2
V13K	7	0	+7	250.0	1.0	31.3	125.0	7.8	62.5	250.0	31.3	49.6	1.0	5.0	1.0
+8	8	0	+8	250.0	1.0	31.3	62.5	7.8	31.3	62.5	31.3	31.3	1.6	8.0	1.6
+9	9	0	+9	<7.8	0.02	7.8	31.3	15.6	15.6	62.5	31.3	22.1	2.2	0.2	0.04
+10	10	0	+10	<7.8	0.02	15.6	15.6	7.8	15.6	31.3	31.3	17.5	2.8	0.2	0.04

[a]*MHC is the maximal peptide concentration that produces no hemolysis of human red blood cells after 18 h in the standard microtiter dilution method. When no detectable hemolytic activity was observed at 1000 μg/ml, a value of 2000 μg/ml was used for calculation of the therapeutic index and fold decreased. When hemolytic activity was still observed at 7.8 μg/ml, a value of 3.9 μg/ml was used for calculation of the therapeutic index and fold decreased.*
[b]*The fold decrease in hemolytic activity compared to that of lead compound, V13K.*
[c]*MIC is minimal inhibitory concentration that inhibited growth of six P. aeruginosa strains in brain heart infusion (BHI) medium at 37 °C after 24h. MIC is given based on three sets of determinations. When no detectable antimicrobial activity was observed at 500 μg/ml, a value of 1000 μg/ml was used for calculation of the therapeutic index and fold increased.*
[d]*GM, geometric mean of the MICs for the six P. aeruginosa clinical isolates.*
[e]*The fold improvement in antimicrobial activity (geometric mean) compared to that of lead compound, V13K.*
[f]*Therapeutic index is the ratio of the MHC value (μg/ml) over the geometric mean MIC value (μg/ml). Large values indicate greater antimicrobial specificity.*
[g]*The fold improvement in therapeutic index compared to that of lead compound, V13K.*

The minimal inhibitory concentrations (MIC) against six different *Pseudomonas aeruginosa* clinical isolates were determined along with the maximal peptide concentration that produces no hemolysis of human red blood cells (MHC) (Table 2).

The number of positively charged residues on the polar face and net charge are important for both antimicrobial activity and hemolytic activity. The number of positively charged residues on the polar face is more important than the net charge. Comparing peptide +4S and peptide +4E, both had the same net charge (+4) but a different number of positively charged residues. Peptide +4E had 7 positively charged residues and had significantly more antimicrobial activity than peptide +4S with 4 positively charged residues.

There is an optimum number of positively charged residues on the polar face for an optimum therapeutic index. Reducing the net charge decreased antimicrobial activity. Increasing the net charge increased both antimicrobial activity and hemolytic activity. For example, increasing the positive charge on the polar face (peptide +4S to peptide +8) decreased hemolytic activity by 2 fold (125μg/ml to 250μg/ml). However, a further increase of one positive charge (peptide +9) dramatically increased hemolytic activity more than 32 fold (250μg/ml to <7.8μg/ml). Thus, there was a 50-fold increase in the therapeutic index by increasing the positive charge from +4 (peptide +4S) to +7 (peptide V13K) and only a 1.6 fold in increasing the positive charge from +7 (peptide V13K) to peptide +8. On the other hand, a further increase of one positive charge on the polar face (peptide +8 to peptide +9) resulted in a decrease in the therapeutic index of 40-130 fold depending on different groups of organisms (40 fold for *Pseudomonas aeruginosa* strains).

References

1. Chen, Y., C. T. Mant, S. W. Farmer, R. E. Hancock, M. L. Vasil and R. S. Hodges. *J Biol Chem* **280** (13), 12316-12329 (2005).

Action of Bauhinia-derivated compounds on *Callosobruchus maculatus* development

Joana T. Sumikawa[1], Marlon V. Brito[1], Ana Paula U. Araújo[2], Maria Ligia R. Macedo[3], Maria Luiza V. Oliva[1] and Antonio Miranda[4]

[1]Department of Biochemistry, UNIFESP, São Paulo, SP; [2]Institute of Physics, USP-São Carlos, São Carlos, SP; [3]Department of Natural Science, UFMS, Três Lagoas, MS; [4]Department of Biophysics, UNIFESP, São Paulo, SP, Brazil

Introduction

Insect pests are a major cause of damage to the world's commercially agricultural crops. For example, a current strategy against cowpea weevil exploits the differences in the activity of plant peptidase inhibitors. Protease inhibitors isolated from the seeds of different species of *Bauhinia* [BbKI (Bauhinia bauhinioides Kallikrein Inhibitor), BbCI (Bauhinia bauhinioides Cruzipain Inhibitor] and BrTI (Bauhinia rufa Trypsin Inhibitor], with 81 percent identity in their primary sequences [1-4], were investigated on *Callosobruchus maculatus* development. Only BrTI [4] that contains **RGD** and **RGE** sequence motifs affected the development of the bruchid. These sequences were shown to be essential for BrTI insecticidal activity since the modified BbKI, rBbKIm, (recombinant form encoding some amino acids residues of BrTI) strongly inhibited the insect development.

In this work, synthetic peptides related to BrTI sequence, YLEPVA**RGD**GGLA-NH_2; YLEPVA**RGE**GGLA-NH_2 and IVYYPD**RGE**TGL-NH_2 were evaluated in their ability to influence on the development of *C. Maculatus* larvae. Peptides were synthesized by solid phase on a MBHA-resin. They were purified by HPLC and characterized by LC/MS and AAA. CD experiments on aqueous solution were also performed. Toxic activity of the peptides on *C. maculatus* was determined by measuring the deleterious effects caused on larvae fed on artificial seeds incorporating different concentrations (w/w) of the peptides. *C. maculatus* were reared on *Vigna unguiculata* (genotype Epace-10) at 29°C and relative humidity of 50% in a controlled environment chamber. Following incubation for 20 days the seeds were opened and the larvae were weighted and the number of their emergence were determined. Effective dose for 50% response [(ED_{50}) = peptide concentration that decreases the weight of larva to 50% from the control] and lethal doses [(LD_{50}) = peptide concentration that decreases the number of larvae emergence to 50% from the control] were also determined.

Results and Discussion

YLEPVA**RGD**GGLA-NH_2 comprising the sequence **RGD** did not interfere on the insect development (see Figure 1). In contrast, YLEPVA**RGE**GGLA-NH_2 presented a great lethality on *C. maculatus*, interfering on emergency and on the surviving of the larva [ED_{50} = 0.16% and LD_{50} = 0.09%] (see Figure 1). These results indicate the importance of **RGE** sequence confirmed by the effect observed for IVYYPD**RGE**TGL-NH_2 (see Figure 1). Removal of the TGL amino acids residues resulted on IVYYPD**RGE**-NH_2 that showed to be more effective (LD_{50} = 0.056% and LD_{50} = 0.085%) than the native protein. Beside, the same peptide inhibited the midgut arginil aminopeptidase from *C. maculatus* larvae.

S. Del Valle et al. (eds.), *Peptides for Youth: The Proceedings of the 20th American Peptide Symposium*, 563
DOI: 10.1007/978-0-387-73657-0_247, © Springer Science+Business Media, LLC 2009

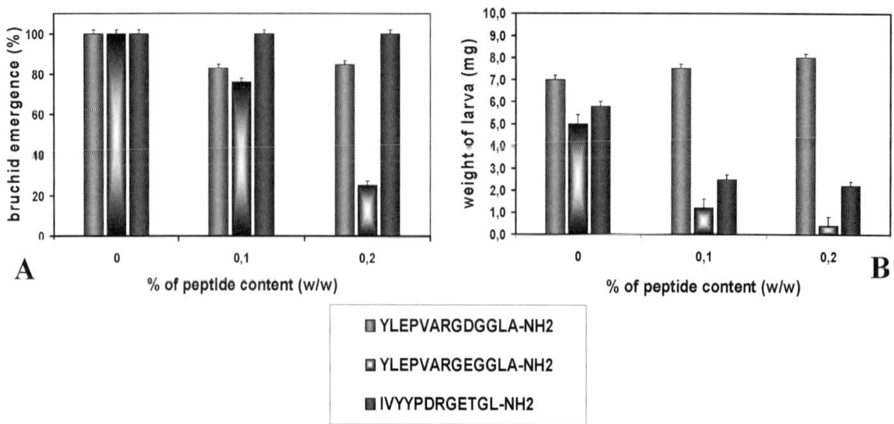

Fig. 1. Effect of YLEPVARGDGGLA-NH_2; YLEPVARGEGGLA-NH_2 and IVYYPDRGETGL-NH_2 on C. maculatus larvae development: (A) mortality (B) weight, using an artificial seed bioassay.

CD spectra of YLEPVA**RGD**GGLA-NH_2 (0.12 mM) and YLEPVA**RGE**GGLA-NH_2 (0.13 mM) did not indicate the existence of any preferential secondary structure in aqueous solution and they were not affected by changes in the peptide concentration within the 0.12 to 1.0 mM range. However, CD spectrum of IVYYPD**RGE**TGL-NH_2 (0.12 mM) presents some content of secondary structure, but it was also not affected by changes in peptide concentration in the same concentration range (data not shown). No clear correlation between CD results and the biological activity of the peptides was found. The feasible toxic mechanism of the peptides may be explored for appliance on crop pest control.

Acknowledgments

This work was supported by CAPES, CNPq, FAPESP and SPDM/FADA.

References

1. Oliva, M. L., Mendes, C. R., Juliano, M. A., Chagas, J. R., Rosa, J. C., Greene, L. J., Sampaio, M. U. and Sampaio, C. A. M. *Immunopharmaco.* **45**, 163-169 (1999).
2. Oliveira, C., Santana, L. A., Carmona, A. K., Cezari, M. H., Sampaio, M. U., Sampaio, C. A. M. and Oliva, M. L. V. *Biol. Chem.* **382**, 847-852 (2001).
3. Araújo, A. P. U., Hansen, D., Vieira, D. F., Oliveira, C., Santana, L. A., Beltramini, L. M., Sampaio, C. A. M., Sampaio, M. U. & Oliva, M. L. V. *Biol. Chem.* **386**, 561-568 (2005).
4. Nakahata, A. M., Bueno, N. R., Rocha, H. A. O., Franco, C. R. C., Chammas, R., Juliano M. A., Nader H. B., Sampaio, M. U. and Oliva, M. L. V. *Int. J. Biol. Macromol.* **40**, 22-29 (2006).

Positively charged gramicidin A based peptides form two types of membrane channels

Sergey I. Kovalchuk[1], Elena A. Kotova[2], Tatyana B. Stoilova[1], Sergey V. Sychev[1], Natalya S. Egorova[1], Andrey Y. Surovoy[1], Yuri N. Antonenko[2] and Vadim T. Ivanov[1]

[1]*Shemyakin-Ovchinnikov Institute of Bioorganic Chemistry Russian Academy of Sciences, Miklukho-Maklaya, 16/10, 117871, Moscow, Russia;* [2]*Belozersky Institute of Physico-Chemical Biology, Moscow State University, 119992, Moscow, Russia*

Introduction

In recent years investigation of new membrane-active antimicrobial peptides attracted much attention. At present at least 5 types of pore structures are proposed: rigid intramolecular tubes formed by head-to-head dimerization of gramicidin A (gA) [1], intramolecular pores resulting from stacking of cyclic peptides [2], intermolecular "barrel-stave" pores formed by bundles of α-helical rods of alamethicin [3], porin-like high-conductance intermolecular pores with a β-barrel structure [4], intermolecular peptide–lipidic pores with pore walls formed both by peptide molecules and lipid headgroups [5–6].

The factors remain unclear for forming pores of different structure by peptide molecules of one sort. Searching of new pore structure types also is of great interest. In the present work, we addressed this question by studying the activity of gramicidin A positively charged derivatives. It was found that peptides having a positively charged moiety conjugated with gramicidin A sequence via a short linker form classical potassium ion selective pores at low concentrations and large unselective pores at high concentrations.

Results and Discussion

Table 1. Sequences of charged gramicidin A analogues.

1	Ac – VGAlAvVvWlWlWlWa - GSGPKKKRKVC
2	Ac – VGAlAvVvWlWlWlWa - GSGEEEEC
3	Ac – VGAlAvVvWlWlWlWa - GSGPKKKRKVG
4	CPKKKRKVGSG - VGAlAvVvWlWlWlWa
5	CKKKKGSG - VGAlAvVvWlWlWlWa
6	CKKKKKKGSG - VGAlAvVvWlWlWlWa
7	CKKKKKKKKGSG - VGAlAvVvWlWlWlWa
8	CKKKKKKKKKKGSG - VGAlAvVvWlWlWlWa

v – D-valine, l – D-leucine, a – β-alanine

Pore-forming activity of charged analogues of gramicidin A (table 1) in planar bilayer lipid membranes and liposomes was studied. Positively charged derivatives (1 and 3 – 8) formed two different types of pores depending on their concentration. At low concentrations they showed classical gramicidin cationic conductance, whereas at high concentrations large unselective pores were observed. The single-channel conductance of the large pore was 320 pS in 100 mM choline chloride as judged from the fluctuation analysis of the multi-channel current. By contrast, a

S. Del Valle et al. (eds.), *Peptides for Youth: The Proceedings of the 20th American Peptide Symposium*,
DOI: 10.1007/978-0-387-73657-0_248, © Springer Science+Business Media, LLC 2009

negatively charged analogue (2) exhibited solely the classical cationic channel activity. CD spectroscopic analysis of the peptides in membranes indicated the presence of both the single-stranded right-handed $\beta^{6.3}$-helical and $\beta^{5.6}$-double helical conformations, with the fraction of the latter being larger than in gramicidin A itself. Formation of unselective pores by the Cys-containing positively charged analogue (1) was strongly influenced by the redox state of the peptide in contrast to the peptide (3) containing Gly residue. In particular, oxidation-induced dimerization led to an increase in the transmembrane current and loss of cationic selectivity of planar bilayer lipid membranes. Besides, it resulted in an increase in the carboxyfluorescein leakage from liposomes. This phenomenon took place at high concentrations of the peptide and was not observed at the single-channel level. The peptide (4) containing the charged moiety at the N-terminus instead of the C-terminus of the gramicidin A sequence didn't show any decrease in the unselective pore-forming activity while absolute loss of classical gramicidin activity took place. Analogues (5 – 8) containing positively charged sequences of different length showed an increase in the carboxyfluorescein leakage from liposomes in direct proportion to the number of lysine residues. Based on these findings, we suggest that the unselective pore represents a "barrel-stave" structure analogous to the one formed by bundles of α-helical rods of alamethicin (fig.1). We suggest that the special feature that converts positively charged gramicidin analogues into pore formers is the ability to form amphipathic-like structures through the intermolecular interaction between the positively charged groups of the hydrophilic domain and the Trp indole groups of gramicidin A. Besides we suggest that there is no functional relation between the classical gramicidin activity and unselective pores. The results obtained provide new ways for design and synthesis of membrane active peptides.

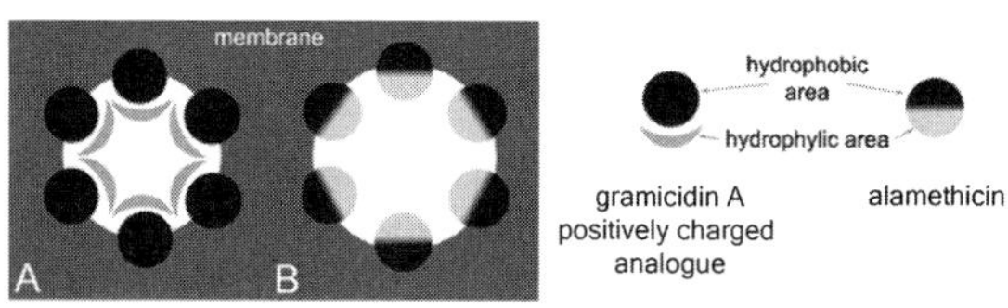

Fig. 1. Scheme of oligomeric pore. A. A hypothetical unselective pore formed by positively charged gramicidin A analogues. B. An oligomeric pore formed by alamethicin.

Acknowledgments

This work was supported in part by a grant from the RFBR: 06-04-48523 (Y.N.A.).

References

1. Koeppe, R.E. and Andersen, O.S. Annu. Rev. Biophys. Biomol. Struct. 25, 231–258 (1996)
2. Fernandez-Lopez, S., Kim, H.S., Choi, E.C., Delgado, M., Granja, J.R., Khasanov, A., Kraehenbuehl, K., Long, G., Weinberger, D.A., Wilcoxen, K.M. and Ghadiri, M.R. Nature 412, 452–455 (2001)
3. Sansom, M.S. Prog. Biophys. Mol. Biol. 55, 139–235 (1991)
4. Thundimadathil, J., Roeske, R.W. and Guo, L. Biochem. Biophys. Res. Commun. 330, 585–590 (2005)
5. Zemel, A., Fattal, D.R. and Ben Shaul, A. Biophys. J. 84, 2242–2255 (2003)
6. Zakharov, S.D., Kotova, E.A., Antonenko, Y.N. and Cramer, W.A. Biochim. Biophys. Acta 1666, 239–249 (2004)

Effects of Tumor-Associated Mutations in the p53 Tetramerization Domain on Oligomerization State and Transcriptional Activity

Rui Kamada, Tomoko Terai, Takao Nomura, Yoshiro Chuman, Toshiaki Imagawa and Kazuyasu Sakaguchi

Department of Chemistry, Faculty of Science, Hokkaido University, Sapporo, Japan

Introduction

Tumor suppressor protein p53 plays an important role to maintain genomic integrity from cellular stress by trans-activating the target genes involved in different cellular functions, such as cell cycle arrest and apoptosis [1]. The tetramer formation of p53 through the C-terminal tetramerization domain (TD) is essential for its activity. The p53 tetramerization domain consists of a β-strand (Glu326-333), a tight turn (Gly334), and an α-helix (Arg335-Gly356) (Fig. 1A). Approximately 50 % of human tumors carry inactivating mutations in the *p53* gene. To date, 41 point mutations have been found in 22 positions among 31 residues of the tetramerization domain (Fig. 1B). It is important to understand the mechanism of malignant transformation by mutations in the tetramerization domain [2, 3].

In this study, we performed thermodynamic analysis of p53 tetramerization domain peptides with mutations found in human tumors in order to clarify the quantitative correlation between stability of oligomeric structure and transcriptional activity for p53.

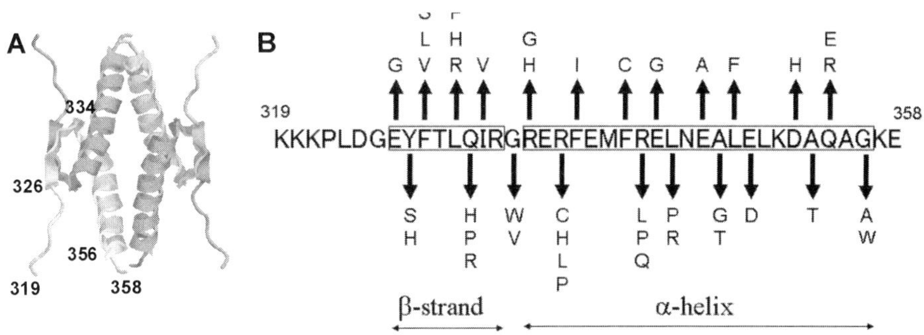

Fig. 1. (A) Structure of the p53TD (B) Sequences of the wild-type and mutant p53TD peptides.

Results and Discussion

The wild-type (WT) and tumor-associated mutant peptides comprising of residues 319-358 were chemically synthesized, and their secondary structure were analyzed by circular dichroism (CD) in sodium phosphate buffer, pH7.5, at 4°C (Fig. 1B). The WT peptides exhibited a characteristic spectrum for the p53 tetramer structure. Several mutant peptides, such as L330R, R342P, and L344P, existed in random-coil structure at 4°C (Fig. 2A). This data indicated that the Pro mutations in α-helix or β-strand region disrupt the secondary structure, and these mutant peptides could not form tetramer even at high peptide concentration and low temperature. Mutation of Leu330 residue to Arg also led destruction of tetrameric structure of p53. Leu330 is located at the center of the hydrophobic core in p53 tetrameric structure.

DOI: 10.1007/978-0-387-73657-0_249, © Springer Science+Business Media, LLC 2009

stingly, G334W showed a β-dominant spectrum after heating and cooling, indicating G334W peptide shifted β-strand rich conformation. Gly334 residues form a sharp turn connecting the β-strand with the α-helix. Our data suggested that non-glycin residues could not adopt a backbone conformation of Gly334 [4]. Other mutant peptides exhibited almost identical spectra as the WT peptide at 10 μM monomer concentration at 4°C.

We analyzed thermodynamic stability of mutant peptides that could form wild type like tetrameric structure under the condition employed. As shown in Fig. 2B, tetrameric structures of mutant peptides were destabilized. Each mutations had a different effect for stability of tetrameric structure. Mutation Phe341 to Cys dramatically destabilized the tetrameric structure (ΔTm = -38.5°C, compared with WT). Phe341 is located in the hydrophobic core in the tetrameric structure, suggesting that the hydrophobicity of Phe341 is critical for stabilization of the p53 tetramer. In addition, substitution of Cyclohexylalanine (Cha) for Phe341 dramatically induced stabilization for the p53 tetrameric structure (data not shown). R337L induced significant destabilization to p53 tetrameric structure (ΔTm = -33.5°C), probably due to loss of salt-bridge between the guanidinium group of Arg337 and Asp352. In contrast, mutations of the residues accessible to solvent, such as G356W, were less effective in destabilization for the tetrameric structures (ΔTm = -0.9°C). It is possible that such a small effect of destabilization could affect the function of cellular p53 protein, since these mutations were also found in human tumors. Taken together, it is probably that the stability of protein structure and functions are stringently regulated in cell.

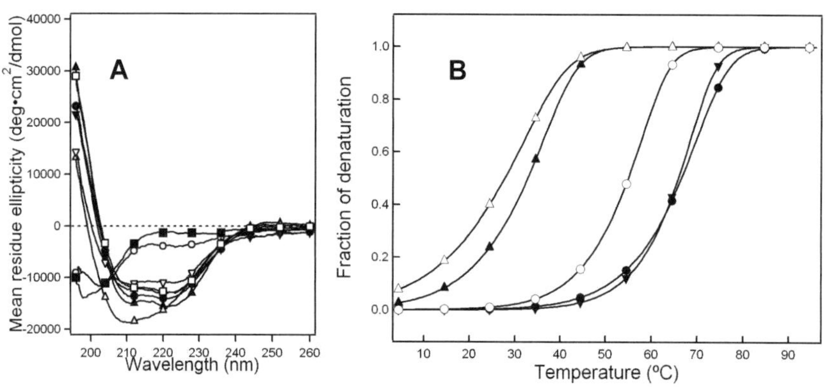

Fig. 2 (A) CD spectra of p53TD peptides in phosphate buffer, pH7.5 at 4°C. ●, wild-type; ○, L330R; ▲, G334W; Δ, G334W after heating; ▼, R337L; ▽,F341C ; ■, L344P; □,G356W. The peptide monomer concentration was 10 μM. (B) Thermal denaturation of peptides was analyzed by measuring the ellipticity at 222 nm for 10 μM, in the range of 4°C to 96°C. ●, wild-type; ○, G334W; ▲, R337L; Δ, F341C; ▼, G356W

References

1. Vogelstein, B., Lane, D., and Levine, A. J., *Nature*, **408**, 307-310 (2000).
2. Clore, G.. M., Ernst, J., Clubb, R., Omichinski, J. G., Sakaguchi, K., Appella, E., and Gronenborn, A. M., *Nat. Strc. Biol.*, **2**, 386-391 (1995).
3. Hollstein, M., *et al. Nucleric Acids Res.*, **24**, 141-146 (1996).
4. Higashimoto, Y., Asanomi, Y., Takakusagi, S., Lewis, M. S., Uosaki, K., Durell, S. R., Anderson. C. W., Appella, E., and Sakaguchi, K. *Biochemistry*, **45**, 1608-1619 (2006).

Self-association regions in the CARD of Bcl-10

M. Ruvo[1], A. Sandomenico[1], D. Marasco[1], S. M. Monti[1], B. Tizzano[1], A. De Capua[1], C. Pedone[1], A. Saporito[1], R .M. Vitale[1], S. Formisano[2], R. Stilo[3] and D. Liguoro[4]

[1]IBB-CNR – via Mezzocannone, 16 – 80134, Napoli-Italy, [2]Università degli Studi di Napoli Federico II, -Via S. Pansini, 5, -80131, Napoli-Italy, [3]Università degli Studi del Sannio di Benevento, Via Port' Arsa, 1, -82100, Benevent-Italy, [4]IEOS-CNR,- Via S. Pansini 5,-80131 Napoli-Italy

Introduction

Many apoptotic signals are mediated by the association of proteins containing homologous domains, such as death domains (DD), death effector domains (DED), and caspase recruitment domains (CARD). Several CARD-containing proteins appear now to be directly involved in modulation of nuclear factor-kB (NF-kB) [1] and in the regulation of inflammatory responses and apoptosis. Among these, Bcl-10 (apoptosis regulator **B**-**c**ell **l**ymphoma **10**) is one of the best studied for its correlation with MALT lymphomas [2, 3] and more recently as a downstream effector of lymphocyte antigen receptors to promote the activation of the IkappaB kinase complex leading to the phosphorylation and degradation of IkappaB [2]. Bcl-10 has a bipartite structure consisting of an N-terminal CARD domain and a C-terminal serine-threonine rich domain acting as a regulatory unit that undergoes multiple phosphorylation by several kinases. Most activities of Bcl-10 are accomplished by a dimeric form of the protein; in lymphomas it has been found that a gene translocation event causes over-expression and oligomerization and that the CARD domain mediates the protein self-association [3]. In order to investigate the Bcl-10 CARD regions responsible of protein-protein interactions, here we followed an approach whereby, after protein proteolysis and peptide fractionation, fragments of the CARD domain are utilized as competitors in a self-association ELISA-like assay. The fragments identified have a nM activity and have been characterized by NMR to determine their structural properties in solution.

Results and Discussion

Bcl-10 CARD domain was expressed as a His-tagged, double Cys-Ser mutated protein of the wild-type human variant. To monitor the CARD self-association, we set up an ELISA assays where the Bcl-10 CARD was adsorbed on the plate surface at different concentrations (0.1-0.6 μM), whereas for binding biotinylated-CARD was utilized (10^{-5}-10^{-7} M) . The concentration of biotinylated-CARD resulting in a 50% of maximum binding was 0.5 μM, and it was taken as an estimation of the K_D value of the self-association equilibrium. A protein aliquot (2 μmoles) was then extensively hydrolyzed with trypsin (1:50 w/w, ON, 37°C) monitoring the reaction progression by LC-MS/MS. The tryptic peptides were thus fractionated by RP-HPLC and the fractions submitted to a competitive ELISA assay under the reported conditions. Several fractions containing overlapping fragments capable to completely abolish the Bcl-10 CARD-CARD interaction, were identified in the C-terminal region of Bcl-10 CARD domain. Fragments [91-98], [78-98], and [68-98], were chemically synthesized by SPPS and tested in a dose-response assay. Notably, while the shortest fragment [91-98] blocked Bcl 10 CARD self-association

S. Del Valle et al. (eds.), *Peptides for Youth: The Proceedings of the 20th American Peptide Symposium*, 569
DOI: 10.1007/978-0-387-73657-0_250, © Springer Science+Business Media, LLC 2009

with an IC_{50} of 34 nM (Fig. 1), the longer fragments were essentially ineffective, suggesting that in these molecules the [91-98] residues are somewhat masked or in a different conformation. CD spectra of the three synthetic fragments showed they have typical α-helical structures, though a partial loss of Cotton effect in the far-UV region, suggestive of aggregation, was recorded over time. Contacting residues were also investigated by preparing double Ala-scan peptides of Bcl-10 [91-98], identifying T91, Q92, I95 and L96 as those mediating the protein-protein interaction.

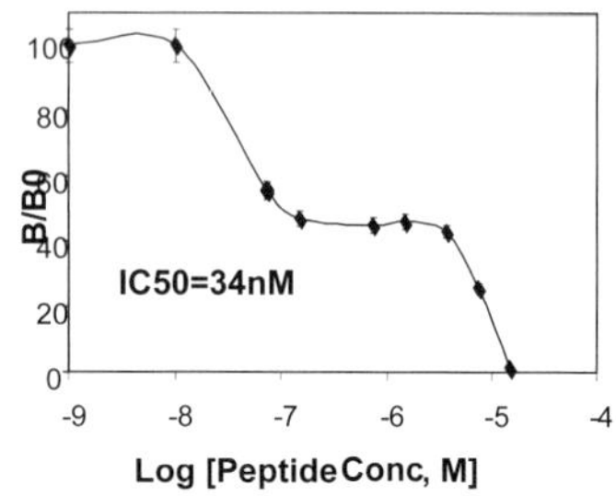

Fig. 1. Competitive ELISA assay with CARD [91-98] synthetic fragment: normalized signals of Absorbance at 490 nm is plotted against Log of Bcl10-CARD [91-98] concentration.

A preliminary solution structure characterization was carried out by NMR on the Bcl-10 [91-98] and [78-98] peptides. Analysis by 1H NMR, at diverse peptide concentrations (mM scale), aqueous buffer solutions, pH and ionic strengths showed broad and unresolved proton signals, suggestive of oligomerization. Notably, the addition of 10-30% (v/v) of CD_3CN to the solution increased the chemical shifts spreading and sharpened signals. A DOSY [4] experiment performed in H_2O and in 30% CD_3CN (v/v) showed that the apparent molecular masses calculated from the diffusion coefficient, were consistent with trimeric structures in water and monomeric structures in 30% CD_3CN. In conclusion, the [91-98] region of Bcl-10 appears strongly involved in the CARD domain-mediated protein dimerization-oligomerization process and key residues for this interaction have also been identified. The reduced activity exhibited by the longer fragments [89-98] and [68-98] can be imputed to the oligomeric state of these peptides that is likely persistent also at the low concentrations where the ELISA assay is carried out.

Acknowledgments

The support of project FIRB RBNE03PX83_005 is acknowledged.

References

1. Park H.H., et al., *Annu. Rev. Immunol.* **25**, 561-86 (2007).
2. Thome M. *Nat. Rev. Immunol.* **4**, 348-59 (2004).
3. Guiet, C. and P. Vito, *J. Cell. Biol.* **148**, 1131-40 (2000).
4. Johnson, J. C. S. *Prog. Nucl. Magn. Reson. Spectrosc.* **34**, 203-256 (1999).

GHRP-2 Attenuates Burn-Induced Dysfunctions in Rodents

Ambikaipakan Balasubramaniam, Rashika Joshi, Lou Ann Friend and J. Howard James

Department of Surgery, University of Cincinnati Medical Center and Shriners Hospital for Children, Cincinnati, OH 45267-0558, USA

Introduction

After major burn injury, enhanced production of catabolic hormones and inflammatory cytokines and reduced levels of anabolic hormones result in prolonged hypermetabolism, loss of lean body mass and muscle wasting. Various treatments, including exercise and anabolic hormones such as insulin, growth hormone (GH), IGF-I or steroids, have been proposed or employed to reduce debilitating body protein loss, but they are far from satisfactory. Recently, we have shown that the newly isolated anti-cachectic peptide, ghrelin, is down regulated in rats after burn injury, and that exogenous ghrelin restores lost body weight by increasing food intake and enhancing adiposity in mice [1, 2]. Furthermore, ghrelin also significantly attenuated skeletal muscle protein breakdown in rats with burn injury by normalizing muscle mRNA expression of E3 ubiquitin ligases, MuRF1 and MAFbx [3, 4]. However, ghrelin's short half-life may limit its clinical application. Moreover, our ultimate goal is to develop longer-acting, lower molecular weight ghrelin analogs. In this regard, a synthetic hexapeptide, growth hormone releasing peptide-2 (GHRP-2), has been shown to be a stable and potent ghrelin receptor agonist. GHRP-2 enhances body weight in rodents, exhibits anti-inflammatory effects in arthritic rats, and improves protein catabolism in critically ill patients [5]. We have therefore investigated the effects of GHRP-2 on body weight, food intake and skeletal muscle protein breakdown in rodents with burn injury.

Results and Discussion

In the first study, we monitored food intake and body-weight daily for 10 days in 3 groups of mice subjected to either 20% surface area dorsal scald burn or a sham procedure (Burn-GHRP-2, Burn-Saline and Sham-Saline, n=12–14 per group). Mice were treated daily with 0.10 ml saline or GHRP-2 (50 μg/kg, *SC*). Compared to the Sham-Saline group, mice in the Burn-Saline group continued to lose body weight from post-burn day-1 (Figure 1), reaching significant levels on day 7-10 after burn. GHRP-2 treatment significantly prevented the burn-induced body weight loss from day 1-7. Although not statistically significant, GHRP-2 treatment also continued the tendency to enhance the body weight of burn-injured mice, compared to the Burn-Saline group, during days 8-10. Comparison of the metabolic rates on day 11-13 after burn revealed that Burn-GHRP-2 mice exhibited a significantly higher respiratory quotient (RQ) relative to the Burn-Saline group during this period. These observations suggest that, as in the case of ghrelin, GHRP-2 induces a preferential metabolism of carbohydrate over fat, thus increasing adiposity. This may, at least in part, account for the GHRP-2-induced body weight gain in burn-injured mice. Both the Burn-Saline and Burn-GHRP-2 groups consumed comparable amounts of food, and this was significantly greater than that consumed by the Sham-Saline group. It appears, therefore, that increased food intake by both groups of burned mice was driven by the demand for more energy due to post-burn hypermetabolism.

S. Del Valle et al. (eds.), *Peptides for Youth: The Proceedings of the 20th American Peptide Symposium*,
DOI: 10.1007/978-0-387-73657-0_251, © Springer Science+Business Media, LLC 2009

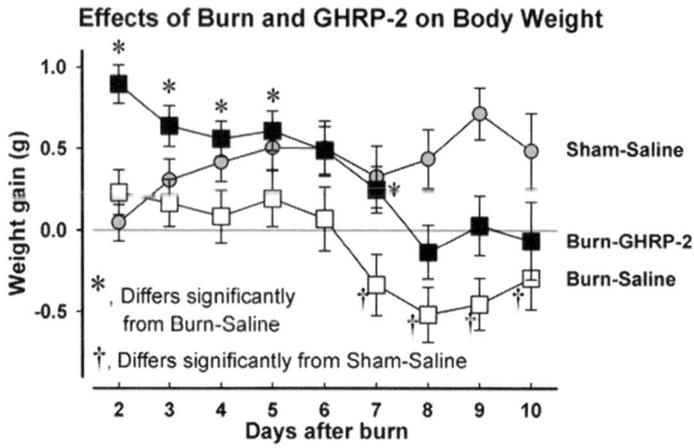

Fig. 1. Effects of GHRP-2 (50 μg/mice/daily, SC) attenuated the burn-induced body weight loss in mice.

In a separate study, the effect of GHRP-2 (10 mg/kg, released SC over 24 h from an osmotic pump) on muscle protein breakdown was determined in extensor digitorum longus (EDL) muscles from young rats (50-70 g, 10-12/group) subjected to either 30% dorsal flame burn or sham procedures. As indicated by the enhanced release of both tyrosine and 3-methylhistine, burn injury increased muscle protein breakdown, which was significantly attenuated by GHRP-2 treatment. Moreover, GHRP-2 treatment significantly suppressed the burn-induced increase in mRNA expression of MuRF1 and MAFbx in the gastrocnemius muscle. It appears, therefore, that GHRP-2 also suppresses burn-induced muscle protein breakdown through attenuating production of E3 ubiquitin ligases.

In summary, these results demonstrate for the first time that GHRP-2 can mimic the beneficial effects of intact ghrelin in attenuating burn-induced body weight loss as well as skeletal muscle protein breakdown. Also, the results suggest that a dose of GHRP-2 higher than what was employed here may be required to completely reverse weight loss as post-burn hypermetabolism becomes progressively more severe.

Acknowledgments

The work was funded by a grant (#8570) from Shriners Hospitals for Children.

References

1. Balasubramaniam, A., *et al. Peptides* **27**, 1624-1631 (2006).
2. Balasubramaniam, A., *et al. J. Burn Care & Res.* **26**, S108 (2005).
3. James, J. H., *et al.. J. Burn Care Rehab.* **26**, S107 (2005).
4. Joshi, R., *et al. J Burn Care & Res* **27**, S164 (2006).
5. Granado, M., *et al. Am. J. Physiol. Endocrinol. Metab.* **288**, 486-492 (2005).

Peptides binding the type E immunoglobulins

Annamaria Sandomenico[1,2], Daniela Marasco[1], Simona M. Monti[1], Michele Saviano[1], Carlo Pedone[1,2], Ettore Benedetti[1,2] and Menotti Ruvo[1]

[1]*Istituto di Biostrutture e Bioimmagini, CNR – via Mezzocannone, 16 – 80134, Napoli - Italy*
[2]*Università di Napoli, via Mezzocannone, 16, 80134, Napoli, Italy.*

Introduction

The human high affinity IgE receptor, hFcεRI, found on the surface of mast cells and basophils, is believed to mediate allergic diseases, anaphylaxis and asthma through binding of IgE itself. hFcεRI contains four distinct polypeptide chains: an α chain, a β chain and a dimer of γ chains. The extracellular portion of the α chain binds with high affinity to the Fc region of the IgE (K_D=10^{-9}M), whereas the β and γ chains are responsible for down-stream signal propagation through phosphorylation of their intracytoplasmatic immunoreceptor tyrosine-based activation motifs (ITAM) [1]. In the effort to obtain a receptor antagonist working by binding the soluble IgE, we have designed a peptide-based receptor mimetic on the basis of the crystallographic structure of the complex between the hFcεRI α chain and the IgE Fc portion [2]. To this aim, fragments containing key residues for Fcε interaction and derived from the D2 domain (interaction site 1, Y129-H134, [2]) and D1-D2 junction of FcεRIα chain (interaction site 2, W110-W113, W87, C151-S162, [2]) have been selected and joined by linkers useful to ensure the proper distance between the two sites (Fig. 1A). Furthermore, by mutating S162 to Cysteine, a disulfide bridge has been introduced on site 2 in order to force this region in a loop. We have thus tested this polypeptide, denoted as IgE-Trap, assessing its capacity to bind IgE through ELISA and BIAcore assays and we have evaluated its specificity for IgE over IgG and IgA. IgE binding capacity of IgE-trap subdomains have been also determined using several synthetic constructs reproducing separate IgE-Trap sites.

Results and Discussion

The IgE-trap polypeptide and sub-domains were produced by SPPS following the Fmoc methodologies and purified by RP-HPLC. Where required, cyclization was achieved by spontaneous oxidation in slightly basic buffers. The final products were characterized by LC-MS. Interaction between the designed polypeptide (in its cyclized as well as reduced form) and the immunoglobulin was initially investigated by SPR (BIACORE 3000) on an IgE-derivatized chip. Solutions of the peptides at increasing concentrations were fluxed through the chip observing dose-dependent binding curves (Fig. 1B) and, by data fitting we observed very similar affinities: K_D = $2.4\pm0.5*10^{-5}$ (cyclized) and $2.1\pm0.5*10^{-5}$M (reduced). The activity of the cyclized IgE-trap was also tested using ELISA assays where the IgE-trap or the immunoglobulin were immobilized on microtiter plates, confirming dose-response and saturable bindings. Given the low affinity of the designed polypeptides for the target IgE, we undertook a structure-activity study, investigating affinities of separate IgE-trap sub-domains. To this aim, peptide fragments reproducing the two separate FcεRI sites were prepared by SPPS and tested by SPR. Remarkably, only regions mimicking the site 1 (denoted as IgE-Trap 1st miniloop) and part of the second site (IgE-Trap 2nd miniloop, see legend of Fig. 1) showed binding capacity

S. Del Valle et al. (eds.), *Peptides for Youth: The Proceedings of the 20th American Peptide Symposium*, 573
DOI: 10.1007/978-0-387-73657-0_252, © Springer Science+Business Media, LLC 2009

for IgE. Importantly K_D values measured for these fragments were $6.5\pm0.5*10^{-4}$ M and $5.7\pm0.5*10^{-4}$ M, respectively, about 20 times lower than the whole polypeptide. The 2nd loop, corresponding to a portion of site 2, instead appeared completely inactive, suggesting that the IgE-trap binds to IgE through residues located on the peptide first loop, while residues from the second loop only have a modest influence on the global affinity.

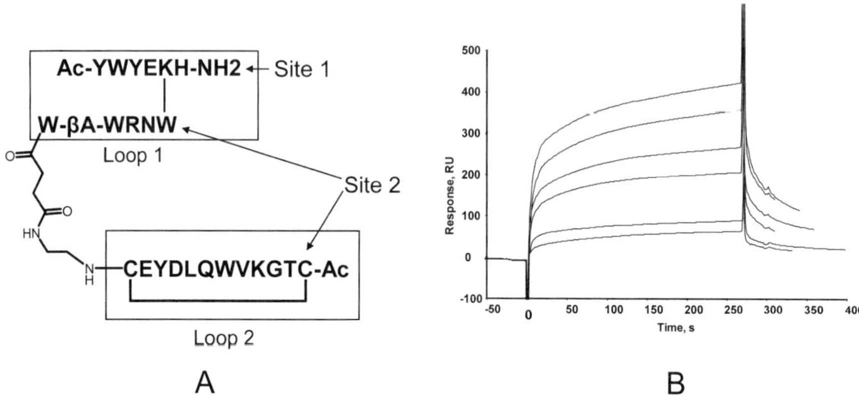

Fig. 1. ***A****: Schematic structure of the designed IgE-Trap; the region denoted as Loop1 and Loop 2 are boxed. The sequences YWYEKH and W-βA-WRNW are denoted as 1*[st] *and 2*[nd] *miniloop, respectively. βA is the amino acid β-Alanine.* ***B****: dose-dependent binding curves by real time BIA to immobilized IgE of the polypeptide IgE-trap ox [2.5 μM – 30μM].*

To assess the specificity of the designed peptides, we performed dose-dependent SPR assays on IgG- and IgA-derivatized sensor chips. Remarkably, no binding was detected in these experiments, suggesting that the interaction for the IgE, though still weak, appears highly specific.

In conclusion, the designed IgE-trap and fragments reproducing site 1 and part of the site 2 (Loop 1, Fig. 1A) have a low affinity for the IgE compared to the receptor, but they exhibit a high specificity for this class of immunoglobulins. Further improvements on peptide structure to increase the affinity are in progress and they will be achieved by introducing in site 2 new structural elements selected from an array of synthetic libraries.

Acknowledgments

Financial support from project FIRB RBNE03PX83_005 is acknowledged.

References

1. Kraft S, Kinet JP., *Nat. Rev. Immunol.* **7**, 365-378 (2007).
2. Garman SC, et al., *Nature* **406**, 259-66 (2000).

Evolution of the Synthetic Process to Prepare the Tripeptide Segment of ABT-510

Calvin L. Becker, Jean Christophe Califano, John C. Tolle, James J. Napier, Lawrence Kolaczkowski, Sou-Jen Chang, Zhenping Tian and Sukamar Manna

GPRD Process Research and Development, Abbott Laboratories, 1401 Sheridan Rd., North Chicago, IL, 60064, USA

Introduction

ABT-510 is a thrombospondin-1 mimetic currently in Phase 2 clinical trials as an angiogenesis inhibitor for treatment of cancer. An efficient synthetic process was developed by the Process R&D group at Abbott and has been scaled to produce several 10 kg batches of the drug substance.

ABT-510

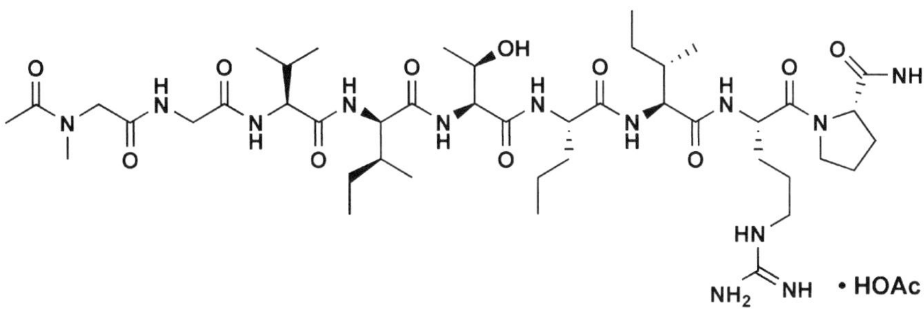

Ac-Sar-Gly-Val-D-allo-Ile-Thr-Nva-Ile-Arg-ProNHEt

Figure 1. The ABT-510 structure.

The process for preparation of the tripeptide segment, Ile-Arg-ProNHEt dihydrochloride salt, underwent significant refinement as the overall process for ABT-510 was being developed. Three distinct generations of the process have led to replacement of $SOCl_2$ with CDI for activation of Z-Pro, replacement of DCC with EDAc for coupling Boc-Arg(NO_2)-OH, and replacement of hydrophobic interaction chromatography for removal of NH_4Cl generated in the hydrogenolysis of the nitro protecting group with reverse osmosis de-salting, which was then subsequently replaced with a simple pH adjustment and evaporative removal of NH_3.

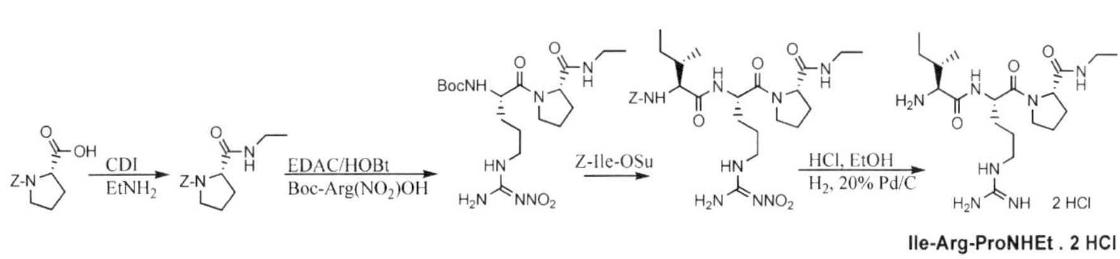

Figure 2. Ile-Arg-ProNHEt dihydrochloride salt synthetic scheme.

S. Del Valle et al. (eds.), *Peptides for Youth: The Proceedings of the 20th American Peptide Symposium*, 575
DOI: 10.1007/978-0-387-73657-0_253, © Springer Science+Business Media, LLC 2009

Figure 2 shows the four steps in the preparation of Ile-Arg-ProNHEt dihydrochloride salt. In the early stages of process development for ABT-510, Z-Pro-OH was activated with thionyl chloride. However, reaction of the acid chloride with 70% ethylamine in water is extremely exothermic, with an adiabatic temperature (ATR) rise of 147°C. Thionyl chloride activation was replaced with 1, 1'-carbonyldiimidazole (CDI) activation. The CDI mediated coupling is safer, with an ATR of 18°C and, in addition, provides a higher yield (87% vs. 70%).

The step 2 coupling was formerly mediated by DCC/HOBt, but removal of DCU from the reaction mixture required a series of filtrations. The switch to EDAC provided a soluble urea byproduct and eliminated the need for filtration during work-up.

Finally, catalytic hydrogenation of Cbz-Ile-Arg(NO_2)-Pro-NHEt was carried out in ethanol, in the presence of HCl (3.6 eq) using 20% palladium on carbon (30-40 psig, 25°C) to afford an ethanol/water solution of the tripeptide H-Ile-Arg-Pro-NHEt.dihydrochloride salt. The nitro protection on the guanidine was reduced under acidic catalytic hydrogenation conditions to produce ammonium chloride (mildly exothermic). Ammonium chloride will lead to amidation of the hexapeptide in the next coupling step of the ABT-510 process and thus must be removed.

The desalting process has undergone three different iterations. Each of these methods was demonstrated at multi-kilo scale. In the first generation the ammonium chloride is removed by hydrophobic interaction chromatography. An aqueous solution of tripeptide dihydrochloride salt and ammonium chloride, adjusted to pH 6 with NaOH, was loaded onto an Amberlite® XAD-16 column. The resin was washed with water to elute NH_4Cl. Product was eluted with 10% methanol. Product fractions were concentrated, chased with ethanol, and crystallized by slow addition to EtOAc. Unfortunately, several mixed fractions were obtained, resulting in an overall yield of 66%. In the second generation process, an aqueous solution of tripeptide dihydrochloride salt and ammonium chloride was diluted with water and desalted by reverse osmosis. The RO process provided a much higher recovery with an overall yield of 86%. However, either of these processes were laborious, time consuming, and required specialized equipment. A third, more efficient process was developed. The pH of the ethanolic hydrogenation mixture was adjusted to pH 8-10 with NaOH to convert the NH_4Cl to NH_3 and NaCl. Distillation of the ethanol/water solution of the tripeptide efficiently reduced the ammonia to < 2 ppm with no chase distillation. The pH of the tripeptide in ethanol was then adjusted to 1-3 by addition of conc. HCl. Thus, the dihydrochloride salt of the product was re-formed, as well as NaCl. Distillation from EtOH reduced the water level to less than 1% and resulted in precipitation of most of the NaCl. Enough ethyl acetate was added to drive the remaining NaCl out of solution, but not the product. NaCl was removed by filtration. The final product, H-Ile-Arg-Pro-NHEt • 2HCl, was crystallized by addition of more ethyl acetate. Following this process, 13 Kg H-Ile-Arg-Pro-NHEt • 2HCl was produced in 89% yield without chromatography or RO.

In the course of ABT-510 process development the 4-step process to produce H-Ile-Arg-Pro-NHEt • 2HCl has been significantly simplified. The overall yield was increased from 35% to 58 %. Chromatography and Reverse Osmosis for removal of NH_4Cl have been eliminated. The process has been demonstrated at > 10Kg scale producing 99.7% pure material.

Retrocyclin-2: a potent anti-HIV θ-defensin that forms a cyclic cystine ladder structural motif

Norelle L. Daly[1], Yi K. Chen[1], K. Johan Rosengren[1], Ute C. Marx[1], Martin L. Phillips[2], Alan J. Waring[3], Wei Wang[3], Robert I. Lehrer[3] and David J. Craik[1]

[1]*Institute for Molecular Bioscience, University of Queensland, Brisbane 4072 Australia;* [2]*Department of Chemistry and Biochemistry, UCLA, Los Angeles, CA, 90095, USA;* [3]*Department of Medicine, David Geffen School of Medicine at UCLA, Los Angeles, CA, 90095, USA.*

Introduction

Retrocyclins are head-to-tail cyclised θ-defensins with potent activity against human immunodeficiency virus, influenza A and herpes simplex virus. Retrocyclins bind carbohydrate-containing surface molecules such as gp120 and CD4 with high affinity (K_d = 10-100 nM) and are localized on the cell membrane. The first θ-defensin discovered, rhesus theta defensin 1 (RTD-1) is a potent antimicrobial peptide found in monkey leucocytes [1]. Human bone marrow expresses mRNA that is homologous to the precursors of RTD-1 [2], but a premature stop codon within the signal sequence domain of the human mRNA transcripts prevents translation. Human θ-defensin sequences, termed retrocyclins, are therefore considered to represent expressed pseudogenes. Synthetic retrocyclins based on the human pseudogene sequences and the processing sites of RTD-1 have been shown to confer protection against HIV-1 infection [2]. Given their potent anti-viral activity, unusual cyclic backbone and unique mode of biosynthesis, determination of the structural and biochemical properties of retrocyclins is of great importance.

The structural features important for activity have yet to be elucidated but here we report the first three-dimensional structure of a retrocyclin, namely one of the most potent forms, retrocyclin-2.

Results and Discussion

NMR spectra of synthetic retrocyclin-2 were recorded under a range of conditions. At millimolar concentrations in aqueous solution broad spectral peaks were observed, consistent with oligomers being present. In the presence of SDS micelles retrocyclin-2 forms a well-defined capped β-hairpin braced by three-disulfide bonds (Figure 1A). We refer to this topological arrangement as a cyclic cystine ladder. A well-defined structure could not be determined in aqueous solution, suggesting that the SDS micelles stabilize the extended cystine ladder conformation of retrocyclin-2.

S. Del Valle et al. (eds.), *Peptides for Youth: The Proceedings of the 20th American Peptide Symposium*, 577
DOI: 10.1007/978-0-387-73657-0_254, © Springer Science+Business Media, LLC 2009

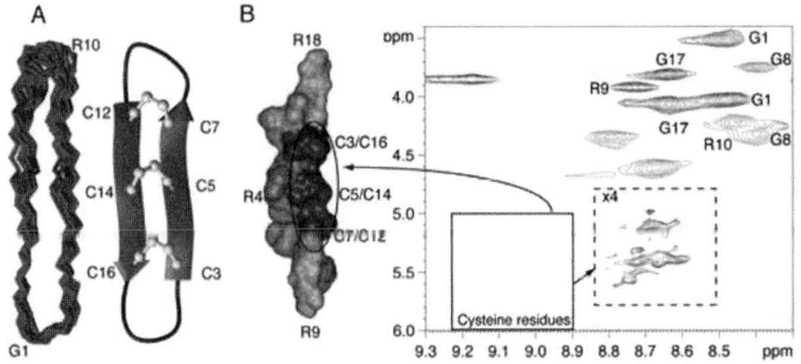

Figure 1. Three-dimensional structure of retrocyclin-2 in 100 mM SDS. (A) A superposition of the 20 lowest energy structures and the lowest energy structure shown with the β-strands as arrows and the disulfide bonds in ball and stick format. the cysteine residues on a surface representation of retrocyclin-2 is highlighted. 546 N. L. Daly et al. *(B) The selective broadening of the cysteine residues highlighted in the TOCSY spectrum recorded at a concentration of 2.6 mM in aqueous solution.*

Analytical ultracentrifugation and NMR diffusion measurements in aqueous buffer indicated that retrocyclin-2 self-associates to form a trimer in a concentration dependent manner. How the monomers interact is an interesting question and the cysteine residues appear to play a crucial role in the self-association process. In addition to the general broadening observed as concentration increased, peaks from the cysteine residues display selective broadening as shown in Figure 1B. A likely explanation is that the cysteine amide protons are at the interface between monomers and are involved in the exchange process between the monomers and the trimer. The most likely association model is where a circular arrangement of the monomers occurs, with the cysteine residues located in the trimer core.

The current study provides information that may be useful in elucidating the mechanism of action of θ-defensins. In particular, the ability of retrocyclins to self-associate may contribute to their high affinity binding for glycoproteins by increasing the valency and enhancing their ability to crosslink cell surface glycoproteins.

Acknowledgments

These studies were supported by grants from the University of Queensland Foundation (NLD), the Australian Research Council (DJC) and the National Institutes of Health (RIL;AI056921). NLD is a NHMRC Industry Fellow. DJC is an ARC Professorial Fellow.

References

1. Tang, Y.-Q., Yuan, J., Ösapay, G., Ösapay, K., Tran, D., Miller, C. J., Ouellette, A. J., and Selsted, M. E. *Science,* **286**, 498-502 (1999).
2. Cole, A. M., Wang, W., Waring, A.J., and Lehrer, R. I.. *Curr. Protein Pept Sci.* **5**, 373-381 (2004).

Isothermal Titration Calorimetry and Inhibition of Platelets Aggregation by [D-Phe/(*Trans*cinnamoyl)-Pro-D-Arg-P1'-CONH2] Peptides Inhibitors of Thrombin

Cristina C. Clement[1], Anna Babinska[2], Elizabeth Kornecki[2] and Manfred Philipp[1]

[1]Department of Chemistry, Lehman College, City University of New York, Bronx, NY, 10468, USA; [2]Department of Anatomy and Cell Biology, SUNY Downstate Medical Center, 450 Clarkson Avenue, Brooklyn, NY, 11203, USA

Introduction

Thrombin is the major product of the plasma coagulation "cascade" of sequential "zymogen-to-protease" steps. As the result of a series of proteolytic cleavages, thrombin converts fibrinogen into fibrin, which deposits at the site of bleeding or thrombosis as the fibrinous portion of a haemostatic plug or thrombotic mass(1). Thrombin also stimulates the platelets through its protein G-coupled receptors PARs (protease activated receptors) (PAR1,4 in humans) being their most potent activator. Thrombin induced-platelets activation plays a critical role in the pathophysiology of thrombosis (1). Activated platelets bind to fibrinogen, causing platelets to aggregate at the site of a cardiovascular injury to form a thrombus that is further stabilized by thrombin-generated fibrin network (1). The discovery of new antithrombotic drugs was primarily focused on finding competitive inhibitors of thrombin, however new approaches target thrombin receptor PAR-1 (such as the PAR-1 structural antagonists) offering additional cardiovascular utility in the treatment of atherosclerosis and restenosis (1).

We previously reported the discovery of new tetrapeptides with the sequence space D-Phe-Pro-D-Arg-P1'-$CONH_2$ as reversible competitive inhibitors of thrombin (2) (were P1' requires small, polar natural or unnatural aminoacids). To further evaluate their potential as antithrombotic drugs and as potential PAR-1 antagonists we synthesized new peptides based on the same sequence space but with D-configuration in P1'. The actual paper presents the structure-activity relationship (SAR) for some L/D tetrapeptides isomers and their evaluation as thrombin inhibitors using biophysical/biochemical approaches (isothermal titration calorimetry and kinetics of thrombin inhibition) and cell-based assays (thrombin mediated platelets aggregation). The synthesis of the D-isomers was performed only for the lead L-peptides (Ki (inhibitory constant) less then 20 μM) (2). Additional tetrapeptides were synthesized with the *trans*cinnamoyl group at P3 position (instead of D-Phe) based on their prediction as high affinity competitive binders of thrombin (2).

Results and Discussion

Synthetic peptide amides were synthesized using Fmoc chemistry on Rink-Amide resins in a 432A Synergy Personal Peptide (ABI) peptide synthesizer or using manual synthetic approaches. The purification of peptides was achieved by RP-HPLC on a C18 column as described earlier (2). Platelets assays were performed by evaluating the ability of synthetic peptides to inhibit platelets aggregation (isolated from healthy human donors) in presence of fibrinogen and thrombin (Figure 1 (B)). A switch from L-Thr into D-Thr in P1' was correlated with a thirteen fold increased inhibitory activity against thrombin and replacement of L-Ala by D-Ala in P1'

S. Del Valle et al. (eds.), *Peptides for Youth: The Proceedings of the 20th American Peptide Symposium*, 579
DOI: 10.1007/978-0-387-73657-0_255, © Springer Science+Business Media, LLC 2009

increased eight fold the affinity of peptide for thrombin (Figure 1(A); P1'-Gly peptide was used as reference). These differences in the binding affinities for P1' L/D isomers were confirmed by ITC. In one typical experiment the titration of peptide L-Ala isomer (P1') released less heat (1.2 kcal/mole of injectant) than the titration of D-Ala (P1') peptide into thrombin (6.5 kcal/mole of injectant) (data not shown). All the peptide D-isomers completely inhibited the aggregation of platelets at concentrations around 200 μM (Figure 1 (B) shows a representative platelets inhibition assay). No inhibition of platelets aggregation was observed with any peptides when ADP or collagen were used to stimulate the platelets, supporting a direct interaction between the peptides and thrombin or PAR-1,4 receptors. The *trans*cinnamoyl-peptides were 2 fold less efficient in the inhibition of platelets aggregation supporting the kinetics data of their two fold less efficiency in inhibition of thrombin *in vitro* (data not shown). Further experiments are under current performance to test this hypothesis. The ability to inhibit the platelets aggregation in presence of thrombin further supports these peptides as new design for anti-thrombotic and anti-atherosclerosis drugs.

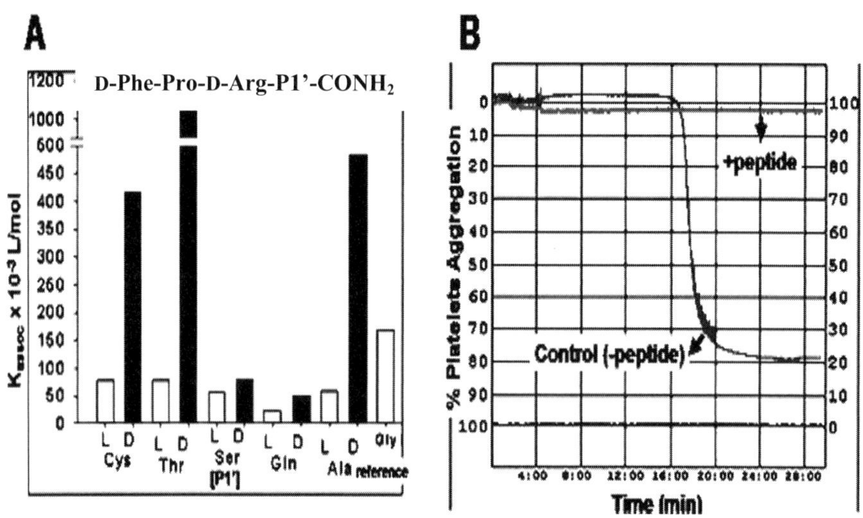

Fig. 1. Structure-activity relationship of L/D-P1'isomers of tetrapeptides D-Phe-Pro-D-Arg-P1'-CONH$_2$ determined by kinetics of thrombin inhibition in vitro (A). Inhibition of thrombin-mediated platelets aggregation by peptide D-Phe-Pro-D-Arg-D-Thr-CONH$_2$ at 220 μM (B).

Acknowledgments

The work was supported by CUNY grants awarded to M. Philipp. We acknowledge the support for conducting ITC experiments in the lab of Professor Richard Magliozzo, Broolyn College, CUNY.

References

1. S. Chackalamannil *et.al. J. Med. Chem.* **48**, 5884-5887 (2005).
2. C.C. Clement and M. Philipp. *Proceeding of American Peptide Society*, Sylvie E. Blondelle (Editor) July 2005.

Trypsin Stability of Lysine-Peptoid Hybrids

Trine S. Ryge[1,2] and Paul R. Hansen[1]

[1]*Department of Natural Sciences, University of Copenhagen, 1871 Copenhagen, Denmark*
[2]*Present address:Zealand Pharma, 2600 Glostrup Denmark*

Introduction

We have previously shown that lysine-peptoid hybrids display low cytotoxicity and potent antibacterial activity against *Staphylococcus aureus* (ATCC 25923), *Escherichia coli* (ATCC 25922) and clinical isolates [1-3]. Trypsin is a serine protease with high specificity for the carboxyl side of arginine and lysine residues. One would expect lysine-peptoid hybrids to be susceptible to enzymatic cleavage by trypsin. However, Miller *et al.* [4] reported that peptoids are stable against several proteolytic enzymes including trypsin.

Here, we investigate the trypsin stability of nine lysine-peptoid hybrids KKYZKYKX-NH_2 (**1**), H-KKYZYKX-NH_2 (**2**), H-YKZKYKXK-NH_2 (**3**), H-YKZKYKXKK-NH_2 (**4**), H-YZYKXKK-NH_2 (**5**), H-YZYKXK-NH_2 (**6**), H-YZKYKXK-NH_2 (**7**), H-KKKKYZYX -NH_2 (**8**) and H-YZYXKKKK- NH_2 (**9**).

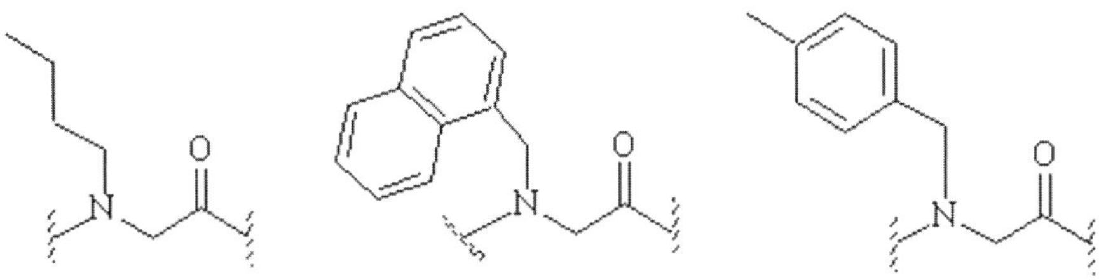

X **Y** **Z**

Figure 1. Structure of building blocks X: [N-(butyl)glycyl]; Y: ([N-(1-naphthalenemethyl) glycyl]; Z: [N-(4-methylbenzyl)glycyl].

Results and Discussion

The lysine-peptoid hybrids were synthesized on a TentaGel S RAM resin by solid-phase synthesis using a combination of Fmoc chemistry and the submonomer approach. Lysine residues were coupled onto resin-bound peptoid residues using HATU (5 eq.) and DIEA (10 eq.) in NMP for 2 hours, followed by a 1 hour recoupling step. The peptoid monomers were coupled by adding 0.6 M bromoacetic acid (10 eq.) in NMP and 3.2 M DIPCDI (12.8 equiv). After 30 minutes, a recoupling was performed. The side chain was introduced by nucleophilic substitution in NMP of the halide with a primary amine (40 equiv.) and agitated for 2h. Cleavage from the resin was performed employing 95:2.5:2.5 TFA/TIS/H_2O. Finally, all lysine-peptoid hybrids were purified by preparative HPLC and characterized by LC-MS.

Stock solutions of the lysine-peptoid hybrids were prepared in 1% DMSO to a concentration of approximately 1 mg/mL. The exact concentration of each stock solution was determined by amino acid analysis [1].

Lysine-peptoid hybrids were incubated at 37°C with TPCK treated trypsin (Sigma, St Louis, Missouri) at a compound:trypsin ratio of 50:1 by weight in

S. Del Valle et al. (eds.), *Peptides for Youth: The Proceedings of the 20th American Peptide Symposium*,
DOI: 10.1007/978-0-387-73657-0_256, © Springer Science+Business Media, LLC 2009

100mM Tris-HCl buffer, pH 8.5. The mixtures were allowed to incubate in 18h to ensure complete digestion.

The lysine-peptoid hybrid digestion fragments were characterized by analytical RP-HPLC and LC-MS.

We found that compounds **1-6** were completely stable to trypsin. Compound **7** was partly cleaved (12%) by trypsin to H-YZK-OH + H-YKXK-OH. As expected, H-KKKKKYZYX-NH_2 and H-YZYXKKKKK-NH_2 were completely digested to H-KKYZYX-NH_2 (97 %) and H-YZYXK-OH (100 %).

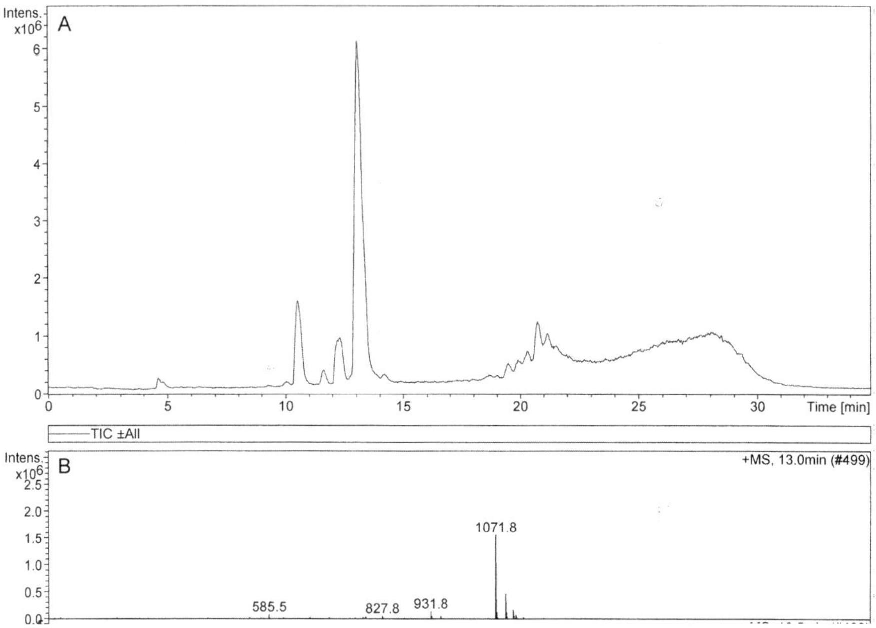

Fig. 2. LC-MS of Trypsin digested YZKYKXK-NH_2. Panel A: UV-trace at 220 nm. The peak at 10.5 min is H-YKXK-OH (calculated mass 606.7 $[M+Na]^+$). The peak at 12.2 min is H-YZK-OH (calculated mass 526.4 $[M+Na]^+$).Panel B: Ion-trace of peak at 13.0 min undigested YZKYKXK-NH_2 (calculated mass 1070.4).

In conclusion, our study indicates that lysine-peptoid amide bonds are resistant to trypsin. This finding combined with the antibacterial properties of lysine-peptoid hybrids make them interesting lead structures for developing future therapeutics.

Acknowledgments

We thank Ms. Jette Petersen for excellent technical assistance. This work was supported by the Danish Research Council grant number 9900234.

References

1. Ryge, T.S. and Hansen, P.R. (2005). *J. Peptide Sci.*, **11**, 727-734.
2. Ryge, T.S. and Hansen, P.R. (2006). *Bioorg. Med. Chem.* **14**, 4444-4451.
3. Ryge, T.S., Frimodt-Møller, N. and Hansen, P.R (2007). Submitted.
4. Miller, S.M. et al., *Bioorg. Med. Chem. Lett,* 1994. **4**, 2657-62.

WORKSHOP: PRODUCTION AND FORMULATION OF PEPTIDE DRUGS

A Comparative Examination of two Fmoc Removal Reagents for Process Improvement to Produce Peptide Drugs

K. Srivastava and M. Davis

Peptide Research and Development, Pharmaceuticals Division, Tyco Healthcare / Mallinckrodt Inc., St. Louis, MO 63147, USA

Abstract

The importance of peptides as therapeutics has been recognized since they were found responsible for a wide variety of biological functions. The recent approval of peptide drugs such as Byetta® (Amylin Pharmaceuticals, Inc.), Fuzeon® (Hoffman-LaRoche Inc.), Integrelin™ (CDR Therapeutics, Inc.), Natrecor® (SCIOS Inc.), Symlin® (Amylin), Teriparatide, and Ziconotide, etc., which demonstrated applications for treatment of such problems as bone metabolism disorders, cardiovascular diseases, diabetes, viral infections and severe chronic pain control, has further endorsed the growing interest in peptides as a potential drug. This growing trend for peptide drugs has drawn our attention for their production in a cost-effective manner. To do so, the improvement in the quality of crude peptides during synthesis, the most critical parameter in the process, is important to prevent yield losses during the more expensive purification step. To accomplish it, we decided to examine the efficacy of the commonly used nucleophilic base piperidine and non-neucleophilic base DBU (1,8-Diazabicyclo[5.4.0]undec-7-ene) for the complete removal of Fmoc group during the synthesis of peptides. According to our investigation, application of piperidine was found more effective than DBU in solid phase synthesis. Details of the investigation will be discussed.

Introduction

Peptides, ubiquitous in living systems, are signal, transport and digestive molecules. As signal molecules, they control biological functions such as cell division, mating, chemotaxis, pain perception, growth and immunity. As transport molecules, they facilitate the passage of ions through cell membranes. While as digestive molecules, they play a critical role in nutrition of both cells and intact organisms. They also act as protective agents such as antibiotics which are excellent antibacterial and antiviral agents. At present, peptides are important commercial entities and have been used in different therapeutic areas like allergy, anti-infection, diagnostics, diabetes, oncology, obesity, arthritis, cardiovascular diseases, etc… As a consequence, the market for peptide drugs surpasses a billion dollars (1).

They have been produced either by solution phase method (small molecues with large volumes) or by solid phase method (large molecules with small volumes) using Boc- and Fmoc-based chemistry (2-5). In the solid-phase approach, the synthesis process is very fast because of its automation capability, but the accumulation of impurities on the resin throughout the assembly of the peptide chain is also a serious problem. Selection of synthesis strategy, coupling method, agitation speed, resin particle size, and incomplete deprotection / coupling reactions all can have impact on the performance of the peptide synthesis and the generation of the various impurities. These factors may not be noticeable for small scale (research scale) synthesis, however, when taken to pilot / production scale they then become a cause

S. Del Valle et al. (eds.), *Peptides for Youth: The Proceedings of the 20th American Peptide Symposium*, 585
DOI: 10.1007/978-0-387-73657-0_257, © Springer Science+Business Media, LLC 2009

for concern. Therefore, it becomes necessary to find a means to solve these problems. Recently, there has been growing interest in the use of Fmoc-chemistry (5) which utilizes a mild base during the synthesis and a mild acid during the last step of the synthesis to avoid by-product formation. The economical cost of Fmoc-amino acids as a raw material, compared to ten to fifteen years ago, is also a factor for its wide application. Incomplete deprotection and coupling reactions are considered the major factors in the generation of impurities. In Fmoc-based peptide synthesis, the most commonly used base for the Fmoc group removal is either piperidine (6, 7) or DBU (8-11).

To improve the quality of peptide during chain assembly and to prevent yield losses during purification, we decided to examine the effectiveness of both bases for the deprotection of Fmoc group during the synthesis of peptides. Our investigation demonstrated that piperidine base was comparatively more effective than DBU in solid phase synthesis.

Experimental Procedures

1. Synthesis of an analogue of LH-RH

Synthesis of an analogue of LH-RH was carried out manually by the solid –phase method, using the Fmoc / t.Butyl synthesis strategy and starting from 10.0g of Rink Amide-MBHA resin (sub=0.47 mmol / g or 4.7 mm / total). The Fmoc protection was removed with 20% piperidine in Dimethylformamide. The synthetic protocol was as follows: 1. DMF wash (3 x 3 min.), 2. Fmoc-deprotection (1 x 5 min., 1 x 20 min.), 3. DMF wash (6-8 x 3 min.), 4. coupling of Fmoc-amino acid (3 hours to overnight), 5. DMF wash (3 x 3 min.), 6. DCM wash (1 x 3 min.), 7. Ninhydrin assay. A 3-fold excess of the respective Fmoc-amino acid was coupled by DIC-HOBT coupling method (12, 13) in DMF + DCM (3:1). The completeness of each coupling reaction during synthesis was monitored by Kaiser test (14) and a second coupling was performed when the test was found positive. Following groups were chosen for protecting side-chain functionalities: t.butyloxycarbonyl (Boc) for Trp, t.butylether (t.Bu) for Ser and Tyr, trityl (Trt) for His, and 2, 2, 4, 6, 7-pentamethyl-dihydrobenzofurane-6-sulfonyl (Pbf) for Arg. The yield of the peptidyl resin was 91% (17.8g) compared to a theory yield of 19.5g. Cleavage of the peptide from the resin (1.0g scale) with side-chain deprotection was performed by treatment with TFA + H_2O + TIS + EDT (92.5:2.5:2.5:2.5) for two to three hours. The cleavage solution after filtration of the cleaved resin was treated with ether to obtain the crude peptide in 97% yield (0.31g) and with a purity of 79.5% or more as determined by analytical HPLC.

A second synthesis of the analogue of LH-RH was performed again on a 5.0g scale of Rink Amide-MBHA resin (0.50 mmol / g capacity) as described above. The removal of the Fmoc group was carried out this time with 2% DBU in Dimethylformamide containing 3% piperidine. The yield of the peptidyl resin was 84.0% (8.4g) compared to a theory yield of 10.0 g. The obtained peptide –bound resin (2.5g) was treated with TFA + H_2O + TIS + EDT for 2 to 3 hours at room temperature. The cleaved resin was removed by filtration and the filtrate was treated with 10-20 volumes of ether, filtered, washed successively with ether and air-dried to yield the crude peptide in 96.4% (0.80g) yield. According to analytical HPLC, the purity of the product was greater than 77.2%.

2. Synthesis of an analogue of α-MSH (4-10):

Two 5.0g scale syntheses of α-Melanocyte Stimulating Hormone were carried out on a Wang resin (sub. = 0.61 mmol / g) using (a) 20% piperidine in DMF and (b) 2% DBU in DMF containing 3% piperidine as deprotection reagents for comparative study, employing 3 mole equivalents of reagents. The yields of the peptidyl resins were 64.6% (6.41g) and 66.8% (6.63g), respectively, compared to a theory yield of 9.9g.

Two 1.0g peptide resin samples (one from each synthesis) were cleaved with TFA + H_2O + TIS + EDT for 3 hours to produce crude peptide in 96.8% (0.31g) yield with a HPLC purity of 56.76% or more and 87.5% (0.28g) yield with a HPLC purity of only 37.6%, respectively.

Results and Discussion

The two most commonly used bases for the removal of Fmoc group in Fmoc-chemistry are piperidine and DBU (1,8-Diazabicyclo[5.4.0]undec-7-ene). A 20% concentration of piperidine base has generally been used from the beginning of the Fmoc-SPPS and later on DBU was introduced in only 2 to 5% concentration. Time to time other bases such as diethylamine, dibutylamine, 4-aminomethylpiperidine, tris(2-aminoethyl) amine (5,16) have also been applied in Fmoc-chemistry. Since the amount of DBU used was very small in comparison to piperidine for the removal of Fmoc group, we therefore, decided to examine both bases from an economical standpoint for scale-up of the synthesis process of peptides. Consequently, two syntheses were carried out on two different types of solid support. One synthesis was carried out on a Wang resin where the c-terminus amino acid was attached to the carrier with an ester bond, while the other synthesis was performed on a Rink Amide-MBHA resin where the c-terminus amino acid was linked to support by an amide bond. The ester bond on Wang resin is generally succeptable to base during the removal of Fmoc group of dipeptide-resin resulting in premature loss of peptide due to diketopiperazine formation, especially when the c-terminus amino acid is either proline or glycine. On the other hand, the amide bond on Rink Amide resin is unaffected under the same condition. According to our investigation, we found that piperidine base was more effective than DBU base and yielded a cleaner product.

Conclusion

The investigation has demonstrated that the synthesis of peptides carried out on two different types of solid support using 20% piperidine for the removal of N-α-Fmoc group was superior in comparison to 2% DBU deprotection reagent.

The HPLC determined purity of the synthesized peptide employing 20% piperidine in the process indicated its advantage for the enhancement of the quality of peptide drugs to be produced.

References

1. M. Ayoub, et al, Chimica Oggi, Chemistry Today, *24 (4)*, 46-48, 2006.
2. R. B. Merrifield, Science, *232*, 341-347, 1986.
3. G. Barany, et al, IJPPR, *30*, 705-739, 1987.
4. S. Sakakibara, Biopolymers (Peptide Science), *37*, 17-28, 1995.
5. G.B. Fields, et al, IJPPR, *35*, 161-214, 1990.
6. V. J. Hruby, et al, J. Peptide Res., *56 (2)*, 70-79, 2000.

7. M. Fridkin, et al, J. Peptide Sci., *11 (1)*, 45-52, 2005.
8. F. Albericio, et al, J. Peptide Res., *56 (2)*, 63-69, 2000.
9. J. D. Wade, et al, Peptide Res., *4*, 194-199, 1991.
10. S. A. Kates, et al, Peptide Res., *9*, 106-113, 1996.
11. I. Szabo, et al, Biopolymers (Peptide Science), *88 (1)*, 20-28, 2007.
12. G. B. Bloomberg, et al, Tetrahedron Lett., *34*, 4709, 1993.
13. D. Hudson, JOC, *53*, 617-624, 1988.
14. E. Kaiser, et al, Anal. Biochem., *34*, 595-598, 1970.
15. L. A. Carpino, et al, JOC, *55*, 1673-1675, 1990.

LH-RH Analogue

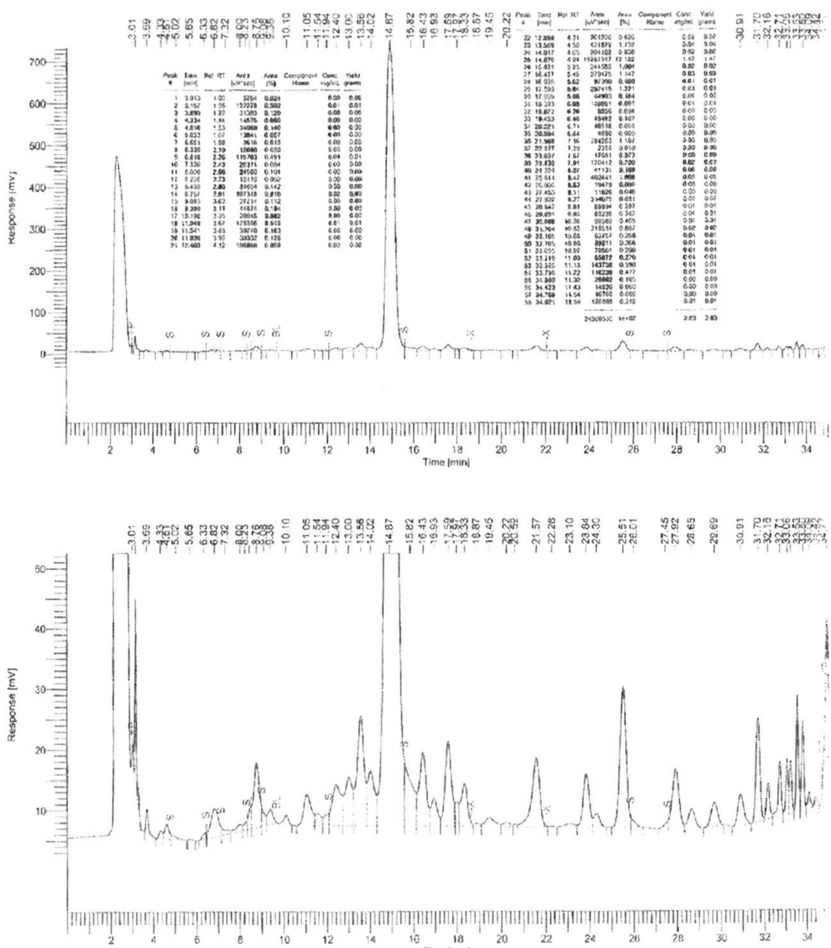

LH-RH Analogue

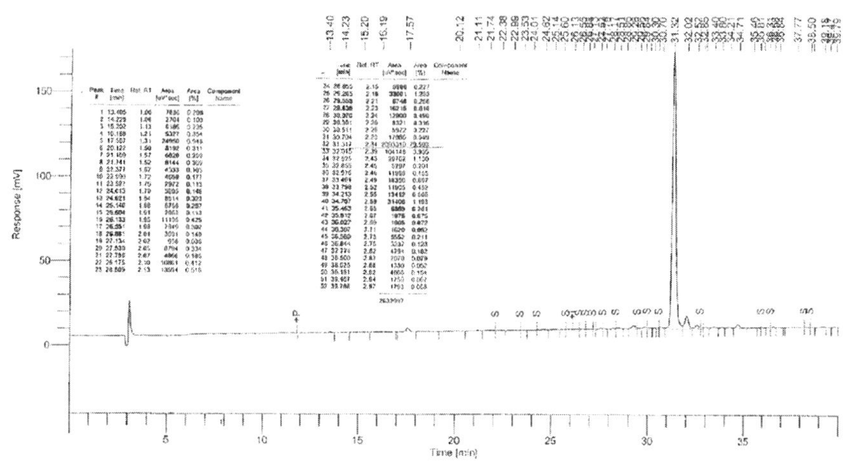

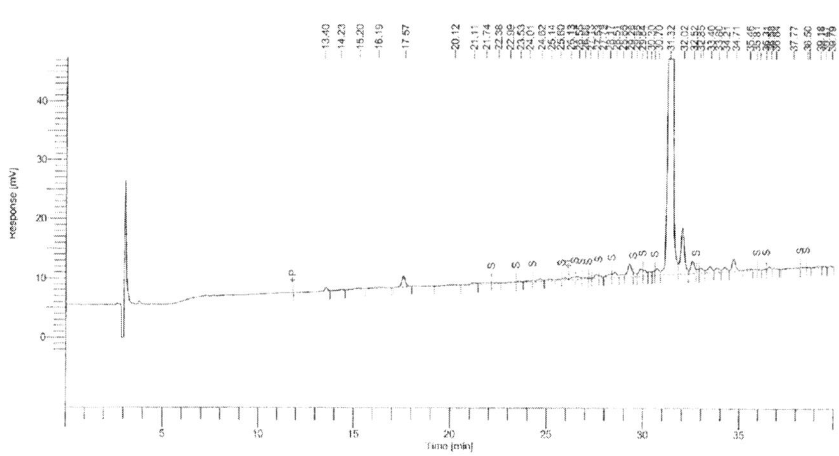

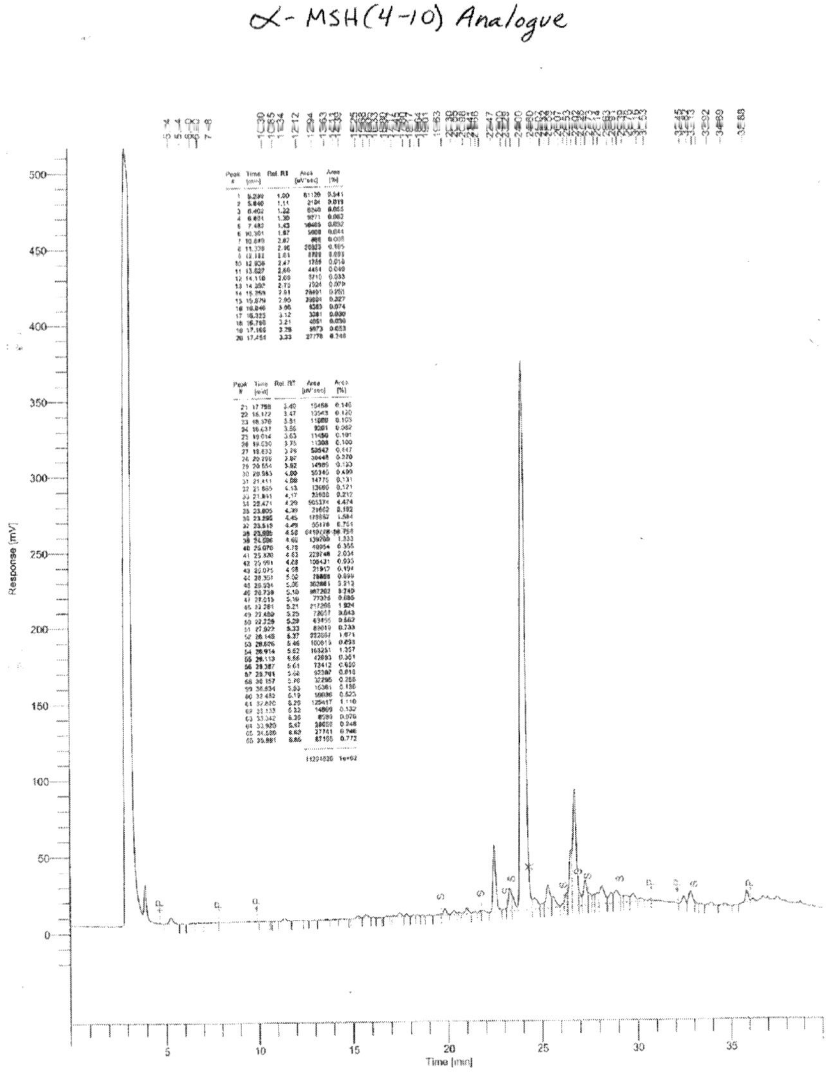
α- MSH(4-10) Analogue
Response [mV]
Time [min]

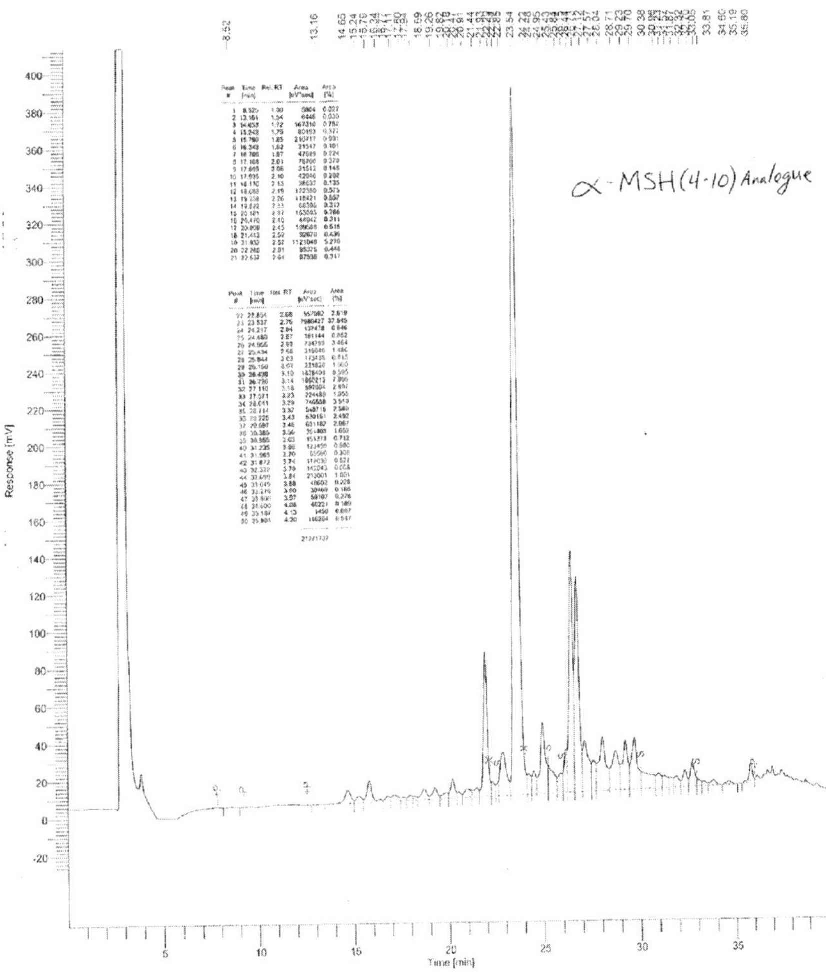
α-MSH(4-10) Analogue
Response [mV]
Time [min]

Bicyclic Organo-Peptides as Models for Carbohydrate Binding Proteins

Claudia James-Johnson and Predrag Cudic

Department of Chemistry and Biochemistry, Florida Atlantic University, 777 Glades Road, Boca Raton, FL 33431, USA

Introduction

Carbohydrate recognition is one of the most sophisticated recognition processes in biological systems.[1] This process mediates many important aspects of cell-cell recognition, such as inflammation, cell differentiation, tumor cell colonization and metastasis. However, due to its complexity and weak binding affinities, carbohydrate recognition by natural systems is still poorly understood. On the other hand, studies on synthetic carbohydrate receptors could make significant contributions to a better understanding of this process and lead to the development of new therapeutics. Given the importance of the carbohydrate recognition, it is not surprising that design and synthesis of artificial receptors for these important biological substrates attracted a great deal of attention in biomimetic chemistry. Current efforts are mainly focused on design of the receptors for complexation of monosacharides or short oligosaccharides.[2] This crude simplification has been based on the structural studies of the lectins, revealing that carbohydrate binding sites are typically shallow binding pockets where only binding of the terminal sugar moieties of complex polysaccharides can occur. Binding of an individual lectin to monosaccharide substrate is extremely weak; the dissociation constant (K_d) for these complexes are typically in the 0.1-10 mM. This is largely due to the solvent-exposed nature of the lectin binding sites, which make few direct contacts with monosaccharide substrate.[3]

Carbohydrate binding protein mimicking molecules using the peptide based systems represent particularly attractive approach for the development of artificial carbohydrate receptors. As a model system we chose a cyclic cationic decapeptide antibiotic polymyxin B, which is known to bind the lipid A moiety of LPS with the high affinity.[4] Using standard Fmoc SPPS we prepared cyclic polymyxin analogue **1** bridged with a bipyridine group in order to introduce structural constraints pertaining to the stabilization of peptide's binding conformation, and to incorporate additional hydrogen-bonding acceptor site. Therefore, this bicyclic organo-peptide receptor molecule possesses the necessary three-dimensional structure, limited flexibility and lipophilic binding pocket where binding can occur through a combination of hydrophobic interaction and possibly hydrogen bonds between receptor molecules and carbohydrate substrates.

Results and Discussion

To demonstrate suitability of bicyclic peptide structures for design of artificial carbohydrate receptors we successfully prepared bicyclic peptide **1** (Figure 1) on Rink-MBHA resin (loading: 0.72 mmol/g, Novabiochem) using standard Fmoc solid-phase chemistry. Our synthetic approach was comprised of the resin attachment of Asp *via* side chain, successful use of combination of four quasi-orthogonal removable protecting groups, stepwise Fmoc solid phase synthesis of a

S. Del Valle et al. (eds.), *Peptides for Youth: The Proceedings of the 20th American Peptide Symposium*, 593
DOI: 10.1007/978-0-387-73657-0_258, © Springer Science+Business Media, LLC 2009

linear precursor, on-resin head-to-tail cyclization, and in the final step on-resin attachment of the bipyridine dicarboxylate moiety. Formation of the final product **1** was confirmed by MALDI-TOF MS (m/z=126.15 calculated; m/z=126.45 observed) and RP-HPLC (R_t=23.2 min., linear gradient 100% A→100% B in 30 min., A=0.1%TFA in H_2O, B=0.1% TFA in MeCN:H_2O=9:1).

The dissociation constants (K_d) for the complexes between receptors **1** with variety of monosaccharides in water were determined by UV/Vis and fluorimetric titration in cacodylate buffer (50 mM, pH=6.5), and the observed values are in the range from ~ 3 to 3.5 mM, similar to the K_d determined for lectin/monosaccharide complexes. The stoichiometry for complexes of **1** with monosaccharides was determined by the mole ratio method (Job plot) to be 1:1. Determined dissociation constants indicate that the selectivity of the receptor molecule with the respect to a specific monosaccharide is moderate. However, with these results we showed that it is possible to construct peptide-based artificial receptor molecules capable of binding monosaccharide substrates in water. Our results also demonstrate that macrobicyclic peptide **1** represent a promising basis for the design of new and more efficient carbohydrate receptors that may have application in analytical or medicinal field. Further optimization of receptor **1** using combinatorial chemistry approach is currently underway in order to improve receptors binding properties and selectivity.

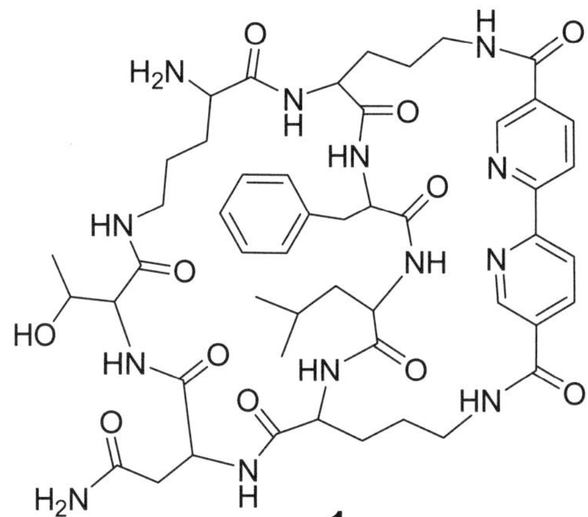

Fig. 1. Peptide based artificial receptor for monosaccharide binding in water.

References

1. (a) Varki, A. *Glycobiology*, **3**, 97-130 (**1993**); (b) Fukuda, M.; Hindsgaul, O. In *Molecular and Cellular Glycobiology*; Hames, B. D., Glover D. M., Eds. University Press: Oxford, 2000.
2. (a) Striegler, S. *Curr. Org. Chem.* 7, 81-102, (**2003**); (b) Davis, A. P., Wareham, R. S. *Angew. Chem., Int. Ed.* **38**, 2978-2996 (**1999**)
3. Sharon, N., Lis, H. The Structural Basis for Carbohydrate Recognition by Lectins, The molecular *Immunology of Complex Carbohydrates-2*, A. M. Wu Ed., Kluwer Academic/ Plenem Publishers, New York, NY, **2001,** 1-16.
4. Vaara, M. *FEMS Microbiol. Lett.* **18**, 117-121 (1983)

Impurity Rejection in the Crystallization of ABT-510 as a Method to Establish Starting Material Specifications

John C. Tolle, Calvin L. Becker, Jean C. Califano, Jane L. Chang, Kevin Gernhardt, James J. Napier, Steven J. Wittenberger and Judy Yuan

Abbott Global Pharmaceutical Process R&D, Dept. R450, Bldg R13
1401 Sheridan Road, North Chicago, IL, 60064-6290, USA

Introduction

ABT-510 [1] is a thrombospondin-1 mimetic currently in Phase 2 clinical trials as an angiogenesis inhibitor for treatment of cancer. An efficient synthetic process was developed by the Process R&D group at Abbott and has been scaled to produce several 10 kg batches of the drug substance. Some key features of the process include:

- 3 Crystalline peptide segments as starting materials
- Crystallization of an intermediate hexapeptide
- Purification and isolation of ABT-510 as a crystalline HCl salt
- Simple ion exchange to the acetate salt
- Crystallization [2] of ABT-510 acetate salt
- No chromatography

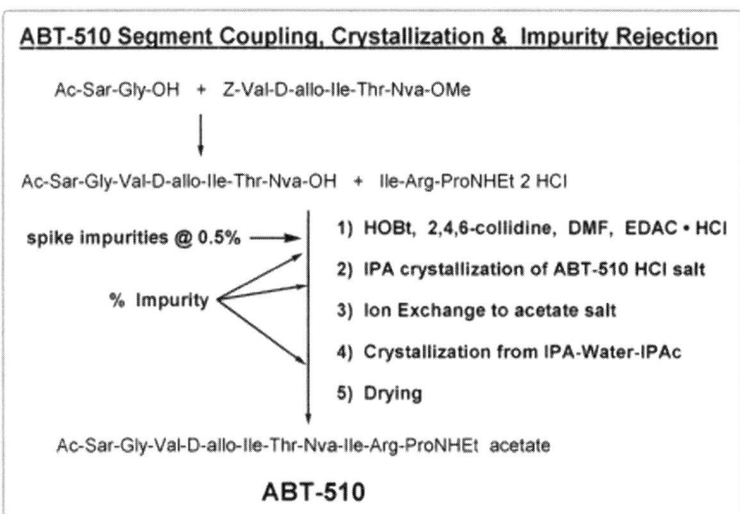

Study Objectives

The level of impurities in batches of ABT-510 made by this process is typically below 0.25% total. As a way to consistently achieve this very high purity in future batches, this study was undertaken. The primary objectives were to 1). Understand the fate of possible impurities that could be present in the three peptide segments as they are carried into the downstream process, 2). Develop analytical specifications for these segments, and 3). Establishing the three segments as regulatory starting materials.

S. Del Valle et al. (eds.), *Peptides for Youth: The Proceedings of the 20th American Peptide Symposium*, 595
DOI: 10.1007/978-0-387-73657-0_259, © Springer Science+Business Media, LLC 2009

Results and Discussion

Over 30 ABT-510 analogs were synthesized, including all 13 point-isomers, all 9 decapeptides due to double-incorporation of an amino acid, and several others that were impurities that had been detected at some point during process development. Up to 5 of these were spiked at 0.5 mole% each into a reaction mixture in which ABT-510 hydrochloride is formed. The level of each impurity relative to ABT-510 was determined by HPLC after spiking and after isolation of the crystalline ABT-510 hydrochloride salt and, finally, after isolation of the crystalline ABT-510 acetate salt. This sequence of steps was repeated until the rejection properties of all of the impurities had been determined. The purity specifications for the three peptide segments, from which these ABT-510 analogs could arise when carried into the ABT-510 process, were established as shown in *Table 1.*

Table 1. Peptide Segment Specifications Based on Impurity Rejection in ABT-510 Crystallization

Impurity Rejection	Impurity Spec'n in Segment	Impurity ID*
80-100%	≤ 0.50%	+D-allo-Ile, L-Ile[4], L-allo-Ile[4], D-Thr, D-allo-Thr, L-allo-Thr, D-Nva, D-allo-Ile[7], D-Ile[7]
65-80%	≤ 0.30%	D-Arg, Gly[3]
50-65%	≤ 0.20%	D-Val, D-Pro
30-50%	≤ 0.15%	+Sar, +Gly, +Val, +Thr, +Arg
0-30%	≤ 0.10%	D-Ile[4], D-Nva[4], Nle[6], Val[7], +Nva, L-allo-Ile[7], +Ile, +Pro

* + = decapeptides derived from double-incorporation of the designated amino acid

Summary

Understanding impurity rejection in a drug substance crystallization process is valuable for establishing purity specifications for the starting materials used in the process. Impurity rejection has been determined for all known ABT-510 impurities and for many of the reasonable & conceivable impurities. Based on this study, a very high purity specification (e.g., > 99.7%) can be set for ABT-510 with a high level of confidence.

References

1. Henkin, J., Haviv, F., Bradley, M., Kalvin, D. and Schneider, A., *JP 2002516342 WO 1999061476 (1999)*
2. Sheikh, A., Pal, A., Viswanath, S. and Tolle, J., *J. Pharm. Sci.,* published online 7 August 2007

Spatially Defined Peptide-Polymer Biotherapeutic Synthesis on Novel Cross-linked Beads with 'Spatially Tunable' and 'Isolated' Functional Sites

M.K. Jacob and K.S. Kumar*

Division of Molecular Medicine, Rajiv Gandhi Centre for Biotechnology, Thycaud P.O., Thiruvananthapuram, Kerala, India. 695 014, kskumar_rgcb@yahoo.com

Introduction

Functionalized crosslinked polymers prepared by either pre or post functionalization strategies do not have any control over the point of functional group incorporation within the polymer network. Synthesis of polypeptide using these polymers can lead to the formation of number of deletion and truncated peptide sequences due to unfavorable site-site interaction between the growing peptide chains and also due to macromolecular network of the polymer. The subsequent conjugation with high molecular weight linear polymers can results in difficulties during purification of the target polymer-bioconjugate. This paper discusses about the synthesis and application of a novel crosslinked polymer in which unfavorable site-site interactions are avoided by the spatial tuning of functional sites to a predefined distance within cross-linked polymers. In addition to 'site isolation' the polymer has been designed in such a way that all functional sites have a homogeneous and amphipathic microenvironment, higher solvation and accessibility, which is not altogether present in any of the commercial polymers[1]. This polymer is used for the synthesis of PEGylated antimicrobial peptides(AMPs) and our studies showed that the property of the PEGylated AMPs varies with the length of PEG and there was a significant reduction of hemolysis in all cases. That is therapeutic index of AMPs can be enhanced by PEGylation.

Results and Discussion

Spatially tunable and functional site isolated polymer synthesis was achieved by introducing a tetra functional PEG molecule, 3,3' PEG Bis(1-(4-vinylphenoxy) propan-2-ol) (bis(VPP)-PEG) (5) to polystrene. The crosslinker was synthesised by the ring opening of PEG bis epoxides (1) with hydroxy styrene (2) as shown in **Scheme 1** (25% yield). The products were characterized using FT-IR, ^{1}H and ^{13}C NMR. The tetra functional PEG (3) at low cross linking percentage (1-6%) was introduced to polystyrene network by free radical aqueous suspension polymerization technique **(Scheme 2).** This polymer having isolated hydroxyl functional groups separated with pre defined distance using PEG backbone is obtained in the bead form. The distance between the hydroxyl functional group can be spatially tuned in a defined manner by altering the length of PEG bisepoxides (1) from 200 Da to 1000 Da. The final loading capacity of the crosslinked polymer beads could be easily controlled from 0.1-0.9 mmole/g as per the synthetic requirement.

The new polymer has been characterized by various spectroscopic techniques and surface morphology was studied by SEM photograph which showed the bead diameter is of 90-130 μm. **(Fig 1).** The incorporation of bis (VPP) PEG to polystyrene network improved the swelling property of the polymer compared to other polystyrene based polymer supports. The presence of phenyl and alkyl ether linkages in the polymer makes the network chemically stable which was studied using FT-IR.

S. Del Valle et al. (eds.), *Peptides for Youth: The Proceedings of the 20th American Peptide Symposium*, 597
DOI: 10.1007/978-0-387-73657-0_260, © Springer Science+Business Media, LLC 2009

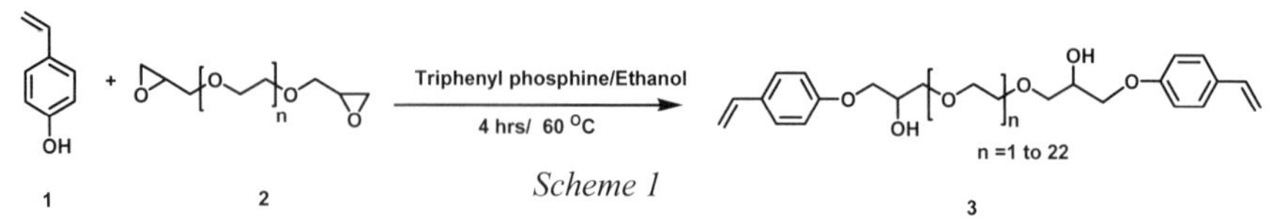

Scheme 1

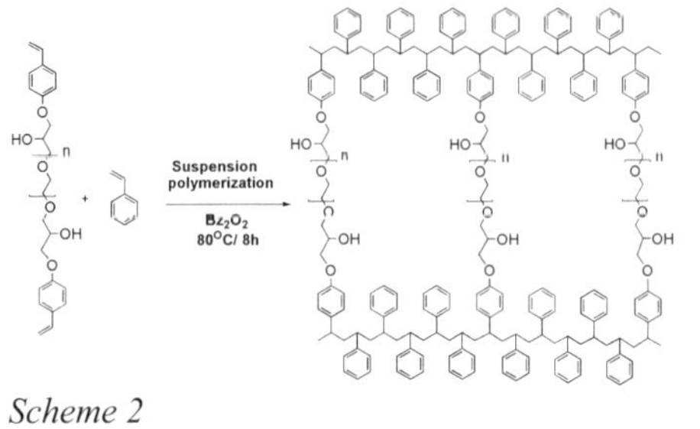

Scheme 2

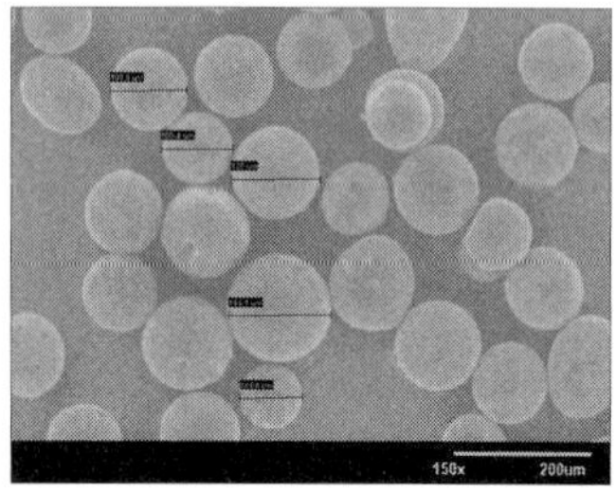

Fig 1. SEM profile of the new polymer

The utility of the new polymer was established by the synthesis of various structural classes of antimicrobial peptide (AMPs) biotherapeutics like PEGylated Tachyplesin, Indolicidin, and Bactenecin using Fmoc SPPS [2]. Tachyplesin-I was synthesized using the new polymer and commercial resins like Tentagel™ and Merrifield ™ resin under the same reaction conditions. The N-terminus of the peptidyl resins was PEGylated using carboxyl mPEG ranging from 500-5000 Da. Quantitative mPEG conjugation could not be achieved in commercial resins [3]. $mPEG_{5000}$-Tachyplesin (PEG-Tach) conjugate was obtained in 80% yield from the new polymer whereas its synthesis using commercial resin resulted in only 30-50% yield as revealed from HPLC analysis **(Fig 4)**. Presence of high quantity of deletion and truncation PEGylated peptides in the cleavage product obtained from commercial resins complicated the purification of the target PEGylated peptide. High purity of PEG-Tach synthesised using the new polymer was evident from the HPLC and MALDI TOF MS **(Fig 3&5)** results. This high quality product established our concept of optimal isolation of functional sites is a prerequisite to achieve maximum synthetic efficiency. Similar results were obtained when PEGylated Indolicidin (PEG-Indol) and Bactenicin (PEG-Bact) were synthesised using the new polymer. The ACP fragment (65-74) which showed aggregation tendency during its synthesis were also succefully synthesised on the new support in very high purity. **(Fig 2)**.

Native peptides and their polymer -bioconjugates were subjected to antibacterial and RBC haemolysis assays **(Table 1).** It showed that antibacterial property varies with the length of PEG but there was significant reduction of hemolysis on PEGylation. That is therapeutic index of AMPs can be enhanced by PEGylation.

The high yield and purity of the PEGylated peptides and the crude ACP fragment stems from the fact that peptides are grown from the isolated functional sites of the tetra functional PEG, which provides an amphipathic and homogeneous reaction environment within the new polymer. Thus it's beyond doubt that the unique feature of 'spatially defined' and 'site isolated' synthesis within the new polymer makes it a novel contribution towards all solid phase synthesis applications.

Table 1 Antibacterial(μM) and haemolysis assay of AMPs and PEG-AMP conjugates

	Tachypl -esin-I	PEG_{1100} -Tach	Indolici -din	PEG_{1100} -Indol	Bactene -cin	PEG_{1100}- Bact
E.coli	> 2	> 2	b/w.50-100	b/w.100-120	>100	>100
P. aeruginosa	b/w 12.5-25	b/w 12.5-25	>120	>120	>100	>100
RBC lysis at MIC	4.4 %	1.2 %	5.7 %	0.89 %	1.93%	1.04%

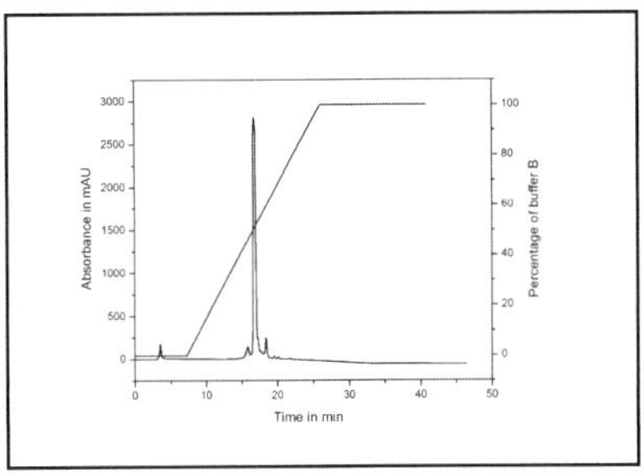

Fig 2 HPLC profile of crude ACP (65-74) from the new polymer

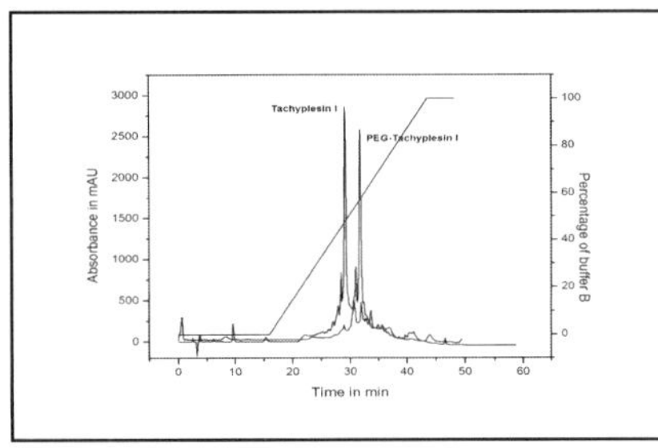

Fig 3 HPLC profile of crude Tachyplesin and PEG1100-Tach from the new polymer

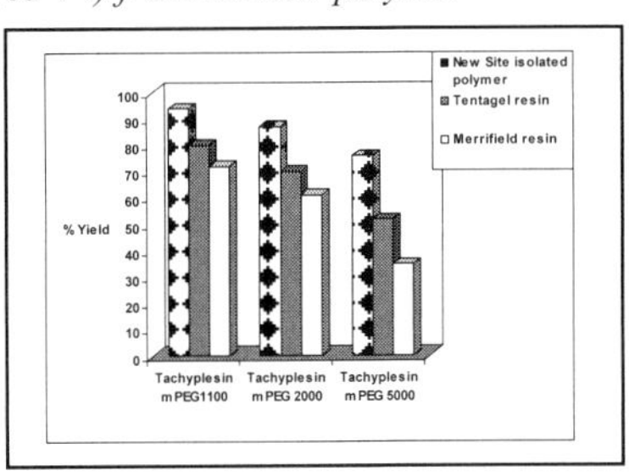

Fig 4 Comparative synthetic yield of PEG-Tach

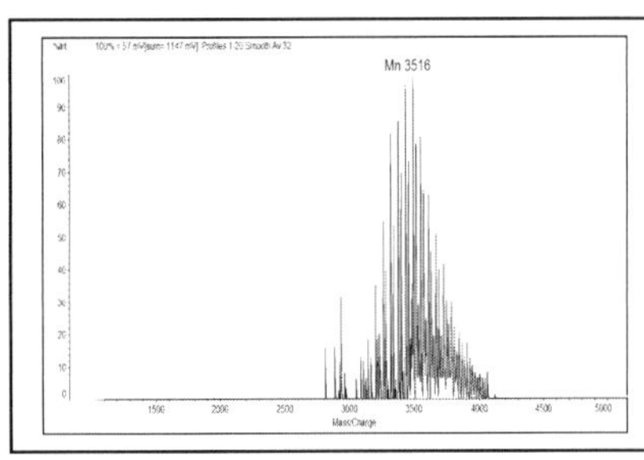

Fig 5 MALDI TOF MS of PEG1100- Tach

Acknowledgments

The authors thank CSIR, New Delhi for the project funding to KSK & research fellowship to MKJ.

References

1. Delgado, M., Janda, K.D. *Current Organic Chemistry*. **6**, 1031-1043 (2002).
2. Fields, G. B. *Methods Enzymol*. **289**, 3-13 (1999).
3. Yi-An Lu. Arthur, M. F. *Reactive Polymers*. **22**, 221-229 (1994).

WORKSHOP: DELIVERY: PEPTIDE DRUG TO PHYSIOLOGICAL TARGET

Supramolecular Aggregates derivatized by CCK8 Peptide as Selective Nanocarriers for Drug Delivery

Antonella Accardo, Diego Tesauro, Giancarlo Morelli *, Luigi Del Pozzo, Carlo Pedone, Anna Lucia Tornesello and Ettore Benedetti

[1]*Department of Biological Sciences & Interuniversitary Center for Research on Bioactive Peptides (CIRPeB) University of Naples Federico II and IBB-CNR Napoli Italy*

Introduction

In last years the supramolecular aggregates became very versatile tools in biology, biochemistry and medicine for a large number of applications, first of all as drugs carriers [1,2]. Unfortunately, many drugs often have a very narrow therapeutic window, meaning that the therapeutic concentration is not much lower than the toxic one. In several cases the toxicity can be reduced or the efficacy enhanced by the use of an appropriate drug carrier which changes the temporal and spatial distribution of the drug, i.e. its pharmacokinetic and biodistribution. A challenging objective would be to deliver to a tissue of interest large quantities of therapeutic agents by using peptide containing supramolecular nanovectors, such as micelles or vesicles [3,4]. The aim of this project is to formulate mixed supramolecular aggregates capable to vehicle drugs on receptors overexpressed in human tumors. The mixed aggregates are formed by two amphipilic monomers. Both molecules have the same two lipophilic C_{18} chains. Therefore, the hydrophilic head is shaped in one case by a peptidic fragment to recognize cellular receptors and in the other case by chelating agent able to coordinate metal ions in order to visualize the tools by nuclear medicine techniques or magnetic resonance imaging (MRI). The selected bioactive peptide is CCK_8, the C-terminal sequence of the cholecystokinin hormone. The peptide provides binding sequence for the cholecystokinin receptor subtypes 1 and 2 (CCK1-R and CCK2-R) overexpressed in many solid tumors [5]. The chelating agent moiety is DTPAGlu (see Figure 1) able to coordinate ^{111}In as radioactive probe and to drive the aggregate formation as anionic surfactant. Here are reported the synthesis, the formulation of aggregates and the biological assays. Moreover, the drug loading capacity and the efficiency using the cytotoxic drug doxorubicin have been studied.

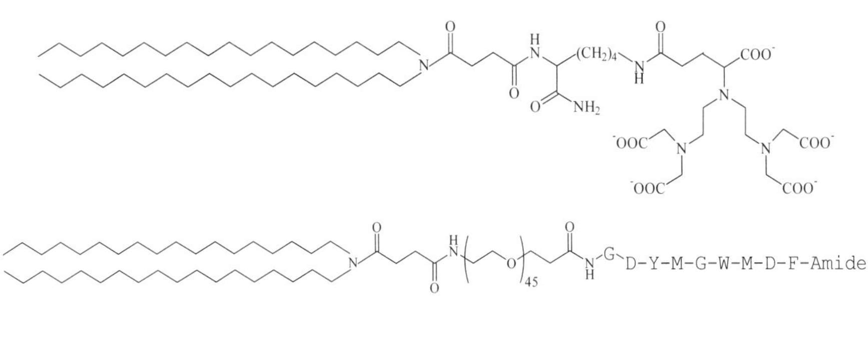

Fig. 1. Schematic representation of the two monomers employed to formulate aggregates.

S. Del Valle et al. (eds.), *Peptides for Youth: The Proceedings of the 20th American Peptide Symposium*, 603
DOI: 10.1007/978-0-387-73657-0_261, © Springer Science+Business Media, LLC 2009

Results and Discussion

Both monomers were synthesized on the solid support Rink-Amide Resin using standard procedures of solid phase peptide synthesis by Fmoc strategy. After HPLC purification, the monomers were dissolved in a small amount of methanol /chloroform (50/50) mixture, and subsequently the solvent was evaporated under a stream of nitrogen. The film was hydrated by phosphate buffer solution (pH 7.4), stirred by vortex and sonicated. Finally the solution was extruded through a polycarbonate membrane with 100 nm size pores in order to obtain aggregates of homogeneous sizes. The supramolecular compounds have been structurally characterized by Small-Angle Neutron Scattering (SANS) and Dynamic Light Scattering (DLS) techniques and by Cryo-Transmission Electron Microscopy. In the experimental conditions we investigated (pH 7.4 and molar ratio between monomers 30/70), there is the presence of: rod-like micelles with a radius of ~ 40 Å and length above 700 Å, open bilayer fragments with thickness ~ 65 Å, and probably vesicles.

Biological characterization of radiolabeled aggregates was carried out binding ^{111}In-labeled aggregates to A431 cells overexpressing the CCK2-R compared to control cells at 37°C and at 4°C. Biodistribution results of experiments were performed in receptor positive and receptor negative xenografts bearing nude mice 18 h after injection. Liver and spleen displayed highest activity retention levels attributed to lipophilic tails. The retention of radiolabel in the receptor expressing xenografts was higher compared to the control xenografts. The incubation of receptor positive and control cells with aggregates filled with doxorubicin showed significantly lower cell survival in receptor expressing cells, with respect to control cells, for samples incubated in the presence of 100 and 200 ng/mL of doxorubicin.

(a) (b)

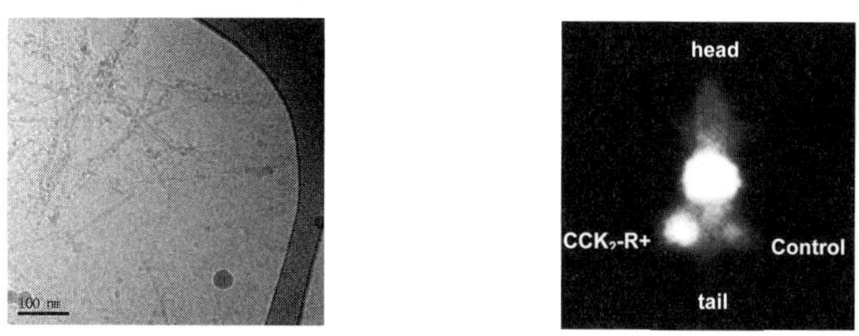

Fig. 2. (a) Cryo-TEM image of aggregates; (b) gamma camera image (dorsal view). The receptor positive xenograft (+, left flank) compared to the control tumor (-, right flank).

Acknowledgments

The authors thank the European Molecular Imaging Laboratories Network (EMIL) for financial support.

References

1. Torchilin, V.P. *Nature Reviews* **2005**, *4*, 145-160.
2. Ferrari, M. *Nature reviews* **2005**, *5*, 161-171.
3. Mulder, W.J.M.; Strijkers, G.J.; van Tilborg, G.A.F.; Griffioen, A.W.; Nicolay, K. *NMR Biomed.* **2006**, *19*, 142-164.
4. Accardo, A.; Tesauro, D.; Roscigno, P.; Gianolio, E.; Paduano, L.; D'Errico, G.; Pedone, C. and Morelli G. *J.Am.Chem. Soc.* **2004**, *126,* 3097-3107.
5. Reubi, J. C.; Schaer, J. C.; Waser, B., Eds. **1997**, *57*, 1377-1386.

The role of Cell Penetrating Peptides (CPPs) in membrane lipid phase behavior: a novel aspect elucidating peptide-mediated delivery

Cinzia Esposito[12], AiWei Tian[1], Mario Scrima[2], Karolina Kieres[2], Pietro Campiglia[2], Anna Maria D'Ursi[1] and Tobias Baumgart[2]

[1]*Department of Chemistry, University of Pennsylvania, Philadelphia, PA, 19104 USA,*
[2]*Dipartimento di Scienze Farmaceutiche, University of Salerno, Fisciano I-84084 Italy*

Introduction

Cell Penetrating Peptides (CPPs) constitute a class of short cationic peptides, which share the ability to deliver therapeutic agents (proteins, oligonucleotides) into target cells. The mechanisms proposed for the CPPs cellular uptake (endocytic, endosomal escape, etc.) are still controversial.

In the hypothesis that membrane lateral heterogeneity generated by multi lipid and protein components in cell membrane could play an important role in CPPs translocation process, we conducted studies on two popular CPPs, penetratin (RQIKIWFQNRRMKWKK) [1] and pep-1 (KETWWETWWTEWSQPKKKRKV) [2], in the presence of giant unilamellar vesicles (GUVs). In particular, we used GUVs containing different quaternary mixtures of 1,2-Dioleoyl-*sn*-Glycero-3-Phosphocholine (DOPC), 1,2-Dipalmitoyl-*sn*-Glycero-3-Phosphocholine (DPPC), Cholesterol (Chol), 1,2-Dipalmitoyl-*sn*-Glycero-3-[Phospho-*rac*-(1-glycerol)] (Sodium Salt) (DPPG) and showing liquid-liquid fluid phase coexistence, which provides an effective approach to characterize biophysical properties of "lipid raft" in cells, i.e. sterol and sphingolipid-enriched microdomains, supposed to be involved in membrane trafficking, signal transduction and protein sorting. Lipid rafts resemble liquid ordered (Lo) domains, which separate from disordered (Ld) domains [3].The addition of DPPG provided a negatively net charged membrane surface, for which especially penetratin shows high affinity. Noteworthy, DPPG partitions in Lo phase. Giant plasma membrane vesicles (GPMVs) were also used to combine biophysical and cell biological approaches. GPMV are vesicles obtained from biological cells by chemically induced plasma membrane vesiculation or "blebbing". They include membrane lipids as well as proteins, providing a realistic biological membrane composition. Liquid-liquid fluid coexistence in GPMVs can be also observed [4].

5-(6) carboxyfluorescein (CF) was chosen to label peptides at the N terminus, while *Texas-Red DPPE* was used to label vesicles; its partitioning in Lo phase, detectable in the red channel, allowed us to simultaneously observe the peptide preferences in the green channel by confocal laser microscopy technique.

Our preliminary results show that both peptides bind the liquid disordered phase in membrane, with important exceptions.

Results and Discussion

Our experiments evidence that both peptides, at different concentrations, bind the Ld phase in different quaternary mixtures of DOPC/DPPC/Chol/DPPG showing liquid-liquid phase coexistence. However, in our statistic observation, we found few vesicles in which penetratin and pep-1 bind Lo phase or sometimes do not

S. Del Valle et al. (eds.), *Peptides for Youth: The Proceedings of the 20th American Peptide Symposium*, 605
DOI: 10.1007/978-0-387-73657-0_262, © Springer Science+Business Media, LLC 2009

discriminate between different domains. This versatile behavior leads us to hypothesize that specific environmental conditions potentially coupled to conformational peptide changes may affect the rearrangement of such peptides in membrane domains (Fig. 1).

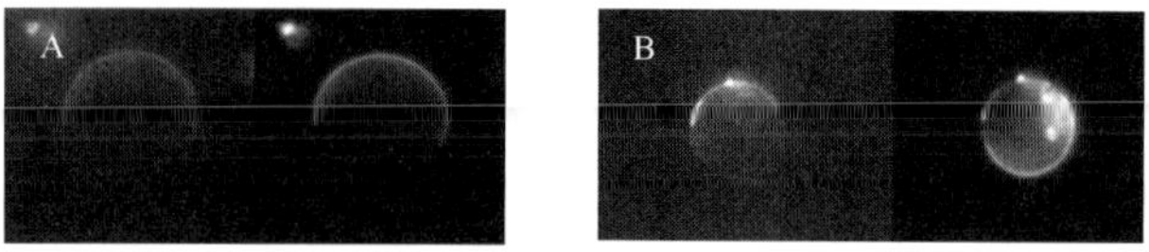

Fig. 1. Confocal fluorescence microscopy of the equatorial section of two different Texas Red-DPPE GUVs showing CF-penetratin (on the left) binding to Ld phase (A) and Lo phase (B) as detected by Texas Red (on the right) which binds Lo phase.

The use of a micropipette allowed us to fix the vesicles and monitor in time the binding kinetics of the peptides injected in real time. Even in this case, we observed opposite behaviors (Fig. 2) for both peptides, potentially indicating compositional differences between vesicles.

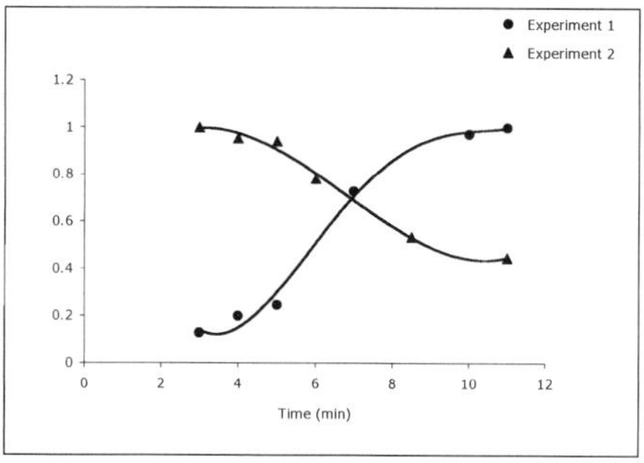

Fig. 2. Quantification of CF-penetratin fluorescence intensity in Lo versus Ld phase in two different GUVs (experiment 1 and 2) with the same composition.

Furthermore, our preliminary findings on GPMVs confirm that both peptides bind Ld phase. However, the brighter CF-peptide fluorescence found inside the GPMVs vesicles compared to GUV ones, suggests that some membrane proteins embodied in vesicles could specifically increase the peptide uptake in cells.

The translocation process of CPPs into cells could be regulate by membrane lateral heterogeneity. Here, we demonstrate by fluorescence microscopy that penetratin and pep-1 bind the Ld phase. However, some specific conditions may alter the common Lo partitioning of such peptides in a time and space-dependent manner, switching them to Lo phase.

Therefore, our future goal will be the identification and the quantification of such factors modulating the peptide partitioning in membrane domains.

References

1. Thoren Per E.G., *et al. FEBS Letters* **482**, 265-268 (2000).
2. Henriques S.T., *et al. Biochim Biophys Acta.* **1669**, 75-86 (2005).
3. Baumgart, T., *et al. Nature* **425**, 821-824 (2003).
4. Baumgart, T., *et al. Proc Natl Acad Sci USA.* **9**, 3165-3170 (2007)

Factors that influence oral bioavailability; A cathepsin K inhibitor for human studies

Daniel F. Veber

Drug Discovery Consultant, 290 Batleson Rd. Ambler, PA 19002

Introduction

The development of drug candidates that demonstrate favorable pharmacokinetic properties has been a major issue limiting the use of peptides as leads in the drug discovery process. Conventional wisdom has placed peptides as intractable leads both because of the presence of the peptide bond and the high molecular weight that is common to any compound that exceeds the tripeptide level. The analysis by Lipinski of the properties common to human drugs [1] that led to his "rule of 5" generally finds peptides as deficient in these properties. A more recent analysis of a more structurally diverse database of the rat oral bioavailability of drug candidates from our laboratory at SmithKline Beecham Pharmaceuticals (now GlaxoSmithKline or GSK) added new insight to the definition of properties that should be considered for achieving good oral bioavailability [2]. In addition to representing examples from nearly all programs at SmithKline Beecham over more than 10 years of research, it included many compounds with low oral bioavailability. Thus our assessments recognize properties both favorable and unfavorable for oral bioavailability. This is a missing aspect of studies that focus only on the human drug database.

Our studies recognized a number of important conclusions that are often not fully appreciated in studies designed to elucidate factors that influence drug oral bioavailability. First we noted a relatively low correlation of oral bioavailability with the log of the permeation rate through a lipid bilayer (r=0.31). It was clear that below a minimal permeation rate (100nm/s) good availability is seldom seen (13%) while at higher permeation rates good oral bioavailability is seen 3 times as often (35%). Notably, even the highest permeation rates are only a low predictor of good oral bioavailability. Thus, the all too common focus on "permeability" misses the key contributors to oral bioavailability which Kwan has carefully detailed [3] as being related to first pass clearance by numerous clearance enzymes which are both metabolic (e.g. P-450's, gluconuridases, proteases, etc.) and transport (e.g. p-glycoproteins) in nature. At the same time, our study highlighted the low correlation of oral bioavailability with in vivo clearance rate, a necessary consequence of the fact that oral bioavailability reflects the difference between first pass and whole body clearance rates.

Considering the extremely diverse and numerous barriers to high oral bioavailability as detailed in ref. 3, it is remarkable that we were able to recognize low molecular flexibility as measured by rotatable bond count (nRot) of <8 and polar surface area (PSA) of $<140A^2$ as being associated with high oral bioavailability independent of whether the molecular weight was above or below 500 Daltons. One can mistakenly argue that the high correlation of molecular weight with nRot and the high correlation of the sum of H-bond donors and acceptors with PSA (r~0.7 and ~0.9 respectively) make these correlated properties interchangeable in any rules created to optimize good oral bioavailability. In fact, it is the divergences from a perfect correlation that offer novel opportunities for drug development in the peptide field. Thus, if a very flexible peptide can be constrained to its bioactive conformation having a real or effective nRot of <8 then the potential for high oral

S. Del Valle et al. (eds.), *Peptides for Youth: The Proceedings of the 20th American Peptide Symposium*, 607
DOI: 10.1007/978-0-387-73657-0_263, © Springer Science+Business Media, LLC 2009

bioavailability should be greatly enhanced. In addition, constraint of peptides through cyclization can enhance internal hydrogen bonding thereby effectively eliminating some H-bond donors and acceptors and reducing the polar surface area.

Exemplifying this line of thinking, one can propose that it is a combination of conformational constraint and hidden polar groups that contribute to the good oral bioavailability of cyclosporine A in spite of its high molecular weight. It is likely that non bonded as well as bonded constraints contribute to the high degree of rigidity that is well recognized in this cyclic peptide.

An additional key point from the studies of reference 2 is that neither the overall rule that recommends nRot<10 and PSA<140A nor the individual components can be expected to serve as absolute filters to assure favorable oral bioavailability. Indeed, compounds having a high degree of flexibility (nRot>10) have a real possibility (15-20%) of showing good oral bioavailability while those with lowest flexibility (nRot<8) are not assured of high oral bioavailability. Nonetheless, they show good oral bioavailability about 70% of the time. Thus, a medicinal chemist will increase the likelihood of success about 3-fold when nRot is less than 8 whether the molecular weight is over or under the magic 500 level.

Rotational freedom is only partially measured by nRot. Non-bonded interactions, particularly steric, can have a major impact on the flexibility of a molecule and correspondingly can also influence oral bioavailability. We have seen a convergence of bonded and non-bonded constraints in the successful development of the cathepsin K inhibitor, relacatib, to be discussed below.

Results and Discussion

The potential application of inhibitors of the cysteine protease, cathepsin K, for the treatment of bone loss diseases has been a goal in our laboratories since it was recognized as specifically expressed in osteoclasts. The background and rationale for this effort has been reviewed [4]. Early, structure-based design of inhibitors resulted in ketones of the type I (fig. 1) which showed *in vitro* inhibitory potency of 1.6nM. Compounds in this class were not satisfactory for *in vivo* studies because of poor pharmacokinetic properties. Specific cyclization using methylene bridges between a carbon adjacent to the ketone carbonyl and the sulfonamide NH gave compounds which showed significantly improved pharmacokinetics. This structural class was optimized as the azepanone, II (fig. 1) which demonstrated Ki=0.16nM and gave oral bioavailability in the rat of 42%. These results have been reviewed and the successful *in vivo* outcome attributed to the conformational constraint imposed by the ring structure [5]. Here we see that nRot has been reduced from 11 to 8 by including the conformational constraint. The studies of reference 2 would predict an increase in the probability of seeing good oral bioavailability and this is indeed the case, even though we have not quite met the nRot level of <8 for optimal probability of success. Indeed, we see that the oral bioavailability of II in the monkey is an unsatisfactory 7% and this compound did not proceed to clinical evaluation in humans.

X-ray crystallographic studies of compound II as bound to cathepsin K revealed that the amino group attached to the azepanone ring was in an axial relationship to the ring while in the unbound form it was seen to be in the energetically more favorable equatorial relationship. This observation led to the preparation of a series of ring methylated, isomeric structures having methyl groups individually placed as indicated by the arrows in fig. 2. It was hoped that these groups would alter the preferred conformations of the 7-membered ring; perhaps in a way that would favor the cathepsin K bound conformation in the ground state. It was also hoped that it might selectively prevent first pass clearance at one of the rate determining points. A large variation in both the potency and the oral bioavailability of the isomeric compounds is seen as indicated in fig. 2. Potency and oral bioavailability did not optimize in the same compound. However, the combination of properties found when the methyl group was placed on position-7 of the azepanone was sufficiently favorable that it was chosen for advancement to animal safety studies and human clinical trials [6]. This compound, now known as relacatib or SB-462795, has good oral bioavailability in the monkey (27.6%) as well as the rat (89%) and has been evaluated in detailed in vivo studies related to bone resorption [7].

The 3 most active azepanones of fig.2 (III, IV and V) have oral bioavailability in the rat of <5% , 89%, and 27% respectively in spite of having identical molecular weight, H-bond acceptors, H-bond donors, cLogP and number of rotatable bonds. They also have a rate of flux through synthetic lipid bilayers that are equivalent to each other and to compound II. Clearly the differences must arise from alteration of some first pass clearance pathways rather than differences in "permeation" or passage through lipid bilayers. Possible reasons for the differences have been discussed in reference 6. I personally favor the view that internal, non-bonded interactions are the source of the differences. Compound V, having the methyl group on position-6 of the azepanone ring is likely to have altered flexibility of the 7-membered ring but is not likely to have altered flexibility of the side chains of positions-1 and 4. Thus, its' net oral bioavailability is similar to the parent, compound II. Methylation of the 5-position (III) should more greatly impact the rotational freedom of the 4-position C-N bond. Apparently it does this in a way that increases the rate of some (unknown) first-pass clearance mechanism even though it is the most potent cathepsin K inhibitor seen. Methylation of the 7-position (IV) can influence the rotational freedom of the 1-position sulfonamide and these non-bonded interactions are likely to impact the 2 rotatable bonds of the sulfonamide (i.e. N-S-C). It appears this rigidification now reduces an interaction with one or more first pass clearance mechanisms, thereby improving oral bioavailability. The unexpected importance of non-bonded interactions by the introduction of a methyl substitution in altering oral bioavailability is an important enhancement of the role of rotational

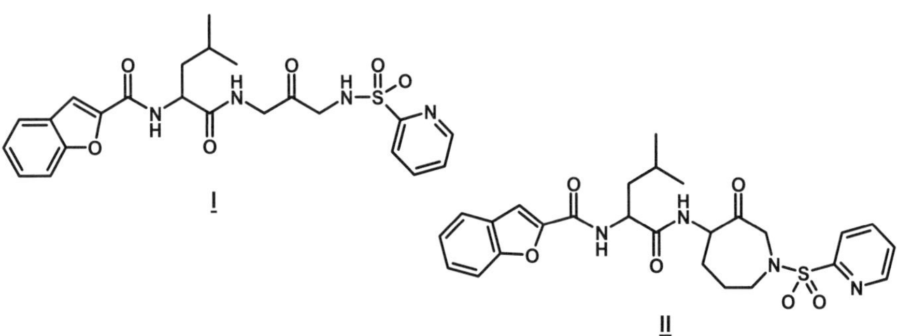

I

II

Fig. 1. Lead cathepsin K inhibitor compound I and the covalently constrained azepanone compound II.

freedom in oral bioavailability. I see the use of methyl substitution which can reduce molecular flexibility while not adding rotatable bonds as likely to prove to be a valuable tool in the search for improved pharmacokinetics. Taken together, bonded and non-bonded conformational constraints can serve as valuable tools for the optimization of peptide-based leads.

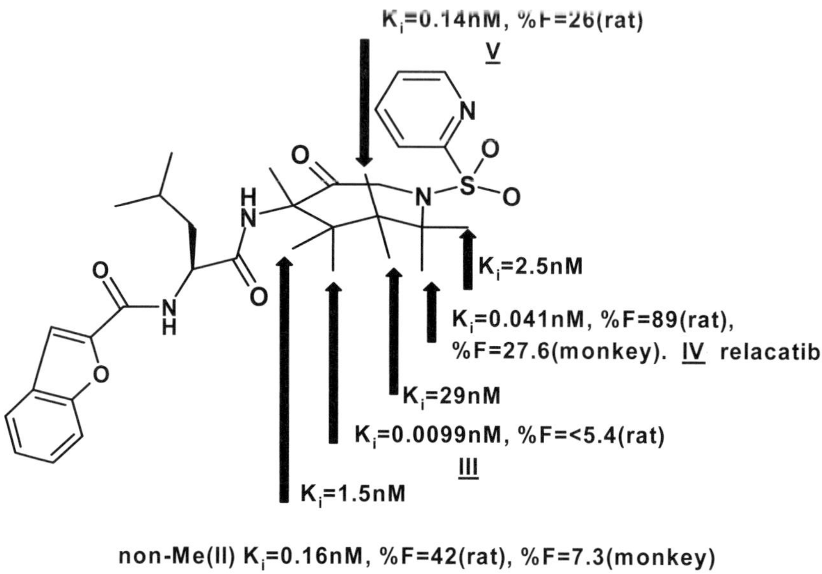

Fig. 2. Cathepsin K inhibitory potency and oral bioavailability (%F) in the species indicated for compounds having methyl substitution individually placed at the positions indicated by the arrows. Data for the unsubstituted azepanone (II) are summarized at the bottom of the figure. The 3 most potent methyl analogs have been designated III, IV and V.

Acknowledgments

I am thankful for the continuing valued discussions on these topics with Dr. Kenneth D. Kopple.

References

1. Lipinski, C. A. *Pharmacol Toxicol. Methods.* **44**, 235-249 (2000).
2. Veber, D. F., Johnson, S. R., Cheng, H-Y., Smith, B. R., Ward, K. W., and Kopple, K. D. *J. Med. Chem.* **45**, 2615-2623 (2002).
3. Kwan, K. C., *Drug Metab. and Disp.* **25**, 1329-1336 (1997).
4. Yamashita, D. S. and Dodds, R. A., *Curr. Pharm. Design.* **6**, 1-24 (2000).
5. Veber, D. F. *et. al.*, *Peptides 2000: Proc. 26th Eur. Pept. Symp*; Martinez, J. and Fehrentz, J.-A., Eds.; Editions EDK; Paris, 2000; pp113-114.
6. Yamashita, D. S. et. al. *J. Med. Chem.* **49**, 1597-1612 (2006).
7. Kumar, S. et. al. *Bone.* **40**, 122-131 (2007).

Injectable Biodegradable Polymer Depots For Minimally Invasive Delivery Of Peptides and Proteins

Li Zhang and Steven P. Schwendeman

Department of Pharmaceutical Sciences, University of Michigan, Ann Arbor, MI 48109-1065, USA

Introduction

With hundreds of biotechnological drugs, mostly peptides and proteins, undergoing clinical trials, there is an urgent need for improved methods to deliver these magnificent biomacromolecules. Unfortunately, when administered to the body by noninvasive routes, most of peptide and protein drugs suffer from significant biological barriers such as poor drug permeability across intestinal mucosa and susceptibility to proteolysis, which in turn, prevents the drug from reaching the systemic circulation and the desired pharmacological target. Therefore, this class of drug molecules is most commonly delivered by injection (e.g., subcutaneously (s.c.) or intravenously (i.v.)) allowing immediate (i.v.) or easy (s.c.) access to the blood circulation. However, even when administered i.v., the body normally eliminates the drug rapidly, requiring the drug be injected repeatedly (e.g., multiple times a day or week). Frequent injections often result in significant pain, poor patient compliance, physiological stress, and peak-and-valley drug levels in the bloodstream.

Several promising methods to overcome difficulties with peptide and protein delivery have been pursued, such as a) altering the chemical structure of drug molecule to increase plasma half-life (e.g., PEGylation), b) utilizing mucosal routes with improved drug bioavailability (e.g., pulmonary and intranasal), and c) extending drug release by microencapsulating the drug in biodegradable polymer depots. Injection frequency can be reduced from daily injections to 1-6 months for current marketed controlled-release peptide formulations, which are prepared from the most common biodegradable polymer for long-term controlled release, poly(lactic-co-glycolic acid) (PLGA). PLGA as a polyester degrades by nonenzymatic hydrolysis to yield nontoxic byproducts while slowly and continuously releasing the medication. PLGA depots have successfully been used to treat a variety of diseases, including those afflicting children and young adults, e.g., central precocious puberty (leuprolide acetate) [1], acromegaly (octreotide acetate) [2], disorders of short stature (human growth hormone) [3], and endometriosis (leuprolide) [1]. In addition to systemic delivery of polypeptides, injectable PLGA depots are also used for drug delivery directly to the target (i.e., so-called "site-specific drug delivery"), e.g., allowing novel growth factors to promote growth of new blood vessels (e.g., basic fibroblast growth factor) or bone (e.g., bone morphogenetic proteins). PLGA depots are not limited to drug delivery and are among the most sought after alternative to aluminum-based adjuvants, which are the only adjuvants approved for human use in the US, and are used to deliver childhood vaccines. Finally, novel peptide antigens, which currently cannot be delivered effectively with aluminum compounds may also benefit from this safe and flexible dosage form.

Despite tremendous progress in development of injectable peptide depots, several issues have impeded their more widespread development. Important obstacles include: a) stability of the peptide/protein when encapsulated in the polymer, b) a difficulty to microencapsulate simply and inexpensively, c) poor

S. Del Valle et al. (eds.), *Peptides for Youth: The Proceedings of the 20th American Peptide Symposium*, 611
DOI: 10.1007/978-0-387-73657-0_264, © Springer Science+Business Media, LLC 2009

control of drug release over the first day following administration (i.e., the initial burst release), and d) a low immune response to PLGA-encapsulated peptide antigens. We summarize here two examples of our approaches to overcome these limitations, namely: a) stabilization of PLGA-encapsulated bFGF to promote blood reperfusion and rescue ischemic SCID murine hindlimbs [4,5], and b) the strong and unexpected enhancement of the antibody response to a birth control peptide antigen encapsulated in pH-modified PLGA [6].

Results and Discussion

During biodegradation of PLGA, the polymer becomes an acid producer as ester bonds are cleaved to yield the corresponding carboxylic acid and alcohol moieties. Whereas an acidic pH (as low as < 3 [7]) in the aqueous peptide-containing PLGA pores is often desirable for stability of many peptide drugs, most protein molecules cannot survive this microenvironment. For example, bovine serum albumin (BSA) undergoes extensive noncovalent aggregation and peptide bond hydrolysis and heparin-stabilized bFGF similarly loses immunoreactivity during one-month release from PLGA 50/50 (D,L-lactic/glycolic acid ratio) [4]. Formulations involving the co-incorporation in PLGA poorly soluble bases (i.e., antacids) such as $Mg(OH)_2$ have been shown to prevent this pH drop homogeneously throughout the polymer and facilitate stabilization and controlled release of BSA and bFGF in vitro [4].

Such protein-stabilized injectable PLGA cylindrical implants (0.8 mm diameter) containing bFGF, $Mg(OH)_2$ and other protein specific stabilizers (e.g., heparin) (stabilized—S group) were prepared by extruding a suspension of solid protein and excipients in PLGA 50/50 (inherent viscosity = 0.60 dL/g)/acetone into silicone rubber tubing before vacuum drying and removal of the solvent. Implants were placed at the site of injury in young male SCID mice whose external iliac and femoral artery and vein had been ligated and cut to create a hindlimb ischemia [5]. Limbs were visually monitored for necrosis and reperfusion of blood flow was measured noninvasively by laser Doppler perfusion imaging (LDPI) using each animal's unimpaired hindlimb as control. Polymer control groups where bFGF was unstable in vitro or absent in PLGA were also evaluated, namely: a) partially stabilized, PS, which had bFGF-specific stabilizers but no $Mg(OH)_2$; unstabilized, US, which had no stabilizer or $Mg(OH)_2$; and blank, B, which had stabilizers and $Mg(OH)_2$ but no bFGF. As shown in Table 1, after 6 weeks of implantation, nearly all ischemic limbs had recovered in the S group but underwent necrosis or exhibited negligible reperfusion in control groups (PS, US, B).

Table 1. [a] Recovery of murine ischemic hindlimbs after 6-weeks implantation of PLGA/bFGF

[b] PLGA formulation	Limb survival	[c] Reperfusion in surviving limbs (%)
Stabilized – S	5 / 5	87 ± 7
Partially stabilized – PS	1 / 4	32
Unstabilized – US	0 / 4	—
Blank – B	1 / 4	21

[a] *Data from [5]*

[b] *All doses were 0.1 μg bFGF and protein loading was ~0.002% w/w. bFGF-specific stabilizers included (w/w): 0.01% heparin, 0.01% EDTA, 2.3% sucrose, 12.7% BSA (S), 15.7% gum arabic (PS); $Mg(OH)_2$ was added at 3% (S); gum arabic was substituted for BSA in PS because BSA aggregates without $Mg(OH)_2$.*

[c] *mean ± SEM (n=5) for S group. PS and US had 1 surviving limb imaged and B had none*

A second example to improve PLGA depot development is the unexpected finding that when an antacid ($MgCO_3$) is co-encapsulated with a human chorionic gonadotropin (hCG)-based peptide antigen to control PLGA microclimate pH, a strong and persistent antibody response is observed. Antibodies against hCG exert antifertility action without disturbance of the normal ovulation function or hormone secretion. Once the antibody level has declined after several months, immunity can be boosted to continue contraception or discontinued to recover fertility.

The chimeric peptide consisting of a universal T-cell epitope from tetanus toxoid (TT2, residues 830-844)) and a B-cell epitope from hCG (C-terminal peptide35, CTP35) was encapsulated in PLGA 50/50 microspheres (1 – 10 μm) by the double emulsion/solvent evaporation method. One mg of antigen was administered to adult, specific pathogen-free New Zealand white rabbits in the following groups: a) a single dose of microencapsulated antigen + or – 1% $MgCO_3$; b) 3 doses of antigen in PBS emulsified in an unsafe oily vehicle (squalene:mannide monooleate (4:1)) and c) 3 doses of soluble antigen in PBS (negative control). As shown in Table 2, peak serum antibody levels, determined by RIA, were as high as the positive control group for microencapsulated antigen + $MgCO_3$, but very low for – $MgCO_3$ or negative controls. Similarly, the duration of the antibody response for microencapsulated antigen + $MgCO_3$ also mimicked the positive control (*data not shown*) [6]. Ongoing efforts are focused at understanding the mechanism of the $MgCO_3$ effect and its potential generality to other peptide antigens.

Table 2. [a] *Peak serum anti-hCG response to PLGA/C-TT2-CTP35 microspheres*

[b] Peptide formulation	Peak serum anti-hCG level (nM)
Soluble antigen (neg. control)	21 ± 6
Antigen in squalene (pos. control)	730 ± 290
Encapsulated antigen + $MgCO_3$	720 ± 190
Encapsulated antigen – $MgCO_3$	15 ± 3

[a] *Data from [6]; experimental values represent mean ± SEM (n=5)*
[b] *Antigen dose was 1 mg; soluble and emulsified antigen groups were boosted at 4 and 10 weeks; positive control group also contained 25 μg nor-MDP*

Acknowledgments
The contributions of many former students (particularly C. Cui and Y. Zhong) and collaborators (particularly V. Stevens, S. R. Mallery, and D. Mooney) in the research described here are gratefully acknowledged.

References

1. http://www.lupron.com/.
2. http://www.sandostatin.com/.
3. http://www.nutropin.com/.
4. Zhu, G., Mallery, S. R. and Schwendeman, S. P. *Nat. Biotech.*, **18**, 52-57 (2000).
5. Zhong, Y., *et al. J. Control. Release*, in press.
6. Cui, C., Stevens, V. C. and Schwendeman, S. P. *Vaccine*, **25**, 500-509 (2007).
7. Ding, A. G., Shenderova, A. and Schwendeman, S. P. *J. Am. Chem. Soc.*, **128**, 5384-5390 (2006).

Author Index

Subject Index

D

I